Handbook of Experimental Pharmacology

Volume 90/I

Catecholamines I

Contributors

H. Bönisch, J. P. M. Finberg, W. W. Fleming, K.-H. Graefe, S. Z. Langer, J. Lehmann, H. Matthaei, P. B. Molinoff, A. Philippu, K. F. Tipton, U. Trendelenburg, D. P. Westfall, H. Winkler, B. B. Wolfe, M. B. H. Youdim

Editors

U. Trendelenburg and N. Weiner

Springer-Verlag
Berlin Heidelberg New York
London Paris Tokyo

U. Trendelenburg, Professor Dr.
Department of Pharmacology and Toxicology
University of Würzburg
Versbacher Str. 9
D – 8700 Würzburg

N. Weiner, Professor Dr.
Department of Pharmacology
Colorado University, Medical Center
4200 East Ninth Street, Denver, CO 802 62, USA

With 43 Figures

ISBN-13: 978-3-642-46627-4 e-ISBN-13: 978-3-642-46625-0
DOI: 10.1007/978-3-642-46625-0

Library of Congress Cataloging-in-Publication Data.
Catecholamines/contributors, H. Bönisch ... [et al.]; editors, U. Trendelenburg and N. Weiner. p.cm. — (Handbook of experimental pharmacology; v. 90/I-) Bibliography: v. 1, p. Includes index. ISBN-13: 978-3-642-46627-4
1. Catecholamines—Metabolism. 2. Catecholamines—Agonists. 3. Catecholamines—Antagonists. 4. Catecholamines—Receptors. I. Bönisch, H. II. Trendelenburg, U. (Ullrich), 1922–. III. Weiner, Norman, 1928–. IV. Series: Handbook of experimental pharmacology; v. 90/I, etc. QP905. H3 vol. 90/1, etc. [QP801.C33] 615'.1 s—dc19 [615'.78]

Typesetting: Interdruck, GDR;
2122/3020 – 54 32 10

List of Contributors

H. BÖNISCH, Institut für Pharmakologie und Toxikologie der Universität Würzburg, Versbacher Str. 9, D – 8700 Würzburg

J. P. M. FINBERG, Technion – Israel Institute of Technology, School of Medicine, The B. Rappaport Family, Medical Sciences Building, Efron Street, P. O. B. 9697, Haifa 31096, Israel

W. W. FLEMING, Department of Pharmacology and Toxicology, West Virginia University, Medical Center, Morgantown, WV 26506, USA

K.-H. GRAEFE, Insitut für Pharmakologie und Toxikologie der Universität Würzburg, Versbacher Str. 9, D – 8700 Würzburg

S. Z. LANGER, Laboratoire d'Etudes et de Recherches Synthélabo (L. E. R. R.), 58, rue de la Glacière, F – 75013 Paris

J. LEHMANN, Ciba Geigy Corporation, Summit, NJ 07901, USA

H. MATTHAEI, Institut für Pharmakodynamik und Toxikologie der Universität Innsbruck, Peter-Mayr-Str. 1, A - 6020 Innsbruck

P. B. MOLINOFF, Department of Pharmacology, University of Pennsylvania, School of Medicine, 36th and Hamilton Walk, Philadelphia, PA 19104–1184, USA

A. PHILIPPU, Institut für Pharmakodynamik und Toxikologie der Universität Innsbruck, Peter-Mayr-Str. 1, A – 6020 Innsbruck

K. F. TIPTON, Department of Biochemistry, Trinity College, Dublin, Ireland

U. TRENDELENBURG, Institut für Pharmakologie und Toxikologie der Universität Würzburg, Versbacher Str. 9, D – 8700 Würzburg

D. P. WESTFALL, Department of Pharmacology and Toxicology, West Virginia University Medical Center, Morgantown, WV 26506, USA

H. WINKLER, Institut für Pharmakologie, Universität Innsbruck, Peter-Mayr-Str. 1, A – 6020 Innsbruck

B. B. WOLFE, Department of Pharmacology, University of Pennsylvania School of Medicine, 36th and Hamilton Walk, Philadelphia, PA 10104-1184, USA

M. B. H. YOUDIM, Technion – Israel Institute of Technology, Faculty of Medicine, The B. Rappaport Family, Medical Sciences Building, Efron Street, P. O. B. 9697, Haifa 31096, Israel

Preface

The two new volumes *Catecholamines* are actually the third "state-of-the-art" report in the series of *Heffter-Heubner's Handbook of Experimental Pharmacology*. The first (*Adrenalin und adrenalinverwandte Substanzen* by P. Trendelenburg, 1924) covered the subject in 163 pages. *Catecholamines* (edited by H. Blaschko and E. Muscholl, 1972) was published in one volume of slightly more than 1000 pages. The succesor, also called *Catecholamines* now appears in two volumes – and with a further increase in the total number of pages.

It is up to future readers to decide whether present-day authors (able to rely on computers) are more verbose than the earlier ones or whether the undeniable explosion of knowledge accounts for the increase in size. The editors prefer the second of these hypotheses. They should like to draw the sceptical reader's attention to Fig. 1 of the preface of Iversen (1967) which illustrates the explosive increase in yearly publications (quoted by the author) between 1955 and 1965. If that figure suggested that the rate of relevant publications approached a maximum (or V_{max}) at the end of this period, one may legitimately ask whether perhaps this V_{max} characterized certain memory stores rather than the rate of relevant publications. Can any individual store the information contained in 150 publications per year (or 0.41 per day!), expecially when it is highly likely that the publication explosion continued unabatedly throughout the 1970s and 1980s? This high rate of publication is one of the reasons why there is an urgent need for critical reviews that enable the non-expert to enter a new field with a minimum waste of time or the expert to check quickly on progress in neighbouring fields.

The present volumes are a successor to, but *not* a new edition of, *Catecholamines* of 1972. Blaschko (1972) had to acknowledge the sad fact that the editors failed to receive this or that promised chapter. The editors of the present volumes were by no means luckier. Irrespective of these missing chapters, it would have been quite impossible to aim at a presentation of *all* areas in which catecholamines are under active study. Hence, in spite of the increased size of Volume 90, *Catecholamines* (1988) resembles the volume of 1972 in presenting "selected topics". We hope to offer a variety of topics that pleases some and offends very few.

As Szekeres (1980/81) edited the volumes *Adrenergic Activators and Inhibitors* (also in this series), the present editors felt entitled to concentrate on "basic mechanisms", with relatively little emphasis on "organ- or system-specific actions".

The editorial board of *Heffter-Heubner's Handbook* selected two persons inexperienced in the formidable task of editing such a wealth of information. Hence (and for reasons of health), fruition of our editorial efforts came later than originally anticipated. This fact is mentioned in order to convince the contributors to these volumes that our thanks for their patience and unflagging enthusiasm are very sincere.

Any selection of topics *must* (to some degree) reflect the editors' bias. Moreover, if such a volume has a long gestation period, topics of very recent interest cannot be included. Within these limitations, we tried to follow the splendid example set by Blaschko and Muscholl with *Catecholamines* (1972): experts were asked to review areas of active research, irrespective of whether they deal with basic mechanisms or clinical problems. It is, of course, in the hospitals that our increasing knowledge of the catecholamines should eventually lead to improved treatment of patients.

Thus, whether researchers or clinicians are in love with central or peripheral catecholamines, we hope to have provided them with interesting and helpful reading material.

As successors to the 1972 volume, the present volumes should be regarded as a unit. Although some attempt was made to arrange the table of contents in a certain order, the publication in *two* volumes arose much more from the publisher's concern about the formidable weight of one single volume than from any wish of the editors to subdivide this area into two separate parts. Moreover, the late arrival of this or that chapter necessitated deviations from an optimal order in the table of contents. In recognition of the fact that the two volumes constitute a unity, the subject index at the end of the second volume covers *both* volumes. The subject index at the end of the first volume is meant to be helpful for those who buy the first volume only.

U. TRENDELENBURG and N. WEINER

References

Blaschko H (1972) Introduction – Catecholamines 1922–1971. In: Blaschko H, Muscholl E (eds) Catecholamines. Springer, Berlin Heidelberg New York, 33:1–15 (Handbook of Experimental Pharmacology Vol 33)

Blaschko H, Muscholl E (1972) Catecholamines. Handbook of Experimental Pharmacology, Vol 33. Springer, Berlin Heidelberg New York

Iversen LL (1967) The uptake and storage of noradrenaline in sympathetic nerves. Cambridge University Press, Cambridge

Szekeres L (1980/81) Adrenergic Activators and Inhibitors. Handbook of Experimental Pharmacology Vol 54/I and 54/II. Springer, Berlin Heidelberg New York

Trendelenburg P (1924) Adrenalin und adrenalinverwandte Substanzen. In: Handbuch der Exp. Pharmakologie Heffter A, (ed.) 2:1130-1293, Springer Berlin Heidelberg

Contents

CHAPTER 1

Transport and Storage of Catecholamines in Vesicles
A. PHILIPPU AND H. MATTHAEI. With 3 Figures 1

CHAPTER 5

The Mechanism of Action of Indirectly Acting Sympathomimetic Amines

CHAPTER 6

The Extraneuronal Uptake and Metabolism of Catecholamines

CHAPTER 7

Catecholamines Receptors
B. WOLFE and B. MOLINOFF. With 4 Figures 321

CHAPTER 8

Presynaptic Receptors on Catecholamine Neurones

List of Abbreviations

ACTH	corticotropin
ADP	adenosine diphosphate
AMP	adenosine monophosphate
ATP	adenosine triphosphate
cAMP	cyclic adenosine monophosphate
cGMP	cyclic guanosine monophosphate
CNS	central nervous system
COMT	catechol-O-methyl transferase
CSF	cerebrospinal fluid
CTP	cytidine triphosphate
DBH	dopamine-β-hydroxylase
DOMA	dihydroxymandelic acid
DOPAC	dihydroxyphenylacetic acid
DOPEG	dihydroxyphenylethylene glycol
DOPET	dihydroxyphenylethanol
EC_{50}	half-maximally effective concentration
EDTA	ethylenediaminetetraacetate
FAD	flavin-adenine dinucleotide
FRL	fractional rate of loss
GABA	gamma-aminobutyric acid
GDP	guanosine diphosphate
GTP	guanosine triphoshate
5-HT	5-hydroxytryptamine, serotonin
HVA	methoxyhydroxyphenylacetic acid (homovanillic acid)
IC_{50}	half-maximally inhibitory concentration
ITP	inosine triphosphate
MAO	monoamine oxidase
MN	metanephrine
MOPEG	methoxyhydroxyphenylethylene glycol
NADH	nicotinamide-adenine dinucleotide
NA/DOPEG	noradrenaline/dihydroxyphenylethylene glycol
NMN	normetanephrine
OMI	3-O-methyl-isoprenaline
PG	prostaglandin
PNMT	phenylethanolamine-N-methyl transferase
UDP	uridine diphosphate
$uptake_1$	neuronal noradrenaline uptake

$uptake_2$	extraneuronal noradrenaline uptake
UTP	uridine triphosphate
VMA	methoxyhydroxymandelic acid (vanillylmandelic acid)

CHAPTER 1

Transport and Storage of Catecholamines in Vesicles

A. PHILIPPU AND H. MATTHAEI

A. Introduction

Transport and storage of catecholamines in vesicles of peripheral nerves and of the central nervous system closely resemble those in chromaffin granules of the adrenal medulla. In fact, a great deal of our knowledge of the properties of vesicles was gained by experiments carried out with chromaffin granules. Hence, properties of catecholamine-storing particles of the adrenal medulla will be discussed together with the properties of storage vesicles of nerves and central neurons. The term "chromaffin granule" will be used for the storage particles of the adrenal medulla.

Since the appearance of the volume Catecholamines in 1972, new information has been published dealing with the biogenesis of chromaffin granules and synaptic vesicles and their constituents. A short outline of biogenesis will appear at the beginning of this chapter.

B. Biogenesis

I. Formation and Types of Vesicles

There is good evidence that chromaffin granules and synaptic vesicles are formed near the Golgi apparatus. Electron microscopic studies show that chromaffin granules are present within the Golgi membranes, or the Golgi-associated endoplasmic reticulum. Similarly, particles resembling synaptic vesicles in their morphology are formed near the Golgi apparatus in perikarya (see COUPLAND 1972; LANE and SWALES 1976; see HOLTZMAN 1977) and transported along the axons to the nerve terminals. Formation of synaptic vesicles also takes place in the axons (PELLEGRINO DE IRALDI and DE ROBERTIS 1970). The vesicles formed in the axons originate from the agranular reticulum, which is comparable to the agranular reticulum and the rough endoplasmic reticulum of the perikaryon (TEICHBERG and HOLTZMAN 1973; DUCROS 1974; DROZ 1975).

At least two types of vesicles may be distinguished in noradrenergic nerves by morphological and biochemical methods: the large dense-core vesicles and the small dense-core vesicles (CHUBB et al. 1970; BISBY and FILLENZ 1971; see BLOOM 1972; NELSON and MOLINOFF 1976). The large dense-core vesicles have a diameter of approximately 80–120 nm and seem to correspond to the heavy vesicles (according to their sedimentation rates after homogenization of the tissue), while the small dense-core vesicles have a diameter of 40–60 nm and seem to be identical with the light vesicles (see SMITH 1972; THURESON-KLEIN et al. 1973). The large vesicles contain DBH, but the presence of the enzyme in the small vesicles is still questionable. BISBY et al. (1973), NELSON and MOLINOFF (1976), DE POTTER and DE SMET (1980), NEUMAN et al. (1984) found DBH activity in both large and small vesicles. Small vesicles with DBH activity were also found in the splenic nerve of the dog. These vesicles seem to derive from the large vesicles after exocytosis and their number increases on nerve stimulation (DE POTTER and CHUBB 1977). On the other hand, according to CHUBB et al. (1970), KIRKSEY et al. (1977), KLEIN et al. (1979), DBH activity in the fraction of small vesicles is either low or absent, and might be due to the presence of large dense-core vesicles rather than to small vesicles (KLEIN et al. 1979). In addition to these nerve vesicles, the bovine splenic nerve contains large dense-core vesicles (300–800 nm) originating from mast cells and some dense-core vesicles (120–140 nm), which bud off from small intensely fluorescent (SIF) cells (THURESON-KLEIN et al. 1979).

Fractionation studies concerning the presence of different types of synaptic vesicles in the central nervous system are scarce. Large dense-core vesicles which contain noradrenaline (THOMAS 1979) and DBH (COYLE and KUHAR 1974; NAGY et al. 1977) have been described in cerebral cortex and hypothalamus. Also found were small dense-core vesicles, which possess low DBH activity (COYLE and KUHAR 1974) and seem to correspond to the small vesicles of peripheral noradrenergic nerves (CHUBB et al. 1970). DBH was found in small agranular vesicles of locus coeruleus and hypothalamus (CIMARUSTI et al. 1979). Other authors (NAGY et al. 1977; THOMAS 1979) were unable to distinguish more than one population of noradrenergic synaptic vesicles in cerebral cortex.

When results concerning the existence of various types of vesicles are interpreted, it should be kept in mind that technical procedures (such as homogenization of tissue, separation of vesicles by gradient centrifugation, fixation for electron microscopic purposes) as well as ageing of the biological material until separation of the vesicles is achieved, may lead to morphological and biochemical alterations (YEN et al. 1973; THURESON-KLEIN et al. 1973; THOMAS 1979).

II. Biogenesis of Proteins, Mucopolysaccharides and Phospholipids

Among the proteins of the chromaffin granules and storage vesicles are the acidic glycoproteins chromogranin A, B and C (BLASCHKO et al. 1967; DE POTTER et al. 1970; LAGERCRANTZ 1971a; see STJÄRNE 1972; WINKLER and SMITH 1975; LAGERCRANTZ 1976; BARTLETT et al. 1976; HUBER et al. 1979; ROSA et al. 1983; FISCHER-COLBRIE and FRISCHENSCHLAGER 1985; FISCHER-COLBRIE et al. 1986a), chromomembrin B (WINKLER 1971; HÖRTNAGL et al. 1971) and proteins with enzymatic properties, such as DBH, ATPase (KIRSHNER 1957; HILLARP 1958a; LEVIN et al. 1960; BANKS 1965; KIRSHNER et al. 1966; STJÄRNE et al. 1967; HÖRTNAGL et al. 1969; DE POTTER et al. 1970; LAGERCRANTZ 1971a; see STJÄRNE 1972), cytochrome b_{561} (SPIRO and BALL 1961; ICHIKAWA and YAM-

ANO 1965; BANKS 1965; APPS et al. 1980a), NADH: (acceptor) oxidoreductase (FLATMARK et al. 1971) and phosphatidylinositol kinase (TRIFARÓ and DWORKIND 1971; PHILLIPS 1973). However, it should be kept in mind that the distribution of chromogranins is by far less specific than it was previously thought to be; chromogranins (synonym: secretogranins) have not only been found in catecholamine-containing chromaffin granules and synaptic vesicles, but also in several types of endocrine cells (ROSA et al. 1981, 1983, 1985; COHN et al. 1982; NOLAN et al. 1985). Cytochrome b_{561} seems to be identical with chromomembrin B (APPS et al. 1980a; FLATMARK and GRØNBERG 1981). Most of the proteins are membrane-bound. Approximately 50 % of chromogranin A is soluble. DBH is 60 % membrane-bound; the remainder is soluble within the matrix of the granules (BJERRUM et al. 1979; WINKLER and WESTHEAD 1980). Immunoelectron microscopic studies suggest that the soluble DBH is not an artifact due to proteolysis (AUNIS et al. 1978). Storage vesicles possess a higher concentration of phospholipids (LAGERCRANTZ 1971b), than do chromaffin granules (BLASCHKO et al., 1967; for details see Chap. 2).

Chromaffin granules also contain opioids, such as enkephalin precursors and dynorphin (LEWIS et al. 1979; STERN et al. 1979; VIVEROS et al. 1979; HOLZ 1980; DAY et al. 1982; LIVETT et al. 1982; KOBAYASHI et al. 1985; FISCHER-COLBRIE et al. 1986b) and enzymes necessary for the production of the peptides from their precursors (LINDBERG et al. 1982, 1984; EVANGELISTA et al. 1982). Opioid peptides were identified in large dense-core vesicles of nerves (FRIED et al. 1981; KLEIN et al. 1982; NEUMAN et al. 1984; KOBAYASHI et al. 1985) and vas deferens (NEUMAN et al. 1984). Substance P was found in large (CUELLO et al. 1977) and small (PICKEL et al. 1979) vesicles of the brain. Neuropeptide Y is localized in large dense-core vesicles of vas deferens (FRIED et al. 1985) and in chromaffin granules (FISCHER-COLBRIE et al. 1986b).

The biogenesis of the secretory proteins (chromogranins and DBH) was studied by "pulse" labelling of the perfused bovine adrenal medulla with ^{3}H-leucine (WINKLER et al. 1972). A short time after the introduction of the label (1.5 min) soluble proteins were labelled in the microsomal fraction, which consists mainly of rough and smooth endoplasmic reticulum. Some hours later, soluble, newly synthesized chromogranins were found in particles which sedimented to a region of the gradient density tube corresponding to the upper part of the mature chromaffin granules. These results indicate that newly formed granules are less dense than mature granules (VIVEROS et al. 1971; BENEDECZKY and SMITH 1972). Very probably, the low density of these granules is due to their low concentration of catecholamines, ATP and other constituents. On the other hand, the membrane-bound chromogranins did not incorporate significant amounts of labelled leucine. Similar results were obtained with labelled fucose, which was used as precursor of glycoproteins; soluble glycoproteins were labelled, while the incorporation into membranes of newly synthesized granules was low. Again, the labelled glycoproteins were located in the immature particles just above the fraction of the mature granules (GEISSLER et al. 1977). Likewise, soluble chromogranins and a proteoglycan were labelled with ^{35}S-sulphate, (FALKENSAMMLER et al. 1985).

^{32}P-Phosphate failed to label the phospholipids, which are almost exclusively located in the membrane of the chromaffin granules (BAUMGARTNER et al. 1974; WINKLER et al. 1972). These results suggest that the concomitant biogenesis of the membranes of the vesicles and their soluble macromolecules is not likely.

However, different results were obtained when immunoprecipitation techniques were used. GAGNON et al. (1976), as well as LEDBETTER et al. (1978) reported labelling of both soluble and membrane-bound DBH with ^{3}H-leucine. The amount of labelled chromogranin was 5-20 times higher than that of radioactive soluble DBH. The amount of chromogranin was also twenty times higher than that of the soluble DBH (LEDBETTER et al. 1978). These findings indicate the simultaneous synthesis of chromogranin, membrane-bound and soluble DBH. Probably, the discrepancies between the results of WINKLER et al. (1972) and those of GAGNON et al. (1976) and LEDBETTER et al. (1978) are due to the different methods used for the determination of the labelled proteins. The question of the specific retrieval of the membranes after exocytosis will be discussed elsewhere (see Chap. 2).

It is pertinent to postulate that the proteins synthesized in the endoplasmic reticulum are transported into the Golgi region to be packaged into chromaffin granules. Indeed, electron radiographic studies revealed a concentration of ^{3}H-leucine within the endoplasmic reticulum 5 min after "pulse" labelling, while by 15 min ^{3}H-leucine was associated with dense granules of the Golgi region (KOBAYASHI et al. 1978). Electron micrographs show fusion of small coated vesicles with presecretory granules (AL-AMI 1969; BENEDECZKY and SMITH 1972). This led to the view that the small coated vesicles undertake a shuttle service to transport newly synthesized macromolecules from the endoplasmic reticulum to the new granules (BENEDECZKY and SMITH 1972).

III. Biogenesis of Catecholamines

The catecholamine content of chromaffin granules of adrenal medulla is approximately 2.5 μmol · mg protein^{-1}, while each chromaffin granule contains 3×10^{6} molecules of catecholamines (see WINKLER, 1976). Vesicles of peripheral noradrenergic nerves contain 70 nmol · mg protein^{-1} (YEN et al. 1976), the noradrenaline content of each vesicle being 6000-15000 molecules (KLEIN et al. 1977). It is now well established that chromaffin granules and vesicles of peripheral noradrenergic nerves take up dopamine, which, in the presence of ascorbate (LEVIN et al. 1960), is converted to noradrenaline by the vesicular DBH (see STJÄRNE and LISHAJKO 1967; EULER and LISHAJKO 1968; LADURON and BELPAIRE 1968; KIRSHNER 1975). In the central nervous system, dopamine is transformed to noradrenaline in the synaptic vesicles of the noradrenergic neurons. In adrenergic neurons of CNS and adrenal medulla, noradrenaline is further converted to adrenaline by PNMT, which is in the axoplasm (KIRSHNER and MC GOODALL 1957; see AXELROD 1962; see AXELROD 1966; LADURON and BELPAIRE 1968). Hence, the synthesized noradrenaline has to be released from the vesicles of granules to be transformed into adrenaline in the axoplasm and the newly formed adrenaline is taken up into, and stored within the vesicles.

However, a discrepancy exists between findings obtained with biochemical and immunohistochemical procedures, because immunohistochemical studies revealed some PNMT in the periphery of chromaffin granules (VAN ORDEN et al. 1977). PNMT was also found within adrenaline-containing granules, as well as on their membranes (NAGATSU et al. 1979). On the other hand, it was recently shown that PNMT is synthesized on free polysomes in the cytosol, suggesting that the enzyme obtains its subcellular localization during its translation (SABBAN and GOLDSTEIN 1984). The latter finding might also indicate that the complex transport of precursors from cytosol to chromaffin granules and vice versa is an indispensable prerequisite for the biosynthesis of adrenaline.

A direct conversion of dopamine to epinine by PNMT in the cytoplasm of chromaffin cells was proposed by LADURON (1972) and LADURON et al. (1974). According to these authors, the newly formed epinine is taken up into the chromaffin granules and converted to adrenaline by DBH. SCHÜMANN and BRODDE (1976) came to different conclusions, since they did not detect any appreciable amounts of epinine in brain and adrenal medulla after inhibition of DBH. The latter results are in good agreement with the finding of PENDLETON and GESSNER (1975) that dopamine is not a substrate for PNMT.

Determination of the DBH activity and of the noradrenaline content of large dense-core vesicles isolated from the splenic nerves revealed 4 (range: 2-9) molecules DBH per vesicle and 30 molecules noradrenaline per molecule DBH, the synthesis rate or noradrenaline being 60-270 molecules $\cdot$ vesicle$^{-1} \cdot s^{-1}$ (KLEIN et al. 1977). In chromaffin granules DBH was estimated to be 136-208 molecules $\cdot$ granule^{-1} (see WINKLER and WESTHEAD 1980).

IV. Biogenesis of ATP

Chromaffin granules and synaptic vesicles contain high amounts of nucleotides (BLASCHKO et al. 1956; FALCK et al. 1956; SCHÜMANN 1958). Besides ATP, chromaffin granules also contain ADP, AMP, GTP, GDP, UTP and UDP (HILLARP and THIEME 1959; GOETZ et al. 1971; VAN DYKE et al. 1977). Small dense-core vesicles found in the sympathetic nerves of vasa deferentia contained ^{3}H-ATP when the tissue was pulse labelled with ^{3}H-adenosine (ABERER et al. 1979). Moreover, the perfused cat adrenal gland phosphorylated labelled adenosine. High amounts of the radioactivity were found in chromaffin granules (STEVENS et al. 1972). The question arises whether catecholamine-storing vesicles are able to synthesize ATP. WINKLER et al. (1972) found the labelled nucleotides concentrated in the mitochondrial fraction of the adrenal gland a few minutes after infusion of ^{32}P-phosphate. After some hours, chromaffin granules had accumulated high amounts of labelled nucleotides in comparison to mitochondria. However, no significant amounts of ^{3}H-AMP and ^{3}H-ADP were found in chromaffin granules. Moreover, inhibitors of mitochondrial oxidative phosphorylation prevented the appearance of ^{3}H-ATP in chromaffin granules. The results suggest that the storage vesicles are unable to synthesize ATP *de novo* (PEER et al. 1976).

C. Uptake

I. Uptake of Catecholamines

1. Dependence on Temperature: Nucleotide and Ionic Requirements

Uptake of radioactive noradrenaline into storage particles was first demonstrated by electron microscopic radiography in the pineal body of the rat (WOLFE et al. 1962). *In vitro* experiments with isolated vesicles of peripheral sympathetic nerves and ganglia revealed that the spontaneous release of endogenous noradrenaline was partly compensated for by uptake of labelled catecholamines added to the incubation medium (EULER and LISHAJKO 1963a, 1963b, 1969; PHILIPPU et al. 1967; see EULER 1972).
Addition of ATP and magnesium to the incubation medium greatly enhanced the uptake of catecholamines into chromaffin granules (CARLSSON et al. 1962; KIRSHNER 1962), vesicles of sympathetic nerves (EULER and LISHAJKO 1963b, 1969) and of ganglia (PHILIPPU et al. 1967) and synaptic vesicles of the brain (PHILIPPU et al. 1968, 1969). The uptake of catecholamines into chromaffin granules was found to be dependent on a magnesium-stimulated ATPase (TAUGNER 1971).

The stimulation of the uptake of catecholamines into chromaffin granules and vesicles by ATP-Mg^{2+} is dependent on temperature and obeys Michaelis-Menten kinetics. In chromaffin granules, the Q_{10} was 2.8, or even 6 in the presence of reserpine (KIRSHNER 1962), in synaptic vesicles of the brain a Q_{10} of 20 was reported (HALARIS and DEMET 1978). The uptake was stimulated by other nucleotides, too (such as ITP, deoxy-ATP, GTP, UTP, CTP, ADP), though to a lesser extent than by ATP (KIRSHNER 1962; CARLSSON et al. 1962, 1963; EULER and LISHAJKO 1963b). Magnesium ions could be replaced by manganese or cobalt ions (KIRSHNER 1962; CARLSSON et al. 1963).

More recently, the characteristics of the uptake of amines into synaptic vesicles of the brain were studied. ATP-Mg^{2+} enhanced the uptake of dopamine and noradrenaline into synaptic vesicles isolated from hypothalamus and caudate nucleus. The ATP-Mg^{2+}-dependent uptake obeyed Michaelis-Menten kinetics, the uptake in the absence of ATP-Mg^{2+} did not. The retention in the absence of ATP-Mg^{2+} was reduced at 0° C, while the retention in the presence of ATP-Mg^{2+} was abolished at this temperature (PHILIPPU et al. 1968, 1969; PHILIPPU and BEYER 1973; LENTZEN and PHILIPPU 1981). At 25° C, the ATP-Mg^{2+}-dependent retention reached a steady state after 10–20 min and remained constant thereafter. At 37° C the ATP-Mg^{2+}-dependent retention was short-lasting; it reached a maximum after approximately 5 min, it then declined and was abolished by 20 min, indicating that the amine taken up was again liberated from the vesicles into the incubation medium (PHILIPPU and BEYER 1973; SEIDLER et al. 1977; TSUDZUKI 1981). The short-lasting retention at 37° C was not due to breakdown of exogenous dopamine or ATP added to the incubation medium, since a renewed addition of these compounds failed to restore the retention of dopamine (PHILIPPU and BEYER

Table 1. K_m and V_{max}-values of the ATP-Mg^{2+}-dependent uptake of monoamines into chromaffin granules and vesicles

Tissue and Species	(−)-NA K_m[a]	(−)-NA V_{max}[b]	(±)-NA K_m	(±)-NA V_{max}	(+)-NA K_m	(+)-NA V_{max}	(−)-A K_m	(−)-A V_{max}	DA K_m	DA V_{max}	TY K_m	TY V_{max}	OA K_m	OA V_{max}	5-HT K_m	5-HT V_{max}	Authors
Adrenal Medulla	8 000																Jonasson et al. 1964
Cattle	18[c]	7 700			70	6 000	16				8	1 100			9	7 000	Phillips 1974
			18.5												4.3		Kanner et al. 1979
	110[c]																Da Prada et al. 1975
	32.5[c]	12 900					35.1[c]	15 200	16.2[c]	14 000					4.7[c]	5 100	Carty et al. 1985
Heart	2														0.5		Bareis and Slotkin 1979
Rat	3.4[d]		10.3[d]		5.7[d]										0.7[d]		Angelides 1980
Whole	0.4																Ferris and Tang 1979
Brain															0.1		Halaris and Demet 1978
Rat	8	4.9	1.5	5.5	17.3	5.1											Matthaei et al. 1980, 1988
									1.6								Tanaka et al. 1976
									0.3								Slotkin et al. 1978b
Caudate			1.3						0.9	12.3			5.3	8.5	0.5	13.3	Matthaei et al. 1976
Nucleus			1.8						0.4		0.1		1.8				Lentzen and Philippu 1981
Pig																	
Corpus Striatum	0.4																Slotkin et al. 1978c
Rat																	
Cerebral Cortex	0.5																Slotkin et al. 1978c
Rat																	

[a] μM; [b] pmol · mg protein^{-1} · min^{-1}; [c] "Ghosts" of chromaffin granules; [d] "Ghosts" of synaptic vesicles

1973). Probably, the retention ceases because of the accelerated ageing of the thermo-labile vesicles at this temperature.

Maximum enhancement of the uptake of noradrenaline into synaptic vesicles of the rat brain was observed at 1 mM each of ATP and magnesium (SEIDLER et al. 1977). The uptake was also stimulated, though to a lesser extent, by GTP and UTP while other nucleotides (ADP, AMP, cAMP, cGMP, CTP) were ineffective. Maximal stimulation of uptake was reached with magnesium, cobalt and manganese, though the affinity constants of the latter two ions were lower than those of magnesium. A slight stimulation was elicited by zinc, nickel and calcium, while barium, chromium and strontium were found to be ineffective. Potassium and sodium also were ineffective, but lithium inhibited the ATP-Mg^{2+}-dependent uptake of noradrenaline (SLOTKIN et al. 1978a). It is tempting to postulate that the inhibition of the vesicular uptake of noradrenaline may account for the therapeutic effects of lithium on psychiatric disorders (SLOTKIN et al. 1978a). Approximately 10 mM lithium is necessary to inhibit the ATP-Mg^{2+}-dependent uptake of noradrenaline by 50 per cent. Since an intraneuronal concentration of lithium takes place (see WESPI 1969), it seems likely that sufficient intraneuronal concentrations are reached during therapy with lithium to elicit a substantial inhibition of the uptake of noradrenaline.

In synaptic vesicles of rat cerebrum, the uptake of dopamine was stimulated by ATP and GTP, but strongly inhibited by the ATP analogue imidodiphosphate. Uptake was stimulated not only by magnesium but also by calcium, the latter ion possessing approximately one third of the effectiveness of magnesium (TANAKA et al. 1976).

K_m-values of ATP-Mg^{2+}-dependent uptake of catecholamines into vesicles and chromaffin granules are given in Table 1. It is noteworthy that affinities for uptake into vesicles are rather similar in peripheral nerves and central nervous system, whereas affinities are higher for transport into vesicles than for transport into chromaffin granules.

Recently, the kinetic parameters for net uptake of some amines into granule "ghosts" were determined by amperometry. The main advantage of this technique is the on-line determination of the amines in the incubation medium (CARTY et al. 1985). The K_m- and V_{max}-values for catecholamines and 5-HT are fairly similar with those described previously (Table 1). The affinity for dopamine is lower than that for 5-HT, but higher than the affinities for noradrenaline and adrenaline. The same order of affinities was found for the ATP-Mg^{2+}-dependent uptake of 5-HT, dopamine and noradrenaline into striatal vesicles (MATTHAEI et al. 1976; LENTZEN and PHILIPPU 1981) (Table 1).

2. Specificity of the Uptake of Catecholamines: Structure-Uptake Relationship

Even chromaffin granules, which store exclusively catecholamines, lack a strictly specific process of uptake. Under the same experimental conditions, chromaffin granules accumulated higher amounts of 5-HT than of the catecholamines dopamine, noradrenaline and adrenaline. The uptake of tyramine was not enhanced by ATP-Mg^{2+} (CARLSSON et al. 1963).

The uptake of various monoamines into chromaffin granules was extensively investigated by SLOTKIN and KIRSHNER (1971). They found that the number and location of hydroxyl groups greatly influence the ability of the amine to be taken up: a p-hydroxyl group enhances uptake, while addition of a β-hydroxyl group or of a second phenolic group reduces the ATP-Mg^{2+}-dependent uptake into the granules.

ATP-Mg^{2+} also stimulated the uptake of noradrenaline into "ghosts" of chromaffin granules (TAUGNER 1971). The affinity of 5-HT for uptake into "ghosts" was higher than that of noradrenaline (Table 1). On the other hand, ATP-Mg^{2+} also increased the uptake of tyramine (PHILLIPS 1974; DA PRADA et al. 1975), phenylethylamine and histamine (DA PRADA et al. 1975) into "ghosts".

Similarly, ATP-Mg^{2+} activated the uptake of 5-HT (MATTHAEI et al. 1976; SLOTKIN et al. 1978b; HALARIS and DEMET 1978), dopamine, noradrenaline and octopamine (PHILIPPU and BEYER 1973; MATTHAEI et al. 1976) into synaptic vesicles of caudate nucleus and whole brain. A first attempt to study the uptake of tyramine into synaptic vesicles of the caudate nucleus revealed that the transport of this indirectly acting sympathomimetic amine was not stimulated by ATP-Mg^{2+} (MATTHAEI et al. 1976). This finding was surprising, because octopamine, with an additional hydroxyl group on the side chain, showed a good affinity for the ATP-Mg^{2+}-dependent system (Table 1). In

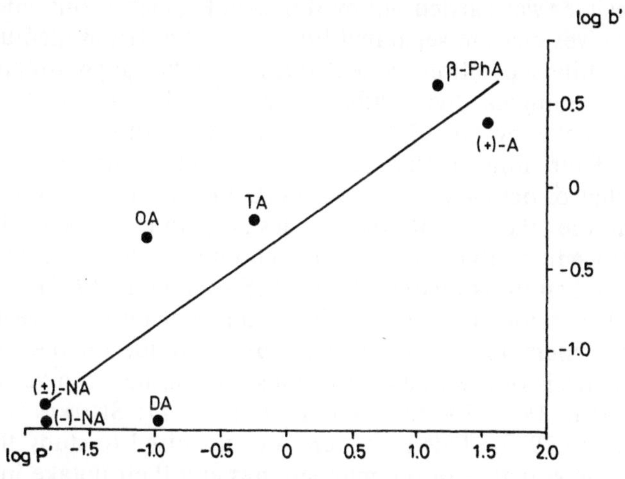

Fig. 1. Relation between the lipophilicity of phenethylamines and their non-saturable, ATP-Mg^{2+}-independent uptake into synaptic vesicles of the caudate nucleus. Ordinate: slopes b' (ratio of the amount of the amine taken up as pmol·mg protein^{-1}·min^{-1}/concentration of the non-protonated amine in the incubation medium as pmol·ml^{-1}); abscissa: log of the partition coefficients of the non-protonated part of the amines, NA: noradrenaline, DA: dopamine, OA: octopamine, TA: tyramine, β-PhA: β-phenethylamine, A: amphetamine. Mean values of 6–13 experiments, $r = 0.905$, $p < 0.01$. LENTZEN and PHILIPPU (1981), with permission

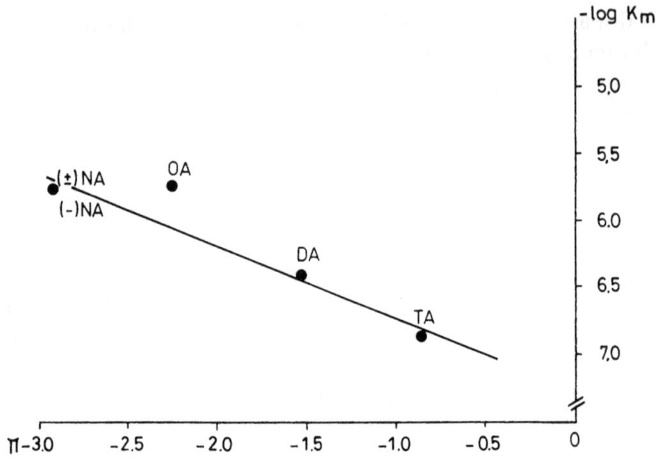

Fig. 2. Relation between substituted group constants and affinities of phenethylamines for uptake into synaptic vesicles of the caudate nucleus in the presence of ATP-Mg^{2+}. Ordinate: $-\log K_m$(M), abscissa: π-values. π-Value of the benzene ring was taken as zero, π-values for the single substituted groups were taken from LEO et al. (1971). For abbreviations see legend to Fig. 1. Mean values of 3-13 experiments. $r = 0.951$, $p < 0.05$. LENTZEN and PHILIPPU (1981), with permission

these studies, the synaptic vesicles were isolated from the medium after incubation by ultracentrifugation. Different results were obtained, when separation of the vesicles was carried out by filtration through membrane filters. By this procedure, vesicles are separated from their incubation medium within a few seconds, while separation by centrifugation takes approximately 90 min. Filtration of the samples after incubation revealed that tyramine is also taken up into the vesicles by an ATP-Mg^{2+}- and temperature-dependent process. The affinity of tyramine to the ATP-Mg^{2+}-dependent uptake site was even higher than that of octopamine (Table 1) (LENTZEN and PHILIPPU 1977). Apparently, tyramine taken up into the vesicles is again liberated during the centrifugation procedure. Experiments with chromaffin granules also revealed that tyramine taken up is readily liberated (KNOTH et al. 1982).

The above mentioned findings indicate that separation of vesicles by filtration through membrane filters is a useful technique for the investigation and kinetic analysis of the transport of biogenic amines into vesicles (KLEIN and LAGERCRANTZ 1971; LENTZEN and PHILIPPU 1977; SLOTKIN et al. 1978b; FERRIS and TANG 1979). This procedure was also used to study the relationship between the structure of the monoamines and their uptake into synaptic vesicles of the caudate nucleus. In the absence of ATP-Mg^{2+}, phenethylamines (noradrenaline, dopamine, octopamine, tyramine, amphetamine, β-phenylethylamine) were taken up into the vesicles, the rates of uptake of the undissociated amines being proportional to their lipophilic character (Fig. 1). ATP-Mg^{2+} enhanced the uptake of amines, provided they possessed at least one hydroxyl-group on the benzene ring. An additional hydroxyl-group on the ring or on the side chain decreased the affinity for uptake (LENTZEN and PHI-

LIPPU 1981). Hence, the structure-uptake relationship in synaptic vesicles of the brain seems to be very similar to that in chromaffin granules (SLOTKIN and KIRSHNER 1971; see above). A correlation was found between K_m-values of the phenethylamines for ATP-Mg^{2+}-dependent uptake and their substituent constants (Fig. 2), indicating that the affinities of the amines are dependent on their lipophilicity (LENTZEN and PHILIPPU 1981).

3. Stereospecificity of the Uptake of Catecholamines

In the absence of ATP-Mg^{2+}, the uptake of noradrenaline into "ghosts" of chromaffin granules (TAUGNER 1972a) and synaptic vesicles of the brain (LENTZEN and PHILIPPU 1981) is not stereospecific.

CARLSSON et al. (1963) and DA PRADA et al. (1975) reported that ATP-Mg^{2+} elicited a preferential uptake of (−)-noradrenaline into chromaffin granules. According to TAUGNER (1972a), the ATP-Mg^{2+}-dependent uptake of (−)-noradrenaline into "ghosts" of chromaffin granules is 15–20% higher than that of the (+)-isomer. When inhibitors of the ATP-Mg^{2+}-dependent uptake of noradrenaline into "ghosts" of chromaffin granules were used, it was found that the affinity of the (−)-isomer to the transport system was approximately 4 times higher than that of the (+)-isomer (PHILLIPS 1974). Stereospecificity for uptake into chromaffin granules seems to be less pronounced than for uptake into synaptic vesicles of sympathetic nerves (Table 1), in which the ratio of the ATP-Mg^{2+}-dependent uptake of (−)-noradrenaline versus that of (+)-noradrenaline is approximately 9 (EULER and LISHAJKO 1967).

In synaptic vesicles of the heart, complete stereospecificity was claimed for the ATP-Mg^{2+}-dependent uptake of noradrenaline, since unlabelled (±)-noradrenaline was half as effective as unlabelled (−)-noradrenaline in inhibiting the uptake of radioactive (−)-noradrenaline (BAREIS and SLOTKIN 1979). However, for the investigation of stereoselectivity (−)- and (+)-isomers should be used rather than (−)- and (±)-compounds. In the latter case, even if a complete stereospecificity exists the (−)-inhibitor will be only twice as effective as the racemate in inhibiting the uptake of the (−)-agonist. Thus, even slight errors may lead to erroneous conclusions concerning the degree of stereospecificity.

The K_m-value for uptake of (−)-noradrenaline into "ghosts" of synaptic vesicles isolated from the heart (Table 1) is one third of the K_m-value for uptake of (±)-noradrenaline (ANGELIDES 1980). The K_m-value for uptake of (−)-noradrenaline into synaptic vesicles of the brain is approximately half that of (+)-noradrenaline, showing that the affinity of the (−)-isomer is twice as high as that of (+)-noradrenaline (MATTHAEI et al. 1980, 1988).

Taken together, these results show that the ATP-Mg^{2+}-dependent uptake of catecholamines is stereospecific, though the degree of stereospecificity varies according to origin and state (intact or lysed) of the vesicles and the procedure used for determination of the stereospecificity as well.

4. Uptake of Catecholamines into Different Synaptic Vesicles of the Brain

Extensive comparative studies of the uptake of catecholamines into synaptic vesicles originating from various regions of the brain are scarce. Vesicles of the caudate nucleus show a similar ratio of uptake dopamine/noradrenline to that of vesicles isolated from cerebral cortex (SLOTKIN et al. 1978c). The K_m-values for ATP-Mg^{2+}-dependent uptake of noradrenaline into vesicles isolated from three brain regions (cerebral cortex, striatum and rest of brain) are almost identical (SLOTKIN et al. 1978b).

Monoamines, which are supposed to be neurotransmitters in the central nervous system, possess similar affinities for the energy-dependent uptake into synaptic vesicles of the brain. As already mentioned, the K_m-values of 5-HT, dopamine and noradrenaline for uptake into synaptic vesicles of the caudate nucleus are comparable (PHILIPPU and BEYER 1973; PHILIPPU and MATTHAEI 1975). These similarities led to the view that the synaptic vesicles of the striatum take up more than one neurotransmitter. Analysis of the ATP-Mg^{2+}-dependent uptake of dopamine and 5-HT by the ABC-test (SCRIVER and WILSON 1964) (analysis of the inhibition of the uptake of two different molecules by a third one) revealed that tyramine competitively inhibited the uptake of dopamine and 5-HT, the K_i-values being similar for the two agonists. Moreover, the uptake of noradrenaline was also inhibited in a competitive way. Since the population of particles used for these experiments contained noradrenergic, dopaminergic and serotoninergic vesicles, it seems that vesicles originating from different neurons possess similar uptake properties and are able to take up more than one neurotransmitter. Furthermore, the results indicate that noradrenaline, dopamine and 5-HT share the same transport system for uptake into the vesicles (MATTHAEI et al. 1976; LENTZEN and PHILIPPU 1977). Similar conclusions were drawn by FERRIS and TANG (1979), who studied the potencies of isomers of amphetamine, methylphenidate and deoxypipradol in inhibiting the uptake of noradrenaline and dopamine into synaptic vesicles of the brain.

In an elegant attempt to characterize those synaptic vesicles of the brain which take up monoamines, SLOTKIN et al. (1978b) used the neurotoxins 6-hydroxydopamine and 5,6-dihydroxytryptamine to destroy catecholaminergic and serotoninergic neurons, respectively. Surprisingly, the intracisternal application of 5,6-dihydroxytryptamine in rats did not affect the ATP-Mg^{2+}-dependent uptake of either noradrenaline or 5-HT, while 6-hydroxydopamine reduced the uptake of both amines into the vesicles. Therefore, noradrenaline and 5-HT seem to be taken up into synaptic vesicles of catecholaminergic neurons rather than into vesicles of serotoninergic neurons. The results also support the view that different monoamines use a common carrier for transport into synaptic vesicles of the brain.

5. Effects of Drugs on the Uptake of Catecholamines

Drugs affecting ATPase activity, electron transfer chain and electrochemical proton gradient will be discussed in Sections C.IV.1, C.IV.3 and C.VI.3, respectively.

a. Inhibition by Drugs which Deplete Catecholamines

As early as 1962, it was demonstrated that reserpine inhibits the uptake of catecholamines into chromaffin granules (KIRSHNER 1962; CARLSSON et al. 1963; JONASSON et al. 1964; see PHILIPPU 1976). Since reserpine and prenylamine elicit a specific inhibition of the uptake of catecholamines without affecting their transport through plasma membranes, these drugs possess a particular interest as pharmacological tools for characterizing uptake into subcellular, amine-storing vesicles. In vitro, reserpine competitively inhibits the ATP-Mg^{2+}-dependent uptake of catecholamines into chromaffin granules (JONASSON et al. 1964). In synaptic vesicles of the hypothalamus and caudate nucleus of the pig it was also shown that reserpine and prenylamine competitively inhibit the uptake of catecholamines, the K_i-values being 10 nM and 20 nM, respectively (PHILIPPU et al. 1969, 1975). A K_i of 3 nM was found for inhibition by reserpine of the uptake of noradrenaline into vesicles of the rat brain (SEIDLER et al. 1978) and of 20 nM for uptake into membranes of chromaffin granules (KANNER et al. 1979). In contrast to this competitive inhibition by reserpine of the ATP-Mg^{2+}-dependent uptake of noradrenaline, pretreatment of rats with reserpine elicits a non-competitive inhibition of the ATP-Mg^{2+}-dependent uptake of noradrenaline into isolated synaptic vesicles of the brain (SEIDLER et al. 1978).

The binding of reserpine to heart microsomes persists for several weeks (GIACHETTI and SHORE 1975), indicating the irreversible character of the drug binding. Similarly, the ATP-Mg^{2+}-dependent uptake of noradrenaline into synpatic vesicles of the brain was found to be reduced for up to three weeks after reserpine treatment (SEIDLER et al. 1978). The action of reserpine on intact chromaffin granules and on isolated membranes of chromaffin granules also seems to be irreversible, since the inhibitory effect of the drug was not lessened by washing (KIRSHNER 1962; KANNER et al. 1979). The reserpine-induced inhibition of the uptake of amines into chromaffin granules was abolished when the washing medium contained phospholipid vesicles. Probably, the particles provide a hydrophobic milieu that removes reserpine from the membranes of the chromaffin granules (KANNER et al. 1979).

The requirement of ATP-Mg^{2+} for amine uptake led WEAVER and DEUPREE (1982) to investigate whether these substances influence the binding of ^{3}H-reserpine to chromaffin granule "ghosts". They found that, in the presence of ATP-Mg^{2+}, reserpine specifically bound to a site which seemed to be associated with the catecholamine carrier (WEAVER and DEUPREE 1982; DEUPREE and WEAVER 1984). SCHERMAN and HENRY (1984) also found that ATP enhanced the binding of reserpine to the membrane, but that the drug is not accumulated inside the granule "ghosts". In the absence as in the presence of ATP, reserpine bound to two classes of binding sites, the high affinity binding site being involved in the monoamine carrier activity. Since the density of this binding site was lower than that of the binding site with low affinity, the authors propose the existence of two forms of the carrier; an active form possessing both binding sites and an inactive form devoid of the high affinity binding site (SCHERMAN and HENRY 1984).

b. Inhibition by Monoamines

Besides tyramine (CARLSSON et al. 1963; EULER and LISHAJKO 1968; MATTHAEI et al. 1976; see PHILIPPU 1976), several monoamines are able to inhibit the ATP-Mg^{2+}-dependent uptake of amines into chromaffin granules and synaptic vesicles. These compounds inhibit the uptake of catecholamines competitively, provided their amino nitrogen is separated from a hydrophobic moiety by at least a 2-carbon bridge. The inhibitory potencies are β-carbolines〉indolealkylamines〉phenylalkylamines〉n-alkylamines. Though similar inhibitory potencies for the different classes of drugs were found in chromaffin granules and in synaptic vesicles of the brain, detailed analysis of the structure-activity relationship for each drug revealed that catecholamine transport of synaptic vesicles differs from that of chromaffin granules. Generally, the uptake sites of the synaptic vesicles of the brain seem to be less selective than those of chromaffin granules (SLOTKIN et al. 1975, 1979a).

Though ATP-Mg^{2+} does not enhance the uptake of (+)-amphetamine into synaptic vesicles of the caudate nucleus (LENTZEN and PHILIPPU 1981), this amine inhibits competitively the ATP-Mg^{2+}-dependent uptake of dopamine and noradrenaline. The K_i value was approximately 2.7 μM for both amines (PHILIPPU and BEYER 1973).

R-(−)-Amphetamine also inhibits the ATP-Mg^{2+}-dependent uptake of dopamine and noradrenaline into synaptic vesicles of the brain; the IC_{50} value of the R-(−)-enantiomer is approximately ten times higher than that of the S-(+)-form. S-(+)-Deoxypipradol and 1R:2R-methylphenidate are approximately as potent as R-(−)-deoxypipradol and 1R:2S-methylphenidate, respectively, in inhibiting the ATP-Mg^{2+}-dependent uptake of the amines. Thus, the receptor site of the amine pump seems to be selective for the stereochemical configuration of the alpha-carbon of amphetamine, but not for that around the analogous carbon of deoxypipradol and methylphenidate (FERRIS and TANG 1979).

6. Ontogenesis of the Uptake of Catecholamines into Synaptic Vesicles

Studies on the postnatal development of the synaptosomes of the rat brain revealed that the number of synaptic vesicles increases from new-born to adult animals (GLEZER 1969; JONES and REVELL 1970; KANERVA et al. 1975, 1976). In contrast to the development of the uptake mechanism of noradrenaline into synaptosomes that reaches its maximum 17–18 days after birth (COYLE and AXELROD 1971, KIRKSEY et al. 1978), the full development of the ATP-Mg^{2+}-dependent uptake of catecholamines into synaptic vesicles is not achieved even at 38 days of age. While the V_{max}-value increased progressively until maturity of the animals, the K_m-value and the inhibition of the uptake by reserpine were very similar in vesicles of 3 days-old and adult rats. Constancy of K_m and steady increase of V_{max} indicate that the vesicles of new-born rats possess a fully functioning uptake system for catecholamines, whereas the number of vesicles increases during development (KIRKSEY et al. 1978). In connection with this observation it is noteworthy that treatment of pregnant or nursing rats with methadone retards brain development, as well as synaptosomal and ATP-Mg^{2+}-dependent uptake of catecholamines and 5-HT (SLOTKIN et al. 1979b).

II. Uptake of Nucleotides

When perfused adrenal glands (STEVENS et al. 1972, 1975) or vasa deferentia (ABERER et al. 1979) were "pulse" labelled with precursors of ATP, a specific labelling of the ATP in the catecholamine-storing particles of the organs was found. Although earlier attempts to demonstrate an uptake of nucleotides under *in vitro* conditions were either unsuccessful (KIRSHNER 1962; BAUMGARTNER et al. 1973) or revealed a slight uptake of labelled ATP (CARLSSON et al. 1963), more recent results indicate that ATP can penetrate the membrane of chromaffin granules, though a considerable part is bound to the membranes (TAUGNER 1975). Experiments with chromaffin granules revealed that isolated storage particles take up not only radioactive ATP (KOSTRON et al. 1977a), but also ADP, AMP, GTP, UTP (ABERER et al. 1978; WEBER and WINKLER 1981), phosphoenolpyruvate, phosphate and sulphate (WEBER et al. 1983). Hence, the nucleotide carrier, like the monoamine carrier, does not possess a high specificity. The uptake of these substances is dependent on temperature and Mg^{2+} and obeys Michaelis-Menten kinetics. The nucleotides were incorporated into the soluble content of the granules, since they were completely released by hypo-osmotic shock. The uptake of nucleotides, phosphoenolpyruvate, PO_4^{3-} and SO_4^{2-} into the vesicles was competitively inhibited by atractyloside (KOSTRON et al. 1977a; ABERER et al. 1978; WEBER et al. 1983), which also blocks the uptake of ATP into mitochondria (KLINGENBERG 1970; VIGNAIS 1976).

An uptake of ATP and other nucleotides into mitochondria contaminating the population of the chromaffin granules is unlikely, because atractyloside inhibits the uptake of nucleotides into mitochondria in a non-competitive way (VIGNAIS et al. 1973). Moreover, the uptake of AMP into chromaffin granules is inhibited by atractyloside, while AMP has no affinity to the mitochondrial carrier of nucleotides (KLINGENBERG 1970; VIGNAIS 1976). The inhibition of the uptake of nucleotides into vesicles by NEM (N-ethylmaleimide) or EDTA and the stimulation by Mg^{2+} indicate that the uptake of nucleotides is dependent on ATPase activity. This view is strengthened by the finding that analogues of ATP (adenyl imidodiphosphate and adenylyl (β, γ-methylene)diphosphonate), which inhibit the ATPase of the granules, reduce the uptake of all three nucleotides. The uptake is also inhibited by uncouplers of the oxidative phosphorylation. Therefore, both catecholamines and nucleotides are taken up into chromaffin granules and synaptic vesicles by energy-dependent and carrier-mediated processes.

According to ABERER et al. (1978), ATP is taken up into chromaffin granules as the intact molecule and its uptake is dependent on ATPase activity. In experiments of TAUGNER et al. (1979) atractyloside failed to inhibit the uptake of nucleotides, which were taken up into the granules at a low rate even in the absence of Mg^{2+} (i.e., when ATPase was inactive). No satisfactory explanation of the differing results can be given now. The authors proposed that ATP passes the membrane of chromaffin granules by a unidirectional transphosphorylating process (TAUGNER et al. 1979; TAUGNER and WUNDERLICH 1979): ATP binds to an enzyme which is on the outside of the membrane and this binding facilitates the binding of ADP to an enzyme on the inner surface of the membrane. Hence, a transfer of the terminal phosphate from ATP outside to ADP inside of the membrane takes place.

III. Uptake of Ascorbate

Chromaffin granules contain high concentrations (13 m*M* to 22 m*M*) of ascorbate so that a concentration gradient of at least 4 exists between intravesicular space and cytosol (TERLAND and FLATMARK 1975; INGEBRETSEN et al. 1980). Ascorbate of the chromaffin granules seems to be of importance as cofactor of DBH for the conversion of dopamine to noradrenaline (LEVIN et al. 1960). Labelled ascorbate is taken up by chromaffin granules, provided cytochrome c is present in the incubation medium. Apparently, in the presence of cytochrome c, dehydroascorbate is formed and taken up by an energy-independent process (TIRRELL and WESTHEAD 1979). In spite of the uptake, the concentration of dehydroascorbate in the chromaffin granules is very low (INGEBRETSEN et al. 1980). It seems that dehydroascorbate taken up into the granules is again reduced to ascorbate, which cannot permeate the granular membrane.

IV. Enzymes Involved in the Uptake of Catecholamines, Nucleotides and Ascorbate

1. ATPase

The amine storing particles of the adrenal medulla contain a specific ATPase (HILLARP 1958a), and in a purified vesicle fraction a Mg^{2+}-dependent ATPase activity was found (BANKS 1965). Since the ATPase activity of the catecholamine storing particles is inhibited by NEM at concentrations which abolish amine uptake (KIRSHNER 1965; TAUGNER and HASSELBACH 1968), it was postulated that uptake of catecholamines into chromaffin granules depends on the function of Mg^{2+}-activated ATPase (TAUGNER 1971). The association of ATPase with chromaffin granules has been questioned (LADURON et al. 1976). However, APPS and REID (1977) and KONINGS and DE POTTER (1980) reported that ATPase activity is associated with the granule membrane. The histochemical evidence for this was presented by BENEDECZKY and CARMICHAEL (1980). Electrophoretic studies revealed that at least 3 types of subunits of the ATPase are present, which have electrophoretic mobilities similar to the major subunits of the mitochondrial ATPase. The enzymes from mitochondria and chromaffin granules differ in their behaviour towards some inhibitors, such as aureovertin and oligomycin (APPS and GLOVER 1978; APPS and SCHATZ 1979).

The ATPase of chromaffin granules can be solubilized by cholate. The activity of the enzyme, which is reduced by removing the lipids, is restored during reconstitution with phospholipids of chromaffin granules or of soybeans (BUCKLAND et al. 1979). Solubilization of membranes of chromaffin granules and subsequent centrifugation revealed the presence of two different ATPase enzymes: a heavier enzyme identical to the mitochondrial ATPase and an ATPase that seems to represent the ATPase activity of chromaffin granules (CIDON and NELSON 1982, 1983). More recently, two ATPases were isolated from membranes of chromaffin granules. ATPase I is involved in the uptake of catecholamines, while the function of ATPase II is still obscure (PERCY et al. 1985).

The affinity of reconstituted phospholipid vesicles for noradrenaline is similar to that of the "ghosts" (ISAMBERT and HENRY 1981). Experiments with highly purified membranes of chromaffin granules revealed that the ATPase has a pH optimum of 7.4 (JOHNSON et al. 1982), thus confirming findings of TAUGNER (1971).

In intact chromaffin granules, ATP-Pi exchange and ATP formation from ADP and Pi were observed. Of these reactions, the first one is independent of Mg^{2+} and has its pH optimum at 7.0, while the second one is Mg^{2+}-dependent and has its pH optimum higher than 8.0. Hence, the ATPase of chromaffin granules seems to be reversible; like the ATPase of mitochondria, it is able to form ATP from ADP and inorganic phosphate, as well as to exchange the phosphate group of ATP with inorganic phosphate (TAUGNER et al. 1980).

2. Phosphoryl Group-Transferring Enzymes

ATP may undergo several transformations during its transport into the granules. As already mentioned, membranes of granules as well as their soluble proteins possess an enzyme, which can transfer the terminal phosphate from ATP to ADP (TAUGNER 1974; TAUGNER and WUNDERLICH 1979; TAUGNER et al. 1979). Like ATPase, this enzyme is also activated by Mg^{2+}, but the activities of the two enzymes are influenced differently by thiol reagents or acetone extraction.

Additionally, chromaffin granules possess a phosphatidylinositol kinase (TRIFARÓ and DWORKIND 1971; PHILLIPS 1973; MULLER and KIRSHNER 1975), which phosphorylates the membrane by an ATP-dependent process. The phosphorylation of the membrane is inhibited by ADP, NEM and gramicidin. It should be pointed out that this kinase is not identical with the phosphoryl group-transferring enzyme, since high concentrations of magnesium (100 m*M*) activate phosphorylation of the membrane by the kinase, but inhibit the activity of the phosphoryl group-transferring enzyme. On the other hand, phosphatidylinositol kinase is different from ATPase, because the latter enzyme is activated by low concentrations (1 m*M*) of magnesium (MULLER and KIRSHNER 1975).

3. Electron-Transferring Enzymes

The membrane of the chromaffin granules contains electron-transferring enzymes such as NADH: (acceptor)oxidoreductase, flavoprotein(s) and cytochrome b_{561} (FLATMARK et al. 1971; TERLAND and FLATMARK 1973, 1980). Indeed, in the presence of ferricyanide as an artificial electron acceptor (FLATMARK et al. 1971) or dopamine (BASHFORD et al. 1976), chromaffin granules oxidize NADH and this reaction is abolished by p-chloromercuribenzoate, potassium cyanide (FLATMARK et al. 1971), antimycin A, oligomycin and rotenone (BASHFORD et al. 1976). It is noteworthy that these compounds, which inhibit electron-transferring enzymes, also inhibit the ATP-Mg^{2+}-dependent uptake of catecholamines into synaptic vesicles (EULER and LISHAJKO 1969) and "ghosts" of chromaffin granules (BASHFORD et al. 1976).

Reducing equivalents seem to be necessary for the synthesis of catecholamines (SKOTLAND and LJONES 1980), as well as to keep them in a reduced state. Furthermore, for the uptake of catecholamines and ascorbate (see Sect. C.III) into granules reducing equivalents are required, but their precise role is still unclear. The reducing equivalents are taken up into the granules by a transmembrane electron carrier driven by the electrochemical gradient (see Sect. C.VI.2) (NJUS et al. 1983).

V. Uptake of Calcium

Chromaffin granules take up $^{45}Ca^{2+}$ by a saturable and temperature-dependent process. These properties of uptake as well as inhibition by strontium indicate that calcium is taken up by a carrier-mediated process, for example by facilitated diffusion (KOSTRON et al. 1977b; GRATZL et al. 1981; KRIEGER-BAUER and GRATZL 1982). The uptake of calcium is also inhibited by sodium, which additionally releases calcium from the vesicles. Hence, a sodium-calcium-exchange seems to operate (GRATZL et al. 1981; PHILIPPS 1981; KRIEGER-BAUER and GRATZL 1982, 1983). ATP-Mg^{2+} did not seem to influence the uptake of calcium into intact chromaffin granules (KOSTRON et al. 1977b; GRATZL et al. 1981; KRIEGER-BAUER and GRATZL 1982) and granule "ghosts" (PHILLIPS 1981). Similarly, the uptake of calcium into synaptic vesicles of the brain did not seem to be dependent on ATP. The slight enhancement of the uptake of calcium by ATP was attributed to contamination of the synaptic vesicles by mitochondria and microsomes (DIAMOND and GOLDBERG 1971; TSUDZUKI 1979).

On the other hand, ATP-dependent uptake of calcium into noradrenaline-storing vesicles of splenic nerves (BURGER and BELLERSHEIM 1976) and into intact chromaffin granules (HÄUSLER et al. 1981; NIEDERMAIER and BURGER 1981) has been reported. Even if the rate of uptake of calcium into vesicles is much lower than that into mitochondria, the uptake does not seem to be due to contamination of the vesicles by these particles. Indeed, incubation with various ions, inhibitors or other drugs influences the uptake of calcium into chromaffin granules and synaptic vesicles in a different way than the uptake into mitochondria and microsomes (BURGER and BELLERSHEIM 1976; NIEDERMAIER and BURGER 1981; HÄUSLER et al. 1981; BURGER et al. 1984).

The ATP-Mg^{2+}-dependent uptake of calcium into chromaffin granules was confirmed by GRAFENSTEIN et al. (1983). They found that, at low concentrations of calcium in the incubation medium ($< 10^{-8}$ *M*), the ATP-Mg^{2+}-dependent uptake of calcium into mitochondria is negligible, while that into chromaffin granules is considerable. Thus, under certain experimental conditions it is possible to distinguish the ATP-Mg^{2+}-dependent uptake of calcium into mitochondria from that into chromaffin granules. These findings show that chromaffin granules and synaptic vesicles take up calcium by an ATP-Mg^{2+}-dependent process. Additionally granules (and possibly vesicles) take up calcium by a process which is ATP-independent, but sodium-dependent.

It is worth mentioning that membranes of chromaffin granules bind magnesium to a higher extent than calcium, while ATP increases the binding of the former ion only (BALZER et al. 1984).

VI. Utilization of Energy Required for Uptake

1. pH of Chromaffin Granules

Substances which are useful tools in the investigation of biochemical properties of mitochondria and chloroplasts (such as 1-anilinonaphthalene-8-sulfonic acid (ANS) and methylamine), as well as nuclear magnetic resonance techniques made it possible to determine the internal pH of the chromaffin granules. When chromaffin granules were incubated with the fluorescent ANS, fluorescence was enhanced after addition of ATP. Presumably, ATP generated a proton gradient (see Sect. C.VI.2.b). The enhancement of fluorescence by ATP was inhibited by N-(3-t-butyl-5-chlorosalicyl)-2-chloro-4-nitroanilide (S-13), bis-hexafluoroacetone and carbonyl cyanide-p-trifluoromethoxyphenylhydrazone (FCCP), which render the membrane permeable to protons. In the presence of these agents, lowering of pH of the suspending buffer below 5.5 enhanced fluorescence of the probe, while at pH above 5.5 fluorescence was reduced. Assuming that the proton ionophores lead to equilibrium of protons between suspending buffer and chromaffin granules, the pH of the intravesicular space should be approximately 5.5 (BASHFORD et al. 1975a). This finding was confirmed when the pH of the vesicles was determined by using labelled methylamine. According to SCHULDINER et al. (1972), only the neutral form of this substance is freely permeable across membranes. Thus, when an equilibrium is reached, the ratio of the intravesicular protonated form to that in the medium will reflect the distribution of the protons between inside and outside of the granules according to $AH^+_{in}/AH^+_{out} = H^+_{in}/H^+_{out}$. Changes in pH of the incubation medium produced equivalent changes in pH gradient between outside and inside of the particles (ΔpH). At pH 5.5 the concentration of labelled methylamine in the medium was equal to that in the granules indicating that the pH of granules is 5.5 (JOHNSON and SCARPA 1976a). In agreement with these observations, the pH of chromaffin granules was found to be 5.65 ± 0.15 by nuclear magnetic resonance and it declines by 0.5 pH units on addition of ATP (see CASEY et al. 1977; NJUS et al. 1978; POLLARD et al. 1979). Probably, the drop of pH by ATP is due to transport of H^+ into the granules.

2. Ion Movements Across the Chromaffin Granule Membrane

a. Ion Permeability of the Membrane

In the presence of chloride, ATP releases the content of the chromaffin granules (OKA et al. 1965; TRIFARÓ and POISNER 1967; LISHAJKO 1969). The influence of various anions on the release of catecholamines from chromaffin granules was investigated by TAUGNER (1972b), who found the following order of effectiveness: $F^- < Cl^- < Br^- < I^- = CH_3CO_2^- < CCl_3CO_2^- < SCN^-$. These ions were also effective in the absence of ATP, though to a lesser extent. ATP is probably liberated from chromaffin granules during incubation, thus producing release of the granule contents (CASEY et al. 1976). The same order of effectiveness of ions was found by DOLAIS-KITABGI and PERLMAN (1975), who used the potassium ionophore valinomycin instead of ATP. They concluded that valinomycin-mediated transport of potassium into the chromaffin granules leads to release of their contents, provided that permeable anions are present. Indeed, the order of effectiveness of the various anions parallels their degree of permeability through membranes (PHILLIPS 1977). Valinomycin was ineffective in the presence of isethionate, which is a large, polar anion and does not readily pass membranes. The concomitant release of catecholamines, proteins and nucleotides indicates that, in the presence of valinomycin and other ionophores, lysis of the chromaffin granules occurs (DOLAIS-KITABGI and PERLMAN 1975; PAPADOPOULOU-DAIFOTIS et al. 1977). In this connection it should be mentioned that the membrane of the chromaffin granules is highly impermeable to phosphate and sulphate ions and to cations, such as H^+, K^+, Na^+, Mg^{2+}, Ca^{2+} (CASEY et al. 1976; JOHNSON and SCARPA 1976b; PHILLIPS 1977).

The dependence of the ATP-induced release on the permeability of the membrane of chromaffin granules to anions as well as the finding that an increase in the osmotic pressure of the incubation medium inhibits the release, indicate that ATP also leads to lysis of the chromaffin granules. Hydrolysis of ATP is needed to translocate protons into the chromaffin granules (see Sect. C.VI.2.b); influx of protons is accompanied by an influx of permeable anions to neutralize the intragranular charge of protons (CASEY et al. 1976). In turn, these anions increase the osmotic pressure inside the granules, thus leading to influx of water and lysis of the particles.

b. Generation of an Electrochemical Gradient Across the Membrane

As already mentioned, a high ΔpH exists across the membrane, while the permeability for protons is very low. Therefore, in the absence of ionophores or ATP, the diffusion potential is too low to be measurable. In the presence of FCCP the membranes of the chromaffin granules become freely permeable to protons (see Sect. C.VI.1). However, FCCP does not affect ΔpH, since efflux of protons establishes an electrical potential which limits their further movement. On the other hand, when chromaffin granules were incubated with FCCP in the presence of potassium and valinomycin, influx of potassium was associated with efflux of protons; thus ΔpH was reduced. Similar effects were elicited by nigericin, which catalyses the electroneutral exchange K^+/H^+ (JOHNSON and SCARPA 1976a).

Incubation of chromaffin granules with ATP in the presence of a permeant anion leads to influx of protons and a drop of intravesicular pH (see Sect. C.VI.1.). Nuclear magnetic resonance studies showed that, on addition of ATP, the ratio "proton translocation/ATP hydrolysis" is approximately 1 (NJUS et al. 1978).

In "ghosts" of chromaffin granules uptake of protons was also observed. The influx of protons was dependent on ATP hydrolysis, since NEM inhibited ATPase activity and uptake of protons to the same extent (FLATMARK and INGEBRETSEN 1977). Therefore, hydrolysis of ATP leads to translocation of protons. In "ghosts" the ratio "proton translocation/ATP hydrolysis" was found to be 1.7 (JOHNSON et al. 1982).

Consequently, incubation of "ghosts" of chromaffin granules with ATP and a permeant anion leads to development of ΔpH. On the other hand, in the absence of a permeant anion in the incubation medium, ATP does not influence the proton gradient. The pH of the intragranular space remains unchanged, but ATP shifts the electrical potential ($\Delta\psi$) of the membrane from -70 mV (according to HOLZ 1979) or -90 mV (according to JOHNSON and SCARPA 1979) to $+50$ mV (HOLZ 1978; JOHNSON et al. 1979). Addition of ammonia (NH_3), which permeates membranes in unchanged form and associates with a proton to form ammonium (NH_4^+), produces a concentration-dependent decrease in the proton gradient. The electrical potential across the membrane of chromaffin granules was increased, suggesting that both ΔpH and $\Delta\psi$ are components of the electrical gradient. Provided that energy is delivered from hydrolysis of ATP, $\Delta\psi$ increases when ΔpH is decreased and vice versa (PHILLIPS and ALLISON 1978; JOHNSON and SCARPA 1979) according to the equation

$$\Delta p = \Delta\psi \frac{2.3\,RT}{\mathrm{F}} \cdot \mathrm{pH}$$

(Δp: protonmotive force, R: gas constant, T: absolute temperature, F: Faraday constant)

3. Bioenergetic Aspects of the Uptake of Catecholamines and ATP

The inhibition of the ATP-Mg^{2+}-dependent uptake of catecholamines into chromaffin granules by agents that uncouple oxidative phosphorylation was shown by EULER and LISHAJKO (1969). More recently, it was found that these substances not only inhibit the uptake of catecholamines into chromaffin granules, they also inhibit the enhancement of ANS fluorescence (BASHFORD et al. 1975b) (see Sect. C.VI.I), while they increase ATPase activity. Dicyclohexycarbodiimide (DCCD), which blocks "proton conducting channels" in mitochondria (MITCHELL 1973, 1974) inhibited ATPase activity, ANS fluorescence and uptake of catecholamines (BASHFORD et al. 1976). Similar results were obtained when the uptake of catecholamines in synaptic vesicles of the brain was studied (TOLL et al. 1977; TOLL and HOWARD 1978; MARON et al.

1979b). Furthermore, it was found that ATP-Mg^{2+}-dependent uptake of 5-HT (MARON et al. 1979a) and glutamate (NAITO and UEDA 1985) into synaptic vesicles was inhibited by nigericin; in the presence of potassium ions, nigericin dissipates the transmembrane proton gradient.
Therefore, ATP-induced uptake of catecholamines seems to be dependent on the existence of a proton gradient. If so, the rate of uptake of catecholamines should be correlated with the magnitude of ΔpH and take place even in the absence of ATP. Indeed, when high concentrations of catecholamines were added to isolated chromaffin granules (suspended in a medium of pH higher than 6) uptake of catecholamines and alkalinization of the intravesicular space were found. On the other hand, uptake of catecholamines was observed only when ΔpH was greater than 0.5 units, and decreased when external pH approached internal pH of the chromaffin granules. As expected, destruction of ΔpH by NH_3 decreased the uptake of catecholamines (JOHNSON et al. 1978).

Furthermore, when ΔpH was imposed artificially (e.g., by preparing "ghosts" of chromaffin granules in acidic milieu and resuspending them in a more alkaline medium), adrenaline (SCHULDINER et al. 1978), 5-HT (PHILLIPS 1978) and dopamine (INGEBRETSEN and FLATMARK 1979) were taken up by a carrier-mediated process. Recently, the catecholamine carrier of chromaffin granules was solubilized by using cholate and incorporated into liposomes. Generation of an artificial ΔpH led to reserpine-sensitive accumulation of 5-HT (MARON et al. 1979b).

Uptake of catecholamines was also found, when a diffusion potential was produced by addition of potassium ions and valinomycin to the incubation medium. The potential-driven uptake of catecholamines had a pH optimum similar to that of ATP-induced uptake and it was inhibited by reserpine (NJUS and RADDA 1979).

The question arises, whether ΔpH or $\Delta\psi$ of the electrochemical gradient (see Sect. C.VI.2.b) is responsible for the uptake of catecholamines and 5-HT into chromaffin granules. When membranes of chromaffin granules were suspended in a medium containing chloride, addition of ATP established a proton concentration gradient, but no transmembrane potential was generated. Dissipation of ΔpH by NH_3 or FCCP inhibited uptake of catecholamines and 5-HT. On the other hand, in a medium containing isethionate addition of ATP caused an electrical potential, but no proton gradient was produced. Destruction of $\Delta\psi$ by SCN^- and FCCP decreased uptake of catecholamines and 5-HT. The most pronounced uptake of catecholamines was found when both ΔpH and $\Delta\psi$ were functionally intact (JOHNSON et al. 1979). Therefore, it seems that uptake of amines is dependent on both components of the electrochemical gradient (JOHNSON et al. 1979; APPS et al. 1980b; KANNER et al. 1980) (Fig. 3). FCCP also abolished the ATP-Mg^{2+}-dependent binding of reserpine to granule "ghosts", indicating that binding of the drug requires the existence of an electrochemical gradient (WEAVER and DEUPREE 1982).

High concentrations of chlorpromazine also inhibit the ATP-Mg^{2+}-dependent uptake of catecholamines into chromaffin granules (JONASSON et al. 1964; PLETSCHER 1977) and dissipate the $\Delta\psi$ as well (DRAKE et al. 1979). How-

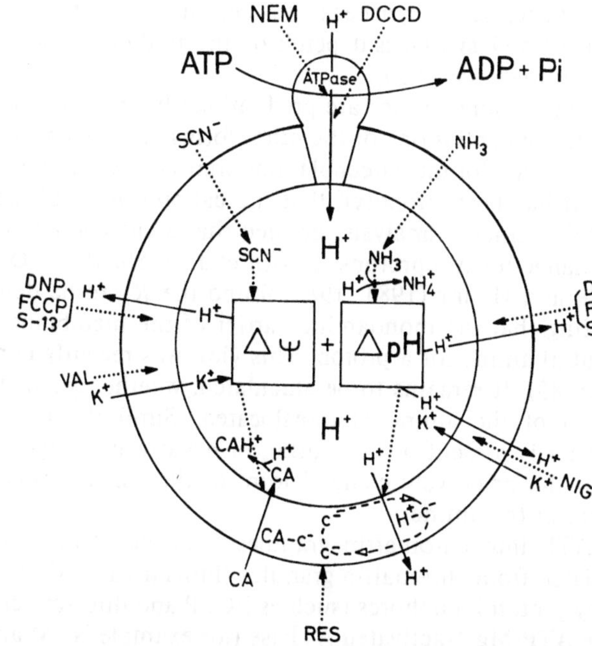

Fig. 3. Schematic representation of bioenergetic processes involved in transport of catecholamines and ATP into vesicles. Arrows: transport of ions or catecholamines, dotted arrows: effects of drugs, $\Delta\psi$ or ΔpH on transport. At the outside of the membrane, ATPase hydrolyses ATP and translocates protons into the vesicles. ATPase is inhibited by NEM or DCCD. Translocation of protons leads to formation of an electrochemical gradient. Ionophores (such as DNP, FCCP, S-13) dissipate the electrochemical gradient. ΔpH is abolished by NH_3, $\Delta\psi$ by permeable anions, such as SCN^-. Valinomycin (VAL) promotes the influx of K^+ (for details see text). Nigericin (NIG) leads to electro-neutral exchange K^+/Na^+ and consequently it dissipates ΔpH. Outside of vesicle, the neutral or negatively charged carrier (c^-) binds an uncharged lipophilic amine; it is still unclear, whether the uncharged amine is present in its neutral, or in its zwitterion form. The complex, which is negatively charged, moves vectorially inward in the presence of a membrane potential (inside positive). At the inner side of the vesicle, dissociation of the amine from the carrier is facilitated by the protonation of the amine. Due to the high concentration of protons inside the vesicle, protonation of the carrier takes place. Thus, each catecholamine is taken up in exchange for one proton. Changes in $\Delta\psi$ and/or ΔpH influence the movement of the carrier and consequently the transport of catecholamines. Reserpine (RES) binds to the carrier, thus inhibiting the transport of catecholamines

ever, the concentrations of the agent needed to elicit these effects are too high to permit any meaningful speculation concerning the mechanism of action of chlorpromazine or other phenothiazines as antipsychotic agents.

$\Delta\psi$ is indispensable for ATP synthesis in membrane ghosts (Roisin et al. 1980). Addition of ADP-Mg^{2+} generates $\Delta\psi$, enhances the rate of ATP synthesis in intact granules and increases the efflux of catecholamines. On the other hand, high concentrations of adrenaline reduce the efflux of catecholamines, dissipate $\Delta\psi$ and decrease the synthesis of ATP. These observations led to the view that the rate of ATP synthesis depends on the gradient of catecholamines (TAUGNER and WUNDERLICH 1984).

Below pH 8.0 the monoamines are predominantly present in the cationic form, whereas the concentrations of the other forms are low and vary according to pH. Since the K_m for the catecholamine transport was found to be independent of pH, it has been proposed that the cationic form of the amine is translocated. Stoichiometric analysis revealed that a catecholamine cation is taken up in exchange for two protons (KNOTH et al. 1980, 1981). On the other hand, SCHERMAN and HENRY (1981; 1983) found the K_m to be dependent on pH thus indicating that the monoamine carrier of the membrane binds and transports a neutral amine for a proton. This view was recently confirmed by KOBOLD et al. (1985). It remains to be elucidated, whether the neutral or the zwitterionic form of the amine is translocated. Similarly, it is not clear whether the carrier is present in its neutral, or negatively charged form. It is conceivable that the negatively charged form of the carrier translocates the zwitterionic form of the amine.

Uptake of ATP into chromaffin granules (ABERER et al. 1978) or into membranes isolated from chromaffin granules (PHILLIPS and MORTON 1978) is also inhibited by proton ionophores (such as FCCP and dinitrophenol) and by inhibitors of the ATP-Mg^{2+}-activated ATPase (for example NEM and DCCD), indicating that the uptake of ATP is also linked to a proton pump. The uptake of ATP was inhibited by SCN^-, while NH_3 and nigericin were ineffective. Hence, in contrast to the uptake of catecholamines, the uptake of ATP into chromaffin granules is driven only by the electrical component of the electrochemical gradient (ABERER et al. 1978; WEBER and WINKLER 1981).

Different carriers seem to be involved in the uptake of catecholamines and ATP, since reserpine does not influence the uptake of the latter into chromaffin granules; moreover, atractyloside, which inhibits uptake of nucleotides (however, see TAUGNER et al. 1979), does not affect the uptake of catecholamines (ABERER et al. 1978; PHILLIPS and MORTON 1978).

The ATP-Mg^{2+}-dependent uptake of calcium seems to be driven by ΔpH because the transport of calcium into intact chromaffin granules is abolished by NH_3 (BURGER et al. 1984).

D. Storage

I. Storage in Chromaffin Granules

Chromaffin granules and catecholamine storing vesicles of peripheral nerves, ganglia, vas deferens and brain contain high amounts of ATP (see STJÄRNE 1972; PHILIPPU 1976). Chromaffin granules contain remarkably high concentrations of catecholamines (> 0.6 M) as well. In spite of these high intragranular concentrations of catecholamines and ATP, the osmolality of the soluble components is surprisingly low (BLASCHKO et al. 1956; FALCK et al. 1956). To explain this puzzling "anomaly" (SHARP and RICHARDS 1977), it was proposed that ATP with its four negative charges (LOHMANN 1932) binds four positively charged catecholamines to form osmotically inert complexes (BLASCHKO et al. 1956). This view was supported by the finding that the molar ratio catecholamines/ATP of chromaffin granules is approximately 7 (BLASCHKO et al. 1956; FALCK et al. 1956), indicating that part of the catecholamines is indeed bound to ATP (BLASCHKO et al. 1956). The problem of the low osmolality seemed to be solved definitively, when molar ratios of 4-5 were found in many laboratories (HILLARP 1958b; BANKS 1965; SMITH and WINKLER 1967; TRIFARÓ and DWORKIND 1970; SEN et al. 1979), suggesting that the catecholamines of the granules may be almost totally bound to ATP. Nevertheless, molar ratios substantially higher than 4-5 have been reported repeatedly and indicate that part of the intragranular catecholamines cannot be bound to ATP. Similar observations led HILLARP (1960) to the conclusion that two separate pools of catecholamines may exist in chromaffin granules. This view is supported by the biphasic efflux of catecholamines from chromaffin granules (SLOTKIN et al. 1971) and from synaptic vesicles of the caudate nucleus (PHILIPPU and BEYER 1973). Recently, TERLAND et al. (1979) determined the molar ratio in populations of chromaffin granules isolated by isopycnic gradient centrifugation. The molar ratio in the total population was slightly higher than 7, thus resembling the molar ratio found by BLASCHKO et al. (1956), FALCK et al. (1956), HILLARP (1960) and VAN DYKE et al. (1977). On the other hand, the molar ratio in noradrenaline-containing granules was close to those reported by HILLARP (1958b), BANKS (1965), SMITH and WINKLER (1967), TRIFARÓ and DWORKIND (1970) and SEN et al. (1979). These results also indicate that the catecholamines of the chromaffin granules cannot be totally bound to ATP.

The detection of calcium and magnesium in the chromaffin granules indicated that these ions may be involved in the storage of the catecholamine-ATP complex (BOROWITZ et al. 1965; COLBURN and MAAS 1965; PHILIPPU and SCHÜMANN 1966; RAJAN et al. 1977). Catecholamines aggregate with ATP, and the formation of aggregates is greatly increased by calcium and magnesium. Moreover, in the presence of calcium, catecholamines and ATP sediment jointly with chromogranins suggesting that chromogranins act as a matrix for the aggregate (BERNEIS et al. 1969, 1973; PLETSCHER et al. 1974; MORRIS et al. 1977). Aggregates have a high molecular weight (2000-10000) and are stable only at low temperatures. At higher temperatures, destruction of the thermolabile complex leads to liberation of its constituents. It is of interest to note that the efflux of catecholamines from chromaffin granules is accompanied by liberation of ATP, magnesium and calcium ions as well (PHILIPPU and SCHÜMANN 1966).

Nuclear magnetic resonance spectroscopy was also used to investigate the storage of catecholamines and ATP in the granules. As early as 1964 (WEINER and JARDETZKY), it was shown that at pH 5.6 (which is close to the pH of intragranular space, see Sect. C.VI.1), one molecule of ATP can interact with three molecules of adrenaline. The interaction seems to involve an ionic bond between the nitrogen of adrenaline and the phosphate moiety of ATP, as well as a hydrogen bond between the beta-hydroxyl-group of the amine and a phosphate oxygen of the nucleotide. The authors postulated that interaction between catecholamines and ATP contributes to the formation of ternary complexes with a protein of the chromaffin granules.

To explain the low osmotic activity, it was proposed that chromogranins form a gel which binds adrenaline and ATP in an osmotically inactive form (DANIELS et al. 1974). More recent nuclear magnetic resonance studies are inconsistent with the view that a gel may exist in the intragranular space and that high molecular weight complexes of catecholamines and ATP are formed; they suggest that chromogranin, which is present as a random-coil polypeptide (SMITH and WINKLER 1967), may form ternary complexes together with noradrenaline and ATP by cross-linking cationic side chains of the protein (SHARP and RICHARDS 1977; GRANOT and ROSENHECK 1978; SHARP and SEN 1978; SEN et al. 1979; SEN and SHARP 1980, 1981).

However, participation of chromogranins in ternary catecholamine-ATP-protein complexes has also been questioned. With their polyelectrolyte properties, chromogranins seem to play only some role in osmotic stabilization (SEN and SHARP 1982; HELLE et al. 1985). Measurements of the osmotic pressure indicate that solutions containing ATP and catecholamines are nonideal, thus partly accounting for the osmotic stabilization within the chromaffin granules (SEN and SHARP 1982; KOPELL and WESTHEAD 1982). In fact, the low osmolality of the aqueous phase of the granules remains to be clarified.

A two-pool storage system has been proposed for granules and synaptic vesicles (UVNÄS and ABORG 1980a,b, 1984). Both pools seem to be localized to the matrix. The non-specific pool 1 has the properties of a cation exchanger, while pool 2 is specific for (−)-noradrenaline and (−)-adrenaline. This two-pool storage system seems to be in agreement with the biphasic efflux of catecholamines from chromaffin granules and synaptic vesicles (SLOTKIN et al. 1971; PHILIPPU and BEYER 1973; YEN et al. 1976). Chemical construction and intermolecular arrangement of the storage system are unknown.

II. Storage in Synaptic Vesicles

The molecular ratio catecholamines/ATP is higher in noradrenaline-storing vesicles than in chromaffin granules. In a mitochondria-poor population of vesicles of splenic nerves the molecular ratio was found to be higher than 7 (DE POTTER et al. 1970; LAGERCRANTZ and STJÄRNE 1974; YEN et al. 1976). Moreover, at 37 °C the rate of spontaneous release of ATP is much lower than that of catecholamines, thus leading to further increase of the molar ratio during incubation (EULER et al. 1963c; STJÄRNE 1964; LAGERCRANTZ and STJÄRNE 1974; YEN et al. 1976). These results indicate that in synaptic vesicles a large part of catecholamines is not stored as a complex with ATP. As already mentioned, the biphasic efflux of catecholamines supports the existence of two pools of catecholamines in synaptic vesicles and chromaffin granules (SLOTKIN et al. 1971; PHILIPPU and BEYER 1973; YEN et al. 1976; UVNÄS and ABORG 1980a,b, 1984). These observations together with the high rate of efflux of catecholamines from synaptic vesicles (see PHILIPPU 1976), may indicate that storage mechanisms in synaptic vesicles are less developed than in chromaffin granules. It is likely that in synaptic vesicles uptake processes play a major role in accumulation and retention of catecholamines.

E. References

Aberer W, Kostron H, Huber E, Winkler H (1978) A characterization of the nucleotide uptake by chromaffin granules of bovine adrenal medulla. Biochem J 172:353-360

Aberer W, Stitzel R, Winkler H, Huber E (1979) Accumulation of (^{3}H)ATP in small dense core vesicles of superfused vasa deferentia. J Neurochem 33:797-801

Al-Lami F (1969) Light and electron microscopy of the adrenal medulla of Macaca mulata monkey. Anat Rec 164:317-332

Angelides KJ (1980) Transport of catecholamines by native and reconstituted rat heart synaptic vesicles. J Neurochem 35:949-962

Apps DK, Glover LA (1978) Isolation and characterization of magnesium adenosine-triphosphatase from the chromaffin granule membrane. FEBS Lett 85:254-258

Apps DK, Reid GA (1977) Adenosine triphosphatase and adenosine diphosphate/adenosine triphosphate isotope-exchange activities of the chromaffin granule membrane. Biochem J 167:279-300

Apps DK, Schatz G (1979) An adenosine triphosphatase isolated from chromaffin-granule membranes is closely similar to F_1-adenosine triphosphatase of mitochondria. Eur J Biochem 100:411-419

Apps DK, Pryde JG, Phillips JH (1980a) Cytochrome b_{561} is identical with chromomembrin B, a major polypeptide of chromaffin granule membranes. Neuroscience 5:2279-2287

Apps DK, Pryde JG, Phillips JH (1980b) Both the transmembrane pH gradient and the membrane potential are important in the accumulation of amines by resealed chromaffin-granule “ghosts”. FEBS Lett 111:386-390

Aunis D, Hesketh J, Devilliers G (1978) Immunohistochemical and immunocytochemical localization of myosin, chromogranin A and dopamine-β-hydroxylase in nerve cells in culture and adrenal glands. J Neurocytol 9:255-274

Axelrod J (1962) Purification and properties of phenylethanolamine-N-methyltransferase. J Biol Chem 237:1656-1660

Axelrod J (1966) Methylation reactions in the formation and metabolism of catecholamines and other biogenic amines. Pharmacol Rev 18:95-113

Balzer H, Khan AR, Ristic-Radivojevic S (1984) Comparative studies on Ca^{2+}- and Mg^{2+}-binding of scarcoplasmic reticulum and chromaffin granule membranes. Biochem Pharmacol 33:21-29

Banks P (1965) The adenosine-triphosphatase activity of adrenal chromaffin granules. Biochem J 95:490-496

Bareis DL, Slotkin TA (1979) Synaptic vesicles isolated from rat heart: 1-(^{3}H)nor epinephrine uptake properties. J Neurochem 32:345-351

Bartlett SF, Lagercrantz H, Smith AD (1976) Gel electrophoresis of soluble and insoluble proteins of noradrenergic vesicles from ox splenic nerve: A comparison with proteins of adrenal chromaffin granules. Neuroscience 1:339-344

Bashford CL, Radda GK, Ritchie GA (1975a) Energy-linked activities of the chromaffin granule membrane. FEBS Lett 50:21-24

Bashford CL, Casey RP, Radda GK, Ritchie GA (1975b) The effect of uncouplers on catecholamine incorporation by vesicles of chromaffin granules. Biochem J 148:153-155

Bashford CL, Casey RP, Radda GK, Ritchie GA (1976) Energy-coupling in adrenal chromaffin granules. Neuroscience 1:399–412
Baumgartner H, Winkler H, Hörtnagl H (1973) Isolated chromaffin granules: Maintenance of ATP content during incubation at 31 °C. Eur J Pharmacol 22:102–104
Baumgartner H, Gibb JW, Hörtnagl H, Snider SR, Winkler H (1974) Labelling of chromaffin granules in adrenal medulla with (^{35}S)sulfate. Mol Pharmacol 10:678–685
Benedeczky I, Carmichael SW (1980) Ultrastructural demonstration of Mg-ATPase activity of the chromaffin granule. Histochem 68:317–320
Benedeczky I, Smith AD (1972) Ultrastructural studies on the adrenal medulla of golden hamster: Origin and fate of secretory granules. Z Zellforsch Mikrosk Anat 124:367–386
Berneis KH, da Prada M, Pletscher A (1969) Micelle formation between 5-hydroxytryptamine and adenosine triphosphate in platelet storage organelles. Science 165:913–914
Berneis KH, Goetz U, da Prada M, Pletscher A (1973) Interaction of aggregated catecholamines and nucleotides with intragranular proteins. Naunyn-Schmiedeberg's Arch Pharmacol 277:291–296
Bisby MA, Fillenz M (1971) The storage of endogenous noradrenaline in sympathetic nerve terminals. J Physiol (Lond) 215:163–179
Bisby MA, Fillenz M, Smith AD (1973) Evidence for the presence of dopamine β-hydroxylase in both populations of noradrenaline storage vesicles in sympathetic nerve terminals of the rat vas deferens. J Neurochem 20:245–248
Bjerrum OJ, Helle KB, Bock E (1979) Immunochemically identical hydrophilic and amphiphilic forms of the bovine adrenomedullary dopamine β-hydroxylase. Biochem J 181:231–237
Blaschko H, Born GVR, d'Iorio A, Eade NR (1956) Observations on the distribution of catecholamines and adenosine triphosphate in the bovine adrenal medulla. J Physiol (Lond) 133:548–557
Blaschko H, Firemark H, Smith AD, Winkler H (1967) Lipids of the adrenal medulla: lysolecithin a characteristic constituent of chromaffin granules. Biochem J 104:545–549
Bloom FE (1972) Electron microscopy of catecholamine containing structures. In: Blaschko H, Muscholl E (eds) Catecholamines. Springer Berlin Heidelberg New York, pp 46–78 (Handbook of Experimental Pharmacology, Vol 33)
Borowitz JL, Fuwa E, Weiner N (1965) Distribution of metals and catecholamines in bovine adrenal medulla subfractions. Nature (Lond) 205:42–43
Buckland RM, Radda GK, Wakefield LM (1979) Reconstitution of the Mg^{2+}-ATPase of the chromaffin granule membrane. FEBS Lett 103:323–327
Burger A, Bellersheim M (1976) Ca^{2+}-uptake into noradrenaline-storing granules of bovine splenic nerves. Naunyn-Schmiedeberg's Arch Pharmacol 296:47–57
Burger A, Niedermaier W, Langer R, Bode U (1984) Further characteristics of the ATP-stimulated uptake of calcium into chromaffin granules. J Neurochem 43:806–815
Carlsson A, Hillarp NA, Waldeck B (1962) A Mg^{2+}-ATP dependent storage mechanism in the amine granules of the adrenal medulla. Med Exp 6:47–53
Carlsson A, Hillarp NA, Waldeck B (1963) Analysis of the Mg^{2+}-ATP dependent storage mechanism in the amine granules of the adrenal medulla. Acta Physiol Scand 59, Suppl 215:1–38
Carty SE, Johnson RG, Vaughan T, Pallant A, Scarpa A (1985) Amine transport into chromaffin ghosts. Kinetic parameters of net uptake of biologically and pharmaco-

logically relevant amines using an on-line amperometric technique. Eur J Biochem 147:447-452

Casey RP, Njus D, Radda GK, Sehr PA (1976) Adenosine triphosphate-evoked catecholamine release in chromaffin granules. Osmotic lysis as a consequence of proton translocation. Biochem J 158:583-588

Casey RP, Njus D, Radda GK, Seeley J, Sehr PA (1977) The biochemistry of the uptake, storage and release of catecholamines. Horiz Biochem Biophys 3:224-256

Chubb IW, de Potter WP, de Schaepdryver AF (1970) Two populations of noradrenaline-containing particles in the spleen. Nature (Lond) 228:1203-1204

Cidon S, Nelson N (1982) Properties of a novel ATPase enzyme in chromaffin granules. J Bioenerg Biomembranes 14:499-512

Cidon S, Nelson N (1983) A novel ATPase in the chromaffin granule membrane. J Biol Chem 258:2892-2896

Cimarusti DL, Saito K, Vaughn JE, Barber R, Roberts E, Thomas PE (1979) Immunocytochemical localization of dopamine-β-hydroxylase in rat locus coeruleus and hypothalamus. Brain Res 162:55-67

Cohn DV, Zangerle R, Fischer-Colbrie R, Chu LLH, Elting JJ, Hamilton JW, Winkler H (1982) Similarity of secretory protein I from parathyroid gland to chromogranin A from adrenal medulla. Proc Natl Acad Sci, USA 79: 6056-6059

Colburn RW, Maas JW (1965) Adenosine triphosphate-metal-norepinephrine ternary complexes and catecholamine binding. Nature (Lond) 208:37-41

Coupland RE (1972) The chromaffin system. In: Blaschko H, Muscholl E (eds) Catecholamines. Springer Berlin Heidelberg New York, pp 16-45 (Handbook of Eperimental Pharmacology, vol 33)

Coyle JT, Axelrod J (1971) Development of the uptake and storage of DL(^{3}H)norepinephrine in rat brain. J Neurochem 18:2061-2075

Coyle JT, Kuhar MJ (1974) Subcellular localization of dopamine β-hydroxylase and endogenous norepinephrine in the rat hypothalamus. Brain Res 65:475-487

Cuello AC, Jessell TM, Kanazawa I, Iversen LL (1977) Substance P: localization in synaptic vesicles in rat central nervous system. J Neurochem 29:747-751

Daniels A, Korda A, Transwell A, Williams RJP, Williams FRS (1974) The internal-structure of the chromaffin granule. Proc R Soc Lond B 187:353-361

Da Prada M, Obrist R, Pletscher A (1975) Discrimination of monoamine uptake by membranes of adrenal chromaffin granules. Br J Pharmacol 53:257-265

Day R, Denis D, Barabe J, St.-Pierre S, Lemaire S (1982) Dynorphin in bovine adrenal medulla. I. Detection in glandular und cellular extracts and secretion from isolated chromaffin cells. Int J Pept Protein Res 19:10-17

De Potter WP, Chubb IW (1977) Biochemical observations on the formation of small noradrenergic vesicles in the splenic nerve of the dog. Neuroscience 2:167-174

De Potter WP, de Smet FH (1980) Biochemical evidence for two types of noradrenaline storage particles in rabbit iris. Experientia 36:1282-1285

De Potter WP, Smith AD, de Schaepdryver AF (1970) Subcellular fractionation of splenic nerve: ATP, chromogranin A and dopamine β-hydroxylase in noradrenergic vesicles. Tissue & Cell 2:529-546

Deupree JD, Weaver JA (1984) Identification and characterization of the catecholamine transporter in bovine chromaffin granules using [^{3}H]reserpine. J Biol Chem 259:10907-10912

Diamond I, Goldberg AL (1971) Uptake and release of ^{45}Ca by brain microsomes, synaptosomes and synaptic vesicles. J Neurochem 18:1419-1431

Dolais-Kitabgi J, Perlman RL (1975) The stimulation of catecholamine release from chromaffin granules by valinomycin. Mol Pharmacol 11:745-750

Drake RAL, Harvey SAK, Njus D, Radda GK (1979) The effect of chlorpromazine on bioenergetic processes in chromaffin granule membranes. Neuroscience 4:853-861

Droz B (1975) Synaptic machinery and axoplasmic transport: maintenance of neuronal connectivity. In: The Nervous System. Raven Press New York

Ducros C (1974) Ultrastructural study of the organization of axonal agranular reticulum in octopus nerve. J Neurocytol 3:513-523

Euler US v (1972) Synthesis, uptake and storage of catecholamines in adrenergic nerves. In: Blaschko H, Muscholl E (eds) Catecholamines. Springer Berlin Heidelberg New York, pp 186-230 (Handbook of Experimental Pharmacology, vol 33)

Euler US v, Lishajko F (1963a) Catecholamine release and uptake in isolated adrenergic nerve granules. Acta Physiol Scand 57:468-480

Euler US v, Lishajko F (1963b) Effect of adenine nucleotides on catecholamine release and uptake in isolated adrenergic nerve granules. Acta Physiol Scand 59:454-461

Euler US v, Lishajko F (1967) Reuptake and net uptake of noradrenaline in adrenergic nerve granules with a note on the affinity for l- and d-isomers. Acta Physiol Scand 71:151-162

Euler US v, Lishajko F (1968) Effect of directly and indirectly acting sympathomimetic amines on adrenergic transmitter granules. Acta Physiol Scand 73:78-92

Euler US v, Lishajko F (1969) Effects of some metabolic cofactors and inhibitors on transmitter release and uptake in isolated adrenergic nerve granules. Acta Physiol Scand 77:298-307

Euler US v, Lishajko F, Stjärne L (1963c) Catecholamines and adenosine triphosphate in isolated adrenergic nerve granules. Acta Physiol Scand 59:495-496

Evangelista R, Ray P, Lewis RV (1982) A "trypsin-like" enzyme in adrenal chromaffin granules: a proenkephalin processing enzyme. Biochem Biophys Res Commun 106:895-902

Falck B, Hillarp NA, Högberg B (1956) Content and intracellular distribution of adenosine triphosphate in cow adrenal medulla. Acta Physiol Scand 36:360-376

Falkensammer G, Fischer-Colbrie R, Winkler H (1985) Biogenesis of chromaffin granules: Incorporation of sulfate into chromogranin B and into a proteglycan

Ferris RM, Tang FLM (1979) Comparison of the effects of the isomers of amphetamine, methylphenidate and deoxypipradol on the uptake of l-(^{3}H)norepinephrine and (^{3}H)dopamine by synaptic vesicles from rat whole brain, striatum and hypothalamus. J Pharmacol Exp Ther 210:422-428

Fischer-Colbrie R, Frischenschlager I (1985) Immunological characterization of secretory proteins of chromaffin granules: Chromogranins A, chromogranins B and enkephalin-containing peptides. J Neurochem 44: 1854-1861

Fischer-Colbrie R, Hagn C, Kilpatrick L, Winkler H (1986a) Chromogranin C: A third component of the acidic proteins in chromaffin granules. J Neurochem 47:318-321

Fischer-Colbrie R, Diez-Guerra J, Emson PC, Winkler H (1986b) Bovine chromaffin granules: Immunological studies with antisera against neuropeptide Y, (Met)enkephalin and bombesin. Neuroscience 18:167-174

Flatmark T, Grønberg M (1981) Cytochrome b-561 of the bovine adrenal chromaffin granules. Molecular weight and hydrodynamic properties in micellar solutions of Triton X-100. Biochem Biophys Res Commun 99:292-301

Flatmark T, Ingebretsen OC (1977) ATP-dependent proton translocation in resealed chromaffin granule ghosts. FEBS Lett 78:53-56

Flatmark T, Lagercrantz H, Terland O, Helle KB, Stjärne L (1971) Electron carriers of the noradrenaline storage vesicles from bovine splenic nerves. Biochim Biophys Acta 245:249-252

Fried G, Lundberg JM, Hökfelt T, Lagercrantz H, Fahrenkrug J, Lundgren G, Holmstedt B, Brodin E, Efendic S, Terenius L (1981) Do peptides coexist with classical transmitters in the same neuronal storage vesicles? In: Chemical neurotransmission. Stjärne L, Hedquist P, Lagercrantz H, Wennmalm A, (eds), pp 105-111 Academie Press New York

Fried G, Terenius L, Hökfelt T, Goldstein M (1985) Evidence for differential localization of noradrenaline and neuropeptide Y in neuronal storage vesicles isolated from rat vas deferens. J Neurosci 5:450-458

Gagnon C, Schatz R, Otten U, Thoenen H (1976) Synthesis, subcellular distribution and turnover of dopamine β-hydroxylase in organ cultures of sympathetic ganglia and adrenal medulla. J Neurochem 27:1083-1084

Geissler D, Martinek A, Margolis RU, Skrivanek JA, Ledeen R, König P, Winkler H (1977) Composition and biogenesis of complex carbohydrates of ox adrenal chromaffin granules. Neuroscience 2:685-693

Giachetti A, Shore PA (1975) On the formation of adrenergic storage granules as measured by reserpine labelling. Naunyn-Schmiedeberg's Arch Pharmacol 288:345-354

Glezer II (1969) On morphogenesis of synaptic apparatus of cerebral cortex on albino rat. Arkh Anat Gistol Embriol 56:77-84

Goetz U, da Prada M, Pletscher A (1971) Adenine-, guanine- and uridine 5′phosphonucleotides in blood platelets and storage organelles of various species. J Pharmacol Exp Ther 178:210-215

Grafenstein HRK v, Neumann E (1983) ATP-stimulated accumulation of calcium by chromaffin granules and mitochondria from the adrenal medulla. Biochem Biophys Res Commun 117:245-251

Granot J, Rosenheck K (1978) On the role of ATP and divalent metal ions in the storage of catecholamines. ^{1}H NMR studies of bovine adrenal chromaffin granules. FEBS Lett 95:45-48

Gratzl M, Krieger-Brauer H, Ekerdt R (1981) Latent acetylcholinesterase in secretory vesicles isolated from adrenal medulla. Biochem Biophys Acta 649:355-366

Halaris AE, Demet EM (1978) Active uptake of (^{3}H)5-HT by synaptic vesicles from rat brain. J Neurochem 31:591-597

Häusler R, Burger A, Niedermaier W (1981) Evidence for an inherent, ATP-stimulated uptake of calcium into chromaffin granules. Naunyn-Schmiedeberg's Arch Pharmacol 315:255-267

Helle KB, Reed RK, Pihl KE, Serck-Hanssen G (1985) Osmotic properties of the chromogranins and relation to osmotic pressure in catecholamine storage granules. Acta Physiol Scand 123:21-33

Hillarp NA (1958a) Enzymic systems involving adenosine phosphate in the adrenaline and noradrenaline containing granules of the adrenal medulla. Acta Physiol Scand 42:144-165

Hillarp NA (1958b) Adenosine phosphates and inorganic phosphate in the adrenaline and noradrenaline containing granules of the adrenal medulla. Acta Physiol Scand 42:321-332

Hillarp NA (1960) Different pools of catecholamines stored in the adrenal medulla. Acta Physiol Scand 50:8-22

Hillarp NA, Thieme G (1959) Nucleotides in the catecholamine granules of the adrenal medulla. Acta Physiol Scand 45:328-338

Hörtnagl H, Hörtnagl H, Winkler H (1969) Bovine splenic nerve: characterization of noradrenaline containing vesicles and other cell organelles by density gradient centrifugation. J Physiol (Lond) 205:103–114

Hörtnagl H, Winkler H, Schöpf JAL, Hohenwaller W (1971) Membranes of chromaffin granules. Isolation and partial characterization of two proteins. Biochem J 122:299–304

Holtzmann E (1977) The origin and fate of secretory packages, especially synaptic vesicles. Neuroscience 2:327–355

Holz RW (1978) Evidence that catecholamine transport into chromaffin vesicles is coupled to vesicle membrane potential. Proc Natl Acad Sci USA 75:5190–5194

Holz RW (1979) Measurement of membrane potential of chromaffin granules by the accumulation of triphenylmethylphosphonium cation. J Biol Chem 254:6703–6709

Holz RW (1980) Osmotic lysis of bovine chromaffin granules in isotonic solutions of salts of weak organic acids. Release of catecholamines, ATP, dopamine β-hydroxylase, and enkephalin-like material. J Biol Chem 255:7751–7755

Huber E, König P, Schuler G, Aberer W, Plattner H, Winkler H (1979) Characterization and topography of the glycoproteins of adrenal chromaffin granules. J Neurochem 32:35–47

Ichikawa Y, Yamano T (1965) Cytochrome b-559 in the microsomes of the adrenal medulla. Biochem Biophys Res Commun 20:263–268

Ingebretsen OC, Flatmark T (1979) Active and passive transport of dopamine in chromaffin granule ghosts isolated from bovine adrenal medulla. J Biol Chem 254:3833–3839

Ingebretsen OC, Terland O, Flatmark T (1980) Subcellular distribution of ascorbate in bovine adrenal medulla. Evidence for accumulation in chromaffin granules against a concentration gradient. Biochim Biophys Acta 628:182–189

Isambert MF, Henry JP (1981) Solubilization and reconstitution of the adenosine 5′-triphosphate dependent noradrenaline uptake system of bovine chromaffin granule membrane. Biochimie 63:211–219

Johnson RG, Scarpa A (1976a) Internal pH of isolated chromaffin vesicles. J Biol Chem 251:2189–2191

Johnson RG, Scarpa A (1976b) Ion permeability of isolated chromaffin granules. J Gen Physiol 68:601–631

Johnson RG, Scarpa A (1979) Protonmotive force and catecholamine transport in isolated chromaffin granules. J Biol Chem 254:3750–3760

Johnson RG, Carlson NJ, Scarpa A (1978) pH and catecholamine distribution in isolated chromaffin granules. J Biol Chem 253:1512–1521

Johnson RG, Pfister D, Carty SE, Scarpa A (1979) Biological amine transport in chromaffin ghosts. Coupling to the transmembrane proton and potential gradients. J Biol Chem 254:10963–10972

Johnson RG, Beers MF, Scarpa A (1982) H^+-ATPase of chromaffin granules. Kinetics, regulation stoichiometry. J Biol Chem 257:10701–10707

Jonasson J, Rosengren E, Waldeck B (1964) Effects of some pharmacologically active amines on the uptake of arylalkylamines by adrenal medullary granules. Acta Physiol Scand 60:136–140

Jones DG, Revell E (1970) The postnatal development of the synapse: a morphological approach utilizing synaptosomes. Z Zellforsch 111:179–194

Kanerva L, Hervonen A, Tissari AH, Suurhasho BVA (1975) Fine structure of synaptosomes from developing rat and human brain. J Ultrastruct Res 50:380

Kanerva L, Hervonen A, Tissari AH (1976) Ultrastructural identification of monoaminergic synaptosomes from one-day-old rat brains. Histochemie 48:233–240

Kanner BI, Fishkes H, Maron R, Sharon I, Schuldiner S (1979) Reserpine as a competitive and reversible inhibitor of the catecholamine transporter of bovine chromaffin granules. FEBS Lett 100:175-178

Kanner BI, Sharon I, Maron R, Schuldiner S (1980) Electrogenic transport of biogenic amines in chromaffin granule membrane vesicles. FEBS Lett 111:83-86

Kirksey DF, Klein RL, Thureson-Klein A, White HB (1977) Latency of dopamine β-hydroxylase in purified noradrenergic vesicles. Neuroscience 2:621-634

Kirksey DF, Seidler FJ, Slotkin TA (1978) Ontogeny of (−)-(^{3}H)-norepinephrine uptake: properties of synaptic storage vesicles of rat brain. Brain Res 150:367-375

Kirshner N (1957) Pathway of noradrenaline formation from dopa. J Biol Chem 226:821-825

Kirshner N (1962) Uptake of catecholamines by a particulate fraction of the adrenal medulla. J Biol Chem 327:2311-2317

Kirshner N (1965) The role of the membrane of the chromaffin granules isolated from the adrenal medulla. In: Koelle GB, Douglas WW, Carlsson A, Tŕcka V (eds) Second international pharmacological meeting. Czechoslovac Medical Press Praha Vol 3, pp 225-233

Kirshner N (1975) Biosynthesis of the catecholamines. In Blaschko H, Sayers G, Smith AG (eds) Adrenal Gland. American Physiological Society Bethesda, pp 341-355 (Handbook of Physiology, Vol 6)

Kirshner N, Goodall MC C (1957) The formation of adrenaline from noradrenaline. Biochim Biophys Acta 24:658-659

Kirshner N, Kirshner AG, Kamin DL (1966) Adenosine triphosphate activity of adrenal medulla catecholamine granules. Biochim Biophys Acta 113:332-335

Klein RL, Lagercrantz H (1971) Unidirectional fluxes in isolated splenic nerve vesicles measured by a millipore filter technique: effects of noradrenaline and competitive reversal of reserpine inhibition. Acta Physiol Scand 83:179-190

Klein RL, Kirksey DF, Rush RA, Goldstein M (1977) Preliminary estimates of the dopamine β-hydroxylase content and activity in purified noradrenergic vesicles. J Neurochem 28:81-86

Klein RL, Thureson-Klein AK, Chen Yen SH, Baggett JMC, Gasparis MS, Kirksey DF (1979) Dopamine β-hydroxylase distribution in density gradients: physiological and artefactural implications. J Neurobiol 10:291-307

Klein RL, Wilson SP, Dzielak DJ, Yang WH (1982) Opioid peptides and noradrenaline co-exist in large dense-cored vesicles from sympathetic nerve. Neuroscience 7:2255-2261

Klingenberg M (1970) Metabolite transport in mitochondria. An example for intracellular membrane function. In: Essays in Biochemistry. London: Academic Press

Knoth J, Handloser K, Njus D (1980) Electrogenic epinephrine transport in chromaffin granule ghosts. Biochem 19:2938-2942

Knoth J, Isaacs JM, Njus D (1981) Amine transport in chromaffin granule ghosts. pH dependence implies cationic form is translocated. J Biol Chem 256:6541-6543

Knoth J, Peabody JO, Huettl P, Njus D (1982) Steady-state tyramine uptake in chromaffin ghosts: a balance between electrogenic transport and electroneutral permeation. Biophys. J 37:218

Kobayashi S, Kent CH, Coupland RE (1978) Observations on the localization of labelled amino acid in mouse adrenal chromaffin cells after the injection of L-(4,5-^{3}H)-leucine. J Endocr 78:21-29

Kobayashi S, Miyabayashi T, Uchida T, Yanaihara N (1985) Metenkephalin-arg^6-gly^7-leu^8 in large dense-cored vesicles of splanchnic nerve terminals innervating guinea pig adrenal chromaffin cells. Neurosci Lett 53:247-252

Kobold G, Langer R, Burger A (1985) Does the carrier of chromaffin granules transport the protonated or the uncharged species of catecholamines? Naunyn-Schmiedeberg's Arch Pharmacol 331:209-219

Konings F, De Potter W (1980) The distribution of Mg^{2+} ATPase and its loss of specific activity upon gradient centrifugation of isolated chromaffin granules. J Neurochem 35:1465-1468

Kopell WN, Westhead EW (1982) Osmotic pressures of solutions of ATP and catecholamines relating to storage in chromaffin granules. J Biol Chem 257:5707-5710

Kostron H, Winkler H, Peer LJ, König P (1977a) Uptake of adenosine triphosphate by isolated adrenal chromaffin granules: a carrier-mediated transport: Neuroscience 2:159-166

Kostron H, Winkler H, Geissler D, König P (1977b) Uptake of calcium by chromaffin granules in vitro. J Neurochem 28:487-493

Krieger-Brauer H, Gratzl M (1982) Uptake of Ca^{2+} by isolated secretory vesicles from adrenal medulla. Biochim Biophys Acta 691:61-70

Krieger-Brauer HI, Gratzl M (1983) Effects of monovalent and divalent cations on Ca^{2+} fluxes across chromaffin secretory membrane vesicles. J Neurochem 41:1269-1276

Laduron P (1972) N-Methylation of dopamine to epinine in adrenal medulla: a new model for the biosynthesis of adrenaline. Arch int Pharmacodyn 195:197-208

Laduron P, Belpaire F (1968) Tissue fractionation and catecholamines. II. Intracellular distribution patterns of tyrosine hydroxylase, dopa decarboxylase, dopamine-β-hydroxylase, phenylethanolamine N-methyltransferase and monoamine oxidase in adrenal medulla. Biochem Pharmacol 17:1127-1140

Laduron P, Van Gompel P, Leysen J, Claeys M (1974) In vivo formation of epinine in adrenal medulla. Naunyn-Schmiedeberg's Arch Pharmacol 286:227-238

Laduron P, Aerts D, De Bie K, Van Gompel P (1976) Tissue fractionation and catecholamines. IV. Adenosine triphosphatase in chromaffin granules: a distribution artefact. Neuroscience 1:219-226

Lagercrantz H (1971a) Isolation and characterization of sympathetic nerve trunk vesicles. Acta Physiol Scand 82 Suppl 336:1-40

Lagercrantz H (1971b) Lipids of the sympathetic nerve trunk vesicles. Comparison with adrenomedullary vesicles. Acta Physiol Scand 82:567-570

Lagercrantz H (1976) On the composition and function of large dense cored vesicles in sympathetic nerves. Neuroscience 1:81-92

Lagercrantz H, Stjärne L (1974) Evidence that a major proportion of the noradrenaline is stored without ATP in sympathetic large dence core vesicles. Nature (Lond) 249:843-845

Lane NJ, Swales LS (1976) Interrelations between Golgi, GERL and synaptic vesicles in the nerve cells of insect and gastroped ganglia. J Cell Sci 22:435-453

Ledbetter FH, Kilpatrick D, Sage HL, Kirshner N (1978) Synthesis of chromogranins and dopamine β-hydroxylase by perfused bovine adrenal glands. Am J Physiol 235:475-486

Lentzen H, Philippu, A (1977) Uptake of tyramine into synaptic vesicles of the caudate nucleus. Naunyn-Schmiedeberg's Arch Pharmacol 300:25-30

Lentzen H, Philippu A (1981) Physico-chemical properties of phenethylamines and their uptake into synaptic vesicles of the caudate nucleus. Biochem Pharmacol 30:1759-1764

Leo A, Hansch C, Elkins D (1971) Partition coefficients and their uses. Chem Rev 71:525-616

Levin EY, Levenberg B, Kaufman S (1960) The enzymatic conversion of 3,4-dihydroxyphenylethylamine to norepinephrine. J Biol Chem 235:2080-2086

Lewis RV, Stern AS, Rossier J, Stein S, Udenfriend S (1979) Putative enkephalin precursors in bovine adrenal medulla. Biochem Biophys Res Commun 89:822-829

Lindberg I, Yang HYT, Costa E (1982) Enzymatic production of Mets- and Leus- enkephalin in adrenal chromaffin granules. Adv Biochem Psychopharmacol 33:183-191

Lindberg I, Yang HYT, Costa E (1984) Further characterization of an enkephalin-generating enzyme from adrenal medullary chromaffin granules. J Neurochem 42:1411-1419

Lishajko F (1969) Influence of chloride ions and ATP-Mg^{2+} on the release of catecholamines from isolated adrenal medullary granules. Acta Physiol Scand 75:255-256

Livett BG, Day R, Elde RP, Howe PR (1982) Co-storage of enkephalins and adrenaline in the bovine adrenal medulla. Neuroscience 7:1323-1332

Lohmann K (1932) Untersuchungen zur Konstitution der Adenylpyrophosphorsäure. Biochem Z 254:381-397

Maron R, Kanner BI, Schuldiner S (1979a) The role of a transmembrane pH gradient in 5-hydroxytryptamine uptake by synaptic vesicles from rat brain. FEBS Lett 98:237-240

Maron R, Fishkes H, Kanner BI, Schuldiner S (1979b) Solubilization and reconstitution of the catecholamine transporter from bovine chromaffin granules. Biochemistry 18:4781-4785

Matthaei H, Lentzen H, Philippu A (1976) Competition of some biogenic amines for uptake into synaptic vesicles of the striatum. Naunyn-Schmiedeberg's Arch Pharmacol 293:89-96

Matthaei H, Pramono S, Philippu A (1980) Stereospecificity of the uptake of noradrenaline into synaptic vesicles of the brain. Naunyn-Schmiedeberg's Arch Pharmacol 311 Suppl R57

Matthaei H, Pramono S, Philippu A (1988) Stereoselectivity of noradrenaline uptake into synaptic vesicles of the rat brain. Naunyn-Schmiedeberg's Arch Pharmacol (in press)

Mitchell P (1973) Cation translocation adenosine triphosphatase models: how direct is the participation of adenosine triphosphate and its hydrolysis products in cation translocation. FEBS Lett 33:267-274

Mitchell P (1974) A chemiosmotic molecular mechanism for proton-translocating adenosine triphosphatases. FEBS Lett 43:189-194

Morris SJ, Schultens HA, Schober R (1977) An osmometer model for changes in the buoyant density of chromaffin granules. Biophys J 20:33-48

Muller TW, Kirshner N (1975) ATPase and phosphatidylinositol kinase activities of adrenal chromaffin vesicles. J Neurochem 24:1155-1161

Nagatsu I, Karasawa N, Kondo Y, Inagaki S (1979) Immunocytochemical localization of tyrosine hydroxylase, dopamine β-hydroxylase and phenylethanolamine-N-methyltransferase in the adrenal glands of the frog and rat by a peroxidase-antiperoxidase method. Histochem 64:131-144

Nagy A, Varady G, Joo F, Rakonczay Z, Pilc A (1977) Separation of acetylcholine and catecholamine containing synaptic vesicles from brain cortex. J Neurochem 29:449-459

Naito S, Ueda T (1985) Characterization of glutamate uptake into synaptic vesicles. J Neurochem 44:99-109

Nelson DL, Molinoff PB (1976) Distribution and properties of adrenergic storage vesicles in nerve terminals. J Pharmacol Exp Ther 196:346–359
Neuman B, Wiedermann CJ, Fischer-Colbrie R, Schober M, Sperk G, Winkler H (1984) Biochemical and functional properties of large and small dense-core vesicles in sympathetic nerves of rat and ox vas deferens. Neuroscience 13:921–931
Niedermaier W, Burger A (1981) Two different ATP-dependent mechanisms for calcium uptake into chromaffin granules and mitochondria. Naunyn-Schmiedeberg's Arch Pharmacol 316:69–80
Njus D, Radda GK (1979) A potassium ion diffusion potential causes adrenal uptake in chromaffin granule "ghosts". Biochem J 180:579–585
Njus D, Sehr PA, Radda GK, Ritchie GA, Seeley PJ (1978) Phosphorus-31 nuclear magnetic resonance studies of active proton translocation in chromaffin granules. Biochemistry 17:4337–4343
Njus D, Knoth J, Cook C, Kelley PM (1983) Electron transfer across the cromaffin granule membrane. J Biol Chem 258:27–30
Nolan JA, Trojanowski JQ, Hogue-Angeletti R (1985) Neurons and neuroendocrine cells contain chromogranin: detection of the molecule in normal bovine tissues by immunochemical and immunohistochemical methods. J Histochem Cytochem 33:791–798
Oka M, Ohuchi I, Yoshida H, Imaizumi R (1965) Effect of adenosine triphosphate and magnesium on the release of catecholamines from adrenal medullary granules. Biochim Biophys Acta 97:170–171
Papadopoulou-Daifotis ZP, Morris SJ, Schober R (1977) Differential lysis of adrenaline- and noradrenaline-containing chromaffin granules promoted by the ionophore Br X537A. Neuroscience 2:609–619
Peer LJ, Winkler H, Snider SR, Gibb JW, Baumgartner H (1976) Synthesis of nucleotides in adrenal medulla and their uptake into chromaffin granules. Biochem Pharmacol 25:311–315
Pellegrino de Iraldi A, de Robertis E (1970) Studies on the origin of the granulated and nongranulated vesicles. In: New aspects of storage and release mechanisms of catecholamines. Schümann HJ, Kroneberg G (eds) pp 4–17. Berlin-Heidelberg-New York: Springer
Pendleton RG, Gessner G (1975) Evidence that dopamine is not a substrate for adrenal phenylethanol amine-N-methyltransferase. Mol Pharmacol 11:232–235
Percy, JM, Pryde JG, Apps DK (1985) Isolation of ATPase I, the proton pump of chromaffin-granule membranes. Biochem J 231:557–564
Philippu A (1976) Transport in intraneuronal storage vesicles. In: Paton DM (ed) The mechanism of neuronal and extraneuronal transport of catecholamines. Raven Press New York pp 215–246
Philippu A, Beyer J (1973) Dopamine and noradrenaline transport into subcellular vesicles of the striatum. Naunyn-Schmiedeberg's Arch Pharmacol 278:387–402
Philippu A, Matthaei H (1975) Uptake of serotonin, gamma-aminobutyric acid and histamine into synaptic vesicles of the pig caudate nucleus. Naunyn-Schmiedeberg's Arch Pharmacol 287:191–204
Philippu A, Schümann HJ (1966) Über die Bedeutung der Calcium- und Magnesiumionen für die Speicherung der Nebennierenmarkhormone. Naunyn-Schmiedeberg's Arch Exp Path Pharmak 252:339–358
Philippu A, Pfeiffer R, Schümann HJ, Lickfeld K (1967) Eigenschaften der Noradrenalin speichernden Granula des sympathischen Ganglion stellatum. Naunyn-Schmiederg's Arch Exp Path Pharmak 258:251–265
Philippu A, Burkat U, Becke H (1968) Uptake of norepinephrine by the isolated hypothalamic vesicles. Life Sci 7:1009–1017

Philippu A, Becke H, Burger A (1969) Effect of drugs on the uptake of noradrenaline by isolated hypothalamic vesicles. Eur J Pharmacol 6:96–101

Philippu A, Matthaei H, Lentzen H (1975) Uptake of dopamine into fractions of pig caudate nucleus homogenates. Naunyn-Schmiedeberg's Arch Pharmacol 287:181–190

Phillips JH (1973) Phosphatidylinositol kinase. A component of the chromaffin-granule membrane. Biochem J 136:579–587

Phillips JH (1974) Transport of catecholamines by resealed chromaffin granule "ghosts". Biochem J 144:311–318

Phillips JH (1977) Passive ion permeability of the chromaffin-granule membrane. Biochem J 168:289–297

Phillips JH (1978) 5-Hydroxytryptamine transport by the bovine chromaffin-granule membrane. Biochem J 170:673–679

Phillips JH (1981) Transport of Ca^{2+} and Na^{+} across the chromaffin-granule membrane. Biochem J 200:99–107

Phillips JH, Allison YP (1978) Proton translocation by the bovine chromaffin-granule membrane. Biochem J 170:661–672

Phillips JH, Morton AG (1978) Adenosine triphosphate in the bovine chromaffin granule. J Physiol (Lond) 74:503–508

Pickel VM, Tong HJ, Reis DJ, Leeman SE, Miller RJ (1979) Electron microscopic localization of substance P and enkephalin in axon terminals related to dendrites of catecholaminergic neurons Brain Res 160:387–400

Pletscher A (1977) Effect of neuroleptics and other drugs on monoamine uptake by membranes of adrenal chromaffin granules. Br J Pharmacol 59:419–424

Pletscher A, da Prada M, Berneis KH, Steffen H, Lütbold B, Weber HG (1974) Molecular organization of amine storage organelles of blood platelets and adrenal medulla. Adv Cytopharmacol 2:257–264

Pollard HB, Shindo H, Creutz CE, Pazoles ChJ, Cohen JS (1979) Internal pH and state of ATP in adrenergic chromaffin granules determined by ^{31}P nuclear magnetic resonance spectroscopy. J Biol Chem 254:1170–1177

Rajan KS, Wiehle RD, Riesen WH, Colburn RW, Davis JM (1977) Effect of metal chelating agents on the storage of norepinephrine in vitro by cerebral vesicles. Biochem Pharmacol 26:1703–1708

Roisin MP, Scherman D, Henry JP (1980) Synthesis of ATP by an artificially imposed electrochemical proton gradient in chromaffin granule ghosts. FEBS Lett 115:143–147

Rosa A, Zanini A (1981) Characterization of adenohypophysial polypeptides by two-dimensional gel electrophoresis. II. Sulfated and glycosylated polypeptides. Mol. Cell. Endocr. 24: 181–193

Rosa P, Zanini A (1983) Purification of a sulfated secretory protein from the adenohypophysis. Immunochemical evidence that similar macromolecules are present in other glands. Eur J Cell Biol 31:94–98

Rosa P, Hille A, Lee RHW, Zanini A, De Camilli P, Huttner WB (1985) Secretogranins I and II: two tyrosine-sulfated secretory proteins common to a variety of cells secreting peptides by the regulated pathway. J Cell Biol: 1999–2011

Sabban EL, Goldstein M (1984) Subcellular site of biosynthesis of the catecholamine biosynthetic enzymes in bovine adrenal medulla. J Neurochem 43:1663–1669

Scherman D, Henry JP (1981) pH-Dependece of the ATP-driven uptake of noradrenaline by bovine chromaffin-granule ghosts. J Biochem 116:535–539

Scherman D, Henry JP (1983) The catecholamine carrier of bovine chromaffin granules. Form of the bound amine. Mol Pharmacol 23:431–436

Scherman D, Henry JP (1984) Reserpine binding to bovine chromaffin granule membranes. Mol Pharmacol 25:113-122

Schuldiner S, Rottenberg H, Avron M (1972) Determination of ΔpH in chloroplasts. 2. Fluorescent amines as a probe for the determination of ΔpH in chloroplasts. Eur J Biochem 25:64-70

Schuldiner S, Fishkes H, Kanner BI (1978) Role of a transmembrane pH gradient in epinephrine transport by chromaffin granule membrane vesicles. Proc Natl Acad Sci USA, 75:3713-3716

Schümann HJ (1958) Über den Noradrenalin- und ATP-Gehalt sympathischer Nerven. Naunyn-Schmiedebergs Arch Exp Path Pharmak 233:296-300

Schümann HJ, Brodde OE (1976) Lack of epinine formation in adrenal medulla and brain of rats during cold exposure and inhibition of dopamine-β-hydroxylase. Naunyn-Schmiedeberg's Arch Pharmacol 293:139-144

Scriver CR, Wilson OH (1964) Possible locations for a common gene product in membrane transport of amino-acids and glycine. Nature (Lond) 202:92-93

Seidler FJ, Kirksey DF, Lau Ch, Whitmore WL, Slotkin TA (1977) Uptake of (−)^{3}H-norepinephrine by storage vesicles prepared from whole rat brain: properties of the uptake system and its inhibition by drugs. Life Sci 21:1075-1086

Seidler FJ, Whitmore WL, Slotkin TA (1978) Age-dependence of inhibition of rat brain synaptic vesicle (^{3}H)-norepinephrine uptake caused by administration of a single dose of reserpine: persistence of inhibition in adults but not in neonates. J Pharmacol Exp Ther 206:652-660

Sen R, Sharp RR (1980) The soluble components of chromaffin granules. A carbon-13 NMR survey. Biochim Biophys Acta 630:447-458

Sen R, Sharp RR (1981) High-molecular-weight catecholaminie-ATP aggregates are absent from the chromaffin granule aqueous phase. Biochem J 195:329-332

Sen R, Sharp RR (1982) Molecular mobilities and the lowered osmolality of the chromaffin granule aqueous phase. Biochim Biophys Acta 721:70-82

Sen R, Sharp RR, Domino LE, Domino EF (1979) Composition of the aqueous phase of chromaffin granules. Biochim Biophys Acta 587:75-88

Sharp RR, Richards EP (1977) Molecular mobilities of soluble components in the aqueous phase of chromaffin granules. Biochim Biophys Acta 497:260-271

Sharp RR, Sen R (1978) Molecular mobilities in chromaffin granules. Magnetic field dependence of proton T_1 relaxation times. Biochim Biophys Acta 538:155-163

Skotland T, Ljones T (1980) Direct spectrophotometric detection of ascorbate free radical formed by dopamine β-monooxygenase and by ascorbate oxidase. Biochim Biophys Acta 630:30-35

Slotkin TA, Kirshner N (1971) Uptake, storage and distribution of amines in bovine adrenal medullary vesicles. Mol Pharmacol 7:581-592

Slotkin TA, Ferris RM, Kirshner N (1971) Compartmental analysis of amine storage in bovine adrenal medullary granules. Mol Pharmacol 7:308-316

Slotkin TA, Anderson TR, Seidler FJ, Lau Ch (1975) Inhibition of epinephrine and metaraminol uptake into adrenal medullary vesicles by aralkylamines and alkylamines. Biochem Pharmacol 24:1413-1419

Slotkin TA, Seidler FJ, Whitmore WL, Salvaggio M, Lau Ch (1978a) Ionic and nucleotide cofactor requirements for uptake of (^{3}H)-norepinephrine by rat brain synaptic vesicle preparations. Mol Pharmacol 14:868-878

Slotkin TA, Seidler FJ, Whitmore WL, Lau Ch, Salvaggio M, Kirksey DF (1978b) Rat brain synaptic vesicles: uptake specificities of (^{3}H)-norepinephrine and (^{3}H)serotonin in preparations from whole brain and brain regions. J Neurochem 31:961-968

Slotkin TA, Salvaggio M, Lau Ch, Kirksey DF (1978c) ^{3}H-dopamine uptake by synaptic storage vesicles of rat whole brain and brain regions. Liefe Sci 22:823-830

Slotkin TA, Salvaggio M, Seidler FJ, Whitmore WL (1979a) Structural characteristics for inhibition of (^{3}H)-norepinephrine uptake into rat brain synaptic vesicles by beta-carboline, indolealkylamine, phenylethylamine and n-alkylamine derivatives. Mol Pharmacol 15:607-619

Slotkin TA, Whitmore WL, Salvaggio M, Seidler FJ (1979b) Perinatal methadone addiction affects brain synaptic development of biogenic amine systems in the rat. Life Sci 24:1223-1230

Smith AD (1972) Subcellular localization of noradrenaline in sympathetic neurons. Pharmacol Rev 24:435-457

Smith AD, Winkler H (1967) Purification and properties of an acidic protein from chromaffin granules of bovine adrenal medulla. Biochem J 103:483-492

Spiro MJ, Ball EG (1961) Studies on the respiratory enzymes of the adrenal gland. I. The medulla. J Biol Chem 236:225-230

Stern AS, Lewis RV, Kimura S, Rossier J, Gerber LD, Brink L, Stein S, Udenfriend S (1979) Isolation of the opioid heptapeptide Met-enkephalin (Arg6, Phe7) from bovine adrenal medullary granules and striatum. Proc Natl Acad Sci USA 76:6680-6683

Stevens P, Robinson RL, van Dyke K, Stitzel R (1972) Studies on the synthesis and release of adenosine triphosphate-8-^{3}H in the isolated perfused cat adrenal gland. J Pharmacol Exp Ther 181:463-471

Stevens P, Robinson RL, van Dyke K, Stitzel R (1975) Synthesis, storage and drug-induced release of ATP-8-^{3}H in the bovine adrenal gland. Pharmacology 13:40-55

Stjärne L (1964) Studies of catecholamine uptake storage and release mechanisms. Acta Physiol Scand 62 Suppl 228:1-60

Stjärne L (1972) The synthesis, uptake and storage of catecholamines in the adrenal medulla. In: Blaschko H, Muscholl E (eds), Catecholamines Springer Berlin Heidelberg New York, pp 231-269 (Handbook of Experimental Pharmacology vol 33)

Stjärne L, Lishajko F (1967) Localization of different steps in noradrenaline synthesis to different fractions of a bovine nerve homogenate. Biochem Pharmacol 16:1719-1728

Stjärne L, Roth RH, Lishajko F (1967) Noradrenaline formation from dopamine in isolated subcellular particles from bovine splenic nerve. Biochem Pharmacol 16:1729-1739

Tanaka R, Asaga H, Takeda M (1976) Nucleoside triphosphate and cation requirement for dopamine uptake by plan synaptic vesicles isolated from rat cerebrum. Brain Res 115:273-283

Taugner G (1971) The membrane of catecholamine storage vesicles of adrenal medulla: Catecholamine fluxes and ATPase activity. Naunyn-Schmiedeberg's Arch Pharmacol 270:392-406

Taugner G (1972a) The membrane of catecholamine storage vesicles of adrenal medulla. Uptake and release of noradrenaline in relation to the pH and the concentration of steric configuration of the amine present in the medium. Naunyn-Schmiedeberg's Arch Pharmacol 274:299-314

Taugner G (1972b) The effects of univalent anions on catecholamine fluxes and adenosine triphosphatase activity in storage vesicles from the adrenal medulla. Biochem J 130:969-972

Taugner G (1974) Enzymatic phosphoryl-group transfer by a fraction of the water solu-

ble proteins of catecholamine storage vesicles. Naunyn-Schmiedeberg's Arch Pharmacol 285 Suppl R82
Taugner G (1975) Incorporation of adenosine nucleotides into storage vesicles of the adrenal medulla. Naunyn-Schmiedeberg's Arch Pharmacol 287 Suppl R3
Taugner G, Hasselbach W (1968) Die Bedeutung der Sulfhydryl-Gruppen für den Catecholamin-Transport der Vesikel des Nebennierenmarks. Naunyn-Schmiedeberg's Arch Exp Path Pharmak 260:58-79
Taugner G, Wunderlich I (1979) Partial characterization of a phosphoryl group transferring enzyme in the membrane of catecholamine storage vesicles. Naunyn-Schmiedeberg's Arch Pharmacol 309:45-58
Taugner G, Wunderlich I (1984) Energy coupling and synthesis of ATP in the catecholamine storage organelle of the adrenal medulla. Biogenic amines 1:1-23
Taugner G, Wunderlich I, John F (1979) Distribution and metabolic fate of adenosine nucleotides in the membrane of storage vesicles from bovine adrenal medulla. Naunyn-Schmiedeberg's Arch Pharmacol 309:29-43
Taugner G, Wunderlich I, Junker D (1980) Reversibility of ATP hydrolysis in catecholamine storage vesicles from bovine adrenal medulla. Naunyn-Schmiedeberg's Arch Pharmacol 315:129-138
Teichberg S, Holtzman E (1973) Axonal agranular reticulum and synaptic vesicles in cultured embryonic chick sympathetic neurones. J Cell Biol 57:88-108
Terland O, Flatmark T (1973) NADH (NADPH): (acceptor) oxidoreductase activities of the bovine adrenal chromaffin granules. Biochim Biophys Acta 305:206-218
Terland O, Flatmark T (1975) Ascorbate as a natural constituent of chromaffin granules from the bovine adrenal medulla. FEBS Lett 59:52-56
Terland O, Flatmark T (1980) Oxidoreductase activities of chromaffin granule ghosts isolated from the bovine adrenal medulla. Biochim Biophys Acta 597:318-330
Terland O, Flatmark T, Kryvi H (1979) Isolation and characterization of noradrenaline storage granules of bovine adrenal medulla. Biochim Biophys Acta 553:460-468
Thomas DV (1979) Subcellular distribution of noradrenaline in guinea-pig cerebral cortex and spleen. J Neurochem 32:1259-1267
Thureson-Klein A, Klein RL, Lagercrantz H (1973) Highly purified splenic nerve vesicles: early post-mortem effects on ultrastructùre. J Neurocytol 2:13-27
Thureson-Klein A, Klein RL, Johansson O (1979) Catecholamine-rich cells and varicosities in bovine splenic nerve, vesicle contents and evidence for exocytosis. J Neurobiol 10:309-342
Tirrell JG, Westhead EW (1979) The uptake of ascorbic acid and dehydroascorbic acid by chromaffin granules of the adrenal medulla. Neuroscience 4:181-186
Toll L, Howard BD (1978) Role of Mg^{2+}-ATPase and a pH gradient in the storage of catecholamines in synaptic vesicles. Biochemistry 17:2517-2523
Toll L, Gundersen CB, Howard BD (1977) Energy utilization in the uptake of catecholamines by synaptic vesicles and adrenal chromaffin granules. Brain Res 136:59-66
Trifaró JM, Dworkind J (1970) A new and simple method for isolation of adrenal chromaffin granules by means of an isotonic density gradient. Analyt Biochem 34:403-412
Trifaró JM, Dworkind J (1971) Phosphorylation of membrane components of adrenal chromaffin granules by adenosine triphosphate. Mol Pharmacol 7:52-65
Trifaró JM, Poisner AM (1967) The role of ATP and ATPase in the release of catecholamines from the adrenal medulla. II. ATP-evoked fall in optical density of isolated chromaffin granules. Mol Pharmacol 3:572-580
Tsudzuki T (1979) Adenosine triphosphate-dependent calcium uptake of synaptic vesicle fraction is largely due to contaminating microsomes. J Biochem 86:777-782

Tsudzuki T (1981) Studies on the uptake of (3 H) dopamine by synaptic vesicle fraction isolated from bovine brain. J Biochem 89:61-69

Uvnäs B, Aborg CH (1980a) In vitro studies on a two-pool storage of adrenaline and noradrenaline in granule material from bovine adrenal medulla. Acta Physiol Scand 109:345-354

Uvnäs B, Aborg CH (1980b) In vitro studies on a cation dependent catecholamine release from a two-compartment storage in bovine adrenal medullary granules. Acta Physiol Scand 109:355-362

Uvnäs, B., Aborg, CH 1984) Cation exchange — a common mechanism in the storage and release of biogenic amines stored in granules (vesicles)? III. A possible role of sodium ions in non-exocytotic fractional release of neurotransmitters. Acta Physiol Scand 120:99-107

Van Dyke K, Robinson R, Urquilla P, Smith D, Taylor M, Trush M, Wilson M (1977) An analysis of nucleotides and catecholamines in bovine metullary granules by anion exchange high pressure liquid chromatography and fluorescence. Evidence that most of the catecholamines in chromaffin granules are stored without associated ATP. Pharmacology 15:377-391

Van Orden LS, Redick BJA, Rybarczyk KE, van Orden DE, Baker HA, Hartman BK (1977) Immunocytochemical evidence for particulate localization of phenylethanolamine-N-methyltransferase in adrenal medulla. Neuropharmacol 16:129-133

Vignais PV (1976) Molecular and physiological aspects of adenine nucleotide transport in mitochondria. Biochim Biophys Acta 456:2-39

Vignais PV, Vignais PM, Defaye G (1973) Adenosine diphosphate translocation in mitochondria. Nature of the receptor site for carboxy atractyloside (gummiferin). Biochemistry 12:1508-1519

Viveros OH, Arqueros L, Kirshner N (1971) Mechanism of secretion from the adrenal medulla.-VII. Effect of insulin administration on the buoyant density, dopamine β-hydroxylase and catecholamine content of adrenal storage vesicles. Mol Pharmacol 7:444-454

Viveros OH, Diliberto EJ, Hazum E, Chang KJ (1979) Opiate-like materials in the adrenal medulla: evidence for storage and secretion with catecholamines. Mol Pharmacol 16:1101-1108

Weaver JA, Deupree JD (1982) Conditions required for reserpine binding to the catecholamine transporter on chromaffin granule ghosts. Eur J Pharmacol 80:437-438

Weber A, Winkler H (1981) Specificity and mechanism of nucleotide uptake by adrenal chromaffin granules. Neurosciene 6:2269-2276

Weber A, Westhead EW, Winkler H (1983) Specificity and properties of the nucleotide carrier in chromaffin granules from bovine adrenal medulla. Biochem J 210:789-794

Weiner N, Jardetzky P (1964) A study of catecholamine nucleotide complexes by nuclear magnetic resonance spectroscopy. Naunyn-Schmiedebergs Arch Exp Path Pharmak 248:308-318

Wespi HH (1969) Active transport and passive fluxes of K, Na, and Li in mammalian non-myelinated nerve fibres. Pflügers Arch ges Physiol 306:262-280

Winkler H (1971) The membrane of the chromaffin granules. Phil Trans R Soc Ser B 261:293-303

Winkler H (1976) Commentary. The composition of adrenal chromaffin granules: an assessment of controversial results. Neuroscience 1:65-80

Winkler H, Smith AD (1975) The chromaffin granule and the storage of catecholamines. In: Blaschko H, Sayers G, Smith AD (eds) Adrenal gland. American Physiological Society Bethesda, pp 321-339 (Handbook of Physiology, vol 6)

Winkler H, Westhead E (1980) The molecular organization of adrenal chromaffin granules. Neuroscience 5:1803-1823
Winkler H, Schöpf JAL, Hörtnagl H, Hörtnagl H (1972) Bovine adrenal medulla: subcellular distribution of newly synthesised catecholamines, nucleotides and chromogranins. Naunyn-Schmiedeberg's Arch Pharmacol 273:43-61
Wolfe DE, Potter LT, Richardson KC, Axelrod J (1962) Localizing tritiated norepinephrine in sympathetic axons by electron microscopic autoradiography. Science 138:440-442
Yen SS, Klein RL, Chen-Yen SH (1973) Highly purified splenic nerve vesicles: early post-mortem effects on norepinephrine content and pools. J Neurocytol 2:1-12
Yen SS, Klein RL, Chen-Yen SH, Thureson-Klein A (1976) Norepinephrine:adenosine-triphosphate ratios in purified adrenergic vesicles. J Neurobiol 7:11-22

CHAPTER 2

Occurrence and Mechanism of Exocytosis in Adrenal Medulla and Sympathetic Nerve

H. WINKLER

A. Introduction

Secretion by exocytosis—a term introduced by DE DUVE (1963)—involves the fusion of the membrane of the catecholamine-storing vesicles with that of the cell, followed by the discharge of the soluble contents of this organelle directly into the extracellular space. When the first Handbook on Catecholamines (BLASCHKO and MUSCHOLL 1972) was published, the chapter on the secretion of catecholamines (SMITH and WINKLER 1972) already presented exocytosis as the well established secretion mechanism for adrenal medulla and more tentatively also for sympathetic nerves. In the meantime many additional papers have been published, which have confirmed and extended this concept (550 of the 622 publications cited in this review appeared after 1972). On the other hand, the molecular mechanisms involved in exocytosis are still a matter of speculation.

This article reviews as completely as possible the established facts and concepts and discusses those questions in detail where clear cut answers remain elusive. For the worker in this field this article should be helpful in avoiding experiments which confirm what does not need to be confirmed further, and in selecting experiments leading to answers for the yet unresolved questions. Several reviews in this field are available (SMITH 1968; DOUGLAS 1968; KIRSHNER and VIVEROS 1972; SMITH and WINKLER 1972; SMITH 1972a; HOLTZMAN et al. 1973; DOUGLAS 1974a; VIVEROS 1975; HELLE and SERCK-HANSSEN 1975; GEFFEN and JARROTT 1977; WINKLER 1977; PERLMAN and CHALFIE 1977; NAGASAWA 1977; HOLTZMAN 1977; TRIFARO 1977; SMITH 1978; PHILLIPS and APPS 1979; POLLARD et al. 1979, 1982a; KLEIN and THURESON-KLEIN 1984; BURGOYNE 1984a; LIVETT 1984; POLLARD et al. 1985; ROSENHECK and LELKES 1987; AUGUSTINE et al. 1987)

B. Evidence for Exocytosis

I. Adrenal Medulla

The first suggestion that hormone release from the adrenal medulla might occur by a process of exoytosis was put forward in early morphological studies (DE ROBERTIS and VAZ FERREIRA 1957; DE ROBERTIS and SABATINI 1960). However, other morphologists (see below and SMITH and WINKLER 1972, for references) did not support this concept, and definite proof of exocytosis awaited the results of biochemical studies.

1. Biochemical Evidence for Exocytosis

This evidence rests on three major findings: (i) the various components of the granule are released together in stoichiometric amounts, (ii) components of the granule membrane are retained in the cell, but become exposed on the cell surface, (iii) soluble constituents of the cytosol are not released. A series of experiments performed by several groups in the late sixties established that the secretory process in adrenal medulla fulfilled these criteria (see SMITH 1968; DOUGLAS 1968; KIRSHNER and VIVEROS 1972; SMITH and WINKLER 1972). The most important early data and the more recent findings will be discussed in the following sections.

a. Secretion of Soluble Proteins from Chromaffin Granules

The soluble proteins of bovine chromaffin granules consist of a family of acidic proteins with chromogranin A, B and C as major components and with DBH and glycoprotein III representing minor components (see WINKLER 1976; WINKLER et al. 1986). The crucial observation that the secretion of catecholamines from the adrenal medulla was accompanied by the release of proteins was made by BANKS and HELLE (1965) on perfused bovine adrenals. The ratio of catecholamines to proteins in perfusates during stimulation (by carbachol) was similar to that of catecholamines to soluble proteins in isolated granules. Furthermore, an antiserum against the soluble proteins of chromaffin granules reacted with the proteins released into the perfusate. In this publication the nature of the soluble granule protein was not further defined. However, this paper is usually quoted for providing evidence for the release of chromogranin A, since the antiserum prepared by HELLE was considered for some years to be directed against chromogranin A (see HELLE and SERCK-HANSSEN 1975). Later studies (HELLE et al. 1978) established that this antiserum reacted only with DBH and therefore BANKS and HELLE (1965) were the first to demonstrate the secretion of this protein. Secretion of bovine chromogranin A was first demonstrated by KIRSHNER and collaborators (KIRSHNER et al. 1966, SAGE et al. 1967; KIRSHNER et al. 1967). The release of this granule-specific protein was elicited by various secretagogues (acetylcholine, nicotine, $BaCl_2$). Secretion was calcium-dependent. By immunological techniques the catecholamine/chromogranin A ratio in perfusates during stimulation was found to be similar to that in isolated chromaffin granules. A cytoplasmic enzyme, PNMT, was not released (KIRSHNER et al. 1967). Similar results for the release of chromogranin A, its calcium dependence and a lack of release of a cytoplasmic enzyme (lactate dehydrogenase) were obtained simultaneously by SCHNEIDER et al. (1967; see also SCHNEIDER 1969a, b). In addition it was shown by electrophoretic techniques that, during stimulation of bovine glands, all the soluble protein components (at least 8) of the granule appeared in the perfusate. In confirmation of the results of BANKS and HELLE (1965) the ratio of catecholamines to secreted proteins in the perfusate was similar to the catecholamine/soluble protein ratio in granules (see also SERCK-HANSSEN 1972). Radioactively labelled chromogranin A (as shown by electrophoresis) was released from glands as early as 30 min after "pulse labelling" of proteins with ^{3}H-leucine (WINKLER et al. 1971). Release of newly synthesized (radioactively labelled) chromogranin A was recently reported for isolated phaeochromocytoma cells (SCHUBERT and KLIER 1977).

Chromogranin A is secreted not only under the rather unphysiological condition of a retrogradely perfused bovine gland but also during stimulation of the splanchnic nerve of a calf under *in vivo* conditions (BLASCHKO et al. 1967a). The presence of chromogranin A in human serum was claimed by YAMADA et al. (1978), based on immunological data. However, the methods employed do not rule out the possibility that the antiserum was directed against DBH, which is by far the most potent antigen of the granule lysate (see SAGE et al. 1967). In a more recent study 57 ng chromogranin A per ml serum were found (O'CONNOR and FRIGON 1984; O'CONNOR and DEFTOS 1986).

Quantitative data for the secretion of DBH were first provided for bovine glands by VIVEROS et al. (1968). The ratio of DBH to catecholamines in the perfusates during stimulation was similar to that in isolated chromaffin granules. When rabbits were treated with insulin to cause a massive secretion of catecholamines from the adrenals, the soluble granule pool of DBH inside the gland decreased in parallel with the catecholamines (VIVEROS et al. 1969b, c; see also SORIMACHI and YOSHIDA 1979). Furthermore, quantities of DBH and catecholamines found in perfusates of stimulated glands were consistent with stoichiometric release of these substances. A decrease of DBH in adrenal medulla caused by stimulation was also observed in perfused bovine glands (SERCK-HANSSEN and HELLE 1972) and in rat glands *in vitro* (BOONYAVIROJ et al. 1977). Similar results were obtained in a semiquantitative, immunohistochemical study of lamb adrenals (FRYDMAN and GEFFEN 1973).

All these studies clearly established that secretion of catecholamines is accompanied by the release of DBH. However, as far as the stoichiometric release of the two components is concerned, several other studies have revealed a rather complicated situation. DIXON et al. (1975) worked with cat adrenals which were stimulated either with acetylcholine or via the splanchnic nerve. Under these conditions catecholamine secretion was accompanied by only a small release of DBH. In fact, the ratio of enzyme to catecholamine in the perfusates during stimulation was less than 10% of the ratio of these two substances in isolated granules. A somewhat better but still too low recovery of DBH was reported by SORIMACHI and YOSHIDA (1979). In experiments on perfused bovine glands (DIXON et al. 1975) the relative enzyme release was much higher, although the enzyme/amine ratio in the perfusate was half that in chromaffin granules. A deficit of DBH release was also observed by SERCK-HANSSEN et al. (1980). On the other hand, JACOBS et al. (1978) reported that, during continuous stimulation of isolated bovine glands with acetylcholine, the adrenaline response declined, after an initial maximum, whereas DBH secretion remained constant (see EDWARDS et al. 1979, for similar results on adrenal glands *in situ* in conscious calves).

In the publications just cited, the authors discuss several possibilities which might help to explain the lack of correlation between DBH release and catecholamine secretion: (i) slower escape of the enzyme from the site of exocytosis into the circulation, (ii) rapid inactivation of the secreted enzyme in the perfusates (actually demonstrated by AUNIS et al. 1978), (iii) heterogeneity of the granule population: during prolonged stimulation release may occur from newly formed granules which possess a relatively low content of catecholamines (see WINKLER 1977; SERCK-HANSSEN et al. 1980). Complications (i) and (ii) can be overcome in experiments with isolated cells. It is therefore reassuring that, in such experiments, the ratio of DBH to catecholamines in the medium during stimulation was identical to that found in isolated granules. This was the case for bovine (FENWICK et al. 1978; LEDBETTER and KIRSHNER 1981; KNIGHT and BAKER 1982), for guinea pig (ITO 1983) or for hamster chromaffin cells (LIANG and PERLMAN 1979).

Release of DBH from the adrenal medulla into the blood stream has also been detected under *in vivo* conditions. Thus, during hemorrhagic shock in dogs, catecholamines and DBH are released from the adrenal into the lumboadrenal vein (CUBEDDU et al. 1977; PINARDI et al. 1979). As a consequence the DBH level in the general circulation also increased (see ARNAIZ et al. 1980, for guinea pigs). However, the contribution of adrenal DBH to the overall plasma levels of this enzyme depends very much on the species investigated. The "releasable" DBH pool in the adrenal of several species varies by a factor of 60, being low in rat and high in rabbits and guinea pigs (ARNAIZ et al. 1978; CUBEDDU et al. 1979). Furthermore, the ratio of total releasable enzyme to total plasma enzyme is low in human and rat, but much higher in dogs and guinea pigs (CUBEDDU et al. 1979). It is therefore not surprising that no significant contribution of the adrenal enzyme to the plasma enzyme levels has been observed in rats and human beings (ARNAIZ et al. 1978; CUBEDDU et al. 1979; see WEINSHILBOUM 1979).

b. Secretion of Other Soluble Constituents of Chromaffin Granules

Evidence for the release of glycoprotein III, carboxypeptidase B-like enzyme, mucopolysaccharides (proteoglycans), enkephalins, neuropeptide Y, chromogranin B, nucleotides and calcium has been obtained.

Chromogranin B (FISCHER-COLBRIE and FRISCHENSCHLAGER 1985), glycoprotein III (FISCHER-COLBRIE et al. 1984) and carboxypeptidase B (HOOK and EIDEN 1985), an enzyme involved in enkephalin synthesis, are characteristic constituents of bovine chromaffin granules and are released upon stimulation. Secretion of ^{35}S-labelled mucopolysaccharides (proteoglycans) from cat adrenals, presumably from the granule compartment, was demonstrated by MARGOLIS et al. (1973). In isolated bovine adrenal glands, prelabelled with ^{45}Ca, a release of this isotope from chromaffin granules occurred during stimulation (BOROWITZ et al. 1975). However, no quantitative data on release are available.

Enkephalins, recently discovered constituents of chromaffin granules (LUNDBERG et al. 1979; VIVEROS et al. 1979), are released together with the catecholamines during stimulation of dog adrenal with acetylcholine or $BaCl_2$. The relative amounts in the perfusates correspond to those in the granule lysate (VIVEROS et al. 1979; for bovine chromaffin cells: KILPATRICK et al. 1980; LIVETT et al. 1981; WILSON et al. 1982; bovine glands: BARRON and HEXUM 1984; for cat glands: CHAMINADE et al. 1983). Neuropeptide Y, which is found in high concentrations in bovine chromaffin granules, is also released concomitantly with catecholamines (KATAOKA et al. 1985; FISCHER-COLBRIE et al. 1986). Neurotensin is secreted from cat glands where it occurs in chromaffin granules (RÖKAEUS et al. 1984).

The first indication of ATP release was obtained by STJÄRNE (1964) who identified an ATP-metabolite, hypoxanthine, in perfusates of ox adrenals stimulated with acetylcholine. In quantitative studies on cat and ox adrenal, two groups of investigators (DOUGLAS et al. 1965; DOUGLAS and POISNER 1966; BANKS 1966a) established that the molar ratio of catecholamines to adenine nucleotides in intact granules was similar to the ratio of amines to ATP (and its catabolites) in the perfusate during stimulation (see also ITO 1983, for isolated cells and ROJAS et al. 1985). When the granule pool of nucleotides within the gland was labelled with either ^{32}P-phosphate (STJÄRNE et al. 1970) or ^{3}H-adenosine (STEVENS et al. 1972), the concomitant release of labelled nucleotides and catecholamines was elicited by stimulation of the gland. Further evidence that during stimulation ATP is released from chromaffin granules was recently obtained by phosphorus-31 nuclear magnetic resonance (BEVINGTON et al. 1984). During stimulation the intragranular ATP pool, as measured by this method, declined.

Stimulation of adrenal glands or chromaffin cells with excess potassium or nicotine leads to the release of acetylcholinesterase (CHUBB and SMITH 1975; MIZOBE and LIVETT 1983; MIZOBE and IWAMOTO 1984). It is not clear whether this secretion occurs from the endoplasmic reticulum and/or from chromaffin granules. Within the chromaffin cell, histochemical studies indicate that most of the enzyme is found in the endoplasmie reticulum (SOMOGYI et al. 1975), but according to results following subcellular fractionation of tissue (GRATZL et al. 1981; BURGUN et al. 1985) some of it is also present in chromaffin granules. More recent data (MIZOBE et al. 1984) indicate that release of catecholamines and of this enzyme occurs from different compartments. Therefore, a parallel release of catecholamines with another component does not necessarly imply that this component also originates from chromaffin granules. This is well illustrated by experiments with ascorbic acid. Newly taken up ascorbic acid is released from chromaffin cells together with catecholamines in a calcium-dependent manner (DANIELS et al. 1983). However, this newly taken up ascorbic acid is not dervied from the granule compartment (LEVINE et al. 1983). Data on the release of endogenous ascorbic acid found in chromaffin granules are not available. Cholinomimetics can release lysosomal enzymes from the adrenal gland (SCHNEIDER 1968; 1970; SMITH 1969).

c. Retention of Membrane Constituents During Secretion

Secretion of catecholamines is not accompanied by release of the granule membrane constituents (see also Sect. B.I.4). Thus, no significant amounts of cholesterol and phospholipids (including lysolecithin) appear in the perfu-

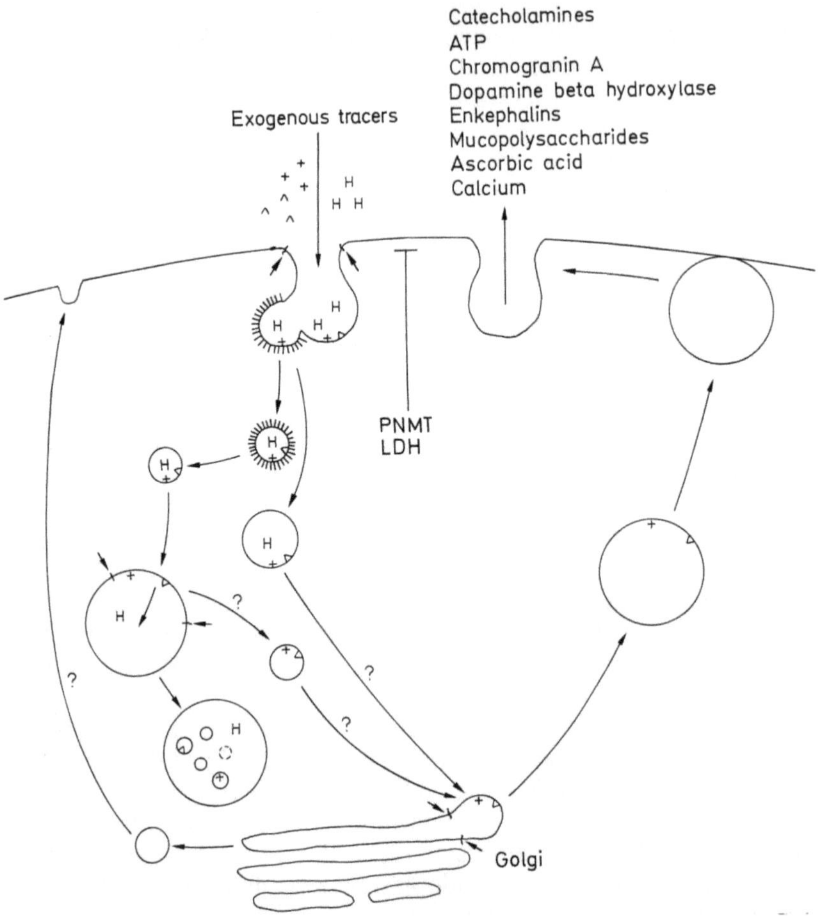

Fig. 1 Exocytosis in adrenal medulla. Chromaffin granules are formed in the Golgi region and attach to the plasma membrane, fuse with it, and the total soluble content is discharged (all or none process). Cytoplasmic enzymes like PNMT or lactate dehydrogenase (LDH) cannot pass through the plasma membrane during this secretory activity. After exocytosis the membranes of chromaffin granule are retrieved in coated vesicles. The regions of membrane fission on the plasma membrane are indicated by small arrows. Subsequently, coated vesicles become smooth vesicles. The size of these vesicles varies from microvesicles (<95 nm) to larger vesicles about the size of chromaffin granules. Some of these vesicles are transported to the Golgi region or directly to new granules and are re-used for another secretory cycle. Other vesicles fuse with lysosomes, discharge their soluble content and then, after fission (at points marked by small arrows), the granule membrane may also reach the Golgi region. Other membranes end up in multivesicular bodies for digestion. The lysosomal pathway has been established with exogenous tracers like horse radish peroxidase (H), which does not bind to the membranes. Recycling of membranes has been demonstrated with tracers like cationized ferritin (+) and specific antibodies (λ) which become bound to the membranes exposed on the cell surface during exocytosis. The figure also suggests that the plasma membrane may be replenished by specific vesicles which are not integrated in the secretory cycle (constitutive pathway). For further details and references see text

sates during catecholamine secretion (SCHNEIDER et al. 1967; TRIFARO et al. 1967). Within the gland the lipid content of the large granule fraction is not decreased when significant catecholamine depletion has occurred (POISNER et al. 1967). Furthermore, the membrane-bound DBH does not decline in the gland during catecholamine secretion, whereas the soluble enzyme (see above) is released (VIVEROS et al. 1969a, c, 1971b). Thus 24 hours after treatment of rabbits with insulin, the catecholamine content was reduced by 80%, the soluble DBH by 60%, whereas the membrane-bound enzyme remained unchanged.

d. Exposure of Membrane Antigens of Chromaffin Granules on Cell Surface During Secretion

During secretion by exocytosis, membrane components of chromaffin granules facing the content of these organelles should become exposed on the surface of chromaffin cells (see Fig. 1). This has actually been demonstrated in several qualitative immunohistochemical studies using an antiserum against DBH (see abstract by FENWICK and LIVETT 1976; WILDMANN et al. 1981; PHILLIPS et al. 1983; DOWD et al. 1983; PATZAK et al. 1984; LOTSHAW et al. 1986). During stimulation an increase of specific immunofluorescence on the cell surface was observed. The exposed DBH molecules appeared to be present in tiny spots and patches. More quantitative data were obtained with antisera against DBH and against another specific glycoprotein of chromaffin granules (GP III: FISCHER-COLBRIE et al. 1982) using a cytotoxicity test, microcomplement fixation (LINGG et al. 1983) and immunofluorescence evaluated with a fluorescence-activated cell sorter (PATZAK et al. 1984). There was a good correlation between the degree of catecholamine secretion and the amount of specific antigen appearing on the cell surface. At the ultrastructural level the exposed patches of chromaffin granules could be labelled by the immunogold technique (PATZAK and WINKLER 1986; PATZAK et al. 1987).

Quite independent evidence for the incorporation of the granule membrane into the plasma membrane was obtained with electrophysiological methods (patch clamp technique). Discrete changes in the electrical capacitance of the plasma membrane were interpreted to represent the addition of granule membrane due to exocytotic activity (NEHER and MARTY 1982).

2. Morphological Evidence for Exoxytosis

The first suggestion that, during secretion, chromaffin granules approach and fuse with the plasma membrane was put forward in the early studies of DE ROBERTIS and VAZ FERREIRA (1957) and DE ROBERTIS and SABATINI (1960). One picture showing continuity of the two interacting membranes was published by COUPLAND (1965) in his paper on the ultrastructure of the rat adrenal. However, the final breakthrough was achieved by DINER (1967; see also GRYNSZPAN-WINOGRAD 1971) who studied the hamster adrenal medulla. She described numerous chromaffin granules which had fused with the plasma membranes and were in various stages of extrusion of their soluble content into the extracellular space. Apparently, this species is best suited for demonstrating the occurrence of exocytotic profiles, even in unstimulated glands (see also BENEDECZKY and SMITH 1972; NAGASAWA and KACHI et al. 1985; DOUGLAS 1972). This may be related to the finding that the secretory response to acetylcholine proceeds in this species for a longer time than in guinea pig or bovine chromaffin cells (LIANG and PERLMAN 1979). In a study on adrenals of several birds, exocytosis was easily observed only in the chaffinch (UNSICKER 1973).

Several morphologists viewed the concept of exocytosis with scepticism (see e.g. discussion by WACKER and FORSSMAN 1972; KAJIHARA et al. 1977) which indicates that it is apparently difficult to capture this fleeting event of exocytosis by conventional electron microscopy. These apparent problems have been overcome by two approaches: (i) stimulation of the adrenal medulla to induce increased secretion, and (ii) the application of freeze-fracturing techniques. Both these approaches were combined in a study on hamster adrenals (SMITH et al. 1973; see also SCHMIDT et al. 1983). In stimulated glands, exocytotic events were observed in large numbers, whereas they were rare during rest (in the absence of calcium). This was confirmed by AUNIS et al. (1979) in a very detailed freeze-fracture study of stimulated and resting bovine glands (see HELLE and SERCK-HANSSEN, 1985 and BROOKS and CARMICHAEL, 1987 for conventional electron microscopy). In cat adrenals induced by acetylcholine to secrete, SMITH and VAN ORDEN (1973) observed a threefold increase in exocytotic events over control. This study and the following ones employed conventional electron microscopy. "Frequent" exocytotic profiles were seen in rat (ABRAHAMS and HOLTZMAN 1973) and hamster adrenals (BENEDECZKY and SOMOGYI 1975) after stimulation by insulin. In pieces of hamster adrenal, incubated with increasing concentrations of calcium, a concomitant increase of exocytotic events was observed (NAGASAWA 1977). In adrenals from mice subjected to stress evidence of exocytotic activity was frequently seen (KOBAYASHI and SERIZAWA 1979).

All these morphological studies have provided convincing evidence that exocytosis occurs in the adrenal medulla of all species investigated. Furthermore, there is a correlation between the frequency of exocytotic events seen morphologically and the release of catecholamines measured biochemically.

After exocytotic release the soluble constituents of chromaffin granules (including proteins of an effective diameter of about 200 Å, see SMITH and WINKLER 1972) have to pass through the capillary wall to reach the circulation. The most likely area of penetration through this barrier is provided by the so called fenestrae, of about 500 Å diameter, possessing a single layered membrane (see ELFVIN 1965). In a freeze-fracture study (RYAN et al. 1975) a surface view of these capillary fenestrae was obtained (for demonstation by scanning electronmicroscopy see CARMICHAEL and ULRICH 1983)

3. Exocytosis: An All or None Release Process

The studies discussed above have established that during secretion chromaffin granules fuse with the plasma membrane and release their soluble contents directly into the extracellular space. However, we have still to deal with the question of whether the total content of a chromaffin granule is discharged during such a fusion event or whether only a partial release takes place. The morphological studies described in the previous section are consistent with total discharge, since the content of the fused granules is very often seen totally released in the extracellular space (see e.g. DINER 1967; BENEDECZKY and SMITH 1972). Evidence that such a total discharge occurs probably in all exocytotic events was provided by a series of biochemical studies (VIVEROS et al. 1969a,b,c,d; 1971a,b). In these experiments on rabbits a severe loss of catecholamines from the adrenals was induced by treatment with insulin. The properties of the subcellular particles in the adrenals after this treatment were characterized by density gradient centrifugation. A population of normal chromaffin granules with their typical high density and their high adrenaline content was still present, although in reduced numbers. Empty vesicle membranes, as measured by their membrane-bound DBH content, were found in the gradient at low density. There was no evidence of chromaffin granules which had lost only part of their content. Such vesicles should have an intermediate density due to the partial loss of soluble constituents.

These results may not have been representative for the total secretory period, since the subcellular fractions were isolated after an extensive release of catecholamines had occurred. In order to overcome this objection SLOTKIN and KIRSHNER (1973) performed additional experiments on rats killed various times after stimulation by insulin. Again, clear evidence for an all or none release of the granule content was obtained.

In a morphometric study on cat adrenal (ECHEVERRIA et al. 1977) the effect of a high dose of reserpine, likely to cause exocytotic release through stimulation of the splanchnic nerve (see SMITH and WINKLER 1972), was investigated. The number of dense-core vesicles decreased significantly. However, the diameter of these granules did not change. It was concluded that smaller chromaffin granules, which had released only part of their content, did not appear; this again supports an all or none release process. In two more recent morphometric studies on stimulated rat adrenals (NORDMANN 1984; THURESON-KLEIN et al. 1984) there was also a parallel decline in the number of chromaffin granules and in the catecholamine content.

4. The Fate of the Granule Membrane After Exocytosis

We have already discussed the evidence that the membranes of chromaffin granules are retained in the cell during exocytosis. We will now deal with several further questions, the first being:

What happens to the granule membrane after it has fused with the plasma membrane? Obviously, either a mechanism of membrane retrieval must exist, or the plasma membrane would enlarge continuously (see PALADE 1959). Morphological evidence indicates that this retrieval occurs in the form of vesicles, since several studies described an increased appearance of empty vesicles in chromaffin cells after stimulation. In addition to several qualitative studies (see SMITH and WINKLER 1972; DOUGLAS 1974, for references) there is one thorough quantitative investigation (KOERKER et al. 1974). In rabbits treated with insulin, the disappearance of dense core chromaffin granules was quantitatively matched by the appearance of electrontranslucent vesicles (empty vesicles). The size of these vesicles (<95 to 320 nm) showed greater variation than that of chromaffin granules (>95 nm to 320 nm), since about half of the vesicles were microvesicles (< 95 nm). Empty vesicles were also encountered in a study employing subcellular fractionation (MALAMED et al. 1968). Perfused adrenal glands were stimulated with acetylcholine. After severe catecholamine depletion (by 77%) the large granule fraction isolated from the gland contained mainly empty vesicles ranging in size from 30 to 370 nm.

Further evidence that these empty vesicles are in fact derived from the plasma membrane, and/or granule membrane fused with it, was obtained in experiments with exogenous tracers. In the adrenals of insulin-treated rats there was a greatly enhanced endocytotic uptake of horseradish peroxidase (ABRAHAMS and HOLTZMAN 1973). The tracer was found in vesicles, tubules, multivesicular bodies and dense bodies. These latter organelles are probably related to lysosmes (see below), but the vesicles are likely to correspond to those seen in the study by KOERKER et al. (1974) discussed above. An endocytotic uptake of exogenous thorium-dioxide into vesicles of hamster adrenal cells was also observed by NAGASAWA and DOUGLAS (1972). A kinetic analysis of endocytotic uptake of horseradish peroxidase following secretion in cultured bovine chromaffin cells was recently provided (VON GRAFENSTEIN et al. 1986). Finally, quite a different method, i.e. the electrical measurement of capacitance changes in chromaffin cells, also indicates that after exocytosis membrane patches are removed from the plasma membrane (NEHER and MARTY 1982).

It seems, therefore, justified to conclude that, after exocytotic release of the granule contents, membrane pieces (possibly including the granule membrane) are retrieved from the plasma membrane in the form of vesicles of varying sizes (see Fig. 1).

How specific is this membrane retrieval? In principle, two pathways are possible: (i) non-specific membrane retrieval: the granule membrane is incorporated into the plasma membrane; subsequently, parts of the mixed membranes are removed; (ii) specific membrane retrieval: the granule membrane is either immediately retrieved at the fusion site, or the granule membrane, after being incorporated into the plasma membrane, is specifically retrieved at a later stage.

The first evidence for an immediate and specific retrieval of granule membranes was provided by DINER (1967; see also GRYNSZPAN-WINOGRAD 1971; BENEDECZKY and SMITH 1972), who reported that a significant proportion (about 30%) of the chromaffin granule membranes already fused at the plasma membrane had coated pits attached to them. HOLTZMAN and DOMINITZ

(1968) had already shown, with horseradish peroxidase as a tracer, that coated pits along the plasma membrane gave rise to coated vesicles in the cytoplasm. More specifically, NAGASAWA and DOUGLAS (1972), using thorium dioxide as a tracer, presented evidence for a sequence of events where the coated pits as seen by DINER took up the marker and subsequently became cytoplasmic coated vesicles and finally smooth vesicles.

The apparently specific and immediate retrieval mechanism suggested by these experiments provides a mechanism for keeping the two membrane compartments separate, which is of course a prerequisite of their specific functions, i.e. carrying acetylcholine receptors, pumping Na^+ and Ca^+ in the case of the plasma membrane and transport of protons, catecholamines and nucleotides (see WINKLER 1977) for the membranes of chromaffin granules. Biochemical and immunological analyses of these two membranes have demonstrated their distinct and specific properties. The plasma membrane fraction isolated by NIJJAR and HAWTHORNE (1974) was characterized by a low content of lysolecithin (4% of total phospholipids) in contrast to the amount (16%) present in chromaffin granules (see WINKLER and CARMICHAEL 1982). This was also the case for the plasma membrane preparations isolated by WILSON and KIRSHNER (1976) and by MALVIYA et al. (1986). These plasma membranes contained only one tenth of the amount of DBH (probably contamination by granule membranes) found in chromaffin granule membranes (see also MEYER and BURGER 1979b). The polypeptide pattern of the two membranes as seen by sodium-dodecyl-sulphate-electrophoresis was significantly different. Similar results were obtained by ZINDER et al. (1978). In addition, it was found that the plasma membrane had a low content of cytochrome b-561. As recently shown (APPS et al. 1980) this protein corresponds to chromomembrin B, one of the major granule membrane proteins (WINKLER 1971, 1976).

These results are in agreement with an immunohistochemical study on rat glands (WINKLER et al. 1974) with an antiserum against chromomembrin B (i.e. cytochrome b-561). A specific reaction (peroxidase-labelled antibody method) was seen on the membranes of chromaffin granules. Plasma membranes were mostly devoid of reaction product; only in some places a positive reaction (artefactual or regions of exocytotic activity?) appeared.

This biochemical (and also functional) specificity of the plasma membrane is best retained if, during secretion, an immediate and specific membrane retrieval of fused granule membrane occurs (see MELDOLESI et al. 1978, for discussion of similar events in exocrine pancreas). With unspecific retrieval, the specificity of the plasma membrane could only be assured if these membranes receive a constant supply of plasma membrane constituents (e.g. by specific "plasma membrane vesicles") that is much faster than granule membrane incorporation; this would constantly "dilute away" the granule membranes incorporated during secretion. There is at present no evidence to support such a fast supply of plasma membranes which should appear in the cell as a vesicle population of significant proportions (compare colonic epithelium: MICHAELS and LEBLOND 1976). This does not exclude the possibility that the plasma membranes receive a supply of specific vesicles from the Golgi region, since specific membrane replenishment cannot occur via the secretory granules (see Fig. 1). In fact recent studies on the pituitary gland have shown that a constitutive pathway of secretion, which is independent of secretory vesicles, provides the plasma membrane with new membrane material (see GUMBINER and KELLY 1982).

More direct evidence for the specific retrieval of the granule membrane has been obtained in immunological studies. As discussed in Sect. B.I.1.d antigens of chromaffin granules become exposed on the cell surface during stimulation. Qualitative evidence that these membrane antigens are again retrieved into chromaffin cells was obtained in an early study (JACOBOWITZ et al. 1975). These authors demonstrated that fluorescent-labelled antibodies against DBH were taken up into chromaffin cells. Similar results were obtained in several more recent studies where careful comparisons between control and stimulated bovine chromaffin cells were made (WILDMANN et al. 1981; DOWD et al. 1983; PHILLIPS et al. 1983). In quantitative experiments (cytotoxicity test: LINGG et al. 1983; immunofluorescence: PATZAK et al. 1984) it was demonstrated that the removal of granule antigens, which had been exposed during stimulation on the cell surface, was quite efficient. Within 30 minutes antigen corresponding to about 20% of the granule population had disappeared. Throughout this process of exposure and retrieval of the granule antigen, the specific immunofluorescence obtained on the cell surface was patchy and never diffuse (PATZAK et al. 1984 see also LOTSKAW et al. 1986). This is the most direct evidence that during exocytosis granule membranes do not "mix" with the plasma membrane but are retrieved specifically and efficiently. This was finally confirmed at the ultrastructural level with immunogold labelling of a specific protein of chromaffin granules (PATZAK and WINKLER 1986; PATZAK et al. 1987).

What is the mechanism of membrane retrieval? As already suggested by early morphological studies (DINER 1967; HOLTZMAN and DOMINITZ 1968; BENEDECZKY and SMITH 1972; NAGASAWA and DOUGLAS 1972) membrane coating seems to be involved in this process. Direct evidence for this was obtained recently. The antigen patches of granule membranes exposed on the cell surface were specifically marked with gold-labelled antibodies. During subsequent incubation these granule patches became coated and finally coated vesicles with the specific immunological label appeared in the cytoplasm (PATZAK and WINKLER 1986). Thus, coated vesicles are responsible for endocytosis of granule membranes incorporated into the plasma membrane. The exact molecular mechanism of this process is still unknown. Coated vesicles have been isolated from several tissues, including the adrenal medulla (PEARSE 1976; see also PEARSE and BRETSCHER 1981). Their coat is made up of a major protein species, called clathrin, of molecular weight 180000 (PEARSE 1976). This protein, together with an additional component of molecular weight 110000, appears in Tris/Cl in the form of filamentous aggregates, but can be reconstituted into basket-like structures (KEEN et al. 1979; SCHOOK et al. 1979). A similar process might occur in the cytoplasm. It is not known whether retrieval by clathrin-coated vesicles involves additional cellular components. In any case retrieval of granule antigens from the cell surface was not inhibited by colchicine, cytochalasin B or trifluoperazine (PATZAK et al. 1984). This would indicate that neither microtubules, nor microfilaments nor calmodulin participate in membrane retrieval. However energy seems required, since a complete block of retrieval was obtained in the presence of sodium azide (PATZAK et al. 1984). Finally the most recent studies on endocytosis in cultured bovine chromaffin cells indicate that this process may not be dependent on Ca^{2+} ions quite in contrast to exocytosis (von GRAFENSTEIN et al. 1986)

What is the final fate of the retrieved granule membranes within the chromaffin cell? In a previous review (WINKLER 1977; see also HOLTZMAN 1977; MELDOLESI et al. 1978; MELDOLESI and CECCARELLI 1981) two possibilities have been discussed in great detail. I will, therefore, concentrate here only on recent studies shedding some further light on this question. We face two alternatives for the final fate of the granule membrane. (i) There may be uptake into lysosomes followed by digestion; in this case the synthesis of both secretory and membrane proteins must occur at the same rate to ensure biogenesis of complete granules. (ii) The retrieved membranes may return to the Golgi region and be re-used for packaging newly synthesized secretory proteins. In this case the rate of synthesis of the granule membranes is expected to be lower than that of the soluble proteins which are lost by secretion (see WINKLER 1971). Since this was actually observed (WINKLER et al. 1972) we favoured this second mechanism (see WINKLER et al. 1974; WINKLER 1977).

On the other hand, the first possibility was supported by early studies with exogenous tracers (horseradish peroxidase, thorium dioxide), which, after having been taken up into the chromaffin cells subsequent to exocytosis, were found in lysosomes (NAGASAWA and DOUGLAS 1972; ABRAHAMS and HOLTZMAN 1973; BENEDECZKY and SOMOGYI 1978). However, more recent studies employing additional tracers (dextran, cationized ferritin) demonstrated that these tracers, after uptake into secretory cells, reach the stacked Golgi-cisternae, condensing vacuoles and newly formed secretory vesicles within a few minutes after exposure of the cells (see FARQUHAR 1983; lacrimal and parotid gland: HERZOG and FARQUHAR 1977; anterior pituitary: FARQUHAR 1978; thyroid cells: HERZOG and MILLER 1979; adrenal medulla: SUCHARD et al. 1981). These observations are consistent with membrane traffic from the plasma membrane to the Golgi complex and into secretory vesicles. The possibility was discussed (see FARQUHAR 1978, 1982) that the membranes, retrieved from the cell surface, first fuse with lysosomes, discharge their content and subsequently reach the Golgi stacks. Such a mechanism might explain why tracers like horseradish peroxidase, not bound to membranes, end up in lysosomes. Thus, these studies have demonstrated that membranes retrieved from the cell surface reach the Golgi region and become incorporated into newly formed secretory vesicles. However, since the tracers did not specifically mark membranes of secretory vesicles, these studies have not yet established that the recycled membranes are those of secretory vesicles, which fused with the plasma membrane during exocytosis.

Experiments relevant for this latter point have been performed on adrenal medulla tissue. In one study (SUCHARD et al. 1981) isolated bovine chromaffin cells were enzymatically iodinated with ^{125}I during carbachol-induced secretion. One hour later, labelled membrane proteins were found in the chromaffin granule fraction, but only those proteins were labelled which face the content of chromaffin granules. The most likely explanation for these findings is that the granule membrane proteins exposed on the cell surface during exocytosis are iodinated and are subsequently retrieved and finally reutilized in the Golgi region for packaging new granules (see Fig. 1). Even more direct support for such a process was obtained recently (PATZAK and WINKLER 1986). In this study patches of granule membranes exposed on the cell surface of stimulated cells were specifically marked with gold-labelled antibodies. During subsequent incubation these labelled membranes appeared in the cytoplasm, in the Golgi region and also in what appeared to be newly formed chromaffin granules (PATZAK and WINKLER 1986). Therefore the original suggestion (WINKLER et al. 1972; WINKLER et al. 1974; WINKLER 1977) that chromaffin granule membranes are reutilized for several secretory cycles has gained significant support. However, we should finally mention one experiment which apparently does not fit into this scheme. LEDBETTER et al. (1978) reported, that after ^{3}H-leucine labelling of bovine adrenal glands, the synthesis rate of soluble DBH was similar to that of the membrane-bound enzyme (see also NOBILETTI and KIRPEKAR 1983). These resultats disagree with a previous study (WINKLER et al. 1972) in which different synthesis rates of total membrane proteins and secretory proteins of chromaffin granules were reported and they do not support re-use of granule membranes. However, one has to consider that membrane DBH may not be representative of the total membrane proteins. In fact, it has been suggested that the membrane-bound enzyme might be a precursor of the soluble one (see BJERRUM et al. 1979) which diminshes its usefulness as a specific marker of membrane proteins.

5. Conclusions

From the foregoing discussion the following conclusions seem justified (see Fig. 1). Secretion of catecholamines from the adrenal medulla occurs by exocytosis, i.e. chromaffin granules fuse with the plasma membrane and discharge their total soluble content (all or none release) into the extracellular space. The membrane of the chromaffin granule is likely to be specifically and completely retrieved from the plasma membrane by a relatively fast and efficient fission process. Some of these retrieved membranes may become digested by lysosomes, but at least a portion of them returns to the Golgi region for reuse.

II. Sympathetic Nerve

Results of early pharmacological studies (VAN ORDEN et al. 1967; HÄGGENDAL and MALMFORS 1969; see SMITH and WINKLER 1972 for further references) had indicated that the cellular source of noradrenaline released from sympathetic nerve is that present in the amine-storing vesicles. These findings were consistent with secretion directly from this storage site and therefore with exocytosis, but did not prove such a concept. As was the case for the adrenal medulla, this proof depended critically on biochemical evidence (see SMITH 1971, 1972a, 1978; DE POTTER et al. 1971a; SMITH and WINKLER 1972; GEFFEN and JARROTT 1977).

1. Biochemical Evidence for Exocytosis

a. Secretion of Chromogranin A, Dopamine β-Hydroxylase and Enkephalin

In the late sixties two groups of workers obtained independent evidence for the release from sympathetic nerves of proteins specific for noradrenaline-storing vesicles. GEFFEN et al. (1969, 1970) published qualitative immunological data indicating the release of chromogranin A and DBH from sheep splenic nerve. For perfused dog and calf spleen DE POTTER, SMITH and coworkers (DE POTTER et al. 1969a,b; SMITH et al. 1970) obtained the first quantitative data which were of critical importance for establishing the concept of exocytosis in sympathetic nerves. Stimulation of the splenic nerve of the calf caused the calcium-dependent release of 82 nmoles of noradrenaline, 460 units of DBH and 0.21 µg of chromogranin A. A cytoplasmic enzyme, i.e. aromatic amino acid decarboxylase did not appear in the perfusates during stimulation (SMITH et al. 1970). The release of DBH during stimulation of the cat splenic nerve was reported in a subsequent study (GEWIRTZ and KOPIN 1970). When these authors (SMITH et al. 1970; GEWIRTZ and KOPIN 1970) attempted to correlate the relative amounts of the constituents found in the perfusate with those present in the storage vesicles inside the sympathetic nerve, great difficulties were encountered and the relationship appeared poor. This, however, is not surprising since the degree of re-uptake of noradrenaline, the composition of the vesicles in the terminal axon and the contribution of the two populations of vesicles (large and small dense core ones) to secretion is not known (see SMITH and WINKLER 1972 for discussion, and below). Therefore the soluble DBH pool in nerves could be estimated only on the basis of several assumptions. WEINSHILBOUM et al. (1971) tried to overcome this problem by measuring the ratio of noradrenaline to soluble DBH in an aqueous extract of total tissue (guinea-pig vas deferens). When this ratio was compared with the amounts of the components released during stimulation in the presence of re-uptake blockers (e.g. phenoxybenzamine) a reasonable agreement was obtained. However, the enzyme in the perfusate seemed to be too low by a factor of two. In the meantime it has been shown (KIRKSEY et al. 1978) that the soluble DBH pool from large dense core vesicles is only insufficiently released by aqueous extraction. Taking this into account, the lack of agreement between the ratios of noradrenaline to soluble enzyme in the tissue and in the perfusates was probably underestimated by WEINSHILBOUM et al. (1971). Unequal diffusion rates of noradrenaline and DBH in the tissue after secretion from the cell may be one reason for this disproportionate release (see the following). There is evidence that significant amounts of DBH released from nerve terminals enter the lymph (ROSS et al. 1974; NGAI et al. 1974).

DBH is also discharged during stimulation of rabbit, guinea-pig and rat hearts (LANGLEY and GARDIER 1977; ALGATE and LEACH 1978) and this is enhanced by cyclic nucleotides (LANGLEY and WEINER 1978) and angiotensin II (ACKERLY et al. 1978). When a low sodium-urea solution is used for perfusing rabbit hearts, the wash-out of DBH released by high potassium is significantly increased (MUSCHOLL et al. 1980). This solution may open capillary pores and facilitate washout. In a further series of experiments (MUSCHOLL and SPIRA 1982; see also MUSCHOLL et al. 1985), washout from the perfused hearts was further improved by collecting "transmyocardial fluid". The released amounts of noradrenaline and DBH were compared with the tissue pools. Stimulation depleted the hearts of noradrenaline and soluble DBH in a stoichiometric manner, provided that noradrenaline synthesis was blocked. Thus, these studies finally establish that there is a concomitant and stoichiometric release of noradrenaline and soluble DBH from the sympathetic nervous system.

There is also evidence that DBH levels decrease in a vesicle fraction isolated from sympathetic tissues after stimulation. Thus, there is a 30% reduction of this enzyme in the vesicle fraction from rat heart when the animals are subjected to cold stress (FILLENZ and WEST 1974; WEST and FILLENZ 1975). On the other hand, only a small (less than 10%) decrease of DBH content was observed after prolonged stimulation of rat salivary glands (HÄGGENDAL 1982).

When isolated brain synaptosomes were stimulated either electrically or with excess potassium, there was a parallel release of DBH and noradrenaline into the incubation medium (WYLLIE and GILBERT 1980). This preparation, like the isolated chromaffin cells from the adrenal medulla, may prove useful for studying exocytosis in nervous tissue.

Numerous studies established the concomitant release of catecholamines and DBH from several tissues under a variety of conditions. Electrical stimulation or depolarization by potassium or veratridine of guinea-pig vas deferens led to analogous results for the secretion of noradrenaline and DBH (THOA et al. 1975). In perfused dog and cat spleens and in guinea-pig vas deferens phenoxybenzamine increased the stimulated outflow of both components (DE POTTER et al. 1971b; JOHNSON et al. 1971; CUBEDDU et al. 1974a), as a result by its blockade of presynaptic inhibitory α-adrenoceptors (see STARKE 1977). The release of these two components from cat spleen and guinea-pig vas deferens is enhanced by cyclic AMP (WOOTEN et al. 1973; CUBEDDU et al. 1975) and papaverine (CUBEDDU et al. 1974b), whereas colchicine, vinblastine or cytochalasin B inhibit their discharge in the guinea-pig vas deferens (THOA et al. 1972).

After reserpine treatment and depletion of the noradrenaline storage vesicles, stimulation of the nerve still leads to the release of DBH with only small amounts of nordarenaline (THOA et al. 1975; CUBEDDU and WEINER 1975a).

Tyramine treatment leads to secretion of noradrenaline without concomitant release of DBH from perfused spleen (CHUBB et al. 1972) and isolated guinea-pig vas deferens (THOA et al. 1975). This is readily explained by the concept, based on several lines of evidence, that tyramine acts by displacing noradrenaline from its storage site without causing exocytosis. On the other hand, the reduction of noradrenaline but not of enzyme release in stimulated vasa deferentia by halothane remains unexplained (ROIZEN et al. 1975).

During stimulation of sympathetic nerves of the bovine vas deferens met-enkephalin is released together with noradrenaline (DE POTTER et al. 1987).

b. Secretion of Other Constituents from Noradrenaline Storage Vesicles

An attempt to demonstrate release of vesicular ATP from perfused cat spleen was undertaken by STJÄRNE et al. (1970). The ATP in the spleen was labelled with ^{32}P-phosphate. During induced secretion of noradrenaline there was no additional outflow of labelled ^{32}P-nucleotides. Since ATP is not a specific constituent of the vesicles it was probably impossible to pick up a small amount of ATP secreted from vesicles against a background of release from other sources. In fact, WESTFALL et al. (1975) have calculated for the vas deferens that the labelled nucleotide pool in storage vesicles amounts to only 0.7% of the total ATP pool (see also FREDHOLM et al. 1982). On the other hand, release of unlabelled ATP from sympathetic nerves in the vas deferens (FRENCH and SCOTT 1983) and from isolated synaptosomes from brain cortex (WYLLIE and GILBERT 1980; see also WHITE 1978) has been demonstrated. The vesicular origin of this released ATP was not proven.

BLASCHKE and UVNÄS (1979) studied the content of noradrenaline, ATP and ^{35}S-labelled mucopolysaccharides in a vesicle fraction (presumably small dense core vesicles) isolated from spleen. A significant decrease of the noradrenaline content was found in these vesicles when they were isolated after nerve stimulation. However, neither ATP nor ^{35}S-labelled mucopolysaccharides had decreased which was taken as evidence against release of these components concomitant with noradrenaline.

However, vesicle fractions isolated from spleen with simple gradient procedures are quite impure (see SMITH 1972b). It cannot, therefore, be excluded that most of the ATP and mucopolysaccharides present in "vesicle fractions" are due to contamination with other cell particles. A decrease of these components in these crude preparations after nerve stimulation might, therefore, not be expected.

c. Conclusions

All the results just discussed have established that secretion of noradrenaline from sympathetic nerves is accompanied by the release of specific constituents of noradrenaline storage vesicles, i.e. chromogranin A and DBH. In the absence of release of cytoplasmic enzymes the discharge of these molecules which are so divergent in size can only be brought about by a mechanism of exocytosis (see SMITH and WINKLER 1972, for a detailed argument). This conclusion still leaves several questions unanswered, e.g. from which vesicle population (large and/or small dense core vesicles?) is this exocytotic release occurring? Is it an all or none release? However, before we try to tackle these questions we will discuss the morphological evidence for exocytotic release in sympathetic nerve.

2. Morphological Evidence for Exocytosis

FILLENZ (1971) provided the first morphological evidence for exocytosis of large dense core vesicles. The published electron micrograph showed such a vesicle which had fused with the neuronal membrane of a sympathetic nerve in rat vas deferens (see also review by FILLENZ 1979). This was confirmed and extended by detailed studies of THURESON-KLEIN and collaborators (THURESON-KLEIN et al. 1976a,b, 1979a,b; THURESON-KLEIN and STJÄRNE 1979; THURESON-KLEIN 1983; ZHU et al. 1986). In one example (THURESON-KLEIN et al. 1979a, Figs. 3, 10), the content (the dense core) of a vesicle already fused with the neuronal membrane of a splenic nerve terminal could still be seen in the extracellular space. The membrane of the fused vesicle was coated. Stimulation of the nerves from human omental veins led to an increase of exocytotic profiles of large dense core vesicles (THURESON-KLEIN et al. 1979b; THURESON-KLEIN et al. 1976a,b).

FILLENZ (1971) also published an electron micrograph showing a small vesicle in the sympathetic nerve of the vas deferens which had apparently fused with the neuronal membrane. THURESON-KLEIN et al. (1976b, 1979a) found only large dense core vesicles involved in exocytosis in unstimulated preparations of bovine splenic nerves and of nerves in human omental veins. However, stimulation of the latter nerves (in the presence of phentolamine) also led to the fusion of small vesicles with the plasma membrane (THURESON-KLEIN et al. 1979b, THURESON-KLEIN and STJÄRNE 1979).

A very thorough study by BASBAUM and HEUSER (1979) of sympathetic nerves of mouse vas deferens added considerable evidence that exocytotic release can occur from small dense core vesicles. The varicosities of these nerves have 51 small vesicles per μm^2, 73 % of them having a dense core. 30 min after electrical stimulation the number of small vesicles had decreased by 60 % and only 31 % of them had dense cores (see DIXON et al. 1978 for similar results on rat heart sympathetic fibres stimulated with high potassium, see also POLLARD et al. 1982b for mouse vas deferens). After a recovery period the number of vesicles returned to control values, whereas the dense core was still present in only 48 %. When the tissue was exposed to an exogenous tracer, i.e. horseradish peroxidase, under appropriate conditions of stimulation, many small vesicles became labelled. Large dense core vesicles, which represent only 7 % of the total vesicles in the tissue, changed neither in number nor in dense core, nor did they participate in horseradish peroxidase uptake. This interesting study is relevant to many questions concerning exocytosis (see the following). In this section it suffices to say that the most likely explanation for these results is that, during stimulation, small dense core vesicles fuse with the neuronal membrane, discharge their contents and finally reform by an endocytotic process (see Fig. 2). The role of large dense core vesicles has not been elucidated further by this study.

3. Immunohistochemical Evidence for Exocytosis

Fluorescein-labelled antibodies to DBH, injected intravenously, are taken up into sympathetic nerve terminals (JACOBOWITZ et al. 1975). These labelled antibodies reach the cell body by retrograde transport (ZIEGLER et al. 1976). Further studies demonstrated the specificity of this uptake into sympathetic nerves (FILLENZ et al. 1976). In the presence of complement, this antibody treatment leads to lysis and degeneration of sympathetic neurons (COSTA et al. 1976; FURNESS et al. 1977, 1979).

These results indicate that DBH becomes exposed extracellularly at sympathetic varicosities, is subsequently retrieved into the terminals and is finally retrogradely transported in the axon (see Fig. 2). An analysis (RUSH et al. 1979) of these events at the ultrastructural level revealed that the DBH antibodies (conjugated with horseradish peroxidase) were taken up into the terminal in small vesicles (less then 35 nm). Up to 30 % of the vesicle population in a given nerve terminal became labelled. The plasma membrane (two hours after stimulation) showed little positive reaction. These results are in agreement with the data (see preceding section) on horseradish peroxidase uptake into small vesicles by BASBAUM and HEUSER (1979) but, due to the specificity of the DBH-antibodies, demonstrate that uptake of this exogenous tracer occurred into small vesicles actually containing this enzyme. All these findings are consistent with exocytosis from vesicles which contain membrane-bound DBH which is present on the inner surface of the vesicle membrane (see Fig. 2).

4. Electrophysiological Evidence for Exocytosis

Early electrophysiological studies of the vas deferens had already made it likely that noradrenaline is released in discrete packets causing spontaneous excitatory junction potentials (see review by HOLMAN 1970). In a study on the guinea-pig vas deferens BLAKELEY and CUNNANE (1979; see also HIRST and NEILD 1986) concluded that excitatory junction potentials recorded intracellularly were made up of "discrete events", which probably represent the release of a single packet of noradrenaline. It was suggested that each stimulus liberated a small number of packets (from 0 to 10) from one varicosity. These electrophysiological results providing evidence for the quantal release of transmitter are consistent with secretion by exocytosis.

5. The Contribution of Large and Small Dense Core Vesicles to Exocytosis and Their Possible Relationship

Before we enter into this question we have to summarize what is known about the composition of these two types of vesicles (for evidence of two types of vesicles: see SMITH 1972b; GEFFEN and LIVETT 1971; THURESON-KLEIN 1982a). The large dense core vesicles of diameter 85 nm (see BISBY and FILLENZ 1971) resemble in composition chromaffin granules (see LAGERCRANTZ 1976; KLEIN 1982). They contain in their soluble core catecholamines, ATP, chromogranin A and B, secretogranin II (for nomenclature see EIDEN et al. 1987), enkephalins and neuropeptide Y (see LAGERCRANTZ 1976; BARTLETT et al. 1976; KLEIN et al. 1982; NEUMANN et al. 1984; FRIED et al. 1985; HAGN et al. 1986; DE POTTER et al. 1987). When these vesicles have reached the terminals by axonal flow their catecholamine/ATP ratio may be at least in the range of 7 to 12 (YEN et al. 1976; LAGERCRANTZ 1976; KLEIN 1982), which is significantly higher than that of chromaffin granules (average about 4; WINKLER 1976). The membranes of large dense core vesicles contain (see LAGERCRANTZ 1976; BARTLETT et al. 1976) DBH, a Mg^{2+}-activated ATP-ase, cytochrome b-561 (i.e. chromomembrin B: APPS et al. 1980) and lipids with only traces of lysolecithin, in contrast to the high amount in chromaffin granules.

Whereas in bovine chromaffin granules 50% of the DBH is membrane-bound, in the large dense core vesicles this percentage may be lower (about 30 to 40%, KIRKSEY et al. 1978; see also KLEIN 1982).

Much less is known about the composition of small dense core vesicles of diameter 50 nm (BISBY and FILLENZ 1971). It seems likely that they contain not only noradrenaline but also ATP. FRIED et al. (1978), working with the vas deferens of castrated rats, found some ATP in density gradient fractions where the small dense core vesicles equilibrate. However, a contribution from other cell organelles could not be completely excluded. With the same preparation, pulse labelled with ^{3}H-adenosine, ABERER et al. (1979) found ^{3}H-ATP in the relevant fractions after density gradient centrifugation (see also FREDHOLM et al. 1982). In a histochemical study (RICHARDS and DA PRADA 1977) both large and small dense core vesicles in sympathetic nerve gave an uranaffin reaction, thought to be specific for ATP.

Are specific proteins, e.g. chromogranin A, present in the content of small vesicles? Morphological studies convincingly demonstrate that large dense core vesicles contain a proteinaceous core which can still be visualized in the electron microscope when the catecholamines are depleted by reserpine or incubation (see JAIM-ETCHEVERRY and ZIEHER 1969; TILL and BANKS 1976; THURESON-KLEIN et al. 1974; THURESON-KLEIN 1982b), whereas in small vesicles depletion of noradrenaline led to a parallel loss of the dense-cores in most experiments (HÖKFELT 1966; FARRELL 1968; JAIM ETCHEVERRY and ZIEHER 1969). In some experiments, however, a matrix component (ATP, protein?) was still observed in small vesicles although their noradrenaline content had been reduced (DIXON et al. 1978; FRIED et al. 1981). Recent biochemical experiments (NEUMAN et al. 1984) have yielded more conclusive results. Analysis of subcellular fractions by density gradient centrifugation revealed that small dense core vesicles do not contain chromogranin A or enkephalin - like material (see also DE POTTER et al. 1987). They are also devoid of neuropeptide Y, which is present in large dense core vesicles (FRIED et al. 1985).

The membranes of small dense core vesicles are likely to contain a Mg-ATP-ase (necessary for catecholamine uptake to occur in these vesicles: BAREIS and SLOTKIN 1979; FRIED 1981), cytochrome b-561 (FRIED 1978) and DBH (BISBY et al. 1973; NELSON and MOLINOFF 1976; FRIED et al. 1978; LUNDBERG et al. 1977; but see the following and DE POTTER and CHUBB 1977; and WILLEMS and DE POTTER 1983 for evidence of a vesicle fraction devoid of DBH).

Let us now try to correlate these results with the studies on release as discussed in previous sections. Noradrenaline release can occur from both vesicle populations since, during increased sympathetic activity, the noradrenaline content of light (small) and dense (large) vesicles declines as shown by a density gradient centrifugation study (FILLENZ and HOWE 1975). It seems justified to conclude that the release of DBH chromogranin A and peptides is from the large dense core vesicles. The release of these components together with the morphological findings is a clear indication that large dense core vesicles secrete their content, i.e. noradrenaline, chromogranin A, DBH and peptides (and in analogy with the adrenal medulla also ATP and further soluble constituents), by a process of exocytosis. Furthermore, the morphological studies discussed above clearly indicate that also small dense core vesicles secrete their content, i.e. noradrenaline (and ATP?), by exocytosis. These statements still leave several questions unanswered.

Is the release from large dense core vesicles an all or none phenomenon? It has been suggested (see e.g. GEFFEN and LIVETT 1971; GEFFEN and JARROTT 1977) that large dense core vesicles fuse several times with the plasma membrane, lose part of their content and of their membrane each time and are thus converted to small dense core vesicles. If such partial release occurs, one would expect that, during stimulation, a population of partially filled vesicles with reduced density (more membrane relative to content) should appear. However, the careful study by DE POTTER and CHUBB (1977) on perfused dog spleen showed that this is not the case. They analysed the vesicle populations by density gradient centrifugation. After stimulation of the splenic nerve the decrease in the large dense core vesicles was matched by an increase in a defined population of small vesicles without any evidence for only partially filled intermediate vesicles; this strongly argues for an all or none release from large dense core vesicles.

What is the relationship between large and small dense core vesicles? SMITH (1971, 1972a, 1978) has suggested that the small dense core vesicles arise in the terminals from the membranes of large dense core vesicles after they had fused with the neuronal membrane (see Fig. 2). During recent years this concept has gained support from several findings. (i) Large vesicles fused with the plasma membrane exhibit a coating indicative of membrane retrieval (e.g. THURESON-KLEIN et al. 1979a). (ii) BASBAUM and HEUSER (1979) have shown that large dense core vesicles do not take up any horseradish peroxidase added from the outside. Since there is good evidence that they undergo exocytosis, the lack of uptake of horseradish peroxidase into large or intermediate vesicles is consistent with the concept that membrane retrieval of these fused large vesicles leads to small vesicles which actually do contain exogenous tracer (BASBAUM and HEUSER 1979).

In this context one wonders why BASBAUM and HEUSER (1979) did not observe a significant decrease (only from 3.73 per μm² to 3.11) in the number of large dense core vesicles per μm² after stimulation of vas deferens, whereas the number of small vesicles declined. A definite answer cannot be offered; however, one has to consider that in the mouse vas deferens, large dense core vesicles constitute only 7% of the total vesicle population. Therefore, assuming that they have to compete with the small dense core vesicles for the same receptor sites, their chance to secrete by exocytosis is low. In the sympathetic nerve terminals of human veins, large dense core vesicles comprise a relatively high percentage, i.e. 30%, of the total vesicles. Stimulation of this tissue leads to an increase of the small vesicles by 30% (THURESON-KLEIN et al. 1979b). This might be an indication that, during stimulation of these nerves, sufficient numbers of large dense core vesicles are converted into small vesicles to permit the detection of this change by quantitative morphology.

(iii) The study by DE POTTER and CHUBB (1977), already mentioned before, provides a balance sheet for membrane-bound DBH during nerve stimulation. In density gradients, this enzyme disappeared from the large dense core vesicle fraction and quantitatively appeared in a fraction corresponding to small dense core vesicles.

All these three lines of evidence support the idea that small dense core vesicles are formed from large ones. However, one crucial problem remains.

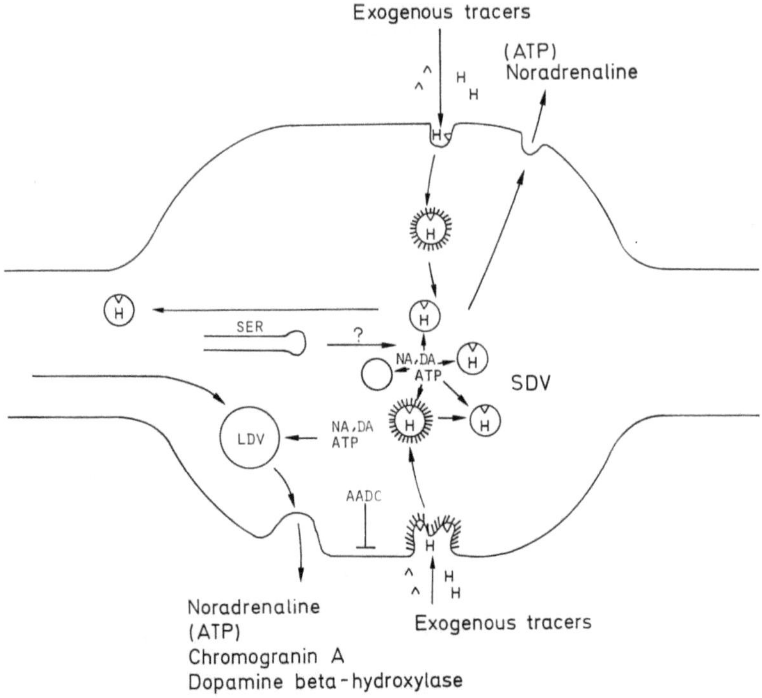

Fig. 2 Exocytosis in sympathetic nerve. Large dense core vesicles (LDV) reach the varicosity by fast axoplasmic flow. They discharge their content by exocytosis (all or none process). Secretion of noradrenaline, chromogranin A, enkephalin and DBH has already been demonstrated. ATP release is likely to occur. Cytoplasmic enzymes (like AADC: aromatic-L-amino acid decarboxylase) are not released during secretory activity. After exocytosis the membranes of LDV are thought to be retrieved, forming small vesicles (SDV). According to the scheme presented here the SDV do not contain any soluble proteins, but have the same membrane as LDV. Both vesicles are thought to contain DBH and, therefore, both types of vesicles can synthesize noradrenaline (NA) from dopamine (DA). A further contribution to the population of small vesicles may come from the smooth endoplasmic reticulum (SER). Such vesicles might have a different composition. SDV can take up noradrenaline and ATP from the cytoplasm. SDV vesicles release their content by exocytosis, are reformed by fission, take up noradrenaline and ATP again and are likely to be re-used for several secretory cycles. Exogenous tracers (H: horseradish peroxidase, $\wedge$: antibodies against DBH) are taken up by vesicles during exocytosis. These markers are found only in small, but not in large vesicles. Some of these vesicles are transported retrogradely to the cell body. For further details and references see text

According to this hypothesis, the membranes of small and large vesicles should be identical. As already pointed out above, the membrane of small vesicles have several proteins in common with those of the large one. However, as far as DBH is concerned, the evidence is controversial. Experiments favouring both the presence of this enzyme in small vesicles (BISBY et al. 1973; NELSON and MOLINOFF 1976; LUNDBERG et al. 1977; FRIED et al. 1978), and also its absence (DE POTTER and CHUBB 1977; WILLEMS and DE POTTER 1983; see also KLEIN and THURESON-KLEIN 1984) have been published. In the most recent study (NEUMAN et al. 1984) of this question additional evidence for the presence of DBH in small dense core vesicles was obtained. Subcellular fractionation was performed with vasa deferentia of two species, i.e. rat and ox. DBH was determined not only enzymatically, but also by studying the biosynthesis of ^{3}H-noradrenaline from ^{3}H-tyrosine. Taking all the available evidence into account we can reach the tentative conclusion, that DBH is present in both types of vesicles. This, however, does not exclude the possibility that an additional small vesicle pool which stores noradrenaline but is devoid of DBH is present (DE POTTER and CHUBB et al. 1977). Such vesicles might be formed from the endpoplasmic reticulum (see DROZ et al. 1975; HOLTZMANN 1977). In this context, we have to discuss recent data on a protein called p38 or synaptophysin. According to immunelectron microscopy, this protein is only found in small synaptic vesicles but is absent from large ones (NAVONE et al. 1986; WIEDENMANN et al. 1986). Analogously, in adrenal medulla synaptophysin was found in a special population of vesicles of unknown function, but was apparently not associated with chromaffin granules (NAVONE et al. 1986). On the other hand, by a biochemical analysis of density gradient fractions we have found that also chromaffin granules contain synaptophysin but only in low concentration (D. OBENDORF and H. WINKLER, to be published), whereas a vesicle unrelated to chromaffin granules contained high concentrations. If, in analogy, large dense core vesicles in nerves contain only small amounts (according to these biochemical results) or no synaptophysin (NAVONE et al. 1986) the small vesicles containing high concentrations of this protein cannot be derived from them. Possibly adrenergic nerves have three types of vesicles: large dense core ones, small dense core ones derived from large ones after their exocytosis and therefore having identical membranes and "synaptic" vesicles containing synaptophysin of unknown function within adrenergic nerves. Only a careful analysis of several membrane antigens will provide a final answer.

Can the small dense core vesicles by reutilized for several cycles of secretion? Several lines of recent evidence favour a re-use of small vesicles (see GEFFEN and LIVETT 1971; SMITH 1972a; DAHLSTRÖM 1973). BASBAUM and HEUSER (1979) have shown that, in stimulated vasa deferentia, small dense core vesicles disappear for some time and subsequently reappear (containing exogenous tracers indicating previous contact with the outside of the nerve). During the post-stimulation period the vesicles which have re-appeared recover a dense core. The most likely explanation for these results is that small vesicles fuse with plasma membrane, discharge their noradrenaline content, reform and slowly fill up again with noradrenaline.

Convincing evidence for re-use has been provided by WAKADE (1979). In the presence of tetraethylammonium chloride stimulation of rat vas deferens leads to near complete depletion of noradrenaline. However, this tissue is still able to take up and retain ^{3}H-noradrenaline in amounts identical to those in control tissues. This newly acquired ^{3}H-noradrenaline can again be released by a calcium-dependent secretory process. The most likely explanation is that the vesicles, having undergone release in the first stimulation period, are still capable of taking up and of releasing noradrenaline in a second one (see GARCIA and KIRPEKAR 1973, for similar experiments and conclusions for cat spleen; see HÄGGENDAL 1982, for salivary gland).

In a further series of experiments on rat hearts (WAKADE and WAKADE 1982a; WAKADE et al. 1982; WAKADE and WAKADE 1984b) these results were confirmed and extended. It was shown that depletion of noradrenaline from the heart was accompanied by a loss of the dense core of small vesicles, whereas these cores reappeared during refilling of the stores (JAIM-ETCHEVERRY and ZIEHER 1983 for sympathetic nerves in pineal gland).

On the other hand, FILLENZ and WEST (1974) suggested that vesicle re-use was limited. After cold treatment of rats, leading to sympathetic stimulation, vesicular pellets (microsomal fractions) from heart contain 30% less DBH than those from control rats. This decrease of DBH appeared to be more than could be accounted for by the loss of soluble enzyme; this seemed to suggest vesicle destruction. However, a redistribution of the vesicle membranes into other subcellular fractions (e.g. by fusion with the neuronal membrane) was not excluded in this study. After prolonged stimulation (24 hours) of the sympathetic nerves by cold exposure, the storage capacity of rat heart vesicles was reduced by 22% (FILLENZ and WEST 1976; see also FILLENZ 1977 for further discussion). This was taken as evidence for vesicle loss without any re-use. However, considering the long period of stimulation, the small decline in storage capacity seems to provide evidence for re-use rather than for significant vesicle disappearance.

Considering all the evidence just discussed, the argument for local re-use of small vesicles seems quite strong. Presumably, after several cycles of release, these vesicles are retrogradely transported to the neuronal body as shown by the retrograde transport of DBH (see BRIMIJOIN and HEILAND 1976) or DBH antibodies, which had previously reacted with this enzyme during exocytosis (see above). Recent evidence suggests that this retrograde transport may involve large dense core vesicles with a reserpine-insensitive transport system (SCHWAB and THOENEN 1983).

6. Conclusions

Large dense core vesicles of sympathetic nerves discharge their total soluble content (at least noradrenaline, chromogranin A, DBH and neuropeptides) by exocytosis (see Fig. 2). It seems likely that the retrieved membranes of these vesicles (containing membrane-bound DBH) give rise to small dense core vesicles. These latter vesicles discharge their content (at least noradrenaline, but not any specific proteins) by exocytosis. They are likely to be re-used for several cycles of release. The significance of small vesicles, possibly formed from endoplasmic reticulum and not containing DBH, is at present difficult to evaluate.

C. Mechanism of Exocytosis

The first step in exocytosis is the attachment of the granule membrane to the plasma membrane; this is followed by the actual fusion process. One essential prerequisite for a molecular understanding of these events is a thorough knowledge of the molecular organization of the two membranes involved. Therefore, an account of the topography of these membranes will be presented first. In this section on the mechanism of exocytosis we will concentrate on adrenal medulla since few data are available for sympathetic nerves.

I. Molecular Organization of Membranes Involved in Exocytosis

Very little is known about the composition and even less about the organization of plasma membranes from chromaffin cells. Several membrane preparations have been isolated (NIJJAR and HAWTHORNE 1974; WILSON and KIRSHNER 1976; ZINDER et al. 1978; MEYER and BURGER 1979b). In contrast to chromaffin granules, their lipid composition is characterized by a low lysolecithin content of 1 to 4% (NIJJAR and HAWTHORNE, 1974; MALVIYA et al. 1986). The proteins have been resolved by sodium dodecylsulphate electrophoresis. About 20 components were seen (ZINDER et al. 1978). The pattern differed significantly from that of granule membrane proteins. A portion of these proteins is exposed on the cytoplasmic surface of the plasma membranes (VAN DER MEULEN et al. 1981).

Much more is known about the granule membranes (for a more detailed discussion see WINKLER and WESTHEAD 1980; WINKLER and CARMICHAEL 1982; WINKLER et al. 1986). The basic structure is a bimolecular lipid leaflet. Detailed data on the localization of phospholipids in one of the two halves of the bimolecular lipid leaflet are available only for lysolecithin (see WINKLER and WESTHEAD 1980). The larger part of this phospholipid (at least 70%) is present in the inner leaflet (VOYTA et al. 1978; DE OLIVEIRA-FILGUEIRAS et al. 1979). The gangliosides are found in the same position (WESTHEAD and WINKLER 1982). Several proteins, or at least part of them, are apparently exposed on the outer face of chromaffin granules (ABBS and PHILLIPS 1980). These proteins include chromomembrin B or cytochrome b-561 (KÖNIG et al. 1976), phosphatidylinositol kinase, NADH: (acceptor) oxidoreductase, the Mg^{2+} activated ATPase (ABBS and PHILLIPS 1980), synaptin (BOCK and HELLE 1977), actin and α-actinin (BADER and AUNIS 1983). Some of these proteins (e.g. cytochrome b-561) span the membrane and are therefore also exposed on the inner side of the granule membrane (ABBS and PHILLIPS 1980; ZAREMBA and HOGUE-ANGELETTI 1985). At the inner surface of the granule membranes are most, if not all, of the glycoproteins (at least their carbohydrate chains) including DBH (HUBER et al. 1979; ABBS and PHILLIPS 1980). The nature of the component binding wheat germ lectin on the outside of chromaffin granules (MEYER and BURGER 1976) has not yet been further characterized. Concanavalin A does not bind to the surface of chromaffin granules (EAGLES et al. 1975; HUBER et al. 1979).

The membrane intercalated particles seen after freeze-etching of isolated chromaffin granules, probably representing proteins, can move translationally (SCHOBER et al. 1977; SCHULER et al. 1978). The nature of these proteins has not yet been identified but they are likely to include the proton-translocating part of the ATPase and carriers for catecholamines and nucleotides (see WINKLER and WESTHEAD 1980; WINKLER et al. 1986).

Chromaffin granules bear a net negative charge on their outside (BANKS 1966b; MATTHEWS et al. 1972; DEAN and MATTHEWS 1974). It is likely that the majority of these charges is contributed by acidic phospholipids, i.e. phosphatidylserine and phosphatidylinositol (see discussion by WINKLER and WESTHEAD 1980).

A distinct class of negative groups has been characterized by MORRIS and SCHOBER (1977). Trivalent cations (like lanthanum) were found to bind to specific sites which slightly protruded from the granule surface and were spaced 52 nm apart. This is in contrast to the majority of negative groups (probably lipids) which are separated by only a few nm. MORRIS and SCHOBER (1977) have suggested that these lanthanum binding sites represents proteins bearing a cluster of acidic groups.

II. Additional Factors Necessary for Exocytosis

In this section five possible factors will be discussed. However, as will become obvious later, exocytosis may depend on quite a number of special conditions and cofactors (see also AUGUSTINE et al. 1987).

1. Metabolic Energy

In the presence of both cyanide and iodoacetate, which block the production of metabolic energy, secretion of catecholamines from perfused bovine glands is abolished (KIRSHNER and SMITH 1969). Similar results were obtained for perfused cat glands (RUBIN 1970a). The demonstration, in "leaky medullary cells", that ATP is necessary for exocytosis is consistent with these results (BAKER and KNIGHT 1981) and demonstrates that energy is needed somewhere after calcium entry. However, we do not know at which step of the sequence of events ATP is required. In rat pheochromocytoma (PC12) cells made leaky by alpha-toxin there is no requirement for ATP (AHNERT-HILGER et al. 1985; AHNERT-HILGER and GRATZL 1987). This might indicate that ATP is required for some early step in exocytosis, e.g. for approach to the plasma membrane and not for the final steps of attachment and fusion, which should be the same in all secretory cells (see also VILMART-SEUWEN et al. 1986 for paramecium cells where ATP seems not to be required for fusion).

2. Calcium

In a series of experiments on adrenal medulla DOUGLAS and coworkers (see DOUGLAS 1968; RUBIN 1970b; DOUGLAS 1974) established, that calcium is essential for catecholamine secretiion, and it was suggested that this ion acts on a critical site within the cell (for references showing the requirement for calcium in release from sympathetic nerve: see BLAUSTEIN 1979). This concept was further strengthened when it was shown that Ca-ionophores (known to transfer calcium across membranes) stimulate catecholamine release from adrenal medulla or pheochromocytoma cells (COCHRANE et al. 1975; GARCIA et al. 1975; PERLMAN et al. 1980; CARVALHO et al. 1982; for references for the sympathetic nervous system see TRIGGLE 1979). Ca^{2+}-loaded liposomes also release catecholamines (GUTMAN et al. 1979) probably by introducing calcium into the cells. The necessity of calcium entry into the cells was also demonstrated with "leaky cells" of adrenal medulla (see below). These cells are permeable to calcium and exocytosis can be elicited by as low a concentration of calcium as 1 to 5 μM (BAKER and KNIGHT 1978; KNIGHT and BAKER 1982). As measured by a calcium indicator the levels of calcium inside intact chromaffin cells rise from the nanomolar to the micromolar range during stimulation (KNIGHT and KESTEVEN 1983; BURGOYNE 1984; KAO and SCHNEIDER 1985, 1986; CHEEK and BURGOYNE 1985; COBBOLD et al. 1987).

Under certain conditions (stimulation by theophylline and salbutamol in Ca^{2+}-free media) release of intracellular calcium may trigger exocytotic release (GUTMAN and BOONYAVIROJ 1979b). Cyclic AMP may be a mediator of this calcium release (see below for further discussion). An increase of intracellular calcium (by a reduction of calcium efflux) may also explain why the introduction of sodium-deficient media is associated with the release of catecholamines in the absence of extracellular calcium (AGUIRRE et al. 1977).

The physiological entry of calcium into chromaffin cells (after nicotinic stimulation) is likely to occur via voltage-dependent Ca^{2+}-channels similar to the slow Ca^{2+}-channels of the squid axon (for references see BAKER and RINK 1975; KIDOKORO et al. 1979; AGUIRRE et al. 1979; WAKADE and WAKADE 1982b; FENWICK et al. 1982a), which are blocked by calcium channel antagonists (PINTO and TRIFARO 1976; AGUIRRE et al. 1979; HIWATARI and TAIRA 1978; ARQUEROS and DANIELS 1978; SCHNEIDER et al. 1981; CORCORAN and KIRSHNER 1983; ARTALEJO et al. 1987) or by gadolinium ions (BOURNE and TRIFARO 1982) or activated by BAY-K-8644 (MONTIEL et al. 1984; HOSHI and SMITH 1987; LADONA et al. 1987b). Acetylcholine has been shown to depolarize chromaffin cells (DOUGLAS et al. 1967; ISCHIKAWA and KANNO 1978, single channel recording: FENWICK et al. 1982b) and to induce propagated action potentials (BIALES et al. 1976; BRANDT et al. 1976; KIDOKORO and RITCHIE 1980; KIDOKORO et al. 1982). Some Ca^{2+} may enter chromaffin cells through the acetylcholine receptor channels (KIDOKORO et al. 1982; KILPATRICK et al. 1982). Stimulation of muscarinic receptors induces in some species catecholamine secretion, whereas in bovine chromaffin cells no response occurs (see HARISH et al. 1987 for references). For these latter cells it has been shown that muscarinic agonists induce a small rise in intracellular calcium which is apparently insufficient to induce secretion (KAO and SCHNEIDER 1985; CHEEK and BURGOYNE 1985; MISBAHUDDIN et al. 1985; COBBOLD et al. 1987). This rise of calcium is apparently independent from extracellular supply and probably is derived from intracellular stores (COBBOLD et al. 1987; KAO and SCHNEIDER 1985; for rat adrenal: see HARISH et al. 1987).

3. Cyclic Nucleotides

Cholinoceptor stimulants can lead to an increase in cyclic AMP levels within the adrenal gland in vivo or during perfusion (JAANUS and RUBIN 1974; GUIDOTTI and COSTA 1973; GUTMAN and BOONYAVIROJ 1979a). In addition cyclic AMP can elicit a modest catecholamine secretion (PEACH 1972; SERCK-HANSSEN 1974; see for further references HIWATARI and TAIRA 1978; GUTMAN and BOONYAVIROJ 1979b). Therefore, the question has arisen whether this nucleotide is an intracellular mediator of catecholamine release induced by various secretagogues. However, the rise in cyclic AMP in perfused cat glands and in isolated bovine cells stimulated by acetylcholine occurs much later than catecholamine release and only at very high acetylcholine concentrations (JAANUS and RUBIN 1974; SCHNEIDER et al. 1979). This makes it unlikely that cyclic AMP is an essential link in physiological stimulus—secretion coupling. This leaves us still with the task to explain how exogenous cyclic AMP can elicit catecholamine secretion.

Most likely it does so by releasing intracellular calcium (see SERCK-HANSSEN 1974; GUTMAN and BOONYAVIROJ 1979a,b). Salbutamol and theophylline induce catecholamine secretion from isolated rat adrenal even in the absence of extracellular calcium. They both increase the intracellular cyclic AMP concentration (GUTMAN and BOONYAVIROJ 1979b). PGE_2 inhibits this cyclic AMP rise and also blocks catecholamine release by these drugs (GUTMANN and BOONYAVIROJ 1979a). This prostaglandin also reduces ^{45}Ca efflux from adrenal medulla. Dantrolene, thought to lower intracellular calcium, reduces the secretory effect of theophylline (COHEN and GUTMAN 1979). All these experiments are best explained by the suggestion that cyclic AMP releases calcium from subcellular stores; this calcium then causes catecholamine release. This cyclic nucleotide may also increase the available calcium by an effect on calcium fluxes at the plasma membrane (MORITA et al. 1987a,b).

The situation seems similar (see discussion by STJÄRNE 1979) for sympathetic nerves where cyclic AMP causes catecholamine release in the absence of extracellular calcium and facilitates secretion of noradrenaline stimulated by other means (WOOTEN et al. 1973; CUBEDDU et al. 1975; but see STJÄRNE 1976).

SCHNEIDER et al. (1979) reported that acetylcholine produced an increase in cyclic GMP levels in isolated bovine adrenal cells. The time scale of this rise was similar to that of the secretory response. However, the acetylcholine concentration raising cyclic GMP was much lower than that causing secretion. This latter finding is in agreement with a study by YANAGIHARA et al. (1979). In addition, these authors found that high potassium led to catecholamine secretion but did not elevate cyclic GMP levels. Thus, this cyclic nucleotide does not appear to be an important link in catecholamine secretion. In "leaky" medullary cells neither cyclic AMP nor cyclic GMP affect secretion, elicited by the addition of Ca^{2+} (BAKER and KNIGHT 1981; BITTNER et al. 1986), which is the most direct evidence against an essential role of these compounds.

4. Protein Kinase C, Polyphosphoinositide Metabolism and GTP-Binding Proteins

Recent studies ascribe a crucial role to this enzyme in signal transduction for various cellular functions (see NISHIZUKA 1984). This enzyme can be activated by diacylglycerol (formed from inositol phospholipids by the action of phospholipase C and calcium). In thrombocytes diacylglycerol stimulates secretion even without elevation of calcium levels (RINK et al. 1983). Calcium and diacylglycerol may act synergistically on protein kinase C (RINK et al. 1983) leading to phosphorylation of endogenous proteins (NISHIZUKA 1984; see also YAMADA et al. 1987). In leaky chromaffin cells phorbol esters, which act like diacylglycerol, increase the affinity of the exocytotic process for calcium (KNIGHT and BAKER 1983) stimulating secretion especially at low calcium levels (POCOTTE et al. 1985; BROCKLEHURST and POLLARD 1985; KNIGHT and BAKER 1985; PEPPERS and HOLZ 1986). In intact chromaffin cells phorbol esters were originally found to be ineffective (KNIGHT and BAKER 1983; BURGOYNE and NORMAN 1984), however later studies described also a synergistic action of phorbol esters on secretion from such cells, but only for some secretagogues (MORITA et al. 1985b; BROCKLEHURST and POLLARD 1986). Protein kinase C which is activated by the above procedure is found in chromaffin cells in the cytosol, but becomes bound to chromaffin granules in the presence of calcium—a feature shared with other so called chromobindins (see below: CREUTZ et al. 1985; SUMMERS and CREUTZ 1985; TERBUSCH and HOLZ 1986).

By what mechanism can protein kinase C influence the calcium sensitivity of exocytosis? This enzyme phosphorylates proteins and therefore we have to consider that phosphorylation is involved in exocytosis (see later section). However the action of this activated enzyme may be complex. At least in intact cell preparations it increases the influx of calcium (WAKADE et al. 1986) and therefore also interferes with the early steps of stimulus/secretion coupling.

The activator of protein kinase C, diacylglycerol, is produced by the breakdown of polyphosphoinositide by a phosphodiesterase (phospholipase C). In addition arachidonic acid is released during this process (FRYE and HOLZ 1985) which, as discussed in a later section, can induce fusion processes. In this connection we have to mention the observation that muscarinic stimulation of bovine adrenal medulla increases phophatidylinositol turnover (see SWILEM et al. 1986, for references) leading to increased levels of diacylglycerol and inositoltriphosphate (SWILEM et al. 1986; FORSBERG et al. 1986; SWILEM et al. 1986). The latter compound can release calcium from intracellular stores (STOEHR et al. 1986) inducing a modest rise in cytoplasmic calcium (see COBBOLD et al. 1987) which, however, in bovine chromaffin cells is insufficient to stimulate secretion (see previous section). Whether the rise in diacylglycerol and the possible activation of protein kinase C plays any role in muscarinic secretion in those species where such secretion occurs is difficult to evaluate (HARISH et al. 1987), thus in rat, for example, muscarinic secretion is not modified by phorbol esters which activate protein kinase C (WAKADE et al. 1986). Finally GTP-binding proteins are related to the phosphatidylinositol metabolism. By reacting with the phosphodiesterase (phospholipase C) they can increase the breakdown of these phospholipids. In mast cells and neutrophils stable GTP analogues even in the absence of calcium can induce secretion (GOMPERTS et al. 1986; BARROWMAN et al. 1986). It has been suggested (BARROWMAN et al. 1986; see also BURGOYNE 1987) that this occurs by two mechanisms: (i) by liberation of diacylglycerol, the activator of protein kinase C, or (ii) by direct interference in the final steps of exocytosis. However in "leaky" chromaffin cells results are controversial. GTP-analogues were shown in one study to inhibit (bovine) or stimulate (chicken) Ca^{2+}-dependent exocytosis. In the absence of calcium this analogue had no effect (KNIGHT and BAKER 1985b). In another study (BITTNER et al. 1986) a GTP analogue induced Ca^{2+} independent secretion. At present it is impossible to reach definite conclusions (see also BAKER and KNIGHT 1986, for discussion) on the participation of the above factors as essential links in exocytosis. These factors may act as modulators of the events between calcium entry and the final fusion. Whether they are essential links remains doubtful since in well defined exocytosis systems like the paramecium cells, for example, inhibitors of protein kinase C have no effect on secretion (LUMPERT et al. 1987). Also in neutrophils very specific inhibitors of this enzyme do not interfere with secretion (WRIGHT and HOFFMAN 1986).

5. Metalloendoprotease

Recent studies suggested that a metalloendoprotease may be involved in exocytosis (MUNDY and STRITTMATTER 1985). Several inhibitors of this enzyme blocked secretion in adrenals, even in leaky cells. However careful subsequent studies showed that these inhibitors did not block secretion by all secretagogues, especially not secretion caused by a Ca^{2+}-ionophore. The most likely effect of these enzyme inhibitors was an interference with the rise of cytoplasmic calcium, but not with the subsequent steps leading to exocytosis (HARRIS et al. 1986; NAKANISHI et al. 1986).

III. The Mechanism of Membrane Attachment

Transport of chromaffin granules from the Golgi region to the plasma membrane may be achieved either by random movement of these organelles or by some directed transport. An optical study of rat adrenal cells with laser light scattering indicates that the relative movement of cytoplasmic organelles is restricted and slower than one would expect, if free diffusion of the vesicles takes place (ENGLERT 1980). This "immobilization" indicates that chromaffin granules are not moving randomly within the chromaffin cells and are possibly attached to "fibrous proteins". However, this has not been demonstrated directly (e.g. in morphological studies, see POISNER and COOKE 1975). Directed transport is, of course, well documented for the sympathetic neurons where axonal flow accomplishes the task of transporting the amine-storing organelles from the Golgi region to the terminal (see DAHLSTRÖM 1971; BRIMIJOIN 1982).

When chromaffin granules arrive near the plasma membrane the final approach bringing the granules close to attachment may again be due to random movements or depend on contractile proteins as discussed in the following section. The final contact of the membranes is likely to depend on the nature and distribution of their charges: this can be influenced (i) by a change of the charge of these membranes or (ii) by ions like calcium (and cofactors) which may neutralize charges or cause bridging of charges. These possibilities will be discussed in subsequent sections.

1. The Role of Contractile Proteins

Let us first discuss the possible role of microtubules. Chromaffin cells contain an abundance of these structures (for electron micrographs and references see POISNER and COOKE 1975). They are arranged radially, but some run parallel with the plasma membrane. No clear indication was found that chromaffin granules were in any way aligned along these structures. After treatment with colchicine or vinblastine the microtubules disappear (POISNER and COOKE 1975). Since these drugs also inhibit catecholamine release elicited by nicotine or acetylcholine in perfused adrenal gland, POISNER and BERNSTEIN (1971) suggested that microtubules participate in catecholamine secretion (for similar results in sympathetic nerve: see THOA et al. 1972). Furthermore, a stabilizer of microtubules, i.e. D_2O, increased catecholamine secretion. However, although colchicine inhibites acetylcholine - stimulated catecholamine release in rabbit gland, it did not block secretion induced by high potassium (DOUGLAS and SORIMACHI 1972a; TRIFARO et al. 1972; SCHNEIDER et al. 1977; MCKAY and SCHNEIDER 1984). Since there is no reason to suggest that the secretory mechanism of these two secretagogues is dissimilar, it seems unlikely that microtubules are a component essential for exocytosis. In "leaky" medullary cells colchicine has no effect on Ca^{2+}-induced secretion (BAKER and KNIGHT 1981).

These results, however, do not exclude the possibility that microtubules play some role in the secretory activity of adrenal medulla. If microtubules are involved in directed transport of chromaffin granules to the plasma membrane (see before), one would not expect that interruption of this transport would have an immediate effect on secretion. The chromaffin cell is replete with granules and the granules already present near the plasma membrane may therefore participate in the secretion process even in the presence of colchicine. A lack of directed transport may cause a disturbance of secretion only after several secretory cycles over a prolonged period of stimulation. We can, therefore, at present conclude that, although microtubules are unlikely to be involved in the exocytotic process per se, they may still be essential for continuous secretory activity of chromaffin cells.

Chromaffin granules possess binding sites for tubulin, but the functional significance of this finding is still obscure (BERNIER-VALENTIN et al. 1983). Chromaffin cells also contain microtubule-associated protein 2 (BURGOYNE and NORMAN 1985).

Microfilaments with actin-like properties form a discontinuous network beneath the plasma membrane of chromaffin cells (COOKE and POISNER 1976). Cytochalasin B, which interferes with microfilaments, has variable effects on catecholamine secretion. In cat adrenal medulla, cytochalasin blocks catecholamine release elicited by acetylcholine, but has less effect on potassium-induced release (DOUGLAS and SORIMACHI 1972b). In isolated bovine chromaffin cells it has no significant effect (SCHNEIDER et al. 1977, 1981) and this is also true for electrically permeabilized cells (BAKER and KNIGHT 1981) but in digitonin-permeabilized cells it enhances secretion (LELKES et al. 1986). Obviously these experiments have not yielded consistent results. Before we return to the question of the function of the cytoskeleton let us discuss further proteins relevant for this system.

Chromaffin cells contain myosin (TRIFARO and ULPIAN 1976; JOHNSON et al. 1977; CREUTZ 1977), which seems bound to the cell membrane (TRIFARO et al. 1978; HESKETH et al. 1978). In addition actin is present. Most of this protein is present in the cytoplasm (PHILLIPS and SLATER 1975), but the granule membrane also contains part of it (PHILLIPS and SLATER 1975; GABBIANI et al. 1976; MEYER and BURGER 1979a; AUNIS et al. 1980; LEE and TRIFARO 1981). α-actinin, which may provide binding sites for actin, was also shown to be a component of the membrane of chromaffin granules (see JOHNSON et al. 1977; BADER and AUNIS 1983). In addition, a protein with cytochalasin-like activity was found in bovine adrenal medulla (GRUMET and LIN 1981).

An early and important indication for the functional significance of actin was provided by FOWLER and POLLARD (1982a,b), who showed, that F-actin interacts with granule membranes in a calcium-dependent manner. Furthermore, recent studies indicate that spectrin-like molecules may represent binding sites for actin on chromaffin granules (AUNIS and PERRIN 1984; KAO and WESTHEAD 1984). Spectrin (α-fodrin) is also found as a continuous ring beneath the plasma membrane of chromaffin cells.

Thus a great number of cytoskeletal proteins are present in chromaffin cells. In accordance cytoplasmic extracts of adrenal medullary cells form in vitro a supramolecular gel containing cytoskeletal elements, i.e. microfilaments, microtubuli and intermediate filament proteins (CHEEK et al. 1986). Participating proteins are filamin, fodrin, myosin, caldesmon (see also BURGOYNE et al. 1986) and tropomyosin (see COTE et al. 1986a). What are the possible functional implications of these proteins? It has been suggested that the cytoskeleton prevents exocytosis by immobilizing the granules in a network (see LELKES et al. 1986; BADER et al. 1986b; BURGOYNE and CHEEK 1987). When calcium enters the cell chromaffin granules are released and can now approach the plasma membrane where fusion occurs. Several lines of evidence support such a concept. α-fodrin, an actin binding protein, is found close to the plasma membrane. During secretion it changes from a continuous ringlike position to a more patchy appearance (PERRIN and AUNIS 1985) indicating a rearrangement of the cytoskeleton. Furthermore, antibodies against fodrin introduced into digitonin-permeabilized cells inhibit secretion (PERRIN et al. 1987). During stimulation with nicotine (although not with high K^+) actin filaments are first disassembled and then reassembled (CHEEK and BURGOYNE 1986). Gelsolin-like proteins might be involved in such a process and in fact such a protein which can fragment actin in a calcium-dependent manner has been found in bovine chromaffin cells (BADER et al. 1986b; ASHINO et al. 1987). On the other hand caldesmon, which is also present in chromaffin cells (BURGOYNE et al. 1986; SOBUE et al. 1985), may be involved as an actin binding protein in assembling actin (maybe together with fodrin). All these quite recent studies provide considerable support for the suggestion that calcium releases chromaffin granules from a cytoskeletal network. Whether this is a direct effect of these ions or necessitates mediation by some of the factors discussed in other sections (calmodulin, GTP binding proteins, protein kinase C) remains to be established. It is interesting to note that in nerve terminals a phosphorylation/dephosphorylation cycle of synapsin is thought to prevent or elicit fusion, maybe also involving cytoskeletal elements like actin (see BÄHLER and GREENGARD 1987).

2. The Role of Changes in the Charge of Granule Membranes

Two reactions have been described by which an alteration in the charge of the membrane of chromaffin granules can be brought about.

a. Phosphorylation of Phosphatidylinositol

Chromaffin granule membranes possess a phosphatidylinositol kinase (see WINKLER 1976) which phosphorylates monophosphatidylinositol to diphosphatidylinositol. This enzyme is apparently localized on the outer surface of the granule membrane (ABBS and PHILLIPS 1980); thus the surface charge of the granule can be altered by its action. In perfused glands stimulated by acetylcholine diphosphatidylinositol is formed in chromaffin granules (TRIFARO and DWORKIND 1975). However, no increase in the calcium-binding capacity of isolated granules was found after phosphorylation by endogenous enzyme (LEFEBVRE et al. 1976) and so the physiological meaning of this phosphorylation is obscure. On the other hand, granules might become phosphorylated in the cytoplasm and may lose this charge before exocytosis occurs, as suggested by LEVEBVRE et al. (1976). However, there is no experimental evidence to support this idea. The function of phosphatidylinositol kinase in granules is, therefore, still unknown.

b. Phosphorylation of Proteins of Granule Membranes

In the brain evidence has been presented that secretion is accompanied by phosphorylation of several proteins present on synaptic vesicles (DE LORENZO et al. 1979; NESTLER and GREENGARD 1982). Two of these proteins (53 kD and 60 kD) have been identified as α- and β-tubulin (BURKE and DE LORENZO 1982). The relevance of this finding for adrenal medulla seems doubtful, since tubulin is not an integral protein of chromaffin granule membranes (BADER et al. 1981). Another protein which is phosphorylated has been named protein I (UEDA et al. 1979), and more recently synapsin I (DE CAMILLI et al. 1983). This protein is found on synaptic vesicles including the small dense core vesicles of sympathetic nerves. Evidence has been provided that phosphorylation of this protein occurs during secretion from nerves. This phosphorylation is thought to dissociate synapsin from vesicles thereby removing a constraint for the release process (LLINAS et al. 1985). This is an attractive model for release from small vesicles in nerve, however, for adrenal medulla it cannot be relevant since chromaffin granules from the adrenal contain negligible amounts if any of synapsin (FRIED et al. 1982). In a previous section on the cytoskeleton we have already discussed some analogous mechanisms possibly occurring in adrenal medulla.

In isolated chromaffin cells the phosphorylation of two proteins (60 kD and 95 kD) is enhanced during secretion elicited by several stimulants (AMY and KIRSHNER 1981). This phosphorylation seems Ca^{2+}-dependent (see also BAKER et al. 1982; for "leaky cells"). The subcellular localization of these proteins is still unknown. In any case, tyrosine hydroxylase (60 kD) becomes phosphorylated during stimulation with acetylcholine (HAYCOCK et al. 1982; NIGGLI et al. 1984) and therefore one of the proteins described by AMY and KIRSHNER (1981) may be this enzyme.

When isolated chromaffin granules are incubated with (γ^{32}P) ATP, calmodulin and calcium stimulate the phosphorylation of several proteins (one component with 43 kD and three components around 58 kD; BURGOYNE and GEISOW 1981; for brain vesicles: see MOSKOWITZ et al. 1983). The protein with molecular weight 43 kD was the major labelled band (see also KONINGS and DE POTTER 1983), however, subsequent studies (BURGOYNE and GEISOW 1982) showed that this protein represented the α-subunit of pyruvate dehydrogenase present in contaminating mitochondria. On the other hand. TREIMAN et al. (1983) found that calcium and calmodulin enhanced the phosphorylation of a smaller protein (30 kD), whereas cyclic AMP stimulated the phosphorylation of a protein identified as tyrosine hydroxylase (60 kD). Chromaffin granule membrane possess a significant $p60^{c\text{-}SRC}$ kinase activity which has been measured by autophosphorylation (PARSONS and CREUTZ 1986).

In two recent studies (COTE et al. 1986b; LEE and HOLZ 1986) the complexity of the phosphorylation occurring in intact chromaffin cells was highlighted. ^{32}P was incorporated into numerous proteins only a few of them with known function (tyrosine hydroxylase, tubulin, α-subunit of dehydrogenase). The phosphorylation appeared to depend on calmodulin (COTE et al. 1986) and on protein kinase C (LEE and HOLZ 1986; see also BROCKLEHURST et al. 1985) as shown by the activation with phorbol esters. Stimulation of the chromaffin cell led to the enhancement of phosphorylation of 22 proteins and to an increased dephosphorylation of a 20.4 kD protein. As far as the time course is concerned phosphorylation of some proteins preceded or accompanied catecholamine secretion (COTE et al. 1986b). One of these phosphorylated proteins belongs to the class of the so called chromobindins (see below), it is a substrate for protein kinase C and its phosphorylation has a time course paralleling that of catecholamine secretion (MICHENER et al. 1986). All the data discussed in this section emphasize the complexity of phosphorylation reactions in isolated granule membranes or intact chromaffin cells, however they do not yet allow any final conclusion as to the relevance of these reactions for exocytosis, and even less as to the involvement of one specific protein in this process. The issue becomes even more complicated by the suggestion that dephosphorylation of specific proteins may be the essential step. Thus a thio-analogue of ATP leading to irreversible phosphorylation of proteins inhibits secretion in permeabilized chromaffin cells (BROOKS et al. 1984). Two proteins were heavily thiophosphorylated, a 54 kD protein in unstimulated cells and a 43 kD protein in calcium-stimulated cells (BROOKS and BROOKS 1985, 1987).

In this context we should mention that in the well-defined exocytosis system of paramecium cells there is considerable evidence to support the concept that dephosphorylation of a 65 kD phosphoprotein is an essential step in secretion (MOMAYEZI et al. 1987). In these cells ATP is thought to be essential to keep the system in a primed state, but is not needed for the actual fusion process (VILMARI-SEUWEN et al. 1986).

c. Methylation of Granule Membrane Components

Bovine adrenal medulla contains a protein carboxymethylase which confers a methyl group to carboxyl side chains of proteins (DILIBERTO et al. 1976). Of the various subcellular organelles, the chromaffin granules have the highest concentration of substrate for this cytoplasmic enzyme. Two membrane proteins appeared preferentially methylated (GAGNON et al. 1978). However, the soluble proteins of granules can also act as substrates (BORCHARDT et al. 1978). It has been suggested (see GAGNON and HEISLER 1979; DILIBERTO 1982) that this methylation reaction on the granule surface leads to a decrease in charge enabling the granule to attach the plasma membrane. For adrenal medulla, a correlation in time course between secretion and methylation has been obtained (see DILIBERTO 1982). On the other hand, in exocrine pancreas blockers of protein methylation did not interfere with the secretory response (POVILAITIS et al. 1981). This argues against an essential role of methylation in exocytosis.

3. The Role of Calcium and Specific Proteins

In a previous section we discussed the evidence that calcium entry into the cell is an essential factor in exocytosis. We will now deal with the possible role of calcium in the attachment process. In the subsequent section the participation of calcium in the events leading to fusion will be discussed.

BANKS (1966b) found that calcium (m*M*) decreased the negative charge on chromaffin granules and agglutinated these organelles. He therefore suggested that this action of calcium may promote the attachment of chromaffin granules to the cell membrane. These results were confirmed and extended by MATTHEWS et al. (1972) and DEAN and MATTHEWS (1974; see also DEAN 1975, for a theoretical analysis).

Several recent studies tested the effect of calcium on the aggregation of isolated chromaffin granules. Such experiments are considered to be relevant to the secretory process occurring *in vivo*, since compound exocytosis (fusion of additional granules with a granule already fused with the plasma membrane) has been observed in secretory tissues (see DAHL et al. 1979).

EDWARDS et al. (1974) reported in a morphological study that isolated chromaffin granules aggregate and possibly fuse (breakage of apposed membrane) at their contact region in the presence of m*M* concentrations of calcium (but also with slightly higher concentrations of Mg^{2+}). When such aggregated granules were studied by freeze-etching, a lateral displacement of the membrane-intercalated particles (see the following for further discussion) from the areas of membrane contact was observed (SCHOBER et al. 1977; see also MORRIS et al. 1983). Treatment with EDTA led to a separation of the aggregated granules. Aggregation of granules in the presence of m*M* concentration of calcium was also observed by intensity fluctuation spectroscopy (GREEN et al. 1978). The biological relevance of all these experiments is questionable (see MORRIS and BRADLEY 1984) since m*M* concentrations of divalent cations were necessary and Mg^{2+} had the same effect as calcium, whereas exocytosis is calcium-specific (see KNIGHT and BAKER 1982). An aggregation of secretory granules at lower calcium concentration has ben reported in morphological studies by GRATZL, DAHL and collaborators (GRATZL et al. 1977; GRATZL and DAHL 1978; DAHL et al. 1979). When chromaffin granules are incubated with increasing calcium concentrations (from 10^{-7} to 10^{-4} *M*), the percentage of aggregated fused vesicles increases. At 10^{-7} *M* calcium only 2% of the vesicle are fused, at 10^{-5} *M* 5%, at 10^{-4} *M* a plateau is reached at 6%. Fusion was usually demonstrated by freeze-etching but could also be visualized by conventional electronmicroscopy (EKERDT et al. 1981). Mg^{2+} or Mn^{2+} were ineffective in this system. The specificity of this aggregation for calcium and the low concentration of calcium (10^{-5} *M*) required for fusing secretory vesicles are consistent with the exocytotic process *in vivo*.

However these studies on the aggregation of isolated chromaffin granules may not be relevant for the interaction of granules with the plasma membrane. Therefore, in vitro studies involving these two components are important. Such experiments were first performed by DAVIS and LAZARUS (1976) who claimed that addition of adrenal plasma membranes (of undetermined purity), of calcium and ATP to isolated granules led to a release of noradrenaline into the incubation medium. It is, however, peculiar that in these experiments chromaffin granules were isolated through hypertonic sucrose but subsequently incubated under isotonic conditions. This should lead to massive adrenaline release, but the authors reported, that when granules were incubated at 37 °C for 10 min without plasma membrane, no release at all occurred. Furthermore, there was no morphological control. In any case, these experiments were taken up again by DE POTTER and coworkers (KONINGS and DE POTTER, 1981a, b; KONINGS et al. 1983; BOHNER et al. 1985). In the presence of Ca^{2+} (10^{-6} to 10^{-5} *M*) incubation of chromaffin granules with plasma membranes led to a temperature-dependent release of catecholamines, ATP and DBH. Percoll gradient centrifugation of the incubation (1 min at 37 °C) mixture indicated, that a portion of granules had lost their total soluble contents (KONINGS et al. 1983). These results were taken as evidence that, under these *in vitro* conditions, chromaffin granules interact with the plasma membrane leading to an exocytotic release of the total granule contents. However, three aspects of these experiments seem puzzling: (i) ATP-Mg^{2+} was not necessary for release, although they are essential in "leaky" cells of adrenal medulla (BAKER and KNIGHT 1981; (ii) the time course of release is very rapid. Within one minute a plateau is reached (25% of total catecholamines released). (iii) As recently shown a major fraction of isolated plasma membrane vesicles is oriented right-side-out thus presenting the „wrong" surface for vesicles to fuse with (ROSENHECK and PLATTNER 1986). In the most recent development it has been reported that catecholamine release observed during incubation of isolated chromaffin granules with plasma membrane was due to an artefact (DE BLOCK and DE POTTER 1987).

What is the substrate for calcium interaction with chromaffin granules? Fusion of isolated granules by calcium is prevented by pretreatment with protease or neuraminidase (DAHL et al. 1979; EKERDT et al. 1981). This would indicate that calcium interacts with carbohydrate chains on proteins and/or on gangliosides. However, according to several studies (HUBER et al. 1979; ABBS and PHILLIPS 1980; WESTHEAD and WINKLER 1982) most if not all of the glycoproteins and gangliosides of chromaffin granules are present on the inner surface of the granule membranes. In this position they can, of course, not interact with calcium on the outside. On the other hand, MEYER and BURGER (1976) found a component, presumably a glycoprotein, on the surface of chromaffin granules, which bound wheatgerm lectin. Unfortunately, no further characterization of this constituent has been performed. A protein, which binds to chromaffin granules, was isolated from plasma membranes of adrenal medulla (MEYER and BURGER, 1979b). This was achieved by running solubilized plasma membranes through a column with attached chromaffin granules. One protein of molecular weight 51 kD was specifically retained. The possible interaction of calcium in the binding of this protein to chromaffin granules has not been investigated.

Are additional factors aiding the reaction of calcium with chromaffin granules? CREUTZ et al. (1978) isolated from adrenal medulla a soluble protein, which was named synexin (47 kD). It significantly lowered the binding constant for Ca^{2+}-promoted aggregation of intact chromaffin granules. This effect was specific for calcium. Synexin is also present in liver. It was suggested (see also CREUTZ et al. 1982; CREUTZ and STERNER 1983) that this protein binds specifically to granule membranes and promotes with calcium an attachment to the plasma membrane during exocytosis. In a subsequent study (CREUTZ et al. 1979) it was shown that the presence of calcium led to a rapid selfassociation of this protein producing rods of a length of 150 Å. MORRIS and HUGHES (1979) reported that synexin not only aggregates chromaffin granules, but also mitochondria, microsomes and phosphatidylserine vesicles. DABROW et al. (1980) disputed this and claimed that the interaction was specific for chromaffin granules and could be blocked by pretreatment of granules with proteases. However, several studies have now shown that synexin can promote the aggregation and even fusion of phospholipid vesicles (MORRIS et al. 1982; HONG et al. 1981, 1982). Synexin can also bind to the inner side of plasma membrane from chromaffin cells (SCOTT et al. 1985). Recently POLLARD and SCOTT (1982) isolated a protein from liver (68 kD), which inhibits synexin-induced chromaffin granule aggregation. It was named synhibin. ODENWALD and MORRIS (1983) have isolated another protein (66 kD) from liver, which has synexin-like properties. Finally WALKER et al. (1983) isolated a protein (32 kD) from the electric organ of Torpedo; it is also found in rat adrenal medulla and it exhibits synexin-like properties by aggregating isolated chromaffin granules.

Phenothiazine drugs, like trifluoperazine and promethazine, inhibit, at low concentration, synexin acitivity *in vitro* and *in vivo* secretion from cultured chromaffin cells (POLLARD et al. 1982). This is consistent with an involvement of synexin in exocytosis. However, more specific tools than the phenothiazines, e.g. antibodies injected into the cell (see below), have to be used before a conclusion can be reached.

Most recently POLLARD et al. (1987) have suggested that the function of synexin is to provide a "hydrophobic bridge" between the secretory granule and the plasma membrane as the first crucial step leading to fusion. It will have to be seen whether this interesting concept can be supported by direct experimental evidence.

Another interesting group of proteins becomes bound to the membrane of isolated granules in the presence of calcium (CREUTZ 1981a; GEISOW and BURGOYNE 1982). These proteins have been named chromobindins (CREUTZ et al. 1983; CREUTZ and RAUCH-HARRISON 1984). They also include synexin as discussed above. Further analyses have revealed that some of these proteins correspond to proteins isolated from other secretory tissues but having similar properties (CREUTZ et al. 1987). Thus chromobindin 4 was identified as a 32 kD protein called calelectrin or endonexin (see also GEISOW et al. 1986). Chromobindin 20 represents a 67 kD variant of calelectrin. It has synhibin activity. Chromobindin 8 and 9 (substrates for protein kinase C) undergo phosphorylation during secretion (see also MICHENER et al. 1986). One chromobindin represents the protein caldesmon (GEISOW and BURGOYNE 1987). A subgroup of these proteins has been called chromobindin A. These proteins are not released from the granule membrane by the removal of calcium unless ATP is present (MARTIN and CREUTZ 1987). Obviously all these proteins have interesting properties. However direct proof of their relevance in the secretory process is still lacking.

A protein known to modulate many functions of calcium is calmodulin (CHEUNG 1980). It has been suggested, that this protein is also involved in stimulus-secretion coupling (see DE LORENZO 1981). In adrenal medulla calmodulin is present (KUO and COFFEE 1976).

An inhibitor of the action of calmodulin is trifluoperazine. This drug inhibits secretion from mast cells (DOUGLAS and NEMETH 1982) and from endocrine or exocrine pancreas (GAGLIARDINO et al. 1980; HEISLER et al. 1981). Antibodies against calmodulin inhibit exocytosis in an *in vitro* system of sea urchin egg membranes (STEINHARDT and ALDERTON 1982). In adrenal medulla, trifluoperazine inhibits catecholamine-secretion (KENIGSBERG et al. 1982; NISHIBE et al. 1983; BROOKS and TREML 1983a; WADA et al. 1983; WAKADE and WAKADE 1984a). KENIGSBERG et al. (1982) showed that this drug did not inhibit calcium entry into these cells; however, WADA et al. (1983a) concluded that its major effect was on calcium entry. It is, therefore, relevant that in "leaky" cells, calcium-dependent exocytosis was inhibited by trifluoperazine (BAKER and KNIGHT 1981). Furthermore, analysis with the patch clamp technique showed that trifluoperazine inhibits calcium currents in chromaffin cells, but also interferes with secretion at a subsequent step (CLAPHAM and NEHER 1984). It is, however, difficult to draw final conclusions from these experiments, since trifluoperazine inhibits acetylcholine-induced secretion much more than secretion elicited by high potassium (BROOKS and TREML 1983a). These authors also showed that higher concentration of trifluoperazine had a detrimental effect on chromaffin cells (BROOKS and TREML 1984). Calmidazolium, another inhibitor of calmodulin, was shown to block secretion after the entry of calcium (BURGOYNE and NORMAN 1984).

In summary, all these data with drugs indicate, but do not prove, that calmodulin is involved in secretion, probably after calcium entry. Recently more direct evidence has become available (KENIGSBERG and TRIFARO 1984). These authors "injected" (by fusion with antibody-filled erythrocytes) antibodies against calmodulin into chromaffin cells. This led to a clear inhibition of catecholamine secretion. This is the most direct evidence that this protein is involved in the calcium promoted exocytotic process. Little can be said about the mechanism by which calmodulin might facilitate exocytosis. Calmodulin binds to chromaffin granules and facilitates binding of cytosolic proteins to granules (GEISOW et al. 1982; GEISOW and BURGOYNE 1983). Apparently specific proteins of the granule membrane are involved (BADER et al. 1985). BURGOYNE et al. (1982) also claimed that *in vivo* trifluoperazine, the blocker of calmodulin, does not inhibit granule attachment to the plasma membranes, but interferes with subsequent fusion. On the other hand (see previous section), calmodulin can also be involved in phosphorylating reactions.

Botulinum toxin inhibits secretion of catecholamine from chromaffin cells, apparently at a step following calcium entry (KNIGHT et al. 1985; KNIGHT 1986; PENNER et al. 1986). Thus this toxin might interfere with a substrate crucial for the final step in exocytosis. It is therefore very interesting that Botulinum toxin leads to the ADP-ribosylation of a 21 kD protein in homogenates of bovine adrenal medulla (OHASHI and NARUMIYA 1987). Since many of such toxin reactive proteins are GTP-binding ones (OHASHI and NARUMIYA 1987), it appears possible that these results provide a connection with the effect of GTP analogues discussed in a previous section. Future results along this line may provide crucial answers.

4. Conclusions

Exocytosis depends on metabolic energy and is initiated by calcium. However, the specific functions of these factors have not been established. Cyclic nucleotides are unlikely to play a significant role in exocytosis. An involvement of contractile proteins has not yet been directly proven but convincing evidence is accumulating. This is also true for calmodulin, whereas participation of protein kinase C appears only possible, and that of a metalloendoprotease rather unlikely.

Numerous hypotheses are available to explain the membrane attachment during exocytosis. Reactions like phosphorylation and methylation of proteins and lipids have been described which can change the charge of the granule membranes. A correlation of some of these changes with exocytosis has been reported. Calcium has been shown to aggregate granule membranes possibly by interacting with specific membrane constituents and aided by proteins like synexin and/or calmodulin. At present a firm conclusion cannot be reached.

IV. Mechanism of Fusion

After the attachment of the chromaffin granules to the plasma membranes, a rearrangement of the two membranes must occur which leads to fusion and continuity of the two membranes. We will first discuss how the major components of the membrane, i.e. lipids and proteins, behave during this process.

1. The Behaviour of Membrane Proteins During Fusion

It has been claimed, for several secretory tissues, that fusion is preceded by the removal of membrane-intercalated particles as seen in freeze-etching (mast cells: CHI et al. 1976; LAWSON et al. 1977; endocrine pancreas: ORCI et al. 1977; adrenal medulla: AUNIS et al. 1979; see PLATTNER 1978). Since these intercalated particles are usually considered to represent proteins, it was assumed that, before fusion occurs, these membrane-spanning proteins are displaced laterally away from a lipid contact zone (see LUCY 1978). However, more recent studies, employing extremely fast freezing techniques, have shown that the particle clearing was probably an artefact caused by the chemical pretreatment used in those other studies (neuromuscular junction: HEUSER et al. 1979; amebocytes: ORNBERG and REESE 1981; mast cells: CHANDLER and HEUSER 1980; adrenal medulla: SCHMIDT et al. 1983). In fact, as demonstrated by these fast freezing techniques, fusion begins in a very limited region, probably with a small pore (SCHMIDT et al. 1983; see also PLATTNER 1987).

Since membrane proteins apparently are not removed from the site of fusion, they may be involved in the fusion process. However, no data are available to prove this point. The isolation of a plasma membrane protein which specifically binds to chromaffin granules (MEYER and BURGER 1979b) has already been mentioned in the previous section.

2. The Role of Lipids in Fusion

Whatever function we may finally ascribe to the membrane proteins, the final fusion step occurs between two bimolecular lipid leaflets. It goes beyond the scope of this review to discuss the numerous studies on lipid vesicles which have attempted to elucidate this mechanism. The very detailed studies by PAPAHADJOPOULOS and coworkers (see PAPAHADJOPOULOS 1978) on liposomes made up of acidic phospholipids or a mixture of acidic and neutral phospholipids have led to the following conclusions. Such vesicles (e.g. containing phosphatidylserine) fuse at 37 °C in the presence of calcium (10^{-2} to 10^{-4} *M*) but not magnesium (PAPAHADJOPOULOS et al. 1976). Calcium is supposed to promote the close apposition (reversible phase of aggregation: see KOTER et al. 1978). In the second step a destabilization of the membrane occurs by crystallization (i.e. condensed acyl chain packing) of acidic phospholipids in some regions, preceded by a lateral phase separation of phospholipids (PAPAHADJOPOULOS et al. 1977; see also PORTIS et al. 1979). The relevance of this hypothesis for chromaffin granules is difficult to evaluate. Acidic phospholipids (including phosphatidylserine) are present in their membrane (see WINKLER 1976; and see WINKLER and WESTHEAD 1982) and they are probably localized on their external surface.

In the model discussed here, crystallization of phospholipids leading to a transient destabilization of the membrane caused by calcium is thought to be of crucial importance (see also NAYAR et al. 1982). However, destabilization may also be caused by increased fluidity or, in the case of the fusogenic agent lysolecithin, by formation of micelles within the aggregated membranes, causing destabilization and finally fusion (see LUCY 1978). When plasma membranes of chromaffin cells are treated with acetylcholine, an increase in the fluidity of the membrane occurs as measured by a hydrophobic fluorescent probe (SCHNEEWEISS et al. 1979; not confirmed by RIMLE et al. 1983).

The fusogenic action of lysolecithin is of course of special interest for the adrenal medulla, since chromaffin granules are known to contain a high concentration of this phopholipid (BLASCHKO et al. 1967b; see WINKLER 1976; for further references: FRISCHENSCHLAGER et al. 1983). When this phospholipid was discovered, it was suggested that this membrane active phospholipid may be essential for exocytosis. However, other secretory organelles do not contain higher concentration of lysolecithin, which argues against a general role in exocytosis (see MELDOLESI et al. 1978). Furthermore, a detailed study of BAKER et al. (1975) of brain synaptic vesicles revealed no increase in lysolecithin formation or turnover during stimulation. On the other hand, more recent studies again raised the question of the formation of lytic phospholipids during exocytosis. Plasma membranes of chromaffin cells pretreated with phospholipase A_2 caused lysis of chromaffin granules suggesting the possibility of lysolecithin formation in plasma membranes in the *in vivo* process of exocytosis (IZUMI et al. 1986). Recent data indicate that chromaffin granule membranes contain a phospholipase A_2 activity (FRISCHENSCHLAGER 1985; HUSEBYE and FLATMARK 1987). In mast cells and leucocytes secretion is accompanied by methylation of lipids, release of arachidonic acid and lysolecithin formation (HIRATA et al. 1979; CREWS et al. 1980). In chromaffin cells inhibitors of phospholipase A_2 block catecholamine secretion (FRYE and HOLZ 1983; SASAKAWA et al. 1983; WADA et al. 1983b). However, at least one site of this action is a blockade of calcium entry (FRYE and HOLZ 1983; WADA et al. 1983b). It would be interesting to see whether these inhibitors also abolish calcium-dependent secretion in "leaky" cells. On the other hand it has also been shown that secretion from adrenal medullary cells is accompanied by the release of (^{3}H)-inositol labelled products indicating on involvement of a calcium activated phospholipase C (WHITAKER 1985).

Finally we have the interesting observation that arachidonic acid induces a temperature-dependent fusion of chromaffin granules, which had been aggregated with synexin (CREUTZ 1981b; CREUTZ et al. 1982). In fact, during stimulation arachidonic acid is released from isolated chromaffin cells (FRYE and HOLZ 1984, 1985). However, other experiments do not at present support a role of these fatty acids in exocytosis, since in pheochromocytoma cells, grown in the presence of a variety of fatty acids, K-induced secretion did not depend on the presence of highly unsaturated fatty acids (WILLIAMS and McGEE 1982).

In conclusion we have many hypotheses concerning the relationship of lipids with membrane fusion, but their relevance for the *in vivo* process has still to be established.

3. Relationship of Fusion and Mg^{2+}-ATP Release Reaction

OKA et al. (1965, 1974) and POISNER and TRIFARO (1967) discovered that isolated chromaffin granules are lysed in the presence of ATP-Mg. It was speculated that this reaction had some relevance for the exocytotic process. Thus the latter authors proposed that Mg-ATP causes a conformational change in the membranes of chromaffin granules when attached to the plasma membrane by calcium. This conformational change was thought to somehow allow "egress of the granule content at the area of increased membrane permeability". Subsequently, LISHAJKO (1969) demonstrated that lysis of the granule caused by ATP-Mg requires the presence of chloride ions. It was also reported that low concentration of calcium (IZUMI et al. 1971) and of a cytoplasmic protein factor (IZUMI et al. 1975, 1977) increased the rate of lysis. FERRIS et al. (1970) found that some drugs (e.g. P-286) inhibit release *in vitro* and secretion *in vivo*. They speculated that the "influx of water after the vesicles have become attached to the plasma membrane may be an important mechanism for the extrusion of the vesicle contents to the exterior of the cell".

The mechanism of this lysis was finally elucidated by CASEY et al. (1976). They demonstrated that chloride entering the granules led to osmotic lysis. Chloride was drawn into the granule by the positive membrane potential provided by the electrochemical proton gradient which is produced by the Mg-activated ATPase. Other anions could substitute for chloride according to their lipid solubility.

In a series of papers POLLARD and collaborators took up the original speculations that this osmotic lysis of isolated granules has relevance for *in vivo* release in adrenal medulla (HOFFMAN et al. 1976; POLLARD et al. 1977; PAZOLES and POLLARD 1978), in platelets (POLLARD et al. 1977) and in parathyroid cells (BROWN et al. 1978). It was supposed that, after attachment of the granules to the plasma membrane, anions would pass through the two membranes via a special anion transport site driven by the positive membrane potential inside chromaffin granules. This would be followed by water and the increased intragranular pressure would finally cause fission at the site of the attached membranes (chemiosmotic hypothesis: see ZINDER and POLLARD 1980). However, more recent results (see also ENGLERT and PERLMAN 1981) seem to argue strongly against this model. For the chemiosmotic hypothesis the presence of an electrochemical gradient across the granule membrane is essential. However, two groups have shown, that exocytosis still occurs, when the electrochemical gradient (either ΔpH and/or $\Delta\psi$ of chromaffin granules) has been abolished (BAKER and KNIGHT 1982; HOLZ et al. 1983; KNIGHT and BAKER 1985a, see also discussion by HOLZ 1986). Therefore an involvement of the electrochemical gradient in secretion seems already excluded. This still leaves us with the question of whether osmotic factors are essential, since secretion from several cells is inhibited by hypertonic solutions (see ZINDER and POLLARD 1980); this was recently confirmed for the adrenal medulla (HAMPTON and HOLZ 1983; POLLARD et al. 1984; PINTO et al. 1985; LANDONA et al. 1987a) but in permeabilized cells inhibition occurs only at a high degree of hyperosmolarity (HOLZ and SENTER 1986; LADONA et al. 1987a). As pointed out above, it was suggested that granule swelling induces fusion of the granules attached to the plasma membranes, thus swelling had to occur before an actual fusion pore was formed. Recent experiments on mast cells however demonstrate the opposite. A small fusion pore is formed before swelling occurs which is apparently caused by the entry of solutes through the pore (ZIMMERBERG et al. 1987; BRECKENRIDGE and ALMERS 1987). These results seem to exclude the possibility that osmotic factors cause the essential steps leading to exocytosis.

V. Experimental Models for Elucidating the Mechanism of Exocytosis in Adrenal Medulla

The most promising models would seem to be those which are close to the physiological system, but do allow a significant modification of the medium surrounding the granule/plasma membrane system. The cell-free system, with the *in vitro* incubation of plasma membranes and chromaffin granules (see KONINGS and DE POTTER 1981a, b, 1982a) has already been discussed. However in this system the original structure of the cell has been destroyed, which may give rise to artefacts. This is not the case in the so called "leaky" chromaffin cells. These cells have been made "leaky" by electric currents which produce small holes allowing the passage of ions, but not of proteins (see BAKER and KNIGHT 1981; KNIGHT and BAKER 1982; KNIGHT and SCRUTTON 1986). The usefulness of this system has already become obvious in the discussion in several preceding sections. A more recent method relies on staphylococcus aureus α-toxin to induce holes in the plasma membrane (AHNERT-HILGER et al. 1985; BADER et al. 1986a; LIND et al. 1987). However, a disadvantage of these systems is that only small molecules can enter the cell.

It seems, therefore, a significant development that larger "holes" can be made into chromaffin cells without loss of secretory function. Such holes should allow introduction of protein molecules (e.g. antibodies) into these cells (in fact the cytoplasmic protein lactate dehydrogenase leaks out of these permeabilized cells). These "holes" can be produced either by digitonin (DUNN and HOLZ 1983; WILSON and KIRSHNER 1983; HOLZ and SENTER 1985) or by saponin treatment (BROOKS and TREML 1983b). On the other hand, antibodies can also be introduced into intact cells by fusing them with antibody filled erythrocytes (KENIGSBERG and TRIFARO 1984). Results with this elegant method have already been discussed before in the discussion on calmodulin.

VI. General Conclusions

In the various sections we have already presented conclusions for each subtopic. It suffices here to state:

(i) Exocytosis has been firmly established as the physiological secretory mechanism for adrenal medulla and sympathetic nerve.

(ii) For adrenal medulla it has been established that the membranes of chromaffin granules are specifically retrieved from the plasma membrane after exocytosis and are reused for several secretory cycles.

(iii) In sympathetic nerves release of catecholamines together with peptides occur from large dense core vesicles. Small vesicles release only catecholamines, but no peptides. They are reused several times for a synthesis/secretion cycle, however, the origin of small vesicles has not yet been definitely established.

(iv) Numerous studies have dealt with the mechanism of exocytosis, in particular the aspects of membrane attachment and membrane fusion. They have not yet led, but may soon lead, to a proven model explaining these events in molecular terms.

Acknowledgement. The original work of the author quoted in this review was supported by the Dr. Legerlotz-Stiftung and by the Fonds zur Förderung der wissenschaftlichen Forschung (Austria). I am grateful to Miss G. Schwaiger, Mrs. M. Nessmann and Mr. Ch. Trawöger for typing the manuscript and preparing the drawings.

D. References

Abbs MT, Phillips JH (1980) Organisation of the proteins of the chromaffin granule membranes. Biochim Biophys Acta 595:200-201

Aberer W, Stitzel R, Winkler H, Huber E (1979) Accumulation of 3 H-ATP in small dense core vesicles of superfused vasa deferentia. J Neurochem 33:797-801

Abrahams SJ, Holtzman E (1973) Secretion and endocytosis in insulin-stimulated medulla cells. J Cell Biol 56:540-558

Ackerly JA, Blumberg AL, Brooker G, Peach JM (1978) Angiotensin II on the release of dopamine β-hydroxylase and arterial cyclic AMP concentrations. Am J Physiol 235:H281-H288

Aguirre J, Pinto JEB, Trifaro JM (1977) Calcium movements during the release of catecholamines from the adrenal medulla: effects of methoxyverapamil and external cations. J Physiol (Lond) 269:371-394

Aguirre J, Falutz J, Pinto JEB, Trifaro JM (1979) Effects of methoxyverapamil on the stimulation by Ca^{2+}, Sr^{2+} and Na^{2+} and on the inhibition by Mg^{2+} of catecholamine release from the adrenal medulla. Br J Pharmacol 66:591-600

Ahnert-Hilger G, Gratzl M (1987) Further characterization of dopamine release by permeabilized PC12 cells. J Neurochem 49:764-770

Ahnert-Hilger G, Bhakdi S, Gratzl M (1985) Minimal requirements for exocytosis. J Biol Chem 260:12730-12734

Algate DR, Leach GDH (1978) Dopamine β-hydroxylase release following acute selective sympathetic nerve stimulation of the heart, spleen and mesentery. J Pharm Pharmacol 30:162-166

Amy CM, Kirshner N (1981) Phosphorylation of adrenal medulla cell proteins in conjunction with stimulation of catecholamine secretion. J Neurochem 36:847-854

Apps DK, Pryde JG, Phillips JH (1980) Cytochrome b-561 is identical with chromomembrin B1 a major polypeptide of chromaffin granule membranes. Neuroscience 5:2279-2287

Arnaiz JM, Garcia AG, Horga JF, Kirpekar SM (1978) Tissue and plasma catecholamines and dopamine β-hydroxylase activity of various animal species after neurogenic sympathetic stimulation. J Physiol (Lond) 285:515-529

Arnatz JM, Garcia AG, Horga JF, Pascual R, Sanchez-Garcia P (1980) Origin of guinea-pig plasma dopamine β-hydroxylase. Br J Pharmacol 69:41-48

Arqueros L, Daniels AJ (1978) Analysis of the inhibitory effect of verapamil on adrenal medullary secretion. Life Sci 23:2415-2422

Artalejo CR, Garcia AG, Aunis D (1987) Chromaffin cell calcium channel kinetics measured isotopically through fast calcium, strontium, and barium fluxes. J Biol Chem 262:915-926

Ashino N, Sobue K, Seino Y, Yabuuchi H (1987) Purification of an 80 kDa Ca^{2+}-dependent actin-modulating protein, which severs actin filaments, from bovine adrenal medulla. J Biochem. 101:609-617

Augustine GJ, Charlton MP, Smith SJ (1987) Calcium action in synaptic transmitter release. Ann Rev Neurosci 10:633-693

Aunis D, Perrin D (1984) Chromaffin granule membrane F-actin interactions and spectrin-like protein of subcellular organelles: a possible relationship. J Neurochem 42:1558-1569

Aunis D, Serck-Hanssen G, Helle KB (1978) Dopamine β-hydroxylase in perfusates of stimulated bovine adrenals. Isolation and characterization of the released enzyme. Gen Pharmacol 9:37-43

Aunis D, Hesketh JE, Devilliers G (1979) Freeze-fracture study of the chromaffin cell during exocytosis: evidence for connections between the plasma membrane and secretory granules and for movements of plasma membrane-associated particles. Cell Tiss Res 197:433-441

Aunis D, Guerold B, Bader MF, Cieselski-Treska J (1980) Immunocytochemical and biochemical demonstration of contractile proteins in chromaffin cells in culture. Neuroscience 5:2261-2277

Bader MF, Aunis D (1983) The 97-kD-α-actinin-like protein in chromaffin granule membranes from adrenal medulla: evidence for localization on the cytoplasmic surface and for binding to actin filaments. Neuroscience 8:165-182

Bader MF, Ciesielski-Treska J, Thierse D, Hesketh JE, Aunis D (1981) Immunocytochemical study of microtubules in chromaffin cells in culture and evidence that tubulin is not an integral protein of the chromaffin granule membrane. J Neurochem 37:917-933

Bader MF, Hikita T, Trifaro JM (1985) Calcium-dependent calmodulin binding to chromaffin granule membranes: presence of a 65-Kilodalton calmodulin-binding protein. J Neurochem 44:526-539

Bader MF, Thierse D, Aunis D (1986a) Characterization of hormone and protein release from α-toxin-permeabilized chromaffin cells in primary culture. J Biol Chem 261:5777-5783

Bader MF, Trifaro JM, Langley OK, Thierse D, Aunis D (1986b) Secretory cell actin-binding proteins: identification of a gelsolin-like protein in chromaffin cells. J Cell Biol 102:636-646

Bähler M, Greengard P (1987) Synpasin I bundles F-actin in a phosphorylation-dependent manner. Nature 326:704-707

Baines AJ (1987) Synapsin I and the cytoskeleton. Nature 326:646

Baker PF, Knight DE (1978) Calcium-dependent exocytosis in bovine adrenal medullary cells with leaky plasma membranes. Nature 276:620-622

Baker PF, Knight DE (1981) Calcium control of exocytosis and endocytosis in bovine adrenal medullary cells. Phil Trans R Soc Lond B 296:83-103

Baker PF, Knight DE (1982) Is the chromaffin granule membrane potential essential for Ca-dependent exocytosis? J Physiol 326:10P
Baker PF, Knight DE (1986) Exocytosis: control by calcium and other factors. Brit Med Bull 42:399–404
Baker PF, Rink TJ (1975) Catecholamine release from bovine adrenal medulla in response to maintained depolarization. J Physiol 253:593–620
Baker RR, Dowdall MJ, Whittaker VP (1975) The involvement of lysophosphoglycerides in neurotransmitter release, the composition and turnover of phospholipids of synaptic vesicles of guinea-pig cerebral cortex and torpedo electric organ and the effect of stimulation. Brain Res 100:629–644
Baker PF, Knight DE, Niggli V (1982) Protein phosphorylation accompanies calcium-dependent exocytosis in "leaky" bovine adrenal medullary cells. J Physiol 332:118P
Banks P (1966a) The release of adenosine triphosphate catabolites during the secretion of catecholamines by bovine adrenal medulla. Biochem J 101:536–541
Banks P (1966b) An interaction between chromaffin granules and calcium ions. Biochem J 101:18C–20C
Banks P, Helle K (1965) The release of protein from the stimulated adrenal medulla. Biochem J 97:40C–41C
Bareis DL, Slotkin TA (1979) Synaptic vesicles isolated from rat heart: ^{3}H-norepinephrine uptake properties. J Neurochem 32:345–351
Barron BA, Hexum TD (1984) Release of catecholamines and (Met^{5}) enkephalin immunoreactive material from perfused bovine adrenal glands. Eur J Pharmacol 106:593–599
Barrowman MM, Cockcroft S, Gomperts BD (1986) Two roles for guanine nucleotides in the stimulus-secretion sequence of neutrophils. Nature 319:504–507
Bartlett SF, Lagercrantz H, Smith AD (1976) Gel electrophoresis of soluble and insoluble proteins of noradrenergic vesicles from ox splenic nerve: a comparison with proteins of adrenal chromaffin granules. Neuroscience 1:339–344
Basbaum CB, Heuser JE (1979) Morphological studies of stimulated adrenergic axon varicosities in the mouse vas deferens. J Cell Biol 80:310–325
Benedeczky I, Smith AD (1972) Ultrastructural studies on the adrenal medulla of golden hamster: origin and fate of secretory granules. Z Zellforsch 124:367–386
Benedeczky I, Somogyi P (1975) Ultrastructure of the adrenal medulla of normal and insulin-treated hamsters. Cell Tiss Res 162:541–550
Benedeczky I, Somogyi P (1978) Cytochemical localization of exogenous peroxidase in adrenal medullary cells of hamster. Acta Biol Acad Sci Hung 29:155–163
Bernier-Valentin F, Aunis D, Rousset B (1983) Evidence for tubulin-binding sites on cellular membranes: Plasma membranes, mitochondrial membranes and secretory granule membranes. J Cell Biol 97:209–216
Bevington A, Briggs RW, Radda GK, Thulborn KR (1984) Phosphorus-31 nuclear magnetic resonance studies of pig adrenal gland. Neuroscience 11:281–286
Biales B, Dichter M, Tischler A (1976) Electrical excitability of cultured adrenal chromaffin cells. J Physiol 262:743–753
Bisby MA, Fillenz M (1971) The storage of endogenous noradrenaline in sympathetic nerve terminals. J Physiol 215:163–179
Bisby M, Fillenz M, Smith AD (1973) Evidence for the presence of dopamine β-hydroxylase in both populations of noradrenaline storage vesicles in sympathetic nerve terminals of the rat vas deferens. J Neurochem 20:245–248
Bittner MA, Holz RW, Neubig RR (1986) Guanine nucleotide effects of catechola-

mine secretion from digitonin-permeabilized adrenal chromaffin cells. J Biol Chem 261:10182-10188

Bjerrum OJ, Helle KB, Bock E (1979) Immunochemically identical hydrophilic and amphiphilic forms of the bovine adrenomedullary dopamine β-hydroxylase. Biochem J 181:231-237

Blakeley AGH, Cunnane TC (1979) The packeted release of transmitter from the sympathetic nerves of the guinea-pig vas deferens: an electrophysiological study. J Physiol (Lond.) 296:85-96

Blaschke E, Uvnäs B (1979) Effect of splenic nerve stimulation on the contents of noradrenaline, ATP and sulphomucopolysaccharides in noradrenergic vesicle fractions from the cat spleen. Acta Physiol Scand 105:496-507

Blaschko H, Muscholl E (1972) Catecholamines, Handbook of Experimental Pharmacology, vol 33. Springer Berlin Heidelberg New York

Blaschko H, Comline RS, Schneider FH, Silver M, Smith AD (1967a) Secretion of a chromaffin granule protein, chromogranin, from the adrenal gland after spanchnic stimulation. Nature 215:58-59

Blaschko H, Firemark H, Smith AD, Winkler H (1967b) Lipids of the adrenal medulla: lysolecithin, a characteristic constituent of chromaffin granules. Biochem J 104:545-549

Blaustein MP (1979) The role of calcium in catecholamine release from adrenergic nerve terminals. In: Paton DM (ed) The release of catecholamines from adrenergic neurons. Pergamon Press Oxford pp 39-58

Bock E, Helle KB (1977) Localization of synaptin on synaptic vesicle membranes, synaptosomal plasma membranes and chromaffin granule membranes. Febs letters 82:175-178

Bohner K, Boons J, Gheuens J, Konings F, de Potter WP (1985) The use of monoclonal antibodies in the study of the interaction between adrenal medullary cell membranes and chromaffin granules. Biochem Biophys Res Comm 133:1006-1012

Boonyaviroj P, Seiden A, Gutman Y (1977) PGE_2, phenylephrine and dopamine β-hydroxylase release from rat adrenal *in vitro*. Biochem Pharmacol 26:351-352

Borchardt RT, Olsen J, Eiden L, Schowen RL, Rutledge CO (1978) The isolation and characterization of the methyl acceptor protein from adrenal chromaffin granules. Biochem Biophys Res Commun 83:970-976

Borowitz JL, Leslie SW, Baugh L (1975) Adrenal catecholamine release: possible termination mechanism. In: Carofoli E, Clementi F, Drabikovski W (eds) North Holland Amsterdam, Oxford pp 227-242

Bourne GW, Trifaro JM (1982) The gadolinium ion: a potent blocker of calcium channels and catecholamine release from cultured chromaffin cells. Neuroscience 7:1615-1622

Brandt BL, Hagiwara S, Kidokoro Y, Miyazaki S (1976) Action potentials in the rat chromaffin cell and effects of acetylcholine. J Physiol (Lond) 263:417-439

Breckenridge LJ, Almers W (1987) Final steps in exocytosis observed in a cell with giant secretory granules. Proc Natl Acad Sci USA 84:1945-1949

Brimijoin S (1982) Microtubules and the capacity of the system for rapid axonal transport. Fed Proc 41:2312-2316

Brimijoin S, Heilland L (1976) Rapid retrograde transport of dopamine β-hydroxylase as examined by the stop flow technique. Brain Res 102:217-228

Brocklehurst KW, Pollard HB (1985) Enhancement of Ca^{2+}-induced catecholamine release by the phorbol ester TPA in digitonin-permeabilized cultured bovine adrenal chromaffin cells. FEBS 183:107-110

Brocklehurst KW, Pollard HB (1986) Synergistic actions of Ca^{2+} and the phorbol ester TPA on K^+-induced catecholamine release from bovine adrenal chromaffin cells. Biochem Biophys Res Comm 140:990-998
Brocklehurst KW, Morita K, Pollard HB (1985) Characterization of protein kinase C and its role in catecholamine secretion from bovine adrenal-medullary cells. Biochem J 228:35-42
Brooks JC, Brooks M (1985) Protein thiophosphorylation associated with secretory inhibition in permeabilized chromaffin cells. Life Sci 37:1869-1875
Brooks JC, Brooks MH (1987) Thiophosphorylation and phosphorylation of saponin-permeabilized cultured chromaffin cells Neurochem Int 1:31-38
Brooks JC, Carmichael SW (1987) Ultrastructural demonstration of exocytosis in intact and saponin-permeabilized cultured bovine chromaffin cells. Am J Anat 178:85-89
Brooks JC, Treml S (1983a) Effect of trifluoperazine on catecholamine secretion by isolated bovine adrenal medullary chromaffin cells. Biochem Pharmacol 32:371-373
Brooks JC, Treml S (1983b) Catecholamine secretion by chemically skinned cultured chromaffin cells. J Neurochem 40:468-473
Brooks JC, Treml S (1984) Effect of trifluoperazine and calmodulin on catecholamine secretion by saponin-skinned cultured chromaffin cells. Life Sciences 34:669-674
Brooks JC, Treml S, Brooks M (1984) Thiophosphorylation prevents catecholamine secretion by chemically skinned chromaffin cells. Life Sci 35:569-574
Brown EM, Pazoles CJ, Creutz CE, Aurbach GD, Pollard HB (1978) Role of anions in parathyroid hormone release from dispersed bovine parathyroid cells. Proc Natl Acad Sci USA 75:876-880
Burgoyne, RD (1984a) Mechanism of secretion from adrenal chromaffin cells. Biochim Biophys Acta 779:201-216
Burgoyne RD (1984b) The relationship between secretion and intracellular free calcium in bovine adrenal chromaffin cells. Biosci Rep 4:605-611
Burgoyne RD (1987) Control of exocytosis. Nature 328:112-113
Burgoyne RD, Cheek TR (1987) Role of fodrin in secretion. Nature 326:448
Burgoyne RD, Geisow MJ (1981) Specific binding of ^{125}I-calmodulin to and protein phosphorylation in adrenal chromaffin granule membranes. FEBS Lett 131:127-131
Burgoyne RD, Geisow MJ (1982) Phosphoproteins of the adrenal chromaffin granule membrane. J Neurochem 39:1387-1396
Burgoyne RD, Norman KM (1984) Effect of calmidazolium and phorbol ester on catecholamine secretion from adrenal chromaffin cells. Biochim Biophys Acta 805:37-43
Burgoyne RD, Norman KM (1985) Presence of microtubule-associated protein 2 in chromaffin cells. Neuroscience 14:955-962
Burgoyne RD, Geisow MJ, Barron J (1982) Dissection of stages in exocytosis in the adrenal chromaffin cell with use of trifluoperazine. Proc R Soc Lond B 216:111-115
Burgoyne RD, Cheek TR, Norman KM (1986) Identification of a secretory granule-binding protein as caldesmon. Nature 319:68-70
Burgun C, Martinez de Munoz D, Aunis D (1985) Osmotic fragility of chromaffin granules prepared under isoosmotic or hyperosmotic conditions and localization of acetylcholinesterase. Biochim Biophys Acta 839:219-227
Burke BE, de Lorenzo RJ (1982) Ca^{2+} and calmodulin-dependent phosphorylation of

endogenous synaptic vesicle tubulin by a vesicle-bound calmodulin kinase system. J Neurochem 38:1205-1218
Carmichael S, Ulrich RG (1983) Scanning electron microscopy of the mammalian adrenal medulla. Mikroskopie 40:53-64
Carvalho MH, Prat JC, Garcia AG, Kirpekar SM (1982) Ionomycin stimulates secretion of catecholamines from cat adrenal gland and spleen. Am J Physiol 242:E137-E145
Casey RP, Njus D, Radda GK, Sehr PA (1976) Adenosine triphosphate-evoked catecholamine release in chromaffin granules. Osmotic lysis as a consequence of proton translocation. Biochem J 158:583-588
Cena V, Nicolas GP, Sanchez-Garcia P, Kirpekar SM, Garcia AG (1983) Pharmacological dissection of receptor-associated and voltage-sensitive ionic channels involved in catecholamine release. Neuroscience 10:1455-1462
Chaminade M, Foutz AS, Rossier J (1983) Co-release of enkephalins and precursors with catecholamines by the perfused cat adrenal in-situ. Life Sciences 33:21-24
Chandler DE, Heuser JE (1980) Arrest of membrane fusion events in mast cells by quick freezing. J Cell Biol 86:666-674
Cheek TR, Burgoyne RD (1985) Effect of activation of muscarinic receptors on intracellular free calcium and secretion in bovine adrenal chromaffin cells. Biochim Biophys Acta 846:167-173
Cheek TR, Burgoyne RD (1986) Nicotine-evoked disassembly of cortical actin filaments in adrenal chromaffin cells. FEBS 207:110-114
Cheek TR, Hesketh JE, Richards RC, Burgoyne RD (1986) Assembly and characterization of a multi-component cytoskeletal gel from adrenal medulla. Biochim Biophys Acta 887:164-172
Cheung WY (1980) Calmodulin plays a pivotal role in cellular regulation. Science 207:19-27
Chi EY, Lagunoff D, Koehler JK (1976) Freeze-fracture study of mast cell secretion. Proc Natl Acad Sci USA 73:2823-2827
Chubb IW, Smith AD (1975) Release of acectylcholinesterase into the perfusate from the ox adrenal gland. Proc R Soc Lond B 191:263-269
Chubb IW, de Potter WP, de Schaepdryver AF (1972) Tyramine does not release noradrenaline from splenic nerve by exocytosis. Naunyn Schmiedeberg's Arch Pharmacol 274:281-286
Clapham DE, Neher E (1984) Trifluoperazine reduces inward ionic currents and secretion by separate mechanism in bovine chromaffin cells J Physiol (Lond) 353:541-564
Cobbold PH, Cheek TR, Cuthbertson KSR, Burgoyne RD (1987) Calcium transients in single adrenal chromaffin cells detected with aequorin. FEBS Lett 211:44-48
Cochrane DE, Douglas WW, Mouri T, Nakazato Y (1975) Calcium and stimulus-secretion coupling in the adrenal medulla: contrasting stimulating effects of the ionophores X-537A and A23187 on catecholamine output. J Physiol 252:363-378
Cohen J, Gutman Y (1979) Effects of verapamil, dantrolene and lanthanum on catecholamine release from rat adrenal medulla. Br J Pharmacol 65:641-645
Cooke P, Poisner AM (1976) Microfilaments in bovine adrenal cells. Cytobiologie 13:442-450
Corcoran JJ, Kirshner N (1983) Inhibition of calcium uptake, sodium uptake and catecholamine secretion by methoxyverapamil (D600) in primary cultures of adrenal medulla cells. J Neurochem 40:1106-1109
Costa M, Rush RA, Furness JB, Geffen LB (1976) Histochemical evidence for the de-

generation of peripheral noradrenergic axons following intravenous injection of antibodies to dopamine β-hydroxylase. Neuroscience Lett 3:201-207

Cote A, Doucet JP, Trifaro JM (1986a) Adrenal medullary tropomyosins: purification and biochemical characterization. J Neurochem 46:1771-1782

Cote A, Doucet JP, Trifaro JM (1986b) Phosphorylation and dephosphorylation of chromaffin cell proteins in response to stimulation. Neurosciene 19:629-645

Coupland RE (1965) Electron microscopic observations on the structure of the rat adrenal medulla. 1. The ultrastructure and organisation of chromaffin cells in the normal adrenal medulla. J Anat (Lond.) 99:231-254

Creutz CE (1977) Isolation, characterization and Localization of bovine adrenal medullary myosin. Cell Tiss Res 178:17-38

Creutz CE (1981a) Secretory vesicle—cytosol interactions in exocytosis: Isolation by Ca^{2+}-dependent affinity chromatography of proteins that bind to the chromaffin granule membrane. Biochem Biophys Res Commun 103:1395-1400

Creutz CE (1981b) cis-Unsaturated fatty acids induce the fusion of chromaffin granules aggregated by synexin. J Cell Biol 91:247-256

Creutz CE, Rauch-Harrison J (1984) Clathrin light chains and secretory vesicle binding proteins are distinct. Nature 308:208-210

Creutz CE, Sterner DC (1983) Calcium dependence of the binding of synexin to isolated chromaffin granules. Biochem Biophys Res Comm 114:355-364

Creutz CE, Pazoles CJ, Pollard HP (1978) Identification and purification of an adrenal medullary protein (synexin) that causes calcium-dependent aggregation of isolated chromaffin granules. J Biol Chem 253:2858-2866

Creutz CE, Pazoles CJ, Pollard HP (1979) Self-association of synexin in the presence of calcium. J Biol Chem 254:553-558

Creutz CE, Scott JH, Pazoles CJ, Pollard HB (1982) Further characterization and fusion of chromaffin granules by synexin as a model for compound exocytosis. J Cell Biochem 18:87-97

Creutz CE, Dowling LG, Sando JJ, Villar-Palasi C, Whipple JH, Zaks WJ (1983) Characterization of the chromobindins J Biol Chem 258:14664-14674

Creutz CE, Dowling LG, Kyger EM, Franson RC (1985) Phosphatidylinositol-specific phospholipase C activity of chromaffin granule-binding proteins. J Biol Chem 260:7171-7173

Creutz CE, Zaks WJ, Hamman HC, Crane S, Martin WH, Gould KL, Oddie KM, Parsons SJ (1987) Identification of chromaffin granule-binding proteins. J Biol Chem 262:1860-1868

Crews FT, Morita Y, Hirata F, Axelrod J, Siraganian RP (1980) Phospholipid methylation affects immunoglobulin E-mediated histamine and arachidonic acid release in rat leukemic basophils. Biochem Biophys Res Commun 93:42-49

Cubeddu LX, Weiner N (1975) Nerve stimulation-mediated overflow of norepinephrine and dopamine β-hydroxylase. III. Effects of norepinephrine depletion on the alpha presynaptic regulation of release. J Pharmacol Exp Ther 192:1-14

Cubeddu LX, Barnes EM, Langer SZ, Weiner N (1974a) Release of norepinephrine and dopamine β-hydroxylase by nerve stimulation. I. Role of neuronal and extraneuronal uptake and of alpha presynaptic receptors. J Pharmacol Exp Ther 190:431-450

Cubeddu LX, Barnes E, Weiner N (1974b) Release of norepinephrine and dopamine β-hydroxylase by nerve stimulation. II. Effects of papaverine. J Pharmacol Exp Ther 191:444-457

Cubeddu LX, Barnes E, Weiner N (1975) Release of norepinephrine and dopamine β-hydroxylase by nerve stimulation. J Pharmacol Exp Ther 193:105-127

Cubeddu LX, Santiago E, Talmaciu R, Pinardi G (1977) Adrenal origin of the increase in plasma dopamine β-hydroxylase and catecholamines induced by hemorrhagic hypotension in dogs. J Pharmacol Exp Ther 203:587-597

Cubeddu LX, Barbella YR, Marrero A, Trifaro J, Israel AS (1979) Circulating pool and adrenal soluble content of dopamine β-hydroxylase (DBH), in rats, guina pigs, dogs and humans: their role in determining acute stress-induced changes and plasma enzyme levels. J Pharmacol Exp Ther 211:271-279

Dabrow M, Zaremba S, Hogue-Angeletti R (1980) Specificity of synexin-induced chromaffin granule aggregation. Biochem Biophys Res Commun 96:1164-1171

Dahl G, Ekerdt R, Gratzl M (1979) Models for exocytotic membrane fusion. In: Secretory Mechanisms (Eds Hopkins CR, Duncan CJ), Soc for Exp Biol Symp vol XXXIII, pp 349-368, Cambridge: Univ.-Press.

Dahlström A (1971) Axoplasmic transport (with particular respect to adrenergic neurons). Phil Trans Roy Soc Lond B 261:325-358

Dahlström A (1973) Aminergic transmission. Introduction and short review. Brain Res 62:441-460

Daniels AJ, Dean G, Viveros OH, Diliberto EJ (1983) Secretion of newly taken up ascorbic acid by adrenomedullary chromaffin cells originates from a compartment different from the catecholamine storage vesicles. Molec Pharm 23:437-444

Davis B, Lazarus NR (1976) An *in vitro* system for studying insulin release caused by secretory granules—plasma membrane interaction: definition of the system. J Physiol (Lond) 256:709-729

Dean PM (1975) Exocytosis modelling: an electrostatic function for calcium in stimulus-secretion coupling. J theor Biol 54:289-308

Dean PM, Matthews EK (1974) Calcium-ion binding to the chromaffin granule surface. Biochem J 142:637-640

De Block J, De Potter W (1987) The cell-free interaction between chromaffin granules and plasma membranes: An *in vitro* model for exocytosis? FEBS 222:358-359

De Camilli P, Cameron R, Greengard P (1983) Synapsin I (protein I), a nerve terminal-specific phosphoprotein. I. Its general distribution in synapses of the central and peripheral nervous system demonstrated by immunofluorescence in frozen and plastic-section. J Cell Biol 96:1337-1354

De Duve C, Cameron MP (1963) Endocytosis. In: De Reuck AVS, (eds) Lysosomes. Ciba Foundation Symposium Churchill London pp 126

De Lorenzo RJ (1981) The calmodulin hypothesis of neurotransmission. Cell Calcium 2:365-385

De Lorenzo RJ, Freedman SD, Yohe WB, Maurer SC (1979) Stimulation of Ca^{2+} dependent neurotransmitter release and presynaptic nerve terminal protein phosphorylation by calmodulin and a calmodulin-like protein isolated from synaptic vesicles. Proc Natl Acad Sci USA 76:1838-1842

De Oliveira-Filgueiras OM, van den Besselar AMHP, van den Bosch H (1979) Localization of lysophosphatidylcholine in bovine chromaffin granules. Biochem Biophys Acta 558:73-84

De Potter WP, Chubb IW (1977) Biochemical observations on the formation of small noradrenergic vesicles in the splenic nerve of the dog. Neuroscience 2:167-174

De Potter WP, Moerman EJ, de Schaepdryver AF, Smith AD (1969a) Release of noradrenaline and dopamine β-hydroxylase upon splenic nerve stimulation. Proc 4th int Congr Pharmac Abstracts. Schwabe & Co, Basel p 146

De Potter WP, de Schaepdryver AF, Moerman EJ, Smith AD (1969b) Evidence for the release of vesicle proteins together with noradrenaline upon stimulation of the splenic nerve. J Physiol (Lond) 204:102P-104P

De Potter WP, Chubb IW, de Schaepdryver AF (1971a) Pharmacological aspects of peripheral noradrenergic transmission. Arch Int Pharmacol 196:258-287

De Potter WP, Chubb IW, Put A, de Schaepdryver AF (1971b) Facilitation of the release of noradrenaline and dopamine β-hydroxylase at low stimulation frequencies by α-blocking agents. Arch Int Pharmacol 193:191-197

De Potter WP, Coen EP, de Potter RW (1987) Evidence for the coexistence and co-release of (met)enkephalin and noradrenaline from sympathetic nerves of the bovine vas deferens. Neuroscience 20:855-866

De Robertis EDP, Sabatini DD (1960) Submicroscopic analysis of the secretory process in the adrenal medulla. Fed Proc 19:70-78

De Robertis EDP, vaz Ferreira A (1957) Electron microscope study of the excretion of catechol-containing droplets in the adrenal medulla. Exp Cell Res 12:568-547

Diliberto EJ (1982) Protein carboxyl methylation: putative role in exocytosis and in the cellular regulation of secretion and chemotaxis. In: Cohn PM (ed) Cellular Regulation of Secretion and Release. Cell Biology: A comprehensive treatise. Academic Press New York pp 147-192

Diliberto EJ, Viveros OH, Axelrod J (1976) Subcellular distribution of protein carboxymethylase and its endogenous substrates in the adrenal medulla: possible role in excitation-secretion coupling. Proc Natl Acad Sci USA 73:4050-4054

Diner, O (1967) L'expulsion des granules de la médullosurrénale chez le hamster. C R Acad Sci (Paris) 265:616-619

Dixon WR, Garcia AG, Kirpekar SM (1975) Release of catecholamines and dopamine β-hydroxylase from the perfused adrenal gland of the cat. J Physiol (Lond) 244:805-824

Dixon JS, Fozard JR, Gosling JA, Muscholl E, Ritzel H (1978) Atrial sympathetic nerves after perfusion of the rabbit heart with low sodium solutions containing potassium chloride or urea—a biochemical, histochemical and fine structural study. Neuroscience 3, 1157-2267

Douglas WW (1968) Stimulus-secretion coupling: the concept and clues from chromaffin and other cells. Brit J Pharmacol 34:451-474

Douglas WW (1974a) Involvement of calcium in exocytosis and the exocytosis vesiculation sequence. Biochem Soc Symp 39:1-28

Douglas WW (1974b) Exocytosis and the exocytosis-vesiculation sequence: with special reference to neurohypophysis, chromaffin and mast cells, calcium and calcium ionophores. In: Thorn NA, Petersen OH (eds) Secretory mechanisms of exocrine glands. Munksgaard Copenhagen pp 116-136

Douglas WW, Nemeth EF (1982) On the calcium receptor activating exocytosis: inhibitory effects of calmodulin-interacting drugs on rat mast cells. J Physiol 323:229-244

Douglas WW, Poisner AM (1966) On the relation between ATP splitting and secretion in the adrenal chromaffin cell: extrusion of ATP (unhydrolysed) during release of catecholamines. J Physiol (Lond) 183:249-256

Douglas WW, Sorimachi M (1972a) Colchicine inhibits adrenal medullary secretion evoked by acetylcholine without affecting that evoked by potassium. Br J Pharmacol 45:129-132

Douglas WW, Sorimachi M (1972b) Effects of cytochalasin B and colchicine on secretion of posterior pituitary and adrenal medullary hormones. Br J Pharmacol 45:143-144P

Douglas WW, Poisner AM, Rubin RP (1965) Efflux of adenine nucleotides from perfused adrenal glands exposed to nicotine and other chromaffin cell stimulants. J Physiol (Lond) 179:130-137

Douglas WW, Kanno T, Sampson SR (1967) Influence of the ionic environment on the membrane potential of adrenal chromaffin cells and on the depolarizating effect of acetylcholine. J Physiol (Lond) 191:107-121

Dowd DJ, Edwards C, Englert D, Mazurkiewicz JE, YE, HZ (1983) Immunofluorescent evidence for exocytosis and internalization of secretory granule membrane in isolated chromaffin cells. Neuroscience 10:1025-1033

Droz B, Rambourg A, Koenig HL (1975) The smooth endoplasmic reticulum: structure and role in the renewal of axonal membrane and synaptic vesicles by fast axonal transport. Brain Res 93:1-13

Dunn LA, Holz RW (1983) Catecholamine secretion from digitonin-treated adrenal medullary chromaffin cells. J Biol Chem 258:4989-4993

Eagles PAM, Johnson LN, van Horn C (1975) The distribution of concanavalin A receptor sites on the membrane of chromaffin granules. J Cell Sci 19:33-54

Echeverría OM, Vázquez-Nin GH, Chávez B (1977) Correlated ultrastructural and biochemical studies on the mechanisms of secretion of catecholamines. Acta anat 98:313-324

Edwards AV, Furness PN, Helle KB (1979) Adrenal medullary responses to stimulation of the peripheral end of the splanchnic nerve in conscious calves. J Physiol (Lond) 300:51P

Edwards W, Phillips SJH, Morris SJ (1974) Structural changes in chromaffin granules induced by divalent cations. Biochim Biophys Acta 356:164-173

Eiden LE, Huttner WB, Mallet J, O'Conner DT, Winkler H, Zanini A (1987) A nomenclature proposal for the chromogranin/secretogranin proteins. Neuroscience 21:1019-1021

Ekerdt R, Dahl G, Gratzl M (1981) Membrane fusion of secretory vesicles and liposomes. Two different types of fusion. Biochem Biophys Acta 646:10-22

Elfvin LG (1965) The ultrastructure of the capillary fenestrae in the adrenal medulla of the rat. J Ultrastruct Res 12:687-704

Englert DF (1980) An optical study of isolated rat adrenal chromaffin cells. Exp Cell Res 125:369-376

Englert DF, Perlman RL (1981) Permeant anions are not required for norepinephrine secretion from phaeochromocytoma cells. Biochem Biophys Acta 674:136-143

Farquhar MG (1978) Recovery of surface membrane in anterior pituitary cells. Variations in traffic detected with anionic and cationic ferritin. J Cell Biol R35-R42

Farquhar MG (1982) Membrane recycling in secretory cells: pathway to Golgi complex. In: Evered D, Collins GM (eds) Membrane Recycling, Ciba Foundation Symposium, vol 92. Pittman Bath Press pp 157-183

Farquhar MG (1983) Multiple pathways of exocytosis, endocytosis and membrane recycling: validation of a Golgi route. Fed Proc 42:2407-2413

Farrell KE (1968) Fine structure of nerve fibres in smooth muscle of the vas deferens in normal and reserpinized rats. Nature 217:279-281

Fenwick EM, Livett BG (1976) Antigens of adrenal medullary vesicles and their fate following exocytosis. Proc Australian Biochem Soc 9:63

Fenwick EM, Fajdiga PB, Howe NBS, Livett BG (1978) Functional and morphological characterization of isolated bovine adrenal medullary cells. J Cell Biol 76:12-30

Fenwick EM, Marty A, Neher E (1982a) Sodium and calcium channels in bovine chromaffin cells. J Physiol (Lond) 331:599-635

Fenwick EM, Marty A, Neher E (1982b) A patch-clamp study of bovine chromaffin cells and of their sensitivity to acetylcholine. J Physiol (Lond) 331:577-597

Ferris RM, Viveros OH, Kirshner N (1970) Effects of various agents on the Mg^{2+}-ATP stimulated incorporation and release of catecholamines by isolated bovine adren-

omedullary storage vesicles and on secretion from the adrenal medulla. Biochem Pharmacol 19:505-514
Fillenz M (1971) Fine structure of noradrenaline storage vesicles in nerve terminals of the rat vas deferens. Phil Trans Roy Soc Lond B 261:319-323
Fillenz M (1977) The factors which provide short-term and long-term control of transmitter release. Prog Neurobiol 8:251-278
Fillenz M (1979) Ultrastructural studies of the mechanism of release. In: (Paton DM) (ed) The release of catecholamines from adrenergic neurons. Pergamon Press Oxford, pp 17-37
Fillenz M, Howe PRC (1975) Depletion of noradrenaline stores in sympathetic nerve terminals. J Neurochem 24:683-688
Fillenz M, West DP (1974) Changes in vesicular dopamine β-hydroxylase resulting from transmitter release. J Neurochem. 23:411-416
Fillenz M, West DP (1976) Fate of noradrenaline storage vesicles after release. Neurosc Lett 2:285-287
Fillenz M, Gagnon C, Stoeckel K, Thoenen H (1976) Selective uptake and retrograde axonal transport of dopamine β-hydroxylase antibodies in peripheral adrenergic neurons. Brain Res 114:293-303
Fischer-Colbrie R, Frischenschlager I (1985) Immunological characterization of secretory proteins of chromaffin granules: chromogranins A, chromogranins B and enkephalin-containing peptides. J Neurochem 44:1854-1861
Fischer-Colbrie R, Schaçhinger M, Zangerle R, Winkler H (1982) Dopamine β-hydroxylase and other glycoproteins from the soluble content and the membranes of adrenal chromaffin granules: Isolation and carbohydrate analysis. J Neurochem 38:725-732
Fischer-Colbrie R, Zangerle R, Frischenschlager I, Weber A, Winkler H (1984) Isolation and immunological characterization of a glycoprotein from adrenal chromaffin granules. J Neurochem 42:1008-1016
Fischer-Colbrie R, Lassmann H, Hagn C, Winkler H (1985) Immunological studies on the distribution of chromogranin A and B in endocrine and nervous tissues. Neuroscience 16:547-555
Fischer-Colbrie R, Guerra DJ, Emson PC, Winkler H (1986) Bovine chromaffin granules: immunological studies with antisera against neuropeptide Y, Met-enkephalin and bombesin. Neuroscience 18:167-174
Forsberg EJ, Rojas E, Pollard HB (1986) Muscarinic receptor enhancement of nicotine-induced catecholamine secretion may be mediated by phosphoinositide metabolism in bovine adrenal chromaffin cells. J Biol Chem 261:4915-4920
Fowler VM, Pollard HB (1982a) In vitro reconstitution of chromaffin granule—cytoskeleton interactions: ionic factors influencing the association of F-actin with purified chromaffin granule membranes. J Cell Biochem 18:295-311
Fowler VM, Pollard HB (1982b) Chromaffin granule membrane F-actin interactions are calcium sensitive. Nature 5847:336-339
Fredholm BB, Fried G, Hedqvist P (1982) Origin of adenosine released from rat vas deferens by nerve stimulation. Eur J Pharmacol 79:233-243
French AM, Scott NC (1983) Evidence to support the hypothesis that ATP is a cotransmitter in rat vas deferens. Experientia 39:264-266
Fried G (1978) Cytochrome b-561 in sympathetic nerve terminal vesicles from rat vas deferens. Biochim Biophys Acta 507:175-177
Fried G (1980) Small noradrenergic storage vesicles isolated from rat vas deferens—biochemical and morphological characterization. Acta Physiol Scand, Suppl 493:1-28

Fried G (1981) Noradrenaline release and uptake in isolated small dense cored vesicles from rat seminal ducts. Acat Physiol Scand 112:41-46

Fried G, Lagercrantz H, Hökfelt T (1978) Improved isolation of small noradrenergic vesicles from rat seminal ducts following castration. A density gradient centrifugation and morphological study. Neuroscience 3:1271-1291

Fried G, Thureson-Klein Å, Lagercrantz H (1981) Noradrenaline content correlated to matrix density in small noradrenergic vesicles from rat seminal ducts. Neuroscience 6:787-800

Fried G, Nestler EJ, de Camilli P, Stjärne L, Olson L, Lundberg JM, Hökfelt T, Quimet CC, Greengard P (1982) Cellular and subcellular localization of protein I in the peripheral nervous system. Proc Natl Acad Sci USA 79:2717-2721

Fried G, Terenius L, Hökfelt T, Goldstein M (1985) Evidence for differential localization of noradrenaline and neuropeptide Y in neuronal storage vesicles isolated from rat vas deferens. J Neuroscience 5:450-458

Frischenschlager I (1985) Lysolecithin und Phospholipase A_2 in den chromaffinen Granula des Nebennierenmarks und Stimulus-Secretion-Coupling. Ph D Thesis, Innsbruck

Frischenschlager I, Schmidt W, Winkler H (1983) Is lysolecithin an in vivo constituent of chromaffin granules? J Neurochem 41:1480-1483

Frydman R, Geffen LB (1973) Depletion and repletion of adrenal dopamine β-hydroxylase after reserpine. Immunohistochemical and fine structural correlates. J Histochem Cytochem 21:166-174

Frye RA, Holz RW (1983) Phospholipase A_2 inhibitors block catecholamine secretion and calcium uptake in cultured bovine adrenal medullary cells. Mol Pharmacol 23:547-550

Freye RA, Holz RW (1984) The relationship between arachidonic acid release and catecholamine secretion from cultured bovine adrenal chromaffin cells. J Neurochem 43:146-150

Frye RA, Holz RW (1985) Arachidonic acid release and catecholamine secretion from digitonin-treated chromaffin cells: effects of micromolar calcium, phorbol ester and protein alkylating agents. J Neurochem 44:265-273

Furness JB, Lewis SY, Rush R, Costa M, Geffen LB (1977) Involvement of complement in degeneration of sympathetic nerves after administration of antiserum to dopamine β-hydroxylase. Brain Res 136:67-75

Furness JB, Costa M, Rush RA, Geffen LB (1979) Noradrenergic transmission in isolated guinea-pig intestine following in vivo administration of antibodies to dopamine β-hydroxylase. Aust J Exp Biol Med Sci 57:203-209

Gabbiani G, da Prada M, Richards G, Pletscher A (1976) Actin associated with membranes of monoamine storage organelles. Proc Soc Expl Biol Med 152:135-138

Gagliardino JJ, Harrison DE, Christie MR, Gagliardino EE, Ashcroft SJH (1980) Evidence for the participation of calmodulin in stimulus—secretion coupling in the pancreatic β-cell. Biochem J 192:919-927

Gagnon C, Heisler S (1979) Protein carboxyl-methylation: role in exocytosis and chemotaxis. Life Sci 25:993-1000

Gagnon C, Viveros OH, Diliberto EJ, Axelrod J (1978) Enzymatic methylation of carboxyl groups of chromaffin granule membrane proteins. J Biol Chem 253:3778-3781

Garcia AG, Kirpekar SM (1973) Release of noradrenaline from the cat spleen by sodium deprivation. Br J Pharmc 47:729-747

Garcia AG, Kirpekar SM, Prat JC (1975) A calcium ionophore stimulating the secretion of catecholamines from the cat adrenal. J Physiol 244:253-262

Geffen LB, Jarrott B (1977) Cellular aspects of catecholamine neurons. In: Blaschko H, Sayers G, Smith A.D. (eds) The nervous system I. American Physiological Society Baltimore, pp 521-571 Handbook of Physiology, vol 15
Geffen LB, Livett BG (1971) Synaptic vesicles in sympathetic neurons. Physiol Rev 51:98-157
Geffen LB, Livett BG, Rush RA (1969) Immunological localization of chromogranins in sheep sympathetic neurons and their release by nerve impulses. J Physiol 204:58-59P
Geffen LB, Livett BG, Rush RA (1970) Immunohistochemical localization of chromogranins in sheep sympathetic neurones and their release by nerve impulses. In: Schühmann HJ, Kronenberg G (eds) New aspects of storage and release mechanism of catecholamines (Bayer Symposium II). Springer Berlin Heidelberg New York, pp 58-72
Geisow MJ, Burgoyne RD (1982) Calcium-dependent binding of cytosolic proteins by chromaffin granules from adrenal medulla. J Neurochem 39:1735-1741
Geisow MJ, Burgoyne RD (1983) Recruitment of cytosolic proteins to a secretory granule membrane depends on Ca^{2+}-calmodulin. Nature 301:432-435
Geisow MJ, Burgoyne RD (1987) An integrated approach to secretion. Ann N Y Acad Sci 493:563-576
Geisow MJ, Burgoyne RD, Harris A (1982) Interaction of calmodulin with adrenal chromaffin granule membranes. FEBS Lett 143:69-72
Geisow MJ, Fritsche U, Hexham JM, Dash B, Johnson T (1986) A consensus aminoacid sequence repeat in torpedo and mammalian Ca^{2+}-dependent membrane-binding proteins. Nature 320:636-638
Gewirtz GP, Kopin IJ (1970) Release of dopamine β-hydroxylase with norepinephrine during cat splenic nerve stimulation. Nature 227:406-407
Gomperts BD, Barrowman MM, Cockcroft S (1986) Dual role for guanine nucleotides in stimulus-secretion coupling. Fed Proc 45:2156-2161
Gratzl M, Dahl G (1978) Fusion of secretory vesicles isolated from rat liver. J Membr Biol 40:343-364
Gratzl M, Dahl G, Russell JT, Thorn NA (1977) Fusion of neurohypophyseal membranes in vitro. Biochim Biophys Acta 470:45-57
Gratzl M, Krieger-Brauer H, Ekerdt R (1981) Latent acetylcholinesterase in secretory vesicles isolated from adrenal medulla. Biochim Biophys Acta 649:355-366
Green DJ, Westhead EW, Langley KH, Sattelle DB (1978) Aggregation and dispersity of isolated chromaffin granules studied by intensity fluctuation spectroscopy. Biochim Biophys Acta 539:364-371
Grumet M, Lin S (1981) Purification and characterization of an inhibitor protein with cytochalasin-like activity from bovine adrenal medulla Biochim Biophys Acta 678:381-387
Grynszpan-Winograd O (1971) Morphological aspects of exocytosis in the adrenal medulla. Phil Trans Roy Soc Lond B 261:291-292
Guidotti A, Costa E (1973) Involvement of adenosine 3′-5′-monophosphate in the activation of tyrosine hydroxylase elicited by drugs. Science 179:902-904
Gumbiner B, Kelly RB (1982) Two distinct intracellular pathways transport secretory and membrane glycoproteins to the surface of pituitary tumor cells. Cell 28:51-59
Gutman, Y, Boonyaviroj P (1979a) Mechanism of PGE inhibition of catecholamine release from adrenal medulla. Eur J Pharmcol 55:129-136
Gutman Y, Boonyaviroj P (1979b) Activation of adrenal medulla adenylate cyclase and catecholamine secretion. Naunyn-Schmiedebergs Arch Pharmacol 307:39-44

Gutman Y, Lichtenberg D, Cohen J, Boonyaviroj P (1979) Increased catecholamine release from adrenal medulla by liposomes loaded with sodium or calcium ions. Biochem Pharmacol 28:1209-1211

Häggendal J (1982) Noradrenaline and dopamine-beta-hydroxylase levels in rat salivary glands after preganglionic nerve stimulation: Evidence for re-use of amine storage granules in transmitter release. J Neural Transmission 53:147-158

Häggendal J, Malmfords T (1969) The effect of nerve stimulation on catecholamines taken up in adrenergic nerves after reserpine pretreatment. Acta Physiol Scand 75:33-38

Hagn C, Klein RL, Fischer-Colbrie R, Douglas BH, Winkler H (1986) An immunological characterization of five common antigens of chromaffin granules and of large dense-cored vesicles of sympathetic nerve. Neurosci Lett 67:295-300

Hampton RY, Holz RW (1983) Effects of changes in osmolatity on the stability and function of cultured chromaffin cells and the possible role of osmotic forces in exocytosis. J Cell Biol 96:1082-1088

Harish OE, Kao LS, Raffaniallo R, Wakade AR, Schneider AS (1987) Calcium dependence of muscarinic receptormediated catecholamine secretion from the perfused rat adrenal medulla. J Neurochem 48:1730-1735

Harris B, Cheek TR, Burgoyne RD (1986) Effects of metalloendoproteinase inhibitors on secretion and intracellular free calcium in bovine adrenal chromaffin cells. Biochim Biophys Acta 889:1-5

Haycock JW, Bennett WF, George RJ, Waymire JC (1982) Multiple site phosphorylation of tyrosine hydroxylase. J Biol Chem 257:13 699-13 703

Heisler S, Chauvelot L, Desjardins D, Noel Ch, Lambert H, Desy-Audet L (1981) Stimulus-secretion coupling in exocrine pancreas: possible role of calmodulin. Can J Physiol Pharmacol 59:994-1001

Helle KB, Serck-Hanssen G (1975) The adrenal medulla: a model for studies of hormonal and neuronal storage and release mechanism. Molec Cell Biochem 6:127-146

Helle KB, Serck-Hanssen G, Bock E (1978) Complexes of chromogranin A and dopamine β-hydroxylase among the chromogranins of the bovine adrenal medulla. Biochim Biophys Acta 533:396-407

Herzog V, Farquhar MG (1977) Luminal membrane retrieved after exocytosis reaches most Golgi cisternae in secretory cells. Proc Natl Acad Sci USA 74:5073-5077

Herzog V, Miller F (1979) Route of luminal plasma membrane retrieved by endocytosis in thyroid follicle cells. Biol Cell 36:163-166

Hesketh JE, Aunis D, Mandel P, Devilliers G (1978) Biochemical and morphological studies of bovine adrenal medullary myosin. Biol Cell 33:199-208

Heuser JE, Reese TS, Dennis MJ, Jan Y, Jan L, Evans L (1979) Synaptic vesicle exocytosis captured by quick freezing and correlated with quantal transmitter release. J Cell Biol 81:275-300

Hirata F, Axelrod J, Crews FT (1979) Concanavalin A stimulates phospholipid methylation and phosphatidylserine decarboxylation in rat mast cells. Proc Natl Acad Sci USA 76:4813-4816

Hirst GDS, Neild TO (1980) Some properties of spontaneous excitatory junction potentials recorded from arterioles of guinea pigs. J Physiol (Lond) 303:43-60

Hiwatari M, Taira N (1978) Differential effects of D600 on release of catecholamines by acetylcholine, histamine, tyramine and by cyclic AMP from canine adrenal medulla. Jap J Pharmacol 28:671-680

Hoffman PG, Zinder O, Bonner WM, Pollard HB (1976) Role of ATP and β-γ-iminoadenosinetriphosphate in the stimulation of epinephrine and protein release from isolated adrenal secretory vesicles. Arch Biochem Biophys 176:375-388

Hökfelt T (1966) The effect of reserpine on the intraneuronal vesicles of rat vas deferens. Experientia 22:56

Holman M (1970) Junction potentials in smooth muscle. In: Bülbring E, Brading A, Jones A, Tomita T, (eds) Smooth Muscle Arnold Ltd. London pp 244-288

Holtzman E (1977) The origin and fate of secretory packages especially synpatic vesicles. Neuroscience 2:327-355

Holtzman E, Dominitz R (1968) Cytochemical studies of lysosomes, Golgi apparatus and endoplasmic reticulum in secretion and protein uptake by adrenal medulla cells of the rat. J Histochem Cytochem 16:320-336

Holtzman E, Teichberg S, Abrahams SJ, Citkowitz E, Crain SM, Kawai N, Peterson ER (1973) Notes on synaptic vesicles and releated structures, endoplasmic reticulum, lysosomes and peroxisomes in nervous tissue and the adrenal medulla. J Histochem Cytochem 21:349-385

Holz RW (1986) The role of osmotic forces in exocytosis from adrenal chromaffin cells. Ann Rev Physiol 48:175-189

Holz RW, Senter RA (1985) Plasma membrane and chromaffin granule characteristics in digitonin-treated chromaffin cells. J Neurochem 45:1548-1557

Holz RW, Senter RA (1986) Effects of osmolality and ionic strength on secretion from adrenal chromaffin cells permeabilized with digitonin. J Neurochem 46:1835-1842

Holz RW, Senter RA, Sharp RR (1983) Evidence that the H electrochemical gradient across membranes of chromaffin granules is not involved in exocytosis. J Biol Chem 258:7506-7513

Hong K, Düzgünes N, Papahadjopoulos D (1981) Role of synexin in membrane fusion. J Biol Chem 256:3641-3644

Hong K, Düzgünes N, Ekerdt R, Papahadjopoulos D (1982) Synexin facilitates fusion of specific phospholipid membranes at divalent cation concentration found intracellulary. Proc Natl Acad Sci USA 79:4642-4644

Hook YHV, Eiden LE (1985) (MET) enkephalin and carboxypeptidase processing enzyme are co-released from chromaffin cells by cholinergic stimulation. Biochem Biophysic Res Comm 128:563-570

Hoshi T, Smith SJ (1987) Large depolarization induces long openings of voltage-dependent calcium channels in adrenal chromaffin cells. J Neurosci 7:571-580

Huber E, König P, Schuler G, Aberer W, Plattner H, Winkler H (1979) Characterization and topography of the glycoproteins of adrenal chromaffin granules. J Neurochem 32:35-47

Husebye ES, Flatmark T (1987) Characterization of phosholipase activities in chromaffin granule ghosts isolated from the bovine adrenal medulla. Biochim Biophys Acta 920:120-130

Ishikawa K, Kanno T (1978) Influences of extracellular calcium and potassium concentrations on adrenaline release and membrane potential in the perfused adrenal medulla of the rat. Jap J Physiol 28:275-289

Ito S (1983) Time course of release of catecholamine and other granular contents from perifused adrenal chromaffin cells of guinea pig. J Physiol 341:153-167

Izumi F, Oka M, Kashimoto T (1971) Role of calcium for magnesium-activated adenosinetriphosphatase activity and adenosinetriphosphatemagnesium stimulated catecholamine release from adrenal medullary granules. Jap J Pharmacol 21:739-746

Izumi F, Oka M, Morita K, Azuma H (1975) Catecholamine releasing factor in bovine adrenal medulla. FEBS Lett 56:73-76

Izumi F, Kashimoto T, Miyashita T, Wada A, Oka M (1977) Involvement of mem-

brane associated protein in ADP-induced lysis of chromaffin granules. FEBS Lett 76:177-180
Izumi F, Yanagihara N, Wada A, Toyohira Y, Kobayashi H (1986) Lysis of chromaffin granules by phospholipase A_2-treated plasma membranes. FEBS Lett 196:349-352
Jaanus SD, Rubin RP (1974) Analysis of the role of cyclic adenosine 3'-5-monophosphate in catecholamine release. J Physiol (Lond) 237:465-476
Jacobowitz DM, Ziegler MG, Thomas JA (1975) In vivo uptake of antibody to dopamine β-hydroxylase into sympathetic elements. Brain Res 91:165-170
Jacobs TP, Henry DP, Johnson DG, Williams RH (1978) Epinephrine and dopamine β-hydroxylase secretion from bovine adrenal. Am J Physiol E600-E605
Jaim-Etcheverry G, Zieher LM (1969) Selective demonstration of a type of synaptic vesicles by phosphotungstic acid staining. J Cell Biol 42:855-860
Jaim-Etcheverry G, Zieher LM (1983) Ultrastructural evidence for monoamine uptake by vesicles of pineal sympathetic nerves immediately after their stimulation. Cell Tiss Res 233:463-469
Johnson DG, Thoa NB, Weinhilboum R, Axelrod J, Kopin IJ (1971) Enhanced release of dopamine β-hydroxylase from sympathetic nerves by calcium and phenoxybenzamine and its reversal by prostaglandins Proc Natl Acad Sci USA 68:2227-2230
Johnson DH, McCubbin WD, Kay CM (1977) Isolation and characterization of a myosin-like protein from bovine adrenal medulla. FEBS Lett 77:69-74
Kachi T, Banerji TK, Quay WB (1985) Quantitative ultrastructural analysis of differences in exocytosis number in adrenomedullary adrenaline cells of golden hamsters related to time of day, pinealectomy, and intracellular region. J Pin Res 2:253-269
Kajihara H, Hirata S, Miyoshi N (1977) Changes in blood catecholamine levels and ultrastructure of dog adrenal medullary cells during hemorrhagic shock. Virchows Arch B Cell Path 23:1-16
Kao LS, Schneider AS (1985) Muscarinic receptors on bovine chromaffin cells mediate a rise in cytosolic calcium that is independent of extracellular calcium. J Biol Chem 260:2019-2022
Kao LS, Schneider AS (1986) Calcium mobilization and catecholamine secretion in adrenal chromaffin cells. J Biol Chem 261:4881-4888
Kao LS, Westhead EW (1984) Binding of action to chromaffin granules and mitochondrial fractions of adrenal medulla. FEBS Lett 173:119-123
Kataoka Y, Majane EA, Yang HYT (1985) Release of NPY-like immunoreactive material from primary cultures of chromaffin cells prepared from bovine adrenal medulla. Neuropharmacol 24:693-700
Keen JH, Willingham MC, Pastan IH (1979) Clathrin-coated vesicles: isolation, dissociation and factor-dependent reassociation of clathrin baskets. Cell 16:303-312
Kenigsberg RL, Trifaro JM (1985) Microinjection of calmodulin antibodies into cultured chromaffin cells blocks catecholamine release in response to stimulation. Neuroscience 14:335-347
Kenigsberg RL, Côte A, Trifaró JM (1982) Trifluoperazine, a calmodulin inhibitor blocks secretion in cultured chromaffin cells at a step distal from calcium entry. Neuroscience 7:2277-2286
Kidokoro Y, Ritchie AK (1980) Chromaffin cell action potentials and their possible role in adrenaline secretion from rat adrenal medulla. J Physiol (Lond) 307:199-216
Kidokoro Y, Ritchie AK, Hagiwara S (1979) Effect of tetrodotoxin on adrenaline secretion in the perfused rat adrenal medulla. Nature 278:63-65

Kidokoro Y, Miyazaki S, Ozawa S (1982) Acetylcholine-induced membrane depolarization and potential fluctuations in the rat adrenal chromaffin cell. J Physiol (Lond) 324:203-220

Kilpatrick DL, Lewis RV, Stein S, Udenfriend S (1980) Release of enkephalins and ekephalin-containing polypeptides from perfused beef adrenal glands. Proc Nat Acad Sci USA 77:7473-7475

Kilpatrick DL, Slepetis RJ, Corcoran JJ, Kirshner N (1982) Calcium uptake and catecholamine secretion by cultured bovine adrenal medulla cells. J Neurochem 38:427-435

Kirksey DF, Klein RL, Bagget JMcC, Gasparis MS (1978) Evidence that most of the dopamine β-hydroxylase is not membrane bound in purified large dense cored noradrenergic vesicles. Neuroscience 3:71-81

Kirshner N, Smith WJ (1969) Metabolic requirements for secretion from the adrenal medulla. Life Sci 8:799-803

Kirshner N, Viveros OH (1972) The secretory cycle in the adrenal medulla. Pharmacol Rev 24:385-398

Kirshner N, Sage HJ, Smith WJ, Kirshner AG (1966) Release of catecholamines and specific protein from adrenal glands. Science 154:529-531

Kirshner N, Sage HJ, Smith WJ, Kirshner Ag (1967) Mechanism of secretion from the adrenal medulla. 2. Release of catecholamines and storage vesicle protein in response to chemical stimulation. Molec Pharmacol 3:254-265

Klein RL (1982) Chemical composition of the large noradrenergic vesicles. In: (Klein RL, Lagercrantz H, Zimmermann H), (eds) Neurotransmitter Vesicles: Composition, Structure and Function, Academic Press New York, pp 133-173

Klein RL, Thureson-Klein AK (1984) Noradrenergic vesicles: Molecular organization and function. In: (Laitha A) (ed) Handbook Neurochemistry Series, Plenum Press New York pp 71-109

Klein RL, Wilson SP, Dzielak DJ, Yang WH, Viveros OH (1982) Opioid peptides and noradrenaline co-exist in large dense cored vesicles from sympathetic nerve. Neuroscience 7:2255-2261

Knight DE (1986) Botulinum toxin types A, B and D inhibit catecholamine secretion from bovine adrenal medullary cells. FEBS 207:222-226

Knight DE, Baker PF (1982) Calcium-dependence of catecholamine release from bovine adrenal medullary cells after exposure to intense electric fields. J Membrane Biol 68:107-140

Knight DE, Baker PF (1983) The phorbol ester TPA increases the affinity of exocytosis for calcium in "leaky" adrenal medullary cells. FEBS Lett 160:98-100

Knight DE, Baker PF (1985a) The chromaffin granule proton pump and calcium-dependent exocytosis in bovine adrenal medullary cells. J Membrane Biol 83:147-156

Knight DE, Baker PF (1985b) Guanine nucleotides and Ca-dependent exocytosis. FEBS 189:345-349

Knight DE, Kesteven NT (1983) Evoked transient intracellular free Ca^{2+} changes and secretion in isolated bovine adrenal medullary cells. Proc R Soc Lond B 218:177-199

Knight DE, Scrutton CS (1986) Gaining access to the cytosol: the technique and some applications of electropermeabilization. Biochem J 234:497-506

Knight DE, Tonge DA, Baker PF (1985) Inhibition of exocytosis in bovine adrenal medullary cells by botulinum toxin type D. Nature 317:719-721

Kobayashi S, Serizawa Y (1979) Stress-induced degranulation accompanied by vesicle formation in the adrenal chromaffin cells of the mouse. Arch Histol Jap 42:375-388

Koerker RL, Hahn WE, Schneider FH (1974) Electron translucent vesicles in adrenal medulla following catecholamine depletion. Eur J Pharmacol 28:350-359

König P, Hörtnagl H, Kostron H, Sapinsky H, Winkler H (1976) The arrangement of dopamine β-hydroxylase (EC 1.14.2.1.) and chromomembrin B in the membrane of chromaffin granules. J Neurochem 27:1539-1541

Konings F, de Potter W (1981a) Calcium-dependent in vitro interaction between bovine adrenal medullary cell membranes and chromaffin granules as a model for exocytosis. FEBS Lett 126:103-106

Konings F, de Potter W (1981b) In vitro interaction between bovine adrenal medullary cell membranes and chromaffin granules: specific control by Ca^{2+}. Naunyn-Schmiedeberg's Arch Pharmacol 317:97-99

Konings F, de Potter W (1982a) The chromaffin granule—plasma membrane interaction as a model for exocytosis: Quantitative release of the soluble granular content. Bioch Biophys Res Commun 104:254-258

Konings F, de Potter W (1982b) A role for sialic acid containing substrates in the exocytosis-like in vitro interactions between adrenal medullary plasma membranes and chromaffin granules. Biochem Biophys Res Commun 106:1191-1195

Konings F, de Potter W (1983) Protein phosphorylation and the exocytosis-like interaction between isolated adrenal medullary plasma membranes and chromaffin granules. Biochem Biophys Res Commun 110:55-60

Konings F, Majchrowicz B, de Potter W (1983) Release of chromaffin granular content on interaction with plasma membranes. Am J Physiol 244:C309-C312

Koter M, de Kruijff B, van Denen LLM (1978) Calcium-induced aggregation and fusion of mixed phosphatidylcholine-phosphatidic acid vesicles as studied by 31P. Biochim Biophys Acta 514:255-263

Kuo ICY, Coffee CJ (1976) Bovine adrenal medulla troponin-C. Demonstration of a calcium-dependent confirmational change. J Biol Chem 251:6315-6319

Ladona MG, Bader MF, Aunis D (1987a) Influence of hypertonic solutions on catecholamine release from intact and permeabilized cultured chromaffin cells. Biochim Biophys Acta 927:18-25

Ladona MG, Aunis D, Gandia L, Garcia AG (1987b) Dihydropyridine modulation of the chromaffin cell secretory response. J Neurochem 48:483-490

Lagercrantz H (1976) On the composition and function of large densel cored vesicles in sympathetic nerves. Neuroscience 1:81-92

Langley AE, Gardier RW (1977) Effect of atropine and acetylcholine on nerve stimulated output of noradrenaline and dopamine beta-hydroxylase from isolated rabbit and guinea pig hearts. Naunyn Schmiedeberg's Arch Pharmacol 297:251-256

Langley AE, Weiner N (1978) Enhanced exocytotic release of norepinephrine consequent to nerve stimulation by low concentrations of cyclic nucleotides in the presence of phenoxybenzamine. J Pharmacol Exp Ther 205:426-437

Lawson D, Raff MC, Comperts B, Fewtrell C, Gilula NB (1977) Molecular events during membrane fusion. A study of exocytosis in rat peritoneal mast cells. J Cell Biol 72:242-259

Ledbetter FH, Kirshner N (1981) Quantitative correlation between secretion and cellular content of catecholamines and dopamine β-hydroxylase in cultures of adrenal medulla cells. Biochem Pharmacol 30:3246-3249

Ledbetter FH, Kilpatrick D, Sage HL, Kirshner N (1978) Synthesis of chromogranins and dopamine β-hydroxylase by perfused bovine adrenal glands. Am J Phys 235:E475-E486

Lee RWH, Trifaro JM (1981) Characterization of anti-actin antibodies and their use in immunocytochemical studies on the localization of actin in adrenal chromaffin cells in culture. Neuroscience 6:2087-2108

Lee SA, Holz RW (1986) Protein phosphorylation and secretion in digitonin-permeabilized adrenal chromaffin cells. J Biol Chem 261:17089-17098

Lefebvre YA, White DA, Hawthorne JN (1976) Diphosphoinositide metabolism in bovine adrenal medulla. Cand J Biochem 54:746-753

Lelkes PI, Friedman JE, Rosenheck K, Oplatka A (1986) Destabilization of actin filaments as a requirement for the secretion of catecholamines from permeabilized chromaffin cells. FEBS 208:357-363

Levine M, Asher A, Pollard H, Zinder O (1983) Ascorbic acid and catecholamine secretion from cultured chromaffin cells. J Biol Chem 258:13111-13115

Liang BT, Perlman RL (1979) Catecholamine secretion by hamster adrenal cells. J Neurochem 32:927-933

Lind I, Ahnert-Hilger G, Fuchs G, Gratzl M (1987) Purification of alpha-toxin from staphylococcus aureus and application to cell permeabilization. Anal Biochem 164:84-89

Lingg G, Fischer-Colbrier R, Schmidt W, Winkler H (1983) Exposure of an antigen of chromaffin granules on cell surface during exocytosis. Nature 301:610-611

Lishajko F (1969) Influence of chloride ions and ATP-Mg^{2+} on the release of catecholamines from isolated adrenal medullary granules. Acta Physiol Scand 75:255-256

Livett BG (1984) The Secretory Process in Adrenal Medullary Cells. In: Cantin M (ed) Cell Biology of the Secretory Process. Karger, Basel pp 304-358

Livett BG, Dean DM, Whelan LG, Udenfriend S, Rossier J (1981) Co-release of enkephalin and catecholamines from cultured adrenal chromaffin cells. Nature 289:317-319

Llinas R, McGuiness TL, Leonard CS, Sugimori M, Greengard P (1985) Intraterminal injection of synapsin I or calcium/calmodulin-dependent protein kinase II alters neurotransmitter release at the squid giant synapse. Proc Nat Acad Sci USA 82:3035-3039

Lotshaw DP, Ye HZ, Edwards C (1986) Endocytosis of surface bound dopamine β-hydroxylase and plasma membrane following catecholamine secretion by bovine adrenal chromaffin cells. Neurochem Int 9:391-399

Lucy JA (1978) Mechanisms of chemically induced cell fusion. In: Poste G, Nicolson GL (eds) Membrane fusion Cell Surface Rev, vol 5 Elsevier North Holland Publ Amsterdam, pp 267-304

Lumpert CJ, Kersken H, Gras U, Plattner H (1987) Screening of enzymatic mechamisms possibly involved in membrane fusion during exocytosis in parmacium cells. Cell Biol Int Rep 11:405-414

Lundberg J, Bylock A, Goldstein M, Hanson HA, Dahlström A (1977) Ultrastructural localization of dopamine β-hydroxylase in nerve terminals of the rat brain. Brain Res. 120:549-552

Lundberg JM, Hamberger B, Schultzberg M, Hökfelt T, Granberg PO, Efendic S, Terenius L, Goldstein M, Luft R (1979) Enkephaline and somatostatin-like immunoreactivity in human adrenal medulla and phaeochromocytoma. Proc Natl Acad Sci USA 76:4079-4083

Malamed S, Poisner AM, Trifaro JM, Douglas WW (1968) The fate of the chromaffin granule during catecholamine release from the adrenal medulla. III. Recovery of a purified fraction of electrontranslucent structures. Biochem Pharmacol 17:241-246

Malviya AN, Gabellec MM, Rebel G (1986) Plasma membrane lipids of bovine adrenal chromaffin cells. Lipis 21:417-419

Margolis RK, Jaanus SD, Margolis RU (1973) Stimulation by acetylcholine of sulfated mucopolysaccharide release from the perfused cat adrenal gland. Mol Pharmacol 9:590-594

Marin WH, Creutz CE (1987) Chromobindin A, a Ca^{2+} and ATP regulated chromaffin granule binding protein. J Biol Chem 262:2803-2810

Matthews EK, Evans RJ, Dean PM (1972) The ionogenic nature of the secretory-granule membrane. Electrokinetic properties of isolated chromaffin granules. Biochem J 130:825-832

McKay DB, Schneider AS (1984) Selective inhibition of cholinergic receptor-mediated $^{45}Ca^{2+}$ uptake and catecholamine secretion from adrenal chromaffin cells by taxol and vinblastine. J Pharmacol Exp Ther 231:102-108

Meldolesi B, Ceccarelli B (1981) Exocytosis and membrane recycling. Phil Trans R Soc Lond B 296:55-65

Meldolesi J, Borgese N, de Camilli P, Ceccarelli B (1978) Cytoplasmic membranes and the secretory process. In: Poste G, Niclolson GL (eds) Membrane fusion. Cell Surface Review, vol 5. North Holland Publ. Amsterdam, Oxford pp 510-627

Meyer DI, Burger MM (1976) The chromaffin granule surface. Localization of carbohydrate on the cytoplasmic surface of an intracellular organelle. Biochim Biophys Acta 443:428-436

Meyer DI, Burger MM (1979a) The chromaffin granule surface: the presence of actin and the nature of its interaction with the membrane. FEBS Lett 101:129-133

Meyer DI, Burger MM (1979b) Isolation of a protein from the plasma membrane of adrenal medulla which binds to secretory vesicles. J Biol Chem 254:9854-9859

Michaels JE, Leblond CP (1976) Transport of glycoprotein from Golgi apparatus to cell surface by means of "carrier" vesicles as shown by radioautography of mouse colonic epithelium after injection of ^{3}H-fucose. J Microscop Biol Cell 25:243-248

Michener ML, Dawson WB, Creutz CE (1986) Phosphorylation of a chromaffin granule-binding protein in stimulated chromaffin cells. J Biol Chem 261:6548-6555

Misbahuddin M, Isosaki M, Houchi H, Oka M (1985) Muscarinic receptor-mediated increase in cytoplasmic free Ca^{2+} in isolated bovine adrenal medullary cells. FEBS 190:25-28

Mizobe F, Iwamoto M (1984) Veratridine-evoked release of intracellular and external acetylcholinesterase from cultured adrenal chromaffin cells. Biomed Res 5:83-88

Mizobe F, Livett BG (1983) Nicotine stimulates secretion of both catecholamines and acetylcholinesterase from cultured adrenal chromaffin cells. J Neuroscience 3:871-876

Mizobe F, Iwamoto M, Livett BG (1984) Parallel but separate release of catecholamines and acetylcholinesterase from stimulated adrenal chromaffin cells in culture. J Neurochem 42:1433-1438

Momayezi M, Lumpert CJ, Kersken H, Gras U, Plattner H, Krinks MH, Klee CB (1987) Exocytosis induction in paramecium tetraurelia cells by exogenous phosphoprotein phosphatase in vivo and in vitro: possible involvement of calcineurin in exocytotic membrane fusion. J Cell Biol 105:181-189

Montiel C, Artalejo AR, Garcia AG (1984) Effects of the novel dihydropyridine BAY-K-8644 on adrenomedullary catecholamine release evoked by calcium reintroduction. Biochem Biophys Res Com 120:851-857

Morita K, Brocklehurst KW, Tomares StM, Pollard HB (1985a) The phorbol ester TPA enhances A23187—but not carbachol—and high K^+-induced catecholamine secretion from cultured bovine adrenal chromaffin cells. Biochem Biophys Res Comm 129:511-516

Morita K, Brocklehurst KW, Tomares SM, Pollard HB (1985b) The phorbol ester TPA enhances A23187—but not carbachol—and high K^+-induced catecholamine secretion from cultured bovine adrenal chromaffin cells. Biochem Biophys Res Comm 129:511-516

Morita K, Dohi T, Kitayama S, Koyama Y, Tsujimoto A (1987a) Enhancement of stimulation-evoked catecholamine release from cultured bovine adrenal chromaffin cells by forskolin. J Neurochem 48:243-247

Morita K, Dohi T, Kitayama S, Koyama Y, Tsujimoto A (1987b) Stimulation-evoked Ca^{2+} fluxes in cultured bovine adrenal chromaffin cells are enhanced by forskolin. J Neurochem 48:248-252

Morris SJ, Bradley D (1984) Calcium-promoted fusion of isolated chromaffin granules detected by resonance energy transfer between labelled lipids embedded in the membrane bilayer. Biochemistry 23:4642-4650

Morris SJ, Hughes JMX (1979) Synexin protein is non-selective in its ability to increase Ca^{2+}-dependent aggregation of biological and artificial membranes. Biochem Biophys Res Commun 91:345-350

Morris SJ, Schober R (1977) Demonstration of binding sites for divalent and trivalent ions on the outer surface of chromaffin granule membranes. Eur J Biochem 75:1-12

Morris SJ, Hughes JMX, Whittaker VP (1982) Purification and mode of action of synexin: A protein enhancing calcium-induced membrane aggregation. J Neurochem 39:529-536

Morris SJ, Costello MJ, Robertson JD, Südhof TC, Odenwald WF, Haynes DH (1983) The chromaffin granule as a model for membrane fusion: implications for exocytosis. J Auton Nerv System 7:19-33

Moskowitz N, Glassman A, Ores Ch, Shook W, Puszkin S (1983) Phosphorylation of brain synaptic and coated vesicle proteins by endogenous Ca^{2+}/calmodulin- and cAMP-dependent protein kinases. J Neurochem 40:711-718

Mundy DI, Strittmatter WJ (1985) Requirement for metalloendoprotease in exocytosis: evidence in mast cells and adrenal chromaffin cells. Cell 40:645-656

Muscholl E, Spira FJ (1982) Kinetic analysis of stimulation-evoked overflow of noradrenaline and dopamine β-hydroxylase from the isolated rabbit heart. The effect of dopa decarboxylase inhibition. Neuroscience 7:3201-3211

Muscholl E, Racke K, Ritzel H (1980) Facilitation by 'low sodium-urea' medium of the washout of dopamine β-hydroxylase released by potassium ions from the perfused rabbit heart. Neuroscience 5:453-457

Muscholl E, Racke K, Spira FJ (1985) Evidence for exocytotic release of dopamine β-hydroxylase from rabbit heart and of vasopressin from rat neurohypophyses during homogenization and fractionation effects of gadolinium ions, cytochalasin B, gallopamil and different temperatures. Neuroscience 14:79-93

Nagasawa J (1977) Exocytosis: The common release mechanism of secretory granules in granular cells, neurosecretory cells, neurons and paraneurons. Arch Histol Jap 40, Suppl 31-47

Nagasawa J, Douglas WW (1972) Thorium dioxide uptake into adrenal medullary cells and the problem of recapture of granule membrane following exocytosis. Brain Res 37:141-145

Nakanishi A, Morita K, Oka M, Katsunuma N (1986) Evidence against a possible involvement of the serine, and thiol proteases in the exocytotic mechanism of catecholamine secretion in cultured bovine adrenal medullary cells. Biochem Int 13:799-807

Navone F, Jahn R, di Gioia G, Stukenbrok H, Greengard P, de Camilli P (1986) Pro-

tein p38: an integral membrane protein specific for small vesicles of neurons and neuroendocrine cells. J Cell Biol 103:2511-2527
Nayar R, Hope MJ, Cullis PR (1982) Phospholipids as adjuncts for calcium ion stimulated release of chromaffin granule contents: Implications for mechanisms of exocytosis. Biochemistry 21:4583-4589
Neher E, Marty A (1982) Discrete changes of cell membrane capacitance observed under conditions of enhanced secretion in bovine adrenal chromaffin cells. Proc Natl Acad Sci USA 79:6712-6716
Nelson DL, Molinoff PB (1976) Distribution and properties of adrenergic storage vesicles in nerve terminals. J Pharmacol Exp Ther 196:346-359
Nestler EJ, Greengard P (1982) Nerve impulse increase the phosphorylation state of protein I in rabbit superior cervical ganglion. Nature 296:452-454
Neuman B, Wiedermann CJ, Fischer-Colbrie R, Schober M, Sperk G, Winkler H (1984) Biochemical and functional properties of large and small dense core vesicles in sympathetic nerves of rat and ox vas deferens. Neuroscience 13:921-931
Ngai SH, Dairman W, Marchelle M, Spector S (1974) Dopamine β-hydroxylase in dog lymph-effect of sympathetic activation. Life Sci 14:2431-2439
Niggli V, Knight DE, Baker PF, Vigny A, Henry JP (1984) Tyrosine hydroxylase in "leaky" adrenal medullary cells: Evidence for in situ phosphorylation by separate Ca^{2+} and cyclic AMP-dependent systems. J Neurochem 43:646-658
Nijjar MS, Hawthorne JN (1974) A plasma membrane fraction from bovine adrenal medulla: Preparation, marker enzyme studies and phospholipid composition. Biochim Biophys Acta 367:190-201
Nishibe S, Ogawa M, Murata A, Nakamura K, Hatanaka T, Kambayashi JI, Kosaki G (1983) Inhibition of catecholamine release from isolated bovine adrenal medullary cells by various inhibitors: possible involvement of protease, calmodulin and arachidonic acid. Life Sci 32:1613-1620
Nishizuka Y (1984) The role of protein kinase C in cell surface signal transduction and tumour promotion. Nature 308:693-698
Nobiletti JB, Kirpekar SM (1983) Nonreutilization of bovine adrenal chromaffin granule membrane in secretion. Fed Proc 42:380
Nordmann JJ (1984) Combined stereological and biochemical analysis of storage and release of catecholamines in the adrenal medulla of the rat. J Neurochem 42:434-437
O'Connor DT, Frigon RP (1984) Chromogranin A, the major catecholamine storage vesicle soluble protein. J Biol Chem 259:3237-3247
O'Connor DT, Deftos LJ (1986) Secretion of chromogranin A by peptide-producing endocrine neoplasms. N Engl J Med 314:1145-1151
O'Conner DT, Frigon, Odenwald WF, Morris SJ (1983) Identification of a second synexin-like adrenal medullary and liver protein that enhances calcium-induced membrane aggregation. Biochem Biphys Res Commun 112:147-154
Ohashi Y, Narumiya S (1987) ADP-ribosylation of a M 21000 membrane protein by type D botulinum toxin. J Biol Chem 262:1430-1433
Oka M, Ohuchi T, Yoshida H, Imaizumi R (1965) Effect of adenosine triphosphate and magnesium on the release of catecholamines from adrenal medullary granules. Biochim Biophys Acta 97:170-171
Oka M, Ohuchi T, Yoshida H, Imaizumi R (1974) Stimulatory effect of adenosine triphosphate and magnesium on the release of catecholamines from adrenal medullary granules. Jap J Pharmacol 17:199-207
Orci L, Perrelet A, Friend DS (1977) Freeze-fracture of membrane fusions during exocytosis in pancreatic B-cells. J Cell Biol 75:23-30

Ornberg RL, Reese TS (1981) Beginning of exocytosis captured by rapid freezing of limulus amebocytes. J Cell Biol 90:40–54
Palade GE (1959) Functional changes in the structure of cell components. In: Hayashi T (ed) Subcellular Particles. Ronald Press New York, pp 64–83
Papahadjopoulos D, Vail WJ, Pangborn WA, Poste G (1976) Studies on the membrane fusion. II. Induction of fusion in pure phospholipid membranes by calcium ions and other divalent metals. Biochim Biophys Acta 448:265–283
Papahadjopoulos D, Vail WJ, Newton C, Nir S, Jacobson K, Poste G, Lazo R (1977) Studies on membrane fusion. III. The role of calcium-induced phase changes. Biochim Biophys Acta 465:579–598
Papahadjopoulos D (1978) Calcium-induced phase changes and fusion in natural and model membranes. In: Poste G, Nicolson GL (eds) Membrane Fusion. Cell Surface Rev, vol. 5. Elsevier North Holland Publ Amsterdam, pp 766–786
Parsons SJ, Creutz CE (1986) $p60^{c\text{-}SRG}$ activity detected in the chromaffin granule membrane. Biochem Biophys Res Comm 134:736–742
Patzak A, Winkler H (1986) Exocytotic exposure and recycling of membrane antigens of chromaffin granules: ultrastructural evaluation after immunolabelling. J Cell Biol 102:510–515
Patzak A, Böck G, Fischer-Colbrie R, Schauenstein K, Schmidt W, Lingg G, Winkler H (1984) Exocytotic exposure and retrieval of membrane antigens of chromaffin granules: Quantitative evaluation of immunofluorescence on the surface of chromaffin cells. J Cell Biol 98:1817–1824
Patzak A, Aunis D, Langley K (1987) Membrane recycling after exocytosis: an ultrastructural study of cultured chromaffin cells. Exp Cell Res 171:346–356
Pazoles CJ, Pollard HB (1978) Evidence for stimulation of anion transport in ATP-evoked transmitter release from isolated secretory vesicles. J Biol Chem 253:3962–3969
Peach MJ (1972) Stimulation and release of adrenal catecholamine by adenosine 3′–5′ cyclic monophosphate and theophylline in the absence of extracellular Ca^{2+}. Proc Natl Acad Sci USA 69:834–836
Pearse BMF (1976) Clathrin: an unique protein associated with intracellular transfer of membrane by coated vesicles. Proc Natl Acad Sci USA 73:1255–1259
Pearse BMF, Bretscher MS (1981) Membrane recycling by coated vesicles. Ann Rev Biochem 50:85–101
Penner R, Neher E, Dreyer F (1986) Intracellularly injected tetanus toxin inhibits exocytosis in bovine adrenal chromaffin cells. Nature 324:76–78
Peppers SC, Holz RW (1986) Catecholamine secretion from digitonin-treated PC12 cells. J Biol Chem 261:14665–14669
Perlman RL, Chalfie M (1977) Catecholamine release from the adrenal medulla. Clinics Endocrin Metabol 6:551–576
Perlman RL, Cossi AF, Role LW (1980) Mechanisms of ionophore-induced catecholamine secretion. J Pharmacol Exp Ther 213:241–246
Perrin D, Aunis D (1985) Reorganization of α-fodrin induced by stimulation in secretory cells. Nature 315:589–592
Perrin D, Langley OK, Aunis D (1987) Anti-α-fodrin inhibits secretion from permeabilized chromaffin cells. Nature 326:498–501
Phillips JH, Apps DK, Tipton KF (1979) Storage and secretion of catecholamines: The adrenal medulla. In (ed) Physiological and Pharmacological Biochemistry, Internat. Review of Biochemistry, vol 26. University Park Press Baltimore, pp 121–178
Phillips JH, Slater A (1975) Actin in the adrenal medulla. FEBS LETT 56:327–331

Phillips JH, Burridge K, Wilson SP, Kirshner N (1983) Visualization of the exocytosis/endocytosis secretory cycle in cultured adrenal chromaffin cells. J Cell Biol 97:1906-1917

Pinardi G, Talmaciu RK, Santiago E, Cubeddu LX (1979) Contribution of adrenal medulla, spleen and lymph, to the plasma levels of dopamine β-hydroxylase and catecholamines induced by hemorrhagic hypotension in dogs. J Pharmacol Exp Ther 209:176-184

Pinto JEB, Trifaro JM (1976) The different effects of D-600 (methoxyverapamil) on the release of adrenal catecholamines induced by acetylcholine, high potassium or sodium deprivation. Br J Pharmacol 57:127-132

Pinto JEB, Viglione PN, KIM MK (1985) Inhibitory effect of hyperosmolality on catecholamine secretion from the bovine adrenal medulla. Arch Int Pharmacodyn 276:236-246

Plattner H (1978) Fusion of cellular membranes. In Silverstein SC (ed) Transport of macromolecules in cellular systems. Berlin Dahlem Konferenzen, pp 465-488

Plattner H (1987) Synchronous Exocytosis in Paramecium Cells. In: Sowers AE (ed) Cell Fusion. Plenum Publ Corp, pp 69-98

Pocotte SL, Frye RA, Senter RA, Terbush DR, Lee SA, Holz RW (1985) Effects of phorbol ester on catecholamine secretion and protein phosphorylation in adrenal medullary cell cultures. Proc Natl Acad Sci USA 82:930-934

Poisner AM, Bernstein J (1971) A possible role of microtubules in catecholamine release from the adrenal medulla: effect of colchicine, vinca alkaloids and deuterium oxide. J Pharmacol Exp Ther 177:102-108

Poisner AM, Cooke P (1975) Microtubules and the adrenal medulla. Ann NY Acad Sci 253:653-669

Poisner AM, Trifaro JM (1967) The role of ATP and ATPase in the release of catecholamines from the adrenal medulla. I. ATP-evoked release of catecholamines, ATP and protein from isolated chromaffin granules. Mol Pharmacol 3:561-571

Poisner AM, Trifaro JM, Douglas WW (1967) The fate of the chromaffin granule during catecholamine release from the adrenal medulla. II. Loss of protein and retention of lipid in subcellular fractions. Biochem Pharmacol 16:2101-2108

Pollard HB, Scott JH (1982) Synhibin: a new calcium-dependent membrane-binding protein that inhibits synexin-induced chromaffin granule aggregation and fusion. FEBS Lett 150:201-206

Pollard HB, Pazoles CJ, Creutz CE, Ramu A, Strott RP, Brown EM, Aurbach GD, Tack-Goldman KM, Shulman NR (1977) A role for anion transport in the regulation of release from chromaffin granules and exocytosis from cells. J Supramol Struct 7:277-285

Pollard HB, Pazoles CJ, Creutz CE, Zinder O (1979) The chromaffin granule and possible mechanisms of exocytosis. Int Rev Cytol 58:159-197

Pollard HB, Scott JH, Creutz CE, Fowler VM, Pazoles CJ (1982a) Regulation of organelle movement, membrane fusion and exocytosis in the chromaffin cell. In: Frazier WA, Glaser L, Gottlieb DI (eds) Cellular recognition, UCLA Symposium on Molecular and Cellular Biology, vol 3. Alan R Liss Inc New York, pp 893-917

Pollard RM, Fillenz M, Kelly P (1982b) Parallel changes in ultrastructure and noradrenaline content of nerve terminals in rat vas deferens following transmitter release. Neuroscience 7:1623-1629

Pollard HB, Scott JH, Creutz CE (1983) Inhibition of synexin activity and exocytosis from chromaffin cells by phenothiazine drugs. Biochem Biophys Res Com 113:908-915

Pollard HB, Pazoles CJ, Creutz CE, Scott JH, Zinder O, Hotchkiss A (1984) An os-

motic mechanism for exocytosis from dissociated chromaffin cells. J Biol Chem 259:1114–1121
Pollard HB, Ornberg R, Levine M, Kelner K, Morita K, Levine R, Forsberg E, Brocklehurst KW, Duong L, Lelkes PI, Heldman E, Youdim M (1985) Hormone secretion by exocytosis with emphasis on information from the chromaffin cell systems. Vit Horm 42:109–195
Pollard HB, Rojas E, Burns AL (1987) Synexin and chromaffin granule membrane fusion. Ann N Y Acad Sci 493:524
Portis A, Newton C, Pangborn W, Papahadjopoulos D (1979) Studies on the mechanism of membrane fusion: Evidence for an intermembrane Ca^{2+} phospholipid complex, synergism with Mg^{2+}, and inhibition by spectrin. Biochemistry 18:780–790
Povilaitis V, Gagnon C, Heisler S (1981) Stimulus-secretion coupling in exocrine pancreas: role of protein carboxyl methylation. Am J Physiol 240:G199–G205
Quatacker J (1981) The axonal reticulum in the neurons of the superior cervical ganglion of the rat as a direct extension of the Golgi apparatus. Histochem J 13:109–124
Quatacker J, de Potter W (1981) Organization and relationship of the axonal reticulum in the cell body of sympathetic ganglion cells. Acta Histochem Suppl XXIV:33–36
Richards JG, da Prada M (1977) Uranaffin reaction: a new cytochemical technique for the localization of adenine nucleotides in organelles storing biogenic amines. J Histochem Cytochem 25:1322–1336
Rimle D, Morse II PD, Njus D (1983) A spin-label study of plasma membranes of adrenal chromaffin cells. Biochim Biophys Acta 728:92–96
Rink TJ, Sanchez A, Hallam TJ (1983) Diacylglycerol and phorbol ester stimulate secretion without raising cytoplasmic free calcium in human platelets. Nature 305:317–319
Roizen MF, Thoa NB, Moss J, Kopin IJ (1975) Inhibition by halothane of release of norepinephrine, but not of dopamine β-hydroxylase from guinea-pig vas deferens. Eur J Pharmacol 31:313–318
Rojas E, Pollard HB, Heldman E (1985) Real-time measurements of acetylcholine-induced release of ATP from bovine medullary chromaffin cells. FEBS 185:323–327
Rökaeus A, Fried G, Lundberg JM (1984) Occurrence, storage and release of neurotensin-like immunoreactivity from the adrenal gland. Acta Physiol Scand 120:373–380
Rosenheck K, Lelkes PI (1987) Stimulus-secretion coupling in chromaffin cells. CRC Press, Boca Raton
Rosenheck K, Plattner H (1986) Ultrastructural and cytochemical characterization of adrenal medullary plasma membrane vesicles and their interaction with chromaffin granules. Biochim Biophys Acta 856:373–382
Ross SB, Eriksson HE, Hellström W (1974) On the fate of dopamine β-hydroxylase after release from the peripheral sympathetic nerves in the cat. Acta Physiol Scand 92:578–580
Rubin RP (1970a) The role of energy metabolism in calcium-evoked secretion from the adrenal medulla. J Physiol 206:181–192
Rubin RP (1970b) The role of calcium in the release of neurotransmitter substances and hormones. Pharmacol Rev 22:389–428
Rush RA, Millar TJ, Chubb IW, Geffen LB (1979) Use of dopamine β-hydroxylase in the study of vesicle dynamics. In: Usdin E, Kopin IJ, Barchas J (eds) Catecholamines: vol 1. Pergamon Press, Oxford New York, pp 331–336

Ryan US, Ryan JW, Smith DS, Winkler H (1975) Fenestrated endothelium of the adrenal gland: Freeze-fracture studies. Tissue Cell 7:181-190

Sage HJ, Smith WJ, Kirshner N (1967) Mechanism of secretion from the adrenal medulla. I. A microquantitative immunologic assay for bovine adrenal catecholamine storage vesicle protein and its application to studies of the secretory process. Molec Pharmacol 3:81-89

Sasakawa N, Yamamoto S, Kumakura K, Kato R (1983) Prevention of catecholamine release from adrenal chromaffin cells by phospholipase A_2- and lipoxygenase-inhibitors. Jap J Pharmacol 33:1077-1080

Schmidt W, Patzak A, Lingg G, Winkler H, Plattner H (1983) Membrane events in adrenal chromaffin cells during exocytosis: a freeze-etching analysis after rapid cryofixation. Eur J Cell Biol 32:31-37

Schneeweiss F, Naquira D, Rosenheck K, Schneider AS (1979) Cholinergic stimulants and excess potassium ion increase the fluidity of plasma membranes isolated from adrenal chromaffin cells. Biochim biophys Acta 555:460-471

Schneider FH (1968) Observations on the release of lysosomal enzymes from the isolated bovine adrenal gland. Biochem Pharmacol 17:848-851

Schneider FH (1969a) Drug-induced release of catecholamines, soluble protein and chromogranin A from the isolated bovine adrenal gland. Biochem Pharmacol 18:101-107

Schneider FH (1969b) Secretion from the cortex-free bovine adrenal medulla. Brit J Pharmacol 37:371-379

Schneider FH (1970) Secretion from the bovine adrenal gland: release of lysosomal enzymes. Biochem Pharmacol 19:833-847

Schneider FH, Smith AD, Winkler H (1967) Secretion from the adrenal medulla: biochemical evidence for exocytosis. Brit J Pharmacol 31:94-104

Schneider AS, Herz R, Rosenheck K (1977) Stimulus-secretion coupling in chromaffin cells isolated from bovine adrenal medulla. Proc Natl Acad Sci USA 74:5036-5040

Schneider AS, Cline HT, Lemaire S (1979) Rapid rise in cyclic GMP accompanies catecholamine secretion in suspensions of isolated adrenal chromaffin cells. Life Sci 24:1389-1394

Schneider AS, Cline HT, Rosenheck K, Sonenberg M (1981) Stimulus-secretion coupling in isolated adrenal chromaffin cells: Calcium channel activation and possible role of cytoskeletal elements. J Neurochem 37:567-575

Schober R, Nitsch C, Rinne U, Morris SJ (1977) Calcium-induced displacement of membrane associated particles upon aggregation of chromaffin granules. Science 195:495-497

Schook W, Puszkin S, Bloom W, Ores C, Kochwa S (1979) Mechanochemical properties of brain clathrin: interactions with actin and α-actinin and polymerization into basketlike structures of filaments. Proc Natl Acad Sci USA 76:116-120

Schubert D, Klier FG (1977) Storage and release of acetylcholine by a clonal cell line. Proc Natl Acad Sci USA 74:5184-5188

Schuler G, Plattner H, Aberer W, Winkler H (1978) Particle segregation in chromaffin granule membranes by forced physical contact. Biochim Biophys Acta 513:244-254

Schwab ME, Thoenen H (1983) Mechanism of uptake and retrograde axonal transport of noradrenaline in sympathetic neurons in culture: reserpine-resistant large dense-core vesicles as transport vehicles. J Cell Biol 96:1538-1547

Scott JH, Creutz CE, Pollard HB, Ornberg R (1985) Synexin binds in a calcium-dependent fashion to oriented chromaffin cell plasma membranes. FEBS 180:17-23

Serck-Hanssen G (1972) The release of protein in the course of catecholamine secretion from bovine adrenals perfused in vitro. Acta Physiol Scand 86:289-298

Serck-Hanssen G (1974) Effects of theophylline and propranolol on acetylcholine—induced release of adrenal medullary catecholamines. Biochem Pharmacol 23:1225-1234

Serck-Hanssen G, Helle KB (1972) Biochemical and morphological characterization of chromaffin granules accumulating during in vitro secretion from perfused adrenal glands. Biochim Biophys Acta 273:199-207

Serck-Hanssen G, Helle KB, Sommersten B, Pihl KE (1980) Quantitative aspects of the acetylcholine-induced release of dopamine β-hydroxylase and catecholamines from the bovine adrenal medulla. Gen Pharmacol 11:243-249

Slotkin T, Kirshner N (1973) All-or-none secretion of adrenal medullary storage vesicle contents in the rat. Biochem Pharmacol 22:205-219

Smith AD (1968) Biochemistry of adrenal chromaffin granules. In: Campbell PN (ed) The interaction of drugs and subcellular components in animal cells. Churchill Ltd London, pp 239-292

Smith AD (1969) Extracellular release of lysosomal phospholipases from the perfused adrenal gland. Biochem J 114:72P

Smith AD (1971) Secretion of proteins (chromogranin A and dopamine β-hydroxylase) from a sympathetic neuron. Phil Trans Roy Soc Lond B 261:363-370

Smith AD (1972a) Mechanisms involved in the release of noradrenaline from sympathetic nerves. Brit Medical Bulletin 29:123-129

Smith AD (1972b) Subcellular localization of noradrenaline in sympathetic neurons. Pharmacol Rev 24:435-457

Smith AD (1978) Biochemical studies of the mechanism of release. In: Paton DM (ed) The release of catecholamines from adrenergic neurons. Pergamon Press Oxford New York, pp 1-15

Smith DJ, van Orden LS III (1973) Ultrastructural evidence for increased incidence of exocytosis in the stimulated adrenal medulla of the cat. Neuropharmacology 12:875-883

Smith AD, Winkler H (1972) Fundamental mechanisms in the release of catecholamines. In: Blaschko H, Muscholl E (eds) Catecholamines. Springer Berlin Heidelberg New York, pp 538-617 (Handbook of Experimental Pharmacology, vol 33)

Smith AD, de Potter WP, Moerman EJ, de Schaepdryver AF (1970) Release of dopamine β-hydroxylase and chromogranin A upon stimulation of the splenic nerve. Tissue and Cell 2:547-568

Smith U, Smith DS, Winkler H, Ryan JW (1973) Exocytosis in the adrenal medulla. Demonstrated by freeze-etching. Science NY 179:79-82

Sobue K, Tanaka T, Kanda K, Ashino N, Kakiuchi S (1985) Purification and characterization of caldesmon$_{77}$: a calmodulin-binding protein that interacts with actin filaments from bovine adrenal medulla. Proc Natl Acad Sci USA 82:5025-5029

Somogyi P, Chubb IW, Smith AD (1975) A possible structural basis for the extracellular release of acetylcholinesterase. Proc R Soc Lond B 191:271-283

Sorimachi M, Yoshida K (1979) Exocytotic release of catecholamines and dopamine β-hydroxylase from the perfused adrenal gland of the rabbit and cat. Br J Pharmacol 65:117-125

Starke K (1977) Regulation of noradrenaline release by presynaptic receptor systems. Rev Physiol Biochem Pharmacol 77:1-124

Steinhardt RA, Alderton JM (1982) Calmodulin confers calcium sensitivity on secretory exocytosis. Nature 295:154-155

Stevens P, Robinson RL, van Dyke K, Stitzel R (1972) Studies on the synthesis and release of adenosine triphosphate 8-3H in the isolated perfused cat adrenal gland. J Pharmacol Exp Ther 181:463-471

Stjärne L (1964) Studies of catecholamine uptake storage and release mechanisms. Acta physiol scand 62:1-60

Stjärne L (1976) Relative importance of calcium and cyclic AMP for noradrenaline secretion from sympathetic nerves of guinea-pig vas deferens and for prostaglandin-induced depression of noradrenaline secretion. Neuroscience 1:19-22

Stjärne L (1979) Role of prostaglandins and cyclic adenosine monophosphate in release. In: Paton DM (ed) The release of catecholamines from adrenergic neurons. Pergamon Press Oxford, pp 111-142

Stjärne L, Hedqvist P, Lagercrantz H (1970) Catecholamines and adenine nucleotide material in effluent from stimulated adrenal medulla and spleen. A study of the exocytosis hypothesis for hormone secretion and neurotransmitter release. Biochem Pharmac 19:1147-1158

Stoehr SJ, Smolen JE, Holz RW, Agranoff BW (1986) Inositol triphosphate mobilizes intracellular calcium in permeabilized adrenal chromaffin cells. J Neurochem 46:637-640

Suchard SJ, Corcoran JJ, Pressman BC, Rubin RW (1981) Evidence for secretory granule membrane recycling in cultured adrenal chromaffin cells. Cell Biol Int Rep 5:953-962

Summers ThA, Creutz CE (1985) Phosphorylation of a chromaffin granule-binding protein by protein kinase C. J Biol Chem 260:2437-2443

Swilem A-MF, Yagisawa H, Hawthorne JN (1986) Muscarinic release of inositol triphosphate without mobilization of calcium in bovine adrenal chromaffin cells. J Physiol Paris 81:246-251

Terbush DR, Holz RW (1986) Effects of phorbol esters, diglyceride, and cholinergic agonists on the subcellular distribution of protein kinase C in intact or digitonin-permeabilized adrenal chromaffin cells. J Biol Chem 261:17099-17106

Thoa NB, Wooten GF, Axelrod J, Kopin IJ (1972) Inhibition of release of dopamine β-hydroxylase and norepinephrine from sympathetic nerves by colchicine, vinblastine or cytochalasin B. Proc Natl Acad Sci USA 69:520-522

Thoa NB, Wooten GF, Axelrod J, Kopin IJ (1975) On the mechanism of release of norepinephrine from sympathetic nerves induced by depolarizing agents and sympathomimetic drugs. Mol Pharmacol 11:10-18

Thureson-Klein Å (1982a) Insights into the functional role of the noradrenergic vesicles. In: Klein RL, Lagercrantz H, Zimmermann H (Eds) Neurotransmitter Vesicles: Composition, Structure and Function. Academic Press Oxford New York, pp 219-239

Thureson-Klein Å (1982b) Fine structure of the isolated noradrenergic vesicles. In Klein RL, Lagercrantz H, Zimmermann H (Eds) Neurotransmitter Vesicles: Composition, Structure and Function. Academic Press Oxford New York, pp 119-132

Thureson-Klein Å (1983) Exocytosis from large and small dense cored vesicles in noradrenergic nerve terminals. Neuroscience 10:245-252

Thureson-Klein Å, Stjärne L (1979) Ultrastructural features of mast cells in human omental veins. Blood Vessels 16:311-319

Thureson-Klein Å, Chen-Yen SH, Klein RL (1974) Retention of matrix density in adrenergic vesicles after extensive norepinephrine depletion. Experientia 30:935-937

Thureson-Klein Å, Stjärne LL, Brundin J (1976a) Effects of field stimulation on nerve terminals in human blood vessels. In: Bailey GW (ed) 34th Ann Proc Electron Microscopy Soc Amer Miami Beach, FL, USA, pp 108

Thureson-Klein Å, Stjärne L, Brundin J (1976b) Ultrastructure of the nerves in veins from human omentum. Neuroscience 1:333-337
Thureson-Klein Å, Klein RL, Johansson O (1979a) Catecholamine-rich cells and varicosities in bovine splenic nerve, vesicle contents and evidence for exocytosis. J Neurobiol 10:309-324
Thureson-Klein Å, Klein RL, Stjärne LL (1979b) Vesicle populations and exocytosis in noradrenergic terminals of human veins. In Catecholamines. Basic and clinical frontiers, (Usdin E, Kopin IJ, Barchas J) vol 1. Pergamon Press New York, pp 262-264
Thureson-Klein Å, Harless S, Klein R (1984) Ultrastructural changes in adrenaline- and SGC-cells after morphine coincide with alterations of adrenaline and dopamine levels. Cell Tiss Res 236: 53-65
Till R, Banks P (1976) Pharmacological and ultrastructural studies on the electron dense cores of the vesicles that accumulate in noradrenergic axons constricted in vitro. Neuroscience 1:49-55
Treiman M, Weber W, Gratzl M (1983) 3′,5′-cyclic adenosine monophosphate -and Ca^{2+}-calmodulin-dependent endogenous protein phosphorylation activity in membranes of the bovine chromaffin secretory vesicles: Identification of two phosphorylated components as tyrosine hydroxylase and protein kinase regulatory subunit type II. J Neurochem 40:661-669
Trifaro JM (1977) Common mechanisms of hormone secretion. Ann Rev Pharmacol Toxicol 17:27-47
Trifaro JM (1978) Contractile proteins in tissues originating in the neural crest. Neuroscience 3:1-24
Trifaro JM, Dworkind J (1975) Phosphorylation of the membrane components of chromaffin granules: Synthesis of diphosphatidylinositol and presence of phosphatidylinositol kinase in granule membranes. Can J Physiol Pharmacol 53:479-492
Trifaro JM, Ulpian C (1976) Isolation and characterization of myosin from the adrenal medulla. Neuroscience 1:483-488
Trifaro JM, Poisner AM, Douglas WW (1967) The fate of the chromaffin granule during catecholamine release from the adrenal medulla. I. Unchanged efflux of phospholipid and cholesterol. Biochem Pharmacol 16:2095-2100
Trifaro JM, Collier B, Lastowecka A, Stern D (1972) Inhibition by colchicine and by vinblastine of acetylcholine-induced catecholamine release from the adrenal gland: an anticholinergic action, not an effect upon microtubules. Molec Pharmacol 8:264-267
Trifaro JM, Ulpian C, Preiksaitis H (1978) Anti-myosin stains chromaffin cells. Experientia 34:1568-1571
Triggle DJ (1979) Release induced by calcium ionophores. In: Paton DM (ed) The release of catecholamines from adrenergic neurons. Pergamon Press Oxford, pp 303-322
Ueda T, Greengard P, Berzins K, Cohen RS, Blomberg F, Grab DJ, Siekevitz P (1979) Subcellular distribution in cerebral cortex of two proteins phosphorylated by a cAMP-dependent kinase. J Cell Biol 83:308-319
Unsicker K (1973) Fine structure and innervation of the avian adrenal gland. I. Fine structure of adrenal chromaffin cells and ganglion cells. Z Zellforsch 145:389-416
Van der Meulen JA, Emerson DM, Grinstein S (1981) Isolation of chromaffin cell plasma membranes on polycationic beads. Biochim Biophys Acta 643:601-615
Van Orden LS, Bensch KG, Giarman NJ (1967) Histochemical and functional rela-

tionships of catecholamines in adrenergic nerve endings. 2. Extravesicular noradrenaline. J Pharmacol Exp Ther 155:428-439
Vilmart-Seuwen J, Kersken H, Stürzl R, Plattner H (1986) ATP keeps exocytosis sites in a primed state but is not required for membrane fusion: an analysis with paramecium cells in vivo and in vitro. J Cell Biol 103:1279-1288
Viveros OH (1975) Mechanism of secretion of catecholamines from adrenal medulla. In: Blaschko H, Sayers G, Smith AD (eds) Handbook of Physiology, vol VI, Adrenal gland. American Physiological Society, Washington, pp 389-426
Viveros OH, Arqueros L, Kirshner N (1968) Release of catecholamines and dopamine β-oxidase from the adrenal medulla. Life Sci 7:609-618
Viveros OH, Arqueros L, Connett RJ, Kirshner N (1969a) Mechanism of secretion from the adrenal medulla. 3. Studies of DBH as a marker for catecholamine storage vesicle membranes in rabbit adrenal glands. Molec Pharmacol 5:60-68
Viveros OH, Arqueros L, Connett RJ, Kirshner N (1969b) Mechanism of secretion of adrenal medulla. IV. The fate of the storage vesicles following insulin and reserpine administration. Mol Pharmacol 6:69-82
Viveros OH, Arqueros L, Kirshner N (1969c) Mechanism of secretion from the adrenal medulla. V. Retention of storage vesicle membrane following release of adrenaline. Mol Pharmacol 5:342-345
Viveros OH, Arqueros L, Kirshner N (1969d) Quantal secretion from adrenal medulla: all-or-none release of storage vesicle content. Science 165:911-913
Viveros OH, Arqueros L, Kirshner N (1971a) Mechanism of secretion from the adrenal medulla. VI. Effect of reserpine on the dopamine β-hydroxylase and catecholamine content and on the buoyant density of adrenal storage vesicles. Mol Pharmacol 7:434-443
Viveros OH, Arqueros L, Kirshner N (1971b) Mechanism of secretion from the adrenal medulla. VII. Effect of insulin administration on the buoyant density, dopamine β-hydroxylase and catecholamine content of adrenal storage vesicles. Mol Pharmacol 7:444-545
Viveros OH, Diliberto EJ, Hazum E, Chang KJ (1979) Opiate-like materials in the adrenal medulla: Evidence for storage and secretion with catecholamines. Mol Pharmacol 16:1101-1108
Von Grafenstein H, Roberts CS, Baker PF (1986) Kinetic analysis of the triggered exocytosis/endocytosis secretory cycle in cultured bovine adrenal medullary cells. J Cell Biol 103:2343-2352
Voyta JC, Slakey LL, Westhead EW (1978) Accessibility of lysolecithin in catecholamine secretory vesicles to acyl CoA: lysolecithin acyl transferase. Biochem Biophys Res Commun 80:413-417
Wacker Ph, Forssmann WG (1972) Immersion and perfusion fixed rat adrenal medulla: The problem of mixed cells, clear cells and the mode of secretion. Z Zellforsch 126:261-277
Wada A, Yanagihara N, Izumi F, Sakurai S, Kobayashi H (1983a) Trifluoperazine inhibits $^{45}Ca^{2+}$ uptake and catecholamine secretion and synthesis in adrenal medullary cells. J Neurochem 40:481-486
Wada A, Sakurai S, Kobayashi H, Yanagihara N, Izumi F (1983b) Suppression by phospholipase A_2 inhibitors of secretion of catecholamines from isolated adrenal medullary cells by suppression of cellular calcium uptake. Biochem Pharmacol 32:1175-1178
Wakade AR (1979) Recycling of noradrenergic storage vesicles of isolated rat vas deferens. Nature 281:374-376
Wakade AR, Wakade TD (1982a) Biochemical evidence for re-use of noradrenergic storage vesicles in the guinea-pig heart. J Physiol 327:337-362

Wakade AR, Wakade TD (1982b) Secretion of catecholamines from adrenal gland by a single electrical shock: Electronic depolarization of medullary cell membrane. Proc Natl Acad Sci USA 79:3071–3074

Wakade AR, Wakade TD (1984a) Effects of desipramine, trifluoperazine and other inhibitors of calmodulin on the secretion of catecholamines from the adrenal medulla and postganglionic sympathetic nerves of the salivary gland. Naunyn Schmiedebergs Arch Pharmacol 325:320–327

Wakade AR, Wakade TD (1984b) Do storage vesicles or peripheral sympathetic nerves have more than one life cycle? In: Usdin E(ed) Catecholamines: Basic and Peripheral Mechanisms. Alan R Liss Inc, New York, pp 89–103

Wakade AR, Wakade TD, Müller Ch, Schwab M (1982) Epinephrine as a tool to investigate the question of recycling of sympathetic storage vesicles in the heart: chemical and morphological studies. J Pharmacol Exp Ther 221:820–827

Wakade AR, Malhotra RK, Wakade TD (1986) Phorbol ester facilitates ^{45}Ca accumulation and catecholamine secretion by nicotine and excess K^+ but not by muscarine in rat adrenal medulla. Nature 321:698–700

Walker JH, Obrocki J, Südhof TC (1983) Calelectrin, a calcium-dependent membrane-binding protein associated with secretory granules in torpedo cholinergic electromotor nerve endings and rat adrenal medulla. J Neurochem 41:139–145

Weinshilboum RM (1979) Serum dopamine β-hydroxylase. Pharmacol Rev 30:133–166

Weinshilboum RM, Thoa NB, Johnson DG, Kopin IJ, Axelrod J (1971) Proportional release of norepinephrine and dopamine β-hydroxylase from sympathetic nerves. Science 174:1349–1351

West DP, Fillenz M (1975) Changes in adrenals and sympathetic nerve terminals following cold exposure. J Neurochem 25:97–99

Westfall DP, Goto K, Stitzel RE, Fedan JS, Fleming WW (1975) Effects of various denervation techniques on the ATP of the rat vas deferens. Eur J Pharmacol 34:397–400

Westhead EW, Winkler H (1982) The topography of gangliosides in the membrane of the chromaffin granule of bovine adrenal medulla. Neuroscience 7:1611–1614

Whitaker M (1985) Polyphosphoinositide hydrolysis is associated with exocytosis in adrenal medullary cells. FEBS 189:137–140

White TD (1978) Release of ATP from a synaptosomal preparation by elevated extracellular K^+ and by veratridine. J Neurochem 30:329–336

Wiedenmann B, Franke WW, Kuhn C, Moll R, Gould VE (1986) Synaptophysin: a marker protein for neuroendocrine cells and neoplasms. Proc Natl Acad Sci USA 83:3500–3504

Wildmann J, Dewair M, Matthaei H (1981) Immunochemical evidence for exocytosis in isolated chromaffin cells after stimulation with depolarizing agents. J Neuroimmunol 1:353–364

Willems M, de Potter W (1983) Isolation of light noradrenaline vesicles from rat vas deferens. Arch Int Pharmacodyn 258:333–334

Williams TP, McGee R (1982) The effects of membrane fatty acid modification of clonal pheochromocytoma cells on depolarization-dependent exocytosis. J Biol Chem 257:3491–3500

Wilson SP, Kirshner N (1976) Isolation and characterization of plasma membranes from the adrenal medulla. J Neurochem 27:1289–1298

Wilson SP, Kirshner N (1983) Calcium-evoked secretion from digitonin-permeabilized adrenal medullary chromaffin cells. J Biol Chem 258:4994–5001

Wilson SP, Chang KJ, Viveros OH (1982) Proportional secretion of opioid peptides

and catecholamines from adrenal chromaffin cells in culture. J Neuroscience 2:1150-1156
Winkler H (1971) The membrane of the chromaffin granules. Phil Trans R Soc B 261:293-303
Winkler H (1976) The composition of adrenal chromaffin granules: an assessment of controversial results. Neuroscience 1:65-80
Winkler H (1977) The biogenesis of adrenal chromaffin granules. Neuroscience 2:657-683
Winkler H, Carmichael SW (1982) The chromaffin granule. In: Poisner, Trifaro J (eds) The Secretory granule. Elsevier Biomed Press, pp 3-79
Winkler H, Westhead EW (1980) The molecular organization of adrenal chromaffin granules. Neuroscience (in press)
Winkler H, Hörtnagl H, Schöpf JAL, Hörtnagl H, Zur Nedden G (1971) Bovine adrenal medulla: Synthesis and secretion of radioactively labelled catecholamines and chromogranins. Naunyn Schmiedeberg's Arch exp Path Pharmacol 271:193-203
Winkler H, Schöpf JAL, Hörtnagl H, Hörtnagl H (1972) Bovine adrenal medulla: Subcellular distribution of newly synthesized catecholamines nucleotides and chromogranins. Naunyn Schmiedeberg's Arch exp Path Pharmacol 273:43-61
Winkler H, Schneider FH, Rufener C, Nakane PK, Hörtnagl H (1974) Membranes of adrenal medulla: Their role in exocytosis. In: Ceccarelli B, Clementi F, Meldolesi J (eds) Cytopharmacology of secretion. Advances in Cytopharmacology, vol 2. Plenum Press New York, pp 127-139
Winkler H, Apps DK, Fischer-Colbrie R (1986) The molecular function of adrenal chromaffin granules: established facts and controversial results. Neuroscience 18:261-290
Wooten GF, Thoa NB, Kopin IJ, Axelrod J (1973) Enhanced release of dopamine β-hydroxylase and norepinephrine from sympathetic nerves by dibutyryl cyclic adenosine 3′,5′-monophosphate and theophylline. Mol Pharmacol 9:178-183
Wright CD, Hoffman MD (1986) The protein kinase C inhibitors H-7 and H-9 fail to inhibit human neutrophil activation. Biochem Biophys Res Comm 135:749-755
Wyllie MG, Gilbert JC (1980) Exocytotic release of noradrenaline from synaptosomes. Biochem. Pharmacol 29:1302-1303
Yamada R, Sato J, Komiya E, Nakai T (1978) A new serological assay of chromogranine. Endocrin Jap 25:37-41
Yamada K, Iwahashi K, Kase H (1987) K252a, a new inhibitor of protein kinase C, concomitantly inhibits 40K protein phosphorylation and serotonin secretion in a phorbol ester-stimulated platelets. Biochem Biophys Res Comm 144:35-40
Yanagihara N, Isosaki M, Ohuchi T, Oka M (1979) Muscarinic receptor-mediated increase in cyclic GMP level in isolated bovine adrenal medullary cells. FEBS LETT 105:296-298
Yen SS, Klein RL, Chen-Yen SH, Thureson-Klein Å (1976) Norepinephrine adenosinetriphosphate ratios in purified adrenergic vesicles. J Neurobiol 7:11-22
Zaremba S, Hogue-Angeletti R (1985) A reliable method for assessing topographical arrangement of proteins in the chromaffin granule membrane. Neurochem Res 10:19-32
Zhu PC, Thureson-Klein Å, Klein RL (1986) Exocytosis from large dense cored vesicles outside the active synaptic zones of terminals within the trigeminal subnucleus caudalis: a possible mechanism for neuropeptide release. Neuroscience 19:43-54
Ziegler MG, Thomas JA, Jacobowith DM (1976) Retrograde axonal transport of antibody to dopamine β-hydroxylase. Brain Res 104:390-395

Zimmerberg J, Curran M, Cohen FS, Brodwick M (1987) Simultaneous electrical and optical measurements show that membrane fusion precedes secretory granule swelling during exocytosis of beige mouse mast cells. Proc Natl Acad Sci USA 84:1585-1589

Zinder O, Pollard HB (1980) The chromaffin granule: Recent studies leading to a functional model for exocytosis. In: Youdim MBH, Lovenberg W, Sharman DF, Lagnado JR (eds) Essays in Neurochemistry and Neuropharmacology, vol 4. Wiley and Sons Ltd Chichester, pp 126-155

Zinder O, Hoffman PG, Bonner WM, Pollard HB (1978) Comparison of chemical properties of purified plasma membranes and secretory vesicle membranes from the bovine adrenal medulla. Cell Tiss Res 188:153-170

CHAPTER 3

Monamine Oxidase

M.B.H. YOUDIM, J.P.M. FINBERG, and K.F. TIPTON

A. Introduction

The enzyme monoamine oxidase (amine: oxygen oxidoreductase (deaminating) EC 1.4.3.4.) (MAO) catalyses the oxidative deamination of amines according to the overall reaction:

$$RCH_2NH_2 + O_2 + H_2O \rightarrow RCHO + NH_3 + H_2O_2$$

It is present in the central nervous system as well as in peripheral tissues, and it functions in the breakdown of neurotransmitter and hormonal amines as well as those aristing from the diet or from bacterial action. The history of its discovery and the earlier work on its properties have been described in detail (see e.g., BLASCHKO 1952, 1963; TIPTON 1975; YOUDIM 1975, KINEMUCHI et al. 1987) and this review will concentrate on the knowledge gained from more recent studies.

B. Classification

Although much has been written on the classification of the amine oxidases, it is still necessary to discuss this aspect briefly, since a great deal of confusion is evident even in some of the more recent publications. The problem is due to the existence of several other amine oxidases, some of which have similar specificities to MAO. ZELLER (1963a, 1979) has carefully considered the distictions between these enzymes and MAO. He introduced the term "monoamine oxidase" in 1938 to distiguish it from the enzyme that catalysed the oxidative deamination of histamine and short-chain aliphatic diamines, which he called diamine oxidase. Since the latter enzyme was inhibited irreversibly by carbonyl reagents, such as semicarbazide, and by cyanide, ZELLER suggested that this inhibitor sensitivity could be used as a criterion for distinguishing between the two enzymes. This still represents the best method for distinguishing MAO from other enzymes with similar activities, although it is important that the carbonyl reagent used should be selected carefully since a number of substituted hydrazine derivatives are specific inhibitors of MAO (see Sect. L). In addition, cyanide has been shown to be a weak reversible inhibitor of MAO (HOUSLAY and TIPTON 1973a) and very high concentrations of semicarbazide may inhibit this enzyme (WIBO and DUONG 1979) and interfere with some commonly used assay methods (LYLES and SHAFFER 1979).

The different inhibitor sensitivities of the enzymes appear to result from their cofactor requirements, since the carbonyl-reagent sensitive enzymes contain a carbonyl-group that is essential for activity (see e.g. YASUNOBU et al. 1976), whereas MAO is a flavoprotein that contains no such essential group (see Sect. D.III).

The nomenclature of the amine oxidases has been a source of great confusion since the carbonyl-reagent sensitive enzymes from several sources are active towards some monoamines and, indeed, some enzymes that would be classified as diamine oxidases by the criteria of ZELLER are unable to catalyze the deamination of common diamines (see BUFFONI 1966; YASUNOBU et al. 1976). In addition, longer-chain aliphatic diamines can serve as substrates for MAO (see Sect. G). These problems have led the Enzyme Commission of the International Union of Biochemistry (see DIXON and WEBB 1979) to recommend that the name monoamine oxidase (MAO) be replaced by amine oxidase (flavin-containing) and that the carbonyl-reagent sensitive enzymes, which also appear to require bound copper ions for activity, be called amine oxidase (copper-containing) (EC 1.4.3.6).

Several other criteria have been used to distinguish between the amine oxidases but they are not as useful as those based on inhibitor sensitivity. For example, MAO is active towards N-methyl-substituted secondary and tertiary amines, whereas diamine oxidase is not (ZELLER 1963a). The potent irreversible inhibition of MAO by acetylenic inhibitors (see Sect. L) can provide a useful means of distinction (see ZELLER 1979). Although one such inhibitor, clorgyline, has been shown to be a reversible inhibitor of the copper-containing amine oxidase from ox plasma (HOUSLAY and TIPTON 1975a) much higher concentrations are required to inhibit the latter enzyme.

Since plasma from many species contains a copper-containing amine oxidase that is active towards simple monoamines, blood contamination can complicate studies of MAO unless the former enzyme is specifically inhibited. It is, however, unsafe to assume that well-washed tissue homogenates will only contain MAO activity, since, in addition to diamine oxidase, which has a substrate specificity quite unlike that of MAO, a carbonyl-reagent-sensitive amine oxidase has been demonstrated in some tissues (LYLES and CALLINGHAM 1975; LEWINSOHN et al. 1980a; WIBO and DUONG 1979).

C. Distribution and Localization

Monoamine oxidase is found within the cells of most mammalian and avian tissues as well as those of some invertebrates (see BLASCHKO 1952; TIPTON 1975, for reviews). It is however absent from erythrocytes. The relatively high activities of the enzyme in liver and intestine suggest important roles for these tissues in the catabolism of exogenous amines. Its presence in the walls of the blood vessels constituting the blood-brain barrier has been suggested to be an important aspect of the function of that system (DE LA TORRE 1972). In the brain the total MAO activity is relatively evenly distributed amongst the different regions, showing no obvious correlation with neurotransmitter amine levels or with any specific type of innervation (see TIPTON 1975; MACKAY et al. 1978). It is present in glial cells as well as in neurons (SILBERSTEIN et al. 1972; SINHA and ROSE 1972) and the presence of extraneuronal MAO has been demonstrated in kidney (BOADLE and BLOOM 1969), artery (DE LA LANDE et al. 1970), heart (HORITA and LOWE 1972) and liver (TIPTON et al. 1976).

Sympathetic denervation studies have revealed that the enzyme is predominantly extrinsic to catecholamine containing neurons in heart (JARROTT 1971; HORITA and LOWE 1972), brain (URETSKY and IVERSEN 1970), liver (TIPTON et al. 1976) and salivary gland (SNYDER et al. 1965; MUELLER et al. 1968). The observation that MAO activity is associated with non-aminergic nerve terminals (WOODS 1970; FURNESS and COSTA 1971) suggests that not all the neuronal enzyme activity will be lost following sympathectomy. A further complication in the interpretation of results from such studies is that a fall in the activity of an extraneuronal enzyme might follow denervation if its activity was dependent on an intact nerve supply (MARSDEN et al. 1971). Such a mechanism might account for the decrease in MAO activity following sympathectomy of liver (KLINGMAN 1966) which is a poorly innervated organ.

Subcellular fractionation studies have shown the enzyme to be largely mitochondrial although some activity has also been reported to be associated with the microsomal fraction (see TIPTON 1975, for review). It has been suggested that this microsomal enzyme might arise as an artefact due to mitochondrial damage occurring during the cell distruption and fractionation procedures (JARROTT and IVERSEN 1968), but histochemical studies (SHANNON et al. 1974; MULLER and DELAGE 1977). have confirmed that a proportion of the enzyme activity is associated with the endoplasmic reticulum within the cell.

Fractionation of isolated mitochondria has shown MAO to be confined to the outer membrane of these organelles (see TIPTON 1975 for review). Since the mitochondrion is apparently unable to synthesize its outer-membrane proteins (NEUPERT et al. 1967), the activity associated with the endoplasmic reticulum might represent recently synthesized enzyme that will ultimately be incorporated into the mitochondria. Such a precursor role for this fraction would be consistent with the observation that the rate of turnover of the enzyme in the microsomal fraction is faster than that in mitochondria (ERWIN and SIMON 1969). However, SAGARA and ITO (1982) have recently demonstrated that the *in vitro* synthesis of rat liver MAO in a rabbit reticulocyte lysate takes place on free but not membrane-bound polysomes. The rate of degradation of irreversibly inhibited MAO is similar to that of other outer-membrane proteins (see BRUNNER and NEUPERT 1968; PLANZ et al. 1972) which suggests that similar factors regulate the turnover of all proteins in this membrane.

The enzyme is tightly bound to membranes and subcellular fractionation studies have revealed no significant amount of soluble activity in most organs from a number of species. In guinea pig liver, however, a significant amount of activity can be readily extracted into the soluble fraction (WEISSBACH et al. 1957).

D. Properties of the Enzyme

The tight association of MAO with membranes necessitates the use of extremely vigorous procedures, such as detergent treatment, sonication, and extraction into organic solvents, to bring it into solution (see TIPTON 1975; YOUDIM 1975). This problem has hampered attempts to obtain highly purified preparations of the enzyme, but several purification procedures have now been reported and a number of the molecular properties of the enzyme have been determined.

I. Molecular Weight

Values reported for the relative mass (M_R) of solubilized MAO preparations from a number of sources have varied from about 100 000 to several million (for review see TIPTON 1975). These large differences appear to be due to a tendency of the extracted enzyme to form aggregates in solution. This may be overcome by the use of detergents, and active species with M_R values in the range 80 000–100 000 have been found after such treatment of MAO preparations from a number of sources (see e.g., TIPTON 1975; WHITE and GLASSMAN 1977; YASUNOBU et al. 1979). A minimum value of about 100 000 has also been obtained from determination of the flavin content (see e.g. YOUDIM and COLLINS 1971, YOUDIM 1975; YASUNOBU et al. 1979; SALACH 1979) and from the binding of specific inhibitors that react with the flavin component of the enzyme (see COLLINS and YOUDIM 1975; ORELAND et al. 1973a and Sect. K).

Polyacrylamide gel electrophoresis in the presence of sodium dodecyl sulphate (SDS-PAGE) has been used to investigate the presence of subunits in the enzyme. YOUDIM and COLLINS (1971) and COLLINS and YOUDIM (1975) found only one type of subunit, corresponding to a relative molecular mass of 75 000. Since this group had calculated that the active enzyme contained one FAD per 150 000, they concluded that this represented a dimer of similar subunits, only one of which contained FAD. Similar studies on the enzyme purified from ox liver have revealed only one type of subunit with a relative molecular mass of 52 000 (MINAMIURA and YASUNOBU 1978) or 61 000 (SALACH 1979). Since the separation of these subunits does not require reducing conditions (YASUNOBU et al. 1979) it appears that disulfide bonds are not involved in their association, the forces holding them together being entirely non-covalent.

After reaction of MAO with radioactively-labelled pargyline, which reacts specifically with the covalently-bound FAD in the enzyme (see Sect. L), SDS-PAGE yielded values between 55 000 and 70 000 for the relative molecular masses of the FAD-containing subunits of MAO from a number of sources.

These studies suggest that the active form of MAO is a dimer of subunits that have similar relative molecular masses and that this dimer is capable of further aggregation. From studies on the enzyme purified from ox liver MINAMIURA and YASUNOBU (1978) have argued that the two subunits are identical, differing only in that one contains covalently bound FAD. They showed that leucine was the only detectable C-terminal amino acid and that the number of peptides that could be separated after tryptic digestion of the enzyme was consistent with it being a dimer composed of two identical subunits of M_R 50000.

It is possible, however, that the FAD-free subunit may not be essential for activity since Aoki et al. (1977) have reported that treatment of the enzyme from ox brain with increasing concentrations of a polyoxyethylene sulphate detergent (Emarl 20c) yielded active forms with M_R values of 89000 and 44000. White and Stine (1982a, b) have reported that treatment of MAO preparations from several sources with low concentrations of sodium dodecyl sulphate or with phospholipase A2 yielded active forms in the M_R range 50000-55000.

The technique of irradiation-inactivation, which can be used to determine the sizes of proteins in particulate preparations (see Mantle 1978, for a discussion of the technique), has been used to study the relative molecular mass of the enzyme in rat liver. Application of this method to freeze-dried preparations of mitochondrial outer membranes from this source gave a value of about 300000 (Tipton and Mantle 1977) but a similar study with freeze-dried whole mitochondria gave values between 140000 and 200000 (Callingham and Parkinson 1979). Thus it appears that, in its membrane-bound form, the enzyme probably exists as an aggregate, but the number of subunits involved is, as yet, uncertain.

II. Cofactors

1. Flavin

As mentioned earlier, MAO is a flavoprotein containing approximately one mole of FAD per 100000 g of protein. Studies on purified MAO from ox liver have shown the 8 position of the isoalloxazine ring of flavin to be covalently bound to a cysteine residue in the enzyme by a thiother bond (see Fig. 3, Kearney et al. 1971; Walker et al. 1971). This specific type of linkage, which has not thus far been found in any other flavoproteins from animal sources, has characteristic absorbance and fluorescence properties which have allowed the demonstration of its presence in MAO preparations from other sources (Salach et al. 1976; Salach and Detmer 1979). After proteolytic digestion of the ox liver enzyme, Walker et al. (1971) were able to isolate a flavin pentapeptide with the following structure:

```
Ser.-Gly-Gly-Cys-Tyr
                |
               FAD
```

A similar flavin peptide has been isolated after proteolytic digestion of pig liver MAO (Oreland et al. 1973b). Recently Watanabe et al. (1980) have reported that the bound flavin in ox liver MAO is relatively labile, in that it can be released from the enzyme by high concentrations of sulfydryl group-containing compounds. This behaviour is unexpected since Singer has claimed that the bound FAD in MAO is very stable and not susceptible to reductive cleavage (see SALACH et al. 1976).

YAGI and NAOI (1982a, b) have recently reported the presence of an amine oxidase in pig liver and brain that does not contain covalently-bound FAD, although it requires this cofactor for activity. These results may account for earlier reports on the presence of non-covalently-bound FAD in MAO preparations (TIPTON 1968a; HARADA et al. 1971; YOUDIM and SOURKES 1972). This enzyme, which has been purified and obtained in crystalline form from pig liver, has been shown to be active towards a range of biogenic amines, including 5-HT, 2-phenylethylamine, tyramine and dopamine, but to be resistant to inhibition by clorgyline and deprenyl (YAGI and NAOI 1982a, b). An evaluation of the significance of this enzyme in the metabolism of the biogenic amines must await the results of further studies.

Direct evidence for the involvement of FAD in MAO activity comes from absorbance and fluorescence studies of the purified enzyme.

FAD is reduced in the presence of substrate (NARA et al. 1966a; ERWIN and HELLERMAN 1967; TIPTON 1968a; HARADA and NAGATSU 1969; ORELAND 1971; YOUDIM and SOURKES 1972). Several specific irreversible inhibitors of MAO have been shown to react directly with the bound flavin resulting in characteristic changes in its absorbance spectrum (see Sect. L).

2. Metal Ions

An early report indicated that beef liver MAO contained copper that was essential for activity (NARA et al. 1966a) but subsequent work has shown that it is possible to obtain highly purified preparations of the enzyme from this (YASUNOBU et al. 1968) and other sources (YOUDIM and SOURKES 1966, ERWIN and HELLERMAN 1967; TIPTON 1968a; ORELAND 1971; NAGATSU et al. 1972) which contain negligible amounts of copper.

Iron is present in preparations of MAO from rat liver (YOUDIM and SOURKES 1966) and pig liver (ORELAND 1971), but the iron content of the enzymes purified from pig brain (TIPTON 1968a) and ox liver (WEYLER and SALACH 1981; ICHINOSE et al. 1982) was found to be insignificant, and ORELAND (1971) found no correlation between the iron content and the activity of the enzyme from pig liver.

III. The Active Site

The activity of MAO is inhibited by reagents that react with sulfhydryl groups (see TIPTON 1975 for a review of the earlier literature) and studies on the kinetics of the reaction of ox liver MAO with a variety of sulfhydryl reagents led GOMES et al. (1976) to conclude that two of the eight sulfhydryl groups per mol. of enzyme (M_R 100 000) were essential for activity' and substrate protection studies suggest these to be at or near the active site of the enzyme (see also YASUNOBU et al. 1979). Further evidence of the involvement of at least one sulfhydryl group comes from the observation that some suicide inhibitors of the enzyme appear to react with such a group (see Sect. K). WHITE and

GLASSMAN (1977) have reported the activity of rat liver MAO towards 5-HT to be more sensitive to inhibition by sulfhydryl reagents than that towards 2-phenylethylamine. Studies on the photooxidation of MAO and its reaction with ethoxyformic acid have suggested that two histidine residues per 100000 g of protein are essential for activity (HIRAMATSU et al. 1975).

An intriguing series of experiments by GORKIN and his co-workers (see GORKIN 1973, 1976 for reviews) has suggested that treatment of MAO from a number of sources with reagents that oxidise sulfhydryl groups results in a loss of activity towards the usual MAO substrates and the appearance of the ability to oxidise amines such as histamine, putrescine, cadaverine, lysine and AMP. A particularly surprising observation from this group is that this treatment also renders the enzyme insensitive to inhibition by compounds such as pargyline and tranylcypromine (see Sect. L) but sensitive to inhibition by carbonyl-group reagents. It is difficult to understand the mechanism of these effects since pargyline inhibition involves reaction with the FAD in MAO (see Sect. K) and the altered sensitivity would imply the appearance of a reactive carbonyl group in the enzyme. GORKIN (1973, 1976) has suggested that this process may be important *in vivo* under conditions, such as vitamin E deficiency, excess of vitamin D_2 and X-irradiation, that give rise to increased levels of lipid peroxides.

Studies with a series of substrates and inhibitors have indicated a hydrophobic region of the active site to be involved in substrate binding (see WILLIAMS 1974; JOHNSON 1976; ZELLER et al. 1979) and that both an electrophilic group (MCEWEN et al. 1968) and a nucleophilic group (see SEVERINA 1973, 1979) may be important for catalysis. Such studies may, ultimately, be of great value in the design of new MAO inhibitors.

E. Kinetics of the Reaction

Initial rate and product inhibition studies have shown MAO from pig brain (TIPTON 1968b), ox liver (OI et al. 1970, 1971) and rat liver (HOUSLAY and TIPTON 1973a, 1975b) to follow a double-displacement, or ping-pong, steady-state mechanism that can be represented by the following simplified scheme:

where S is the amine substrate and P represents the products derived from it. The reaction can be regarded as taking place in two separate steps. In the first of these the enzyme reacts with the amine to give the aldehyde plus ammonia and a reduced form of the enzyme (H_2E). This reduced enzyme then reacts with oxygen in the second step, to produce hydrogen peroxide and the starting form of the enzyme (E). Direct evidence for this mechanism, and evidence for the relative molecular mass of the active form of the enzyme (see Sect. D.I), has come from studies in the absence of oxygen, when only the first of these steps can occur. Under these conditions the enzyme from pig brain has been shown to form a maximum of one mole of the aldehyde product per 100 000 g of protein (TIPTON 1968b). Similar results have been reported for the production of the aldehyde and ammonia by the enzyme from ox liver (OI et al. 1970; OI and YASUNOBU 1973).

In the case of the enzyme from rat liver it appears that only the aldehyde product is released during the first step of the reaction. Ammonia does not dissociate until after oxygen is bound (HOUSLAY and TIPTON 1973a, 1975b). In contrast, FISCHER et al. (1968) suggested that only ammonia was released from the ox thyroid enzyme before the binding of oxygen. These variations on the basic double-displacement mechanism may result from differences in the assay conditions (see OI et al. 1971), the substrate (see LEFFLER and GRUNWALD 1963) or the nature of the enzyme preparation used (see HOUSLAY and TIPTON 1975b). Kinetic data with MAO from human brain (ROTH 1979a) and platelet (FOWLER et al. 1979) have also suggested that the enzymes from these sources obey a double-displacement mechanism.

The pre-steady state kinetics of purified ox liver MAO have been studied by HUSAIN et al. (1982). With 2-phenylethylamine as the substrate the results of stopped-flow experiments were consistent with a simple double-displacement mechanism, but with benzylamine these workers found the maximum rate of the reaction to be faster than the rate of reoxidation of the enzyme that had been reduced by prior treatment with either benzylamine or dithionite. HUSAIN et al. (1982) interpreted the results with this substrate in terms of a sequential mechanism in which reoxidation of the enzyme precedes dissociation of the products. Such data would, however, also be consistent with a double-displacement mechanism if only one of the products of the first step (aldehyde and ammonia) were to dissociate before the binding of oxygen as, for example, appears to be the case with rat liver MAO (HOUSLAY and TIPTON 1973a, 1975b).

BERNHEIM (1931) first proposed that the reaction catalysed by MAO involved dehydrogenation at the carbon-nitrogen bond to form the corresponding imine, which was then subsequently hydrolysed to give the aldehyde plus ammonia. Such a mechanism has become widely accepted and indirect evidence has been presented in its support (TIPTON and SPIRES 1971). The evidence discussed in Sect. D.III indicates that the flavin component of the enzyme becomes reduced during the reaction and thus the overall reaction can probably be represented by the sequence:

$$\begin{aligned}
&\text{E-FAD} + \text{RCH}_2\text{NH}_2 \rightleftharpoons \text{E-FADH}_2 + \text{RCH}=\text{NH}\\
&\text{RCH}=\text{NH} + \text{H}_2\text{O} \rightleftharpoons \text{RCHO} + \text{NH}_3\\
&\text{E-FADH}_2 + \text{O}_2 \rightleftharpoons \text{E-FAD} + \text{H}_2\text{O}_2
\end{aligned}$$

The imine could be a product of the reaction itself, being hydrolysed after its release, or it could be hydrolysed on the enzyme with subsequent release of the aldehyde and ammonia. Both indirect (TIPTON and SPIRES 1972) and direct (HOUSLAY and TIPTON 1973a, 1975b) kinetic evidence favours the latter alternative in the case of rat liver MAO.

The kinetic mechanism obeyed by MAO will give rise to an initialate equation of the form:

$$\underline{v} = \frac{V}{1 + \frac{K_m^A}{[\text{Amine}]} + \frac{K_m^0}{[O_2]}}$$

where $\underline{v}$ and V are the initial and maximal velocities, respectively, the square brackets represent concentrations and K_m^A, K_m^0 are, respectively, the concentrations of oxygen and amine that will give an initial velocity of one half V at saturating concentrations of the other substrate. Values of the kinetic constants determined for MAO preparations from several sources are shown in Table 1.

The K_m values for oxygen are close to the solubility of oxygen in air-saturated water and thus it might seem that the enzyme would not be saturated with oxygen under normal cellular conditions (see BLASCHKO 1952; VON KORFF 1977).

This might suggest that the MAO activity would be severely affected by competition from other mitochondrial oxidase activities for the available oxygen. TIPTON (1972, 1980) has, however, pointed out that a specific consequence of the double-displacement mechanism is that the K_m value for one substrate will decrease as the concentration of the other is lowered, and this will lead to the activity of MAO being relatively insensitive to fluctuations in the levels of oxygen at relatively low concentrations of the amine substrate.

Inhibition of MAO by high concentrations of the amine substrate has been reported by a number of workers but in many cases this may have resulted from contamination of the substrates with aldehydes which are themselves inhibitory (HOUSLAY and TIPTON 1973a) or from failure to determine the true initial rate of the reaction (KINEMUCHI et al. 1982). The unprotonated form of the amine appears to be the true substrate (see Sect. F) and high substrate inhibition at lower pH values may involve the binding of the protonated form producing an unreactive complex (GABAY et al. 1976; ROTH 1979). It is

Table 1. Kinetic constants of monoamine oxidase

Source	Amine Substrate	pH	k_m(μ*M*)		Reference
			Amine	Oxygen	
Pig Brain	Tyramine	7.0	240	234	TIPTON 1986b; 1972
Ox Liver	Benzylamine	7.4	175	125	OI et al. 1970
Ox Thyroid	Tyramine	7.4	160	120	FISCHER et al. 1968
Rat Liver[a]	Benzylamine	7.2	475 (571)	156 (113)	HOUSLAY and TIPTON 1973a, 1975b

The k_m values are as defined in the text.
[a] Kinetic data for a membrane-bound preparation of the enzyme, data obtained with a solubilized preparation are shown in brackets.

unlikely, however, that such inhibition has any physiological significance since in studies with perfused liver preparations BEN-HARARI et al. (1982) could not detect any inhibition of deamination by high amine concentrations in the perfusing medium.

Inhibition by the products of the reaction might have some role in regulating the activity of the enzyme (see TIPTON and MANTLE 1977, for discussion). Ammonia is a relatively poor inhibitor and it is thus unlikely that it will have a significant effect on the activity of the enzyme *in vivo*. The high efficiency of the systems that remove hydrogen peroxide (see SINET et al. 1980) suggests that inhibition by this product will also not be of significance. The relatively high sensitivity of the enzyme to inhibition by the aldehyde products has led to the suggestion that these might have regulatory importance under some conditions (TURNER et al. 1974), but in the absence of any reliable data on the intracellular concentrations of these compounds this must remain speculative. Competition between different amines could affect the activity towards any one of them. Some possible effects of such competition in terms of the behaviour of MAO *in vivo* have been discussed (TIPTON et al. 1976; TIPTON and MANTLE 1977).

F. Reaction Mechanism

The primary step in amine oxidation is the formation of the imine intermediate (see Sect. E) and RICHTER (1937) suggested that it would be necessary for the amine group to be protonated in order to account for the oxidation of tertiary amines such as hordenine (N,N-dimethyltryptamine) by MAO.

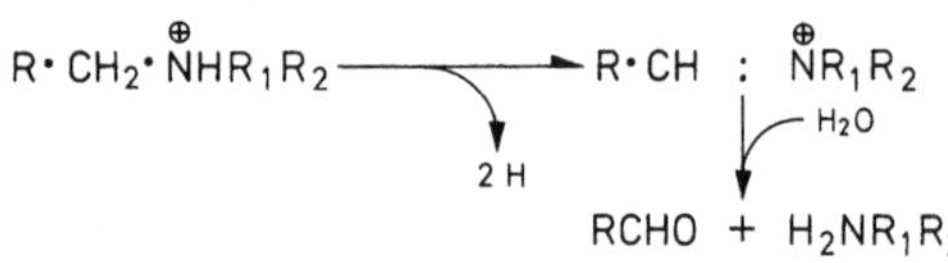

Studies on the effects of pH on the activities of MAO preparations from human liver have, however, shown that the pH-dependence of the K_m value for the amine substrate is as would be expected if only the unprotonated form of the amine were the true substrate (MCEWEN et al. 1968, 1969). Similar results have been obtained for the enzyme from pig brain (WILLIAMS 1974) and other evidence supports the view that the amine functions as a substrate only in its unprotonated form (SMITH et al. 1962; BELLEAU and MORAN 1963; BROWN and HAMILTON 1970; ROTH 1979). However in studies on the effects of pH on the activity of the ox liver enzyme, using benzylamine (pK 9.2 — WILLIAMS 1974) as the substrate, OI et al. (1971) found no evidence for any group with a pK value between 7.0 and 10.2 affecting either the K_m value or the maximal velocity and similar results were obtained over narrower ranges of pH values for the enzymes from rat liver and brain (COQ and BARON 1968; HUSZTI 1972).

WILLIAMS (1974) has proposed a rather elaborate mechanism in which the initial abstraction of a proton from the α-carbon was followed by nucleophilic attack by the amine group of a lysine residue in the enzyme to form a covalent intermediate with the release of ammonia. Loss of a second proton from the intermediate would then result in the formation of a Schiff base which could be hydrolysed to form the aldehyde product without the involvement of an imine as an intermediate in the reaction. There is no direct evidence that a lysine residue in MAO is directly involved in its activity. If such a mechanism were involved one would expect that reduction of the enzyme-substrate mixture by sodium borohydride would lead to inactivation with the formation of a stable compound between the substrate and an enzyme-bound lysine residue. This has, however, not been demonstrated Thus, although the initial stage in the oxidative process may involve the abstraction of a proton from the α-carbon atom, forming a carbanion intermediate, there is no evidence in support of the subsequent steps proposed in this mechanism.

SMITH et al. (1962) showed that N-oxides could act as substrates for MAO and that their metabolism appeared to proceed by way of imine intermediates, indicating that N-oxides themselves are unlikely to serve as alternative intermediates in the oxidation of amines. They proposed a mechanism to account for the oxidation of dimethyltryptamine and its N-oxide which involved the removal of a hydride ion from the unprotonated substrate:

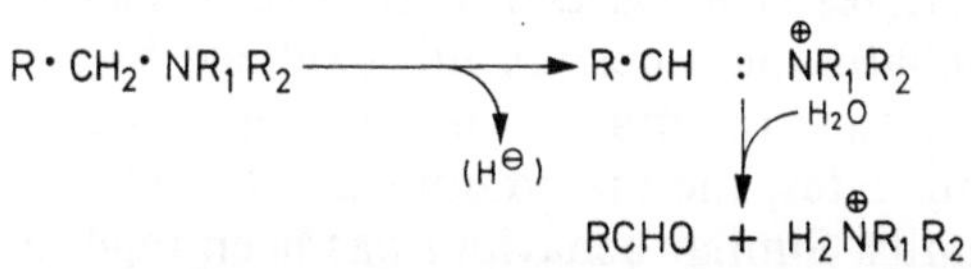

From the results of studies with inhibitors (See Sect. K) and with non-enzymic model reactions, SILVERMAN et al. (1980) have proposed that MAO-catalysed amine oxidation proceeds by two successive one-electron transfers with the formation of a radical cation intermediate. Their proposed mechanism, which avoids the removal of nonacidic protons, which would be necessary in a carbanionic mechanism, is shown in outline below, where F represents the oxidised FAD:

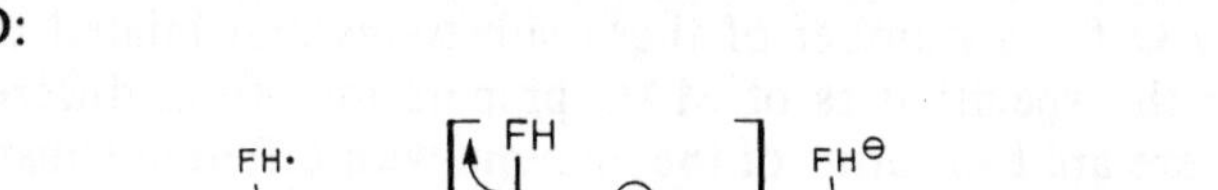

An alternative mechanism for the final step in this reaction involving radical combination followed by two electron transfer to the flavin, was also envisaged by these authors:

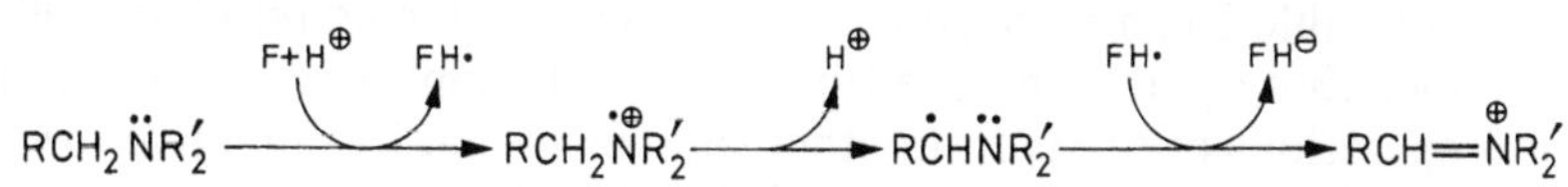

G. Specificity

The substrate specificity of MAO has been reviewed in detail (Blaschko 1952, 1963; Tipton 1975; Youdim 1975) and the data collected in these reviews summarised briefly here. The enzyme is active towards primary, secondary and tertiary amines with the general structure:

$$R \cdot CH_2 \cdot NR_1R_2$$

where R_1 and R_2 may be either hydrogen ions or methyl groups. N-oxides, such as N,N-dimethyltryptamine-N-oxide, are also substrates for the enzymes (Smith et al. 1962). The amino group must be attached to an unsubstituted methylene group. Aniline, amphetamine and other α-methyl-substituted amines are, for example, not substrates (Belleau and Moran 1963; Mantle et al. 1976). Methylamine is not a substrate, although ethylamine may be in some species. The optimum chain-length for activity in alkylamines has been reported to be 5–6 carbon atoms, but this information was not necessarily obtained under conditions that would give maximal velocities in each case and it did not take account of the relative insolubility of the longer-chain aliphatic amines. Branched-chain aliphatic amines such as isoamylamine and isobutylamine are also substrates (see e.g. Oreland 1972; Peers et al. 1980).

Short-chain aliphatic diamines are not oxidized by MAO. Longer chain compounds, in which the two amine groups are separated by 7 or more methylene groups, are substrates; the maximum rate of oxidation is reached when there are 13 such units. Similar behaviour has been reported for ω-amino acids and for quaternary compounds with the general formula

$$H_2N(CH_2)_n\overset{+}{N}(CH_3)_3$$

Recently it has been reported that monoacyl derivatives of cadaverine are also substrates for MAO (Suzuki et al. 1980).

The natural endogenous substrates for MAO are the catecholamines and indolealkylamines and their derivatives. Table 2 shows the kinetic parameters of rat liver MAO for a number of these substrates and related compounds. Differences in the specificities of MAO preparations from different sources suggest that there are two forms of the enzyme with different substrate specificities. It should be borne in mind that, since the proportions of these two forms can vary greatly between different organs and species, a number of apparent discrepancies between earlier results on the specificity of the enzyme could simply have resulted from differences in the proportion of the forms present (see Sect. J). The enzyme from rat liver mitochondria appears to contain approximately equal amounts of the two forms of MAO and thus the results shown in Table 2 can be regarded as representing the activities of both of them. The specificity of the enzyme from ox thyroid appears to be very different from preparations of MAO from other sources in that it is inactive towards benzylamine, adrenaline and noradrenaline although tyramine is a good substrate (Fischer et al. 1968).

Monoamine oxidase appears to show some stereospecificity although it is not absolute. The (−)-isomers of adrenaline and noradrenaline have been

Table 2. The specificity of rat liver monoamine oxidase

Substrate	k_m(μ*M*)	Relative Maximum Velocity[a]
Benzylamine	245	100
Dopamine	405	112
Tyramine	282	200
Tryptamine	18.5	81
2-Phenylethylamine	20.8	118
5-HT	187	124
Vanillylamine	201	68
p-Hydroxybenzylamine	2500	114
Adrenaline	400	70
Noradrenaline	416	71
Octopamine	625	149
m-O-Methylnoradrenaline	200	54
m-O-Methyladrenaline	267	65

[a] Expressed as a percentage of that given by benzylamine. All values were determined at pH 7.2 and at 30 °C with an airsaturating concentration of oxygen. Taken from HOUSLAY and TIPTON 1974.

shown to be oxidised more rapidly than the (+)-isomers (PRATESI and BLASCHKO 1959; GIACHETTI and SHORE 1966). In the case of the enantiomers of phenylethanolamine, WILLIAMS (1977) found that the (+)-isomer was oxidised more rapidly than the (−)-isomer. Inhibitors such as amphetamine and other α-methyl-substituted amines also show some degree of stereospecificity; for example (+)-amphetamine is a more potent inhibitor of the enzyme than the (−)-isomer (MANTLE et al. 1976; FULLER and HEMRICK 1979; ZELLER et al. 1979). When enantiomers of tyramine were produced by replacing one of the α-methylene hydrogens by deuterium, stereospecificity was again seen (BELLEAU and BURBA 1960; BELLEAU et al. 1960) and this and subsequent work has shown that MAO specifically removes the pro-R hydrogen from the α-methylene group (see also STAUNTON and SUMMERS 1978).

The purified enzyme appears to be specific for oxygen as the hydrogen acceptor. MAO from ox liver has been found incapable of catalysing the substrate-dependent reduction of 2,6-dichlorophenolindophenol, cytochrome C or tetrazolium salts (YASUNOBU et al. 1968). Indeed tetrazolium derivatives have been found to be inhibitors of MAO from rat liver (LAGNADO et al. 1971; YOUDIM and LAGNADO 1972) and these compounds and phenazine methosulfate were also found to inhibit the enzyme from rabbit liver and brain (VON KORFF 1977). These results appear to be in conflict with the common use of tetrazolium salts to stain for MAO activity in histological studies (see e.g. GLENNER et al. 1960; BOADLE and BLOOM 1969) and after electrophoretic separation (see e.g. COLLINS et al. 1968; KIM and D'IORIO 1968). The reduction of tetrazolium salts by MAO in the presence of tryptamine has, however, been shown to be due to reaction of the product, indoleacetaldehyde, with the tetrazolium salt rather than from a direct electron transport involving MAO (GLENNER et al. 1960), although further reactions of the aldehyde product could be involved as an additional enzyme may catalyse this process (WEISSBACH et al. 1957; VON KORFF 1977).

Anaerobic reduction of tetrazolium salts by crude preparations of rabbit liver in the presence of tyramine has been reported by BLASCHKO (1952). The most likely explanations for this would be either that MAO can, in fact, use electron acceptors other than oxygen in the presence of additional cellular components or that there is another enzyme present in the cell that can catalyse this reaction. At present there is insufficient evidence to decide between these two alternatives, although GUHA and his co-workers have described a monoamine dehydrogenase system that is able to catalyse the reduction of tetrazolium derivatives in the presence of some MAO substrates under anaerobic conditions. The properties of this system appear to be rather different from those of MAO (GUHA and GOSH 1970; GHOSH and GUHA 1978).

H. The Influence of Membrane Environment

As discussed in Sects. C and D, MAO is tightly bound to the mitochondrial outer membrane. The properties of the enzyme are, to some extent, affected by its membrane environment. The activity of MAO has been shown to be affected by temperature-induced phase-transitions of the membrane lipids (HUANG and EIDUSON 1977; SAWYER and GREENAWALT 1979; HUANG 1980; KWATRA and SOURKES 1981). Removal of lipid material has been shown to decrease the thermal stabilities of MAO from pig liver (ORELAND and EKSTEDT 1972) and rat liver (HOUSLAY and TIPTON 1973b; BAKER and HEMSWORTH 1978; YU 1979).

As might be expected for a membrane-bound enzyme, the activity of MAO has been shown to be affected by a number of lipophilic compounds including local anaesthetics (see eg FOWLER et al. 1980a), naturally-occurring lipids (HOUSLAY 1978; YU and BOULTON 1979) and detergents (WHITE and GLASSMAN 1977; KANDASWAMI and D'IORIO 1978; FOWLER et al. 1980a; ACHEE and GABAY 1981). The observation that extraction of MAO from its membrane environment affects its sensitivity to some inhibitors (see eg HOUSLAY et al. 1976; SCHURR et al. 1978; EIDUSON and BUCKMAN 1979) may be explained by their lipophilicity (HOUSLAY 1977). WOJTCZAK and NALECZ (1979) have, however, shown rat liver MAO to be activated by anionic detergents but inhibited by cationic detergents. Since both these effects were reversed by surfactants, these workers suggested that the surface charge on the membranes is important for MAO activity. The activity of MAO can be directly affected by dietary factors that affect the lipid composition of its environment (DEMISCH et al. 1979; KANDASWAMI and D'IORIO 1979). Further work will, however, be necessary to determine whether such effects could be physiologically significant.

J. Multiple Forms

I. Electrophoretic Studies

Solubilized preparations of MAO from a number of sources can be separated into five bands of activity by electrophoresis on polyacrylamide gels (COLLINS et al. 1968, 1970; SHIH and EIDUSON 1969; YOUDIM et al. 1969; SIERENS and D'IORIO 1970; YOUDIM 1972) or cellulose acetate strips (KIM and D'IORIO 1968) and a great deal of work has been done on the relative activities, inhibitor sensitivities and tissue distributions of these forms. The electrophoretically-separated forms of MAO from rat liver (TIPTON 1972), ox adrenal medulla (TIPTON et al. 1972) and cat brain (YOUDIM and HOLMAN 1975) were found to have widely different phospholipid contents and electrophoresis of the rat liver preparation in the presence of a non-ionic detergent prevented the separation of multiple bands (TIPTON 1972). Treatment of soluble preparations of rat liver MAO with agents that disrupt hydrophobic lipid-protein interactions also resulted in the appearance of a single band of activity following electrophoresis (HOUSLAY and TIPTON 1973b). Thus it appears that these electrophoretically-separable forms of the enzyme are artefacts of the procedures used rather than genuine isoenzymes and this conclusion is further supported by the observations that their presence appears to depend on the method used to render the enzyme soluble (JAIN and SANDS 1974; KANDASWAMI et al. 1977).

II. Selective Inhibitors

JOHNSTON (1968) showed that the sensitivity of rat brain MAO to irreversible inhibition by clorgyline (see Sect. L) is dependent on the substrate used to de-

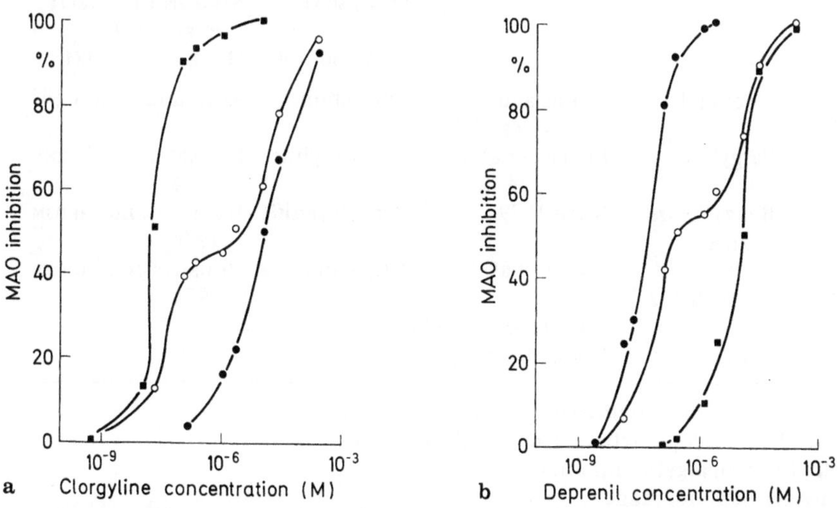

Fig. 1. Clorgyline **(a)** and deprenyl **(b)** inhibition of the oxidation of tyramine (○), 5-HT (■) and benzylamine (●) catalysed by rat liver monoamine oxidase (Taken from FOWLER et al. 1978)

termine its activity. With 5-HT as the substrate, considerably lower concentrations were required to cause inhibition than was the case with benzylamine as the substrate. When the activity was determined using tyramine, about half the activity was inhibited at the concentrations necessary to inhibit 5-HT oxidation, whereas inhibition of the remaining enzyme activity required concentrations of clorgyline similar to those required to inhibit benzylamine oxidation (see Fig. 1a). JOHNSTON interpreted these results as indicating that there are two forms of MAO: one, which he termed the A-form, is sensitive to inhibition by clorgyline and oxidatively deaminates 5-HT; and the other, the B-form, is relatively insensitive to clorgyline and catalyzes the oxidation of benzylamine. He also proposed that tyramine is a substrate for both forms of the enzyme. The irreversible inhibitor deprenyl (see Sect. L) is a more potent inhibitor of the B-form of the enzyme (KNOLL and MAGYAR 1972), giving rise to inhibition curves of the type shown in Fig. 1b.

Table 3. Some selective inhibitors of MAO

	Irreversible	Reference	Reversible	Reference
Selective for A-form	Clorgyline	JOHNSTON 1968	Amphetamine	MANTLE et al. 1976
	Lilly 51641[a]	FULLER 1968	Harmine	GORKIN and TAT'YENENKO 1967
	PCO[b]	MANTLE et al. 1975b	Harmaline	FULLER 1972
			Mexiletine	CALLINGHAM 1977
			α-Methyltryptamine	MANTLE et al. 1976
			Tryptolines	YOUDIM and OPPENHEIM 1980
			MD 780515[f]	STROLIN BENEDETTI et al. 1979.
			FLA 336(+)[g]	OGREN et al. 1981
Selective for B-form	Deprenyl	KNOLL and MAGYAR 1972	Imipramine	ROTH and GILLIS 1974
	Pargyline	FULLER et al. 1970	Amitriptyline	EDWARDS and BURNS 1974
	Benzylhydrazine	ROTH 1979	Benzylcyanide	HOUSLAY and TIPTON 1974
	J 508 (AGN 1133)[c]	KNOLL et al. 1978	4-Cyanophenol	HOUSLAY and TIPTON 1974
	AGN 1135[d]	FINBERG et al. 1980		
	U 1424[e]	KNOLL 1978		

[a] N-2-[O-(chlorophenoxyl)ethyl]ethyl]cyclopropylamine
[b] 5-phenyl-3-(N-cyclopropyl)ethylamine-1,2,4-oxadiazole
[c] N-methyl-N-propargyl-1-aminoindane
[d] N-propargyl-1-aminoindane
[e] N-methyl-N-propargyl-(2-furyl-1-methyl)-ethyl-ammonium. HCl.
[f] 3-[4-(3-cyanophenylmethoxy)phenyl-5-(methoxymethyl)-2-oxazolidinone
[g] (+)-4-dimethylamino-2-methyl-α-methyl-phenethylamine

Several other irreversible inhibitors exhibit a greater potency towards one of the enzyme forms (see Table 3). Kinetic studies also indicate that a number of reversible inhibitors show different Ki values for the two forms and some of these are listed in Table 3.

Selective inhibitors have been used to determine the relative proportions of the two forms of MAO in different tissues and species (see e.g. HALL et al. 1969; SQUIRES 1972; TIPTON et al. 1976). The proportions of the two forms have been estimated from the plateau values of the inhibition curves when tyramine is used as the substrate (see Fig. 1). The relative proportions of the two activities in different tissues have been found to vary considerably (Table 4). The values determined in this way are only approximate since they depend on the assumption that the K_m values of the two forms for tyramine are similar (see Sect. J.III). Ox liver (HALL et al. 1969; SALACH et al. 1979; TIPTON and DELLA CORTE 1979), pig liver and brain (HALL et al. 1969; TIPTON

Table 4. Proportion of the A-form of monoamine oxidase in adult rat and human tissues.

Tissue	A-form activity (Percent of Total A + B)	
	Rat	Human
Kidney	70	57
Liver	40 (55)	54 (73) 45*
Spleen	>95	34
Intestine	70	—
Lung	50	92 55*
Testis	90	—
Vas Deferens	50	—
Brain	55 (74)	38 (53) 25*
Superior Cervical Ganglion	90	—
Pineal Gland	15	—
Heart	(98)	43
Aorta	—	48
Adrenal	—	60
Tongue	—	62
Oesophagus	—	72
Stomach	—	50
Ileum	(66)	47 (89) 75*
Colon	—	53
Pancreas	—	90
Skeletal Muscle	—	28
Diaphragm	—	39
Skin	—	31

Data taken from LEWINSOHN et al. (1980b), TIPTON et al. (1976). The values in brackets are from WHITE and TANSIK (1979), and those indicated by * from RIEDERER et al. 1981.

1971; SQUIRES 1972; but cf. EKSTEDT and ORELAND 1976), human blood platelet (MURPHY and DONELLY 1974; EDWARDS and CHANG 1975) and rabbit liver (HALL et al. 1969) contain essentially only the B-form of the enzyme. In contrast, human placenta has been shown to contain only the A-form of the enzyme (POWELL and CRAIG 1977; SALACH and DETMER 1979).

The existence of a third form of MAO, type C, has been postulated in rat brain (WILLIAMS et al. 1975a, b). This was based on the detection of a form of MAO that was insensitive to inhibition by clorgyline in histochemical studies in which 5-HT was used as the substrate. Subsequent careful studies have, however, shown that these results were an artefact of the methods used and that, like other tissues, rat brain contains only the A- and B-forms of the enzyme (TOYOSHIMA et al. 1979).

III. Substrate Specificities

Selective inhibitors have been used to determine the substrate specificities of the two forms in a number of tissues (see Table 5 for rat liver). The two forms from rat brain and human liver and brain appear to have generally similar specificities, but there are some differences with catecholamines as substrates. GLOVER et al. (1977) reported that dopamine was a substrate only for the B-form of the enzyme in human brain but other workers have found it to be deaminated by both forms (WHITE and GLASSMAN 1977; ROTH and FEOR 1978; O'CARROLL et al. 1983). Noradrenaline appears to be a substrate for both forms of the enzyme from human brain (see WHITE and GLASSMAN 1977 and Table 6) but only for the A-form of MAO from rat brain (GORIDIS and NEFF 1971) and rat liver (Table 5).

Comparison of published data on the specificities of the two forms in different tissues and animal species has revealed a number of departures from the simple classification shown in Table 5 (see FOWLER et al. 1978). These differences have been found to be due, at least in part, to the lack of absolute

Table 5. Substrate specificities of the two forms of rat liver monoamine oxidase

Substrates for the A-form	Substrates for both forms	Substrates for the B-form
Adrenaline	Dopamine	Benzylamine
Noradrenaline	Tyramine	Phenethylamine
3-O-Methyladrenaline	3-Methoxytyramine	Phenylethanolamine
3-O-Methylnoradrenaline	p-Synephrine	*o*-Tyramine
		1,4-Methylhistamine
5-HT		N-*tele*-Methylhistamine

Data from: HOUSLAY and TIPTON (1974): SUZUKI et al. (1979b); WALDMEIER et al. (1977); WILLIAMS (1977).
As discussed in the text there is some controversy about some of the assignments and the specificities of the two forms is dependent on substrate concentration in some cases.

Table 6. Kinetic parameters of the two forms of MAO

Substrate	Source	MAO-A $K_M(\mu M)$	MAO-A $V_{max}(\%)$	MAO-B $K_m(\mu M)$	MAO-B $V_{max}(\%)$
5-HT	rat liver	160	90	2000	10
5-HT	rat brain	180	89	1200	11
Tyramine	rat liver	110	34	580	66
2-Phenylethylamine	rat liver	280	13	20	87
Dopamine	human cerebral cortex	212	49	220	51
Noradrenaline	human cerebral cortex	284	66	238	34

The activity of each form of MAO was determined at 37°C and pH 7.2 after inhibition of one of the forms by treatment with clorgyline or (−)-deprenyl. Data from FOWLER and TIPTON (1981; 1982); TIPTON et al. (1982); O'CARROLL et al. (1983).

specificities of the two forms. Thus 2-phenylethylamine is a substrate for MAO-A although the K_m value is much higher and the maximal velocity is much lower than is the case with MAO-B (Table 6). Earlier work indicated MAO-B to be inhibited by high concentrations of 2-phenylethylamine such that only the activity of MAO-A could be detected at high concentrations of this substrate (KINEMUCHI et al. 1979, 1980; SUZUKI et al. 1979a, b). However this apparent high substrate inhibition has been shown to be due to failure to measure the true initial rate of the reaction since high concentrations of 2-phenylethylamine have been shown to inhibit the MAO-B in a time-dependent reaction (KINEMUCHI et al. 1982). When initial rate measurements were made it was found that at saturating 2-phenylethylamine concentrations the A-form of rat liver MAO would contribute about 13 % of the total activity towards this substrate (TIPTON et al. 1982).

Similar studies with other substrates have given the results shown in Table 6. These data indicate that 5-HT can only be regarded as a specific substrate for MAO-A at relatively low substrate concentrations. Benzylamine, however, appears to be a relatively specific substrate for MAO-B with only a small percentage, if any, of the activity being due to the A-form of the enzyme in rat liver even at very high substrate concentrations (FOWLER and ORELAND 1980; TIPTON and MANTLE 1981). The results of studies on the competition between benzylamine, tyramine and dopamine for oxidation by rat liver MAO suggest that benzylamine is able to bind to both forms of the enzyme although only the B-form possesses significant catalytic activity towards this substrate (HOUSLAY et al. 1974; HOUSLAY and TIPTON 1975c).

As shown in Table 6 the two forms of rat liver MAO have somewhat different K_m values towards tyramine. This indicates that the relative contributions of the two forms to the oxidation of this substrate will depend upon its concentration. Thus FOWLER and TIPTON (1981) have calculated that, with rat liver mitochondria, only 32 % of the total activity towards 50 μM tyramine will be due to MAO-B, but that this proportion will rise to 57 % at a tyramine concentration of 1 mM. These considerations indicate that attempts to estimate the relative proportions of the two forms present in a tissue preparation from the plateau values of clorgyline or deprenyl inhibition curves when tyramine is used as the substrate (see Sect. J.II) will depend on the substrate concentration used. This probably accounts for the differences in published values of proportions of the two forms in any given tissue (see e.g. Table 4).

Recently FOWLER et al. (1980b) and FOWLER and ORELAND (1980) have attempted to determine the molar concentrations of MAO-A and MAO-B present in tissue preparations by determining the concentrations of selective irreversible inhibitors necessary to cause complete inhibition of each of the forms after long incubation times under conditions that minimise any nonspecific binding of the inhibitors (see e.g. FOWLER and CALLINGHAM 1978, 1980). The results of these studies indicated that in human brain and rat liver both forms of the enzyme had similar turnover numbers with tyramine as the substrate. In contrast EGASIRA et al. (1976) and BROWN et al. (1982) have produced evidence to indicate that the turnover number of the A-form in these tissues is considerably greater than that of the B-form.

A number of investigators have used the relative activites towards different substrates as a method for estimating the proportions of the two forms of the enzyme in different tissues (se e.g. EDWARDS and MALSBURY 1978). The factors dicussed above show that great care must be taken in interpreting the activity towards 2-phenylethylamine as being due only to the B-form of the enzyme, and to a lesser extent, that towards 5-HT being only a measure of the A-form activity, unless account is taken of concentration-dependent effects. In particular such effects could lead to considerable errors in studies of tissues where one of the forms predominates.

As discussed in Sect L, the acetylenic inhibitors, including clorgyline, deprenyl and pargyline, act as suicide (K_{cat}) inhibitors of MAO. The enzyme first forms a non-covalent complex with the inhibitor and subsequent reactions within this complex result in covalent bond formation giving rise to irreversible inhibition. In the case of pargyline the selectivity towards MAO-B appears to be due solely to that form of the enzyme having a higher affinity for non-covalent binding of the inhibitor with the subsequent reactions within the non-covalent complexes occurring at similar rates with both forms of the enzyme. Clorgyline, however, has a higher affinity for non-covalent binding to MAO-A and this selectivity is further enhanced by a faster reaction within this complex. Similary the MAO-B selectivity of (−)-deprenyl is due to a greater affinity for non-covalent binding to that form and a faster rate of reaction within that complex than that formed with MAO-A (FOWLER et al. 1980b). HOUSLAY (1977) has shown that clorgyline will partition much more readily into non-polar solvents than will deprenyl (see also WILLIAMS and LAWSON 1975). This might suggest that the different affinities of the two forms for these inhibitors might be due to differences in the hydrophobicity in their binding regions.

IV. Other Evidence for Multiple Forms

Several workers have reported that the optimum pH for MAO depended on the substrate used to assay the enzyme (see HOUSLAY et al. 1976; ACHEE et al. 1977 for reviews) and have interpreted such results as indicating the presence of more than one enzyme having different pH optima. Such arguments are invalid since they take no account of the different p*K* values of the amines which are probably only substrates for the enzyme in the unprotonated form (see Sect. F). Thus, as the pH is varied the concentration of the active form of the substrate, and hence the degree of saturation of the enzyme, will change and this will be different for different amine substrates. When allowances have been made for the different p*K* values of the amines the responses of the ox brain enzyme to pH have been shown to be similar for a number of substrates that showed widely different behaviour if substrate ionization was not taken into account (GABAY et al. 1976).

Differences in the thermal stabilities of the activities of MAO preparations towards different substrates have been found in many studies and have often been interpreted in terms of multiple forms of the enzyme. The relative stabilities of the two forms appear to vary between different species and organs (compare e.g. SQUIRES 1972; HOUSLAY and TIPTON 1973b; GABAY et al. 1976; FUENTES and NEFF 1977; WHITE and GLASSMAN 1977) and there is also evidence that the B-form of the enzyme may be heterogenous by this criterion (SQUIRES 1972; LYLES and GREENAWALT 1978). These differences may be due to different effects of the membrane environment on the stability of the enzyme (see Sect. H).

Antibodies have been prepared to purified preparations of MAO in order to investigate the relationship between the two forms. Studies with antibodies prepared to the purified enzyme from ox liver have indicated that the enzyme in crude preparations from this organ, which has been shown to contain only the B-form of the enzyme (see Sect. K.II), is homogenous by this criterion (HARTMAN et al. 1971). At a concentration sufficient to precipitate all the enzyme from ox liver this antibody was found to precipitate only about 80 percent of the activity in ox brain (HARTMAN 1972; HARTMAN and UDENFRIEND 1972). MCCAULEY and RACKER (1973) claimed that this immunoprecipitable fraction from ox brain corresponded to the A-form of the enzyme whereas the remainder was the B-form. It is difficult to justify this conclusion, however, since they found both these forms to be active towards benzylamine and HARTMAN (1972) had shown them both to be active against 5-HT as well. In addition, studies on the inhibition of the ox brain enzyme by clorgyline, using tyramine as the substrate, have shown that there are approximately equal amounts of the two forms of the enzyme in that organ (TIPTON and MANTLE 1977). A possible explanation of these results is that the portion of the enzyme that was resistant to precipitation was in a different membrane environment from the remainder.

MAYER and his coworkers (DENNICK and MAYER 1977; RUSSELL et al. 1979b) prepared antibodies to the purified B-form of the enzyme from rat and human liver mitochondria and, under conditions designed to minimize interference from binding to membrane lipids (see also RUSSELL et al. 1979c), found the A- and B-forms of the enzyme from several sources to be immunologically identical. Using antibodies raised against MAO-A from human placenta and MAO-B from human platelets, however, CRAIG and his co-workers (BROWN et al. 1982; CRAIG et al. 1982) found that, although there was extensive cross-reaction, the placental enzyme appeared to possess antigenic determinants that were not exhibited by the enzyme from platelets. These placental MAO antigenic determinants were not, however, found in either forms of MAO from human brain or liver. These results might suggest that the specific determinants of the A-form from human placenta were not shared by the A-form in other human tissues. BROWN et al. (1982) however argued that the specific activity of MAO-A in human brain and liver was much greater than that of MAO-B and thus the molar concentration of MAO-A in those tissues was too low to allow its antigenic differences to be detected. This interpretation would, however, be at variance with the results obtained from the titration of MAO activity from human brain with irreversible inhibitors which suggested rather similar proportions of the two forms to be present (FOWLER et al. 1980b).

DENNEY et al. (1982a, b) have prepared a monoclonal antibody towards pargyline-inhibited MAO-B from human platelets. They showed this antibody to react with MAO-B, but not MAO-A, from human liver. When an immunoaffinity column was prepared using this antibody it was found to adsorb the MAO-B activity from soluble preparations from human liver without affecting the MAO-A activity. This important study represents the first clear separation of the two forms from a tissue containing them both. Unfortunately the conditions necessary to elute the bound B-form resulted in loss of activity, but a single polypeptide with a relative molecular mass of 59000 (see Sect. D.I and J.V), as determined by polyacrylamide gel electrophoresis in the presence of sodium dodecyl sulphate, was obtained in this way.

The antibodies prepared in the studies of BROWN et al. (1982), CRAIG et al. (1982) and DENNEY et al. (1982a, b) were not directed against the active-site of the enzyme and, in fact, also reacted with the inactive pargyline-labelled enzyme. Those prepared in the earlier studies of MAYER and his co-workers were partially inhibitory and thus presumably reacted close to the active site of the enzyme. The failure of these antibodies to reveal any difference between the two forms would be consistent with their active site regions being similar.

Genetic analysis has shown that the structural gene locus for the human MAO-A flavin-containing polypetide is on the X-chromosome (PINTAR et al. 1981a). Experiments with somatic cell hybrids between rat hepatoma cells possessing both MAO-A and -B activities and mouse neuroblastoma lacking both these activities gave results that were consistent with MAO-B also being encoded by the X-chromosome (PINTAR et al. 1981b). NAGATSU et al. (1982) also carried out hybridization studies, using neuroblastoma cells, and obtained evidence suggesting that the two forms of the enzyme might be coded by different gene loci.

A number of workers have shown that the mitochondria populations of some organs are heterogeneous and can be separated into fractions of different sizes by density-gradient centrifugation. Studies on the MAO activities of these fractions indicate that mitochondria with a relatively higher activity towards 5-HT can be separated from those that are more active towards 2-phenylethylamine (KROON and VELDSTRA 1972; NEFF et al. 1974; YOUDIM 1974; BOURNE et al. 1975; OWEN et al. 1977). In rat brain OWEN et al. (1977) found a higher proportion of the A-form in mitochondria derived from the synaptosomal fraction, in agreement with the evidence that this form of the enzyme may be enriched in nerve endings (see Sect. I.VI). NELSON et al. (1979), in contrast, could not find evidence for any significant enrichment of A-form activity in mitochondria from the synaptosomal fractions of rat and ox brains. These workers also determined the amounts of the A-form of MAO in mitochondria from these cources by measuring the binding of radioactively labelled harmaline and confirmed that the mitochondria derived from the synaptosomal fractions did not appear to be enriched in this form of the enzyme. Further studies on the distributions of the two forms will be discussed in Sect. I.

V. The Nature of the Two Forms

There has been considerable speculation about the nature of the two forms of MAO (see NEFF and YANG 1974; HOUSLAY et al. 1976; ACHEE et al. 1977; JAIN 1977; MANTLE and TIPTON 1982; SCHURR 1982 for reviews). It has been suggested that they may be two separate proteins (MCCAYLEY and RACKER 1973), conformational isomers of a single enzyme (YOUDIM and COLLINS 1971), a single enzyme that may possess two active sites with different specificities (MANTLE et al. 1975a, b; WHITE and WU 1975) or a single enzyme that is modified in different ways by its membrane environment (HOUSLAY and TIPTON 1973b). Despite a great deal of work, convincing proof that the two forms arise in any one of these ways is still lacking and it may be that a combination of factors is involved.

The suggestion that the two forms of the enzyme may represent a single species that is modified to different extents by its membrane lipid environment receives some support from observations that the stability, acitivity, substrate specificity and inhibitor sensitivity of MAO are affected by the lipid environment of the enzyme (see Sect. H), although the vigour of the procedures used or the presence of detergents could themselves exert a direct effect on the activities of the enzyme (TIPTON et al. 1976). The observations that the differential inhibitor sensitivities of the two forms could be abolished by treatment with chaotropic agents that weaken hydrophobic lipid-protein interactions (HOUSLAY and TIPTON 1973b; TIPTON et al. 1973) or by lipid substitution with dimyristoyl phosphatidyl choline (HOUSLAY 1980) and that the inhibitor sensitivities of the electrophoretically separated acitivities of cat brain MAO varied with their phospholipid contents (YOUDIM and HOLMAN 1975) indicate that the sensitivities of the two forms to these inhibitors are, at least to some extent, dependent on their membrane environments. This could, however, be a reflection of the different lipid-solubilities of clorgyline and deprenyl (see Sect. I.III) rather than indicating identity of the two forms.

The results of several studies have shown the activity and inhibitor sensitivity of MAO-A to be generally more sensitive to lipid extraction and manipulation than those of MAO-B (for review see MANTLE and TIPTON 1982). Studies on the interaction of rat brain MAO with electron-spin resonance probes (HUANG and EIDUSON 1977; HUANG 1980) and the effects of phospholipase digestion and the reconstitution of the enzyme into defined lipid vesicles (HUANG and FAULKNER 1980; HUANG 1981) have led HUANG (1981) to conclude that the active site of MAO-A is buried in the hydrophobic core of the membrane whereas that of MAO-B is closer to, or partly in, the hydrophilic exterior. Immunological studies have shown the A form of MAO to be located on the inner side of the mitochondrial outer-membrane and the B-form to be located on the outer side (RUSSELL et al. 1979a, b, c). This would imply that the two forms will be in different lipid environments *in situ*. It should also be remembered that it is not known whether the two forms can coexist in the same mitochondrion.

The suggestion that the two forms of MAO may be due to two different sites on a single enzyme molecule originally arose from the results of kinetic experiments on the competition between substrates, but subsequent analysis has shown that these results would also be consistent with the involvement of two separate enzyme forms (see MANTLE and TIPTON 1982, for review). SEVERINA (1973, 1979) suggested that MAO might contain a single active site but that differences in substrate structures would affect the ways in which they could orient in this site, giving rise to an apparent duality of forms. Such a model has been supported by ZELLER et al. (1979) on the basis of structure-activity studies. However it is difficult to reconcile this model with the observation that the two forms are on different sides of the mitochondrial outer membrane.

Many of the properties of the two forms of MAO are similar. In rat liver the two forms turn over at similar rates *in vivo* (DELLA CORTE and TIPTON 1980). WHITE and GLASSMAN (1977) showed the two forms in detergent-soluble preparations from human brain and rat liver to have indistinguishable isoelectric points and relative molecular masses and the latter observation has also been made by others (e.g. HOUSLAY and TIPTON 1973b; YOUDIM and COLLINS 1971; COSTA and BREAKEFIELD 1980). Using the technique of irradiation-inactivation (see Sect. D.I) TIPTON and MANTLE (1977) could find no significant difference between the relative molecular masses of the two forms in rat liver mitochondria, but CALLINGHAM and PARKINSON (1979) reported a M_R value of 143 000 for MAO-A and 202 500 for MAO-B from the same source. The FAD cofactor is bound to both forms of the enzyme in the same way and the amino-acid sequences around the binding regions are similar (SALACH et al. 1979; SALACH and DETMER 1979; NAGY and SALACH 1981). The same flavin adduct is also formed on reaction of the two forms with acetylenic inhibitors (SALACH et al. 1979). COSTA and BREAKEFIELD (1980) have, however, shown that the adduct formed by reaction of MAO-A with pargyline is somewhat less stable than that formed with MAO-B.

The relative molecular masses of the FAD-containing subunits have been determined by SDS-PAGE (see Sect. D.I) after reaction with radioactively-labelled pargyline. By selectively protecting one of the two forms of MAO by prior reaction with clorgyline or deprenyl or by using sources enriched in one of them, it has been possible to compare their M_R values. Several workers have reported that there is no difference between the sizes of the subunits derived from the A- and B-forms of MAO (COLLINS and YOUDIM 1975; EDWARDS and PAK 1979; PINTAR et al. 1979; COSTA and BREAKEFIELD 1980). CALLINGHAM and PARKINSON (1979), however, have reported that, for MAO from rat liver, the A-form had a subunit relative molecular mass of 60000 whereas that of the B-form was 55000. Such a difference was also observed by CAWTHON and BREAKEFIELD (1979) but this group later argued that this was an artefact of the solubilization procedure and that the FAD-containing subunits of both forms had the same M_R of 57000 ± 3000 (COSTA and BREAKEFIELD 1980). They also showed that these subunits from MAO-A and -B had identical isoelectric points. BROWN et al. (1981) have reported M_R values of 63000 and 67000 in the FAD-containing subunits of human platelet MAO-B and human placental MAO-A, respectively, and CAWTHON et al. (1981) have determined values of 60000 and 63000, respectively, for the MAO-B and -A subunits from several human sources. In contrast STADT et al. (1982) found the flavin-containing subunits of the two forms of purified rat liver MAO both to have M_R values of 60000. Values reported for the subunit sizes of the B-form of the enzyme purified from ox liver are 61000 (SALACH 1979) and 52000 (MINAMIURA and YASUNOBU 1978). This difference is rather large to be explained on the experimental error likely to be encountered in this type of experiment, and it may be that some form of degradation could have taken place during the purification procedures.

The immunological evidence that the two forms of MAO may not be identical (see Sect. I.IV) is supported by the results of studies on the limited proteolytic digestion of the separated pargyline-labelled subunits. Differences between the peptides released from these subunits of MAO-A and -B (CAWTHON and BREAKEFIELD 1979; PINTAR et al. 1979) suggest that the two forms differ in their amino-acid sequences or in their susceptibility to proteolysis. The possibility that the difference results from a posttranslational modification of one of the forms cannot be excluded. Pig liver MAO has been reported to be glycosylated (ORELAND 1971) and HOUSLAY and MARCHMONT (1980) have shown that treatment of rat liver mitochondrial outer membranes with neuroaminidase results in selective inhibition of MAO-A. Thus it is possible that the two forms may differ in their contents of sialic acid residues or in their association with glycolipid.

The failure of many attempts to separate the two forms of MAO from a single organ such as rat liver (see WHITE and GLASSMAN 1977; STADT et al. 1982) has led several workers to attempt comparisons between the properties of the enzyme forms purified from different sources that contain only one of them. Most published studies on the properties of the purified enzyme have been with the B-form but, more recently, human placenta has been used as the source of the A-form by SALACH and DETMER (1979) and ZELLER et al. (1979). These groups have obtained preparations that are apparently homogeneous, although probably not completely lipid-free, that still retain the characteristics of MAO-A. Comparisons of the properties of such preparations with those of MAO-B purified from other sources may be complicated by species and organ differences but they should nevertheless provide valuable information on the differences between the two forms. The use of immobilized antibodies to separate the two forms (DENNEY et al. 1982a, b) and the recently reported affinity chromatographic procedure for separating MAO-A and MAO-B from ox heart (NAOI and YAGI 1980) should, however, allow a direct evaluation of the differences between them.

VI. Evidence for an Association of Type A Activity with Neurones

The presence of MAO in extraneuronal as well as neuronal tissue has been demonstrated by histochemical techniques, by its occurrence in non-innervated tissue, e.g. platelets, and non-neuronal cells in tissue culture (HAWKINS and BREAKEFIELD 1978; EDELSTEIN et al. 1978; HAWKINS et al. 1979), and by denervation experiments (see Sect. C).

Substrate specificity studies showed that MAO associated with sympathetic ganglia was predominatly type A, exhibiting highest affinity towards neurotransmitters such as noradrenaline and 5-HT (GORIDIS and NEFF 1971). Similarly, denervation experiments in peripheral organs, e.g. pineal gland (NEFF and GORIDIS 1972), vas deferens (JARROTT 1971), nictitating membrane (JARROTT and LANGER 1971) and arteries (COQUIL et al. 1973) showed that the enzyme activity disappearing after denervation was mainly the A type.

In certain tissues of the pig (SQUIRES 1972) nearly all MAO activity may be type B. However, the proportion of the total MAO activity in neuronal tissue (which may contain the type A enzyme) may be too small in relation to the total MAO activity to permit differentiation.

When the MAO activity in various cell types in culture has been examined, cells of both neural (e.g. neuroblastoma) and non-neural (e.g. rodent melanoma, sarcoma, human skin fibroblasts) origin were found to contain predominantly A-type enzyme (DONNELLY et al. 1976; HAWKINS and BREAKEFIELD 1978; EDELSTEIN et al. 1978). Hepatoma cells (HAWKINS and BREAKEFIELD 1978) as well as isolated rat liver parenchymal cells (TIPTON et al. 1976) contained both A and B types whereas HeLa cells (HAWKINS et al. 1979) contained mainly B type enzyme.

In brain tissue, mitochondrial MAO localised in the synaptosomal fraction also shows a relative enrichment of the A type enzyme (BOURNE et al. 1975; STUDENT and EDWARDS 1977—see Sect. I.IV). Rat glioma cells in culture, however, contained exclusively A-type activity (MURPHY et al. 1976), although these cells may differ in developmental characteristics from mature glia, as well as possibly representing a different cell type (HAWKINS and BREAKEFIELD 1978). Similarly, MAO-B is absent in the neonatal rat brain (MANTLE et al. 1976) but develops in the first month after birth. Hemitransection of rat brains led to an increase in the proportion of B-type activity in the operated side, which may be related to an increased glial cell content (ORELAND et al. 1980). Certain nuclei of the rat hypothalamus have a particularly high content of MAO type B (KUNIMOTO et al. 1979). The relation of this finding to a localisation in glial cells or neurones is unknown. Lesion of the nigrostriatal pathway with 6-hydroxydopamine led to a substantial fall in "dopamine-oxidase" activity of rat striatum (AGID et al. 1973), which was shown to be due to a selective reduction of MAO-A (DEMAREST and AZZARO 1979).

The experimental data available at present are suggestive of a predominant occurrence of type A MAO in neurones. However, proof of such selective localisation awaits development of histochemical techniques capable of differentiating the enzyme subtypes.

K. Multiple Forms as an *In Vivo* Reality and Their Function

I. Evidence from Animal Studies

The evidence presented in foregoing sections points to the presence of at least two forms of MAO in mammalian tissue whose activity *in vitro* can be inhibited separately with drugs. The thousand-fold difference in the concentration of clorgyline required to inactivate A and B forms of MAO *in vitro* suggested that it might be possible to inhibit these enzymes selectively *in vivo*. Assay of MAO activity in homogenates or mitochondrial fractions from different tissues of cats, rats and mice pretreated with varying doses of either clorgyline or (−)-deprenyl has shown that substrate-selective inhibition can, indeed, be produced *in vivo* (JOHNSTON 1968, JARROTT and IVERSEN 1971; YOUDIM et al. 1971; NEFF and GORIDIS 1972; NEFF et al. 1973, 1974; YANG and NEFF 1973, 1974; YOUDIM and HOLMAN 1975; GREEN and YOUDIM 1975; SHARMAN 1976; DELLA CORTE and TIPTON 1980; WALDMEIER et al. 1981).

The interest for the pharmacologist is in the observation that the putative neurotransmitters adrenaline, noradrenaline and 5-HT are deaminated by type A MAO and that dopamine is deaminated by both enzyme forms. The sympathomimetic amines tyramine and 2-phenylethylamine are also present in the brain (YOUDIM 1977; PHILIPS 1978; SABELLI et al. 1978; BOULTON 1979). The former is a substrate for MAO-A and B whereas the latter is only oxidised by the B-form (see Sect. J.III). Therefore, by administering MAO inhibitors which preferentially inactivate one enzyme form, the deamination and function of these amines may be selectively modified. NEFF and his colleaugues (1973, 1974) were the first to show that, following administration of clorgyline to rats, the brain concentrations of noradrenaline, 5-HT and dopamine are increased, but that of 2-phenylethylamine remained constant. On the other hand, deprenyl administration resulted in an increase in dopamine and 2-phenylethylamine levels in brain without an apparent change in 5-HT or noradrenaline. Dopamine apparently is deaminated by both enzymes in rat brain (NEFF et al. 1973, 1974; YANG and NEFF 1973, 1974; GREEN et al. 1977). Other reports have largely confirmed these results (GREEN and YOUDIM 1975; MAITRE et al. 1976; SHARMAN 1976; GREEN et al. 1977; WALDMEIER et al. 1981) and further shown the selectivity of enzyme inhibition and amine deamination *in vivo* to be less pronounced than *in vitro*. Thus, the apparent increases in rat brain 5-HT, noradrenaline, or 2-phenylethylamine after administration of either clorgyline or (−)-deprenyl are not as great as those produced when a non-selective inhibitor (e.g. tranylcypromine) or combined clorgyline plus (−)-deprenyl is used. These results have led Green and Youdim (1975) and Green et al. (1977) to conclude that while *in vivo* 5-HT or noradrenaline is deaminated by the A form of the enzyme, when this enzyme form is selectively inhibited by clorgyline, brain 5-HT or noradrenaline concentration rises and the amine becomes deaminated by MAO type B. Analogous results are seen with brain 2-phenylethylamine and MAO-B (Braestrup et al. 1975). These observations are consistent with the results of *in vitro* studies discussed in Sect. I.III.

The differential inhibition of amine deamination *in vivo* is also evident from the concentrations of the acidic and alcoholic metabolites of 5-HT, noradrenaline, dopamine, and 2-phenylethylamine in rat brain and urine (Christmas et al. 1972; Yang and Neff 1974; Neff et al. 1974; Sandler et al. 1974; Sharman 1976; Maitre et al. 1976; Waldmeier et al. 1981). Significantly lower amounts of 5-hydroxyindoleacetic acid, MOPEG, DOPAC and HVA were found following clorgyline as compared to deprenyl treatment. In contrast, only DOPAC and HVA concentrations declined after deprenyl (Neff et al. 1974; Braestrup et al. 1975; Sharman 1976; Maitre et al. 1976; Waldmeier et al. 1981).

In vitro and *in vivo* studies (Roth and Gillis 1975; Bakhle and Youdim 1976, 1979; Gillis and Roth 1977; Youdim et al. 1979) have shown that the properties of lung MAO are fully compatible with criteria already established for A and B types of the enzyme. Another assessment of MAO A and B activity *in situ* can be made using isolated perfused organs, where the cellular and subcellular structure and permeability barriers which influence the access of the substrate to the enzyme are retained. Although isolated perfused organ systems are not ideal, since they lack neuronal and hormonal input, they represent much closer systems to the *in vivo* condition (Youdim and Woods 1975). In one important feature, namely the response of MAO to selective inhibitors, the results in isolated perfused lung resemble the *in vitro* experiments, using homogenates or mitochondrial preparations. Thus, clorgyline but not (−)-deprenyl is a potent inhibitor of 5-HT deamination, while deprenyl is a selective inhibitor of 2-phenylethylamine metabolism (Bakhle and Youdim 1976, 1979).

Substrate inhibition and competition between MAO A and B substrates observed *in vitro*, are factors thought to be relevant to regulation *in vivo* of biogenic amine catabolism (Houslay et al. 1974; Houslay and Tipton 1975a, b; Ekstedt 1976; White and Wu 1975; White and Tansik 1979). The results with isolated perfused lung, where 5-HT and 2-phenylethylamine were infused together either at the same concentrations or where the concentration of one amine far exceeded (two hundred-fold) that of the other, do not show competition between these two substrates (see also Sect. I.III). The selectivity of MAO A and B inhibitors for the enzyme forms and the absence of substrate competition were demonstrated in the perfusate either by bioassay (Bakhle and Youdim 1979) or by measuring deaminated metabolites derived from infused ^{14}C-labelled 5-HT and 2-phenylethylamine. Recently White and Tansik

(1979) using human cortex mitochondrial MAO have reported similar data and pointed out that, as long as the initial concentrations of 5-HT and 2-phenylethylamine are close to their apparent K_m values for MAO-A and B respectively, these substrates do not compete with each other.

These results, together with those previously discussed, demonstrate clearly that the functional division into MAO A and B first defined *in vitro* and then *in vivo*, is also exhibited in perfused organs i.e. in organized tissue (Gillis and Roth 1979; Youdim et al. 1979a, 1980) and strengthens the evidence that multiple forms of MAO which can be selectively inactivated with drugs exist *in vivo*.

II. Evidence from Human Studies

The difficulty of human *in vivo* studies has been the availability of human material, since the selective MAO inhibitors clorgyline and (−)-deprenyl are not in regular clinical use. However, the few results that are available largely confirm those observed in animal experiments. MAO activity has been mea-

Table 7. Post-mortem human brain MAO activity in subjects treated with clorgyline, deprenyl and tranylcypromine

	% Inhibition					
	Clorgyline			Tranylcypromine		
	Tryptamine	Dopamine	Kynuramine	Tryptamine	Dopamine	Kynuramine
Cerebral Cortex	35 ± 15	0	25 ± 5	55 ± 17	77 ± 9	39 ± 11
Cerebellum	50 ± 20	5 ± 2	15 ± 10	47 ± 31	47 ± 31	68 ± 15
Centrum Ovalae	20 ± 12	0	22 ± 13	60 ± 19	60 ± 19	77 ± 9
Basal ganglion	25 ± 15	7 ± 8	27 ± 9	30 ± 35	30 ± 35	62 ± 12

	(−)-Deprenyl	
	5-HT	Dopamine
Caudate	66 ± 5	89 ± 2
Putamen	61 ± 6	88 ± 3
G. Pallidus	69 ± 4	93 ± 3
Thalamus	50 ± 7	86 ± 4
Hypothalamus	61 ± 3	90 ± 5
S. Nigra	67 ± 10	94 ± 2

The daily doses of clorgyline, tranylcypromine and (-)-deprenyl were 30 mg, 30 mg, and 10 mg respectively. The results are means of MAO activity as compared to control brains and are from brains of 12 each clorgyline and tranylcypromine treated (Youdim et al. 1972b) or 7 deprenyl treated (Riederer et al. 1978b) subjects. Brains which were obtained 6–12 hours post-mortem were frozen at −70°C until analysis.

sured in brain samples (including the basal ganglia) obtained at autopsy from groups of geriatric subjects, with terminal diseases, who had been treated for their depression with either tranylcypromine, isocarboxazid or clorgyline (Youdim et al. 1972b). While the non-selective inhibitors tranylcypromine and isocarboxazid inhibited (75–100%) MAO activity, clorgyline at equivalent doses (30 mg) was almost without effect (Table 7), except in the pineal body.

Human brain dopamine and 5-HT concentrations measured in the same regions were significantly raised in those subjects treated with tranylcypromine or isocarboxazid but not clorgyline (Bevan Jones et al. 1972). However, 2-phenylethylamine levels were not measured. These data together with the *in vitro* inhibition studies of Squires (1972) and Riederer et al. (1978b) point to the human brain MAO being largely type B. Confirmation has come from autopsy brain samples of Parkinsonian subjects treated with (−)-deprenyl. At doses far below those of clorgyline used in geriatric subjects (Table 7), (−)-deprenyl was shown to selectively inhibit MAO B. The brain dopamine and 2-phenylethylamine concentrations were significantly increased but that of 5-HT remained unchanged (Riederer et al. 1978a, 1981; Reynolds et al. 1978a). Measurements of urinary and CSF dopamine, noradrenaline and 5-HT as well as their deaminated metabolites have indicated that during (−)-deprenyl treatment, only HVA and MOPEG are decreased and not 5-hydroxyindoleacetic acid (Eisler et al. 1979), confirming results obtained in the rat and mouse (Neff et al. 1974; Sharman 1976; Braestrup et al. 1975).

L. Inhibitors

I. Classification and Mechanism of Action

A wide variety of compounds possess MAO inhibitory properties. The most important of these are: (1) competitive inhibitors, bearing structural resemblance to monoamine substrates but resistant to oxidation by the enzyme (e.g. β-carbolines, aryl- or indolealkylamines), and (2) irreversible inhibitors. Examples of the latter class include hydrazines, cyclopropylamines and propargylamines. The latter compounds are site-directed, suicide inactivators in which an initial phase of the reaction involves competitive, reversible binding with the enzyme receptor site (McEwen et al. 1969; Tipton and Mantle 1981), following which the inhibitor is oxidised to a form which binds covalently to the enzyme.

1. Hydrazines

The detailed pharmacology and structure-activity relationships of these inhibitors have been reviewed previously (Pletscher et al. 1960; Zeller 1963b; Biel et al. 1964; Biel 1970; Ho 1972). Oxidation of the hydrazine to an active intermediate was originally suggested by Eberson and Persson (1962). The intermediate compound was found to be an aryl diazene (Patek and

HELLERMAN 1974) which combines with the C4a position of the enzyme FAD (KENNEY et al. 1979). A serious disadvantage of these drugs is their hepatotoxicity (BERGER 1977).

2. Cyclopropylamines

Compounds containing the cyclopropylamine inactivating group have been developed, some of which show substrate selectivity (see e.g. FULLER 1968; FULLER et al. 1979; MURPHY et al. 1978). The mechanism of the irreversible enzyme inactivation produced by tranylcypromine was recently elucidated by PAECH et al. (1979). The initial stage in the reaction may be oxidation to an imine which reacts with an -SH group at the substrate site of the enzyme forming a thioaminoketal or thiohemiketal (PAECH et al. 1979). In the case of tranylcypromine, attachment of the inhibitor to the FAD group was shown not to occur. The related N-cyclopropyl-N-arylalkyl amines (e.g. Lilly 51641), however, appear to react differently and a scheme involving their attachment to the N-5 position of the flavin coenzyme was suggested by SILVERMAN and HOFFMAN (1979). Recovery of a derivatised form of 2-phenylcyclopropanone from the labelled adduct showed that the cyclopropyl ring did not open during reaction with the enzyme (PAECH et al. 1979), although in one case of tranylcypromine overdose, both methamphetamine and amphetamine were detected in the plasma (YOUDIM et al. 1979b). These compounds are among the most active of the MAO inhibitors, and tranylcypromine has been widely used clinically. Although tranylcypromine is generally considered a non-selective type of inhibitor, its (+)-isomer is a selective inhibitor of the type B enzyme in man, as reflected by elevated plasma phenylethylamine levels, and inhibition of platelet MAO (RIEDERER et al. 1981).

3. Propargylamines

Pargyline was the first of this class of clinically useful MAO inhibitors to be described (SWETT et al. 1963). The mechanism of irreversible enzyme inactivation produced by these acetylenic compounds has been examined in detail (MAYCOCK et al. 1976; YOUDIM 1976; SALACH et al. 1979; SINGER 1979). The propargylamine group itself, as in N,N-dimethylpropargylamine, possesses MAO inhibitory activity (MAYCOCK et al. 1976), while combination of this group with an aromatic moiety *via* a side chain can yield compounds with selectivity for MAO type A, e.g. clorgyline (JOHNSTON 1968), M & B 9303 [N-methyl-N-propargyl-2-2(2,4-dichlorphenylthio) ethylamine, JARROTT 1971], or type B, e.g. deprenyl (KNOLL and MAGYAR 1972), pargyline (NEFF and YANG 1974), AGN 1133 and AGN 1135 (N-methyl-N-propargyl-l-aminoindane and N-propargyl-l-aminoindane, FINBERG et al. 1980) (see also Table 3). Structure acitivity relationships for MAO inhibitory potency in the pargyline series were discussed by SWEET et al. (1963), and structural determinants for A or B selectivity were enumerated by KNOLL et al. (1978), KNOLL (1979) and KALIR et al. (1981). Chain length between the aromatic group and N atom, as well as substitution in the aromatic ring, appear to be of critical importance in determin-

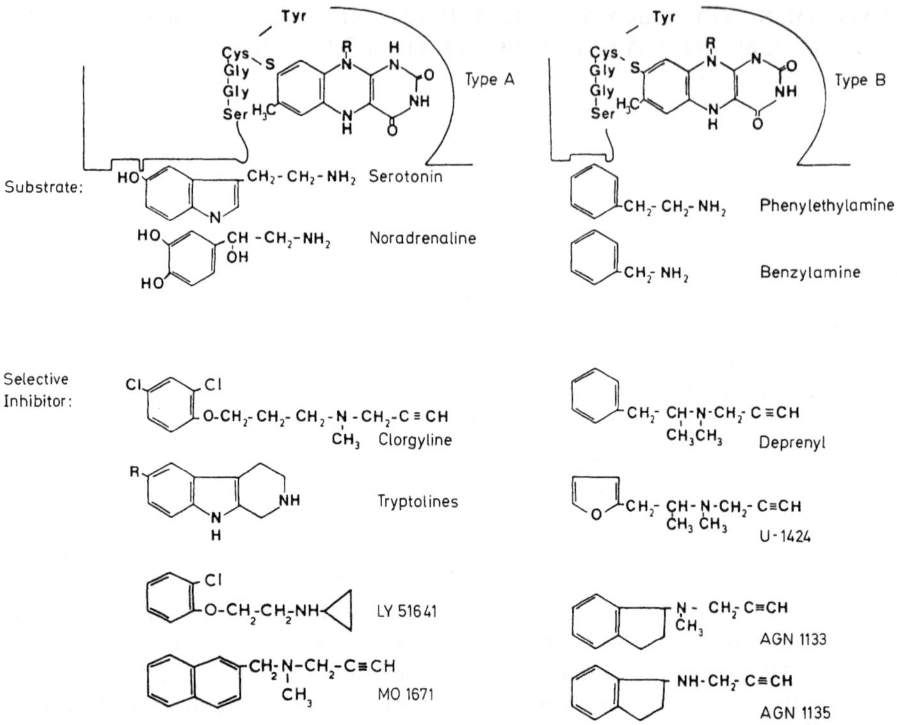

Fig. 2. Schematic representation of the active sites of MAO types A and B, and binding sites of their selective substrates and inhibitors. The flavin combining site is shown as being identical for both enzyme types, selectivity being conferred by an associated site, which, in the case of the type A enzyme, can accept larger carbon chains or substituents between the aromatic center and the nitrogen than in the type B enzyme (Adapted from Youdim and Finberg 1980)

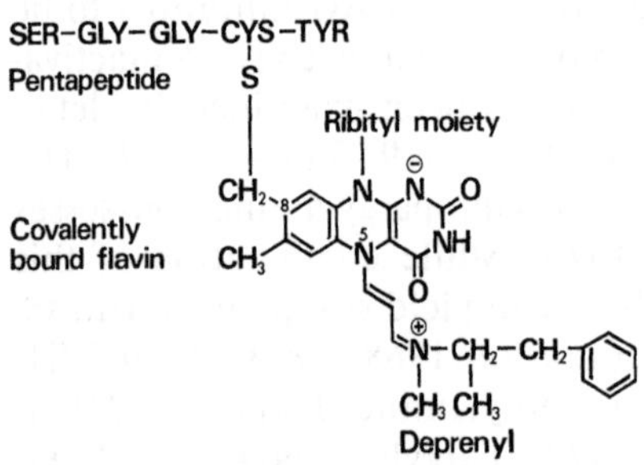

Fig. 3. Structure of cysteinyl-flavin adduct formed when MAO is inactivated by deprenyl. (Taken from Youdim 1978)

ing selectivity (see Fig. 2); however, this relationship does not hold in other chemical classes of inhibitors.

Interaction of inhibitor with MAO leads to reduction of the enzyme-bound FAD and the oxidised inhibitor then reacts covalently with FAD at the N-5 position (Maycock et al. 1976; Salach et al. 1979) (see Fig. 3).

In vivo metabolism of such compounds can result in cleavage of the inhibitor molecule releasing the parent amine. Thus deprenyl is metabolised to methamphetamine and amphetamine (Reynolds et al. 1978a, b) and pargyline to N-methyl benzylamine and benzylamine (Oates et al. 1963; Durden et al. 1976).

4. Reversible Inhibitors

Many compounds from the β-carboline series, particularly harmine and harmaline, have been described which show selectivity for MAO type A, both *in vivo* and *in vitro* (Fuller 1972; Elliott and Holman 1977). Besides being selective reversible inhibitors of MAO-A, the β-carbolines are selective inhibitors of 5-HT uptake into brain synaptosomes and platelets (Elliott and Holman 1977; Youdim and Oppenheim 1980). Other recently described reversible inhibitors of MAO type A are shown in Table 3.

Some drugs have been shown to possess competitive, reversible MAO inhibitory properties in addition to their major therapeutic effect, e.g. furazolidone (Stern et al. 1967), nitroglycerin (Ogawa et al. 1967), amphetamine (Mantle et al. 1976), tricyclic antidepressants (Roth 1978; Achee et al. 1977), phenothiazine neuroleptics (Roth et al. 1979; Youdim and Hefetz 1980) and debrisoquin (Pettinger et al. 1969). Debrisoquin may selectively inhibit neuronal MAO because of its selective concentration in sympathetic neurones (Giachetti and Shore 1967).

II. Selectivity of MAO Inhibition—Acute vs. Chronic Studies

If selective inhibition of one enzyme type or the other is to be maintained *in vivo*, correct dosage schedules are critically important, since all selective MAO inhibitors described so far lack selectivity at high doses. This effect is particularly obvious with type B inhibitors. When rats were treated daily with 1.0 mg/kg deprenyl, substantial inhibition of both A and B MAO in the brain was seen after 14 days treatment, whereas doses of 1.0 and 10 mg/kg harmaline caused selective inhibition of type A enzyme only in brain and liver throughout the 14 day period (Waldmeier and Felner 1978). At a daily s.c. dose of 0.05 to 0.25 mg/kg (−)-deprenyl, however, selective inhibition of rat brain MAO-B was maintained for 3 weeks (Ekstedt et al. 1978; Knoll 1978). In the human a daily oral dose of 10 mg (−)-deprenyl for more than 4 years produced a similar degree of selective MAO-B inhibition in the brain as was seen after one week treatment (Riederer and Reynolds 1980; Riederer et al. 1981). Ogren et al. (1981) claim that reversible inhibitors have therapeutic advantages over irreversible inhibitors since the former may maintain greater selectivity on long term treatment, and may have lower potential toxicity.

III. Pharmacological Actions of MAO Inhibitors

1. Interaction with Centrally-Acting Drugs

Experimentally, little detectable change in gross animal behaviour is seen following acute MAO inhibition, although the combination of an MAO inhibitor with a drug producing blockade of amine reuptake, or with monoamine precursors (see Sect. M) may result in marked central stimulatory effects. Similar reactions occur with other centrally-acting drugs e.g. narcotic analgesics (reviewed by MARLEY 1977).

Evidence has been presented that MAO inhibitors may reduce the metabolic degradation rate of pethidine (CLARK and THOMPSON 1972) and barbiturates (LAROCHE and BRODIE 1960) as well as causing some depression of microsomal drug metabolising enzymes in man (SMITH et al. 1980). Such an effect on metabolism would be expected to prolong and intensify the action of the drug concerned, as is seen with barbiturates; however interaction with elevated levels of central amine neurotransmitters is more likely to explain the rapidly developing toxic reaction seen after pethidine (ROGERS and THORNTON 1969). Apomorphine also produced a hyperthermic response in MAO inhibitor-pretreated rabbits, which could be antagonised by neuroleptic drugs or p-chlorophenylalanine (FJÄLLAND 1979). Interactions between MAO inhibitors and a variety of other psychotropic agents have been reviewed previously (SJÖQVIST 1965; STOCKLEY 1974; HARTSHORN 1976).

2. Potentiation of Peripheral Effects of Sympathomimetic Amines

MAO inhibitor treatment results in potentiation of the intensity and duration of the pressor response to orally or intravenously administered, indirectly acting sympathomimetic amines, such as tyramine, amphetamine, ephedrine, phenylephrine and phenylpropanolamine in man and laboratory animals (for reviews see GOLDBERG 1964; SJÖQVIST 1965; MARLEY and BLACKWELL 1970; STOCKLEY 1974; HARTSHORN 1976; MARLEY 1977). The pressor effects of dopamine (HORWITZ et al. 1960) and L-dopa (HUNTER et al. 1970; CALNE and REID 1973) are also potentiated. However, the pressor response to noradrenaline is generally unchanged (ELIS et al. 1967; PETTINGER and OATES 1968; RAND and TRINKER 1968; BOAKES et al. 1973). HORWITZ et al. (1960) found that the noradrenaline pressor response in man was potentiated only after several weeks of treatment with pheniprazine, at a time when orthostatic hypotension had developed. TRENDELENBURG (1972a) summarized evidence from isolated tissue preparations that some potentiation of the effects of noradrenaline is seen after MAO inhibition, and related this phenomenon to reduction of neuronal net uptake of noradrenaline.

Recent clinical and experimental data with selective MAO inhibitors have shown that the effects of tyramine are potentiated following clorgyline administration (LADER et al. 1970; SIMPSON 1978a; FINBERG et al. 1981) but not after the selective MAO-B inhibitors (−)-deprenyl and AGN 1135 (KNOLL and MAGYAR 1972; KNOLL et al. 1968, 1972; KNOLL 1978; ELSWORTH et al. 1978; FINBERG et al. 1980, 1981). The actions of 2-phenylethylamine however, are potentiated by the selective MAO-B inhibitors. Since tyramine is a substrate for both types of MAO, equivalent potentiation by selective MAO-A and MAO-B inhibitors might have been anticipated. The greater tyramine-potentiating effect of MAO type-A inhibitors may be explained by one or both of the following factors: a) intestinal MAO in man (and rat) is mainly type A (SQUIRES 1972; YOUDIM 1977), and b) neuronal MAO is mainly type A (see Sect. I.VI). The influence of these factors in potentiation of sympathomimetic amines will now be summarized.

a. Effects of Inhibition of Extraneuronal MAO on the Actions of Indirect Sympathomimetic Amines

For indirectly acting amines which are substrates for MAO, the gastrointestinal tract mucosa and liver present the initial obstacle to entry into the general circulation following oral ingestion. Lack of inhibition of gastrointestinal MAO by (−)-deprenyl may be one factor involved in the absence of tyramine potentiation during treatment with this drug (KNOLL 1976; ELSWORTH et al. 1978; BIRKMAYER et al. 1977). In the absence of MAO inhibition, adminstration of tyramine into the portal vein produces a much smaller pressor effect than that following infusion into the systemic circulation (RAND and TRINKER 1968). GARCHA et al. (1979) found a substantial amount of unmetabolized tyramine in arterial blood following intraduodenal instillation of ^{14}C-tyramine in the cat. In man, about 200 to 400 mg of tyramine produces a pressor response following oral administration, although an equivalent pressor effect is obtained after only about 12 mg following MAO inhibitor treatment (ELSWORTH et al. 1978). Evisceration, or hepatic circulatory bypass, produces considerable potentiation of the pressor response to tyramine and β-phenylethylamine in cats (RAND and TRINKER 1968; KNOLL 1976). MAO inhibition also markedly reduces the rate of metabolism of tyramine in dogs (FARAJ et al. 1977). However, although elevated blood levels of tyramine (and other indirectly-acting amines which are substrates for MAO) result from MAO inhibition, this is not the only reason for the potentiation of the sympathomimetric effects of these substances, since the effects of amines which are not substrates for MAO, e.g. ephedrine (ELIS et al. 1967) and amphetamine (SIMPSON 1978a), are also potentiated.

b. Effects of Inhibition of Neuronal MAO on the Actions of Indirect Sympathomimetic Amines

An early explanation for potentiation of sympathomimetic amines by MAO inhibitors was that the latter produce an elevation of neuronal noradrenaline content in the transmitter pool available for release (SJÖQVIST 1965). Although early work showed that tyramine releases noradrenaline from isolated vesicles on a stoichiometric basis (SCHÜMANN and PHILIPPU 1962), the present state of knowledge on the mode of action of tyramine and similar amines does not permit acceptance or rejection of this hypothesis (see TRENDELENBURG 1979, for review on mode of action of phenylethylamines; see also Chap. 5).

Another suggestion (TRENDELENBURG 1972b), that MAO inhibitors prevent deamination of noradrenaline subsequent to its release from the vesicles, thereby presenting a higher concentration of neurotransmitter for transport out of the nerve ending, is more tenable, since a proportion of the noradrenaline released by tyramine is metabolized by deamination (BRANDAO et al. 1980).

Inhibition of neuronal MAO would, therefore, be expected to potentiate tyramine's effects by at least two mechanisms: a) inhibition of the degradation of tyramine (or its product, octopamine) within the neurone, and b) inhibition of noradrenaline deamination subsequent to intraneuronal release. The observation that clorgyline potentiates the effects of tyramine, whereas deprenyl does not, is compatible with an effect on sympathetic nerves, since neuronal MAO is largely type A (see Sect. I.VI).

In isolated tissue preparations (−)-deprenyl reversibly antagonises the effects of tyramine without producing an equivalent inhibition of the noradrenaline response (KNOLL 1976, 1978). This effect was attributed by KNOLL to inhibition of neuronal uptake of tyramine. However, (−)-deprenyl possesses only a weak inhibitory effect on neuronal amine uptake (BRAESTRUP et al. 1975; SIMPSON 1978b). Thus, significant reduction of amine uptake may only appear at doses above those optimal for MAO inhibition. Using the isolated rat vas deferens, FINBERG et al. (1981) concluded that the tyramine antagonistic effect was only seen at doses high enough to inhibit both forms of MAO, and may result from a combination of amine uptake and noradrenaline efflux inhibition. Recent work with another, selective irreversible inhibitor of MAO type B (AGN 1135; FINBERG et al. 1981) has shown that at doses which selectively inhibit MAO type B in brain and peripheral tissues tyramine responses are not potentiated, although the effects of 2-phenylethylamine on cat blood pressure and nictitating membrane are enhanced.

The implication of these results is that any effect of MAO-B inhibition on *in vivo* degradation rate of tyramine is minor, and lack of inhibition of neuronal MAO (type A) is the main reason why the sympathomimetic effects of tyramine are not enhanced by the selective MAO-B inhibitors. Despite the tyramine-potentiating properties of selective MAO-A inhibitors so far described, however, it is quite conceivable that an inhibitor could be developed which possesses both uptake-inhibitor and MAO-A-inhibitor properties, and thus not suffer from tyramine potentiation. In addition, it may be that the relative freedom from side-effects of combined MAO-inhibitors and tricyclic antidepressant therapy (PARE 1976) could partly result from inhibition of tyramine uptake by the tricyclic drug.

3. Effects on Blood Pressure

Reduction of systemic blood pressure, accompanied by orthostatic hypotension, is a common side-effect of MAO-inhibitor therapy. This effect has been described with all chemical classes of MAO inhibitors used clinically, including hydrazines (BIEL et al. 1964), propargylamines (MARONDE and HAYWARD 1963) and cyclopropylamines (ZIRKLE and KAISER 1964). Pargyline reduced blood pressure of hypertensive subjects in the supine as well as upright position, by reducing both peripheral vascular resistance (ONESTI et al. 1964), and cardiac output (SANNERSTEDT 1967). Both central and peripheral mechanisms have been implicated in the antihypertensive effect of MAO inhibitors. Ganglion-blocking and sympathetic neurone blocking actions are seen following injections of phenelzine or pargyline (daily for three days) in the cat (GESSA et al. 1963). Chronic pargyline treatment in cats increased noradrenaline content of autonomic ganglia while reducing that in spleen and heart (SCHOEPKE and WIEGAND 1963; PUIG et al. 1972), leading to the suggestion that ganglion blockade results from an increased inhibitory effect of catecholamines on ganglionic transmission. Given acutely, pargyline elevated the noradrenaline content of rat vascular tissue while reducing synthesis rate (BERKOWITZ et al. 1974). The elevated cytoplasmic noradrenaline concentration following MAO inhibition reduces tyrosine hydroxylase activity by product inhibition (SPECTOR et al. 1967). A role of false neurotransmitter accumulation in production of the adrenergic neuronal blockade is also a possibility, as suggested by KOPIN et al. (1964).

Two studies in spontaneously hypertensive rats (YAMORI et al. 1972; FUENTES et al. 1979) suggest that the antihypertensive effect of pargyline is correlated with an elevated brain stem noradrenaline content, and may be evoked by activation of central α-adrenoceptors. The pargyline-induced fall in blood pressure is blocked by intracerebroventricular injection of 6-hydroxydopamine or phentolamine (FUENTES et al. 1979) demonstrating the importance of a central noradrenergic pathway in mediation of this effect.

Both clorgyline (MURPHY et al. 1981) and Lilly 51641 (FULLER et al. 1979) have been reported to cause orthostatic hypotension in man. This effect appears to be less frequent with (−)-deprenyl therapy (BIRKMAYER et al. 1977; YAHR 1978; MANN and GERSHON 1980). Clorgyline was also more effective than (−)-deprenyl in reducing the blood pressure of salt-sensitive, hypertensive rats (SHALITA and DIKSTEIN 1979).

In addition to their antihypertensive effects, certain MAO inhibitors may cause hypertensive crises not related to tyramine ingestion. Both tranylcypromine (ZIRKLE and KAISER 1964; RAO et al. 1979) and phenelzine (BIEL et al. 1964) can produce a hypertensive response on intravenous injection. The pronounced pressor response to tranylcypromine is not related to inhibition of vasodilatory prostaglandin synthesis (RAO et al. 1979) but is probably mediated by catecholamine release, and, thereforce, can be potentiated by inhibition of MAO; in this way, tranylcypromine, phenelzine and mebanazine can potentiate their own pressor effects (SJÖQVIST 1965; MARLEY 1977). (−)-Deprenyl also possesses intrinsic sympathomimetic activity, and doses of 1 mg/kg and above intravenously increased rat blood pressure (SIMPSON 1978b) and caused contraction of the cat nictitating membrane (FINBERG et al. 1981). Contractile effects on the isolated cat nictitating membrane were also seen with phenelzine and tranylcypromine (TSAI and FLEMING 1965).

4. Other Pharmacological Effects of MAO Inhibitors

In addition to the effects of (−)-deprenyl on amine uptake, certain other MAO inhibitors have been shown to inhibit neuronal amine high affinity uptake (HENDLEY and SNYDER 1968). Inhibition of neuronal MAO reduces net amine uptake by elevating cytoplasmic noradrenaline levels (TRENDELENBURG et al. 1972, 1976); however the effect of deprenyl on amine uptake is believed to be independent of MAO inhibition, since it was detected only at concentrations 100 to 1000 times greater than those which effectively inhibited MAO.

MAO inhibitors may also interfere with reproductive hormones in males and females (KNOLL 1980). (−)-Deprenyl, pargyline and clorgyline elevated plasma levels of prolactin in man (MENDLEWICZ and YOUDIM 1977; MURPHY et al. 1979), although other workers using the rat found that elevated prolactin levels induced by reserpine were reduced by MAO type A inhibitors (KEANE et al. 1979).

M. Physiological Role and Functional Activity of MAO; Biochemical and Behavioural Correlates

The recognition that MAO plays an important role in the degradation of biologically active monoamines has led to the following conclusions: (a) in the CNS MAO is involved in the intra- and extraneuronal inactivation and regulation of the catecholamines, indoleamines and non-catecholic phenylethylamines (BLASCHKO 1952, 1974; BLOOM and GIARMAN 1968; SANDLER and YOUDIM 1972; TIPTON 1975; YOUDIM 1975, 1977; ACHEE et al. 1977; KNOLL 1980), (b) intestinal MAO may be responsible for the inactivation of vasoactive monoamines, e.g. tyramine, hordenine, dopamine and 5-HT, of dietary origin (YOUDIM 1977; MARLEY 1977; KNOLL 1980), (c) liver MAO may control blood levels of pressor amines which have escaped deamination in the gut and by the platelet (BRODIE et al. 1957; DAVISON 1958; YOUDIM and HEFEZ 1980) and (d) platelet, blood vessel and lung MAO could have a crucial role in protecting the organs from toxic effects of circulating active monoamines (SPECTOR et al. 1972; YOUDIM et al. 1980; YOUDIM and HEFEZ 1980). Thus, in different organs MAO may have a variety of functions.

More is known about the functional role of MAO in the CNS than other tissues. The catecholamines and 5-HT, which are synthesized in specific neurons, are thought to move into one of two compartments; either into the storage granular pool or, when that is fully saturated, into the "functional" free pool (NEFF and TOZER 1968; GLOWINSKI et al. 1972; GRAHAME-SMITH 1974). There is a dynamic balance between free and vesicle bound amines in the nerve endings. It has been suggested that the amount of the presynaptic "functional" free pool of these amines is regulated by the accessibility of the amines to MAO, by the granular uptake process and by release into the synaptic cleft (TRENDELENBURG et al. 1972, 1976; GREEN and GRAHAME-SMITH 1975). The exact manner in which this regulation occurs is not known. The process of presynaptic re-uptake is now considered to be a primary system of inactivation of released catecholamines and 5-HT (IVERSEN 1973; GREEN and GRAHAME-SMITH 1975). The function of intraneuronal MAO may, therefore, be to inactivate neurotransmitter monoamines after re-uptake into the nerve endings. However, the extent of monoamine transport by the reuptake system is thought to be related to the activity of MAO in the nerve terminals (TRENDELENBURG et al. 1972, 1976), which regulates the relative concentration of free intra- and extraneuronal amines. Analysis of the available data (TIPTON 1973) has suggested that the concentration of free noradrenaline in the nerve may be about 100 μM, which is slightly lower than the K_m of MAO for this amine. The results suggest, therefore, that MAO is essential for maintaining low levels of cytoplasmic catecholamines and 5-HT. For 5-HT, the main pathway of chemical inactivation is oxidative deamination. The catecholamines can be methylated by COMT prior to deamination. Thus MAO inhibitors have a considerably greater effect on 5-HT containing neurons than on those containing catecholamines (SPECTOR et al. 1963; KOPIN and AXELROD 1963; NEFF and TOZER 1968; FELDSTEIN and WILLIAMSON 1968). In some instances the O-methylated products, e.g. normetanephrine, are more efficiently oxidized than the parent amine (TIPTON 1972; HOUSLAY and TIPTON 1973 a).

The concentrations of dopamine and 5-HT in the rat brain can be increased significantly without any overt behavioural changes occurring in this animal by giving either L-dopa or L-tryptophan, respectively (EVERETT et al. 1963; EVERETT 1966; GRAHAME-SMITH 1971), whereas a characteristic syndrome of hyperactivity and stereotypy occurs if rats are pretreated with a non-selective irreversible MAO inhibitor, e.g. phenelzine or tranylcypromine prior to giving the amino acids. These results were interpreted as demonstrating the ability of intraneuronal MAO to inactivate the increased dopamine or 5-HT synthesized, thereby preventing their release into the synaptic cleft, stimulation of post-synaptic receptors and production of the behavioural changes. When a non-selective MAO inhibitor is given, this intraneuronal metabolism is prevented, the increased amine accumulates in the functional pool and is apparently released on to the receptors, resulting in the hyperactivity syndrome. GREEN et al. (1977) have shown that L-dopa administration to rats pretreated with either clorgyline or deprenyl alone, even at high doses, does not result in increased brain noradrenaline and dopamine levels or the appearance of the hyperactivity syndrome, all of which were observed after administration of both inhibitors together. The results suggest that *in vivo*, dopamine is metabolized by both forms A and B, and when only one form is fully inhibited the amine substrate can still be oxidised by the other form. The assumption is that the amounts of MAO present in the brain are grossly in excess of normal requirements and there is a need to inhibit total brain MAO by 85% or more in order to increase dopamine concentrations and dopamine "functional activity", as judged by the appearance of the hyperactivity syndrome (GREEN et al. 1977). Analogously, although 5-HT, noradrenaline or 2-phenylethylamine may be predominantly deaminated *in vivo* by type A or type B respectively, when each enzyme form is fully inhibited selectively by clorgyline or (−)-deprenyl, the other MAO form can continue to act on the amine (GREEN and YOUDIM 1975). Thus, 5-HT or 2-phenylethylamine brain concentration rises to a much lower extent after the selective inhibitors than after the non-specific inhibitors, and the hyperactivity phenomenon is not observed (GREEN and YOUDIM 1976). The *in vitro* results of EKSTEDT (1976), ACHEE and GABAY (1977), MITRA and GUHA (1980), TIPTON and MANTLE (1981) and FOWLER and TIPTON (1982) support this hypothesis. Similar results have been reported in experiments where the 'functional activity' of 5-HT in terms of behaviour was examined in rats pretreated with a MAO inhibitor and L-tryptophan (GREEN and YOUDIM 1975; SQUIRES and BUUS LASSEN 1975; SQUIRES 1975).

In the brain, as in the periphery, the MAO activity in monoamine neurons may be largely type A (see Sect. I.VI), and it may be this enzyme activity which is important in regulation of amine neurotransmitter functional activity. Thus, for reversal of the reserpine syndrome, selective inhibition of MAO type A is effective, whereas, in order to elicit hyperactivity from L-trytophan or L-dopa, inhibition of types A and B enzyme is required. In both these instances, a minimum of 85% inhibition of enzyme activity must be achieved. The extraneuronal MAO could function as the inactivator of amines taken up by non-neuronal cells, e.g. glial cells.

Another way in which synaptic levels of amines can be elevated is by blocking re-uptake. Thus the combination of MAO inhibitor or amine precursor with an amine uptake blocker results in behavioural hyperactivity similar to that observed with MAO inhibitors plus amine precursors (MODIGH 1973; GREEN and GRAHAME-SMITH 1975; HOLMAN et al. 1976; ASHKENAZI et al. 1983). In instances where MAO alone is inhibited, neuronal re-uptake is adequate to remove excess synaptic amines, but when an MAO inhibitor is combined with an amine precursor, the amounts of amine formed are such that neuronal uptake is unable to maintain the concentration of amines in the synaptic cleft at physiological levels.

The physiological role of type B enzyme is unknown. This enzyme form is specific for some trace amines (e.g. 2-phenylethylamine, tryptamine, 5-methoxytryptamine, 4-methoxy-phenylethylamine, N-methylhistamine) in most species (USDIN and SANDLER 1976), and it is also active towards tyramine and dopamine. The main function of MAO B may be protection against exogenous dietary amines, but such a role appears to be unlikely since the MAO in the small intestine is predominantly type A, and both A and B amine substrates originate in the diet. The importance of the MAO B species probably will remain obscure until the role of trace amines is established.

The presence of 2-phenylethylamine in the rat and human brain and the similarity of its pharmacological activity to amphetamine has led some authors to implicate the increased urinary excretion of 2-phenylethylamine and low platelet MAO activity in the pathogenesis of schizophrenia (USDIN and SANDLER 1976; WYATT et al. 1977; PHILIPS 1978; SABELLI et al. 1978).

An assumption implicit in any study using the platelet is that the MAO activity in this cell reflects the activity of an enzymic form located in the brain, where biochemical variations related to mood would be expected to occur. A comparison of substrate specificity and selective inhibitor sensitivity of MAO in platelets and brain tissue from rabbit and man has shown that the predominant MAO in the rabbit brain and platelet is type A whereas that in man is type B (COLLINS et al. 1970; YOUDIM et al. 1972b; EDWARDS 1976; GLOVER et al. 1977, 1980; HOLZBAUER and YOUDIM 1977; RIEDERER et al. 1978a, b). It is speculated that the lower platelet MAO activity which was reported in chronic schizophrenics on long term treatment with neuroleptics (MURPHY and WYATT 1972) might be associated with a similar reduction in the brain enzyme. Of the more than 30 studies on this subject, the results are equally divided between those in support of and those against low platelet MAO activity in schizophrenia (see BOULLIN 1978; YOUDIM 1979; AGARWAL et al. 1979). The factors which may account for these discrepancies include the MAO inhibitory action of neuroleptics (ROTH and GILLIS 1975; ROTH et al. 1979; YOUDIM and HEFEZ 1980), action of circulating hormones (MURPHY 1976), diet (BROCKINGTON et al. 1976), serum iron (YOUDIM et al. 1975) and the concomitant presence of diseases which are known to affect platelet MAO activity (YOUDIM and HOLZBAUER 1976a; YOUDIM and HEFEZ 1980). Until now, no difference in either MAO-A or MAO-B activities have been found in brain samples of schizophrenics as compared to normal controls (ULENA et al. 1968; DOMINO et al. 1973; SCHWARTZ et al. 1974; ORELAND 1979).

Cyclic variations in the brain tissue concentration of catecholamines, 5-HT and MAO A and B have been reported on several occasions (SANDLER 1968; BIEGON et al. 1979; HOLZBAUER and YOUDIM 1977; GREENGRASS and TONGE 1971). Such variations might be linked with a possible interaction between brain amines and the release of pituitary hormones or with behavioural changes that occur during the oestrous cycle. The modulatory mechanism involved in this process has been suggested to depend on the acitivity of MAO (SANDLER 1968; GREENGRASS and TONGE 1971; HOLZBAUER and YOUDIM 1973; BIEGON et al. 1979).

In aged rats showing a disrupted oestrous cycle, L-dopa administration reinitiates the oestrous cycle (QUADRI et al. 1973). In addition copulatory behaviour of sluggish aged male rats with receptive females was stimulated by L-dopa (TAGLIAMONTE et al. 1974) or low doses of (−)-deprenyl (KNOLL 1981) which elevate brain dopamine (NEFF et al. 1973). Apparently gonadotropins and increased brain MAO B activity due to aging are involved in these reactions. This is supported by the evidence that (a) the tuberoinfundibular dopaminergic neuron system appears to play a role in the regulation of gonadotropin secretion (FUXE et al. 1970) and (b) the MAO of the anterior hypophysis in rats is involved in the process of sexual differentiation of the brain (GAZIRI and LADOSKY 1973; WILSON et al. 1979).

The deaminated products of some amine neurotransmitters activate the oxidation of glucose in the anterior pituitary. The aldehydes formed from the reaction of MAO are substrates for the NADPH-dependent aldehyde reductases (as well as for the NAD^+-dependent aldehyde dehydrogenase) and the increased levels of $NADP^+$ could result in stimulation of the pentose-phosphate pathway in certain circumstances (UNGAR et al. 1973; TURNER et al. 1974; TIPTON et al. 1977). The aldehydes derived from catecholamines and 5-HT may also possess pharmacological activity in the brain (JOUVET 1969, 1974; HOLMAN et al. 1977) and a role has been assigned to them in the mechanism of sleep (JOUVET 1974; SABELLI et al. 1973, 1978; TIPTON et al. 1977). These aldehydes have different but complementary effects in the regulation of slow wave and paradoxical sleep phases. MAO inhibitors are known to block the appearance of normal REM sleep (HOLMAN et al. 1977; CARMAN et al. 1978). Thus, as ACHEE et al. (1977) have suggested, it may be an over-simplification to ascribe all the pharmacological actions of MAO inhibitors to their effects on the increased levels of the biogenic amines alone.

Biogenic monoamines have been shown to have an action on glycogen breakdown (ROBINSON et al. 1972; HIMMS-HAGEN 1972), and adrenaline and noradrenaline are potent glycogenolytic agents. A role for MAO has been assigned in this process since selective MAO-A and B inhibitors enhance glycogenolysis in the liver and brain and increase blood glucose (ISHMAHAN et al. 1978). The results are not unexpected because these inhibitors elevate the catecholamine content of the adrenal gland and brain. There is a possibility that the selective inhibitors interfere with the action of insulin due to increased circulating adrenaline and noradrenaline arising from adrenal glands. These amines inhibit the release of insulin, an effect which may be produced by activation of α-adrenoceptors on the insulin producing β-cells (MONTAGUE and HOWELL 1975) leading to inhibition of adenylate-cyclase and lower intracellular cyclic-AMP (HOWELL and MONTAGUE 1973). Such an inhibitory effect of catecholamines on insulin release might be physiologically important in maintaining elevated blood glucose concentrations arising from catecholamine-induced glycogenolysis during MAO inhibition. The exact mechanism by which selective MAO inhibitors induce glycogenolysis in the brain and liver and elevate blood glucose remains speculative. Involvement of MAO in the mechanism of insulin release is a possibility (ALEYASSINE and GARDINER 1975; HUANG and SCHULZ 1972).

N. Psychiatric and Neurological Disorders: MAO Activity and MAO Inhibitors as Drugs

I. Depressive Illness

The MAO inhibitor antidepressants have been in clinical use for the past 30 years. The older drugs iproniazid, phenelzine, nialamide, isocarboxazide and tranylcypromine are generally regarded as effective antidepressants. Following the recognition of their side effects, particularly the interaction with sympathomimetic amines and amine precursors e.g. tyramine and L-dopa (see Sect. L.III), the clinical use of these drugs became severely curtailed.

However, in the last few years there has been a resurgence of interest in the clinical application of MAO inhibitors (NIES et al. 1975; PARE 1976; MURPHY et al. 1979; see YOUDIM and PAYKEL 1981 for review).

A number of uncontrolled investigations have suggested that the MAO type B inhibitors, pargyline and (−)-deprenyl possess antidepressant properties (DUNLOP 1963; KLINE 1963; VARGA and TRINGER 1967; MENDLEWICZ and YOUDIM 1978; MANN and GERSHON 1980; MENDLEWICZ and YOUDIM 1983).

But in only one of three double blind trials (MENDLEWICZ and YOUDIM 1981; MENDIS et al. 1981; MURPHY et al. 1979), was there a positive antidepressant effect. In contrast, in three double blind studies with selective type A inhibitors, clorgyline and Lilly 51641 (see Table 3) were shown to be as effective as imipramine and superior to amitriptyline (HERD 1969; WHEATLEY 1970; CARMAN et al. 1978; MURPHY et al. 1979).

Both 5-HT and noradrenaline have been implicated in the pathogenesis of depression (VAN PRAAG 1979; MASS 1979). Drugs with antidepressant effects are believed to increase the amount of neurotransmitter amines available at the post synaptic 5-HT or noradrenaline receptors. The classical antidepressants are believed to potentiate the action of 5-HT or noradrenaline; tricyclic compounds through inhibition of uptake and MAO inhibitors through inhibition of amine degradation.

A simple relation between MAO inhibition and increased brain amines cannot explain the therapeutic effectiveness of these drugs. Although MAO inhibition occurs within minutes or hours following administration of the drug, as judged by the enzyme activity in platelets, the antidepressant response is not observed until three to four weeks later (see YOUDIM and PAYKEL 1981). SULSER (1978) demonstrated that all antidepressant modalities lead to an eventual reduction in reactivity of β-adrenoceptors. Chronic administration of tricyclic antidepressants or MAO inhibitors results in a reduced density of β-adrenoceptors (SARAI et al. 1978; CAMPBELL et al. 1979). SULSER (1978) speculated that in depression, β-adrenoceptors are hypersensitive and a reduction of β-adrenoceptor responsiveness may be a crucial element in the recovery process which takes place simultaneously with β-adrenoceptor desensitisation. In addition to this effect on β-adrenoceptors, chronic treatment with MAO inhibitors reduces the number of α_2-adrenoceptors and ^{3}H-5-HT binding sites in the cerebral cortex of rats (SAVAGE et al. 1980; COHEN et al. 1982). The effects of chronic treatment on central β-adrenoceptor number are found after administration of type A but not type B inhibitors (COHEN et al. 1982). For recent reviews of this subject, see YOUDIM and FINBERG (1982, 1985).

II. Parkinson's Disease

The deficiency of dopamine in the striatum (caudate nucleus) of subjects with Parkinson's disease and the efficacy of L-dopa therapy as dopamine replacement are well established (BIRKMAYER and HORNYKIEWICZ 1976; POIRIER et al. 1979). However, the initial beneficial responses to L-dopa are lost within two to three years after the start of the drug treatment (BIRKMAYER et al. 1979). One obvious approach to enhancing the therapeutic effect of L-dopa would be to inhibit the enzyme MAO, thus preventing the deamination of dopamine formed. The early suggestions that this combined regimen might be successful were based on the observations that L-dopa reversal of reserpine-induced sedation in rats is potentiated by MAO inhibitors (CARLSSON et al. 1958). Furthermore, when rats pretreated with pargyline were given L-dopa, they exhibited increased motor activity (EVERETT et al. 1963). The initial trials of BIRKMAYER and HORNYKIEWICZ (1962) with hydrazine MAO inhibitors and L-dopa were abandoned largely because of the induction of hypertension. The knowledge that MAO exists in two functional forms (discussed earlier) together with (a) the discovery of selective inhibitors for each enzyme form, (b) the suggestion by SQUIRES (1972) and YOUDIM et al. (1972a, b) that the major component of human brain MAO is type B and (c) the development of a selective inhibitor of MAO-B, (−)-deprenyl, which does not induce the cheese effect (KNOLL and MAGYAR 1972), gave rise to the possibility of achieving differential brain MAO-B inhibition without incurring undesirable side effects. In 1975 and 1977, BIRKMAYER et al. were the first to show that (−)-deprenyl is beneficial in Parkinson's disease when used as a supplement to L-dopa therapy, especially in patients displaying 'on-off' phenomena (akinesia) and in those who had lost their response to L-dopa (BIRKMAYER et al. 1979). Of the more than 600 patients thus far treated, none has developed hypertension and the side effects are no more than those observed with L-dopa alone (BIRKMAYER et al. 1977; LEES et al. 1977; YAHR 1978; RINNE et al. 1978).

The anti-Parkinson action of (−)-deprenyl is presumed to be related to its inhibition of brain MAO type B, thus making more dopamine available for neurotransmission. Platelet MAO activity, which is considered to be entirely of the B form, was almost completely inhibited by the daily dose (10 mg) of deprenyl in such patients (BIRKMAYER et al. 1977; ELSWORTH et al. 1978). GLOVER et al. (1977), in their initial *in vitro* studies, reported that dopamine, like 2-phenylethylamine, is deaminated predominantly by MAO-B in the human striatum as well as in the platelets. The confirmation of selective MAO-B inhibition has come from brain autopsy studies of Parkinsonian subjects treated with (−)-deprenyl. In seventeen specific areas (including caudate nucleus, putamen, globus pallidus and thalamus), dopamine (MAO-B) and 5-HT (MAO-A) deamination were inhibited by 85–90% and 65% respectively (Table 7) (RIEDERER et al. 1978a, 1981, RIEDERER and YOUDIM 1986). The high degree of brain MAO inhibition which is required to increase the brain level and the functional activity of dopamine and 5-HT (about 85%—see above) may explain why 5-HT was not elevated in the striatum of Parkinsonian patients, although striatal and CSF dopamine levels were increased and urinary excretion of dopamine metabolites was decreased (EISLER et al. 1979; RIEDERER et al. 1981).

The limitations of L-dopa in long-term treatment and management of Parkinsonian patients have led to the use of numerous drugs capable of enhancing striatal dopamine functional activity, by their action as uptake blockers or as releasers and/or as dopamine postsynaptic receptor agonists (FUXE and CALNE 1979). There has been little success with these compounds (see POIRIER et al. 1979). The availability of (−)-deprenyl, which can selectively inhibit the brain enzyme without inducing the untoward potentiation of the actions of sympathomimetic amines such as tyramine and dopamine, has led to its clinical use, and all seven major studies have reported good to excellent results (see BIRKMAYER and YAHR 1978; BIRKMAYER et al. 1982 for reviews). An equally good effect is observed in first-time L-dopa-treated patients as in those who have been on L-dopa for some years. A 40% reduction in L-dopa dosage can be achieved without altering its therapeutic effectiveness (BIRK-

MAYER et al. 1977; YAHR 1978; YOUDIM 1980). Biochemical and pharmacological evidence is available which indicates that (−)-deprenyl is capable of augmenting CNS dopaminergic activity without inducing the untoward effects of other MAO inhibitors. In rats with unilateral nigrostriatal lesions induced by 6-hydroxydopamine, administration of (−)-deprenyl produces a mild ipsilateral rotational response of the type seen with amphetamine. However, when given concomitantly with L-dopa, (−)-deprenyl potentiates the dopamine effect by enhancing and prolonging contralateral rotation (YAHR 1978). This may explain why (−)-deprenyl is effective in reducing the emergence of 'off' periods.

However, it is worth noting that (−)-deprenyl has amphetamine-like properties (SIMPSON 1978a, b; GREEN and YOUDIM 1975) and increases human brain levels of 2-phenylethylamine in (−)-deprenyl-L-dopa treated Parkinsonian subjects (REYNOLDS et al 1978a). 2-Phenylethylamine, an amphetamine-like metabolite of phenylalanine (WYATT et al. 1977; SABELLI et al. 1978) with central stimulatory effects, may have a modulating function (ANTELMAN et al. 1977) on the dopamine formed. Deprenyl by itself is ineffective in the treatment of Parkinson's disease (BIRKMAYER et al. 1977).

In man, (−)-deprenyl is metabolized to methamphetamine and amphetamine (REYNOLDS et al. 1978a, b). Small amounts of (−)-amphetamine, equivalent to those found after a dose of 10 mg of (−)-amphetamine, were found in brains of Parkinsonian subjects treated with (−)-deprenyl (REYNOLDS et al. 1978a, b). It is well known that amphetamine has a central stimulating action mediated by the release of dopamine (MOORE 1977). The (−)-isomer is less active than the (+)-isomer. This effect probably cannot explain the efficacy of (−)-deprenyl in Parkinson's disease, since (+)- or (−)-amphetamine (50 mg/d) has little beneficial effect in Parkinsonian subjects (PARKES et al. 1975). It will be of interest to determine whether other MAO-B selective inhibitors without amphetamine-like activity, e.g. AGN 1135, discussed earlier, possess similar anti-Parkinsonian effects.

O. Future Perspectives

There is little doubt that multiple forms of MAO exist *in vivo* and that these enzyme forms play an important function in the catabolism of biogenic and trace amines. Whether MAO type A and type B represent different molecular forms of the enzyme or a single enzyme in different mitochondrial lipid environments, seems to be of minor physiological importance. What matters is that they mediate distinct functions. For example, their selective inhibition in brain can result in differential increases in 5-HT, noradrenaline, dopamine or 2-phenylethylamine levels. Their activity can be affected independently by hormonal manipulation and there are differences in the rates of their postnatal development.

Wheter the therapeutic activity of drugs which inactivate the enzyme *in vivo* is related to the increase in brain levels of amine transmitter substances is still a matter of conjecture, as are the possible abnormal MAO activity and amine metabolism in psychiatric disorders. What is becoming clear is that chronic treatment with MAO inhibitors may have other effects e.g. a reduction in β-adrenoceptor binding sites, besides MAO inactivation. Such effects might be crucial in determining the antidepressant action of MAO inhibitors. These modalities need to be examined with the newly developed selective MAO type A and type B inhibitors. Selective inhibitors do not yet have an established place in antidepressant therapy, but they offer considerable promise for the future as pharmacological and therapeutic tools and for the treatment of Parkinson's disease.

Development of new techniques for selective localization of the different MAO forms in the brain is an urgent need for further elucidation of the role of MAO in brain amine metabolism.

P. Adddendum

The completion of this chapter by the beginning of 1982 does not do justice to the wealth of information obtained on the biochemistry, pharmacology and function of monoamine oxidases since then. Thus the appreciation and understanding of the mechanism of action of this important regulatory enzyme has greatly changed.

It is now apparent that the lipid in the immediate environment of mitochondrial MAO does not fully explain the substrate specificities and inhibitor selectivities of type A and type B enzyme. Recent studies strongly point to the existence of at least two distinct enzyme proteins under the control of different gene loci (WESTLUND et al. 1985; KOCHERSPERGER et al. 1986) as identified by induction of monoclonal antibodies to the enzyme forms. The use of these antibodies has permitted the study of distribution and location of MAO A and B in rat and human brain. While MAO A is present in catecholaminergic neurones and extraneuronally in glial cells, it is highly concentrated in the former neurones. By contrast MAO B is present in serotonergic neurones of raphe and glial cells (LEVITT et al. 1982). The fact that MAO A has not been identified in serotonergic neurones is odd, since 5-HT is type A substrate and MAO type A inhibitor increases its concentration in the brain. It is more than likely that the monoclonal antibody library is not adequate to identify a possible subtype of MAO A in the serotonergic neurones.

On occasions it has been reported that (−)-deprenyl has antidepressant activity alone (MENDLEWICZ and YOUDIM 1983) or in combination with 5-hydroxytryptophan (MENDLEWICZ and YOUDIM 1979). The localization of MAO B in the raphe nucleus, as discussed above, together with the ability of (−)-deprenyl in increasing 5-HT concentrations in these neurons may explain the antidepressant action of (−)-deprenyl (KUHN et al. 1985; WOLF et al. 1985).

Pharmacologically the enzyme continues to attract much attention. A large number of selective inhibitors of both enzyme forms have now been described (YOUDIM and FINBERG 1982; TIPTON et al. 1984b; TIPTON and YOUDIM 1984; CALLINGHAM 1984; YOUDIM and FINBERG 1985; MONDOVI 1985). Among these are the newly developed reversible inhibitors of MAO A, with minimum tyramine pressor potentiation ("cheese effect"). The rationale being that tyramine, a substrate of MAO A, would displace the inhibitor from its binding site and itself be metabolized by the enzyme A. This procedure was expected to limit the extensive release of noradrenaline by tyramine (see YOUDIM and FINBERG 1985). Although these reversible MAO A inhibitors are an improvement over the irreversible ones, they still possess the "cheese effect" property (TIPTON et al. 1984b). By contrast animal and human studies with selective MAO B inhibitors have shown that this class of inhibitors is devoid of tyramine pressor potentiation (YOUDIM and FINBERG 1982). The reversible inhibitors of MAO A are thought to possess antidepressant activity (TIPTON et al. 1984a, b).

There is no question that selective MAO B inhibitors represent an important class of anti-Parkinson drugs (RINNE 1983), since therapeutic doses of deprenyl have been shown to selectively inhibit MAO B in brains obtained at autopsy from Parkinson subjects, and increase dopamine and phenylethylamine without altering 5-HT or 5-hydroxy-indoleacetic acid in caudate, putamen, substantia nigra and pallidum (RIEDERER and YOUDIM 1986). Dopaminergic neurotoxin MPTP (N-methyl-4-phenyl(1,2,3,6)-tetrahydropyridine) is a substrate for MAO B and induces Parkinsonism in human subjects, primates and rodents (see MARKEY et al. 1986, KINEMUCHI et al. 1987 for reviews). The neurotoxic effects of MPTP can be prevented by selective MAO B inhibitors, suggesting the neurotoxic property resides in the MAO B-derived oxidative product MPP^+ (N-methyl-4-phenylpyridine) (HEIKKILA et al. 1984, 1985). This has led to the suggestion that Parkinsonism may be a neurotoxic event. The longevity noted in Parkinsonian subjects on long term deprenyl treatment has been interpreted as representing the prevention by deprenyl of dopaminergic neurone degeneration (BIRKMAYER et al. 1985). Thus MAO B inhibitors may be of therapeutic use not only as agents to potentiate L-Dopa effects, but also to induce retardation of degeneration of dopamine neurones as a resultant of free radical formation from hydrogen peroxide or an endogenous MPTP like neurotoxin generated by MAO reaction (YOUDIM and RIEDERER 1986).

Acknowledgement. This work was supported by grants from Israel National Council for Research and Development (M.B.H.Y. and J.P.M.F.) and the Medical Research Council of Ireland (K.F.T.). We are grateful to Mrs. Marie Harvey and Mrs. Joyce Carp for their help in the preparation of this manuscript.

Q. References

Achee FM, Gabay S (1977) Studies on monoamine oxidase: inhibition of bovine brain MAO in intact mitochondria by selective inhibitors. Biochem Pharmacol 26:1637-1644

Achee FM, Gabay S (1981) Studies of monoamine oxidases. Effect of Triton X-100 and bile salts on monoamine oxidase in brain mitochondria. Biochem Pharmacol 30:3151-3157

Achee FM, Gabay S, Tipton KF (1977) Some aspects of monoamine oxidase activity in brain. Progr Neurobiol 8:325-348

Agarwal DP, Goedde HW, Schrappe O (1979) Blood platelet monoamine oxidase activity in schizophrenia, affective disorders and alcoholism. In: Singer TP, von Korff RW, Murphy DL (eds) Monoamine Oxidase: Structure, Function and Altered Functions. Academic Press New York, pp 397-402

Agid Y, Javoy F, Youdim MBH (1973) Monoamine oxidase and aldehyde dehydrogenase activity in the striatum of rats after 6-hydroxy dopamine lesion of the nigrostriatal pathway. Br J Pharmacol 48:175-179

Aleyassine H, Gardiner RJ (1975) Dual action of anti-depressant drugs (MAO inhibitors) on insulin release. Endocrinology 96:702-710

Antelman SM, Edwards DJ, Lin M (1977) Phenylethylamine: evidence for a direct postsynaptic dopamine receptor stimulating action. Brain Res 127:317-324

Aoki S, Manabe T, Okuyama T (1977) Molecular weight estimation of bovine brain mitochondrial monoamine oxidase. J Biochem 82:1533-1539

Ashkenazi R, Finberg JPM, Youdim MBH (1983) Behavioural hyperactivity in rats treated with selective monoamine oxidase inhibitors and LM 5008, a selective 5-hydroxytryptamine uptake blocker. Br J Pharmacol 79:765-770

Baker SP, Hemsworth BA (1978) Effect of phospholipid depletion by phospholipases on the properties and formation of the multiple monoamine oxidase forms in the rat liver. Eur J Biochem 92:165-174

Bakhle YS, Youdim MBH (1976) Metabolism of phenylethylamine in rat isolated perfused lung: evidence for monoamine oxidase 'type B' in lung. Br J Pharmacol 56:125-127

Bakhle YS, Youdim MBH (1979) The metabolism of 5-hydroxytryptamine and phenylethylamine in perfused rat lung and *in vitro*. Br J Pharmacol 65:147-154

Belleau B, Burba J (1960) Stereochemistry of the enzymic decarboxylation of amino acids. J Am Chem Soc 82:5751-5752

Belleau B, Moran J (1963) Deuterium isotope effects in relation to the chemical mechanism of monoamine oxidase. Ann N Y Acad Sci 107:822-839

Belleau B, Fang M, Burba J, Moran J (1960) The absolute optical specificity of monoamine oxidase. J Am Chem Soc 82:5752-5754

Ben-Harari RR, Pelleg R, Youdim MBH (1982) Uptake and deamination of non-polar amines by isolated perfused liver and lung. Br J Pharmacol 76:77-85

Berger PA (1977) Antidepressant medications and the treatment of depression. In: Barchas JD, Berger PA, Ciaranello RD, Elliott GR (eds) Psychopharmacology—from Theory to Practice. Oxford University Press New York, pp 174-207

Berkowitz BA, Tarver JJ, Spector S (1974) Control of norepinephrine synthesis in blood vessels and the effects of monoamine oxidase inhibition. J Pharmac Exp Ther 190:21-29

Bernheim MLC (1931) Tyramine oxidase II. The course of the oxidation. J Biol Chem 299-309

Bevan Jones AB, Pare CMB, Nicholson WJ, Price K, Stacey RS (1972) Brain amine concentration after monoamine oxidase inhibitor administration. Br Med J i 17-19

Biegon A, Segal M, Samuel D (1979) Sex differences in behavioural and thermal responses to pargyline and tryptophan. Psychopharmacologia 61:77-80

Biel JH (1970) Monoamine oxidase inhibitor antidepressants. Structure activity relationships. In: Clark WG, del Gundice J (eds) Principles of Psychopharmacology. Academic Press New York, pp 279-287

Biel JH, Horita A, Drukker AE (1964) Monoamine oxidase inhibitor (hydrazines). In: Gordon M (eds) Psychopharmacological Agents Academic Press New York, pp 359-445

Birkmayer W, Hornykiewicz O (1962) Der L-dioxyphenylalanin (L-dopa)-Effekt beim Parkinson-Syndrom des Menschen. Arch Psychiat Nervenkr 203:560-574

Birkmayer W, Hornykiewicz O (1976) Advances in Parkinsonism. Basel: Editions Roche

Birkmayer W, Yahr M (1978) Deprenyl, an inhibitor of MAO-type B in the treatment of Parkinsonism. J Neural Transmis 43:177-286

Birkmayer W, Riederer P, Youdim MBH, Linauer W (1975) Potentiation of anti-akinetic effect after L-dopa treatment by an inhibitor of MAO-B—deprenyl. J Neural Transmis 36:303-323

Birkmayer W, Riederer P, Ambrozi L, Youdim MBH (1977) Implications of combined treatment with 'Madopar' and L-deprenyl in Parkinson's disease. A long term therapy Lancet 1:439-444

Birkmayer W, Riederer P, Youdim MBH (1979) Distrinction between benign and malignant type of Parkinson's disease Clin Neurol Neurosurg 81-3:158-164

Birkmayer W, Riederer P, Youdim MBH (1982) Deprenyl in the treatment of Parkinson's disease. Clin Neuropharmacol. 5:195-230

Birkmayer W, Knoll J, Riederer P, Youdim MBH, Hars V, Marton J (1985) Increased life expectancy resulting from addition of l-deprenyl to Madopar treatment in Parkinson's disease: A longterm study. J Neural Transm is 64:113-127

Blaschko H (1952) Amine oxidase and amine metabolism. Pharmac Rev 4:415-453

Blaschko H (1963) Amine oxidase. In: Boyer PD, Lardy H, Myrbäck K (eds) The Enzymes. 2nd ed vol 8 Academic Press New York, pp 337-351

Blaschko H (1974) The natural history of amine oxidases. Rev Physiol Biochem Pharmacol 70:84-148

Bloom FE, Giarman NJ (1968) Physiologic and pharmacologic consideration of biogenic amines in the nervous system. Ann Rev Pharmacol 8:229-247

Boadle MC, Bloom FE (1969) A method for the fine structural localization of monoamine oxidase. J Histochem Cytochem 17:331-340

Boakes AJ, Laurence DR, Teoh PC, Barar FSK, Benedikter LT, Prichard BNC (1973) Interactions between sympathomimetic amines and antidepressant agents in man. Br Med J i 311-315

Boullin DG (1978) Biochemical indicators of central serotonin function. In: Serotonin and Mental Abnormalities. Wiley, Chichester

Boulton AA (1979) Trace amines in the central nervous system. Int Rev Biochem 26:179-206

Bourne R, Lai JCK, Owen F (1975) Monoamine oxidase activity in distinct populations of rat brain mitochondria. Br J Pharmacol 55:298P

Braestrup C, Andersen H, Randrup A (1975) The monoamine oxidase B inhibitor deprenyl potentiates phenylethylamine behaviour in rats without inhibition of catecholamine metabolite formation. Eur J Pharmacol 34:181-187

Brandao F, Rodrigues-Pereira E, Guilherme Monteiro J, Osswald W (1980) Characteristics of tyramine induced release of noradrenaline: mode of action of tyramine and metabolic fate of the transmitter. Naunyn-Schmiedeborg's Arch Pharmacol 311:9-15

Brockington I, Crow TJ, Johnstone EC, Owen F (1976) An investigation of platelet monoamine oxidase activity in schizophrenia and schizoaffective psychosis. Ciba Found. Symp. 39 (New Ser) Amsterdam: Elsevier Amsterdam

Brodie BB, Tomich EG, Kuntzman R, Shore PA (1957) On the mechanism of action of reserpine. Effects of reserpine on capacity of tissue to bind serotonin. J Pharmacol Exp Ther 119:461-465

Brown GK, Powell JF, Craig IW (1981) Molecular weight differences between human platelet and placental monoamine oxidase. Biochem Pharmacol 29:2595-2603

Brown GK, Powell JF, Graig IW (1982) Immunological studies of human monoamine oxidase. J Neurochem 39:1266-1270

Brown LE, Hamilton GA (1970) Some model reactions and a general mechanism for flavoenzyme-catalyzed dehydrogenations. J Amer Chem Soc 92:7225-7227

Brunner G, Neupert W (1968) Turnover of outer and inner membrane proteins of rat liver mitochondria. FEBS Lett 1:153-155

Buffoni F (1966) Histaminase and related amine oxidases. Pharmacol Rev 18:1163-1199

Callingham BA (1984) In vitro pharmacology of reversible inhibitors of monoamine oxidase. In: Paton W, Mitchell J, Turner P (eds) IUPHAR 9th Inter Cong. Pharmac vol 2 Macmillan London, pp 211-218

Callingham BA (1977) Substrate selective inhibition of monoamine oxidase by mexiletine. Br J Pharmacol 61:118-119P

Callingham BA, Parkinson D (1979) Tritiated pargyline binding to rat liver mitochondrial MAO. In: Singer TP, von Korff RW, Murphy DL (eds) Monoamine Oxidase: Structure, Function and Altered Functions, Academic Press New York, pp 81-86

Calne DB, Reid JL (1973) Actions of levodopa on the blood pressure of conscious rabbits. Br J Pharmacol 48:194-197

Campbell I, Murphy DL, Callagher DW, Tallman JW, Marshall FE (1979) Neurotransmitter-related adaption in the central nervous system following chronic monoamine oxidase inhibition. In: Singer TP, von Korff RW, Murphy DL (eds) Monoamine Oxidase; Structure Function and Altered Functions, Academic Press New York, pp 517-530

Carlsson A, Lindqvist M, Magnusson T, Waldeck B (1958) On the presence of 3-hydroxytyramine in brain. Science 127:471-472

Carman JS, Gillin JC, Murphy DL, Weinberger DR, Kleiman JE, Bigelow LB, Wyatt RJ (1978) Effects of a selective inhibitor of type A monoamine oxidase (Lilly 51461) on behaviour, sleep and circadian rhythms in depressed and schizophrenic patients. Commun Psycharmacol 2:513-524

Cawthon RM, Breakefield XO (1979) Differences in A and B forms for monoamine oxidase revealed by limited proteolysis and peptide mapping. Nature 281:692-694

Cawthon RH, Pintar JE, Heseltine JP, Breakefield XO (1981) Differences in the structure of A and B forms of human monoamine oxidase. J Neurochem 37:363-372

Christmas AJ, Coulson CJ, Maxwell DR, Riddell D (1972) A comparison of the pharmacological and biochemical properties of substrate-selective monoamine oxidase inhibitors. Br J Pharmacol 45:490-503

Clark B, Thompson JW (1972) Analysis of the inhibition of pethidine N-demethylation by monoamine oxidase inhibitors and some other drugs with special reference to drug interactions in man. Br J Pharmacol 44:89-99

Cohen RM, Campbell IC, Dauphin M, Tallman JF, Murphy DL (1982) Changes in α- and β-receptor densities in rat brain as a result of treatment with monoamine oxidase inhibiting antidepressants. Neuropharmacol 21:293-298

Collins GGS, Youdim MBH (1975) The binding of [^{14}C] phenethylhydrazine to rat liver monoamine oxidase. Biochem Pharmacol 24:703-706

Collins GGS, Youdim MBH, Sandler M (1968) Isoenzymes of human and rat liver monoamine oxidase. FEBS Lett 1:215-218

Collins GGS, Sandler M, Williams ED, Youdim MBH (1970) Multiple forms of human brain mitochondrial monoamine oxidase. Nature 225:817-820

Coq H, Baron C (1968) Etude cinetique de l'activité monoamineoxydasique de mitochondries de foie de rat. Bull Soc Chim Biol 50:163-178

Coquil JF, Goridis C, Mack G, Neff NH (1973) Monoamine oxidase in rat arteries: evidence for different forms and selective localization. Br J Pharmacol 48:590-599

Costa MRC, Breakefield XO (1980) Electrophoretic analysis of ^{3}H-pargyline-labelled monoamine oxidases A and B from human and rat cells. Mol Pharmacol 17:199-205

Craig IW, Powell JF, Brown GK, Summers KM (1982) Studies on human monoamine oxidase. In: Kamijo K, Usdin E, Nagatsu T (eds) Monoamine Oxidase: Basic and Clinical Frontiers. Excerpta Medica Amsterdam, pp 18-27

Davison AN (1958) Physiological role of monoamine oxidase. Physiol Rev 38:729-747

Della Corte L, Tipton KF (1980) The turnover of the A- and B-forms of monoamine oxidase in rat liver. Biochem Pharmacol 29:891-895

De la Lande IS, Hill BD, Jellet LB, McNell JM (1970) The role of monoamine oxidase in the response of the isolated central artery of the rabbit ear to tyramine. Br J Pharmacol 40:249-256

De la Torre JC (1972) Dynamics of brain monoamines. Plenum Press New York, pp 87-106

Demarest KT, Azzaro AJ (1979) The association of type A monoamine oxidase with the nigrostriatal dopamine neuron. In: Singer TP, von Korff RW, Murphy DL (eds) Monoamine Oxidase: Structure, Function and Altered Functions, Academic Press New York, pp 423-430

Demisch L, Demisch K, Seiler N (1979) Factors altering platelet monoamine oxidase. The influence of oral glucose intake. Metabolism 28:144-150

Denney RM, Fritz RR, Patel NT, Abell CW (1982a) Human liver MAO A and B separated by immunoaffinity chromatography with MAO-B-specific monoclonal antibody. Science 215:1400-1403

Denney RM, Patel NT, Fritz RR, Abell CW (1982b) A monoclonal antibody elicited to human platelet monoamine oxidase. Isolation and specificity for human monoamine oxidase B but not A. Mol Pharmacol 22:500-508

Dennick RG, Mayer RJ (1977) Purification and immunochemical characterisation of monoamine oxidase from rat and human liver. Biochem J 161:167-174

Dixon M, Webb EC (1979) Enzymes 3rd ed. Longman London

Domino EF, Krause RR, Bowers M (1973) Various enzymes involved with putatitive neurotransmitters. Arch Gen Psychiat 29:195-201

Donnelly CH, Richelson E, Murphy DL (1976) Properties of monoamine oxidase in mouse neuroblastoma NIE-115 cells. Biochem Pharmacol 25:1639-1643

Dunlop E (1963) Antidepressant effects of MAO inhibitors. Ann NY Acad Sci 107:1107-1117

Durden DA, Philips SR, Boulton AA (1976) Identification and distribution of benzylamine in tissue extracts isolated from rats pretreated with pargyline. Biochem Pharmacol 25:858-859

Eberson LE, Persson K (1962) Studies on monoamine oxidase inhibitors: the autoxidation of β-phenylisopropyl hydrazine as a model reaction for irreversible monoamine oxidase inhibition. J Med Pharm Chem 5:738-752

Edelstein SB, Castiglione CM, Breakefield XO (1978) Monoamine oxidase activity in normal and Lesch-Nyhan fibroblasts. J Neurochem 31:1247-1254

Edwards DJ (1976) Monoamine oxidases in brain and platelets: implications for the role of trace amines and drug action. In: Usdin E, Sandler M (eds) Trace Amines in the Brain. Dekker New York

Edwards DJ, Burns MO (1974) Effects of tricyclic antidepressants on human platelet monoamine oxidase. Life Sci 15:2045-2058
Edwards DJ, Chang S-S (1975) Multiple forms of monoamine oxidase in rabbit platets. Life Sci 17:1127-1134
Edwards DJ, Malsbury CW (1978) Characteristics of monoamine oxidase in brain and other organs of the golden hamster. Biochem Pharmacol 27:959-963
Edwards DJ, Pak KY (1979) Selective radiochemical labelling of type A and B active sites of rat liver monoamine oxidase. Biochem Biophys Res Commun 86:350-357
Egashira T, Ekstedt B, Kinemuchi H, Wiberg A, Oreland L (1976) Molecular turnover numbers of different forms of mitochondrial monoamine oxidase in rat. Med Biol 54:272-277
Eiduson S, Buckman T (1979) Studies on MAO using spin-labelled probes. In: Singer TP, von Korff RW, Murphy DL (eds) Monoamine Oxidase: Structure, Function and Altered Functions, Academic Press New York, pp 213-231
Eisler T, Calne DB, Ebert MH, Kopin IJ, Zeigler MG, Levine R, Murphy DL (1979) Biochemical measurements during (−)-deprenyl treatment of Parkinsonism. In: Singer TP, von Korff RW, Murphy DL (eds) Monoamine Oxidase: Structure, Function and Altered Functions, Academic Press New York, pp 497-505
Ekstedt B (1976) Substrate specificity of the different forms of monoamine oxidase in rat liver mitochondria. Biochem Pharmacol 25:1133-1138
Ekstedt B, Oreland L (1976) Heterogeneity of pig liver and pig brain mitochondrial monoamine oxidase. Arch Int Pharmacodyn Ther 222:157-165
Ekstedt B, Magyar K, Knoll J (1978) Does the B form selective monoamine oxidase inhibitor deprenyl lose selectivity by long term treatment? Biochem Pharmacol 28:919-923
Elis J, Laurence DR, Mattie H, Prichard BNC (1967) Modification by monoamine oxidase inhibitors of the effect of some sympathomimetics on blood pressure. Br Med J 2:75-78
Elliot GR, Holman RB (1977) Tryptolines as potential modulators of serotonergic function. In: Usdin E, Hamberg D, Barchas J (eds) Neuroregulators and Psychiatric Disorders, Oxford University Press New York, pp 220-228
Elsworth JD, Glover V, Reynolds GP, Sandler M, Lees AJ,. Phuapradit P, Shaw KM, Stern GM, Kumar P (1978) Deprenyl administration in man. A selective MAO-B inhibitor without the 'cheese effect'. Psychopharmacology 57:33-38
Erwin VG, Hellerman L (1967) Mitochondrial monoamine oxidase I. Purification and characterization of the bovine kidney enzyme. J Biol Chem 242:4230-4238
Erwin VG, Simon RJ (1969) Occurrence of newly synthesized monoamine oxidase in subcellular fractions of rat liver. J Pharm Sci 58:1033-1035
Everett G (1966) The dopa response potentiation test and its use in screening for antidepressant drugs. Excerpta Med Int Cong Series 122:164-167
Everett GM, Wiegand RG, Rinaldi FU (1963) Pharmacologic studies of some nonhydrazine MAO inhibitors. Ann NY Acad Sci 107:1068-1080
Faraj BA, Dayton PE, Camp PM, Wilson JP, Malveaux EJ, Schlant RC (1977) Studies on the fate of tyramine in dogs: the effect of monoamine oxidase inhibition, portafemoral shunt and coronary artery ligation on the kinetics of tyramine. J Pharmacol Exp Ther 200:384-393
Feldstein A, Williamson O (1968) Serotonin metabolism in pineal homogenates. Adv Pharmacol 6:91-96
Finberg JPM, Sabbagh A, Youdim MBH (1980) Pharmacology of selective propargyl "suicide" inhibitors of monoamine oxidase. In: Usdin E, Sourkes TL, Youdim

MBH (eds) Enzymes and Neurotransmitters in Mental Disease. Wiley Chichester, pp 205-219

Finberg JPM, Tenne M, Youdim MBH (1981) Selective irreversible propargyl derivative inhibitors of monoamine oxidase (MAO) without the cheese effect. In: Youdim MBH, Paykel ES (eds) Monoamine Oxidase Inhibitors: State of the Art. Wiley Chichester, pp 31-43

Fischer AG, Schulz AR, Oliner L (1968) Thyroidal biosynthesis of iodothyronines. II. General characteristics and purification of mitochondrial monoamine oxidase. Biochim Biophys Acta 159:460-471

Fjälland B (1979) Antagonism of apomorphine-induced hyperthermia in MAOI-pretreated rabbits as a sensitive model of neuroleptic activity. Psychopharmacology 63:119-123

Fowler CJ, Callingham BA (1978) The effect of age on the number of monoamine oxidase active centres in the rat heart. Biochem Soc Trans 6:955

Fowler CJ, Callingham BA (1980) The inhibition of rat heart type A monoamine oxidase by clorgyline as a method for the estimation of enzyme active centers. Mol Pharmacol 16:546-555

Fowler CJ, Oreland L (1980) The nature of the substrate-selective interaction between rat liver mitochondrial monoamine oxidase and oxygen. Biochem Pharmacol 29:2225-2233

Fowler CJ, Tipton KF (1981) Concentration dependence of the oxidation of tyramine by the two forms of rat liver mitochondrial monoamine oxidase. Biochem Pharmacol 30:3329-3332

Fowler CJ, Tipton KF (1982) Deamination of 5-hydroxytryptamine by both forms of monoamine oxidase in the rat brain J Neurochem 38:733-736

Fowler CJ, Callingham BA, Mantle TJ, Tipton KF (1978) Monoamine oxidase A and B: a useful concept? Biochem Pharmacol 27:97-101

Fowler CJ, Ekstedt B, Egashira T, Kinemuchi H, Oreland L (1979) The interaction between human platelet monoamine oxidase, its monoamine substrates and oxygen. Biochem Pharmacol 28:3063-3068

Fowler CJ, Callingham BA, Mantle TJ, Tipton KF (1980a) The effect of lipophilic compounds upon the activity of rat liver mitochondrial monoamine oxidase-A and B. Biochem Pharmacol 29:1177-1183

Fowler CJ, Oreland L, Marcusson J, Winblad B (1980b) Titration of human brain monoamine oxidase-A and -B by clorgyline and L-deprenil. Naunyn-Schmiedeberg's Arch Pharmacol 311:263-272

Fuentes JA, Neff NH(1977) Inhibition by pargyline of cardiovascular amine oxidase activity. Biochem Pharmacol 26:2107-2112

Fuentes JA, Ordaz A, Neff NH (1979) Central mediation of the antihypertensive effect of pargyline in spontaneously hypertensive rats. Eur J Pharmacol 57:21-27

Fuller RW (1968) Kinetic studies and *in vivo* effects of a new monoamine oxidase inhibitor N-(2[*o*-chlorophenoxy]-ethyl)-cyclopropylamine. Biochem Pharmacol 17:2097-2106

Fuller RW (1972) Selective inhibition of monoamine oxidase. Adv Biochem Psychopharmac 5:339-354

Fuller RW, Hemrick SK (1979) Stereoselective inhibition of monoamine oxidase. In: Singer TP, von Korff RW, Murphy DL (eds) Monoamine Oxidase: Structure, Function, and Altered Functions. Academic Press New York, pp 245-250

Fuller RW, Warren BJ, Molloy BB (1970) Selective inhibition of monoamine oxidase in rat brain mitochondria. Biochem Pharmacol 19:2934-2936

Fuller RW, Slater IH, Mills J (1979) The development of N-cyclopropyl-arylalkylamines as monoamine oxidase inhibitors. In: Singer TP, von Korff RW, Murphy DL (eds) Monoamine Oxidase: Structure, Function and Altered Functions. Academic Press New York, pp 317-333

Furness JB, Costa M (1971) Monoamine oxidase histochemistry of enteric neurones in guinea pig. Histochemie 28:324-336

Fuxe K, Calne DB (1979) Dopaminergic Ergot Derivatives and motor functions. Pergamon Press Oxford

Fuxe K, Goldstein M, Ljungdahl A (1970) Anti-Parkinsonian drugs and central dopamine neurons. Life Sci 9:811-824

Gabay S, Achee FM, Mentes G (1976) Some parameters affecting the activity of monoamine oxidase in purified bovine brain mitochondria. J Neurochem 27:415-424

Garcha G, Imrie PR, Marley E, Thomas DV (1979) Effects of monoamine oxidase inhibitor (MAOI) pretreatment on the fate of intraduodenally instilled [^{14}C]-tyramine. Brit J Pharmacol 67:454P

Gaziri LCJ, Ladowsky W (1973) Monoamine oxidase variation during sexual differentiation. Neuroendocrinology 12:249-256

Gessa GL, Cuenca E, Costa E (1963) On the mechanism of hypotensive effects of MAO inhibitors. Ann NY Acad Sci 107:935-941

Ghosh SK, Guha SR (1978) Further studies on the inhibition of monoamine oxidation by monoamine oxidase inhibitors. Biochem Pharmacol 27:112-114

Giachetti A, Shore PA (1966) Optical specificity of monoamine oxidase. Life Sci 5:1373-1378

Giachetti A, Shore PA (1967) Monoamine oxidase inhibition in the adrenergic neuron by bretylium, debrisoquin, and other adrenergic neuronal blocking agents. Biochem Pharmacol 16:237-238

Gillis CN, Roth JA (1977) The fate of monoamines in perfused rabbit lung. Brit J Pharmacol 59:585-590

Glenner GC, Weissbach H, Redfield BG (1960) The histochemical demonstration of enzymatic activity by a nonenzymatic redox reaction. Reduction of tetrazolium salts by indolyl-3-acetaldehyde. J Histochem Cytochem 8:258-261

Glover V, Sandler M, Owen F, Riley GJ (1977) Dopamine is a monoamine oxidase B substrate in man. Nature 265:80-81

Glover V, Elsworth JD, Sandler M (1980) Dopamine oxidation and its inhibition by (−)-deprenyl. J Neural Transm Suppl 16:163-171

Glowinski J, Hamon M, Javoy F, Morot-Gaudry Y (1972) Rapid effects of monoamine oxidase inhibitors on synthesis and release of central monoamines. Adv Biochem Psychopharmacology 5:423-440

Goldberg LI (1964) Monoamine oxidase inhibitors. J Am Med Assn 190:456-462

Gomes B, Kloepfer HG, Oi S, Yasunobu KT (1976) The reaction of sulphydryl reagents with bovine hepatic monoamine oxidase. Evidence for the presence of two cysteine residues essential for activity. Biochim Biophys Acta 483:347-357

Goridis C, Neff NH (1971) Monoamine oxidase in sympathetic nerves: a transmitter specific enzyme type. Br J Pharmacol 43:814-818

Gorkin VZ (1973) Monoamine oxidase: versatility of catalytic properties and possible biological functions. Adv Pharmacol Chemother 11:1-50

Gorkin VZ (1976) Monoamine oxidase inhibitors and the transformation of monoamine oxidase. Ciba Foundation Symposium 39 (New Ser), Elsevier Amsterdam pp 61-68

Gorkin VZ, Tat'yanenko LV (1967) On the inhibition by harmine of oxidative deamination of biogenic amines. Life Sci 6:791-795

Grahame-Smith DB (1971) Studies *in vitro* on the relationship between brain tryptophan, brain 5-HT synthesis and hyperactivity in rats treated with a monoamine oxidase inhibitor and L-tryptophan. J Neurochem 18:1053-1066

Grahame-Smith DG (1974) How important is the synthesis of brain 5-hydroxytryptamine in the physiological control of its central function? Adv Biochem Psychopharmacology 10:83-91

Green AR, Grahame-Smith DG (1975) 5-Hydroxytryptamine and other indoles in the central nervous system. In: Plenum New York (Handbook of Psychopharmacology, vol 3)

Green AR, Youdim MBH (1975) Effects of monoamine oxidase inhibition by clorgyline, deprenyl and tranylcypromine on 5-hydroxytryptamine concentration in rat brain and hyperactivity following tryptophan administration. Br J Pharmacol 55:415-422

Green AR, Youdim MBH (1976) Use of behavioural model to study the action of monoamine oxidase inhibition *in vivo*. Ciba Foundation Symposium 39 Elsevier Amsterdam, pp 231-240

Green AR, Mitchell BD, Tordoff FC, Youdim MBH (1977) The evidence for dopamine deamination by both type A and type B monoamine oxidase in rat brain *in vivo* and for degree of inhibition of enzyme necessary for increased functional activity of dopamine and 5-hydroxyptamine. Br J Pharmacol 60:343-349

Greengrass PM, Tonge SR (1971) Changes in brain monoamine concentrations during the oestrous cycle in the mouse: possible pharmacological implications J Pharm Pharmacol 23:897-898

Guha SR, Ghosh SK (1970) Inhibition of monoamine oxidation in brain by monoamine oxidase inhibitors. Biochem Pharmacol 19:2929-2932

Hall DWR, Logan BW, Parsons GH (1969) Further studies on the inhibition of monoamine oxidase by M & B 9302 (Clorgyline). I. Substrate specificity in various mammalian species. Biochem Pharmacol 18:1447-1454

Harada M, Nagatsu T (1969) Identification of the flavin in the purified beef brain mitochondrial monoamine oxidase. Experientia 25:583-584

Harada M, Mizutani K, Nagatsu T (1971) Purification and properties of mitochondrial monoamine oxidase in beef brain. J Neurochem 18:559-569

Hartman BK (1972) The discovery and isolation of a new monoamine oxidase from brain. Biol Psychiat 4:147-155

Hartman BK, Udenfriend S (1972) The use of immunological technique for the characterization of bovine monoamine oxidase from liver and brain. Adv Biochem Psychopharmac 5:119-128

Hartman BK, Yasunobu KT, Udenfriend S (1971) Immunological identity of the multiple forms of beef liver mitochondrial monoamine oxidase. Arch Biochem Biophys 147:797-804

Hartshorn EA (1976) Interactions of CNS antidepressant psychotherapeutic agents. In: (eds) Handbook of Drug Interactions 3rd edition. Drug Intelligence Publications Hamilton Illinois

Hawkins M, Breakefield XO (1978) Monoamine oxidase A and B in cultured cells. J Neurochem 30:1391-1397

Hawkins M, Costa MR, Breakefield XO (1979) Distinct forms of monoamine oxidase expressed in hepatoma and Hela cells in culture. Biochem Pharmacol 28:525-528

Heikkila RE, Manzio L, Cabbat FS, Duvoisin RC (1984) Protection against the dopaminergic neurotoxicity of 1-methyl-4-phenyl 1,2,3,6-tetrahydropyridine by monoamine oxidase inhibitors. Nature 311:467-469

Heikkila RE, Duvoisin RC, Finberg JPM, Youdim MBH (1985) Prevention of MPTP-induced neurotoxicity by AGN 1133 and AGN 1135, selective inhibitors of monoamine oxidase B. Eur J Pharmacol 116:313-397

Hendley ED, Snyder SH (1968) Relationship between the action of monoamine oxidase inhibitors on the noradrenaline uptake system and their antidepressant efficacy. Nature 220:1330-1331

Herd JA (1969) A new antidepressant M & B 9302. A pilot study and a double blind controlled trial. Clin Trials 6:119-126

Himms-Hagen J (1972) Effects of catecholamines on metabolism. In: Blaschko H, Muscholl E (eds) Catecholamines. Springer Berlin Heidelberg New York, pp 363-441 (Handbook of Experimental Pharmacology, vol 33)

Hiramatsu A, Tsurushkiin S, Yasunobu KT (1975) Evidence for essential histidine residues in bovine-liver mitochondrial monoamine oxidase. Eur J Biochem 57:587-593

Ho BT (1972) Monoamine oxidase inhibitors. J Pharm Sci 61:821-837

Holman RB, Seagraves E, Elliot GR, Barchas JD (1976) Stereotyped hyperactivity in rats treated with tranylcypromine and specific inhibitor of 5-HT reuptake. Behav. Biol 16:507-514

Holman RB, Dement WC, Guilleminoult C (1977) Sleep disorders and neuroregulators. In: (eds) Neuroregulators and Psychiatric Disorders Oxford University Press New York

Holzbauer M, Youdim MBH (1973) The oestrous cycle and monoamine oxidase activity. Br J Pharmacol 48:600-608

Holzbauer M, Youdim MBH (1977) Physiological control of monoamine oxidase. In: Usdin E, Weiner N, Youdim MBH (eds) Structure and Function of Monoamine Enzymes, Dekker New York, pp 601-627

Horita A, Lowe MC (1972) On the extraneuronal nature of cardiac monoamine oxidase in the rat. Adv Biochem Psychopharmacology 5:227-242

Horwitz D, Goldberg LI, Sjoerdsma A (1960) Increased blood pressure responses to dopamine and norepinephrine produced by monoamine oxidase inhibitors in man. J Lab Clin Med 56:747-753

Houslay MD (1977) A model for the selective mode of action of the irreversible monoamine oxidase inhibitors clorgyline and deprenyl, based on studies of their ability to activate a Ca^{2+}-Mg^{2+} ATPase in defined lipid environments. J Pharm Pharmacol 29:664-669

Houslay MD (1978) Lysolecithin is a selective reversible inhibitor of mitochondrial monoamine oxidase. Biochem Pharmacol 27:1287-1288

Houslay MD (1980) Lipid substitution of mitochondrial monoamine oxidase can lead to the abolition of clorgyline selective inhibition without alteration in the A/B ratio assessed by substrate utilisation. Biochem Pharmacol 29:3211-3213

Houslay MD, Marchmont RJ (1980) Exposure of mitochondrial outer membranes to neuroaminidase selectively destroys monoamine oxidase A activity. J Pharm Pharmacol 32:65-66

Houslay MD, Tipton KF (1973a) The reaction pathway of membrane bound rat liver mitochondrial monoamine oxidase. Biochem J 135:735-750

Houslay MD, Tipton KF (1973b) The nature of the electrophoretically separable multiple forms of rat liver monoamine oxidase. Biochem J 135:173-186

Houslay MD, Tipton KF (1974) A kinetic evaluation of monoamine oxidase activity in rat liver mitochondrial outer membranes. Biochem J 139:645-652

Houslay MD, Tipton KF (1975a) Inhibition of beef plasma amine oxidase by clorgyline. Biochem Pharmacol 24:429-431

Houslay MD, Tipton KF (1975b) Rat liver mitochondrial monoamine oxidase. A change in the reaction mechanism on solubilization. Biochem J 145:311-321

Houslay MD, Tipton KF (1975c) Amine competition for oxidation by rat liver mitochondrial monoamine oxidase. Biochem Pharmacol 24:627-631

Houslay MD, Garrett NJ, Tipton KF (1974) Mixed substrate experiments with human brain monoamine oxidase. Biochem Pharmacol 23:1937-1944

Houslay MD, Tipton KF, Youdim MBH (1976) Multiple form of monoamine oxidase: fact and artefact. Life Sci 19:467-473

Howell SL, Montague W (1973) Adenylcyclase activity in isolated rat islets of Langerhans. Biochem Biophys Acta 320:44-52

Huang CL, Schulz AR (1972) The effect of inhibitors of thyroid MAO on the incorporation of iodide into thyroid slice protein. Life Sci 11:975-982

Huang RH (1980) Lipid-protein interactions in the multiple forms of monoamine oxidase. Enzymatic and esr studies with purified intact rat brain mitochondria. Mol Pharmacol 17:192-198

Huang RH (1981) Topology and lipid protein association of MAO-A and MAO-B. In: Usdin E, Weiner N, Youdim MBH (eds) Function and Regulation of Monoamine Enzymes: Basic and Clinical Aspects. MacMillan London, pp 489-501

Huang RH, Eiduson S (1977) Signification of multiple forms of brain monoamine oxidase *in situ* as probed by electron spin resonance. J Biol Chem 252:284-290

Huang RH, Faulkner R (1980) Lipid-protein interactions in the multiple forms of monoamine oxidases: lipases as probes using purified intact rat brain mitochondria. Mol Pharmacol 18:267-273

Hunter KR, Boakes AJ, Laurence DR, Stern GM (1970) Monoamine oxidase inhibitors and L-dopa. Br Med J 3:388-398

Husain M, Edmondson DE, Singer TP (1982) Kinetic studies on the catalytic mechanism of liver monoamine oxidase. Biochemistry 21:595-600

Huszti Z (1972) Kinetic studies on rat brain monoamine oxidase. Mol Pharmacol 8:385-397

Ichinose M, Gomes B, Sanemori H, Yasunobu KT (1982) Bovine liver mitochondrial monoamine oxidase is not an iron-dependent enzyme. J Biol Chem 257:887-888

Ishmahan G, Parvez H, Parvez S, Youdim MBH (1978) The effect of selective monoamine oxidase inhibitors clorgyline and deprenyl upon tissue glycogen stores and blood glucose levels. In: Bellinger K, Klatzo I, Riederer P (eds) Neurotransmitters in Cerebral Coma and Stroke. Springer Wien

Iversen LL (1973) Catecholamine uptake processes. Br Med Bull 29:130-135

Jain M (1977) Monoamine oxidase: examination of multiple forms. Life Sci 20:1925-1934

Jain M, Sands FL (1974) Electrophoretic homogeneity of solubilized human brain monoamine oxidase. J Neurochem 23:1291-1293

Jarrott B (1971) Occurrence and properties of monoamine oxidase in adrenergic neurones. J Neurochem 18:7-16

Jarrott B, Iversen LL (1968) Subcellular distribution of monoamine oxidase activity in rat liver and vas deferens. Biochem Pharmacol 17:1619-1625

Jarrott B, Iversen LL (1971) Noradrenaline metabolizing enzymes in normal and sympathetically denervated vas deferens. J Neurochem 18:1-6

Jarrott B, Langer SZ (1971) Changes in monoamine oxidase and catechol-O-methyl transferase activities after denervation of the nictitating membrane of the cat. J Physiol (Lond) 212:549-559

Johnson CL (1976) Quantititive structure-acitivity studies on monoamine oxidase inhibitors. J Med Chem 19:600-605

Johnston JP (1968) Some observations upon a new inhibitor of monoamine oxidase in brain tissue. Biochem Pharmacol 17:1285-1297

Jouvet M (1969) Biogenic amines and the state of sleep. Science 163:32-41

Jouvet N (1974) The role of monoaminergic neurons in the regulation and function of sleep. In: (eds) Basic Sleep Mechanism. Academic Press New York

Kalir A, Sabbagh A, Youdim MBH (1981) Selective acetylenic "suicide" and reversible inhibitors of monoamine oxidase type A and B. Br J Pharmacol 73:55-64

Kandaswami C, d'Iorio A (1978) On rat liver mitochondrial monoamine oxidase activity with lipids. Arch Biochem Biophys 190:847-849

Kandaswami C, d'Iorio A (1979) On hepatic mitochondrial monoamine oxidase activity in lipid deficiency. Can J Biochem 57:588-594

Kandaswami C, Diaz Borges JM, d'Iorio A (1977) Studies on the fractionation of monoamine oxidase from rat liver mitochondria. Arch Biochem Biophys 183:273-280

Keane PE, Chanoine F, Strolin-Benedetti M (1979) The effects of specific type A and B MAO inhibitors on reserpine-induced changes in brain dopamine and serum prolactin levels in the rat. In: Singer TP, von Korff RW, Murphy DL (eds) Monoamine Oxidase: Structure, Function and Altered Functions. Academic Press New York, pp 341-346

Kearney EB, Salach JI, Walker WH, Seng RL, Kenney WC, Zeszotek E, Singer TP (1971) The covalently bound flavin of hepatic monoamine oxidase. I. Isolation and sequence of a flavin peptide and evidence for binding at the 8 α position. Eur J Biochem 24:321-327

Kenney WC, Nagy J, Salach JI, Singer TP (1979) Structure of the covalent phenylhydrazine adduct of monoamine oxidase. In: Singer TP, von Korff RW, Murphy DL (eds) Monoamine Oxidase: Structure, Function and Altered Functions, Academic Press New York, pp 25-37

Kim HC, d'Iorio A (1968) Possible isoenzymes of monoamine oxidase in rat tissues. Can J Biochem 46:295-297

Kinemuchi H, Wakui Y, Toyoshima Y, Hayashi N, Kamijo K (1979) β-Phenylethylamine (PEA), a concentration-dependent preferential substrate for multiple forms of MAO. In: Singer TP, von Korff RW, Murphy DL (eds) Monoamine Oxidase: Structure, Function and Altered Functions, Academic Press New York, pp 205-212

Kinemuchi H, Wakui W, Kamijo K (1980) Substrate selectivity of type A and type B monoamine oxidase in rat brain. J Neurochem 35:109-115

Kinemuchi H, Arai Y, Oreland L, Tipton KF, Fowler CJ (1982) Time-dependent inhibition of monoamine oxidase by β-phenethylamine. Biochem Pharmacol 31:959-964

Kinemuchi H, Fowler C-T, Tipton KF (1987) The Neurotoxicity of 1-methyl-4-phenyl-1,2,3,6-tetrahydropyridine (MPTP) and its relevance to Parkinson's Disease. Neurochem Inter 11:359-382

Kline NS (1963) Use of pargyline (Eutonyl) in private practice Ann NY Acad Sci 107:1090-1106

Klingman GI (1966) Monoamine oxidase activity of peripheral organs and sympathetic ganglia of rat after sympathectomy. Biochem Pharmacol 15:1729-1736

Knoll J (1976) Analysis of the pharmacological effects of selective monoamine oxidase inhibitors. Ciba Foundation Symposium 39 (New Ser.), Elsevier Amsterdam pp 135-155

Knoll J (1978) The possible mechanisms of action of (−)-deprenyl in Parkinson's disease. J Neural Transmis 43:177-198

Knoll J (1979) Structure-activity relationships of the selective inhibitors of MAO-B. In: Singer TP, von Korff RW, Murphy DL (eds) Monoamine Oxidase: Structure Function and Altered Functions, Academic Press New York, pp 431-446

Knoll J (1980) Monoamine oxidase chemistry and pharmacology. In: Sandler M (ed) Enzyme Inhibitors on Drugs, MacMillan London, pp 151-171

Knoll J (1981) The pharmacology of selective MAO inhibitors. In: Youdim MPH, Paykel ES (eds) Monoamine Oxidase Inhibition; The State of the Art. Chichester: Wiley. pp 43-64.

Knoll J, Magyar K (1972) Some puzzling pharmacological effects of monoamine oxidase inhibitors. Adv Biochem Psychopharmacol 5:393-408

Knoll J, Vizi ES, Somogyi E (1968) Phenylisopropylmethylpropinylamine (E-250), a monoamine oxidase inhibitor antagonising the effects of tyramine. Arzneim-Forsch 18:109-112

Knoll J, Vizi ES, Magyar K (1972) Pharmacological studies on some central effects of amphetamines. In: Lissale K (ed) Recent Developments of Neurobiology in Hungary III. Results in Neuroanatomy, Neurophysiology, Neuropathophysiology and Neuropharmacology. Publishing House of the Hungarian Academy of Sciences, Budapest, pp 167-217

Knoll J, Ecsery Z, Magyar K, Satory E (1978) Novel (−)-deprenyl-derived selective inhibitors of B-type monoamine oxidase. The relation of structure to their action. Biochem Pharmacol 27:1739-1747

Kochersperger L, Waguespack A, Patterson J, Hsieh C, Weyler W, Salach J, Denney R (1986) Immunological uniqueness of human monoamine oxidases A and B. New evidence from studies with monoclonal antibody to human, MAO A. J Neurosci In press.

Kopin IJ, Axelrod J (1963) The role of monoamine oxidase in the release and metabolism of norepinephrine. Ann NY Acad Sci 107:848-855

Kopin IJ, Fischer JE, Musacchio J, Horst WD (1964) Evidence for a false neurochemical transmitter as a mechanism for the hypotensive effect of monoamine oxidase inhibitors. Biochemistry 52:716-721

Kroon MC, Veldstra H (1972) Forms of rat brain mitochondrial monoamine oxidase. Subcellular fractionation. FEBS Lett 24:173-176

Kuhn DM, Wolf WA, Youdim MBH (1985) 5-Hydroxytryptamine release in vivo from a cytoplasmic pool; studies on the 5-HT behavioural syndrome in reserpinized rats. Br J Pharmacol 84:121-129

Kunimoto N, Hazama H, Kamase H (1979) Regional distribution of type B MAO activity towards β-phenylethylamine in the individual rat hypothalamic nuclei. Brain Res 176:175-179

Kwatra MM, Sourkes TL (1981) Substrate-dependent activation energy of the reaction catalysed by monoamine oxidase. Arch Biochem Biophys 210:531-536

Lader MH, Sakalis G, Tansella M (1970) Interactions between sympathomimetic amines and a new monoamine oxidase inhibitor. Psychopharmacologia 18:118-123

Lagnado JR, Okamoto M, Youdim MBH (1971) The effect of tetrazolium salts on monoamine oxidase activity. FEBS Lett 17:117-120

Laroche MJ, Brodie BB (1960) Lack of relationship between inhibition of monoamine oxidase and potentiation of hexobarbital hypnosis. J Pharmacol Exp Ther 130:134-137

Lees AJ, Shaw KM, Kohout LJ, Stern GM, Elsworth JD, Sandler M, Youdim MBH (1977) Deprenil in Parkinson's disease Lancet 11:791-795

Leffler JE, Grunwald E (1963) Rates and equilibria of organic reactions Wiley New York

Levitt P, Pintar J, Breakefield X (1982) Immunocyto-chemical demonstration of monoamine oxidase B in brain astrocytes and serotonergic neurons. Proc Natl Acad Sci USA 79:6385-6389

Lewinsohn R, Glover V, Sandler M (1980a) β-Phenylethylamine and benzylamine as substrates for human monoamine oxidase A: a source of some anomalies? Biochem Pharmacol 29:777-781

Lewinsohn R, Glover V, Sandler M (1980b) Development of benzylamine oxidase and monoamine oxidase A and B in man. Biochem Pharmacol 29:1221-1230

Lyles GA, Callingham BA (1975) Evidence for a clorgyline resistant monoamine metabolizing activity in the rat heart. J Pharm Pharmacol 27:682-691

Lyles GA, Greenawalt JW (1978) Possible heterogeneity of type B monoamine oxidase in pig heart mitochondria. Biochem Pharmacol 27:923-935

Lyles GA, Shaffer CJ (1979) Substrate specificity and inhibitor sensitivity of monoamine oxidase in rat kidney mitochondria. Biochem Pharmacol 28:1099-1106

Mackay AVP, Davies P, Dewar AJ, Yates CM (1978) Regional distribution of enzymes associated with neurotransmission by monoamines, acteylcholine and GABA in the human brain. J Neurochem 30:827-839

Maitre L, Delini-Stula A, Waldmeier PC (1976) Relations between the degree of monoamine oxidase inhibition and some psychopharmacological responses to monoamine oxidase inhibitors in rats. Ciba Foundation Symposion 39 (New Ser) Elsevier Amsterdam, pp 247-267

Mann J, Gershon S (1980) L-Deprenyl, a selective monoamine oxidase type B inhibitor in endogenous depression. Life Sci 26:877-882

Mantle TJ (1978) Molecular weight determination, density gradient centrifugation, electrophoresis and irradiation inactivation. In: Kornberg HL, Metcalfe JC, Northcote DH, Pogson CI, Tipton KF (eds) Techniques in the Life Sciences: Biochemistry vol B1/1, Elsevier Amsterdam, pp B1056

Mantle TJ, Tipton KF (1982) Monoamine oxidase A and B: time for re-evaluation? In: (eds) Trends in Autonomic Pharmacology, vol 2. Urban & Schwarzenberg, Baltimore München

Mantle TJ, Wilson K, Long RF (1975a) Studies on the selective inhibition of membrane-bound rat liver monoamine oxidase. Biochem Pharmacol 24:2031-2038

Mantle TJ, Wilson K, Long RF (1975b) Kinetic studies of membrane bound rat liver monoamine oxidase. Biochem Pharmacol 24:2039-2046

Mantle TJ, Tipton KF, Garrett NJ (1976) Inhibition of monoamine oxidase by amphetamine and related compounds. Biochem Pharmacol 25:2073-2077

Markey SP, Castagnoli N, Trevor AJ, Kopin IJ (1986) Editors. MPTP: a Neurotoxin Producing a Parkinsonian Syndrome. London: Academic Press

Marley E (1977) Monoamine oxidase inhibitors and drug interactions. In: Grahame-Smith DG (ed) Drug Interactions. MacMillan London

Marley E, Blackwell B (1970) Interactions of monoamine oxidase inhibitors, amines and foodstuffs. Adv Pharmacol Chemother 8:186-239

Maronde RF, Haywood LJ (1963) Evaluation of the monoamine oxidase inhibitor, pargyline, as an antihypertensive agent: A. Clinical results. Ann NY Acad Sci 107:975-979

Marsden CA, Brock OJ, Guldberg HC (1971) Catechol-O-methyl transferase and monoamine oxidase activities in rat submaxillary gland: effects of ligation, sympathectomy and some drugs. Eur J Pharmacol 15:335-342

Mass JW (1979) Catecholamines and the affective disorders. In: Aromatic Amino Acid Hydroylase and Mental Disease. Wiley London

Maycock AL, Abeles RH, Salach JI, Singer TP (1976) The action of acetylenic inhibitors on mitochondrial monoamine oxidase: structure of the flavin site in the inhibited enzyme. Ciba Foundation Symposium 39 (New Ser) Elsevier Amsterdam, pp 37-47

McCauley R, Racker E (1973) Separation of two monoamine oxidases from bovine brain. Mol Cell Biochem 1:73-81

McEwen CM, Sasaki G, Lenz WR (1968) Human liver mitochondrial monoamine oxidase. I. Kinetic studies of model interactions. J Biol Chem 243:5217-5225

McEwen CM, Sasaki G, Jones DC (1969) Human liver mitochondrial monoamine oxidase. II. Determinants of substrate and inhibitor specificities. Biochemistry 8:3952-3962

Mendis M, Pare CMB, Sandler M, Glover V, Stern G (1981) (−)-Deprenyl in the treatment of depression. In: (eds) Monoamine Oxidase Inhibitors: The State of the Art. Wiley Chichester, pp 171-176

Mendlewicz J, Youdim MBH (1977) Monoamine oxidase inhibitors and prolactin secretion. Lancet i:507

Mendlewicz J, Youdim MBH (1978) Anti-depressant potentiation of 5-hydroxytryptophan by L-deprenyl, an MAO 'type B' inhibitor. J Neural Tansmis 43:279-286

Mendlewicz J, Youdim MBH (1979) Antidepressant potentiation of 5-hydroxytryptophan by l-deprenyl in affective illness. J Aff Disord 2:137-146

Mendlewicz J, Youdim MBH (1981) A selective MAO-B inhibitor (l-deprenil) and 5-HTP as anti-depressant therapy. In: (eds) Monoamine Oxidase Inhibitors: The State of the Art. Wiley Chichester, pp 177-188

Mendlewicz J, Youdim MBH (1983) l-Deprenyl, a selective monoamine oxidase type B inhibitor in the treatment of depression. A double blind evaluation. Br J Psychiat 142:507-511

Minamiura N, Yasunobu KT (1978) Bovine liver monoamine oxidase. A modified purification procedure and preliminary evidence for two subunits and one FAD. Arch Biochem Biophys 189:481-482

Mitra C, Guha SR (1980) Serotonin oxidation by type B MAO of rat brain. Biochem Pharmacol 29:1213-1217

Montague W, Howell SL (1975) Cyclic AMP and the physiology of the islets of Langerhans. In: (eds) Advances in Cyclic Nucleotide Research, vol 6. Raven Press New York

Modigh K (1973) Effects of chloroimipramine and protriptyline on the hyperactivity induced by 5-hydroxytryptophan after peripheral decarboxylase inhibition in mice. J Neural Transmis 34:101-109

Mondovi B (1985) Editor. Structure and Functions of Amine Oxidases. Boca Raton: CRC Press

Moore KE (1977) The action of amphetamine on neurotransmitters. Biol Psychiat 12:451-462

Mueller RA, de Champlain J, Axelrod J (1968) Increased monoamine oxidase activity in isoproterenol-stimulated submaxillary glands. Biochem Pharmacol 17:2455-2461

Muller J, Delage C (1977) Ultracytochemical demonstration of monoamine oxidase activity in nervous and non-nervous tissue of the rat. J Histochem Cytochem 25:337-348

Murphy DL (1976) Clinical, genetic, hormonal and drug influences on the activity of human platelet monoamine oxidase. Ciba Foundation Symposium 39 (New Ser) Amsterdam Elsevier, pp 341-351

Murphy DL, Donnelly CH (1974) Monoamine oxidase in man: Enzyme characteristics in platelets, plasma and other human tissues. Adv Biochem Psychopharmacol 12:71-86

Murphy DL, Wyatt RJ (1972) Reduced MAO activity in blood platelets from schizophrenic patients. Nature 238:225-226

Murphy DL, Donelly CH, Richelson E (1976) Substrate and inhibitor related characteristics of monoamine oxidase in C6 rat glial cells. J Neurochem 26:1231-1235

Murphy DL, Donnelly CH, Richelson E, Fuller RW (1978) N-substitued cyclopropylamines as inhibitors of MAO-A and -B forms. Biochem Pharmacol 27:176-1769

Murphy DL, Lipper S, Campbell IC, Major LF, Slater S, Buchsbaum MS (1979) Comparative studies of MAO-A and MAO-B inhibitors in man. In: Singer TP, von Korff RW, Murphy DL (eds) Monoamine Oxidase: Structure, Function and Altered Functions. Academic Press New York, pp 457-475

Murphy DL, Roy B, Pickar D, Lipper S, Cohen RM, Jimerson P, Lake CR, Muscetlola G, Saavedra J, Kopin I (1981) Cardiovascular changes accompanying MAO inhibition in man. In: Usdin E, Weiner N, Youdim MBH (eds) Function and Regulation of Monoamine Enzymes, MacMillan London, pp 549-560

Nagatsu T, Nakano G, Mizutani K, Harada M (1972) Purification and properties of amine oxidases in brain and connective tissue (dental pulp). Adv Biochem Psychopharmacol 5:25-36

Nagatsu T, Nakano T, Kato T, Kano-Tanaka K, Higashida H (1982) Expression of monoamine oxidase A and B types in hybrid cells of neuroblastoma. In: Kamijo K, Usdin E, Nagatsu T (eds) Monoamine Oxidase: Basic and Clinical Frontiers, Excerpta Medica Amsterdam, pp 297-288

Nagy J, Salach JI (1981) Identity of the active site flavin peptide fragments from the human "A"-form of monoamine oxidase and the bovine "B"-form of monoamine oxidase. Arch Biochem Biophys 208:388-394

Naoi M, Yagi K (1980) Effect of phospholipids on beef heart mitochondrial monoamine oxidase. Arch Biochem Biophys 205:18-26

Nara S, Gomes B, Yasunobu KT (1966a) Amine oxidase. VII. Beef liver mitochondrial monoamine oxidase, a copper containing protein. J Biol Chem 241:2774-2780

Neff NH, Goridis C (1972) Neuronal monamine oxidase: specific enzyme types and their rate of formation. Adv Biochem Psychopharmac 5:307-323

Neff NH, Tozer TN (1968) *In vivo* measurement of brain serotonin turnover. Adv Pharmacol 6A:97-109

Neff NH, Yang HYT (1974) Another look at the monoamone oxidases and the MAO inhibitor drugs. Life Sci 14:2061-2074

Neff NH, Yang HYT, Goridis C (1973) Degradation of the transmitter amines by specific types of monoamine oxidases. In: Usdin E, Snyder SH (eds) Frontiers in Catecholamine Research. Pergamon Press New York

Neff NH, Yang HYT, Fuentes JA (1974) The use of selective monoamine oxidase inhibitor drugs to modify amine metabolism in brain. Adv Biochem Psychopharmacol 12:49-57

Nelson DL, Herbet A, Pétillot Y, Pichat L, Glowinski J, Hamon M (1979) [^{3}H]Harmaline as a specific ligand of MAO-A. I. Properties of the active site of MAO A from rat and bovine brains. J Neurochem 32:1817-1827

Neupert W, Brdiczka D, Bucher T (1967) Incorporation of amino acids into the outer and inner membrane of isolated rat liver mitochondria. Biochem Biophys Res Commun 27:488-493

Nies A, Robinson DS, Ravaris CL, Ives JO (1975) The efficacy of the monoamine oxidase inhibitor phenelzine: dose effects and prediction of response. In: (eds) Neuropsychopharmacology. Excerpta Medica Amsterdam

Oates JA, Nirenberg PZ, Jepson JB, Sjoerdsma A, Udenfriend S (1963) Conversion of phenylalanine to phenyl ethylamine in patients with phenylketonuria. Proc Soc Exp Biol Med 112:1078-1081

O'Carroll A-M, Fowler CJ, Phillips JP, Tobia I, Tipton KF (1983) The deamination of dopamine by human brain monoamine oxidase: specificity for the two enzyme forms in seven brain regions. Naunyn-Schmiedeberg's Arch Pharmacol 322:198-202

Ogawa K, Gudbjarnason S, Bing RJ (1967) Nitroglycerin (glyceryl trinitrate) as a monoamine oxidase inhibitor. J Pharmacol Exp Ther 155:449-455

Ogren SO, Ask AJ, Holm AC, Florvall L, Lindbom LO, Lundstrom J, Ross JB (1981) Biochemical and pharmacological properties of a new selective and reversible monoamine oxidase inhibitor, FLA 336 (+). In: Youdim MBH, Paykel ES (eds) Monoamine Oxidase inhibitor — The State of the Art. Wiley Chichester

Oi S, Yasunobu KT (1973) Mechanistic aspects of the oxidation of amines by monoamine oxidase. Biochem Biophys Res Commun 53:631-637

Oi S, Shimada K, Inamasu M, Yasunobu KT (1970) Mechanistic studies of beef liver mitochondrial amine oxidase. XVII. Amine oxidase. Arch Biochem Biophys 139:28-37

Oi S, Yasunobu KT, Westley J (1971) The effect of pH on the kinetic parameters and mechanism of beef liver monoamine oxidase. Arch Biochem Biophys 145:557-564

Onesti G, Novack P, Ramirez O, Brest AN, Moyer JH (1964) Hemodynamic effects of pargyline in hypertensive patients. Circulation 30:830-835

Oreland L (1971) Purification and properties of pig liver mitochondrial monoamine oxidase. Arch Biochem Biophys 146:410-421

Oreland L (1972) Some properties of pig liver mitochondrial monoamine oxidase. Adv Biochem Psychopharmac 5:37-43

Oreland L (1979) The activity of human brain and thrombocyte monoamine oxidase (MAO) in relation to various psychiatric disorders. I. MAO activity in some disease states. In: Singer TP, von Korff RW, Murphy DL (eds) Academic Press New York, pp 379-387

Oreland L, Ekstedt B (1972) Soluble and membrane-bound pig liver mitochondrial oxidase: thermostability, tryptic digestibility and kinetic properties. Biochem Pharmacol 21:2479-2488

Oreland L, Kinemuchi H, Stigbrand T (1973a) Pig liver monoamine oxidase: studies on the subunit structure. Arch Biochem Biophys 159:854-860

Oreland L, Kinemuchi H, Yoo BY (1973b) The mechanism of action of the monoamine oxidase inhibitor pargyline. Life Sci 13:1533-1541

Oreland L, Fowler CJ, Carlsson A, Magnusson T (1980) Monoamine oxidase-A and -B activity in the rat brain after hemitransection. Life Sci 26:139-146

Owen F, Bourne RC, Lai JCK, Williams R (1977) The heterogeneity of monoamine oxidase in distinct populations of rat brain mitochondria. Biochem Pharmacol 26:289-292

Paech C, Salach JI, Singer TP (1979) Suicide inactivation of monoamine oxidase by *trans*-phenylcyclopropylamine In: Singer TP, von Korff RW, Murphy DL (eds) Monoamine oxidase: Structure, Function and Altered Functions. Academic Press New York, pp 39-50

Pare CMB (1976) Introduction to clinical aspects of monoamine oxidase inhibitors in the treatment of depression. Ciba Foundation Symposium 39 (New Ser.) Elsevier Amsterdam, pp 271-280

Parkes JD, Tarsy D, Marsden CD, Bovil KT, Phipps JA, Rose P, Asselman P (1975) Amphetamines in the treatment of Parkinsons's disease. J Neurol Neurosurg Psychiat 38:323-237

Patek DR, Hellerman L (1974) Mitochondrial monoamine oxidase. Mechanism of inhibition by phenylhydrazine and by aralkylhydrazines. Role of enzymatic oxidation. J Biol Chem 249:2372-2380

Peers EM, Lyles GA, Callingham BA (1980) The deamination of isoamylamine by monoamine oxidase in mitochondrial preparations from rat liver and heart: a comparison with phenylethylamine. Biochem Pharmacol 29:1097-1102

Pettinger WA, Oates JA (1968) Supersensitivity to tyramine during monoamine oxidase inhibition in man. Clin Pharmacol Ther 9:341-344

Pettinger WA, Korn A, Spiegel H, Solomon HM, Pocelinko R, Abrams WB (1969) Debrisoquin, a selective inhibitor of intraneuronal monoamine oxidase in man. Clin Pharmacol Ther 10:667-674

Philips SR (1976) β-Phenylethylamine: a metabolically and pharmacologically active amine. In: (eds) Non-Catecholic Phenylethylamines. Dekker New York

Pintar JE, Cawthon RM, Costa MCC, Breakefield XO (1979) A search for structural differences in MAO: electrophoretic analysis of ^{3}H-pargyline labelled proteins. In: Singer TP, von Korff RW, Murphy DL (eds) Monoamine Oxidase: Structure, Function, and Altered Functions, Academic Press New York, pp 185-196

Pintar JE, Barbosa J, Francke U, Castiglione CM, Hawkins M, Breakefield XO (1981a) Gene for monoamine oxidase type A assigned to the human X chromosome. J Neurosci 1:166-175

Pintar JE, Cawthon RM, Hawkins M, Castiglione CM, Breakefield XO (1981b) Biochemical and genetic analysis of MAO-A and B. In: Usdin E, Weiner N, Youdim MBH (eds) Function and Regulation of Monoamine Enzymes. MacMillan London, pp 855-863

Planz G, Quiring K, Palm D (1972) Rates of recovery of irreversibly inhibited monoamine oxidase: a measure of enzyme protein turnover. Naunyn-Schmiedeberg's Arch Pharmacol 273:27-42

Pletscher A, Gey KF, Zeller P (1960) Monoaminoxidase Hemmer. Arzneim-Forschung 2:417-590

Poirier LJ, Sourkes TL, Bedard P (1979) The extrapyramidal system and its disorders. Raven Press New York

Powell JF, Craig IW (1977) Biochemical and immunological studies of the monoamine oxidase activities of cultured human cells. Biochem Soc Trans 5:180-182

Pratesi P, Blaschko H (1959) Specificity of amine oxidase for optically active substrates and inhibitors. Brit J Pharmacol 14:256-260

Puig M, Wakade AR, Kirpekar SM (1972) Effect on the sympathetic nervous system of chronic treatment with pargyline and L-dopa. J Pharmacol Exp Ther 182:130-134

Quadri SK, Kledzik GS, Meites J (1973) Reinitiation of estrus cycle in old constant estrus rats by centrally acting drugs. Neuroendocrinology 11:248-255

Rand MJ, Trinker FR (1968) The mechanism of the augmentation of responses to indirectly acting sympathomimetic amines by monoamine oxidase inhibitors. Brit J Pharmacol 33:287-303

Rao GHR, Einzig S, Redd KR, White JG (1979) Tranylcypromine induced hypertension is not mediated by the inhibition of prostacyclin synthesis. Prostaglandins and Medicine 3:201-210

Reynolds GP, Riederer P, Sandler M, Jellinger K, Seeman D (1978a) Amphetamine and 2-phenylethylamine in post-mortem parkinsonian brain after deprenyl administration. J Neural Transmis 43:271-278

Reynolds GP, Elsworth JD, Blou K, Sandler M, Lees AJ, Stern G (1978b) Deprenyl is metabolized to methamphetamine and amphetamine in man. Brit J Clin Pharmacol 6:542-544

Richter D (1937) Adrenaline and monoamine oxidase. Biochem J 31:2022-2028
Riederer P, Reynolds GP (1980) Deprenyl is a selective inhibitor of brain MAO-B in the long term treatment of Parkinson's disease. Br J Clin Pharmacol 9:98-99
Riederer P, Youdim MBH (1986) Monoamine oxidase activity and monoamine metabolism in brains of Parkinson patients treated with l-deprenyl. J Neurochem 46:1359-1365
Riederer P, Youdim MBH, Birkmayer W, Jellinger K (1978a) Monoamine oxidase activity during (−) deprenil therapy: human brain post mortem studies. Adv Biochem Psychopharmacology 19:377-382
Riederer P, Youdim MBH, Rausch WD, Birkmayer W, Jellinger K, Seemann D (1978b) On the mode of action of L-deprenyl in the human central nervous system. J Neural Transmis 43:217-226
Riederer R, Reynolds GP, Youdim MBH (1981) Selectivity of MAO inhibitors in human brain and their clinical consequences. In: Youdim MBH, Paykel ES (eds) Monoamine Oxidase Inhibitors: The State of the Art. Wiley Chichister
Richter D (1937) Adrenaline and monoamine oxidase. Biochem J 31:2022-2028
Rinne UK (Ed) (1983) A new approch to the treatment of Parkinson's disease. Acta Neurol Scand No 95:7-144
Rinne UK, Sirrtola T, Sonninen V (1978) (−)-Deprenyl treatment of 'on-off' phenomena in Parkinson's disease. J Neural Transmis 43:253-262
Robinson GA, Butcher RW, Sutherland EW (1972) The catecholamines In: (eds) Biochemical Actions of Hormones. vol II. Academic Press New York
Rogers KJ, Thornton JA (1969) The interaction between monoamine oxidase inhibitors and narcotic analysis in mice. Brit J Pharmacol 36:470-480
Roth JA (1978) Inhibition of human brain type B monoamine oxidase by tricyclic psychoactive drugs. Mol Pharmacol 14:164-171
Roth JA (1979) Effect of drugs on inhibition of oxidized and reduced form of MAO. In: Singer TP, von Korff RW, Murphy DL (eds) Monoamine Oxidase: Structure Function and Altered Functions, Academic Press New York, pp 153-168
Roth JA, Feor K (1978) Deamination of dopamine and its 3-O-methylated derivative by human brain monoamine oxidase. Biochem Pharmacol 27:1606-1608
Roth JA, Gillis CN (1974) Deamination of β-phenylethylamine by monoamine oxidase. Inhibition by imipramine. Biochem Pharmacol 23:2537-2545
Roth JA, Gillis CM (1975) Multiple forms of amine oxidase in perfused rabbit lung. J Pharmac Exp Ther 194:537-544
Roth J, Whittmore R, Shakarjian M, Eddy B (1979) Inhibition of human brain type A and type B monoamine oxidase by chlorpromazine and metabolites. Commun Psychopharmacol 3:236-244
Russell SM, Davey J, Mayer RJ (1979a) Vectorial orientation of monoamine oxidase in the mitochondrial outer membrane. Biochem J 181:7-14
Russell SM, Davey J, Mayer RJ (1979b) Immunochemical characterization of monoamine oxidase from human liver, placenta, platelets and brain cortex. Biochem J 181:15-20
Russell SM, Davey J, Mayer RJ (1979c) The topography and turnover of mitochondrial monoamine oxidase. In: Singer TP, von Korff RW, Murphy DL (eds) Monoamine Oxidase: Structure, Function and Altered Functions, Academic Press New York, pp 265-272
Sabelli H-C, Giardina WJ, Mosnaim AD, Sabelli NH (1973) A comparison of the function roles of NE, DA and PEA in the central nervous system. Acta Physiol Pol 24:33-40
Sabelli H-C, Borison RL, Diamon BI, Havdala HS, Narasimhachori N (1978) Phenylethylamine and brain function. Biochem Pharmacol 27:1729-1730

Sagara Y, Ito A (1982) In vitro synthesis of monoamine oxidase of rat liver outer mitochondrial membrane. Biochem Biophys Res Commun 109:1102-1107

Salach JI (1979) Monoamine oxidase from beef liver mitochondria; simplified isolation procedure, properties and determination of its cysteinyl flavin content. Arch Biochem Biophys 192:128-137

Salach JI, Detmer K (1979) Chemical characterization of monoamine oxidase from human placental mitochondria. In: Singer TP, von Korff RW, Murphy DL (eds) Monoamine Oxidase: Structure, Function and Altered Functions Academic Press New York, pp 121-128

Salach JI, Singer TP, Yasunobu KT, Minamiura N, Youdim MBH (1976) Cysteinyl flavin in monoamine oxidase from the central nervous system. Ciba Foundation Sympsoum 39 (New Ser) Elsevier Amsterdam, pp 49-50

Salach JI, Detmer K, Youdim MBH (1979) The reaction of bovine and rat liver monoamine oxidase with [^{14}C]-clorgyline and [^{14}C]-deprenyl. Mol Pharmacol 16:234-241

Sandler M, Youdim MBH (1972) Multiple forms of monoamine oxidase: functional significance. Pharmac Rev 24:331-348

Sandler M, Bonham Carter S, Goodwin BL, Ruthven CRJ, Youdim MBH, Hanington E, Cuthbert MF, Pare CMB (1974) Multiple forms of monoamine oxidase: some *in vivo* correlations. In: (eds) Neuropharmacology of Monoamines and their Regulatory Enzymes. Raven Press New York

Sandler R (1968) Concentration of norepinephrine in the hypothalamus of the rat in relation to the estrus cycle. Endocrinology 83:1383-1386

Sannerstedt R (1967) Hemodynamic effects of pargyline hydrochloride at rest and during exercise in hypertension. Acta Med Scand 181:699-706

Sarai K, Frazer A, Brunswick D, Mendels J (1978) Desmethylimipramine-induced decrease of β-adrenergic receptor binding in rat cerebral cortex. Biochem. Pharmacol 27:2179-2181

Savage DD, Mendels J, Frazer A (1980) Decrease in (^{3}H)-serotonin binding in rat brain produced by the repeated administration of either monoamine oxidase inhibitors or centrally acting serotonin agonists. Neuropharmacol 19:1063-1071

Sawyer ST, Greenawalt JW (1979) Association of monoamine oxidase with lipid. A comparative study of mitochondria from Novikoff hepatoma and rat liver. Biochem Pharmacol 28:1735-1744

Schoepke HG, Wiegand RG (1963) Relation between norepinephrine accumulation or depletion and blood pressure responses in the cat and rat following pargyline administration. Ann NY Acad Sci 107:924-934

Schümann HJ, Philippu A (1962) Release of catecholamines from isolated medullary granules by sympathomimetic amines. Nature 193:890-891

Schurr A (1982) Monoamine oxidase — to B or not to B? Life Sci 30:1059-1063

Schurr A, Poráth O, Krup M, Livne A (1978) The effects of hashish components and their mode of action on monoamine oxidase from the brain. Biochem Pharmacol 27:2513-2517

Schwartz MA, Wyatt RJ, Yang H-YT, Neff NH (1974) Multiple forms of brain monoamine oxidase in schizophrenic and normal individuals. Arch Gen Psychiat 31:557-560

Severina IS (1973) On the substrate-binding sites of the active centre of mitochondrial monoamine oxidase. Eur J Biochem 38:239-246

Severina IS (1979) Mechanism of selective inhibition by clorgyline and deprenyl of the activity of mitochondrial monoamine oxidase and the possible nature of its forms A and B. In: Singer TP, von Korff RW, Murphy DL (eds) Monoamine Oxidase: Structure, Function and Altered Functions. Academic Press New York, pp 169-163

Shalita B, Dikstein S (1979) D-tyrosine prevents hypertension in DOCA-saline treated uninephrectomised rats. Arch 379:245-250

Shannon WA, Wasserking HL, Selingman AM (1974) The ultrastructural localization of monoamine oxidase (MAO) with tryptamine and a new tetrazolium salt, (2-(2'benzothiazolyl)-5-styryl-3-(4'phthalhydrazidyl) tetrazolium chloride (BSPT). J Histochem Cytochem 22:170-182

Sharman DF (1976) Can the intra and extra-homoneuronal metabolism of catecholamines be distinguished in the mammalian central nervous system? Ciba Found Symp 39, (New Ser.) Elsevier Amsterdam, pp 203-216

Shih JC, Eiduson S (1969) Multiple forms of monoamine oxidase in the developing brain. Nature 224:1309-1310

Sierens L, d'Iorio A (1970) Multiple monoamine oxidases in rat liver mitochondria. Can J Biochem 48:659-663

Silberstein SD, Shein HM, Berv KR (1972) Catechol-*O*-methyl transferase and monoamine oxidase activity in cultured rodent astrocytoma cells. Brain Res 41:245-248

Silverman RB, Hoffman SJ (1979) Mechanism of inactivation of monoamine oxidase by N-cydopropyl-N-arylalkyl amines. In: Singer TP, von Korff RW, Murphy DL (eds) Monoamine oxidase: Structure, Function and Altered Functions, Academic Press New York pp 71-79

Silverman RB, Hoffman SJ, Catus WB (1980) A mechanism for mitochondrial monoamine oxidase catalysed amine oxidation. J Amer Chem Soc 102:7126-7128

Simpson LL (1978a) Mechanism of the adverse interaction between monoamine oxidase inhibitors and amphetamine. J Pharmacol Exp Ther 205:392-399

Simpson LL (1978b) Evidence that deprenyl, a type B monoamine oxidase inhibitor, is an indirectly acting sympathomimetic amine. Biochem Pharmacol 27:1591-1595

Sinet PM, Heikkila RE, Cohen G (1980) Hydrogen peroxide production by rat brain *in vivo*. J Neurochem 34:1421-1428

Singer TP (1979) Active site-directed, irreversible inhibitor of monoamine oxidase. In: Singer TP, von Korff RW, Murphy DL (eds) Monoamine Oxidase: Structure Function and Altered Functions. Academic Press New York, pp 7-24

Sinha AK, Rose SPR (1972) Monoamine oxidase and cholinesterase activity in neurons and neuropil from rat cerebral cortex. J Neurochm 19:1607-1610

Sjöqvist F (1965) Psychotropic drugs 2. Interaction between monoamine oxidase (MAO) inhibitor and other substrates. Proc Roy Soc Med 205:967-978

Smith SE, Lambourn J, Typer PJ (1980) Antipyrine elimination by patients under treatment with monoamine oxidase inhibitors. Brit J Clin Pharmacol 9:21-25

Smith TE, Weissbach H, Udenfriend S (1962) Studies on the mechanism of action of monoamine oxidase: metabolism of N,N-dimethyltryptamine and N,N-dimethyltryptamine-N-oxide. Biochemistry 1:137-143

Snyder SH, Fischer J, Axelrod J (1965) Evidence for the presence of monoamine oxidase in sympathetic nerve endings. Biochem Pharmacol 14:363-365

Spector S, Hirsch CW, Brodie BB (1963) Association of behavioural effect of pargyline, a non hydrazine MAO inhibitor with increase in brain norepinephrine. Int J Neuropharmacol 2:81-93

Spector S, Gordon R, Sjoerdsma A, Udenfriend S (1967) End-product inhibition of tyrosine hydroxylase as a possible mechanism for regulation of norepinephrine synthesis. Mol Pharmacol 3:549-555

Spector S, Tarver S, Berkowitz B (1972) Effect of drugs and physiological factors in disposition of catecholamines in blood vessels. Pharmac Rev 24:191-202

Squires RF (1972) Multiple forms of monoamine oxidase in intact mitochondria as characterized by selective inhibitors and thermal stability: a comparison of eight mammalian species. Adv Biochem Psychopharmacol 5:355–370

Squires RF (1975) Evidence that 5-methoxy, N,N-dimethyltryptamine is a specific substrate for MAO-A in the rat: implications for the indoleamine dependent behavioural syndrome. J Neurochem 24:47–50

Squires RF, Buus Lassen J (1975) The inhibition of A and B forms of MAO in the production of a characteristic behavioural syndrome in rats after L-tryptophan loading. Psychopharmacologia 41:145–151

Stadt MA, Banks PA, Kobes RD (1982) Purification of rat liver monoamine oxidase by octyl glucoside extraction and reconstitution. Arch Biochem Biophys 214:223–230

Staunton J, Summers MC (1978) Stereochemical studies on enzymic reactions. In: Kornberg HL, Metcalfe JC, Northcote DH, Pogson CI, Tripton KF (eds) Techniques in the Life Sciences: Biochemistry, vol Bl/11. Elsevier Amsterdam, pp B116:1–33

Stern IJ, Hollifield RD, Wilk S, Buzard J (1967) The anti-monoamine oxidase effects of furazolidone. J Pharmacol Exp Ther 156:492–499

Stockley I (1974) Monoamine oxidase inhibitors, Part 1: Interactions with sympathomimetic amines. In: Drug (eds) Interactions. Pharmaceutical Press London

Strolin Benedetti M, Kan J-P, Keane PE (1979) A new specific reversible type A monoamine oxidase inhibitor: MD 780515. In: Singer TP, von Korff RW, Murphy DL (eds) Monoamine Oxidase: Structure, Function, and Altered Functions. Academic Press New York, pp 335–340

Student AK, Edwards DJ (1977) Subcellular localization of types A and B monoamine oxidase in rat brain. Biochem Pharmacol 26:2337–2342

Sulser F (1978) Functional aspects of the norepinephrine receptor coupled adenylcyclase system in the limbic forebrain and its modification by drugs which precipitate or alleviate depression: molecular approach to an understanding of affective disorders. Pharmakopsychiatrie 11:43–52

Suzuki O, Hattori H, Oya M, Katsumata Y, Matsumoto T (1979a) Oxidation of β-phenylethylamine by both types of monoamine oxidase: effects of substrate concentration and pH. Life Sci 25:1843–1850

Suzuki O, Katsumata Y, Oya M (1979b) Characterization of some biogenic monoamines as substrates for type A and type B monoamine oxidase. In: Singer TP, von Korff RW, Murphy DL (eds) Monoamine Oxidase: Structure Function, and Altered Functions. Academic Press New York, pp 197–204

Suzuki O, Matsumoto T, Oya M, Katsumata M, Stepitaklauco M (1980) Monocylcadaverines as substrates for both monoamine oxidase and diamine oxidase. Low rates of activity. Experientia 36:535–537

Swett LR, Martin WB, Taylor JD, Everett GM, Wykes AA, Gladish YC (1963) Structure-activity relations in the pargyline series. Ann NY Acad Sci 107:891–898

Tagliamonte A, Fratta W, del Fiacco M, Gessa GL (1974) Possible stimulatory role of brain dopamine in the copulatory behaviour of male rats. Pharmac Biochem Behav 2:257–260

Tipton KF (1968a) The prosthetic groups of pig brain mitochondrial monoamine oxidase. Biochem Biophys Acta 159:451–459

Tipton KF (1968b) The reaction pathway of pig brain mitochondrial monoamine oxidase. Eur J Biochem 5:316–320

Tipton KF (1971) Monoamine oxidases and their inhibitors. In: Aldridge WN (ed) Mechanisms of Toxicity. MacMillan London, pp 13–27

Tipton KF (1972) Some properties of monoamine oxidase. Adv Biochem Psychopharmacol 5:11-24
Tipton KF (1973) Biochemical aspects of monoamine oxidase. Brit Med Bull 29:116-119
Tipton KF (1975) Monoamine oxidase. In: Smith AD, Blaschko H (eds) Handbook of Physiology, section 7, vol 2, American Physiological Society Washington, pp 667-691
Tipton KF (1980) Kinetic mechanism and enzyme function. Biochem Soc Trans 8:242-245
Tipton KF, Della Corte L (1979) Problems concerning the two forms of monoamine oxidase. In: Singer TP, von Korff RW, Murphy DL (eds) Monoamine oxidase: Structure, Function and Altered Functions. Academic Press New York, pp 87-99
Tipton KF, Mantle TJ (1977) Dynamic properties of monoamine oxidase. In: Usdin E, Weiner N, Youdim MBH (eds) Structure and Function of Monoamine Enzymes, Dekker New York, pp 559-585
Tipton KF, Mantle TJ (1981) Inhibition of rat liver monoamine oxidase by clorgyline and deprenyl. In: Youdim MBH, Paykel ES (eds) Monoamine Oxidase Inhibitors: The State of the Art. Wiley Chichester
Tipton KF, Spires IPC (1971) The kinetics of 2-phenylethylhydrazine oxidation by monoamine oxidase. Biochem J 125:521-524
Tipton KF, Spires IPC (1972) Oxidation of 2-phenylethylhydrazine by monoamine oxidase. Biochem Pharmacol 21:268-270
Tipton KF, Youdim MBH, Spires IPC (1972) Beef adrenal medulla monoamine oxidase. Biochem Pharmacol 21:2197-2204
Tipton KF, Youdim MBH (1984) The assay of monoamine oxidase activity. In: Parvez S, Nagatsu T, Nagatsu I, Parvez H (eds) Methods in Biogenic Amine Research. Elsevier Amsterdam, pp 441-466
Tipton KF, Houslay MD, Garrett NJ (1973) Allotopic properties of human brain monoamine oxidase. Nature 246:213-214
Tipton KF, Houslay MD, Mantle TJ (1976) The nature and locations of the multiple forms of monoamine oxidase. Ciba Foundation Symposium 39 (New Ser) Elsevier Amsterdam, pp 5-16
Tipton KF, Houslay MD, Turner AJ (1977) Metabolism of aldehydes in brain. In: (eds) Essays in Neurochemistry and Neuropharmacology, vol 1. Wiley Chichester, pp 103-138
Tipton KF, Fowler CJ, Houslay MD (1982) Specificites of the two forms of monoamine oxidase. In: Kamijo K, Usdin E, Nagatsu T (eds) Monoamine Oxidase: Basic and Clinical Frontiers. Excerpta Medica Amsterdam, pp 87-99
Tipton KF, O'Carroll A-M, Hasan F (1984a) Enzymological and pharmacological aspects of monoamine oxidase. In: Paton W, Mitchell J, Turner P (eds) IUPHAR 9th Inter Cong Pharmac vol 2. MacMillan London, pp 179-185
Tipton KF, Dostert P, Strolin-Benedetti M (1984b) Eds. Monoamine Oxidase and Disease. Academic Press London
Toyoshima Y, Kinemuchi H, Kamijo K (1979) Nonexistence of a type C monoamine oxidase in rat brain. J Neurochem 32:1183-1189
Trendelenburg U (1972a) Factors influencing the concentration of catecholamines at the receptors. In: Blaschko H, Muscholl E (eds) Catecholamines, Springer Berlin Heidelberg New York, pp 726-761 (Handbook of Experimental Pharmacology, vol 33)
Trendelenburg U (1972b) Classification of sympathomimetic amines. In: Blaschko H,

Muscholl E (eds) Catecholamines Springer Berlin Heidelberg New York, pp 336-362 (Handbook of Experimental Pharmacology, vol 33)
Trendelenburg U (1979) Release induced by phenethylamines. In: Paton DM (ed) The Release of Catecholamines from Adrenergic Neurones, Pergamon Press New York, pp 333-359
Trendelenburg U, Draskoczy PR, Graefe KH (1972) The influence of intraneuronal monoamine oxidase on neuronal net uptake of noradrenaline and on sensitivity to noradrenaline. Adv Biochem Psychopharmacol 5:371-378
Trendelenburg U, Graefe KH, Henseling M (1976) The part played by monoamine oxidase in the inactivation of catecholamines in intact tissues. Ciba Foundation Symposium 39 (New Ser.) Elsevier Amsterdam, pp 181-195
Tsai TH, Fleming WH (1965) Sympathomimetic actions of monoamine oxidase inhibitors in the isolated nictitating membrane of the cat. Biochem Pharmacol 14:369-371
Turner AJ, Illingworth JA, Tipton KF (1974) Simulation of biogenic amine metabolism in the brain. Biochem J 144:353-360
Ulena H, Kanamura H, Suda S, Nakamura R, Machiyma Y, Takahashi R (1968) Studies on the regional distribution of the monoamine oxidase activity in the brains of schizophrenic patients. Proc Jap Acad 44:1078-1083
Ungar F, Tabakoff B, Alivisatos S (1973) Inhibition of binding of aldehydes of biogenic amines in tissues. Biochem Pharmacol 22:1905-1913
Uretskiy NJ, Iversen LL (1970) Effects of 6-hydroxydopamine on catecholamine-containing neurones in the rat brain. J Neurochem 17:269-278
Usdin E, Sandler M (1976) Trace amines and the brain. Dekker New York
Van Praag HM (1979) Serotonin and pathogenesis of affective disorders. In: Aromatic Amino Acid Hydroxylases and Mental Disease. Wiley London
Varga E, Tringer L (1967) Clinical trial of a new type promptly acting psychoenergetic agent (phenyl-isoprophylmethyl-propynylamine-HCl, E-250). Acta Med Acad Sci Hung 23:289-295
Von Korff RW (1977) Characteristics of monoamine oxidase of mitochondria isolated from rabbit brain and liver. Biokhimiya 42:396-402
Waldmeier PC, Felner AE (1978) Deprenil: loss of selectivity for inhibition of B-type MAO after repeated treatment. Biochem Pharmacol 27:801-802
Waldmeier PC, Feldtrauer JJ, Maitre L (1977) Methylhistamine: evidence for selective deamination by MAO B in the rat brain *in vivo*. J Neurochem 29, 785-790
Waldmeier PC, Felner AE, Maitre L (1981) Long term effects of selective MAO inhibitors on MAO activity and amine metabolism. In: Youdim MBH, Paykel ES (eds). Monoamine Oxidase Inhibitors. The State of Art. Wiley Chichester pp 85-97
Walker WH, Kearney EB, Seng RL, Singer TP (1971) The covalently bound flavin of hepatic monoamine oxidase. 2. Identification and properties of cysteinyl flavin. Eur J Biochem 24:328-331
Watanabe K, Minamiura N, Yasunobu KT (1980) Thiols liberate covalently bonded flavin from monoamine oxidase. Biochem Biophys Res Commun 94:579-585
Weissbach H, Redfield BG, Udenfriend S (1957) Soluble monoamine oxidase; its properties and actions on serotonin. J Biol Chem 229:953-963
Westlund K, Denney R, Kochersperger L, Rose R, Abell C (1985) Distinct monoamine oxidase A und B population in primate brain. Science 230:180-183
Weyler W, Salach JI (1981) Iron content and spectral properties of highly purified bovine liver monoamine oxidase. Arch Biochem Biophys 212:147-153
Wheatley D (1970) Comparative trial of a new monoamine oxidase inhibitor in depression J Psychiat 117:573-574

White HL, Glassman AT (1977) Multiple binding sites of human brain and liver monoamine oxidase: substrate specificities, selective inhibitions, and attempts to separate enzyme forms. J Neurochem 29:987-997

White HL, Stine DK (1982a) Characterization of active MAO-A and B sites by various biochemical techniques. In Monoamine Oxidase: Basic and Clinical Frontiers, edited by Kamijo K, Usdin E and Nagatsu T, Excerpta Medica, pp 62-73

White HL, Stine DK (1982b) Monoamine oxidase A and B as components of a membrane complex. J Neurochem 38:1429-1436

White HL, Transik RL (1979) Characterization of multiple substrate binding sites of MAO. In: Singer TP, von Korff RW, Murphy DL (eds) Monoamine Oxidase: Academic Press New York, pp 129-144

White HL, Wu JC (1975) Multiple binding sites of human brain monoamine oxidase as indicated by substrate competition. J Neurochem 25:21-26

Wibo M, Duong AT (1979) Semicarbazide-sensitive amine oxidase in rat aorta: a plasma-membrane enzyme. Arch Internat Physiol Biochem 87:868-869

Williams CH (1974) Monoamine oxidase. I. Specificity of some substrates and inhibitors. Biochem Pharmacol 23:615-628

Williams CH (1977) Beta-phenylethanolamine as a substrate for monoamine oxidase. Biochem Soc Trans 5:1770-1771

Williams CH, Lawson J (1975) Monoamine oxidase. III. Futher studies of inhibition by propargylamines. Biochem Phamacol 24:1889-1891

Williams D, Gascoigne JG, Williams ED (1975a) A tetrazolium technique for the histochemical demonstration of multiple forms of rat brain monoamine oxidase. Histochem J 7:585-597

Williams D, Gascoigne JE, Williams ED (1975b) A specific form of rat brain monoamine oxidase in circumventricular structures. Brain Res 100:231-235

Wilson WE, Agrawal AK, Zeller EA (1979) Androgen modulation of brain monoamine oxidase types A and B in the preweanling rat. In: Singer TP, von Korff RW, Murphy DL (eds) Monoamine Oxidase: Structure, Function and Altered Functions. Academic Press New York, pp 309-315

Wolf WA, Youdim MBH, Kuhn DM (1985) Does brain 5-HIAA indicate serotonin release or monoamine oxidase activity? Eur J Pharmac 109:381-387

Wojtczak L, Nalecz MJ (1979) Surface charge of biological membranes as a possible regulator of membrane-bound enzymes. Eur J Biochem 94:99-107

Woods RI (1970) The innervation of frog's heart. I. An examination of the autonomic postganglionic nerve fibres and a comparison of autonomic and sensory ganglionic cells. Proc Roy Soc B176:43-54

Wyatt RJ, Gillin JC, Stoff DM, Moja EA, Tinkelberg JR (1979) Phenylethylamine and the neuropsychiatric disturbances. In: Usdin E, Barchas ID, Hamburg D (eds) Neuroregulators and Psychiatric Disorders. Oxford University Press New York

Yagi K, Naoi M (1982a) Crystalline pig liver mitochondrial monoamine oxidase. In: Kamijo K, Usdin E, Nagatsu T (eds) Monoamine Oxidase: Basic and Clinical Frontiers. Academic Press New Yourk, pp 1-10

Yagi K, Naoi M (1982b) Crystallization of a monoamine oxidase purified from pig liver mitochondria. Biochem Int 4:457-463

Yahr M (1978) Overview of present day treatment of Parkinson's disease. J Neural Transmiss 43:227-238

Yamori Y, de Jong W, Yamabe H, Lovenberg W, Sjoerdsma A (1972) Effects of L-dopa and inhibitors of decarboxylase and monoamine oxidase on brain noradrenaline levels and blood pressure in spontaneously hypertensive rats. J Pharm Pharmacol 24:690-695

Yang H-YT, Neff NH (1973) β-Phenylethylamine: a specific substrate for type B monoamine oxidase of brain. J Pharmacol Exp Ther 187:365-371

Yang H-YT, Neff NH (1974) The monoamine oxidases of brain: selective inhibition with drugs and the consequences on the metabolism of the biogenic amines. J Pharmacol Exp Ther 189:733-740

Yasunobu KT, Igaue I, Gomes B (1968) The purification and properties of beef liver mitochondrial monoamine oxidase. Adv Pharmacol 6A:43-59

Yasunobu KT, Ishizaki M, Minamiura N (1976) The molecular, mechanistic and immunological properties of amine oxidases. Mol Cell Biochem 13:3-29

Yasunobu KT, Watanabe K, Zeidan H (1979) Monoamine oxidase: some new findings. In: Singer TP, von Korff RW, Murphy DL (eds) Monoamine Oxidase: Structure, Function and Altered Functions. Academic Press New York, pp 251-263

Youdim MBH (1972) Multiple forms of monoamine oxidase and their properties. Adv Biochem Psychopharmacol 5:67-77

Youdim MBH (1974) Heterogeneity of rat brain mitochondrial monoamine oxidase. Adv Biochem Psychopharmacol 11:59-63

Youdim MBH (1975) Monoamine deaminating system in mammalian brain. In: Blaschko H (ed) M.T.P. International Review of Science, Biochemistry Section, vol 12. MTP London, pp 169-209

Youdim MBH (1976) Rat liver mitochondrial monoamine oxidase—an iron-requiring flavoprotein. In: Singer, TP (eds) Flavins and Flavoproteins. Elsevier Amsterdam

Youdim MBH (1977) Tyramine and Psychiatric disorders. In: Usdin E, Barchas TD, Hamburg D (eds) Neuroregulators and Psychiatric Disorders. Oxford University Press New York

Youdim MBH (1978) Requirement of iron for monoamine oxidase activity. In: Usdin E, Weiner N, Youdim MBH (eds) Structure and Fuction of Monoamine Enzymes. Dekker New York, pp 587-599

Youdim MBH (1979) Functional activity of brain monoamine oxidase. In: Usdin E, Sourkes TL, Youdim MBH (eds) Neuro-Psychopharmacology. Pergamon Press New York

Youdim MBH (1980) : The use of selctive monoamine oxidase type B inhibitors in the treatment of Parkinson's disease. In: Usdin E, Sourkes TL, Youdim MBH (eds) Enzymes and Neurotransmitters in Mental Disease. Wiley London

Youdim MBH, Collins GGS (1971) Dissociation and reassociation of rat liver mitochondrial monoamine oxidase. Eur J Biochem 18:73-78

Youdim MBH, Finberg JPM (1980) Sites of action of monoamine oxidase inhibitors. In: Littauer UZ, Dudai Y, Silman I, Teichberg VI, Vogel Z (eds) Neurotransmitters and their Receptors. Wiley Chichester, pp 73-87

Youdim MBH, Finberg JPM (1982) Monoamine oxidase inhibitor antidepressants. In: Grahame-Smith DG, Cowen PJ (eds) Psychopharmacology I, part 1: Preclinical Psychopharmacology. Excerpta Medica Amsterdam, pp 38-70

Youdim MBH, Finberg JPM (1985) Monoamine oxidase inhibitor antidepressants. In: Grahame-Smith DG (ed) Psychopharmacology 2/1. Excerpta Medical Amsterdam, pp 35-70

Youdim MBH, Hefez A (1980) Platelet function and MAO activity in psychiatric disorders. In: Rothman A (eds) Platelets: Cellular Response Mechanism and Their Biological Significance. Wiley Chester

Youdim MBH, Holman B (1975) The nature of inhibition of cat brain monoamine oxidase by clorgyline. J Neural Transmis 37:11-24

Youdim MBH, Holzbauer M (1976a) Physiological and pathological changes in tissue monoamine oxidase activity. J Neural Transmis 38:193-230
Youdim MBH, Lagnado JR (1972) The effects of tetrazolium salts on monoamine oxidase activity. Adv Biochem Psychopharmac 5:289-292
Youdim MBH, Oppenheim B (1981) The effect of 1,2,3,4-tetrahydro-β-carbolines on monoamine metabolism in the human platelet and platelet aggregation. Neuroscience. 6:801-810
Youdim MBH, Sourkes TL (1966) Properties of purified soluble monoamine oxidase. Can J Biochem 44:1397-1400
Youdim MBH, Paykel ES (1981) Monoamine oxidase inhibitors: The State of the Art. Wiley Chichester
Youdim MBH, Riederer P (1986) MAO type B: its relation to MPTP induced and classical Parkinsonism. In: Markey SP, Castagnoli N, Trevor AJ, Kopin IJ (eds) A Neurotoxin Producing a Parkinsonism Syndrome. Academic Press London, pp 203-213
Youdim MBH, Sourkes TL (1972) The flavin prosthetic groups of purified rat liver mitochondrial monoamine oxidase Adv Biochem Psychopharmacol 5:45-43
Youdim MBH, Woods HF (1975) The influence of tissue environments on the roles of meteabolic processes and the properties of enzymes. Biochem Pharmacol 24:317-323
Youdim MBH, Collins GGS, Sandler M (1969) Multiple forms of rat brain monoamine oxidase. Nature 223:626-628
Youdim MBH, Collins GGS, Sandler M (1971) Monoamine oxidase, multiple forms and selective inhibitors. Biochem J 121:134-136
Youdim MBH, Collins GGS, Sandler M (1972c) Isoenzymes of soluble mitochondrial monoamine oxidase. In: Shugar D (ed) Enzymes and Isoenzymes. Structure, Properties and Function. Academic Press New York, pp 281-289
Youdim MBH, Collins GGS, Sandler M, Bevan Jones AB, Pare CMB, Nicholson WJ (1972b) Human brain monoamine oxidase: multiple forms and selective inhibitors. Nature 236:225-228
Youdim MBH, Woods HF, Mitchell B, Grahame-Smith DG (1975) Human platelet monoamine oxidase activity in iron-deficiency anaemia. Clin Sci Mol Med 48:289-295
Youdim MBH, Ben-Harari RR, Bakhle YS (1979a) Comparison of monoamine oxidase activity in perfused organ and *in vitro*. In: Singer TP, von Korff RW, Murphy DL (eds) Monoamine Oxidase: Structure, Function and Altered Functions. Academic Press New York pp 361-377
Youdim MBH, Aronson JK, Blau K, GreeN AR, Grahame-Smith DG (1979b) Tranylcypromine ('Parnate') overdose: measurement of tranylcypromine concentrations and MAO inhibitory activity and identification of amphetamines in plasma. Psychological Medicine 9:377-382
Youdim MBH, Ben-Harari RR, Bakhle YS (1980) Inactivation of monoamines by the lung. In: Porter R, Knight T (eds) Metabolic Activities of the Lung. Ciba Foundation Symposium 78 (New series). Elsevier Amsterdam pp 105-128
Yu PH (1979) Effect of lipid-depletion on type-A and type-B monoamine oxidase in rat heart and bovine liver mitochondria. In: Singer TP, von Korff RW, Murphy DL (eds) Monoamine Oxidase: Structure, Function, and Altered Functions. Academic Press, pp 233-244
You PH, Boulton AA (1979) Activation of platelet monoamine oxidase by plasma in the human. Life Sci 25:31-36

Zeller EA (1938) Über den enzymatischen Abbau von Histamin und Diaminen. Helv Chim Acta 21:880-890

Zeller EA (1963a) Diamine oxidase. In: Boyer PD, Lardy H, Myrbäck K (eds) The Enzymes 2nd Ed, vol 8. Academic Press New York, pp 313-335

Zeller EA (1963b) A new approach to the analysis of the interaction between monoamine oxidase and its substrates and inhibitors. Ann NY Acad Sci 107:811-820

Zeller EA (1979) Classification and nomenclature of monoamine oxidase and other amine oxidases. In: Singer TP, von Korff RW, Murphy DL (eds) Monoamine Oxidase: Structure, Function, and Altered Functions. Academic Press New York, pp 531-537

Zeller EA, Arora KL, Gürne DH, Huprikar SA (1979) On the topochemistry of the active site of monoamine oxidases types A und B. In: Singer TP, von Korff RW, Murphy DL (es) Monoamine Oxidase: Structure, Function and Altered Functions. Academic Press New York, pp 101-120

Zirkle CL, Kaiser C (1964) Monoamine oxidase inhibitors (nonhydrazines). In: Gordon M (ed) Psychopharmacological Agents, vol 1. Academic Press New York, pp 445-554

CHAPTER 4

The Transport of Amines Across the Axonal Membranes of Noradrenergic and Dopaminergic Neurones

K.-H. GRAEFE and H. BÖNISCH

A. Introduction

In contrast to acetylcholine, which is removed at cholinergic junctions by enzymic degradation, the transmitter noradrenaline is primarily inactivated by membrane transport. Two membrane transport processes are capable of removing noradrenaline and other catecholamines from the extracellular fluid: neuronal and extraneuronal uptake. While the mechanisms related to the latter will be dealt with elsewhere in this volume by TRENDELENBURG (see Chap. 6), the present chapter is intended to cover mechanisms related to the neuronal fate of catecholamines. It will concentrate on recent results and consider aspects that have not been extensively treated in earlier reviews and books (e.g. IVERSEN 1967, 1975; von EULER 1972; HERTTING and SUKO 1972; PATON 1976a). The main emphasis is given to the amine carrier system associated with the axonal membrane of noradrenergic neurones.

Although neuronal uptake is primarily responsible for the ability of noradrenergic and dopaminergic neurones to eliminate catecholamines from the extracellular fluid, the effectiveness of the neurone as a whole in inactivating catecholamines is greatly strengthened by vesicular uptake and MAO which both constitute intraneuronal "sink" mechanisms. They are highly effective in keeping the axoplasmic transmitter concentration low, thus enabling the neurone to function as a compartment which virtually irreversibly inactivates the amine. While it will be necessary to consider the neuronal metabolism of catecholamines, the vesicular uptake system is touched on here only to the extent required for a coherent account (for review see STJÄRNE 1972; PHILIPPU 1976 and this volume, Chap. 7).

Since the functional aspects of neuronal uptake have been comprehensively reviewed by TRENDELENBURG (1972), there is no need to discuss the evidence for the widely accepted view that the neuronal uptake system represents the major mechanism for terminating the action of noradrenaline in the biophase (i.e., the extracellular space in the immediate neighbourhood of the adrenoceptors) of sympathetically innervated organs. The view is based on the uptake theory of denervation and cocaine supersensitivity and it is this hypothesis which states that neuronal uptake is capable of generating a concentration gradient, with the noradrenaline concentration in the biophase being considerably lower than that in the incubation medium. It is quite obvious that this hypothesis may have implications for noradrenaline transport studies. If it is assumed that the biophase is not far away from what may be termed the neurophase (the extracellular fluid adjacent to the neuronal uptake sites), it may well be erroneous to equate the substrate concentration in the biophase with the amine concentration in the incubation medium. Another rather obvious point, which has not always received the attention it deserves, is that the fractional size of the tissue compartment involved in neuronal uptake is very small indeed; in peripheral organs it is likely to be smaller than 0.01, even if tissues exhibiting a rich sympathetic innervation are considered. In certain synaptosomal preparations of brain tissue, the fraction of tissue occupied by noradrenergic nerve terminals was found to be larger than 0.01 (IVERSEN and SCHON 1973).

These factors explain why (in contrast to other carrier-mediated membrane transport systems) important pieces of information as to the mechanism of neuronal uptake are still missing. Recently, a clonal cell line derived from the rat phaeochromocytoma (cultured PC-12 cells) has proven useful in neuronal uptake studies, since they possess the noradrenaline carrier system (BÖNISCH 1984; BÖNISCH et al. 1984; HARDER and BÖNISCH 1985; FRIEDRICH and BÖNISCH 1986). Such cells show outgrowth of nerve fibres when exposed to nerve growth factor and exhibit many other properties of noradrenergic neurones (GREENE and TISCHLER 1982). These cells are homogeneous in nature; they offer the possibility of characterizing the noradrenaline carrier system in terms of the underlying mechanism and even of isolating the carrier protein for reconstitution experiments.

B. Neuronal Uptake

I. Terminology and Methodology

1. Definition of Terms

Before a description of the properties of neuronal uptake is presented, it may be useful to define some terms which are widely used to describe noradrenaline (or dopamine) uptake. The term *uptake* refers strictly to the transmembrane flux of the amine substrate into the neurone; it comes about at first unidirectionally at a rate called *initial rate of uptake*. Thereafter, net inward transport continues at progressively decreasing rates. This is due to efflux of the translocated amine out of the neurone (GRAEFE et al. 1971). Hence, one speaks of *net uptake* when measurements of uptake are taken at a time at which the unidirectional influx is partially counterbalanced by back flux. To estimate the saturation constants for amine uptake (K_m and V_{max}), initial rates should be determined, because the Michaelis-Menten equation generally applies to the rates of unidirectional transport. The mathematical description of net transport is usually much more complex than that of unidirectional transport (WILBRANDT und ROSENBERG 1961; STEIN 1967).

To measure net uptake of an amine, the fate of this amine subsequent to uptake has to be taken into consideration. Neuronal uptake may lead either to *accumulation* of the unchanged amine in the neurone, to amine metabolism (by MAO) or to both, depending on the experimental conditions and the amine used in the study. Except when initial rates of accumulation are being determined, measurements of accumulation usually reflect *net accumulation*; this term takes into account the occurrence of both back flux and metabolism of the transported substrate. Net accumulation will take place only when the rate of net uptake exceeds the rate of metabolism. Under conditions of inhibition of vesicular uptake (after pretreatment with reserpine), for example, net uptake of noradrenaline is unaffected (LINDMAR and MUSCHOLL 1964; HAMBERGER et al. 1964), whereas net accumulation of noradrenaline is greatly reduced (MUSCHOLL 1960; AXELROD et al. 1961; DENGLER et al. 1962; KOPIN et al. 1962). Therefore, it may be incorrect to equate net accumulation with net uptake. The same reasoning applies to studies in which ^{3}H-noradrenaline is being used. The net accumulation of total radioactivity (i.e., unchanged amine plus metabolites) in the tissue usually does not reflect net uptake, since the main neuronal metabolite of noradrenaline (i.e., DOPEG) is highly lipid soluble and leaves the tissue with ease (see Sect. C.I).

Many early studies aimed at measurements of the neuronal *retention* of noradrenaline. This is the case whenever the incubation with noradrenaline is followed by a 5 to 10-min exposure to amine-free medium. This method has the disadvantage of being heavily dependent on the function of the vesicular uptake and storage mechanism. Since, in addition, back flux and metabolism of the transported amine continue during the wash-out period, measurements of retention must be regarded as being inferior to measurements of net accumulation for determination of uptake.

From these considerations it follows that the effect of drugs on neuronal uptake of noradrenaline should not be examined merely by measuring accumulation or retention, since any effect of the drug under study on either vesicular uptake or MAO is then bound to obscure the results. However, when MAO is inhibited or non-metabolizable noradrenaline analogues (e.g., metaraminol) are used as substrates, the net accumulation of amine may well serve to measure neuronal net uptake. An alternative procedure is to measure the *removal* or *net removal* of an amine from the external medium. This method was first described by LINDMAR and MUSCHOLL (1964). It measures the amount of amine taken up by the tissue and offers the advantage of being largely independent of the intraneuronal fate of the substrate. It is clear that in studies with ^{3}H-noradrenaline, net removal equals net accumulation of the unchanged amine plus metabolite formation.

2. Methodological Considerations

In studies of neuronal uptake, both incubated tissue slices and perfused isolated organs have been used. In tissue slices (or strips prepared from thin-walled organs) it may be difficult to ensure that all parts of the tissue are adequately equilibrated with the external medium. Slices or strips should be prepared as uniformly as possible and thin enough so as to avoid difficulties that may arise from both the slow penetration of amines into tissues and any pronounced variation in diffusion distances. Problems of this kind can be disregarded, however, when conditions are selected in which the amount of amine taken up is linearly related to the weight (or the protein content) of the tissue sample. Such conditions are usually met in studies with isolated neuronal cells (HARDER and BÖNISCH 1984) and homogenates of brain tissue or synaptosomes (SNYDER et al. 1970; KUHAR 1973). Another possibility to overcome diffusion problems is the use of perfused isolated organs such as the spleen (GILLESPIE and KIRPEKAR 1965) or the heart (IVERSEN 1963; LINDMAR and MUSCHOLL 1964; GILLIS and PATON 1966; JARROTT 1970).

Although neuronal uptake of amines is easily demonstrated *in vitro*, one must beware of neglecting uptake into non-neuronal cells. Extraneuronal uptake, when existent, takes place in parallel with neuronal uptake (LIGHTMAN and IVERSEN 1969; HERMANN and GRAEFE 1977; FIEBIG and TRENDELENBURG 1978). To make allowance for this, it is possible to determine, in a parallel set of tissue preparations, the non-neuronal amine distribution after inhibition of neuronal uptake (ROSS and RENYI 1964; HERMANN and GRAEFE 1977; KELLER and GRAEFE 1979; HENSELING 1983). An alternative procedure is to examine the uptake of amines in the presence of an inhibitor of extraneuronal uptake which is without effect on neuronal uptake (GRAEFE et al. 1978). In brain tissue preparations extraneuronal uptake is by no means absent (HENDLEY 1976). Moreover, the experimenter is faced with the added uncertainty that „nonspecific“ neuronal uptake sites exist in brain. For instance, apart from the transport into noradrenergic and dopaminergic neurones, there can be amine uptake into other neurones or cells (SNYDER and COYLE 1969; FERRIS and STOCKS 1972; KIMELBERG 1986). Problems of this kind can be circumvented by the use of isolated neuronal cells (e.g. PC-12 cells).

Initial rates of neuronal uptake can be determined by measuring either amine accumulation (IVERSEN 1963; IVERSEN et al. 1965) or removal (GRAEFE et al. 1978). In any case, the time course of uptake has to be established. It is the inspection of the time course which tells us whether or not a plot of the log rate of uptake against time is required to estimate initial rates by extrapolation to zero time. When the time course is initially linear, a more direct determination of the initial rate is possible. It is a characteristic feature of the neuronal carrier system that it can maintain initial rates for much longer when exposed to low rather than to high substrate concentrations (IVERSEN 1963; GRAEFE et al. 1978).

It is clear that in all studies of accumulation or removal allowance must be made for distribution into extracellular spaces. As extracellular tissue compartments equilibrate relatively quickly with the external medium, it is obvious that the amount of substrate present in the extracellular fluid contributes considerably to total uptake when the period of exposure to the substrate is short. It is because of this reasoning that, in a number of studies, the uptake of amines was determined in the presence of a marker of the extracellular fluid (HERMANN and GRAEFE 1977; SANCHEZ-ARMASS and ORREGO 1978; GRAEFE et al. 1978).

As far as initial rates of uptake are concerned, the formal description of transport kinetics is equivalent to that of enzyme kinetics. Therefore, the apparent Michaelis-Menten constants, K_m and V_{max}, for the neuronal uptake of an amine can be estimated from the saturation curve (relating initial rates to substrate concentrations) with the help of one of the various graphic methods available (SEGEL 1975) or by the method described by WILKINSON (1961). The latter technique fits the Michaelis-Menten equation directly to the experimental data using the least-squares method. Any distortion of the saturation curve caused by additional overlapping processes with first-order kinetics (e.g., extracellular distribution, cellular uptake by simple diffusion, extraneuronal uptake far below saturation) leads to overestimation of both K_m and V_{max}. On the other hand, underestimates of the two parameters will be obtained whenever initial rates of uptake are being underestimated (especially at high substrate concentrations where "initial rates" easily happen to be rates of net uptake). In studies with tissue slices, conditions will sometimes be met in which the diffusion of substrate into the tissue is rate-limiting. Such conditions cause a distortion of the saturation curve which does not affect the determination of V_{max}, but does give rise to overestimation of the apparent K_m value (GREEN 1976).

II. Characteristics of Neuronal Uptake

1. Basic Properties

The physicochemical properties of noradrenaline and dopamine (lipid solubility, degree of ionisation at physiological pH) are such that the ability of these amines to penetrate cell membranes by simple diffusion must be considered rather poor indeed (PATON 1976a; MACK and BÖNISCH 1979). Therefore, it is perhaps not surprising that some carrier system is likely to be involved whenever these amines cross cell membranes. Mechanisms responsible for carrier-mediated transport have the property of becoming saturated as the substrate concentration is increased. This has in fact been found for the mechanism underlying neuronal uptake. Numerous studies have characterized the uptake of various sympathomimetic amines into noradrenergic and dopaminergic neurones as being saturable; they have shown that the initial uptake process is governed by saturation kinetics of the Michaelis-Menten type. The results reported in most of these studies show that the apparent K_m of the system for a number of substrates is in the micromolar range (see Tables 1 and 2). The value of K_m is defined as the concentration of substrate at which the initial rate of uptake has reached half of its maximum value (V_{max}). It should not be taken to mean the dissociation constant of the carrier-substrate complex (K_D). Equating K_m with K_D requires the assumption that the translocation of the carrier-substrate complex across the axonal membrane is rate-limiting in the whole transport cycle. The available evidence indicates that this assumption is not valid. This will be discussed in more detail in Sect. E.

When determined in incubated tissue slices, rates of uptake depend not only on the substrate concentration, but also on such factors as the surface area to volume ratio of the tissue sample. In studies using rat brain tissue it was noted that, at any given amine concentration in the bath, the initial rate of ^{3}H-(±)-noradrenaline accumulation was much higher in homogenates than in slices (SNYDER et al. 1968; SNYDER and COYLE 1969). Despite these differences, both slices and homogenates gave virtually identical K_m values for ^{3}H-(±)-noradrenaline uptake. Hence, although there can be sequestration of part of the uptake sites in tissue slices, the diffusion of the substrate towards the accessible sites does not, at least in brain tissue, appear to be rate-limiting in neuronal uptake (for discussion see GREEN 1976).

In perfused organs the rate of perfusion is of some importance. Lowering of the perfusion rate was shown to reduce the initial rate of uptake in the rat and rabbit heart (IVERSEN and KRAVITZ 1966; GRAEFE and BÖNISCH 1978). Moreover, in a study using rabbit hearts perfused with (−)-noradrenaline, DRASKÓCZY and TRENDELENBURG (1968) found the difference between the arterial and venous concentration of noradrenaline (indicating net removal of the amine due to neuronal uptake) to be inversely related to the perfusion rate. The effect of changes of the perfusion rate on the saturation kinetics of neuronal uptake was examined with ^{3}H-(−)-noradrenaline as substrate in rabbit hearts by GRAEFE and BÖNISCH (1978). While the value of V_{max} increased with increasing rates of perfusion, there was no flow-dependent change in the apparent K_m. If uptake sites were arranged in series, the arterio-venous difference of the ^{3}H-(−)-noradrenaline concentration would reflect a substrate gradient within the tissue, and an increase in perfusion rate (which decreases the gradient) would then cause the apparent K_m to decrease. As the K_m remained unaltered, it was concluded that the neuronal uptake sites in the perfused rabbit heart are operationally arranged in parallel. The flow-dependent changes in V_{max} indicate that the availability of carrier sites is limited by the rate of perfusion (GRAEFE and BÖNISCH 1978).

Besides the property of being saturable, the neuronal carrier system brings about transport against considerable substrate gradients. On perfusion of rat and rabbit hearts with 0.06 to 0.12 μmol/l noradrenaline, for instance, tissue to perfusate (tissue/medium) ratios of 60 to 120 (calculated with reference to the wet weight of the heart) have been observed after 30 to 100 min of perfusion (IVERSEN 1963; GRAEFE et al. 1971; see also Fig. 1). The substrate gradients reported for dopaminergic neurones (synaptosomes of rat striatum) are even greater (SNYDER and COYLE 1969; HOLZ and COYLE 1974). Considering the fact that the tissue compartment involved is very small indeed (in the perfused heart perhaps 1% or less), the tissue/medium ratios mentioned above (60–120) must have been in the order of 6000 to 12000 (for further discussion, see Sect. D.I). This would hardly be possible if the mechanism underlying neuronal uptake were passively mediated transport (sometimes termed "facilitated diffusion").

One should not neglect, however, vesicular uptake and its contribution to the reported large tissue/medium ratios. Vesicular uptake appears to contribute a great deal when the exposure of the tissue to an amine is long lasting. BÖNISCH and GRAEFE (1976) determined steady-state accumulation ratios in rabbit hearts perfused with 0.3 μmol/l ^{3}H-(−)-noradrenaline. MAO was inhibited in these experiments. While hearts with intact storage vesicles attained an average tissue/medium ratio of 100, a value of 44 was found when hearts from reserpine-pretreated animals were used. Thus, the large net accumulation usually observed in normal tissues (vesicles operating) is not due to neuronal uptake alone, but is partly a result of vesicular uptake and storage.

As reported by IVERSEN (1963), neuronal uptake leads to some exchange of exogenous with endogenous noradrenaline. After having been taken up by the neurone, many foreign amines structurally related to noradrenaline also exchange with endogenous noradrenaline (MUSCHOLL 1972). However, it is clear that such exchange phenomena do not play any part in the initial uptake process. One would expect, therefore, K_m and V_{max} for noradrenaline transport to remain unchanged after depletion of the endogenous noradrenaline pool. This is what actually has been found (GRAEFE et al. 1978).

In tissues of reserpine-pretreated animals a pronounced accumulation of noradrenaline takes place in the axoplasm of noradrenergic neurones when MAO is inhibited. It is, of course, always possible to show axoplasmic accumulation for amines which are not metabolized by MAO (e.g. α-methylnoradrenaline and metaraminol; CARLSSON et al. 1967; GIACHETTI and SHORE 1966; HAMBERGER 1967). As the capacity of the neurone to accumulate noradrenaline within the axoplasm is quite pronounced, it is not surprising that there has been a good deal of uncertainty about the possible existence of axoplasmic binding and/or trapping sites. Such sites have often been postulated (LINDMAR and LÖFFELHOLZ 1974a; ENNA and SHORE 1974, GIACHETTI and HOLLENBECK 1976), but to prove or disprove their existence experimentally is a difficult task.

After inhibition of vesicular uptake and MAO, the neurone can be described as a one-compartment system if the assumption is made that there is no vesicular trapping and/or axoplasmic binding. The efflux out of this compartment could come about either by a first-order process or by a carrier-mediated process obeying Michaelis-Menten kinetics. If such a system is exposed to a constant concentration of noradrenaline, the time course of net accumulation of the amine in the compartment can be determined. It yields two parameters: the half-time required for the approach of the system to a steady state of net accumulation and the tissue/medium concentration ratio attained at steady state. Depending on the mechanism responsible for efflux, the two parameters may or may not change when the substrate concentration imposed on the system is raised. As to this point the following predictions can be made on theorethical grounds (STEIN 1967). If efflux were to follow first-order kinetics, the half-time for approach to steady state would be independent of the substrate concentration, whereas the steady-state tissue/medium ratio would be inversely related to the substrate concentration. If, on the other hand, efflux were to be mediated by a saturable process, the half-time would increase with increasing substrate concentrations, while the tissue/medium ratio would be independent of the substrate concentration. Neither of these predictions agreed with experimental results. When rabbit hearts (in which vesicular uptake and MAO had been inhibited) were perfused with various concentrations of noradrenaline, both the half-time and the steady-state tissue/medium ratio decreased as the perfusion concentration of noradrenaline was increased (LINDMAR and LÖFFELHOLZ 1974a; BÖNISCH and GRAEFE 1976;

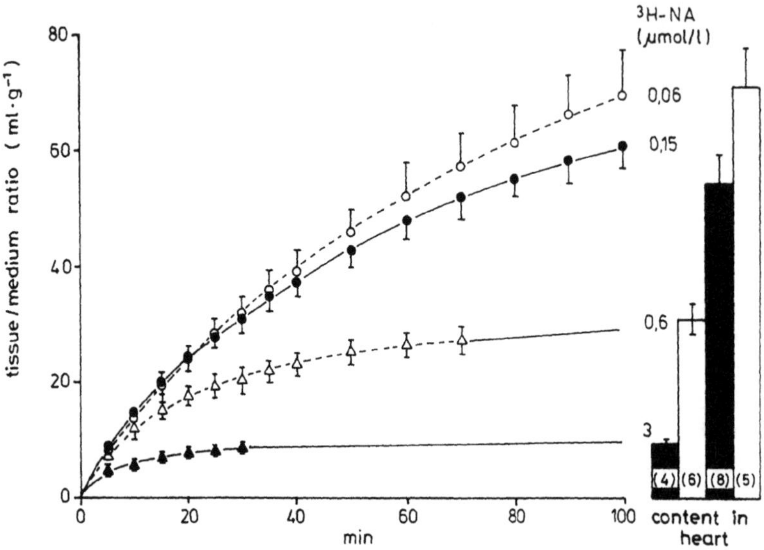

Fig. 1. Dependence of the tissue/medium ratio of ^{3}H-noradrenaline on the perfusion concentration of the ^{3}H-amine in isolated rabbit hearts. Hearts were obtained from reserpine- and pargyline-pretreated rabbits and perfused at a constant rate with Tyrode's solution containing 0.06–3.0 µmol/l ^{3}H-(−)-noradrenaline. From the arterio-venous difference of the amine concentration, tissue/medium ratios of the ^{3}H-noradrenaline concentration were calculated and plotted against time. Removal was determined only until the arterio-venous difference had reached a value of 2 % of the arterial concentration. After the end of perfusion (100 min), the ^{3}H-noradrenaline content of the heart was determined and expressed in terms of tissue/medium ratios in ml/g (see columns on the right). Shown are means (±SEM) of at least 4 observations each. Note that at low amine concentrations in the medium a steady state of ^{3}H-noradrenaline accumulation in the heart was not reached within the period of observation. In those instances the amine content of the heart at steady state (see Fig. 2) was calculated: the amount of amine not yet accumulated in the tissue by the end of perfusion was estimated from the ratio "rate of removal at the end of perfusion/k" (where k is the rate constant characterizing the terminal slope of the log removal rate vs. time curve) and added to the amount of amine present in the tissue at the end of perfusion. Taken from BÖNISCH (1982)

Fig. 1). This pattern of results is inconsistent with the assumption that the noradrenergic neurone (after pretreatment with reserpine) is a one-compartment system. Independent evidence against this assumption is shown in Fig. 2: the saturation curve relating steady-state accumulation to substrate concentration involves more than one (probably two) saturable component(s). This finding supports the notion that, in spite of inhibition of vesicular uptake, the neurone behaves like a two- rather than a one-compartement system.

After pretreatment with reserpine and inhibition of MAO, intraneuronal binding and/or trapping may contribute considerably to the ability of the neu-

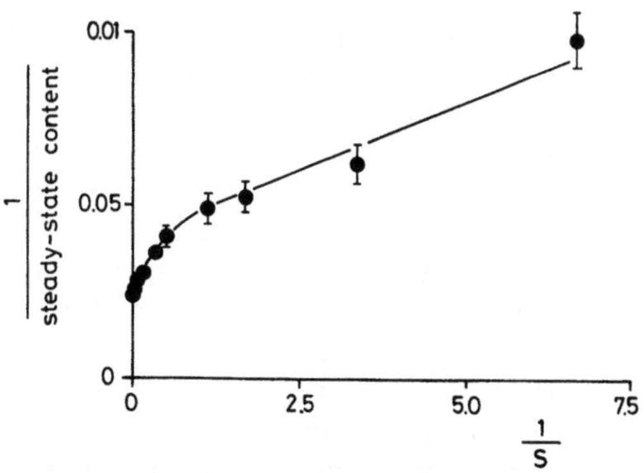

Fig. 2. Saturability of the neuronal steady-state accumulation of ^{3}H-noradrenaline in the perfused isolated rabbit heart. Ordinate: reciprocal of the ^{3}H-noradrenaline content of the heart at the steady state of amine accumulation in $nmol^{-1} \cdot g$ (corrected for distribution into the total tissue water space which was estimated to be $0.8\ ml \cdot g^{-1}$). Abscissa: reciprocal of the ^{3}H-(−)-noradrenaline concentration in the perfusion fluid (S) in $\mu mol^{-1} \cdot l$. For details, see legend to Fig. 1. Amine accumulation due to extraneuronal uptake, if any, is small enough to be neglected (GRAEFE et al. 1978; GRAEFE 1981). Shown are means (±SEM) of at least 3 observations each. The solid line was drawn to fit the following Eq.: 1/steady-state content $= (1/A_1 + B_1/A_1S) + (1/A_2 + B_2/A_2S)$ (where the terms on the right represent two saturable components obeying the Michaelis-Menten equation; A_1 and A_2 = maximum steady-state content at infinitely high substrate concentrations for component 1 and 2, respectively; B_1 and B_2 = substrate concentrations at which A_1 and A_2, respectively, are half-maximal). $A_1 = 22.4$ nmol/g, $B_1 = 0.17$ µmol/l, $A_2 = 26.8$ nmol/g, $B_2 = 9.4$ µmol/l. Taken from BÖNISCH (1982)

rone to accumulate noradrenaline. While there is no convincing evidence to support the existence of axoplasmic binding sites, the evidence for amine trapping within the storage vesicles via a reserpine-resistant mechanism is fairly strong. It is now well established that storage vesicles generate a proton gradient (inside acid) across their membranes (JOHNSON and SCARPA 1976) and that this gradient is not affected by reserpine (HOLZ 1978). Hence, in spite of pretreatment with reserpine, the acid interior of storage vesicles serves as an "amine trap": depending on their lipophilicity, amines cross the vesicular membrane in their non-ionized form by simple diffusion and are then trapped inside in their ionized form (JOHNSON et al. 1982; KNOTH et al. 1984). Amines trapped in this way can be released from the neurone by exocytosis (FARNEBO 1971; BÖNISCH and RODRIGUES-PEREIRA 1983).

2. Structural Requirements

Many compounds, more or less structurally related to noradrenaline, have been shown to be substrates for neuronal uptake. Cocaine, a highly selective inhibitor of the system, is a powerful tool to demonstrate neuronal uptake. A compound is commonly assumed to be a substrate when uptake and accumulation of this compound are susceptible to inhibition by cocaine. Saturability of uptake, dependence of uptake on the presence of extracellular Na^+ and the observation that the efflux of the compound from previously incubated tissues is markedly accelerated in response to sympathomimetic amines, constitute additional experimental evidence that the compound under study is a substrate. However, one has to be aware of the limitations of the methods. For instance, compounds that cause release or acceleration of efflux of noradrenaline are not necessarily substrates of the carrier, even if these effects are inhibited by cocaine. Both, ouabain (PATON 1973) and veratridine (ROSS and KELDER 1976; BÖNISCH et al. 1983), accelerate the efflux of axoplasmic ^3H-noradrenaline, but no one would claim these two substances to be substrates of the carrier. Needless to say, drugs which competitively inhibit noradrenaline uptake need not themselves be substrates; they may well act as "pure" (non-transported) inhibitors.

The structural requirements for the neuronal amine carrier have been judged by many authors from a comparison of the potencies of various phenylethylamines for inhibition of neuronal uptake. This involved determination of either IC_{50} or K_i values. Using β-phenylethylamine (phenethylamine) as reference compound, studies of this kind indicate that the presence of phenolic hydroxyls or α-methylation of the side chain decreases, while β-hydroxylation, N-alkylation and O-methylation (in that order) increase the IC_{50} of the amines at both noradrenaline (BURGEN and IVERSEN 1965; MUSCHOLL and WEBER 1965; IVERSEN 1967) and dopamine uptake sites (HORN 1973; RAITERI et al. 1977a). The amine carriers of noradrenergic and dopaminergic neurones appear to differ with regard to their ability to distinguish between phenylethylamines having an asymmetric centre either at the α- or β-carbon of the side chain. In the noradrenergic system, the order of potencies for inhibition of uptake is $R(-) > S(+)$ for β-hydroxylated (BURGEN and IVERSEN 1965; COYLE and SNYDER 1969) and $R(-) \leqq S(+)$ for α-methylated phenylethylamines (HOLMES and RUTLEDGE 1976; see also Table 3). In the dopaminergic system, on the other hand, the rank order is $R(-) = S(+)$ in the case of β-hydroxylated (COYLE and SNYDER 1969) and $R(-) < S(+)$ in the case of α-methylated phenylethylamines (HOLMES and RUTLEDGE 1976; see also Table 3). The optimal configuration for those phenylethylamines in which both the α- and β-carbon are asymmetric appears to be the $\alpha S{:}\beta R$ configuration (HENDLEY et al. 1972). Studies of the inhibitory potencies of conformationally rigid analogues of amphetamine have shown that the preferred conformation at the uptake sites of dopaminergic and noradrenergic neurones is probably the *anti* conformation, i.e., the conformation with the sides chain fully extended and the amino group above the plane of the benzene ring (HORN and SNYDER 1972).

The competitive nature of transport inhibition by phenylethylamines has been demonstrated for adrenaline (IVERSEN 1965), amphetamine (BÖNISCH 1984) and metaraminol (ZEITNER and GRAEFE 1986) in the noradrenergic system, and for amphetamine (HORN et al. 1971) in the dopaminergic system. On theoretical grounds, all substrates should be competitive inhibitors and their values of K_i should be identical with their K_m values. This was found to be true for noradrenaline (IVERSEN 1967), adrenaline (IVERSEN 1965) and tyramine (BÖNISCH and RODRIGUES-PEREIRA 1983) in the noradrenaline carrier system, and for noradrenaline and dopamine (SNYDER and COYLE 1969; COYLE and SNYDER 1969) as well as for p-hydroxyamphetamine (FISCHER and CHO 1979) in the dopamine carrier system.

The structural specificity of neuronal uptake can also be judged from the ability of the carrier system to translocate a given substrate. The parameter best suited for estimating the effectiveness of transport is the ratio of V_{max}/K_{max} (i.e., the rate constant for uptake; k_{uptake})[1]. Since the rate of uptake (v) at very low substrate concentrations (S) is proportional to this ratio, values of k_{uptake} can likewise be determined from measurements of the ratio of v/S obtained at substrate concentrations far below K_m. To be comparable, values of k_{uptake} should be determined in the same tissue or organ, since the V_{max} is subject to marked organ differences (HERMANN and GRAEFE 1977). For this and other reasons, the following statements about the structural requirements for the uptake of phenylethylamines are mainly based on results reported for the same tissue and by the same group of authors. Despite the limited amount of information available (some of which is summarized in Tables 1 and 2), the following conclusions can be drawn:

a) Alkylation of the primary amino group reduces the effectiveness of transport; noradrenaline is more effectively transported than adrenaline (Table 1), and isoprenaline is hardly taken up at all (HERTTING 1964; CALLINGHAM and BURGEN 1966).

b) α-Methylation does not affect the transport effectiveness. α-Methylnoradrenaline is transported as effectively as (or even better than) noradrenaline (LINDMAR and MUSCHOLL 1965). Many other α-methylated phenylethylamines are good substrates, notably metaraminol (Table 1) and α-methyltyramine (IVERSEN 1966). The latter amine (also known as p-hydroxyamphetamine) is also transported into dopaminergic neurones (FISCHER and CHO 1979).

c) The presence of the β-hydroxyl group is not essential for uptake. The carrier systems associated with noradrenergic and dopaminergic neurones differ somewhat with respect to the effect of β-hydroxylation. Dopaminergic neurones take up dopamine at higher rates than ($\pm$)-noradrenaline (Table 2). For noradrenergic neurones, on the other hand, dopamine appears to be just as good a substrate as ($\pm$)-noradrenaline (Table 1), and β-hydroxylation of tyramine to octopamine does not lead to a pronounced change in k_{uptake} (Table 1). Also, while the dopamine carrier does not distinguish between the enantiomers of noradrenaline (Table 2), some stereoselectivity in favour of the (−)-enantiomer has been observed in the noradrenergic system (Table 1).

d) The presence of the two phenolic hydroxyl groups does not appear to be an absolute requirement. Many phenylethylamines with one phenolic hydroxyl (e.g. metaraminol, tyramine, octopamine) are good substrates (Table 1). It should be emphasized that measurements of uptake of phenylethylamines which lack one phenolic hydroxyl are often fraught with experimental difficulties. In experiments with amines such as metaraminol (HERMANN and

[1] The Michaelis-Menten equation simplifies to $v=(V_{max}/K_{max})\cdot S=k_{uptake}\cdot S$ when the substrate concentration (S) is very low relative to the K_m of the substrate.

Table 1. Apparent kinetic constants for amine uptake into noradrenergic neurones of various tissues

Substrate	K_m (μmol/l)	V_{max} (nmol/g · min)	V_{max}/K_m (min^{-1})	Organ or tissue	Reference
Dopamine	0.7	1.45	2.07	rat heart (perfused)	HELLMANN et al. 1971
(±)-Noradrenaline	0.7	1.36	1.94	rat heart (perfused)	IVERSEN 1963
(−)-Noradrenaline	0.3	1.18	3.93	rat heart (perfused)	IVERSEN 1963
(+)-Noradrenaline	1.4	1.72	1.23	rat heart (perfused)	IVERSEN 1963
(±)-Adrenaline	1.4	1.03	0.74	rat heart (perfused)	IVERSEN 1965
(−)-Noradrenaline	0.3	0.03	0.10	rat atrium	ROSS and GOSZTONYI 1975
Bretylium	3.0	0.10	0.03	rat atrium	ROSS and GOSZTONYI 1975
(±)-Noradrenaline	1.0	1.04[b]	1.04	rat vas deferens	IVERSEN and LANGER 1969
(−)-Noradrenaline	3.2	3.30	1.03	rat vas deferens	SAMMET and GRAEFE 1979
(±)-Metaraminol	1.7	3.30	1.94	rat vas deferens	HERMANN and GRAEFE 1977
(−)-Noradrenaline	1.0	0.30	0.30	rat vas deferens	ROSS 1976a
Bretylium	5.0	0.60	0.12	rat vas deferens	ROSS 1976a
(±)-Noradrenaline	1.6	0.58[b]	0.36	rat immature uterus	GREEN and MILLER 1966
(±)-Adrenaline	1.6	0,75[b]	0.47	rat immature uterus	GREEN and MILLER 1966
(±)-Noradrenaline	1.7	0.97[b]	0.57	rat iris	JONSSON et al. 1969
Dopamine	0.1	0.66[a]	6.60	rat hypothalamus (synaptosomes)	SNYDER and COYLE 1969
(±)-Noradrenaline	0.4	1.00[a]	2.50	rat hypothalamus (synaptosomes)	SNYDER and COYLE 1969
(−)-Noradrenaline	0.2	0.65	3.25	rat hypothalamus (synaptosomes)	IVERSEN et al. 1971
(+)-Noradrenaline	0.8	1.08	1.35	rat hypothalamus (synaptosomes)	IVERSEN et al. 1971
(±)-Noradrenaline	0.3	0.37	1.23	rat hypothalamus (slices)	SHASKAN and SNYDER 1970
5-HT	8.6	8.55	0.99	rat hypothalamus (slices)	SHASKAN and SNYDER 1970
(−)-Noradrenaline	3.3	5.65	1.71	rabbit heart (perfused)	GRAEFE et al. 1978
(+)-Noradrenaline	3.9	4.30	1.10	rabbit heart (perfused)	GRAEFE et al. 1978
(±)-Metaraminol	5.9	7.85	1.33	rabbit heart (perfused)	GRAEFE et al. 1978
Tyramine	1.0	1.82	1.82	rabbit heart (perfused)	BÖNISCH and RODRIGUES-PEREIRA 1983
Amezinium	0.9	1.20	1.33	rabbit heart (perfused)	STEPPELER and STARKE 1982

[a] V_{max} expressed as nmol per g pellet per min
[b] V_{max} calculated from the data given by authors

Table 1 (continued)

Substrate	K_m (μmol/l)	V_{max} (nmol/g · min)	V_{max}/K_m (min^{-1})	Organ or tissue	Reference
(−)-Noradrenaline	2.3	0.53	0.23	rabbit aortic strips	HENSELING 1983
(−)-Noradrenaline	2.2	3.00	1.36	cat heart (perfused)	GRAEFE et al. 1978
Dopamine	0.7	0.11	0.16	mouse cerebral cortex (slices)	ROSS and RENYI 1966b
(±)-Noradrenaline	0.4	0.11	0.28	mouse cerebral cortex (slices)	ROSS and RENYI 1964
(±)-Octopamine	0.7	0.10	0.14	mouse cerebral cortex (slices)	ROSS et al. 1968
Tyramine	0.4	0.04	0.10	mouse cerebral cortex (slices)	ROSS and RENYI 1966b

GRAEFE 1977), octopamine (ROSS et al. 1968), tyramine (ROSS and RENYI 1966a; BÖNISCH and RODRIGUES-PEREIRA 1983) and phenylephrine (RAWLOW et al. 1980) a considerable non-saturable, cocaine-resistant uptake overshadows the carrier-mediated component of transport. The cocaine-resistant tissue uptake is even more pronounced for non-phenolic phenylethylamines and may, hence, be a consequence of the high lipid solubility of the amines (absence of polar groups). This probably explains why several studies failed to demonstrate the neuronal uptake of amphetamine, β-phenylethanolamine, norephedrine and ephedrine (ROSS and RENYI 1966a; THOENEN et al. 1968; ROSS et al. 1968; GOLKO and PATON 1976; BÖNISCH and RODRIGUES-PEREIRA 1983). However, neuronal uptake of amphetamine was clearly demonstrated to occur in both synaptosomes of rat brain (RUTLEDGE and VOLLMER 1979) and cul-

Table 2. Apparent kinetic constants for amine uptake into dopaminergic neurones of rat striatum

Substrate	K_m (μmol/l)	V_{max} (nmol/g · min)	V_{max}/K_m (min^{-1})	Preparation	Reference
Dopamine	0.2	5.26	26.3	synaptosomes	HARRIS and BALDESSARINI 1973
Dopamine	0.4	20.00[a]	50.0	synaptosomes	SNYDER and COYLE 1969
(±)-Noradrenaline	2.0	20.00[a]	10.0	synaptosomes	SNYDER and COYLE 1969
(−)-Noradrenaline	2.0	2.34	1.17	synaptosomes	IVERSEN et al. 1971
(+)-Noradrenaline	1.9	2.80	1.45	synaptosomes	IVERSEN et al. 1971
(±)-Noradrenaline	1.75	4.05	2.31	slices	SHASKAN and SNYDER 1970
5-HT	8.50	29.8	3.97	slices	SHASKAN and SNYDER 1970

[a] V_{max} expressed as nmol per g pellet per min

tured PC-12 cells (BÖNISCH 1984). A quaternary ammonium derivative of amphetamine was also shown to be a substrate (ROSS 1976a).

e) Methylation of phenolic hydroxyl groups reduces transport effectiveness drastically. For example, normetanephrine does not seem to be taken up by noradrenergic neurones (SIMMONDS and GILLIS 1968; HENDLEY et al. 1970).

Hence, the structural requirements for the transport of phenylethylamines (i.e., for the value of k_{uptake}) are strikingly similar to those described above for the potency of phenylethylamines for inhibition of uptake (i.e., for the values of K_i or K_m). However, changes in the phenylethylamine structure produced by the introduction or alteration of substituents bring about changes in K_i or K_m (IVERSEN 1967) which are much more pronounced than those in k_{uptake} (Tables 1 and 2). The obvious reason for this is that changes in K_m go hand in hand with similar changes in V_{max}, so that differences between the ratios of V_{max}/K_m (i.e., k_{uptake}) for various substrates are not very pronounced. This is already evident from the results summarized in Table 1 (especially from those given for the perfused rabbit heart) and even more so from a recent report by LANGELOH et al. (1987). Using the rat vas deferens, these authors have determined values of K_m and V_{max} for various substrates of the neuronal noradrenaline carrier and found a correlation between both parameters (Fig. 3). These results indicate that large substrate-dependent changes in K_m (range: 0.9 to 44.3 µmol/l) are accompanied by small substrate-dependent changes in k_{uptake} (range: 0.05 to 0.35 min^{-1}). This observation is in agreement with the results in Tables 1 and 2 where the maximum difference between the values of k_{uptake} within comparable groups of data is not very marked (i.e. 4 to 5-fold).

Since most of the studies mentioned above were carried out with radioactively labelled amines, it is pertinent to note that increases in the mass of a given amine molecule by introduction of a label may affect rates of uptake. For instance, tritium labelling of the noradrenaline molecule reduces transport effectiveness. This reduction is dependent on the number of tritium substituents per molecule and probably also on the position of tritium in the noradrenaline molecule (tritium labelling at the α- and β-carbon > tritium labelling at the β-carbon). This was shown in the rat vas deferens and the rabbit perfused heart in which values of k_{uptake} were determined for unlabelled noradrenaline and for noradrenaline tritiated in different positions and to various degrees (TRENDELENBURG et al. 1983).

Besides the amines closely related to noradrenaline and dopamine, many other compounds unrelated to the phenylethylamine structure are substrates of the neuronal amine carrier systems (for review see ROSS 1976a). These include tryptamine analogues such as 5-HT (Tables 1 and 2; Fig. 3), adrenergic neurone blocking agents such as bretylium (Table 1; Fig. 3), guanethidine (GIACHETTI and HOLLENBECK 1976; Fig. 3) and bethanidine (Fig. 3) as well as other compounds such as amezinium (Table 1; Fig. 3). Recent evidence suggests that even as simple a molecule as choline$^+$ may serve as a substrate of the noradrenaline carrier (UNGELL et al. 1986). Hence, the overall specificity of the neuronal amine carrier systems is rather poor. The only common denominator in the structures of all known substrates appears to be the presence of an ionizable nitrogen not incorporated into an aromatic system.

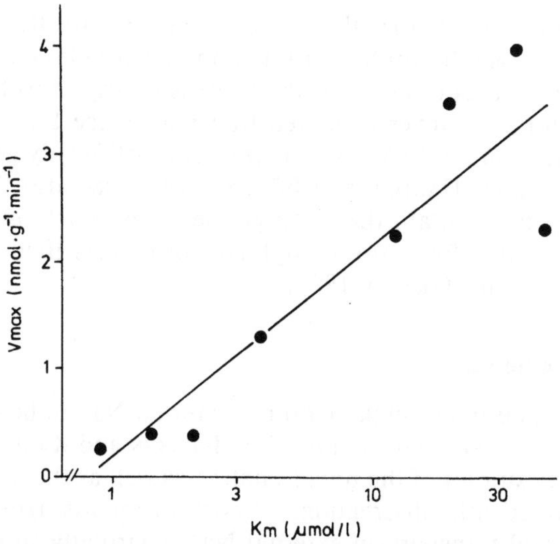

Fig. 3. Relationship between V_{max} and K_m of various substrates for neuronal uptake. Ordinate: V_{max} in nmol · g^{-1} · min^{-1}. Abscissa (log scale): K_m in µmol/l. Results were obtained in vasa deferentia of reserpine-pretreated rats; MAO and COMT were inhibited. Symbols indicate results for the following substrates (from left to right): amezinium, tyramine, dopamine, (−)-noradrenaline, guanethidine, bretylium, bethanidine and 5-HT. Shown are means taken from LANGELOH et al. (1987)

3. Temperature-Dependence and Metabolic Requirements

The neuronal accumulation of amines is highly dependent on temperature (DENGLER et al. 1962; GIACHETTI and SHORE 1966; HAMBERGER 1967; SACHS 1970; HARRIS and BALDESSARINI 1973). That is why the uptake observed at 0–2 °C was often used as a blank value. A high temperature-sensitivity of the noradrenaline carrier was also noted in plasma membrane vesicles obtained from cultured PC-12 cells (HARDER and BÖNISCH 1984). The Q_{10} of noradrenaline uptake has been reported to be 2.2 (GREEN and MILLER 1966). HOLZ and COYLE (1974) studied the influence of temperature on the dopamine uptake system of rat striatal synaptosomes. The Arrhenius plot of their results shows a bend at about 30 °C, the value of Q_{10} being 1.7 above and 4.5 below this temperature.

Several studies dealt with the question of whether the neuronal amine accumulation is dependent on free energy derived from cell metabolism. Anoxia, or the presence of an inhibitor of either anaerobic glycolysis (iodoacetate) or oxidative metabolism (dinitrophenol) may or may not affect net accumulation of amines. However, treatment of tissues *in vitro* with a combination of metabolic inhibitors (iodoacetate plus dinitrophenol) does impair amine accumulation (HAMBERGER 1967; WAKADE and FURCHGOTT 1968; PATON 1968).

Though indicating that the expenditure of metabolic energy is needed for net accumulation of amines to occur, the studies mentioned above do not answer the question of whether the carrier system is directly coupled to free energy sources (e.g. ATP). Studies with membrane vesicles ("ghosts") derived from plasma membranes of cultured PC-12 cells clearly show that ATP *per se* is *not* required for the carrier-mediated transport, since the amine transport into these vesicles can be driven by ion gradients artificially generated across the vesicle membrane (HARDER and BÖNISCH 1984; see also Sect. B.II.4). As ATP is necessary to maintain these ion gradients in intact neurones, the carrier system responsible for neuronal uptake can be classified as a secondary active transport system (GRAEFE 1976).

4. Ionic Requirements

The finding that neuronal uptake requires external Na^+ to be operative dates back to 1966 (BOGDANSKI and BRODIE 1966; IVERSEN and KRAVITZ 1966). Since then a great many studies of the noradrenaline as well as the dopamine uptake system corroborated this observation. Almost all reports agreed that the dependence of inward transport on external Na^+ is virtually complete (IVERSEN and KRAVITZ 1966; BOGDANSKI et al. 1968, 1970b; HORST et al. 1968; BOGDANSKI and BRODIE 1969; SUGRUE and SHORE 1969a; HARRIS and BALDESSARINI 1973; HOLZ and COYLE 1974; SANCHEZ-ARMASS and ORREGO 1978; KELLER and GAEFE 1979). Many other monovalent cations have been tested (Li^+, K^+, Rb^+, Cs^+, $Tris^+$, $choline^+$ and NH_4^+), but none of them are able to substitute for Na^+ in bringing about amine transport (PATON 1968, 1971; BOGDANSKI and BRODIE 1966; KELLER and GRAEFE 1979). Thus, the requirement for Na^+ appears to be absolute. This conclusion was recently confirmed in studies with cultured PC-12 cells and membrane vesicles derived from these cells (BÖNISCH et al. 1984).

When determined at a constant amine concentration, the relation between initial rate of uptake and external Na^+ concentration was found to be linear (BOGDANSKI et al. 1968; WHITE and PATON 1972; HARRIS and BALDESSARINI 1973; HOLZ and COYLE 1974), hyperbolic (IVERSEN and KRAVITZ 1966; HORST et al. 1968; BOGDANSKI and BRODIE 1969; SUGRUE and SHORE 1969a; BOGDANSKI et al. 1970b; KELLER and GRAEFE 1979; FRIEDRICH and BÖNISCH 1986) or biphasically hyperbolic (SANCHEZ-ARMASS and ORREGO 1978). The substances used to maintain isosmolarity in Na^+-deficient media were sucrose (IVERSEN and KRAVITZ 1966; BOGDANSKI et al. 1968, 1970b; BOGDANSKI and BRODIE 1969; SUGRUE and SHORE 1969a; WHITE and PATON 1972; HARRIS and BALDESSARINI 1973; HOLZ and COYLE 1974), Li^+ (HORST et al. 1968), $choline^+$ (HOLZ and COYLE 1974; SANCHEZ-ARMASS and ORREGO 1978) or $Tris^+$ (KELLER and GRAEFE 1979; FRIEDRICH and BÖNISCH 1986).

The progressive activation of amine transport by increasing Na^+ concentrations may theoretically come about by a Na^+-dependent decrease in the K_m for amine substrates, a Na^+-dependent increase in V_{max} or by both. The finding that inward transport is completely dependent on the presence of external Na^+ would strongly suggest Na^+-dependent changes in V_{max} which, in addition, may or may not be accompanied by changes in K_m. While some authors have shown that reducing the external Na^+ concentration decreases V_{max} without affecting K_m (SUGRUE and SHORE 1969a; WHITE and PATON 1972; HOLZ and COYLE 1974), others reported that low-Na^+ medium leaves V_{max} unchanged and increases K_m (BOGDANSKI et al. 1970b; SANCHEZ-ARMASS and ORREGO 1978).

Thus, although it is generally agreed that neuronal uptake is Na^+-dependent, disagreemant prevails as regards the kinetics of interaction between amine substrates and Na^+ at the uptake site. The following considerations may help to clarify this problem. *First*, the substances commonly used to replace Na^+ (or NaCl) in Na^+-deficient media (namely sucrose, Li^+ or choline$^+$) were later shown to have inhibitory effects above and beyond those resulting solely from the lack of Na^+. When comparing several NaCl substitutes, KELLER and GRAEFE (1979) regarded Tris HCl as a "pure" replacement substance, because the activation of neuronal uptake by low Na^+ concentrations was most pronounced in the presence of this substitute. All other substitutes tested in this study (including sucrose, LiCl and choline chloride) were inhibitory in comparison with Tris HCl. Sucrose-containing low-Na^+ medium is inevitably Cl^--deficient; it impairs neuronal uptake, presumably as a consequence of the carrier system being dependent not only on Na^+ but also on Cl^- (SANCHEZ-ARMASS and ORREGO 1977; FRIEDRICH and BÖNISCH 1986). *Second*, most of the studies were carried out with "normal" tissue preparations in which the endogenous noradrenaline stores had not been depleted. It has been shown that the use of normal tissues, as opposed to those obtained from reserpine-pretreated animals, can lead to distortion of results (SANCHEZ-ARMASS and ORREGO 1978), probably because the endogenous noradrenaline discharged from the neurone after exposure to low-Na^+ medium (GARCIA and KIRPEKAR 1973) interferes with the uptake of the exogenous ^3H-amine. *Third*, extraneuronal distribution of substrates (which is independent of external Na^+) was not considered in all studies. This may also have contributed to the discrepancies in the literature outlined above.

All these points were taken into consideration when SAMMET and GRAEFE (1979) reexamined the effects of Na^+ on the neuronal uptake of ^3H-(−)-noradrenaline. As illustrated in Fig. 4, neuronal uptake of noradrenaline followed Michaelis-Menten kinetics irrespective of whether noradrenaline or Na^+ was taken as the variable substrate. Moreover, increasing fixed concentrations of Na^+ (or noradrenaline) progressively decreased the apparent K_m of the system for noradrenaline (or Na^+) and gradually increased the apparent V_{max}. These results were confirmed by FRIEDRICH and BÖNISCH (1986) in a study with cultured PC-12 cells. They show that the noradrenaline carrier system exhibits the kinetic properties of a two-substrate sequential enzyme reaction in which both noradrenaline and Na^+ (1:1) must bind to the carrier for amine transport to occur. They also represent kinetic evidence for the view that noradrenaline and Na^+ act as mutually cooperative co-substrates.

As already mentioned, the inward transport of noradrenaline is also dependent on external Cl^- (SANCHEZ-ARMASS and ORREGO 1977). In cultured PC-12 cells exposed to a fixed high Na^+ concentration and various Cl^- concentrations, the K_m for noradrenaline uptake decreased and the V_{max} increased with gradually increasing Cl^- concentrations (FRIEDRICH and BÖNISCH 1986).

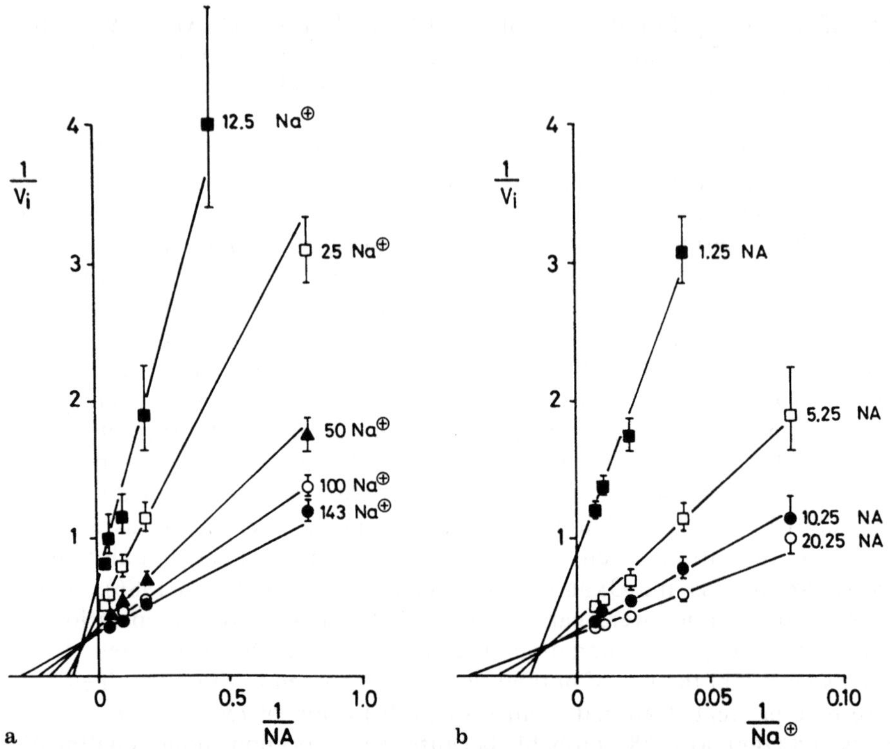

Fig. 4a, b. Lineweaver-Burk analysis of the kinetics of interaction between noradrenaline and Na^+ in neuronal uptake. Ordinates: $1/v_i$ (v_i = initial rate of neuronal uptake of ^3H-noradrenaline in $nmol \cdot g^{-1} \cdot min^{-1}$). Abscissae: **a** 1/NA (NA = concentration of ^3H-noradrenaline in the incubation medium in μmol/l); **b** $1/Na^+$ (Na^+ = concentration of Na^+ in the incubation medium in mmol/l). Rat vasa deferentia were obtained from reserpine-pretreated animals and incubated for 1 min at 37 °C in medium containing various concentrations of ^3H-(−)-noradrenaline (1.25–30.25 μmol/l) and Na^+ (12.5–143 mmol/l). MAO and COMT were inhibited. In Na^+-deficient media, Na^+ was replaced by isosmolar quantities of $Tris^+$. Neuronal uptake is defined here as the difference of tritium uptake observed in the absence and presence of 100 μmol/l cocaine. Shown are means ±SEM of at least 6 observations each. Taken from SAMMET and GRAEFE (1979)

In contrast to the fact that no other cation can effectively substitute for Na^+ to bring about transport, the action of external Cl^- is partially mimicked by Br^- and less so by SCN^-. Replacement of all external Cl^- by SO_4^{2-}, NO_3^-, $CH_3SO_4^-$ or isethionate$^-$ virtually abolishes neuronal uptake (SANCHEZ-ARMASS and ORREGO 1977; FRIEDRICH and BÖNISCH 1986).

Plasma membrane vesicles of cultured PC-12 cells are capable of taking up and accumulating noradrenaline up to a concentration gradient of about 500, provided that inwardly directed Na^+/Cl^- and outwardly directed K^+ gradients are generated across the plasma membrane during the preparation procedure (HARDER and BÖNISCH 1984). The uptake of noradrenaline by these membrane vesicles is again strictly dependent on the concentrations of external Na^+ and Cl^-. In addition, it was found that replacement of internal K^+ by Li^+ drastically reduces noradrenaline transport into the vesicle and that internal K^+ can be effectively substituted by Rb^+ (HARDER and BÖNISCH 1985). Hence, inward transport brought about by the neuronal carrier is dependent not only on external Na^+ and Cl^- but also on internal K^+.

III. Inhibitors of Neuronal Uptake

As has already been discussed (Sect. B. II.2), phenylethylamines and other substrates of the neuronal amine carrier systems are competitive inhibitors of neuronal uptake, because they simply compete with noradrenaline or dopamine for common carrier sites. Since neuronal uptake is Na^+-dependent, it is not surprising that the inhibition of uptake produced by alternative substrates is also Na^+-dependent. This has been shown by ZEITNER and GRAEFE (1986) who reported that the values of K_i for the (+)-amphetamine- and (−)-metaraminol-induced inhibition of ^{3}H-noradrenaline uptake increase with decreasing extracellular Na^+ concentrations. The results of Fig. 5a provide independent evidence for the Na^+-dependence of the inhibitory actions of (+)-amphetamine and (−)-metaraminol; they show that the uptake inhibition produced by these phenylethylamines is more pronounced at high than at low Na^+ concentrations (i.e., is mixed-type with respect to Na^+). Recent evidence also suggests that the potency of (−)-metaraminol at inhibiting ^{3}H-noradrenaline uptake is not affected by increases of the extracellular K^+ concentration (UNGELL and GRAEFE 1987).

All other inhibitors that will be described here are probably "pure" inhibitors. The available evidence presented below indicates that drugs such as cocaine and desipramine are dead-end inhibitors of the noradrenaline carrier.

1. Cocaine

Cocaine is one of the most specific inhibitors of neuronal uptake. The inhibition produced by cocaine is competitive with respect to amine substrates (MUSCHOLL 1961; IVERSEN 1963; BERTI and SHORE 1967; MAXWELL et al. 1970; GRAEFE et al. 1978). The K_i for cocaine at the noradrenaline carrier ranges from 0.4 to 2.9 μmol/l (IVERSEN 1967; BERTI and SHORE 1967; IVERSEN and LANGER 1969; GRAEFE et al. 1978; BÖNISCH 1984) and shows little species variability (JARROTT 1970). Cocaine appears to be more potent at the noradrenaline than at the dopamine carrier (Table 3). Of the two enantiomers of cocaine the (−)-form is much more potent than the (+)-form (KRELL and PATIL 1972; KOE 1976).

The inhibition by cocaine of noradrenaline uptake is Na^+-dependent in that the K_i for cocaine decreases with decreasing external Na^+ concentration (ZEITNER and GRAEFE 1986). In addition, cocaine is competitive not only with respect to noradrenaline as substrate (see above), but also with respect to Na^+ (Fig. 5c). Thus, the magnitude of uptake inhibition produced by cocaine is

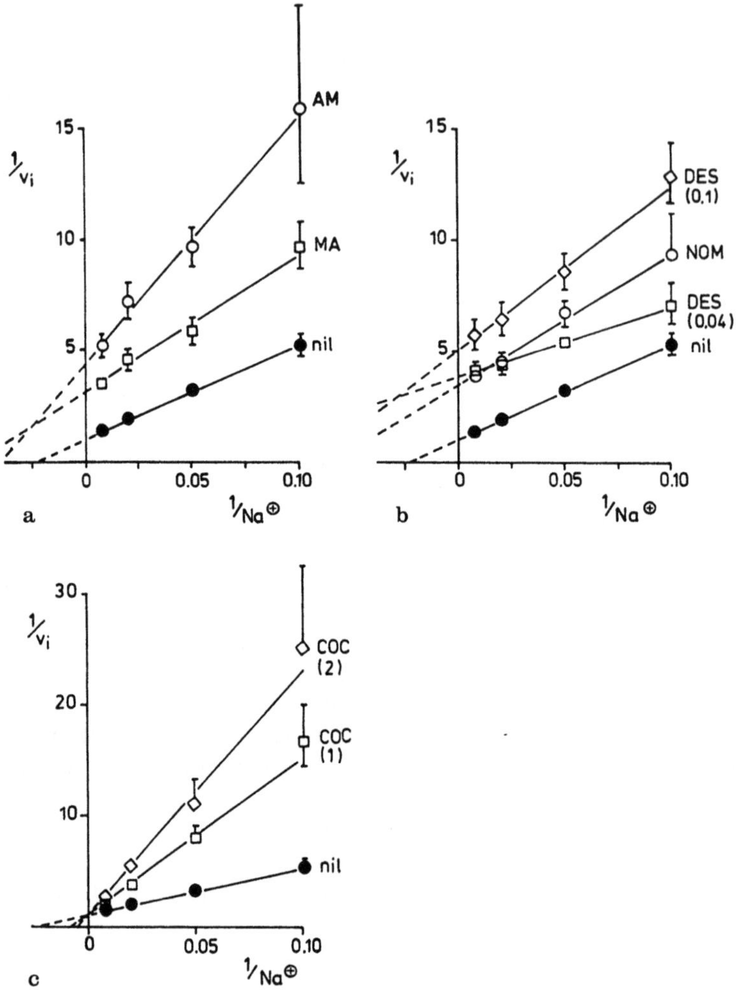

Fig. 5 a–c. Effects of some uptake inhibitors on the activation by Na^+ of the neuronal uptake of ^3H-noradrenaline; rat vas deferens, 37 °C. Ordinates: $1/v_i$ (v_i = initial rate of the neuronal ^3H-noradrenaline uptake in $nmol \cdot g^{-1} \cdot min^{-1}$). Abscissae: $1/Na^+$ (Na^+ = concentration of Na^+ in the incubation medium in mmol/l). Vasa deferentia obtained from reserpine-pretreated rats were exposed (MAO and COMT inhibited) to 1.2 μmol/l ^3H-(−)-noradrenaline and various concentrations of Na^+ (10–140 mmol/l; isosmolarity maintained with $Tris^+$), and the neuronal uptake of ^3H-noradrenaline was determined after 1 min of incubation either in the absence (nil) or presence of (+)-amphetamine (AM; 1 μmol/l), (−)-metaraminol (MA; 2 μmol/l), nomifensine (NOM; 0.1 μmol/l), desipramine (DES; 0.04 or 0.1 μmol/l) or cocaine (COC; 1 or 2 μmol/l). Shown are means (±SEM) of 20 (nil) and 6–8 (in the presence of inhibitors) observations each. Taken from ZEITNER and GRAEFE (1986)

Table 3. Comparison of the sensitivity to various drugs of the noradrenergic (A) and dopaminergic (B) neurone uptake system.

Compound	IC_{50} (μmol/l)[a]		
	A	B	Reference
(+)-Amphetamine	0.3	0.5	FERRIS et al. 1972
	0.4	1.5	KOE 1976
(−)-Amphetamine	0.4	2.1	FERRIS et al. 1972
	0.7	5.7	KOE 1976
Cocaine	0.7	3.4	ROSS and RENYI 1978
	0.3	1.7	KOE 1976
Benztropine	2.7	0.1	HORN et al. 1971
	0.5	0.5	KOE 1976
Imipramine	1.0	8.0	HORN et al. 1971
	0.07	20.0	KOE 1976
Desipramine	0.05	50.0	HORN et al. 1971
	0.006	21.0	KOE 1976
Nomifensine	0.03	0.1	SCHACHT and HEPTNER 1974
	0.02	0.4	KOE 1976
Nisoxetine	0.001	0.4	WONG and BYMASTER 1976
	0.008	2.9	KOE 1976
GBR-12935[b]	0.04	0.0016	VAN DER ZEE et al. 1980

[a] IC_{50} values have been determined at substrate concentrations of 0.1 or 0.2 μmol/l; the subtrates were labelled noradrenaline (A) and dopamine (B), respectively
[b] 1-[2-(diphenylmethoxy)ethyl]-4-(3-phenylpropyl)-piperazine

less pronounced at high than at low Na^+ concentrations. The potency of cocaine in inhibiting neuronal uptake is not affected by increases in the extracellular K^+ concentration (UNGELL and GRAEFE 1987).

2. Tricyclic Antidepressants and Related Compounds

Most antidepressant agents are powerful inhibitors of noradrenaline uptake, especially the secondary amine derivatives of the tricyclic antidepressants (e.g., desipramine, nortriptyline and protriptyline) and some newer drugs which are structurally unrelated to the conventional tricyclics (e.g., nomifensine, nisoxetine, oxaprotiline). Reviews on antidepressant drugs have been published (MAXWELL and WHITE 1978; IVERSEN and MACKAY 1979) and structure-activity relationships for the inhibition by these drugs of noradrenaline and dopamine uptake described (MAXWELL et al. 1969; SALAMA et al. 1971; KOE 1976; DE PAULIS et al. 1978; VAN DER ZEE et al. 1980; WALDMEIER et al. 1982). The tricyclic antidepressants inhibit uptake by noradrenergic much more potently than uptake by dopaminergic neurones (Table 3). The same is true for nisoxetine and, to a lesser degree, for nomifensine, whereas GBR-12935 is more selective for the dopamine than for the noradrenaline carrier (Table 3). Benztropine, an anticholinergic agent, was originally described as being a selective inhibitor of the dopamine carrier (Table 3), but subsequent studies failed to confirm this finding (Table 3; ORLANSKY and HEIKKILA 1974; HUNT et al. 1979). The inhibition by tricyclic antidepressants of the noradrenaline carrier system is competitive with respect to amine substrates (MAXWELL et al. 1969; BERTI and SHORE 1967; IVERSEN and LANGER 1969; HORN et al. 1971; BÖNISCH 1984). By contrast, amine uptake by dopaminergic neurones appears to be non-competitively antagonized by these drugs (HORN et al. 1971). Similarly, benztropine inhibits amine uptake competitively in the hypothalamus and non-competitively in the striatum (HORN et al. 1971). Nomifensine, on the other hand, appears to act as a competitive inhibitor in both regions (SCHACHT and HEPTNER 1974; TUOMISTO 1977).

The inhibition of uptake by both desipramine and nomifensine observed in noradrenergic neurones is dependent on the extracellular Na^+ concentration (ZEITNER and GRAEFE 1986). In contrast to cocaine, however, desipramine and nomifensine behave as mixed-type inhibitors with respect to Na^+ and impede noradrenaline uptake to a greater extent at high than at low Na^+ concentrations (Fig. 5b). Increases in the extracellular K^+ concentration do not affect the potency of desipramine for inhibition of noradrenaline uptake (UNGELL and GRAEFE 1987).

3. Irreversible Uptake Inhibitors

Certain nitrogen mustards such as phenoxybenzamine (IVERSEN and LANGER 1969; SANCHEZ-GARCIA et al. 1977), N-2-chloroethyl-N-ethyl-2-bromobenzylamine hydrochloride (DSP4; ROSS 1976b; ROSS and RENYI 1976) and N-2-chloroethyl-N-ethyl-2-methylbenzylamine (xylamine; CHO et al. 1980; FISCHER et al. 1983) have been shown to act as irreversible inhibitors of the neuronal amine carrier. In contrast to the poor selectivity of phenoxybenzamine, DSP4 and xylamine are both fairly selective inhibitors of the noradrenaline carrier. The three compounds are chemically related in that they all are β-chloroethylbenzylamines which, in aqueous solution, undergo intramolecular cyclization to yield aziridinium ions. As the alkylating properties of aziridinium ions probably mediate the irreversible action of these drugs on the neuronal carrier system (RANSOM et al. 1982), their time-dependent formation goes hand in hand with a time-dependent inhibition of transport. The interaction of these drugs with the amine transporter is Na^+-dependent (CHO et al. 1980) and appears to involve inward transport followed by covalent binding of the active principle to the carrier protein (FISCHER et al. 1983). During the initial phase of drug exposure, the carrier sites can be protected from irreversible inhibition by high concentrations of substrates or reversible inhibitors of the carrier system.

4. Monovalent Cations

As will be discussed in more detail in Sect. E, low-K^+ or K^+-free medium impairs neuronal uptake because the Na^+ pump is inhibited under these conditions. A characteristic feature of this type of uptake impairment is that it occurs only after a certain time lag. High K^+ concentrations also antagonize

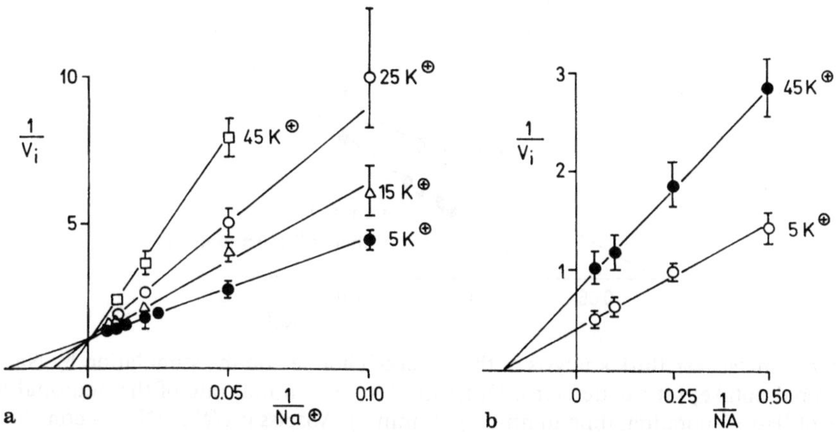

Fig. 6 a, b. Lineweaver-Burk analysis of the effects of K^+ on the stimulation by Na^+ of the neuronal ^3H-noradrenaline uptake (a) and on the saturation kinetics of the neuronal ^3H-noradrenaline uptake (b). Ordinates: reciprocal of the initial rate of ^3H-noradrenaline uptake in $nmol^{-1} \cdot g \cdot min$ (**a** and **b**). Abscissae: reciprocal of the Na^+ concentration in the incubation medium in $mmol^{-1} \cdot l$ (**a**); reciprocal of the ^3H-noradrenaline concentration in the incubation medium in $\mu mol^{-1} \cdot l$ (**b**). Results were obtained in vasa deferentia of reserpine-pretreated rats. MAO and COMT were inhibited. The ^3H-(−)-noradrenaline concentration was 1.1 µmol/l in **a** and 2–20 µmol/l in **b**. The Na^+ concentration was 10–100 mmol/l in **a** and 50 mmol/l in **b**. The K^+ concentration used in these experiments is indicated on the graphs (in mmol/l). Isosmolarity in incubation media was maintained with $Tris^+$. Taken from KELLER and GRAEFE (1979) **(a)** and UNGELL and GRAEFE (1987) **(b)**

neuronal uptake (COLBURN et al. 1968; HARRIS and BALDESSARINI 1973), but this inhibitory effect manifests itself without any delay (KELLER and GRAEFE 1979). The inhibitory action of high K^+ was originally thought to be due to K^+ competing with Na^+ for common sites on the noradrenaline carrier, because K^+ was shown to behave like a competitive inhibitor of the transport-stimulating action of Na^+ (BOGDANSKI and BRODIE 1969; SUGRUE and SHORE 1969b; Fig. 6a). If K^+ were to interfere competitively with the action of Na^+, high K^+ concentrations should increase the K_m and decrease the V_{max} for noradrenaline transport, since both kinetic parameters are known to be Na^+-dependent (see Sect. B.II.4). However, high K^+ concentrations produce a noncompetitive type of inhibition of noradrenaline transport; i.e. they decrease V_{max} without any change in K_m (SUGRUE and SHORE 1969b; Fig. 6b). Hence, the K^+-induced inhibition of noradrenaline transport is probably due to membrane depolarization rather than a consequence of K^+ hindering the action of Na^+ (for further discussion, see UNGELL and GRAEFE 1987).

Several other monovalent cations (e.g., Li^+, choline$^+$, Cs^+, Rb^+ and NH_4^+) also have antagonistic effects on noradrenaline uptake (KELLER and GRAEFE 1979). The degree of uptake inhibition observed in the presence of choline$^+$ increases with increasing external Na^+ concentration (Fig. 7). Moreover, recent evidence suggests that choline$^+$ is a substrate of the noradrenaline carrier

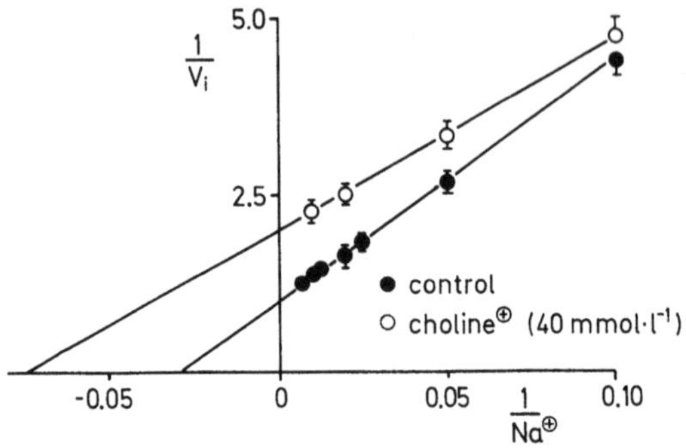

Fig. 7. Lineweaver-Burk analysis of the effect of choline$^+$ on the stimulation by Na^+ of neuronal uptake; rat vas deferens. Ordinate: $1/v_i$ (v_i = initial rate of the neuronal uptake of ^{3}H-(−)-noradrenaline in nmol·g^{-1}·min^{-1}). Abscissa: $1/Na^+$ (Na^+ = concentration of Na^+ in the incubation medium in mmol/l). The ^{3}H-(−)-noradrenaline concentration was 1.1 µmol/l. Shown are means ±SEM taken from KELLER and GRAEFE (1979). Note that the inhibition of uptake by choline$^+$ was the more pronounced, the higher the Na^+ concentration

(UNGELL et al. 1986). Li^+ appears to inhibit noradrenaline uptake by competing with Na^+ (apparent K_i = 112 mmol/l; KELLER and GRAEFE 1979). The uptake of noradrenaline and dopamine by synaptosomes of rat brain is also inhibited by Li^+ (BALDESSARINI and YORK 1970; STEFANINI et al. 1976).

5. Binding Studies with Inhibitors of Uptake

The recent discovery of specific, high-affinity binding sites for ^{3}H-imipramine and their association with the 5-HT carrier system located in the plasma membranes of platelets and serotoninergic neurones (for review, see LANGER and BRILEY 1981) has prompted binding studies with ^{3}H-desipramine. By analogy, this ligand was considered to be a suitable label of the neuronal noradrenaline carrier. High-affinity binding of ^{3}H-desipramine was indeed demonstrated in peripheral tissues, in the central nervous system and in PC-12 cells (REHAVI et al. 1981; HRDINA 1981; LEE and SNYDER 1981; LANGER et al. 1981; BÖNISCH and HARDER 1986). Several lines of evidence support the view that, after correction for non-specific binding, ^{3}H-desipramine labels a site on plasma membranes of noradrenergic neurones which is involved in noradrenaline transport (for review, see LANGER 1984). *First*, sympathectomy brought about by either surgical or chemical means results in a pronounced decrease in the number of ^{3}H-desipramine binding sites. *Second*, the densitiy of noradrenergic innervation parallels the density of ^{3}H-desipramine binding sites (i.e. the value of B_{max}). *Third*, the potencies of various drugs for inhibition of noradrenaline transport closely correlate with the potency values of these drugs for inhibition of ^{3}H-desipramine binding (with the notable exception of substrates of the noradrenaline carrier; see below). *Fourth*, the binding of ^{3}H-desipramine and the transport of noradrenaline are stereoselectively antagonized by oxaprotiline, cocaine and amphetamine. *Fifth*, the binding of ^{3}H-desipramine is Na^+-dependent, just as the inhibition by desipramine of noradrenaline transport is Na^+-dependent; increasing Na^+ concentrations gradually increase the affinity of the binding site for desipramine ($1/K_D$) without changing B_{max}. *Finally*, intact or solubilized plasma membranes derived from PC-12 cells (i.e. pure neuronal plasma membranes) also contain a single, non-interacting population of binding sites to which ^{3}H-desipramine is bound with high affinity (Bönisch and Harder 1986; Schömig and Bönisch 1986).

The observation that substrates of the noradrenaline carrier are relatively weak inhibitors of ^{3}H-desipramine binding was taken to indicate that the ^{3}H-desipramine and the substrate binding sites on the carrier are not identical (Raisman et al. 1982). It was also hypothesized that ^{3}H-desipramine binds near the substrate recognition site to a site that modulates noradrenaline transport through an allosteric mechanism (Langer et al. 1984). While this is one possible explanation of the poor ability of substrates to inhibit ^{3}H-desipramine binding, there is an alternative one (see Sect.E).

Recently, aryl-1,4-dialk(en)yl-piperazine derivatives such as GBR-12935 [1-(2-(diphenylmethoxyl)ethyl)-4(3-phenylpropyl)-piperazine] have been found to be very potent and rather selective inhibitors of dopamine uptake in the rat corpus striatum (van der Zee et al. 1980). Tritiated GBR-12935 was also used as ligand for labelling the dopamine carrier in the striatum (Berger et al. 1985; Janowsky et al. 1986). The specific binding of this ligand was of high affinity (K_D = 1–2 nmol/l), Na^+-dependent and stereoselectively antagonized by the diastereoisomers of methylphenidate. Surgical denervation of the striatum by section of the medial forebrain bundle reduced both the uptake of dopamine and the specific binding of another aryl-1,4-dialkenyl-piperazine analogue (^{3}H-GBR-12783) (Bonnet et al. 1986).

C. Neuronal Amine Metabolism

I. Formation of Metabolites Through MAO Activity

MAO plays a pivotal role in the neuronal metabolism of noradrenaline (Blaschko 1972; Kopin 1972; Graefe and Henseling 1983). The neuronal compartment of MAO may represent but a small fraction of the tissue's MAO activity (Lowe and Horita 1970). However, its intracellular localisation, together with the fact that the availability of substrates for this enzyme is enhanced in noradrenergic and dopaminergic neurones by a highly effective uptake system, explains why the neuronal pool of MAO is so important in densely innervated tissues.

In most of the reports published in the 1960s the "deaminated metabolites" of ^{3}H-noradrenaline were not further specified. More recent studies have made use of refined chromatographic procedures (Levin 1973; Graefe et al. 1973; Cubeddu et al. 1979a) to separate the various metabolites of radioactively labelled catecholamines. The neuronal formation of deaminated metabolites resulting from exposure to low concentrations of ^{3}H-noradrenaline (^{3}H-DOPEG and ^{3}H-DOMA) or ^{3}H-dopamine (^{3}H-DOPET and ^{3}H-DOPAC) was assessed under three experimental conditions: *first*, during the spontaneous outflow from tissues previously labelled with these catecholamines (LANGER 1974; LANGER et al. 1975; CUBEDDU et al. 1979a); *second*, during and after amine release elicited by nerve stimulation (LANGER 1974; LANGER et al. 1975; FARAH et al. 1977; CUBEDDU et al. 1979b); *third*, during the exposure of tissues to these catecholamines (LEVIN 1974; GRAEFE et al. 1977; HENSELING and TRENDELENBURG 1978; PAIVA and GUIMARAES 1978; FIEBIG and TRENDELENBURG 1978; BÖNISCH 1980; GRAEFE 1981; HENSELING 1983). Under all three experimental conditions, ^{3}H-DOPEG was formed from ^{3}H-noradrenaline at higher rates than ^{3}H-DOMA, while with ^{3}H-dopamine as the parent amine ^{3}H-DOPAC was the main metabolite. The formation of these metabolites during spontaneous outflow is affected little, if at all, by cocaine, indicating that under these conditions MAO acts upon the transmitter amines after their leakage from storage vesicles into the axoplasm. When, on the other hand, oxidative deamination occurs subsequent to neuronal uptake, metabolite formation is greatly reduced in the presence of cocaine.

Studies with ^{3}H-(−)-, ^{3}H-(±)- or ^{14}C-(+)-noradrenaline as substrates have indicated that there are two types of stereoselectivity (LEVIN 1974; GRAEFE et al. 1977; HENSELING and TRENDELENBURG 1978; GRAEFE and HENSELING 1983). After inhibition of vesicular uptake by pretreatment with reserpine, the total amount of deaminated metabolites formed in the noradrenergic neurone is higher for ^{3}H-(−)- than for ^{3}H(±)- or ^{14}C-(+)-noradrenaline (stereoselectivity of MAO). Moreover, ^{3}H-(−)-noradrenaline is transformed to DOPEG at higher rates than ^{3}H-(±)- or ^{14}C-(+)-noradrenaline, and the ratio "rate of formation of DOPEG/rate of formation of DOMA" is higher for the ^{3}H-(−)- than for the ^{14}C-(+)-enantiomer (or the ^{3}H-racemate). The latter type of stereoselectivity cannot be ascribed to MAO; it probably resides in the aldehyde reductase and (or) aldehyde dehydrogenase which both are in series with MAO and form DOPEG and DOMA, respectively.

Although ^{3}H-DOPEG is usually found to be a more predominant neuronal metabolite of ^{3}H-(−)-noradrenaline than ^{3}H-DOMA, it is pertinent to note that the rate of formation of ^{3}H-DOPEG relative to that of ^{3}H-DOMA decreases when tissues are exposed to increasing concentrations of ^{3}H-(−)-noradrenaline (GRAEFE and HENSELING 1983). In rabbit aortic strips, for instance, the ratio of the rates of formation of ^{3}H-DOPEG to ^{3}H-DOMA was 1.75 during incubation with 0.12 μmol/l and 0.72 during incubation with 1.2 ^{3}H-(−)-noradrenaline (HENSELING and TRENDELENBURG 1978). Similarly, the ratio of ^{3}H-DOPEG/^{3}H-DOMA decreased from 11.7 to 2.22 when, in the cat perfused heart, the perfusion concentration of ^{3}H-(−)-noradrenaline was increased from 0.003 to 0.3 μmol/l (GRAEFE and HENSELING 1983). Hence, when compared with the DOMA-forming enzyme, the enzyme responsible for the production of DOPEG has the lower K_m.

The kinetics of the formation of ^{3}H-DOPEG and ^{3}H-DOMA have been studied in more detail in perfused isolated hearts of the cat and rabbit (GRAEFE 1981). In both spezies there is, first of all, virtually no extraneuronal formation of ^{3}H-DOPEG and ^{3}H-DOMA, nor are there any pronounced species differences with regard to the kinetic results of this study. On perfusion of the hearts at a constant rate with 0.3 μmol/l ^{3}H-(−)-noradrenaline, the neuronal compartment of MAO equilibrates very quickly with the perfusion medium, provided hearts of reserpine-pretreated animals are being used. In normal hearts, on the other hand, the formation of ^{3}H-DOPEG and ^{3}H-DOMA attains a steady rate only after a considerable time lag, the duration of which is similar for both metabolites. Apparently, the presence of intact storage vesicles prolongs the time required for the axoplasmic concentration of ^{3}H-noradrenaline to reach a steady level. The appearance of the two metabolites in the perfusion fluid reaches a constant rate later than does their formation. This discrepancy in time course must be ascribed to the time taken for distribution of the metabolites in the tissue: the concentration of the metabolites in the tissue rises progressively until a rate of efflux from the tissue is generated which matches the rate of metabolite formation. Within the time of observation (120 min), constant rates of appearance of ^{3}H-DOPEG and ^{3}H-DOMA in the perfusion fluid are attained in hearts of reserpine-pretreated animals (after about 70 min), but not in normal hearts. This steady state is accompanied by a steady level of ^{3}H-noradrenaline net removal from the perfusion fluid. The rate of net removal of the parent amine observed during steady state is fully accounted for by the rates of the neuronal formation of ^{3}H-DOPEG and ^{3}H-DOMA.

In the same experiments (GRAEFE 1981), it was found that the steady-state rate of metabolite appearance in the perfusion fluid (^{3}H-DOPEG > ^{3}H-DOMA) is approached more quickly for ^{3}H-DOPEG than for ^{3}H-DOMA, and that the steady-state tissue levels are much lower for ^{3}H-DOPEG than for ^{3}H-DOMA. The relation between rate of appearance and tissue level at steady state is linear for both ^{3}H-DOPEG and ^{3}H-DOMA, indicating first-order distribution (or efflux) kinetics. The rate constant for efflux is considerably higher for ^{3}H-DOPEG (0.54 min^{-1} in the cat and 0.72 min^{-1} in the rabbit heart) than for ^{3}H-DOMA (0.15 min^{-1} and 0.06 min^{-1}, respectively). Thus, the ability to penetrate cell membranes appears to be pronounced for ^{3}H-DOPEG and poor for ^{3}H-DOMA. These results are in keeping with several other reports in which the kinetics of efflux have been examined for a number of monoamine metabolites originating from both neuronal and extraneuronal compartments (for review, see TRENDELENBURG et al. 1979). They also support the notion that metabolite concentrations in tissues cannot be regarded as a measure of metabolite formation (LANGER et al. 1975; FARAH et al. 1977).

Recent evidence indicates that tritium labelling of the noradrenaline molecule may impair its oxidative deamination by MAO (STARKE et al. 1980; TRENDELENBURG et al. 1983). While (−)-noradrenaline-7-^{3}H (with the label present at the β-carbon of the side chain) is a slightly better substrate of MAO than unlabelled (−)-noradrenaline, (−)-noradrenaline-7,8-^{3}H (with varying proportions of the label also present at the α-carbon) and (−)-noradrenaline-2,5,6-^{3}H (ring-labelled noradrenaline with high specific activity) are clearly inferior substrates. Tritium substitution at the α-carbon not only hinders the deamination of noradrenaline by MAO, it also leads to a MAO-induced formation of tritium water which falsifies the results of established column chromatographic procedures (STARKE et al. 1980). As a consequence of the reduced rates of intraneuronal deamination by MAO, the net accumulation of ^{3}H-noradrenaline inside the neurone is more pronounced after exposure to (−)-noradrenaline-7,8^{3}H or (−)-noradrenaline-2,5,6-^{3}H than after exposure to (−)-noradrenaline-7-^{3}H. When compared with the latter ^{3}H-amine, preloading of tissues with one of the former two ^{3}H-amines also gives rise to higher ratios of ^{3}H-noradrenaline/^{3}H-DOPEG in the spontaneous outflow of radioactivity from neuronal stores (for further discussion, see TRENDELENBURG et al. 1983).

The metabolism of endogenous and unlabelled exogenous noradrenaline was studied by STARKE and his co-workers (STARKE et al. 1981; MAJEWSKI et al. 1982). They used high pressure liquid chromatography with electrochemical detection and determined the spontaneous and induced outflow of noradrenaline, DOPEG and DOMA from incubated guinea-pig atria and rabbit perfused hearts. According to their results, DOPEG is the only deaminated metabolite of endogenous noradrenaline; DOMA was detected neither in the outflow nor in tissues. However, during and after exposure of the tissues to unlabelled (−)- or (+)-noradrenaline there was formation of DOPEG *and* DOMA (DOPEG > DOMA) and the ratio of the neuronal formation of DOMA to that of DOPEG was higher after exposure to (+)- than after exposure to (−)-noradrenaline. Moreover, during and after exposure of tissues to increasing concentrations of (−)-noradrenaline (0.17 to 17 μmol/l) increasing amounts of DOMA relative to DOPEG were formed intraneuronally. Hence, as already indicated by the results of experiments with radioactively labelled noradrenaline, the neuronal formation of DOPEG and DOMA is stereoselective, and DOMA formation comes into play whenever the DOPEG-forming enzyme is saturated by exposure to outside (−)-noradrenaline concentrations higher than about 0.2 μmol/l.

II. Is There any Intraneuronal Formation of O-Methylated Metabolites?

The existence of a neuronal compartment of COMT has been inferred from decreases in COMT activity observed after sympathectomy (JARROTT and IVERSEN 1971; JARROTT and LANGER 1971). However, the lack of convincing evidence for O-methylated metabolites being of neuronal origin casts some doubt on the view that a neuronal COMT is involved in the metabolism of noradrenaline. The finding that the transformation of ^{3}H-noradrenaline to ^{3}H-MOPEG is often only partially reduced or even unaffected after inhibition of extraneuronal uptake is by no means unequivocal proof for neuronal COMT activity. As DOPEG is highly lipid soluble (MACK and BÖNISCH 1979) and penetrates cell membranes with ease by simple diffusion, ^{3}H-MOPEG may well be formed from neuronal ^{3}H-DOPEG by extraneuronal COMT. In support of this view, several studies have shown that changes in the formation of ^{3}H-DOPEG go hand in hand with similar changes in the formation of ^{3}H-MOPEG or ^{3}H-O-methylated-deaminated metabolites (i.e. ^{3}H-MOPEG plus ^{3}H-

VMA) (LANGER et al. 1972; GRAEFE et al. 1977; FIEBIG and TRENDELENBURG 1978; GRAEFE 1981).

While the intraneuronal O-methylation of ^{3}H-DOPEG to ^{3}H-MOPEG remains doubtful, attempts to demonstrate presynaptic formation of ^{3}H-NMN have either failed or turned out to be not very impressive. Neither in the rat heart (FIEBIG and TRENDELENBURG 1978) nor in the cat heart (GRAEFE and HENSELING 1983) or rabbit aorta (HENSELING 1983) were there indications of any intraneuronal transformation of ^{3}H-(−)-noradrenaline to ^{3}H-NMN. Large amounts of unchanged ^{3}H-(−)-noradrenaline accumulate in sympathetic nerve terminals of the perfused heart of reserpine-pretreated cats when MAO is blocked; even so, a cocaine-sensitive formation of ^{3}H-NMN could not be detected (GRAEFE and HENSELING 1983). The spontaneous outflow from noradrenergic nerve endings preloaded with ^{3}H-noradrenaline usually contains small amounts of ^{3}H-NMN. This formation of ^{3}H-NMN is hardly susceptible to inhibition by inhibitors of extraneuronal uptake (LUCHELLI-FORTIS and LANGER 1975; GRAEFE et al. 1977). This is the only observation that points towards the possible existence of a small functional pool of neuronal COMT.

In support of the view that the decreases in COMT activity observed after surgical sympathectomy (see above) do not reflect losses of neuronal COMT, BRANCO et al. (1984) have shown that the extraneuronal O-methylation of ^{3}H-isoprenaline is greatly reduced in sympathetically denervated blood vessels.

D. Neuronal Efflux

I. Terminology and Methodology

The discharge of ^{3}H-catecholamines from previously labelled neurones evoked by various means has always been described as amine release. There are basically two mechanisms by which amine release can be elicited: one is dependent on, and the other independent of, the presence of external Ca^{2+} (PATON 1973, 1976b; THOA et al. 1975; ROSS and KELDER 1976, 1979; RAITERI et al. 1979). The Ca^{2+}-dependent type of release, which is known to come about by exocytosis, is not affected by the presence of inhibitors of the neuronal amine transporter. The Ca^{2+}-independent release, on the other hand, is highly susceptible to inhibition by low concentrations of cocaine or desipramine and therefore thought to be mediated by the amine carrier operating in the outward direction. This latter type of evoked amine release would be better termed "carrier-mediated efflux" or "outward transport". There is of course also a spontaneous amine efflux which is also Ca^{2+}-independent and due mainly to passive diffusion of the non-ionized amine species across the axonal membrane.

The spontaneous efflux of radioactivity from tissues previously loaded with, e.g., ^{3}H-noradrenaline and then washed out with amine-free medium declines rapidly at first and slowly thereafter. In the late component of efflux, the efflux declines monoexponentially with time. When, at any given time during washout, the rate of tritium efflux is expressed as a fraction of the amount of ^{3}H-amine remaining in the tissue, the "fractional rate of loss" is obtained (PATON 1973; GRAEFE et al. 1977). This parameter is a measure of the instantaneous rate constant for efflux; it falls with time initially and remains virtually unchanged during the late component of efflux. This efflux was shown to be of neuronal origin (PATON 1973; LINDMAR and LÖFFELHOLZ 1974a; GRAEFE and FUCHS 1979). It usually accounts for most of the ^{3}H-amine previously taken up by the tissue and has always been referred to as neuronal efflux.

The spontaneous neuronal efflux of tritium from normal tissues labelled with ^{3}H-noradrenaline or ^{3}H-dopamine appears to be mainly a consequence of the leakage of the ^{3}H-amine from storage vesicles followed by MAO-induced amine metabolism within the axoplasm, since under these circumstances only a small proportion (10% or less) of the radioactivity leaving the tissue is composed of the parent ^{3}H-amine (LANGER et al. 1975; CUBEDDU et al. 1979a). To study the neuronal efflux of unchanged amines under these experimental conditions, the formation of metabolites, the vesicular uptake and storage and possible effects of drugs under study on MAO and/or storage vesicles have to be taken into consideration. Therefore, many efflux studies have been carried out after inhibition of both vesicular uptake and MAO (PATON 1973; LINDMAR and LÖFFELHOLZ 1974a and b; ROSS and KELDER 1976, 1979; RAITERI et al. 1977b; GRAEFE and FUCHS 1979). The following discussion is based mainly on work in which this approach was used.

When vesicular uptake and MAO are inhibited (by pretreatment with reserpine and pargyline, respectively) very high axoplasmic noradrenaline concentrations can be achieved. STEFANO and TRENDELENBURG (1984) have demonstrated that saturation of the small remainder of neuronal MAO occurs in vasa deferentia obtained from rats pretreated with reserpine and pargyline and previously exposed to 2 μmol/l ^{3}H-(−)-noradrenaline. If the assumption is made that the pretreatment with pargyline (an irreversible inhibitor of MAO) does not affect the K_m of that part of the enzyme that escaped inhibition, then the axoplasmic concentration of ^{3}H-noradrenaline must have been in the range of 1 to 2 mmol/l (i.e. concentrations of noradrenaline likely to saturate MAO). The build-up of concentrations of this magnitude indicates that the axonal membrane is able to prevent very effectively the outward movement of noradrenaline. One must also be aware of the fact that the remainder of neuronal MAO activity is then exposed to very high substrate concentrations. Hence, in spite of a high degree of MAO inhibition by pargyline, it is almost impossible to abolish the formation of ^{3}H-DOPEG in the spontaneous neuronal efflux of ^{3}H-noradrenaline (STEFANO and TRENDELENBURG 1984).

II. Spontaneous Neuronal Efflux

The axonal membrane appears to function as a highly effective barrier to the efflux of unchanged axoplasmic noradrenaline or dopamine when the Na^+- and K^+-gradients across the membrane are normal. This explains why, under conditions of intact or incompletely inhibited MAO, in normal and reserpine-pretreated tissues (previously loaded with ^{3}H-noradrenaline or ^{3}H-dopamine) most of the radioactivity that leaves the neurone consists of the relatively lipophilic deaminated metabolites ^{3}H-DOPEG or ^{3}H-DOPAC. It also explains why, after inhibition of MAO and vesicular uptake, the spontaneous efflux of the unchanged amines exhibits a very low apparent rate constant. The half life for axoplasmic ^{3}H-noradrenaline determined under these conditions probably depends mainly on the degree of MAO inhibition and ranges from 1 to 2.5 h (PATON 1973; HOLBACH and LÖFFELHOLZ 1975; HENSELING et al. 1976; MEKANONTCHAI and TRENDELENBURG 1979; STEFANO and TRENDELENBURG 1984). The half-life remains virtually unchanged when the amount of amine initially accumulated in the axoplasm is varied over a wide range, indicating that the spontaneous neuronal efflux cannot be easily saturated by increases of the internal amine concentration (HENSELING et al. 1976; STEFANO and TRENDELENBURG 1984).

Since active membrane transport systems usually operate in both directions (though at differing rates), it is conceivable that the carrier responsible for neuronal uptake contributes to neuronal efflux. The available evidence indicates that the spontaneous amine efflux is perhaps both carrier-mediated and due to passive diffusion. Inhibition of the carrier by either cocaine or desipramine moderately enhances the efflux of axoplasmic ^{3}H-noradrenaline in most intact tissues. However, under experimental conditions under which re-uptake of the amine is largely prevented (i.e. in superfused synaptosomes), both cocaine and desipramine have the opposite effect: they antagonize the spontaneous efflux of ^{3}H-noradrenaline (RAITERI et al. 1974, 1977b). This apparent conflict of results can be explained if cocaine and desipramine are non-transported inhibitors of the carrier; they both arrest the carrier at the outer face of the membrane and inhibit mediated transport in both directions. The cocaine-induced enhancement of efflux commonly observed in intact tissues is then the result of two opposite effects of the drug: while the efflux via the amine carrier is inhibited, that part of efflux which comes about by simple diffusion is "enhanced" because re-uptake is likewise inhibited. Since the net effect of cocaine is usually "enhancement of efflux", it can be concluded that part of the spontaneous efflux of ^{3}H-noradrenaline is due to simple diffusion.

Two other observations support the view that both carrier-mediated transport and passive diffusion contribute to the spontaneous neuronal efflux. Firstly, the neuronal retention of ^{3}H-metaraminol is more dependent on neuronal re-uptake than is the retention of ^{3}H-noradrenaline (CARLSSON and WALDECK 1968), and the enhancement of the neuronal efflux produced by cocaine is much more pronounced for ^{3}H-metaraminol as substrate than for ^{3}H-noradrenaline (PATON 1981). These findings can readily be explained by the fact that metaraminol is much more lipid soluble than is noradrenaline (MACK and BÖNISCH 1979). Consequently, the relative contribution by simple diffusion to the spontaneous neuronal efflux is expected to be higher for metaraminol than for noradrenaline. Secondly, the spontaneous efflux of noradrenaline and dopamine is markedly temperature-dependent; Q_{10} values ranging from 1.8

to 5.0 have been reported (LINDMAR and LÖFFELHOLZ 1972; PATON 1973; FISCHER and CHO 1979). Hence, it is reasonable to conclude that the spontaneous neuronal efflux is in part carrier-mediated.

III. Carrier-Mediated Neuronal Efflux

1. Efflux Induced by Phenylethylamines

It is well known that indirectly acting sympathomimetic amines release endogenous noradrenaline or dopamine through a mechanism independent of Ca^{2+} and sensitive to cocaine or desipramine. No doubt, this type of induced transmitter efflux has all the characteristics of being mediated by the neuronal amine carrier. Compounds that are able to evoke outward transport of noradrenaline from noradrenergic neurones without interfering with the potential and/or ion gradients across the axonal membrane are all substrates of the noradrenaline carrier. For this reason, PATON (1973) described the phenomenon of coupling between the inward transport of the "inducing" amine and the outward transport of noradrenaline as "accelerative exchange diffusion". This counterflow phenomenon is known to occur in other membrane transport systems and is usually taken to indicate mobility of the carrier site (STEIN 1967). The evidence in favour of the "carrier hypothesis" of the phenylethylamine-induced release of noradrenaline is discussed in more detail elsewhere (BÖNISCH and TRENDELENBURG this volume, Chap. 5)

2. Efflux Induced by Changes in Transmembrane Ion Gradients

The efflux of tritium from the axoplasm of noradrenergic and dopaminergic neurones (previously loaded with ^{3}H-noradrenaline and ^{3}H-dopamine, respectively) is greatly accelerated in response to a low-Na^{+} medium (PATON 1973; RAITERI et al. 1979). This acceleration of tritium efflux is exclusively due to an acceleration of efflux of the unchanged ^{3}H-amine (GRAEFE and FUCHS 1979). On exposure to Na^{+}-free medium, rabbit perfused hearts show a 38-fold increase in the rate of noradrenaline efflux (LINDMAR and LÖFFELHOLZ 1974b) and rat vasa deferentia a 40-fold increase in the rate of ^{3}H-noradrenaline efflux (GRAEFE and FUCHS 1979). This enormous increment in efflux is not just a consequence of the blockade of neuronal re-uptake; it reflects true acceleration of carrier-mediated efflux, because it is not observed at low ambient temperature (ROSS and KELDER 1979) and is largely prevented in the presence of cocaine or desipramine or nomifensine (PATON 1973; RAITERI et al. 1977b; 1979; ROSS and KELDER 1979; GRAEFE and FUCHS 1979). The observed inhibition by cocaine and desipramine is concentration-dependent: IC_{50} values of 2.2 µmol/l (cocaine) and 0.35 µmol/l (desipramine) have been obtained in studies with the rat vas deferens (GRAEFE and FUCHS 1979; ROSS and KELDER 1979). Moreover, changes of the degree of neuronal filling with ^{3}H-noradrenaline did not alter the potency of cocaine as inhibitor of the carrier-mediated efflux (GRAEFE and FUCHS 1979). Hence, cocaine (a competitive inhibitor of noradrenaline inward transport) appears to be a noncompetitive inhibitor of noradrenaline outward transport. This kind of result would be expected if cocaine were to act as a non-transported inhibitor. Interestingly enough, substrates of the carrier such as metaraminol do not antagonize the acceleration of efflux evoked by Na^+-free medium (GRAEFE et al. 1984).

Studies in which the effects of various external Na^+ concentrations were examined have shown that the rate of amine efflux is inversely proportional to the concentration of Na^+ in the medium (PATON 1973; RAITERI et al. 1979). This inverse relationship between the external Na^+ concentration and the rate of carrier-mediated efflux was analysed in more detail in the rat vas deferens labelled with ^3H-noradrenaline (GRAEFE and FUCHS 1979; GRAEFE et al. 1984). The results show that extracellular Na^+ inhibits the carrier-mediated neuronal efflux in a concentration-dependent manner; 50% inhibition of ^3H-noradrenaline outward transport was observed at a Na^+ concentration of about 10 mmol/l. This observation was interpreted to mean that the binding of Na^+ to the free carrier results in a complex which is unable to cross the membrane, so that its formation on the outside of the membrane limits the availability of carrier sites on the inside (for further discussion, see Sect. E.).

In the efflux experiments of GRAEFE and FUCHS (1979) $Tris^+$ was used to maintain isosmolarity in Na^+-deficient media. When, on the other hand, Li^+ was the replacement cation, Na^+ was less potent in inhibiting the outward transport of ^3H-noradrenaline; in other words, the increase in efflux elicited by low external Na^+ concentrations was more pronounced in Li^+- than in $Tris^+$-containing media (GRAEFE and FUCHS 1979; BÖNISCH and LANGELOH 1986). Thus, external Li^+ appears to antagonize the binding of Na^+ to the carrier. In doing so, it accelerates the outward transport in response to low-Na^+ medium. The same line of reasoning applies to the effects of high K^+ concentrations in Na^+-deficient media. High external K^+ concentrations accelerate the efflux of axoplasmic ^3H-noradrenaline (in the absence of Ca^{2+}!) more than would result solely from the lack of external Na^+ (WAKADE and KIRPEKAR 1974; PATON 1976b; ROSS and KELDER 1979). Hence, external K^+ may compete with Na^+ for carrier sites through which external Na^+ hinders carrier-mediated efflux. However, it is more likely that K^+ accelerates efflux by membrane depolarization (see Sect. B.III.4).

Other studies in which the *internal* Na^+ concentration was increased have added further weight to the notion that the neuronal amine carrier can bring about outward transport. The experimental tools commonly used to elevate the intracellular Na^+ concentration (inhibition of Na^+ pumping by either K^+-free medium or ouabain; opening of the fast Na^+ channel by veratridine) all cause marked acceleration of carrier-mediated neuronal efflux (PATON 1973; LINDMAR and LÖFFELHOLZ 1974b; ROSS and KELDER 1976, 1979; RAITERI et al. 1979; BÖNISCH et al. 1983; STUTE and TRENDELENBURG 1984). Thus, neuronal efflux, like neuronal uptake, is dependent on the presence of Na^+ on the appropriate side of the axonal membrane.

The question of whether the saturability of the outward transport of noradrenaline is dependent on the inside Na^+ concentration was investigated in rat vasa deferentia previously loaded with a wide range of ^3H-noradrenaline concentrations. When the carrier-mediated efflux of ^3H-noradrenaline was evoked by Na^+-free medium, the observed peak rates of efflux were linearly

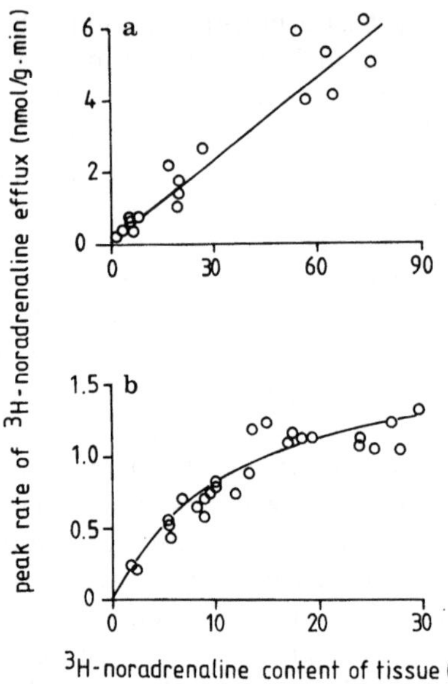

Fig. 8 a, b. Relationship between the rate of ^{3}H-noradrenaline outward transport and the ^{3}H-noradrenaline content in tissue; rat vas deferens. Ordinates: peak rates of the carrier-mediated ^{3}H-noradrenaline efflux observed after exposure to Na^{+}-free medium **(a)** or 100 µmol/l veratridine **(b)** in nmol · g^{-1} · min^{-1}. Abscissae: ^{3}H-noradrenaline content in tissue in nmol · g^{-1}. Vasa deferentia (the MAO and COMT of which were inhibited) were obtained from reserpine-pretreated rats and incubated for 30 min with various concentrations of ^{3}H-(−)-noradrenaline (0.3–100 µmol/l). Thereafter, they were washed with amine-free medium until the fractional rate of tritium loss from the tissue remained virtually constant (i.e., after 100 min of washout). Tissues were then exposed for 10 min to either Na^{+}-free medium (Na^{+} replaced by Li^{+}) or 100 µmol/l veratridine and rates of tritium efflux determined at 1-min intervals. Unchanged ^{3}H-noradrenaline accounted for more than 90% of the radioactivity in tissues and washout media. Taken from BÖNISCH et al. (1986)

related to the ^{3}H-amine content of the tissue (Fig. 8 a). On the other hand, the peak efflux rates observed after exposure to veratridine approached saturation as the ^{3}H-noradrenaline content in the tissue was raised (Fig. 8 b). Since the achievable degree of neuronal filling with ^{3}H-noradrenaline is limited by the saturability of neuronal uptake, the linear relationship in Fig. 8 a does not exclude saturability of the outward transport at much higher degrees of neuronal filling than those that can be achieved. Therefore, the results of Fig. 8 indicate that the K_m for the outward transport of ^{3}H-noradrenaline is much lower when the efflux is elicited by veratridine than when the efflux is elicited by Na^{+}-free medium. It is well known that veratridine, but not Na^{+}-free medium, produces an increase in the intracellular Na^{+} concentration (consequent upon opening of the fast Na^{+} channel). Hence, the results of Fig. 8 suggest that the K_m for the outward transport of noradrenaline is inversely related to the internal Na^{+} concentration, just as the K_m for the inward transport of noradrenaline is inversely related to the external Na^{+} concentration (see Fig. 4).

E. Proposed Mechanism of Neuronal Uptake: A Summing up

I. General Considerations

A simple model for the carrier(C)-mediated transport of a substrate (S) across the axonal membrane, which can be applied to any carrier system, is presented in Fig. 9. A slightly revised version of this model was recently analysed mathematically and tested experimentally (SCHÖMIG et al. 1988). In this model, one transport cycle from out to in consists of four steps: the *association* step, i.e. the formation of the carrier-substrate complex (SC) at the outer face of the membrane (step 1), the *translocation* of SC from the outer to the inner face of the membrane (step 2), the *dissociation* of CS at the inner face of the membrane (step 3) and the *recycling* of free C from in to out (step 4). Initial rates of uptake (i.e. rates of unidirectional inward transport) will take place in this model whenever the steps 2, 3 and 4 proceed virtually only counterclockwise, and after the distribution of the available carrier in the positions a, b, c and d (Fig. 9) has reached a steady state. The relative steady-state concentrations of carrier sites in these four positions will depend on the relative magnitude of the first-order rate constants characterizing the steps 2, 3, and 4. If step 2 is rate-limiting in the whole transport cycle, most of the carrier will distribute in the positions a and b, and it is only under this condition that K_m (the substrate concentration sufficient to produce $V_{max}/2$) approximate K_D (the equilibrium dissociation constant for SC) and that V_{max} is the same for all substrates. If, on the other hand, step 3 is rate-limiting, carrier will be "trapped" in position c, whereas "trapping" of carrier in position d will occur if step 4 is rate-limiting. Under the latter two conditions K_m will not approximate K_D, and V_{max} will vary with the substrate. In fact, the value of K_m relative to that of K_D will then be the lower, the lower the fractional availabilty of carrier on the outer face of the membrane. The fractional availabilty of carrier in the positions a and b will limit not only K_m but also V_{max}. In addition, however, the overall first-order rate constant for the steps 2, 3 and 4 (i.e. the total carrier "conductance") will also be an important determinant of V_{max} (SCHÖMIG et al. 1988).

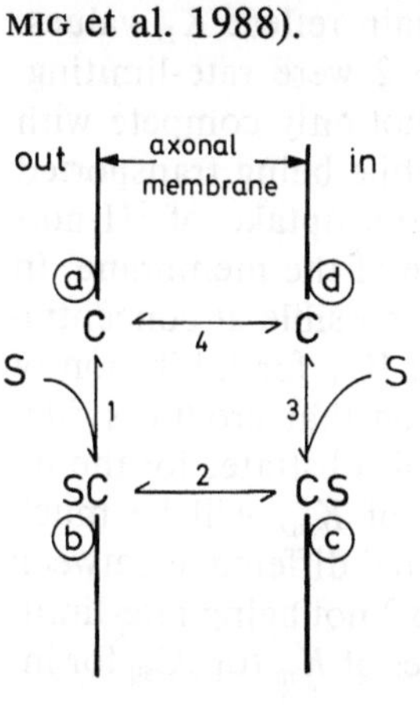

Fig. 9. Simple model for the transport of substrates (S) through the carrier system (C) associated with the axonal membrane of the noradrenergic neurone. One transport cycle for the uptake of S by the neurone consists of the association step (1), the translocation step (2), the dissociation step (3) and the recycling of the carrier (4). The four positions in which the carrier distributes under conditions of steady state (i.e., when uptake proceeds at initial rates) are indicated by a, b, c and d

As far as the noradrenaline carrier system is concerned, the available evidence is not compatible with the view that step 2 is rate-limiting. The results presented in Fig. 3 show that V_{max} is substrate-dependent and tends to increase with K_m. To explain these observations with the help of Fig. 9, the following kinetic considerations are pertinent (for discussion, see SCHÖMIG et al. 1988). *First*, if step 4 is rate-limiting (with carrier being trapped in position d) and the rate constant for step 3 substrate-dependent, K_m will be the lower and V_{max} the higher, the higher the rate constant for step 3. *Second*, if step 3 is rate-limiting (with carrier being trapped in position c) and its rate constant substrate-dependent, K_m and V_{max} will be the higher, the higher the rate constant for step 3. *Third*, if step 3 or 4 is rate-limiting and the rate constant for step 2 substrate-dependent, K_m will be the lower and V_{max} the higher, the higher the rate constant for step 2. Hence, the results of Fig. 3 can be reconciled with the model in Fig. 9 only, when the assumption is made that step 3 is rate-limiting and its rate constant substrate-dependent.

However, if step 2 has to be subdivided into sequential steps 2a, 2b, 2c etc., then the results of Fig. 3 would also be obtained if step 2b or 2c were rate-limiting and substrate-dependent.

The results of the ^{3}H-desipramine binding studies (see Sect. B.III.5) can also be discussed on the basis of the model in Fig. 9. Several reports show that non-transported (dead-end) inhibitors of the noradrenaline carrier (e.g. desipramine, nisoxetine and cocaine) are about equipotent in inhibiting both ^{3}H-desipramine binding and ^{3}H-noradrenaline uptake, while substrates of the carrier (e.g. metaraminol, tyramine and amphetamine) are much less potent in inhibiting ^{3}H-desipramine binding than in inhibiting ^{3}H-noradrenaline uptake (RAISMAN et al. 1982; LEE et al. 1982; LANGER et al. 1984; BÖNISCH and HARDER 1986). To account for these observations, it must be realized that, under the conditions of ^{3}H-desipramine binding studies (membrane preparations), the carrier can distribute only in positions a and b, so that the potencies of both substrates and inhibitors simply reflect their values of K_D. Under conditions of ^{3}H-noradrenaline uptake studies (intact tissues), on the other hand, substrates (while competing with ^{3}H-noradrenaline) will enter the transport cycle, whereas inhibitors will not. This is the reason why the potencies of inhibitors in antagonizing ^{3}H-noradrenaline uptake again reflect K_D values. The same reasoning would apply to substrates if step 2 were rate-limiting. However, if step 2 is not rate-limiting, substrates will not only compete with ^{3}H-noradrenaline for common carrier sites, but also (while being transported themselves) limit the availability of carrier sites for the uptake of ^{3}H-noradrenaline by "trapping" part of them on the inner face of the membrane. In this way, 50% inhibition of ^{3}H-noradrenaline uptake is possible at concentrations of the competing substrate that are far below the IC_{50} for inhibition of ^{3}H-desipramine binding (i.e. the substrate concentration that produces 50% occupancy of carrier sites). Hence, the potency values of substrates for the inhibition of ^{3}H-noradrenaline uptake (i.e. their values of K_m) will be much lower than their values of K_D. In other words, the reported difference between K_m and K_D values for substrates speaks in favour of step 2 not being rate-limiting in the noradrenaline carrier system. Therefore, values of K_m (or IC_{50} for inhibition of ^{3}H-noradrenaline uptake) for substrates do not reflect dissociation (or affinity) constants for the carrier-substrate complex.

If it is accepted that ^{3}H-desipramine labels selectively the noradrenaline carrier and the assumption is made that there is a 1:1 interaction between ^{3}H-desipramine and the carrier, then the turnover number of the carrier can be calculated from the ratio of V_{max} of ^{3}H-noradrenaline transport to B_{max} of ^{3}H-desipramine binding. A V_{max} of 3.6 nmol $\cdot$ g^{-1} $\cdot$ min (SAMMET and GRAEFE 1979) and a B_{max} of 25.4 pmol $\cdot$ g^{-1} (WETZEL et al. 1981; RAISMAN et al. 1982) were found in the rat vas deferens. From these results a turnover number of 142 noradrenaline molecules per min is obtained, indicating a duration of each noradrenaline transport cycle of 400 ms. A cycle length of very similar duration (397 ms) was found for the ^{3}H-noradrenaline transport in PC-12 cells (BÖNISCH and HARDER 1986). As the V_{max} varies with the substrate, the duration of one transport cycle is also substrate-dependent. Nevertheless, the cycle length for noradrenaline transport is surprisingly long. This also means that, on outside exposure to a substrate, the carrier system will not be able to provide initial rates of uptake instantaneously;—some time will elapse before the distribution of the carrier in the four positions of the model in Fig. 9 has reached a steady state (i.e. before uptake proceeds at initial rates). In perfused rabbit hearts, in which rates of the neuronal uptake of ^{3}H-noradrenaline have been measured at intervals of 5 s, a transient acceleration of uptake was observed before uptake took place at initial rates (GRAEFE et al. 1978; GRAEFE and BÖNISCH 1978). Although interpreted differently by the authors, this transient acceleration of uptake may well have the meaning of a pre-steady state.

II. Models Accounting for the Ion-Dependence

There is abundant evidence that the uptake of amines by noradrenergic and dopaminergic neurones is absolutely dependent on the presence of external Na^+. For this reason and in analogy to other Na^+-dependent membrane transport systems, it has been suggested that the energy inherent in the transmembrane Na^+ concentration gradient is utilized by the neuronal carrier system to bring about uphill amine transport. Serveral models have been proposed to account for the neuronal uptake of amines (BOGDANSKI and BRODIE 1969; WHITE and PATON 1972; WHITE 1976; SANCHEZ-ARMASS and ORREGO 1979; SAMMET and GRAEFE 1979; BÖNISCH et al. 1984). Although based in part on distinctly differing experimental results, they all have as a component the basic concept underlying the "Na^+-gradient hypothesis" of secondary active membrane transport (CRANE 1965; SCHULZ and CURRAN 1970): downhill transport of Na^+ and uphill transport of amines are coupled through the operation of a carrier system which brings about co-transport of Na^+ and amine substrates. This working hypothesis is strongly supported by the finding that ouabain (DENGLER et al. 1962; GIACHETTI and SHORE 1966) as well as K^+-free medium (GILLIS and PATON 1967; COLBURN et al. 1968) impair neuronal uptake. Both procedures are known to cause a running down of the transmembrane Na^+ gradient, since they inhibit the Na^+ pump (Na^+K^+-ATPase). Their inhibitory effect is therefore considered indirect; it occurs only after a certain lag period (TISSARI et al. 1969; SUGRUE and SHORE 1969b), is temporally correlated with changes in the Na^+ concentration gradient (WHITE 1976) and appears to be dependent on the external Na^+ concentration (BOGDANSKI et al. 1970a). As a consequence of their mechanism of action, both ouabain (BERTI and SHORE 1967; HORN et al. 1971) and K^+-free medium (WHITE and PATON 1972; HOLZ and COYLE 1974) produce noncompetitive inhibition of amine transport.

Direct evidence for the coupling of amine transport with the transport of Na^+ is still lacking. However, the findings of SAMMET and GRAEFE (1979) as well as FRIEDRICH and BÖNISCH (1986) strongly support the view that noradrenaline and Na^+ serve as co-substrates of the carrier; they show that the carrier system exhibits the kinetic properties of a two-substrate sequential enzyme reaction in which the formation of a ternary complex (carrier plus Na^+ plus noradrenaline; CNaNA) is necessary for amine transport to take place. To account for their results, SAMMET and GRAEFE (1979) proposed a simple model (model A; Fig. 10). It shows the two ligands noradrenaline and Na^+ combining with the carrier in an obligatory order, i.e. Na^+ adds first and noradrenaline second. This model includes translocation of the ternary complex (CNaNA) and mobility of the free (unloaded) carrier, but does not include translocation of the binary complex CNa. The formation of CNa and its inability to traverse the membrane can be inferred from the observation that the carrier-mediated efflux of noradrenaline into low-Na^+ medium is antagonized by external Na^+ in a concentration-dependent manner. As this inhibitory action of Na^+ is confined to that side of the membrane towards which the amine flux is directed, it probably reflects *trans*-inhibition of noradrenaline transport (for discussion, see GRAEFE 1976). The drop in inward transport observed under conditions of elevated intracellular Na^+ concentrations (e.g. inhibition of Na^+ pumping) may likewise be interpreted as reflecting *trans*-inhibition by Na^+ of amine transport. These *trans*-inhibitory effects of Na^+ suggest formation of CNa on the *trans*-side of the membrane and also immobility of CNa; Na^+ sequesters the carrier on the *trans*-side of the membrane in an inactive

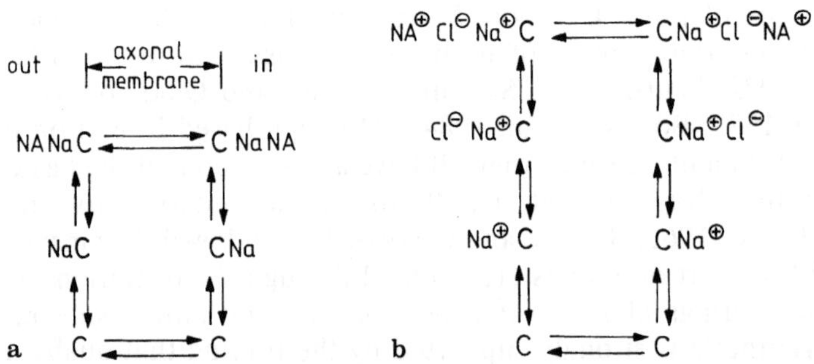

Fig. 10. Compulsory order mechanism of the co-transport of **(a)** noradrenaline (NA) and sodium (Na) or **(b)** noradrenaline (NA^+), sodium (Na^+) and chloride (Cl^-) across the axonal membrane of noradrenergic neurones. C= neuronal noradrenaline carrier

form and limits the availability of carrier sites for amine transport towards that side. If CNa represents an immobile state of the carrier, its distribution between the two membrane surfaces will always reflect the Na^+ concentration gradient across the membrane. In other words, pooling of carrier sites occurs on that side of the membrane at which the Na^+ concentration is higher.

If it is accepted that the ternary complex is the only complexed carrier form in model A that is translocated, then the other possible binary complex CNA (if formed at all) should also be immobile and its formation should likewise manifest itself in *trans*-inhibition of amine transport. This possibility was tested with the help of efflux experiments carried out in the rat vas deferens (GRAEFE et al. 1984). It was found that high concentrations of extracellular (−)-metaraminol do not antagonize the carrier-mediated efflux of ^3H-noradrenaline elicited by Na^+-free medium. Hence, a Na^+-independent combination of amine substrates with the carrier does not appear to take place. Therefore, the noradrenaline carrier system is best described with Na^+ being the leading substrate in an ordered reaction, and model A does not include any formation of CNA.

The differential effect of Na^+ on the ability of inhibitors to antagonize the inward transport of noradrenaline (see Sect. B.III) can also be explained on the basis of model A. For example, the inhibition of noradrenaline uptake by desipramine is competitive with respect to noradrenaline and mixed-type with respect to Na^+. This indicates that desipramine has the ability to combine with the CNa complex, but is unable to bind to the free carrier. It behaves like amine substrates in this respect, but in contrast to substrates, it is not transported by the carrier. The observation that the specific binding of ^3H-desipramine is critically dependent on the presence of Na^+ clearly supports the notion that desipramine binds to the CNa complex and not to the free carrier. Cocaine, on the other hand, competitively interacts with both Na^+ and noradrenaline. This shows that the site on (or the state of) the noradrenaline carrier which combines with cocaine differs from that mediating the inhibitory effects of desipramine. Cocaine resembles desipramine in being a dead-end inhibitor. Though circumstantial, the evidence for this contention is fairly strong: while substrates of the carrier evoke marked acceleration of the outward transport of ^3H-noradrenaline when added to the external medium (i.e. they produce *trans*-acceleration of outward transport), cocaine and desipramine clearly produce *trans*-inhibition of outward transport.

Although model A in Fig. 10 is sufficient to account for the Na^+-dependence of neuronal uptake, there are several reasons for extending this simple model. *First* of all, the dependence of inward transport on external Cl^- (see Sect. B. II. 4) has to enter the picture. In analogy to the Na^+-dependence, the kinetics of the effect of Cl^- on ^3H-noradrenaline transport (examined at saturating Na^+ concentrations; FRIEDRICH and BÖNISCH 1986) suggest co-transport of Cl^-. *Second*, in addition to the Na^+ concentration gradient, the electrical potential gradient across the membrane (i.e. the resting membrane potential) also appears to constitute a driving force for inward transport. This was recently shown by HARDER and BÖNISCH (1985) who examined the ^3H-noradrenaline transport into plasma membrane vesicles derived form PC-12 cells under conditions of high external Na^+ and high internal K^+ concentrations. They found a transient acceleration of ^{3}H-noradrenaline uptake in response to a transient hyperpolarization of the membrane potential (inside negative) elicited by the addition of valinomycin. This is direct evidence for the membrane potential contributing to the driving force for neuronal uptake. It also indicates that the translocated carrier species is positively charged. This is shown in the extended model in Fig. 10 (model B) which also includes the role of Cl^- as co-substrate. Since the charge introduced by Na^+ is neutralized by the co-transport of Cl^-, the translocated amine is likely to be responsible for the net positive charge of the translocated quaternary complex in model B. Noradrenaline is predominantly ionized at physiological pH and exists mainly as a cation in aqueous solution. The fact that completely ionized compounds, which have a quaternary nitrogen (e.g. bretylium, the quaternary ammonium derivative of amphetamine and choline$^+$), are substrates of the carrier, speaks in favour of the ionized form of noradrenaline accounting for the flow of positive charge. The observed dependence of inward transport on internal K^+ (see Sect. B.II.4) shows that this ion also plays an important role in neuronal uptake. However, this is probably due to the fact that high internal K^+ concentrations are necessary to maintain the resting membrane potential which is mainly a K^+ diffusion potential (BLAUSTEIN and GOLDRING 1975).

According to the models in Fig. 10, the carrier system has the potential to bring about amine transport in either direction. However, under conditions of normal transmembrane electrochemical gradients for Na^+ and K^+, amine transport takes place almost exclusively in one direction, namely the inward direction. This is the consequence of two asymmetries present at the axonal membrane under normal conditions: the inward downhill Na^+ concentration gradient and the electrical potential gradient (inside negative). These asymmetries depend ultimately upon the energy-dependent translocation of Na^+ and K^+ brought about through the action of the Na^+K^+-ATPase. The Na^+ concentration gradient leads to pooling of carrier sites on the outside of the membrane and brings about much higher values of V_{max} and lower values of K_m for inward than for outward transport. The resting membrane potential accelerates the inward and hinders the outward translocation of the carrier-substrate complex (i.e. it increases the V_{max} for inward transport). Hence, these asymmetries constitute driving forces for inward transport and represent effective "brakes" for outward transport. In addition, intraneuronal sink mechanisms (MAO and vesicular uptake) also support the inward transport of noradrenaline, since they keep the substrate concentration for outward transport very low. These properties of the system explain why neuronal uptake is so efficient and able to terminate the action of neurally released noradrenaline.

As far as the amine transporter associated with dopaminergic neurones is concerned, the mechanism of amine transport there is likely to be similar to that described above for noradrenergic neurones. No doubt, the dopamine carrier system has been less extensively investigated than the noradrenaline carrier system. However, the available evidence indicates that the dopamine transport, like the noradrenaline transport, is coupled to the electrochemical Na^+ gradient across the axonal membrane and that dopamine, like noradrenaline, is translocated in its protonated form. The latter contention is based on the observation that the neurotoxic cationic compound MPP(+) (N-methyl-4-phenylpyridine) is taken up by dopaminergic neurones at high rates (JAVITCH and SNYDER 1985).

Acknowledgements. The authours' results described in this review originate from work supported by the Deutsche Forschungsgemeinschaft (Gr 490, Bo 521, SFB 176).

F. References

Axelrod J, Whitby LG, Hertting G (1961) Effect of psychotropic drugs on the uptake of ^{3}H-noropinephrine by tissues. Science 133:383-384

Baldessarini RJ, York C (1970) Effects of lithium and pH on synaptosomal metabolism of noradrenaline. Nature (London) 228:1301-1303

Berger P, Janowsky A, Vocci V, Skolnick P, Schweri MM, Paul SM (1985) (^{3}H)GBR-12935: A specific high affinity ligand for labeling the dopamine transport complex. Eur J Pharmacol 107:289-290

Berti F, Shore PA (1967) A kinetic analysis of drugs that inhibit the adrenergic neuronal membrane amine pump. Biochem Pharmacol 16:2091-2094

Blaschko H (1972) Introduction. Catecholamines 1922-1971. In: Blaschko H, Muscholl E (eds) Catecholamines. Springer, Berlin Heidelberg New York, pp 1-15 (Handbook of Experimental Pharmacology, vol 33)

Blaustein MP, Goldring JM (1975) Membrane potentials in pinched-off presynaptic nerve terminals monitored with a fluorescent probe: evidence that synaptosomes have potassium diffusion potentials. J Physiol (London) 247:589-615

Bönisch H (1980) The rate constants for the efflux of deaminated metabolites of ^{3}H-dopamine from the perfused rat heart. Naunyn-Schmiedeberg's Arch Pharmacol 314:231-235

Bönisch H (1982) Mechanismus der Freisetzung von Noradrenalin durch indirekt wirkende Sympathomimetika. Habilitationsschrift, Universität Würzburg

Bönisch H (1984) The transport of (+)amphetamine by the neuronal noradrenaline carrier. Naunyn-Schmiedeberg's Arch Pharmacol 327:267-272

Bönisch H, Graefe K-H (1976) Distribution kinetics of ^{3}H-(−)-noradrenaline (NA) and ^{3}H-(±)-metaraminol (MA) in the perfused rabbit heart. Naunyn-Schmiedeberg's Arch Pharmacol 293, R4

Bönisch H, Harder R (1986) Binding of ^{3}H-desipramine to the neuronal noradrenaline carrier of rat phaeochromocytoma cells (PC-12 cells). Naunyn-Schmiedeberg's Arch Pharmacol 334:403-411

Bönisch H, Langeloh A (1986) Neuronal efflux of noradrenaline induced by Tris or lithium as substitutes for extracellular sodium. Naunyn-Schmiedeberg's Arch Pharmacol 333:13-16

Bönisch H, Rodrigues-Pereira E (1983) Uptake of ^{14}C-tyramine and release of extravesicular ^{3}H-noradrenaline in isolated perfused rabbit hearts. Naunyn-Schmiedeberg's Arch. Pharmacol 323:233-244

Bönisch H, Graefe K-H, Keller B (1983) Tetrodotoxin-sensitive and -resistant effects of veratridine on the noradrenergic neurone of the rat vas deferens. Naunyn-Schmiedeberg's Arch Pharmacol 324:264-270

Bönisch H, Friedrich U, Fritsch H, Harder R (1984) Transport of noradrenaline across the cell membrane of isolated neuronal cells with special reference to PC-12 cells.

In: Fleming WW, Graefe K-H, Langer SZ, Weiner N (eds) Neuronal and Extraneuronal Events in Autonomic Pharmacology. Raven Press, New York, pp 63-74
Bönisch H, Fuchs G, Graefe K-H (1986) Sodium-dependence of the saturability of carrier-mediated noradrenaline efflux from noradrenergic neurones in the rat vas deferens. Naunyn-Schmiedeberg's Arch Pharmacol 332:131-134
Bogdanski DF, Brodie BB (1966) Role of sodium and potassium ions in storage of norepinephrine by synaptic nerve endings. Life Sci 5:1563-1569
Bogdanski DF, Brodie BB (1969) The effects of inorganic ions on the storage and uptake of ^{3}H-norepinephrine by rat heart slices. J Pharmacol Exp Ther 165:181-189
Bogdanski DF, Tissari AH, Brodie BB (1968) Role of sodium, potassium, ouabain, and reserpine in uptake, storage and metabolism of biogenic amines in synaptosomes. Life Sci 7:419-428
Bogdanski DF, Blaszkowski TP, Tissari AH (1970a) Mechanisms of biogenic amine transport and storage. IV. Relationship between K^+ and the Na^+ requirement for transport and storage of 5-hydroxytryptamine and norepinephrine in synaptosomes. Biochim Biophys Acta 211:521-532
Bogdanski DF, Tissari AH, Brodie BB (1970b) Mechanism of transport and storage of biogenic amines. III. Effects of sodium and potassium on kinetics of 5-hydroxytryptamine and norepinephrine transport by rabbit synaptosomes. Biochim Biophys Acta 219:189-199
Bonnet J-J, Protais P, Chagraoui A, Costentin J (1986) High-affinity ^{3}H-GBR 12783 binding to a specific site associated with the neuronal dopamine uptake complex in the central nervous system. Europ J Pharmacol 126:211-222
Branco D, Teixeira AA, Azevedo J, Osswald W (1984) Structural and functional alterations caused at the extraneuronal level by sympathetic denervation of blood vessels. Naunyn-Schmiedeberg's Arch Pharmacol 326:302-312
Burgen ASV, Iversen LL (1965) The inhibition of noradrenaline uptake by sympathomimetic amines in the rat isolated heart. Br J Pharmacol 25:34-49
Callingham BA, Burgen ASV (1966) The uptake of isoprenaline and noradrenaline by the perfused rat heart. Mol Pharmacol 2:37-42
Carlsson A, Waldeck B (1968) Different mechanism of drug-induced release of noradrenaline and its congeners α-methylnoradrenaline and metaraminol. Europ J Pharmacol 3:165-168
Carlsson A, Lundborg P, Stitzel R, Waldeck B (1967) Uptake, storage and release of ^{3}H-α-methylnorepinephrine. J Pharmacol Exp Ther 158:175-182
Cho AK, Ransom RW, Fischer JB, Kammerer RC (1980) The effects of xylamine, a nitrogen mustard, on (^{3}H)-norepinephrine accumulation in rabbit aorta. J Pharmacol Exp Ther 214:324-327
Colburn RW, Goodwin FK, Murphy DL, Bunney WE, Davis JM (1968) Quantitative studies of norepinephrine uptake by synaptosomes. Biochem Pharmacol 17:957-964
Coyle JT, Snyder SH (1969) Catecholamine uptake by synaptosomes in homogenates of rat brain: stereoselectivity in different areas. J Pharmacol Exp Ther 170:221-231
Crane RK (1965) Na^+-dependent transport in the intestine and other animal tissues. Fed Proc 24:1000-1006
Cubeddu LX, Hoffmann IS, Ferrari GB (1979a) Metabolism and efflux of ^{3}H-dopamine in rat striatum: presynaptic origin of 3,4-^{3}H-dihydroxyphenylacetic acid. J Pharmacol Exp Ther 209:165-175

Cubeddu LX, Hoffmann IS, Paris VB (1979b) Effects of papaverine on the release and metabolism of dopamine in rat striatum. J Pharmacol Exp Ther 209:73-78
Dengler HJ, Michaelson IA, Spiegel HE, Titus EO (1962) The uptake of labeled norepinephrine by isolated brain and other tissues of the cat. Int J Neuropharmacol 1:23-28
Draskoczy PR, Trendelenburg U (1968) The uptake of l- and d-norepinephrine by the isolated perfused rabbit heart in relation to the stereoselectivity of the sensitizing action of cocaine. J Pharmacol Exp Ther 159:66-73
Enna SJ, Shore PA (1974) On the nature of the adrenergic neuron extragranular amine binding site. J Neural Transmission 35:125-135
Euler US von (1972) Synthesis, uptake and storage of catecholamines in adrenergic nerves, the effect of drugs. In: Blaschko H, Muscholl E (eds) Catecholamines. Springer, Berlin Heidelberg New York, pp 186-230 (Handbook of Experimental Pharmacology, vol 33)
Farah MB, Adler-Graschinsky E, Langer SZ (1977) Possible physiological significance of the initial step in the catabolism of noradrenaline in the central nervous system of the rat. Naunyn-Schmiedeberg's Arch Pharmacol 297:119-131
Farnebo LO (1971) Effects of reserpine on release of ^{3}H-noradrenaline, ^{3}H-dopamine, and ^{3}H-metaraminol from field stimulated rat iris. Biochem Pharmacol 20:2715-2726
Ferris RM, Stocks BD (1972) Kinetic analysis of ^{3}H-dl-norepinephrine and ^{3}H-dopamine uptake into homogenates of rat striatum and hypothalamus and purified synaptosomes of rat whole brain. Abstracts, 5th Intern. Congress on Pharmacol, San Francisco, p 68
Ferris RM, Tang FLM, Maxwell RA (1972) A comparison of the capacities of isomers of amphetamine, deoxypipradrol and methylphenidate to inhibit the uptake of tritiated catecholamines into rat cerebral cortex slices, synaptosomal preparations of rat cerebral cortex, hypothalamus and striatum and into adrenergic nerves of rabbit aorta. J Pharmacol Exp Ther 181:407-416
Fiebig ER, Trendelenburg U (1978) The neuronal and extraneuronal uptake and metabolism of ^{3}H-(−)-noradrenaline in the perfused rat heart. Naunyn-Schmiedeberg's Arch Pharmacol 303:21-35
Fischer JF, Cho AK (1979) Chemical release of dopamine from striatal homogenates: evidence for an exchange diffusion model. J Pharmacol Exp Ther 208:203-209
Fischer J, Waggaman LA, Ransom RW, Cho AK (1983) Xylamine, an irreversible inhibitor of norepinephrine uptake, is transported by this same uptake mechanism in cultured rat superior cervical ganglia. J Pharmacol Exp Ther 226:650-655
Friedrich U, Bönisch H (1986) The neuronal noradrenaline transport system of PC-12 cells: kinetic analysis of the interaction between noradrenaline, Na^{+} and Cl^{-} in transport. Naunyn-Schmiedeberg's Arch Pharmacol 333:246-252
Garcia AG, Kirpekar SM (1973) Release of noradrenaline from the cat spleen by sodium deprivation. Br J Pharmacol 47:729-747
Giachetti A, Hollenbeck RA (1976) Extra-vesicular binding of noradrenaline and guanethidine in the adrenergic neurones of the rat heart: a proposed site of action of adrenergic neurone blocking agents. Br J Pharmacol 58:497-504
Giachetti A, Shore PA (1966) Studies in vitro of amine uptake mechanisms in heart. Biochem Pharmacol 15:607-614
Gillespie JS, Kirpekar SM (1965) The inactivation of infused noradrenaline by the cat spleen. J Physiol (London) 176:205-227
Gillis CN, Paton DM (1966) Effects of hypothermia and anoxia on retention of noradrenaline by the cat perfused heart. Br J Pharmacol 26:426-434

Gillis CN, Paton DM (1967) Cation dependence of sympathetic transmitter retention by slices of rat ventricles. Br J Pharmacol 29:309-318

Golko DS, Paton DM (1976) Characteristics of accumulation of ephedrine in rabbit atria. Can J Physiol Pharmacol 54:93-100

Graefe K-H (1976) Methodology of catecholamine transport studies: definitions of terms. In: Paton DM (ed) The mechanism of neuronal and extraneuronal transport of catecholamines. Raven Press, New York, pp 7-35

Graefe K-H (1981) The disposition of ^{3}H-(−)-noradrenaline in the perfused cat and rabbit heart. Naunyn-Schmiedeberg's Arch Pharmacol 318:71-82

Graefe K-H, Bönisch H (1978) The influence of the rate of perfusion on the kinetics of neuronal uptake in the rabbit isolated heart. Naunyn-Schmiedeberg's Arch Pharmacol 302:275-283

Graefe K-H, Fuchs G (1979) On the mechanism of neuronal efflux of axoplasmic ^{3}H-(−)-noradrenaline. In: Usdin E, Kopin IJ, Barchas J (eds) Catecholamines: basic and clinical frontiers, vol 1. Pergamon Press, New York Oxford Toronto Sydney Frankfurt Paris, pp 268-270

Graefe K-H, Henseling M (1983) Neuronal and extraneuronal uptake and metabolism of catecholamines. Gen Pharmacol 14:27-33

Graefe K-H, Bönisch H, Trendelenburg U (1971) Time-dependent changes in neuronal net uptake of noradrenaline after pretreatment with pargyline and/or reserpine. Naunyn-Schmiedeberg's Arch Pharmacol 271:1-28

Graefe K-H, Stefano FJE, Langer SZ (1973) Preferential metabolism of ^{3}H-(−)-norepinephrine through the deaminated glycol in the rat vas deferens. Biochem Pharmacol 22:1147-1160

Graefe K-H, Stefano FJE, Langer SZ (1977) Stereoselectivity in the metabolism of ^{3}H-noradrenaline during uptake into and efflux from the isolated rat vas deferens. Naunyn-Schmiedeberg's Arch Pharmacol 299:225-238

Graefe K-H, Bönisch H, Keller B (1978) Saturation kinetics of the adrenergic neurone uptake system in the perfused rabbit heart. A new method for determination of initial rates of amine uptake. Naunyn-Schmiedeberg's Arch Pharmacol 302:263-273

Graefe K-H, Zeitner C-J, Fuchs G, Keller B (1984) Role played by sodium in the membrane transport of ^{3}H-noradrenaline across the axonal membrane of noradrenergic neurones. In: Fleming WW, Graefe K-H, Langer SZ, Weiner N (eds) Neuronal and Extraneuronal Events in Autonomic Pharmacology. Raven Press, New York, pp 51-62

Green AL (1976) The kinetics of enzyme action and inhibition in intact tissues and tissue slices, with special reference to cholinesterase. J Pharm Pharmacol 28:265-274

Green RD, Miller JW (1966) Evidence for the active transport of epinephrine and norepinephrine by the uterus of the rat. J Pharmacol Exp Ther 152:42-50

Greene LA, Tischler AS (1982) PC-12 pheochromocytoma cultures in neurobiological research. Adv Cell Neurobiol 3:373-414

Hamberger B (1967) Reserpine-resistant uptake of catecholamines in isolated tissues of the rat. Acta physiol scand 71, Suppl 295:1-56

Hamberger B, Malmfors T, Norberg K-A, Sachs C (1964) Uptake and accumulation of catecholamines in peripheral adrenergic neurons of reserpinized animals, studied with a histochemical method. Biochem Pharmacol 13:841-844

Harder R, Bönisch H (1984) Large-scale preparation of plasma membrane vesicles from PC-12 pheochromocytoma cells and their use in noradrenaline transport studies. Biochim Biophys Acta 775:95-104

Harder R, Bönisch H (1985) Effects of monovalent ions on the transport of noradrenaline across the plasma membrane of neuronal cells (PC-12 cells). J Neurochem 45:1154-1162

Harris JE, Baldessarini RJ (1973) The uptake of ^{3}H-dopamine by homogenates of rat corpus striatum: effects of cations. Life Sci 13:303-312

Hellmann G, Hertting G, Peskar B (1971) Uptake kinetics and metabolism of 7-^{3}H-dopamine in the isolated perfused rat heart. Br J Pharmacol 41:256-269

Hendley ED (1976) The mechanism of extraneuronal transport of catecholamines in the central nervous system. In: Paton DM (ed) The mechanism of neuronal and extraneuronal transport of catecholamines. Raven Press, New York, pp 313-324

Hendley ED, Taylor KM, Snyder SH (1970) ^{3}H-Normetanephrine uptake in rat brain slices. Relationship to extraneuronal accumulation of norepinephrine. Europ J Pharmacol 12:167-179

Hendley ED, Snyder SH, Fauley JJ, La Pidus JB (1972) Stereoselectivity of catecholamine uptake by brain synaptosomes: studies with ephedrine, methylphenidate and phenyl-2-piperidyl carbinol. J Pharmacol Exp Ther 183:103-116

Henseling M (1983) Kinetic constants for uptake and metabolism of ^{3}H-(−)-noradrenaline in rabbit aorta. Naunyn-Schmiedeberg's Arch Pharmacol 323:12-23

Henseling M, Trendelenburg U (1978) Stereoselectivity of the accumulation and metabolism of noradrenaline in rabbit aortic strips. Naunyn-Schmiedeberg's Arch Pharmacol 302:195-206

Henseling M, Eckert E, Trendelenburg U (1976) The distribution of ^{3}H-(±)-noradrenaline in rabbit aortic strips after inhibition of the noradrenaline-metabolizing enzymes. Naunyn-Schmiedeberg's Arch Pharmacol 292:205-217

Hermann W, Graefe K-H (1977) Relationship between the uptake of ^{3}H-(±)-metaraminol and the density of adrenergic innervation in isolated rat tissues. Naunyn-Schmiedeberg's Arch Pharmacol 296:99-110

Hertting G (1964) The fate of ^{3}H-isoproterenol in the rat. Biochem Pharmacol 13:1119-1128

Hertting G, Suko J (1972) Influence of neuronal and extraneuronal uptake on disposition, metabolism and potency of catecholamines. In: Snyder SH (ed) Perspectives in neuropharmacology. Oxford University Press, New York London Toronto, pp 267-300

Holbach H-J, Löffelholz K (1975) Differences between noradrenaline release and enhancement of noradrenaline efflux evoked by DMPP in the rabbit heart. Naunyn-Schmiedeberg's Arch Pharmacol 287:R6

Holmes JC, Rutledge CO (1976) Effects of the d- and l-isomers of amphetamine on uptake, release and catabolism of norepinephrine, dopamine and 5-hydroxytryptamine in several regions of rat brain. Biochem Pharmacol 25:447-451

Holz RW (1978) Evidence that catecholamine transport into chromaffin vesicles is coupled to vesicle membrane potential. Proc Natl Acad Sci USA 75:5190-5194

Holz RW, Coyle JT (1974) The effects of various salts, temperature, and the alkaloids veratridine and batrachotoxin on the uptake of ^{3}H-dopamine into synaptosomes from rat striatum. Mol Pharmacol 10:746-758

Horn AS (1973) Structure-activity relations for the inhibition of catecholamine uptake into synaptosomes from noradrenergic and dopaminergic neurones in rat brain homogenates. Br J Pharmacol 47:332-338

Horn AS, Snyder SH (1972) Steric requirements for catecholamine uptake by rat brain synaptosomes: studies with rigid analogues of amphetamine. J Pharmacol Exp Ther 180:523-530

Horn AS, Coyle JT, Snyder SH (1971) Catecholamine uptake by synaptosomes from rat brain: structure-activity relationships of drugs with differential effects on dopamine and norepinephrine neurons. Mol Pharmacol 7:66–80

Horst WD, Kopin IJ, Ramey ER (1968) Influence of sodium and calcium on norepinephrine uptake by isolated perfused rat hearts. Am J Physiol 215:817–822

Hrdina PD (1981) Pharmacological characterization of ^{3}H-desipramine binding in rat cerebral cortex. Prog Neuro-Psychopharmacol 5:553–557

Hunt P, Raynaud J-P, Leven M, Schacht U (1979) Dopamine uptake inhibitors and releasing agents differentiated by the use of synaptosomes and field-stimulated brain slices in vitro. Biochem Pharmacol 28:2011–2016

Iversen LL (1963) The uptake of noradrenaline by the isolated perfused rat heart. Br J Pharmacol 21:523–537

Iversen LL (1965) The uptake of adrenaline by the rat isolated heart. Br J Pharmacol 24:387–394

Iversen LL (1966) Accumulation of α-methyltyramine by the noradrenaline uptake process in the isolated rat heart. J Pharm Pharmacol 18:481–484

Iversen LL (1967) The uptake and storage of noradrenaline in sympathetic nerves. Cambridge Univ Press, Cambridge

Iversen LL (1975) Uptake mechanisms for biogenic amines. In: Iversen LL, Iversen SD, Snyder SH (eds) Handbook of Psychopharmacology, vol 3. Plenum Press, New York London, pp 381–442

Iversen LL, Kravitz EA (1966) Sodium dependence of transmitter uptake at adrenergic nerve terminals. Mol Pharmacol 2:360–362

Iversen LL, Langer SZ (1969) Effects of phenoxybenzamine on the uptake and metabolism of noradrenaline in the rat heart and vas deferens. Br J Pharmacol 37:627–637

Iversen LL, MacKay AVP (1979) Pharmacodynamics of antidepressants and antimanic drugs. In: Paykel ES, Coppen A (eds) Psychopharmacology of affective disorders. Oxford University Press, Oxford New York Toronto, pp 60–90

Iversen LL, Schon FE (1973) The use of autoradiographic techniques for the identification and mapping of transmitter-specific neurones in CNS. In: Mandel AJ (ed) New concepts in neurotransmitter regulation. Plenum Press, London New York, pp 153–193

Iversen LL, Glowinski J, Axelrod J (1965) The uptake and storage of norepinephrine in the reserpine-pretreated rat heart. J Pharmacol Exp Ther 150:173–183

Iversen LL, Jarrott B, Simmonds MA (1971) Differences in the uptake, storage and metabolism of (+)- and (−)-noradrenaline. Br J Pharmacol 43:845–855

Janowsky A, Berger P, Vocci F, Labarca R, Skolnick P, Paul SM (1986) Characterization of sodium-dependent ^{3}H-GBR-12935 binding in brain: a radioligand for selective labelling of the dopamine transport complex. J Neurochem 46:1272–1276

Jarrott B (1970) Uptake and metabolism of ^{3}H-noradrenaline by the perfused hearts of various species. Br J Pharmacol 38:810–821

Jarrott B, Iversen LL (1971) Noradrenaline metabolizing enzymes in normal and sympathetically denervated vas deferens. J Neurochem 18:1–6

Jarrott B, Langer SZ (1971) Changes in monoamine oxidase and catechol-O-methyl transferase activities after denervation of the nictitating membrane of the cat. J Physiol (London) 212:549–559

Javitch JA, Snyder S (1985) Uptake of MPP(+) by dopamine neurons explains selectivity of parkinsonism-inducing neurotoxin, MPTP. Europ J Pharmacol 106: 455–456

Johnson RG, Scarpa A (1976) Internal pH of isolated chromaffin vesicles. J Biol Chem 251:2189-2191

Johnson RG, Sally EC, Hayflick S, Scarpa A (1982) Mechanismus of accumulation of tyramine, metaraminol, and isoproterenol in isolated chromaffin granules and ghosts. Biochem Pharmacol 31:815-823

Jonsson G, Hamberger B, Malmfors T, Sachs C (1969) Uptake and accumulation of ^{3}H-noradrenaline in adrenergic nerves of rat iris. Effect of reserpine, monoamine oxidase and tyrosine hydroxylase inhibition. Europ J Pharmacol 8:58-72

Keller B, Graefe K-H (1979) The inhibitory effect of some monovalent cations on the stimulation by Na^{+} of the neuronal uptake of noradrenaline. Naunyn-Schmiedeberg's Arch Pharmacol 309:89-97

Kimelberg HK (1986) Occurrence and functional significance of serotonin and catecholamine uptake by astrocytes. Biochem Pharmacol 35:2273-2281

Knoth J, Peabody JO, Huettl P, Njus D (1984) Kinetics of tyramine transport and permeation across chromaffin-vesicle membranes. Biochem 23:2011-2016

Koe BK (1976) Molecular geometry of inhibitors of the uptake of catecholamines and serotonin in synaptosomal preparations of rat brain. J Pharmacol Exp Ther 199:649-661

Kopin IJ (1972) Metabolic degradation of catecholamines. The relative importance of different pathways under physiological conditions and after administration of drugs. In: Blaschko H, Muscholl E (eds) Catecholamines. Springer, Berlin Heidelberg New York Tokio, pp 270-282 (Handbook of Experimental Pharmacology, vol 33)

Kopin IJ, Hertting G, Gordon EK (1962) Fate of norepinephrine-^{3}H in the isolated perfused rat heart. J Pharmacol Exp Ther 138:34-40

Krell RD, Patil PN (1972) Steric aspects of adrenergic drugs. XX. Accumulation of (−)- and (+)-norepinephrine-^{14}C by peripheral tissues of the rat. J Pharmacol Exp Ther 182:273-283

Kuhar MJ (1973) Neurotransmitter uptake: a tool in identifying neurotransmitter-specific pathways. Life Sci 13:1623-1634

Langeloh A, Bönisch H, Trendelenburg U (1987) The mechanism of the ^{3}H-noradrenaline-releasing effect of various substrates of uptake$_1$: multifactorial induction of outward transport. Naunyn-Schmiedeberg's Arch Pharmacol 336:602-610

Langer SZ (1974) Selective metabolic pathways for noradrenaline in the peripheral and in the central nervous system. Med Biol 52:372-383

Langer SZ (1984) ^{3}H-Imipramine and ^{3}H-desipramine binding: non-specific displaceable sites or physiologically relevant sites associated with the uptake of serotonin and noradrenaline? Trends Pharmacol Sci 5:51-52

Langer Sz, Briley M (1981) High-affinity ^{3}H-imipramine binding: a new biological tool for studies in depression. Trends Neurosci 4:28-31

Langer SZ, Stefano FJE, Enero MA (1972) Pre- and postsynaptic origin of the norepinephrine metabolites formed during transmitter release elicited by nerve stimulation. J Pharmacol Exp Ther 183:90-102

Langer SZ, Farah MB, Luchelli-Fortis MA, Adler-Graschinsky E, Filinger EJ (1975) Metabolism of endogenous noradrenaline. In: Tuomisto J, Paasonen MK (eds) Proc. of the 6th International Congress of Pharmacology, vol 2. Helsinki, pp 17-31

Langer SZ, Raisman R, Briley M (1981) High-affinity ^{3}H-DMI binding is associated with neuronal noradrenaline uptake in the periphery and the central nervous system. Europ J Pharmacol 72:423-424

Langer SZ, Tahraoui L, Raisman R, Arbilla S, Najar M, Dedek J (1984) ^{3}H-Desipramine labels a site associated with the neuronal uptake of noradrenaline in the peripheral and central nervous system. In: Fleming WW, Graefe K-H, Langer SZ, Weiner N (eds) Neuronal and Extraneuronal Events in Autonomic Pharmacology. Raven Press, New York, pp 37–49

Lee C-M, Snyder SH (1981) Norepinephrine neuronal uptake binding sites in rat brain membranes labeled with ^{3}H-desipramine. Proc Natl Acad Sci USA 78:5250–5254

Lee C-M, Javitch JA, Snyder SH (1982) Characterization of ^{3}H-desipramine binding associated with neuronal norepinephrine uptake sites in rat brain membranes. J Neurosci 2:1515–1525

Levin JA (1973) Paper chromatographic assay of ^{3}H-norepinephrine and its five major metabolites. Anal Biochem 51:42–60

Levin JA (1974) The uptake and metabolism of ^{3}H-l- and ^{3}H-dl-norepinephrine by intact rabbit aorta and by isolated adventitia and media. J Pharmacol Exp Ther 190:210–226

Lightman S, Iversen LL (1969) Role of uptake$_2$ in the extraneuronal uptake and metabolism of catecholamines in the isolated rat heart. Br J Pharmacol 37:638–649

Lindmar R, Löffelholz K (1972) Differential effects of hypothermia on neuronal efflux, release and uptake of noradrenaline. Naunyn-Schmiedeberg's Arch Pharmacol 274:410–414

Lindmar R, Löffelholz K (1974a) Neuronal and extraneuronal uptake and efflux of catecholamines in the isolated rabbit heart. Naunyn-Schmiedeberg's Arch Pharmacol 284:63–92

Lindmar R, Löffelholz K (1974b) The neuronal efflux of noradrenaline: dependency on sodium and facilitation by ouabain. Naunyn-Schmiedeberg's Arch Pharmacol 284:93–100

Lindmar R, Muscholl E (1964) Die Wirkung von Pharmaka auf die Elimination von Noradrenalin aus der Perfusionsflüssigkeit und die Noradrenalinaufnahme in das isolierte Herz. Naunyn-Schmiedeberg's Arch Pharmacol 247:469–492

Lindmar R, Muscholl E (1965) Die Aufnahme von α-Methylnoradrenalin in das isolierte Kaninchenherz und seine Freisetzung durch Reserpin und Guanethidin in vivo. Naunyn-Schmiedeberg's Arch Pharmacol 249:529–548

Lowe MC, Horita A (1970) Cardiac monoamine oxidase: stability of activity after chemical sympathectomy with 6-hydroxydopamine. Nature (London) 228:175–176

Luchelli-Fortis MA, Langer SZ (1975) Selective inhibition by hydrocortisone of ^{3}H-normetanephrine formation during ^{3}H-transmitter release elicited by nerve stimulation in the isolated nerve-muscle preparation of the cat nictitating membrane. Naunyn-Schmiedeberg's Arch Pharmacol 287:261–275

Mack F, Bönisch H (1979) Dissociation constants and lipophilicity of catecholamines and related compounds. Naunyn-Schmiedeberg's Arch Pharmacol 310:1–9

Majewski H, Hedler L, Steppeler A, Starke K (1982) Metabolism of endogenous and exogenous noradrenaline in the rabbit perfused heart. Naunyn-Schmiedeberg's Arch Pharmacol 319:125–129

Maxwell RA, White HL (1978) Tricyclic and monoamine oxidase inhibitor antidepressants: structure-activity relationships. In: Iversen LL, Iversen SD, Snyder SH (eds) Handbook of Psychopharmacolgy, vol 14. Plenum Press, New York London, pp 85–155

Maxwell RA, Keenan DP, Chaplin E, Roth B, Eckhardt SB (1969) Molecular features affecting the potency of tricyclic antidepressants and structurally related com-

pounds as inhibitors of the uptake of tritiated norepinephrine by rabbit aortic strips. J Pharmacol Exp Ther 166:320-329
Maxwell RA, Chaplin E, Eckhardt SB, Soares JR, Hite G (1970) Conformational similarities between molecular models of phenethylamine and of potent inhibitors of the uptake of tritiated norepinephrine by adrenergic nerves in rabbit aorta. J Pharmacol Exp Ther 173:158-165
Mekanontchai R, Trendelenburg U (1979) The neuronal and extraneuronal distribution of ^{3}H-(−)-noradrenaline in the perfused rat heart. Naunyn-Schmiedeberg's Arch Pharmacol 308:199-210
Muscholl E (1960) Die Hemmung der Noradrenalin-Aufnahme des Herzen durch Reserpin und die Wirkung von Tyramin. Naunyn-Schmiedeberg's Arch Pharmacol 240:234-241
Muscholl E (1961) Effect of cocaine and related drugs on the uptake of noradrenaline by heart and spleen. Br J Pharmacol 16:352-359
Muscholl E (1972) Adrenergic false transmitters. In: Blaschko H, Muscholl E (eds) Catecholamines. Springer, Berlin Heidelberg New York Tokio, pp 618-660 (Handbook of Experimental Pharmacology, vol 33)
Muscholl E, Weber E (1965) Die Hemmung der Aufnahme von α-Methylnoradrenalin in das Herz durch sympathomimetische Amine. Naunyn-Schmiedeberg's Arch Pharmacol 252:134-143
Orlansky H, Heikkila E (1974) An evaluation of various antiparkinsonian agents as releasing agents and uptake inhibitiors for ^{3}H-dopamine in slices of rat neostriatum. Europ J Pharmacol 29:284-291
Paiva MQ, Guimaraes S (1978) A comparative study of the uptake and metabolism of noradrenaline and adrenaline by the isolated saphenous vein of the dog. Naunyn-Schmiedeberg's Arch Pharmacol 303:221-228
Paton DM (1968) Cation and metabolic requirements for retention of metaraminol by rat uterine horns. Br J Pharmacol 33:277-286
Paton DM (1971) The effects of Na^{+} and K^{+} on the uptake of metaraminol by rabbit ventricular slices. Br J Pharmacol 41:65-75
Paton DM (1973) Mechanism of efflux of noradrenaline from adrenergic nerves in rabbit atria. Br J Pharmacol 49:614-627
Paton DM (1976a) Characteristics of uptake of noradrenaline by adrenergic neurons. In: Paton DM (ed) The mechanism of neuronal and extraneuronal transport of catechoamines. Raven Press, New York, pp 49-66
Paton DM (1976b) Characteristics of efflux of noradrenaline from adrenergic neurons. In: Paton DM (ed) The mechanism of neuronal and extraneuronal transport of catecholamines. Raven Press, New York, pp 155-174
Paton DM (1981) Effect of cocaine on the efflux of noradrenaline, octopamine and metaraminol in rabbit atria. IRCS Med Sci 9:128
De Paulis T, Kelder D, Ross SB, Stjernström NE (1978) On the topology of the norepinephrine transport carrier in rat hypothalamus. Mol Pharmacol 14:596-606
Philippu A (1976) Transport in intraneuronal storage vesicles. In: Paton DM (ed) The mechanism of neuronal and extraneuronal transport of catecholamines. Raven Press, New York, pp 215-246
Raisman R, Sette M, Pimoule C, Briley M, Langer SZ (1982) High-affinity ^{3}H-desipramine binding in the peripheral and central nervous system: a specific site associated with the neuronal uptake of noradrenaline. Europ J Pharmacol 78:345-352
Raiteri M, Levi G, Federico R (1974) d-Amphetamine and the release of ^{3}H-norepinephrine from synaptosomes. Europ J Pharmacol 28:237-240

Raiteri M, Del Carmine R, Bertollini A, Levi G (1977a) Effect of sympathomimetic amines on the synaptosomal transport of noradrenaline, dopamine and 5-hydroxytryptamine. Europ J Pharmacol 41:133-143

Raiteri M, Del Carmine R, Bertollini A, Levi G (1977b) Effect of desmethylimipramine on the release of ^{3}H-norepinephrine induced by various agents in hypothalamic synaptosomes. Mol Pharmacol 13:746-758

Raiteri M, Cerrito F, Cervoni AM, Levi G (1979) Dopamine can be released by two mechanisms differentially affected by the dopamine transport inhibitor nomifensine. J Pharmacol Exp Ther 208:195-202

Ransom RW, Kammerer RC, Cho AK (1982) Chemical transformations of xylamine (N-2′-chloroethyl-N-ethyl-2-methylbenzylamine) in solution. Mol Pharmacol 21:380-386

Rawlow A, Fleig H, Kurahashi K, Trendelenburg U (1980) The neuronal and extraneuronal uptake and deamination of ^{3}H-(−)-phenylephrine in the perfused rat heart. Naunyn-Schmiedeberg's Arch Pharmacol 314:237-247

Rehavi M, Skolnick P, Hulihan B, Paul SM (1981) "High affinity" binding of (^{3}H)-desipramine to rat cerebral cortex: relationship to tricyclic antidepressant-induced inhibition of norepinephrine uptake. Europ J Pharmacol 70:597-599

Ross SB (1976a) Structural requirements for uptake into catecholamine neurons. In: Paton DM (ed) The mechanism of neuronal and extraneuronal transport of catecholamines. Raven Press, New York, pp 67-93

Ross SB (1976b) Long-term effects of N-2-chloroethyl-N-ethyl-2-bromobenzylamine hydrochloride on noradrenergic neurones in the rat brain and heart. Br J Pharmacol 58:521-527

Ross SB, Gosztonyi T (1975) On the mechanism of the accumulation of ^{3}H-bretylium in peripheral sympathetic nerves. Naunyn-Schmiedeberg's Arch Pharmacol 288:283-293

Ross SB, Kelder D (1976) Effect of veratridine on the fluxes of ^{3}H-noradrenaline and ^{3}H-bretylium in the rat vas deferens in vitro. Naunyn-Schmiedeberg's Arch Pharmacol 295:183-189

Ross SB, Kelder D (1979) Release of ^{3}H-noradrenaline from the rat vas deferens under various in vitro conditions. Acta physiol scand 105:338-349

Ross SB, Renyi AL (1964) Blocking action of sympathomimetic amines on the uptake of tritiated noradrenaline by mouse cerebral cortex tissue in vitro. Acta pharmacol toxicol 21:226-239

Ross SB, Renyi AL (1966a) Uptake of tritiated tyramine and (+)-amphetamine by mouse heart slices. J Pharm Pharmacol 18:756-757

Ross SB, Renyi AL (1966b) Uptake of some tritiated sympathomimetic amines by mouse brain cortex slices in vitro. Acta pharmacol toxicol 24:297-309

Ross SB, Renyi AL (1976) On the long-lasting inhibitory effect of N-(2-chloroethyl)-N-ethyl-2-bromobenzylamine (DSP 4) on the active uptake of noradrenaline. J Pharm Pharmacol 28:458-459

Ross SB, Renyi AL (1978) Effect of (+)-amphetamine on the retention of ^{3}H-catecholamines in slices of normal and reserpinized rat brain and heart. Acta pharmacol toxicol 42:328-336

Ross SB, Renyi AL, Brunfelter B (1968) Cocaine-sensitive uptake of sympathomimetic amines in nerve tissue. J Pharm Pharmacol 20:282-288

Rutledge CO, Vollmer S (1979) Evidence for carrier mediated efflux of norepinephrine displaced by amphetamine. In: Usdin E, Kopin IJ, Barchas J (eds) Catecholamines: basic and clinical frontiers, vol 1. Pergamon Press, New York Oxford Toronto Sydney Frankfurt Paris, pp 304-306

Sachs C (1970) Noradrenaline uptake mechanisms in the mouse atrium. Acta physiol scand 79, Suppl 341:1-67
Salama AI, Insalaco JR, Maxwell RA (1971) Concerning the molecular requirements for the inhibition of the uptake of racemic ^{3}H-norepinephrine into rat cerebral cortex slices by tricyclic antidepressants and related compounds. J Pharmacol Exp Ther 178:474-481
Sammet S, Graefe K-H (1979) Kinetic analysis of the interaction between noradrenaline and Na^{+} in neuronal uptake: kinetic evidence for co-transport. Naunyn-Schmiedeberg's Arch Pharmacol 309:99-107
Sanchez-Armass S, Orrego F (1977) A major role for chloride in ^{3}H-noradrenaline transport by rat heart adrenergic nerves. Life Sci 20:1829-1838
Sanchez-Armass S, Orrego F (1978) Noradrenaline transport by rat heart sympathetic nerves: a re-examination of the role of sodium ions. Naunyn-Schmiedeberg's Arch Pharmacol 302:355-261
Sanchez-Armass S, Orrego F (1979) Plasma membrane noradrenaline transport: a rotary liquid pore model. In: Usdin E, Kopin IJ, Barchas J (eds) Catecholamines: basic and clinial frontiers, vol 1. Pergamon Press, New York Oxford Toronto Sydney Frankfurt Paris, pp 307-309
Sanchez-Garcia P, Garcia AG, Matinez-Sierra R, Velasco-Martin A (1977) Inhibition of norepinephrine uptake by phenoxybenzamine and desmethylimipramine in the isolated guinea-pig atrium. J Pharmacol Exp Ther 201:192-198
Schacht U, Heptner W (1974) Effect of nomifensine (HOE 984), a new antidepressant, on uptake of noradrenaline and serotonin and on release of noradrenaline in rat brain synaptosomes. Biochem Pharmacol 23:3413-3422
Schömig E, Bönisch H (1986) Solubilization and characterization of the ^{3}H-desipramine binding site of rat phaeochromocytoma cells (PC-12 cells). Naunyn-Schmiedeberg's Arch Pharmacol 334:412-417
Schömig E, Körber M, Bönisch H (1988) Kinetic evidence for a common binding site for substrate and inhibitors of the neuronal noradrenaline carrier. Naunyn-Schmiedeberg's Arch Pharmacol (in press)
Schulz SG, Curran PF (1970) Coupled transport of sodium and organic solutes. Physiol Rev 50:637-718
Segel IH (1975) Enzyme kinetics. Behaviour and analysis of rapid equilibrium and steady-state enzyme systems. Wiley and Sons, New York London Sydney Toronto
Shaskan EG, Snyder SH (1970) Kinetics of serotonin accumulation into slices from rat brain: relationship to catecholamine uptake. J Pharmacol Exp Ther 175:404-418
Simmonds MA, Gillis CN (1968) Uptake of normetanephrine and norepinephrine by cocaine-treated rat hearts. J Pharmacol Exp Ther 159:283-289
Snyder SH, Coyle JT (1969) Regional differences in ^{3}H-norepinephrine and ^{3}H-dopamine uptake into rat brain homogenates. J Pharmacol Exp Ther 165:78-86
Snyder SH, Green AI, Hendley ED (1968) Kinetics of ^{3}H-norepinephrine accumulation into slices from different regions of the rat brain. J Pharmacol Exp Ther 164:90-102
Snyder SH, Kuhar MJ, Green AI, Coyle JT, Shaskan EG (1970) Uptake and subcellular localisation of neurotransmitters in brain. Int Rev Neurobiol 13:127-158
Starke K, Steppeler A, Zumstein A, Henseling M, Trendelenburg U (1980) False labelling of commercially available ^{3}H-catecholamines. Naunyn-Schmiedeberg's Arch Pharmacol 311:109-112
Starke K, Hedler L, Steppeler A (1981) Metabolism of endogenous and exogenous nor-

adrenaline in guinea-pig atria. Naunyn-Schmiedeberg's Arch Pharmacol 317:193-198
Stefanini E, Argiolas A, Gessa GL (1976) Effect of lithium on dopamine uptake by brain synaptosomes. J Neurochem 27:1237-1239
Stefano FJE, Trendelenburg U (1984) Saturation of monoamine oxidase by intraneuronal noradrenaline accumulation. Naunyn-Schmiedeberg's Arch Pharmacol 328:135-141
Stein WD (1967) The movement of molecules across cell membranes. Acacemic Press, New York, London
Steppeler A, Starke K (1982) Fate of (^{3}H)-amezinium in sympathetically innervated rabbit tissues. Biochem Pharmacol 31:1075-1080
Stjärne L (1972) The synthesis, uptake and storage of catecholamines in the adrenal medulla. The effect of drugs. In: Blaschko H, Muscholl E (eds) Catecholamines. Springer, Berlin Heidelberg New York, pp 231-269 (Handbook of Experimental Pharmacology, vol 33)
Stute N, Trendelenburg U (1984) The outward transport of axoplasmic noradrenaline induced by a rise of the sodium concentration in the adrenergic nerve endings of the rat vas deferens. Naunyn-Schmiedeberg's Arch Pharmacol 327:124-132
Sugrue MF, Shore PA (1969a) The mode of sodium dependency of the adrenergic neuron amine carrier. Evidence for a second, sodium-dependent, optically specific and reserpine-sensitive system. J Pharmacol Exp Ther 170:239-245
Sugrue MF, Shore PA (1969b) The mode of potassium action on the adrenergic neuron amine transport system. Life Sci 8:1337-1341
Thoa NB, Wooten GF, Axelrod J, Kopin IJ (1975) On the mechanism of release of norephinephrine from sympathetic nerves induced by depolarizing agents and sympathomimetic drugs. Mol Pharmacol 11:10-18
Thoenen H, Hürlimann A, Haefely W (1968) Mechanism of amphetamine accumulation in the isolated perfused heart of the rat. J Pharm Pharmacol 20:1-11
Tissari AH, Schönhöfer PS, Bogdanski DF, Brodie BB (1969) Mechanism of biogenic amine transport. II. Relationship between sodium and the mechanism of ouabain blockade of the accumulation of serotonin and norepinephrine by synaptosomes. Mol Pharmacol 5:593-604
Trendelenburg U (1972) Factors influencing the concentration of catecholamines at the receptors. In: Blaschko H, Muscholl E (eds) Catecholamines. Springer, Berlin Heidelberg New York Tokyo, pp 726-761 (Handbook of Experimental Pharmacology, vol 33)
Trendelenburg U, Bönisch H, Graefe K-H, Henseling M (1979) The rate constants for the efflux of metabolites of catecholamines and phenethylamines. Pharmacol Rev 31:179-203
Trendelenburg U, Stefano FJE, Grohmann M (1983) The isotope effect of tritium in ^{3}H-noradrenaline. Naunyn-Schmiedeberg's Arch Pharmacol 323:128-140
Tuomisto J (1977) Nomifensine and its derivatives as possible tools for studying amine uptake. Europ J Pharmacol 42:101-106
Ungell A-L, Graefe K-H (1987) Failure of K^+ to affect the potency of inhibitors of the neuronal noradrenaline carrier in the rat vas deferens. Naunyn-Schmiedeberg's Arch Pharmacol 335:250-254
Ungell A-L, Bönisch H, Graefe K-H (1986) Choline$^+$: a substrate of the neuronal noradrenaline carrier in the rat vas deferens. Naunyn-Schmiedeberg's Arch Pharmacol 334:223-227
Van der Zee P, Koger HS, Gootjes J, Hespe W (1980) Aryl-1,4-di-alk(en)ylpiperazines as selective and very potent inhibitors of dopamine uptake. Eur J Med Chem 15:363-370

Wakade AR, Furchgott RF (1968) Metabolic requirements for the uptake and storage of norepinephrine by the isolated left atrium of the guinea pig. J Pharmacol Exp Ther 163:123-135

Wakade AR, Kirpekar SM (1974) Calcium-independent release of ^{3}H-norepinephrine from reserpine-pretreated guinea-pig vas deferens and seminal vesicles. J Pharmacol Exp Ther 190:451-458

Waldmeier PC, Baumann PA, Hauser K, Maitre L, Storni A (1982) Oxaprotiline, a noradrenaline uptake inhibitor with an active and an inactive enantiomer. Biochem Pharmacol 31:2169-2176

Wetzel HW, Briley MS, Langer SZ (1981) ^{3}H-WB 4101 binding in the rat vas deferens: effects of chronic treatment with desipramine and prazosin. Naunyn-Schmiedeberg's Arch Pharmacol 317:187-192

White TD (1976) Models for neuronal noradrenaline uptake. In: Paton DM (ed) The mechanism of neuronal and extraneuronal transport of catecholamines. Raven Press, New York, pp 175-193

White TD, Paton DM (1972) Effects of external Na^{+} and K^{+} on the initial rates of noradrenaline uptake by synaptosomes prepared from rat brain. Biochim Biophys Acta 266:116-127

Wilbrandt W, Rosenberg T (1961) The concept of carrier transport and its corollaries in pharmacology. Pharmacol Rev 13:109-183

Wilkinson GN (1961) Statistical estimation in enzyme kinetics. Biochem J 80:324-332

Wong DT, Bymaster FP (1976) Effect of nisoxetine on uptake of catecholamines in synaptosomes isolated from discrete regions of rat brain. Biochem Pharmacol 25:1979-1983

Zeitner C-J, Graefe K-H (1986) Sodium-dependence of the potency of inhibitors of the neuronal noradrenaline carrier in the rat vas deferens. Naunyn-Schmiedeberg's Arch Pharmacol 334:397-402

CHAPTER 5

The Mechanism of Action of Indirectly Acting Sympathomimetic Amines

H. Bönisch and U. Trendelenburg

A. Introduction

Since 1959 it has been established that indirectly acting amines exert their effects by releasing endogenous noradrenaline from adrenergic nerve endings (Burn and Rand 1958). However, the mechanism by which noradrenaline is able to leave the adrenergic nerve endings has remained a subject of debate. About eight years ago a review (Trendelenburg 1978) posed a dilemma regarding the actions of indirect sympathomimetic amines which we were then unable to resolve satisfactorily. The noradrenaline-releasing effects of the indirectly acting sympathomimetic amines would be easy to explain if they were all substrates of uptake$_1$ (i.e. of cocaine-sensitive, sodium-dependent neuronal uptake). However, for the highly lipophilic (+)-amphetamine evidence for its being a substrate of uptake$_1$ was missing, at least for peripheral, adrenergically innervated organs; indeed, the then available evidence seemed to indicate that (+)-amphetamine was a nontransported inhibitor of uptake$_1$ (Thoenen et al. 1968; see also Bönisch and Rodrigues-Pereira 1983). Raiteri et al. (1974) were unable to obtain evidence for any carrier-mediated uptake of (+)-amphetamine into brain synaptosomes but Azzaro et al. (1974) reported that, for a low concentration of (+)-amphetamine (0.1 μmol/l), uptake was sensitive to desipramine.

If there is only one indirectly acting amine which is not a substrate of uptake$_1$, we continue to face the mystery regarding how the process of "diffusion through the axonal membrane, followed by release of vesicular noradrenaline into the axoplasm" can permit enough noradrenaline to leave the varicosity to activate α- and β-adrenoceptors. After all, the highly polar noradrenaline molecule cannot easily diffuse through the axonal membrane. Only two processes are presently known to induce a release of endogenous noradrenaline from adrenergic nerve endings: calcium-dependent exocytotic release and calcium-independent carrier-mediated outward transport. Since the noradrenaline-releasing effect of indirectly acting amines is independent of the presence of calcium (Lindmar et al. 1967; Thoenen et al. 1969; Thoa et al. 1975; Ross and Kelder 1979), only the second mechanism remains. However, if (+)-amphetamine is not transported into the adrenergic nerve ending, it cannot possibly induce any carrier-mediated outward transport of noradrenaline (Trendelenburg 1978).

Recently, Bönisch (1984) succeeded in demonstrating that (+)-amphetamine is a substrate of uptake$_1$ (see Sect. D for details). In view of this new finding, we are now permitted to base this review on the assumption that *all* indirectly acting amines are substrates of uptake$_1$. The review will concentrate on the mechanisms of action of these agents. Other aspects of this general area (i.e. especially the classification of sympathomimetic amines as direct or indirect) have been dealt with in earlier reviews (Muscholl 1966; Trendelenburg 1972).

B. The Adrenergic Nerve Ending

The reader is referred to the chapter by Graefe and Bönisch (this volume Chap. 4) for a more detailed account of various functions of adrenergic nerve endings.

I. Inward Transport by Uptake$_1$

According to the proposal of Sammet and Graefe (1979) the sodium gradient influences the "normal" direction of this transport by two mechanisms. While the unloaded carrier is freely mobile, it is immobilized as soon as sodium binds to it. Since the sodium concentration is much higher on the outside than on the inside of the axonal membrane, most carrier molecules face the extracellular space and only a small fraction faces the axoplasm. Morever, the concentration gradient also determines the affinity of noradrenaline to the carrier: it is high on the outside and low on the inside of the axonal membrane. Noradrenaline is bound to the carrier only subseqent to the binding of sodium (and of chloride; see Sanchez-Armass and Orrego 1977, Friedrich and Bönisch 1986). The fully loaded carrier regains mobility and inward transport takes place. On the inside of the axonal membrane, noradrenaline, sodium and chloride dissociate from the carrier, and the unloaded carrier can return to the outside, since it is again fully mobile (see above).

One consequence of the asymmetry of the system is that, normally, there is very little outward transport of noradrenaline: very few carrier molecules are available on the inside of the axonal membrane, and the affinity of axoplasmic noradrenaline to the carrier is low.

II. The Outward Transport of Noradrenaline

Any decrease of the normal sodium gradient is able to induce carrier-mediated outward transport of noradrenaline, since the mobile unloaded carrier redistributes itself according to the change of the sodium gradient. The sodium gradient is reduced either by a lowering of the outside sodium concentration or by raising the inside sodium concentration. The latter result is achieved either by inhibition of the membrane ATPase (Na^+, K^+-ATPase) (by a cardiac glycoside or by reduction of the outside K^+ concentration) or by the opening of the fast sodium channels (by veratridine). If experiments are carried out in calcium-free solution (to prevent any exocytotic release of noradrenaline in response to the depolarization of the varicosities), these procedures are well known to induce brisk outward transport of noradrenaline (Paton 1973; Lindmar and Löffelholz 1974; Ross and Kelder 1976; and many others).

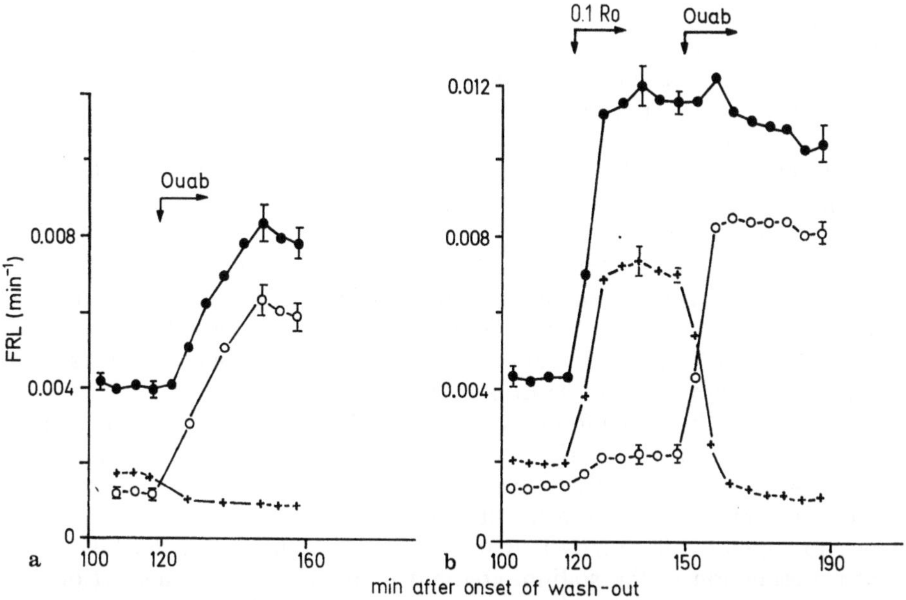

Fig. 1 a, b. The effect of ouabain on the efflux of tritium, ^{3}H-noradrenaline and ^{3}H-DOPEG either in the absence **(a)** or in the presence of a reserpine-like compound **(b)**. Rat vasa deferentia in calcium-free medium, first loaded with ^{3}H-noradrenaline and then washed out with amine-free solution. Ordinates: FRL (rate of efflux/tissue tritium content) (min^{-1}). Abscissae: time after onset of wash-out (min). Shown are means ±S.E. of 4 experiments each: ● — tritium; ○ — noradrenaline; + — DOPEG. DOMA and O-methylated metabolites are not shown, since their contribution to total efflux is very small. **a** From the 120th min of wash-out onwards, 3 mmol/l ouabain was present in the incubation medium. **b** From the 120th min of wash-out onwards, 0.1 µmol/l Ro 4-1284 was present in the incubation medium, from the 150th min onwards additionally 3 mmol/l ouabain. (Drawn from the results of STUTE and TRENDELENBURG 1984)

When intact adrenergic nerve endings are loaded with ^{3}H-noradrenaline, and when the tissue is then washed out with amine-free solution for about 100 min, tritium leaves the tissue at a fractional rate of loss (= rate of efflux of tritium/simultaneously determined tritium content of tissue) that is constant with time (see Fig. 1, from 100 to 120 min). This spontaneous efflux of tritium is composed primarily of ^{3}H-DOPEG and relatively little ^{3}H-noradrenaline (i.e. the ratio NA/DOPEG is < 1). The spontaneous efflux of tritium reflects the leakage of ^{3}H-noradrenaline from the storage vesicles. As the highly polar ^{3}H-noradrenaline cannot easily diffuse through the axonal membrane, it is largely deaminated by neuronal MAO, a process that results in the production of ^{3}H-DOPEG, a highly lipophilic metabolite (MACK and BÖNISCH 1979) that can easily leave the adrenergic nerve ending. For a discussion of the spontaneous efflux of tritium from adrenergic nerve endings, see STUTE and TRENDELENBURG (1984).

PATON (1981) loaded rabbit atria (obtained from reserpine- and pargyline-pretreated animals) with ^{3}H-noradrenaline, ^{3}H-octopamine and ^{3}H-metaraminol, respectively. On exposure to 30 µmol/l cocaine, the spontaneous efflux of tritium was increased by factors of 2.0, 2.3 and 4.5, respectively. As cocaine inhibits not only neuronal re-uptake but also the outward transport of these substrates of $uptake_1$, the results indicate that the unidirectional outward diffusion is ^{3}H-noradrenaline < ^{3}H-octopamine < ^{3}H-metaraminol. This finding is consistent with the view that outward diffusion is dependent on the lipid solubility of the various amines, and that outward diffusion of the highly polar noradrenaline should be low.

On inhibition of the membrane ATPase (Fig. 1a), there is a brisk increase in the efflux of ^{3}H-noradrenaline as a consequence of outward transport. Simultaneously, the efflux of ^{3}H-DOPEG declines. As shown by Stute and Trendelenburg (1984), this outward transport of noradrenaline always goes hand in hand with an increase in the ratio NA/DOPEG.

It is noteworthy that inhibition of the membrane ATPase also increases the rate of efflux of tritium (Fig. 1a). On the one hand, it is conceivable that the rise of the internal sodium concentration increases the gross rate of leakage of ^{3}H-noradrenaline from the storage vesicles; however, there is no evidence to support this proposal. A more likely explanation is the following. When a reduction of the sodium gradient induces outward transport of ^{3}H-noradrenaline, the axoplasmic concentration of ^{3}H-noradrenaline declines. As a consequence, the rate of formation of ^{3}H-DOPEG falls. As a second consequence, the rate of re-uptake of axoplasmic ^{3}H-noradrenaline into the storage vesicles also declines. Even if the gross rate of leakage of ^{3}H-noradrenaline is unaltered by the change in the sodium gradient, the *net* rate of leakage from the storage vesicles it increased when re-uptake into the storage vesicles declines. Thus, we face the possibility that—in intact adrenergic nerve endings—the rate of tritium efflux directly reflects the rate of net leakage of ^{3}H-noradrenaline from the storage vesicles.

Up to this point, the discussion dealt with outward transport of ^{3}H-noradrenaline induced by a decrease of the sodium gradient. For many carrier systems a second mechanism is well known, by which outward transport can be induced: if the labelled substrate A is inside the cells, any exposure of the tissue to the unlabelled substrate B is able to increase the efflux of substrate A, i.e. to induce facilitated exchange diffusion. This phenomenon is the consequence of the inward transport of substrate B, which makes carrier available (on the inside of the cell membrane) for the outward transport of substrate A. For adrenergic nerve endings, Paton (1973) was the first to propose that facilitated exchange diffusion can take place. In this review, an attempt will be made to show that indirectly acting sympathomimetic amines are all able to induce facilitated exchange diffusion.

III. The Effect of Reserpine-like Drugs

Reserpine and reserpine-like drugs (like Ro 4-1284; 2-hydroxy-2-ethyl-3-isobutyl-9,10-dimethoxy-1,2,3,4,6,7-hexahydro-11b-H-benzo(a)-quinolizine) increase the rate of net leakage of ^{3}H-noradrenaline from the storage vesicles. As illustrated in Figure 1b, they induce a pronounced increase in the rate of efflux of tritium and of ^{3}H-DOPEG. On the other hand, these agents cause only a minor increase in the rate of efflux of ^{3}H-noradrenaline (Fig. 1b). As reserpine-like agents are not transported into the adrenergic nerve ending, and as they do not affect the sodium gradient, they are unable to increase the availability of the carrier on the inside of the axonal membrane. Moreover, in the absence of any pronounced outward transport of ^{3}H-noradrenaline, the increased levels of free axoplasmic ^{3}H-noradrenaline lead to a pronounced increase in the rate of formation of ^{3}H-DOPEG. Figure 1b illustrates the typical events elicited by "depleting agents" which fail to cause any substantial release of ^{3}H-noradrenaline into the extracellular space.

Figure 1b also shows the effects of ouabain in the continued presence of the reserpine-like compound. As soon as the change in the sodium gradient increases the availability of the carrier on the inside of the axonal membrane, pronounced outward transport of ^{3}H-noradrenaline takes place, simultaneously with a steep decline of the efflux (and formation) of ^{3}H-DOPEG.

The results presented in Fig. 1 illustrate the following points: **1)** Any increase in the availability of the carrier on the inside of the axonal membrane leads to an outward transport of ^{3}H-noradrenaline that goes hand in hand with a decline in the rate of neuronal deamination of the amine; **2)** pronounced increases in the axoplasmic concentration of ^{3}H-noradrenaline fail to cause any pronounced efflux of ^{3}H-noradrenaline from the adrenergic nerve ending if the increase in vesicular net leakage of ^{3}H-noradrenaline is not accompanied by an increased availability of the carrier on the inside of the axonal membrane; **3)** however, if high axoplasmic levels of ^{3}H-noradrenaline are combined with an increased availability of the carrier for outward transport (Fig. 1b; in the presence of Ro 4-1284 and ouabain), then ^{3}H-noradrenaline is transported out of the nerve ending at high rates. For further evidence, the reader is referred to Bönisch et al. (1983) and to Stute and Trendelenburg (1984).

It should be added here that ^{3}H-DOPEG is the preferred neuronal deaminated metabolite. Under the experimental conditions of Fig. 1, ^{3}H-DOMA amounts to less than 20 % of ^{3}H-DOPEG. Moreover, in this figure (and in all others presented in this review) the extraneuronal enzyme, COMT, was inhibited by the presence of 10 μmol/l U-0521 (3,4-dihydroxy-2-methyl propiophenone). Hence, hardly any O-methylated metabolites were detected.

C. The Experimental Models

Three different models are used in the analysis of the mechanism of action of indirectly acting amines. They are presented in descending order of "pharmacological intervention".

I. Adrenergic Nerve Endings After Inhibition of Vesicular Uptake and of MAO

The loading of the adrenergic nerve endings of animals pretreated with reserpine and an irreversible inhibitor of MAO results in rather high axoplasmic levels of ^{3}H-noradrenaline. However, it would be false to regard this model as a "one-compartment system". Although the vesicular uptake mechanism is inhibited by reserpine, some ^{3}H-noradrenaline enters the storage vesicles by diffusion. As the storage vesicles maintain their low inside pH (Johnson and Scarpa 1976) even after pretreatment with reserpine, ^{3}H-noradrenaline combines with a proton inside the vesicle and is unable to diffuse across the lipophilic chromaffin granule membrane. This trapping mechanism can lead to a considerable accumulation of ^{3}H-noradrenaline inside the storage vesicles (Johnson et al. 1982; Phillipps 1982; Bönisch and Rodrigues-Pereira 1983). Thus, under these experimental conditions, there are two distribution compartments for ^{3}H-noradrenaline, and they are arranged in series.

When the rate of efflux is expressed as a fractional rate of loss (FRL = rate of efflux/tritium content of tissue), any increase in the rate of efflux from a one-compartment system should increase the FRL from a low to a higher plateau. Fig. 2a illustrates that this is not the case for the efflux (induced by tyramine) from adrenergic nerve endings of the vas deferens of rats pretreated with reserpine and pargyline. Efflux curves are then characterized by the early appearance of a peak, followed by a gradual decline of the FRL. The peak is due mainly to the efflux of ^{3}H-noradrenaline from the axoplasmic compartment; however, as the duration of the incubation increases, efflux of ^{3}H-noradrenaline from the storage vesicles into the axoplasm plays an increasing role in determining the shape of the efflux curve. The transient response to indirectly acting amines is not due to any interference of low concentrations of the indirectly acting amine with the outward transport of ^{3}H-noradrenaline (after the uptake of the indirectly acting amine into the adrenergic nerve ending), since similar efflux curves are generated by exposure of the tissue to veratridine (which induces an outward transport of ^{3}H-noradrenaline through an

---

▶

Fig. 2a–c. Tyramine-induced release of tritium from rat vasa deferentia under different experimental conditions (calcium-free medium, COMT inhibited in all experiments). **a** after pretreatment with reserpine and pargyline (to inhibit vesicular uptake and MAO); **b** after pretreatment with pargyline (to inhibit MAO); **c** no pretreatment. All tissues were loaded with 1 µmol/l ^{3}H-noradrenaline for 30 min and then washed out with amine-free solution. From the 110th min of wash-out onwards tyramine (6, 60 or 600 µmol/l) was added to the incubation medium. (Ordinates: FRL (min^{-1}); abscissae: time (in min) after onset of wash-out. Shown are means (and some S.E.) of 3 experiments each. Note that the efflux of tritium consisted mainly of ^{3}H-noradrenaline, even when there was no pretreatment (c; see also Fig. 6). At the 110th min of wash-out tritium in the tissue amounted to 7.70 + 1.01 **(a)**, 7.66 ± 0.48 **(b)** and 6.84 ± 0.68 **(c)** nmol/g (6 experiments each). Peak rates of tritium efflux induced by 60 µmol/l tyramine were: 450 ± 58 **(a)**, 179 ± 14 **(b)** and 94.2 ± 20.0 **(c)** pmol · g^{-1} · min^{-1} (3 experiments each). (Bönisch, unpublished observations)

increase of the intracellular sodium concentration; BÖNISCH et al. 1983; BÖNISCH and TRENDELENBURG 1987).

By comparing the outward transport of ^{3}H-noradrenaline induced by 100 µmol/l veratridine plus 1 mmol/l ouabain (in calcium-free solution to prevent any exocytotic release) with that simulated by a mathematical two-compartment system (with the compartments arranged in series), SCHÖMIG and

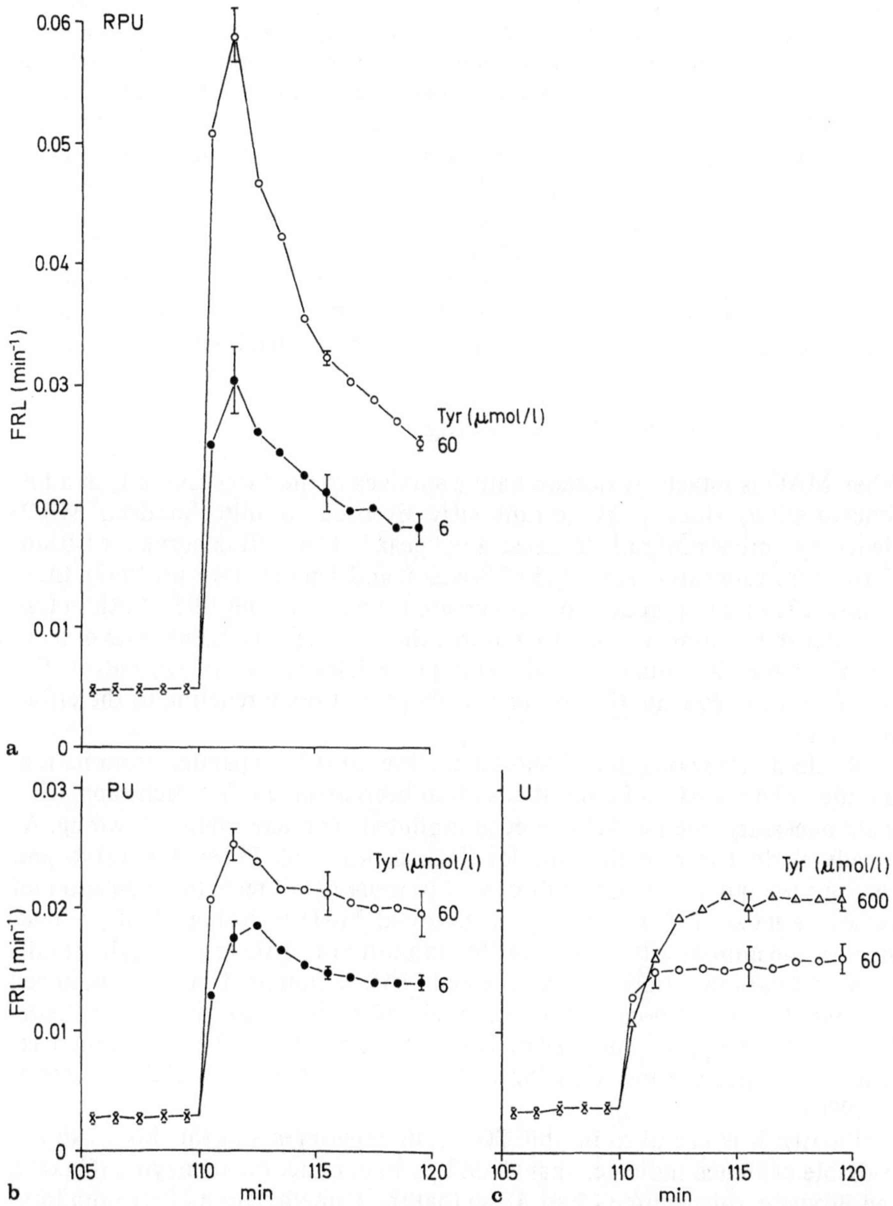

TRENDELENBURG 1987 proposed that about 45 % of "neuronal ^{3}H-noradrenaline" are distributed into the axoplasm, and 55 % into the storage vesicles (under experimental conditions like those in Fig. 2a).

II. Adrenergic Nerve Endings After Inhibition of MAO

Again we deal with a "two compartment system", but in the absence of any pretreatment with reserpine the vesicular compartment is considerably greater than in model (I). As a consequence, the peak of the efflux curve is attenuated, while the tendency to reach a late plateau is increased (Fig. 2b; see also Fig. 4 of BÖNISCH and TRENDELENBURG 1987). It should also be noted that the "peak FRL" is considerably lower in Fig. 2b than in Fig. 2a. As indicated in the legend to Fig. 2, just before the administration of tyramine the tritium content of the tissue (after identical loading) was as high in these tissues as in those of Fig. 2a. Hence, the accumulation of ^{3}H-noradrenaline in the axoplasmic compartment must be considerably smaller in the tissues of Fig. 2b than in those of Fig. 2a. This is also borne out by the finding that the omission of the pretreatment with reserpine lowered not only the peak FRL (compare Fig. 2a and b) but also peak rates (in pmol·g^{-1}·min^{-1}; see legend to Fig. 2).

III. Intact Adrenergic Nerve Endings

When MAO is intact, ^{3}H-noradrenaline survives in the axoplasm only at a low concentration, since it is continuously exposed to mitochondrial MAO. Hence, tyramine now fails to cause any "peak" of the efflux curve for tritium (Fig. 2c; for veratridine see Fig. 5 of BÖNISCH and TRENDELENBURG 1987), there is only a smooth approach to an increased plateau of the FRL. Both before and after the administration of tyramine, the FRL reflects the net rate of leakage of ^{3}H-noradrenaline from the storage vesicles (into the axoplasm). See Sect. G, for the contribution of the metabolites of noradrenaline to the efflux of tritium.

While it is obvious that ^{3}H-noradrenaline must be separated from its metabolites when MAO is intact, it has often been assumed that such separation is not necessary when MAO had been inhibited. This assumption is wrong. As already evident from earlier studies (HENSELING et al. 1978; HENSELING and TRENDELENBURG 1978) and as discussed in more detail recently (STEFANO and TRENDELENBURG 1984), the V_{max} of neuronal MAO is so high that pretreatment of the animal with an irreversible inhibitor of MAO (e.g. pargyline) fails to cause total inhibition. When neuronal MAO is inhibited, such pronounced accumulation of ^{3}H-noradrenaline can take place in the axoplasm that the reduction of the V_{max} of the enzyme is largely compensated for by the increase in the substrate concentration (STEFANO and TRENDELENBURG 1984; CASSIS et al. 1986).

Finally, it is useful to inhibit COMT in experiments of this kind. All the available evidence indicates that COMT is an extraneuronal enzyme (GRAEFE and BÖNISCH, this volume Chap. 4), so that the O-methylation of ^{3}H-noradrenaline, ^{3}H-DOPEG or ^{3}H-DOMA takes place subsequent to the efflux of these compounds from the nerve endings. As these extraneuronal events are irrelevant for any analysis of the mechanism of action of indirectly acting amines, for ease of analysis, COMT should be inhibited.

D. Carrier-Mediated Uptake of (+)-Amphetamine

While a cocaine-sensitive, sodium-dependent uptake of noradrenaline is easily demonstrated in a large variety of adrenergically innervated organs, analogous experiments with ^{3}H-(+)-amphetamine yielded negative results (THOENEN et al. 1968; BÖNISCH and RODRIGUES-PEREIRA 1983). However, it should be realized that the adrenergic nerve endings of an isolated organ represent a very small percentage of total tissue mass. Hence, it is conceivable that the active uptake of (+)-amphetamine into a very small percentage of the tissue is masked by the diffusional entry of the highly lipophilic (+)-amphetamine (MACK and BÖNISCH 1979) into *all* cells of this tissue.

To test this hypothesis, the uptake of ^{3}H-(+)-amphetamine was studied in cultures of PC-12-cells (rat phaeochromocytoma cells) which are free of any admixture of other cell types. The uptake of ^{3}H-noradrenaline into PC-12-cells is saturable, characterized by a K_m that is typical for uptake$_1$, sensitive to inhibition by cocaine and desipramine, and absolutely dependent on the presence of extracellular sodium and chloride (BÖNISCH 1984; FRIEDRICH and BÖNISCH 1986). Thus, PC-12-cells have uptake$_1$.

In these cells, there is a cocaine- and desipramine-sensitive uptake of ^{3}H-(+)-amphetamine that is absolutely dependent on the presence of extracellular sodium and chloride. However, as one would expect for an amine with very high lipophilicity, there is also a cocaine-resistant and sodium-independent entry of ^{3}H-(+)-amphetamine into PC-12-cells that is considerably greater than that for ^{3}H-noradrenaline.

Unfortunately, even in PC-12-cells the diffusional influx of ^{3}H-(+)-amphetamine is so high that a K_m for uptake$_1$ of ^{3}H-(+)-amphetamine cannot be determined (BÖNISCH 1984). However, the K_m should be equal to that concentration of (+)-amphetamine that halves uptake$_1$ of a tracer concentration of ^{3}H-noradrenaline (= 0.2 µmol/l).

It is of interest to note that there have been earlier hints of an uptake of ^{3}H-(+)-amphetamine by uptake$_1$. AZZARO et al. (1974) reported that the uptake of 0.1 µmol/l of this amine into synaptosomes (prepared from rat brain) was reduced by cocaine or desipramine. Moreover, RUTLEDGE and VOLLMER (1979) determined the uptake of 6.25 to 25 µmol/l ^{3}H-(+)-amphetamine into synaptosomes prepared from rat cerebral cortex, at 20° C. Analysis of the results yielded a K_m of 58 µmol/l. As the K_m exceeded the highest substrate concentration more than two-fold, this must be regarded as a rough guess. But even so, the K_m-value is surprisingly high. However, it should be realized that—as in PC-12-cells—it is virtually impossible to measure "initial rates of uptake of ^{3}H-(+)-amphetamine", since efflux of ^{3}H-(+)-amphetamine begins very soon after the onset of the incubation; again, this is the unavoidable disadvantage of the very high lipophilicity of (+)-amphetamine. For these reasons, it is much more likely that the K_m for uptake$_1$ of amphetamine is considerably lower, as indicated by the IC_{50} (see above).

E. The Release of ^{3}H-Noradrenaline After Inhibition of MAO and of Vesicular Uptake

If we select that experimental model which ensures a good (but not exclusive) distribution of the neuronal ^{3}H-noradrenaline into the axoplasm (model I of Sect. C), we can test the hypothesis that indirectly acting amines elicit facilitated exchange diffusion of preloaded ^{3}H-noradrenaline. There should be a strict correlation between the K_m for uptake$_1$ of various unlabelled (indirectly acting) amines and the EC_{50} of their ^{3}H-noradrenaline releasing effect. Moreover, if there should be differences between amines with respect to the V_{max} for uptake$_1$, this might be reflected by corresponding differences between the maxima of their ^{3}H-noradrenaline-releasing effects.

Table 1 presents the K_m-values for uptake$_1$, while Fig. 3a and b shows the concentration-response curves for the ^{3}H-noradrenaline releasing effect of var-

Table 1. K_m and V_{max} for uptake$_1$ and "equieffective concentrations" (for the release of ^{3}H-noradrenaline) for 12 amines; vasa deferentia of reserpine- and pargyline-pretreated rats, COMT inhibited. Shown are means. Taken from LANGELOH et a. (1987)

Amine	K_m (μmol/l)[a]	EC FRL 0.03 min^{-1}(μmol/l)[b]
(+)-amphetamine	0.35	1.41
amezinium	0.90	[c]
(−)-metaraminol	1.3	17.4
tyramine	1.4	10.4
dopamine	2.0	14.7
(−)-noradrenaline	3.7	29.9
debrisoquin	3.7	[c]
(±)-phenylephrine	7.9	51.4
guanethidine	12.0	128
bretylium	19.2	116
bethanidine	34.8	307
5-HT	44.3	392

[a] $K_m = IC_{50}$, i.e. that concentration of unlabelled amine which halved the initial rate of neuronal uptake of a very low concentration of ^{3}H-(−)-noradrenaline

[b] EC FRL 0.03 min^{-1}: equieffective concentrations determined at the level indicated by the horizontal line in Fig. 3

[c] Because of the low maximum effect of this amine, no EC FRL 0.03 min^{-1} was calculated.

ious substrates of uptake$_1$. Most importantly, there is a positive correlation between the EC_{50} and the K_m for uptake$_1$ (LANGELOH et al. 1987).

From the concentration-response curves of Fig. 3a and b equieffective concentrations were calculated at the level of Δ FRL of 0.03 min^{-1} (see horizontal lines in Fig. 3a and b). These values appear in Table 1. They indicate a high degree of correlation between the K_m for uptake$_1$ and equieffective ^{3}H-noradrenaline-releasing concentrations.

Fig. 3a and b shows phenomena that require an explanation. For instance, from a comparison of Fig. 3a and b with Table 1 it is obvious that the EC_{50} (for release of ^{3}H-noradrenaline) exceeds K_m for uptake$_1$. This will be discussed in Sect. F.

For amines with low K_m (see Table 1), concentration-response curves tend to have low maxima, and they tend to be "bell-shaped" (see Fig. 3a and b). The low maximum effects observed for amines with low K_m are due to the fact that K_m and V_{max} for uptake$_1$ were found to be positively correlated: the lower the K_m, the lower the V_{max} (see Chap. 4 by GRAEFE and BÖNISCH, this volume). This correlation was obtained, when the V_{max} was determined for those eight substrates of Table 1 that were available as labelled compounds. Hence, it was not surprising to see low maximum effects (Fig. 3) for amines characterized by low maximum rates of inward transport. The descending limb of the concentration-response curves requires a different explanation. As illustrated in Fig. 3, the phenomena of a descending limb of the concentration-response curve can be ranked: (+)-amphetamine > tyramine > (±)-metaraminol > amezinium. This sequence supports the view that there are two determinants, the lipophilicity of the amine and its K_m for uptake$_1$. The higher the lipophilicity, the higher the tendency towards a pronounced descending limb. However, even for a highly hydrophilic amine a descending limb is evident, provided its K_m for uptake$_1$ is low (e.g. amezinium). Hence, it is likely that a very pronounced inward diffusion across the axonal membrane is responsible for the descending limb of the concentration-response curve—either at rather low outside concentration (when the amine is highly lipophilic, e.g. (+)-amphetamine), or at very high outside concentration (when the amine is very hydrophilic; e.g. amezinium). Conceivably, this very pronounced inward diffusion generates such high axoplasmic amine levels that the unlabelled amine is able to compete with ^{3}H-noradrenaline for outward transport. However, this proposal is possible only if the inward transport of substrates of uptake$_1$ lowers the K_m for outward transport, since otherwise a successful competition is impossible: if there is no increase in the sodium concentration on the inside of the axonal membrane, outward transport is not saturated even by much higher axoplasmic concentrations than those attained in these experiments (BÖNISCH et al. 1986). As will be discussed in Sect. F.III, the cotransport of sodium and chloride is responsible for this lowering of the K_m for the outward transport of ^{3}H-noradrenaline.

It should be noted that a bell-shaped concentration-response curve was already observed by ROSS and KELDER (1979) in experiments in which unlabelled (+)-amphetamine released ^{3}H-bretylium from adrenergic nerve endings.

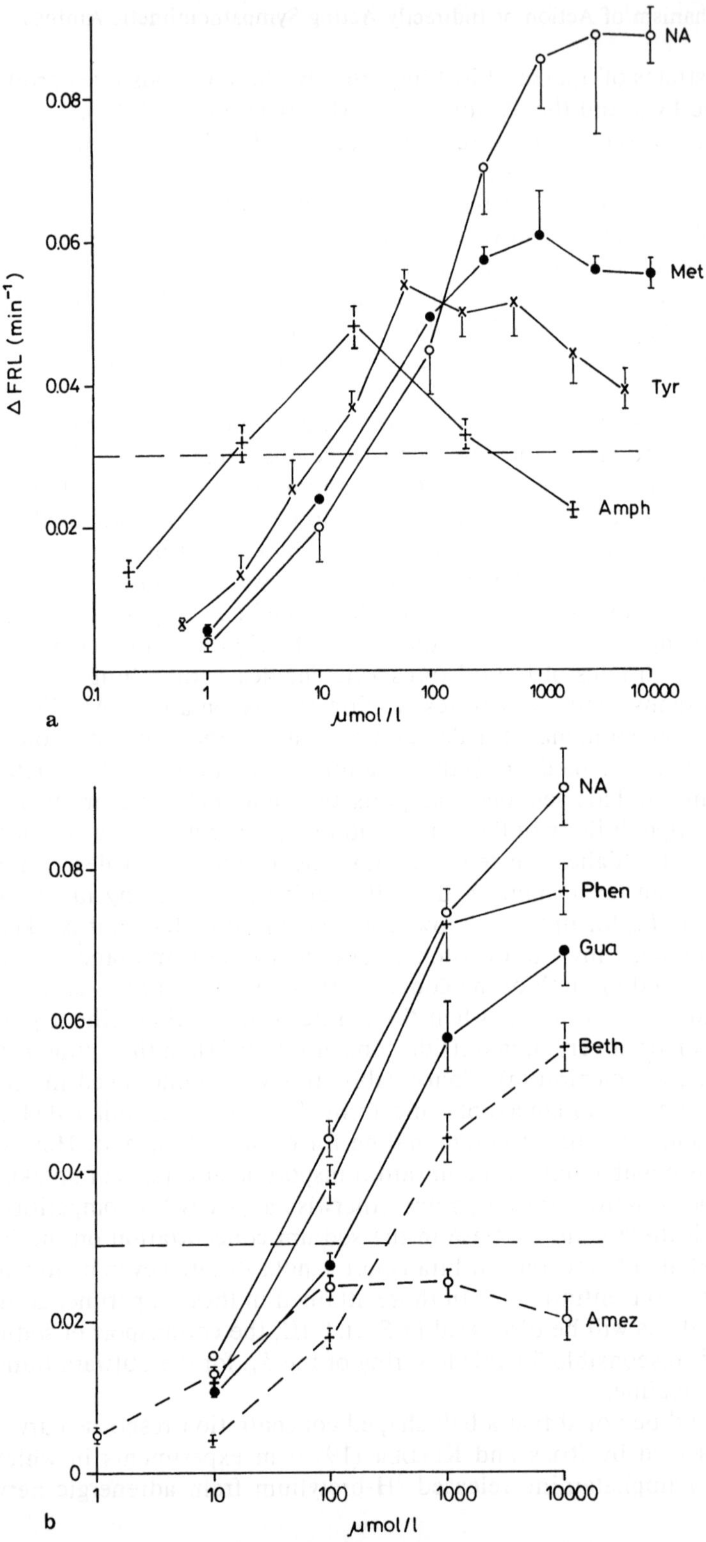

Δ FRL (min^{-1})
0.08
0.06
0.04
0.02
0.1
1
10
100
1000
10000
µmol/l
NA
Met
Tyr
Amph
a
0.08
0.06
0.04
0.02
0
1
10
100
1000
10000
µmol/l
NA
Phen
Gua
Beth
Amez
b

F. Factors Involved in the Release of Axoplasmic ^{3}H-Noradrenaline

I. Facilitated Exchange Diffusion

As pointed out in the Introduction, any inward transport of a substrate of uptake$_1$ should increase the availability of the carrier (for outward transport) on the inside of the axonal membrane. Or, in other words, the inward transport of a substrate of uptake$_1$ should increase the V_{max} for outward transport.

II. The Co-transport of Sodium

SAMMET and GRAEFE (1979) presented kinetic evidence indicating that the inward transport of a substrate of uptake$_1$ is accompanied by an inward co-transport of sodium. Any consequent rise of the sodium concentration at the inside of the axonal membrane would result in a decrease of the K_m for the outward transport of ^{3}H-noradrenaline. Evidence favouring this possibility was obtained in experiments in which the Na^+, K^+-ATPase was inhibited (by ouabain or by omission of K^+ from the medium). The consequent decrease of the Na^+ concentration gradient should decrease the uptake of tyramine and (+)-amphetamine, hence also their ^{3}H-noradrenaline-releasing effect. However, when the exposure of the tissue to tyramine [or (+)-amphetamine] was simultaneous with inhibition of the sodium pump, the release of ^{3}H-noradrenaline was transiently enhanced (Fig. 4; BÖNISCH 1986). A similar phenomenon was observed by LEVI et al. (1976) who determined the efflux of ^{3}H-noradrenaline from rat brain synaptosomes exposed to unlabelled noradrenaline either in the absence or in the presence of ouabain, and by LIANG and RUTLEDGE (1982) who measured the efflux of ^{3}H-dopamine from corpus striatum slices exposed to (+)-amphetamine either in the absence or in the presence of ouabain.

◄

Fig. 3a, b. ^{3}H-noradrenaline-releasing effect of various substrates of uptake$_1$; rat vas deferens, calcium-free medium, vesicular uptake, MAO and COMT inhibited (by pretreatment with reserpine and pargyline, and by the presence of 10 µmol/l U-0521, respectively). Tissues were loaded with 1 µmol/l ^{3}H-noradrenaline for 30 min in **a**, or with 0.2 µmol/l ^{3}H-noradrenaline for 60 min in **b**; they were then washed out with amine-free solution. From the 100th min of wash-out onwards, incubation media were changed once per min. From the 110th min of wash-out onwards, various indirectly acting amines were present in the medium (for 10 min). Ordinates: Δ FRL for ^{3}H-noradrenaline (peak FRL — basal FRL) (min^{-1}); abscissae: concentration of unlabelled amines (µmol/l; log scale). Shown are means ±S.E. of 3 to 7 experiments per point. Panel **a**: NA — (−)-noradrenaline; Met — (±)-metaraminol; Tyr — tyramine; Amph — (+)-amphetamine. Panel **b**: NA — (−)-noradrenaline; Phen — (−)-phenylephrine; Gua — guanethidine; Beth — bethanidine; Amez — amezinium. Note that the experiments of panels a and b were carried out in two series (at different times); hence each panel shows (−)-noradrenaline as a reference substance. Horizontal lines indicate the level at which equieffective concentrations (Table 1) were determined. Taken from LANGELOH et al. (1987)

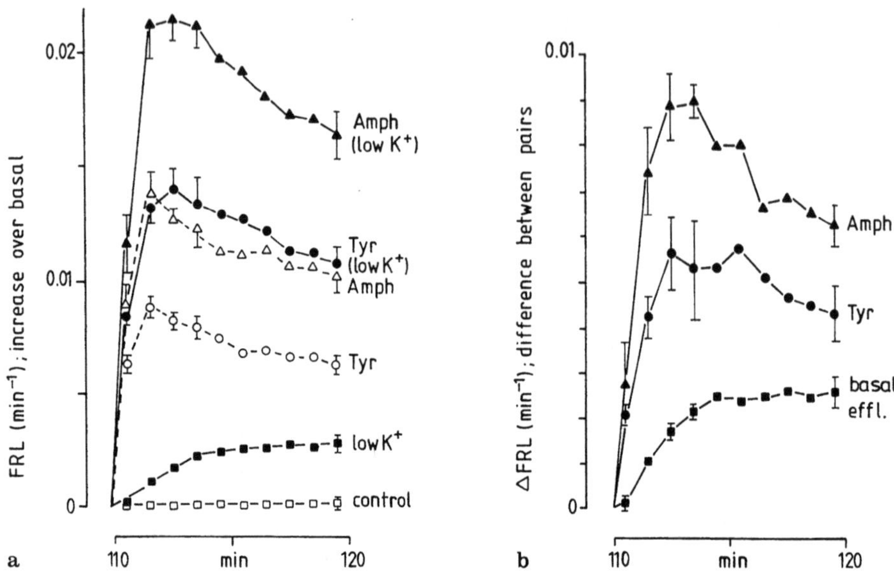

Fig. 4a, b. Potentiation of the ³H-noradrenaline-releasing effect of tyramine and (+)-amphetamine by inhibition of the membrane ATPase. Rat vasa deferentia (vesicular uptake, MAO and COMT inhibited; calcium-free medium), first loaded with 1 μmol/l ³H-noradrenaline for 30 min and then washed out with amine-free solution. From the 110th min onwards there was either no change (control) or the membrane ATPase was inhibited by reduction of the K^+ concentration of the medium (low K^+); parallel strips were exposed to 0.2 μmol/l (+)-amphetamine (Amph) or 0.6 μmol/l tyramine (Tyr), either in normal medium or in medium with reduced K^+ (low K^+). **a** increase in the FRL for tritium over the basal level (min^{-1}). **b** effect of low K^+ on basal efflux of radioactivity and on tyramine- and (+)-amphetamine-induced release of tritium (as calculated from the paired experiments shown in **a**. Taken from BÖNISCH (1986)

III. The Co-transport of Chloride

A recent analysis of the dependence of uptake$_1$ on the presence of Cl^- in the extracellular medium (FRIEDRICH and BÖNISCH 1986), gave results that were strikingly similar to those obtained earlier with Na^+ (SAMMET and GRAEFE 1979): there is an absolute necessity for the presence of Cl^- in the extracellular fluid, and any reduction of the extracellular Cl^- concentration increases K_m and decreases V_{max} of uptake$_1$. Hence, it is likely that Cl^- is cotransported together with Na^+, and such co-transport should further decrease the K_m for ³H-noradrenaline on the inside of the axonal membrane (i.e. the K_m for outward transport).

In the experiments discussed here, the tissue content (at the time of administration of unlabelled substrates) amounted to about 3 nmol/g. However, in one series of experiments tissues were loaded with a high concentration of ³H-noradrenaline, and the tissue content amounted to about 55 nmol/g. Irrespective of whether outward transport was induced by (+)-amphetamine,

5-HT or veratridine, the peak FRL for outward transport was lower for heavily loaded than for lightly loaded tissues. As shown by BÖNISCH et al. (1986) for outward transport in the presence of veratridine, saturability of outward transport is then demonstrable.

IV. Inhibition of Neuronal Re-uptake

Especially when $EC_{50} > K_m$, it is obvious that the ^3H-noradrenaline-releasing substrates of $uptake_1$ must impair the re-uptake of ^3H-noradrenaline across the axonal membrane (subsequent to outward transport). Although inhibition of neuronal re-uptake would not affect unidirectional outward transport, it should increase the net outward movement.

Thus, we face the possibility that up to four different factors are involved in the type of ^3H-noradrenaline release discussed here. It is of critical importance to realize that the relative contribution of each of the four factors to total release is directly dependent on K_m for $uptake_1$. As discussed in more detail by LANGELOH et al. (1987), EC_{50} must exceed K_m, if multiple factors (the relative contribution of each one of which depends on K_m) determine the rate of outward transport of ^3H-noradrenaline.

Fig. 5 b shows the concentration-dependence of the ^3H-noradrenaline-releasing effect of (−)-noradrenaline after "normalization" of the concentration. Normalization was achieved by the conversion of concentrations to fractions of V_{max} (where $K_m = 0.5$). Fig. 5 b shows that the experimental points fall onto a

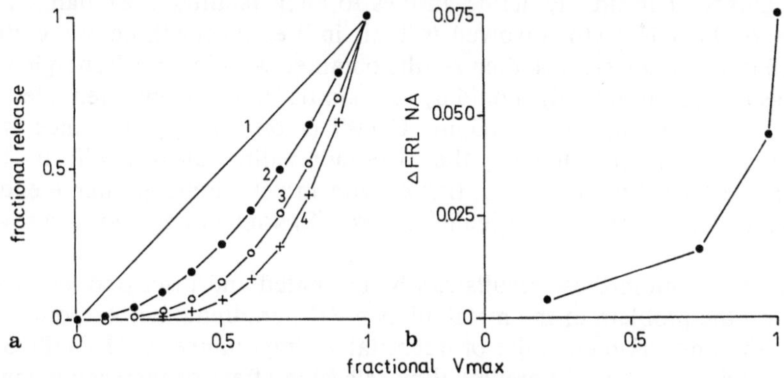

Fig. 5 a, b. Dependence of the fractional rate of outward transport of ^3H-noradrenaline on fractional V_{max} for the releasing amine. Shown are theoretical curves **(a)** and results of experiments with (−)-noradrenaline **b**. Ordinates: fractional release of ^3H-noradrenaline (maximum = 1) in a; rate of release (Δ FRL, see Fig. 1a) in **b**. Abscissae: fractional V_{max}, calculated (for $V_{max} = 1$) from K_m (Table 1) and concentration of (−)-noradrenaline, with the help of the Michaelis-Menten equation. As explained in text, curve 1 was calculated for the direct dependence of fractional release on fractional V_{max} ($y = x$); for curves 2, 3 and 4 $y = x^2$, x^3 and x^4, respectively, illustrating the assumption that 2, 3 or 4 multiplicative factors all induce release of ^3H-noradrenaline, each one in dependence on fractional V_{max}. Results in **b** taken from LANGELOH et al. (1987)

curve that has a high degree of upward concavity. It should be noted that corresponding points calculated for the ^{3}H-noradrenaline-releasing effect of the other 11 substrates were very close to the points of Fig. 5b; or in other words, upward concave curves were obtained for *all* substrates of uptake$_1$. Fig. 5a shows theoretical curves for $y = x^1$, x^2, x^3 and x^4, respectively. As the curve presented in Fig. 5b is very similar to that curve in Fig. 5a for which $y = x^4$, it is likely that the shape of the experimental curve (Fig. 5b) is determined by a) the fact that outward transport of ^{3}H-noradrenaline is induced not only by facilitated exchange diffusion but by additional three other factors (see above), and b) the fact that the relative magnitude of the contribution by each of the four factors is directly dependent on K_m. Hence, the four factors are multiplicative.

These considerations lead to the conclusion that the postulate "indirectly acting amines release ^{3}H-noradrenaline through induction of facilitated exchange diffusion" is too simple. Also involved in this process are affinity changes (as a consequence of the co-transport of Na^+ and Cl^-) as well as an inhibition of re-uptake of the labelled amine across the axonal membrane.

G. The Release of ^{3}H-Noradrenaline from Adrenergic Nerve Endings with Intact Storage Vesicles and Intact MAO (COMT Inhibited)

If the ability of indirectly acting amines to elicit facilitated exchange diffusion were the sole factor involved in their indirect sympathomimetic effects exerted onto *normal* tissues, then results obtained with intact adrenergic nerve endings (COMT inhibited) should by very similar to those obtained after inhibition of vesicular uptake, MAO and COMT (shown in Fig. 3). Hence, it was of considerable interest to study the ^{3}H-noradrenaline-releasing effects of various indirectly acting amines in tissues with intact adrenergic nerve endings (i.e. with intact MAO and vesicular uptake) (LANGELOH and TRENDELENBURG 1987).

However, before these results can be presented and discussed, one has to deal with the problem of the metabolites of ^{3}H-noradrenaline. As pointed out above, the only ^{3}H-metabolite of quantitative importance is ^{3}H-DOPEG. Figure 6 illustrates the ^{3}H-noradrenaline-releasing effect of increasing concentrations of tyramine (Fig. 6a) with the simultaneous efflux of ^{3}H-DOPEG from the tissue (Fig. 6b). Concentrations of tyramine increasing from 0.1 to 100 µmol/l cause increasing degrees of ^{3}H-noradrenaline release with a maximum at 1000 µmol/l tyramine (Fig. 6a). As far as ^{3}H-DOPEG is concerned, a totally different relation to the concentration of tyramine was observed (Fig. 6 b). Only at the very lowest concentration of tyramine (0.1 µmol/l) was the efflux of ^{3}H-DOPEG comparable to that of ^{3}H-noradrenaline. At all higher concentrations, the efflux of ^{3}H-DOPEG became progressively smaller (in relation to that of ^{3}H-noradrenaline). It is evident from Fig. 6 that the ratio NA/DOPEG increases with increasing concentrations of tyramine. This obser-

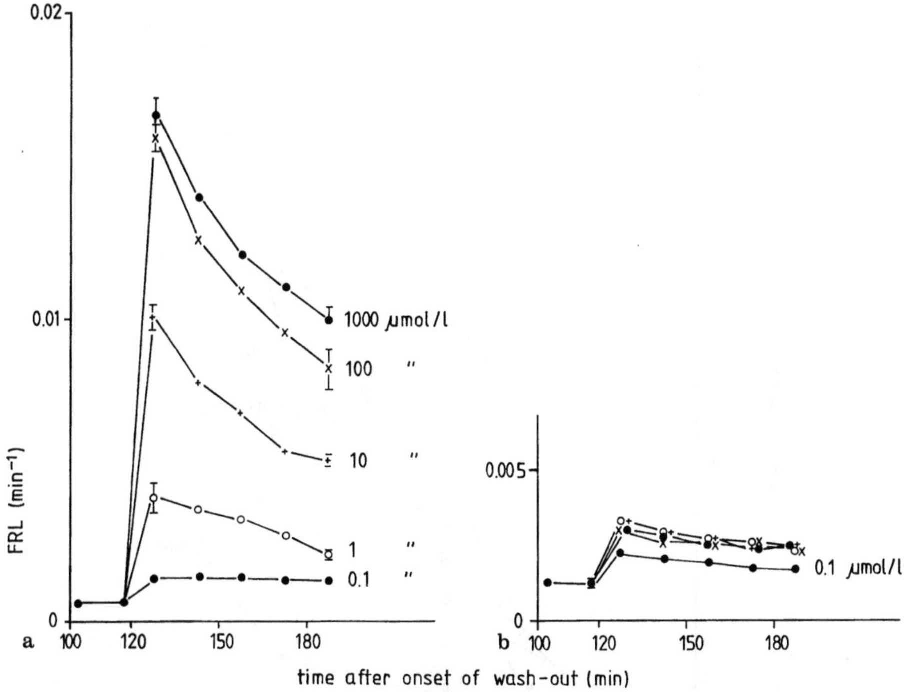

Fig. 6a, b. The release of ^{3}H-noradrenaline **(a)** and ^{3}H-DOPEG **(b)** by tyramine; rat vas deferens (COMT inhibited; calcium-free medium) first loaded with 0.1 µmol/l ^{3}H-noradrenaline for 60 min, then washed out with amine-free solution. From the 120th min of wash-out onwards, tyramine was present in various concentrations. Ordinates: FRL (min^{-1}) (= rate of efflux of ^{3}H-noradrenaline **(a)** or ^{3}H-DOPEG **(b)**/tritium content of tissue at the same time). Abscissae: time (in min) after onset of wash-out. Shown are means (with some S.E.) of 5 to 7 experiments per point. (LANGELOH and TRENDELENBURG 1987)

vation is in agreement with the proposal of STUTE and TRENDELENBURG (1984) that any outward transport of ^{3}H-noradrenaline lowers the axoplasmic ^{3}H-noradrenaline concentration and, hence, the rate of formation of ^{3}H-DOPEG. In the case of tyramine, the efflux of ^{3}H-DOPEG is not indicative of any lowering of the axoplasmic ^{3}H-noradrenaline concentration, but the results of Fig. 6 b can be interpreted as indicating that the outward transport of ^{3}H-noradrenaline (induced by tyramine) minimizes the rise in axoplasmic ^{3}H-noradrenaline in proportion to the ^{3}H-noradrenaline-releasing effect of tyramine.

Basically, similar observations were made with *all* indirectly acting amines studied here. For further discussion, see the following.

Figure 7 shows the concentration-response curves for the indirectly acting amines presented in Fig. 3. Before the results of Figs. 7 and 3 are compared, some details of Fig. 7 have to be discussed. For instance, sigmoidal concentration-response relationships were obtained for virtually all indirectly acting amines for concentrations up to 1 mmol/l. However, for some amines the re-

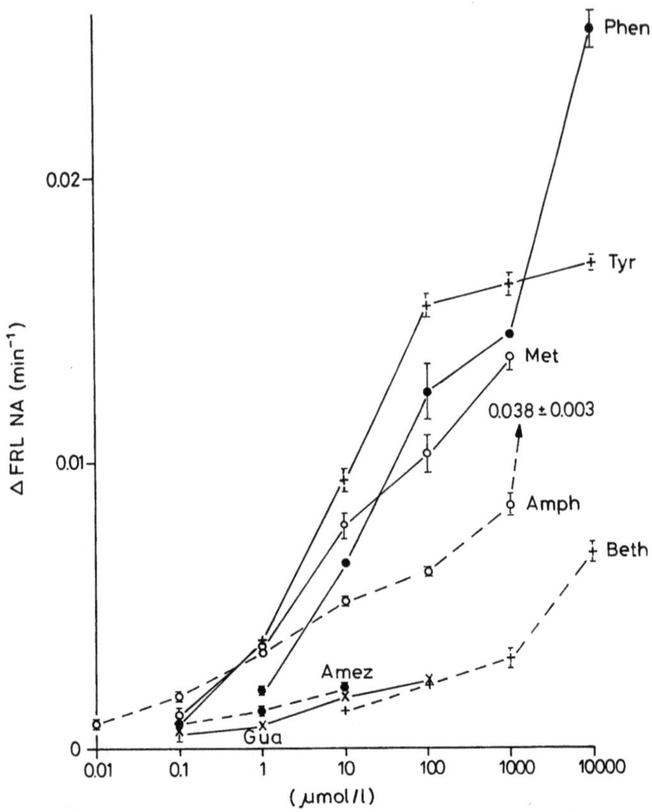

Fig. 7. The ^{3}H-noradrenaline-releasing effect of various indirectly acting amines; rat vas deferens, calcium-free medium, only COMT is inhibited (by the presence of 10 μmol/l U-0521). Tissues were loaded with ^{3}H-noradrenaline and then washed out with amine-free solution. Incubation media were changed every 5 min. From the 120th min onwards the various amines were present in the incubation media for 40 to 60 min. Ordinates: Δ FRL for ^{3}H-noradrenaline (peak FRL – basal FRL); abscissa: concentration (μmol/l). Shown are means ±S.E. of 3 to 6 experiments each. Note that the effect of 10 mmol/l (+)-amphetamine exceeds the scale of the ordinate. Abbreviations: Phen — (−)-phenylephrine; Tyr — tyramine; Met — (±)-metaraminol; Amph — (+)-amphetamine; Beth — bethanidine; Amez — amezinium; Gua — guanethidine. (LANGELOH and TRENDELENBURG 1987)

sponse to 10 mmol/l deviated from this relationship, most clearly for (+)-amphetamine. Since (+)-amphetamine is the most lipophilic of the amines studied here, the following explanation is proposed. At this very high concentration of (+)-amphetamine there is such a massive passive diffusion through the axonal membrane that the storage vesicles are exposed to about 10 mmol/l (+)-amphetamine. As demonstrated by JOHNSON et al. (1982), very high concentrations of highly lipophilic amines are able to enter the storage vesicles in such large amounts that they decrease the normal pH-gradient across the vesicular membrane (inside < outside). If that happens, the proportion of ^{3}H-noradrenaline in the protonated form is reduced, and a massive efflux of ^{3}H-noradrenaline (into the axoplasm) takes place, which, in turn, leads to an outward transport of ^{3}H-noradrenaline at very high rates.

If the occasional atypical effects of 10 mmol/l concentrations are disregarded, the concentration-response curves of Fig. 7 should reflect the release of ^{3}H-noradrenaline as a consequence of the inward transport of the indirectly acting amines.

Comparison of Figs. 3 and 7 reveals pronounced differences, especially with regard to the relative maxima of the concentration-response curves. For instance, while guanethidine, bretylium and bethanidine were quite effective in eliciting facilitated exchange diffusion (Fig. 3), they were very poor releasers of ^{3}H-noradrenaline in intact adrenergic nerve endings (Fig. 7). Moreover, rather similar maximal effects of (+)-amphetamine and tyramine in Fig. 3 stand in contrast to clear differences in Fig. 7.

When vesicular uptake and MAO are intact (Fig. 7), the concentration of axoplasmic ^{3}H-noradrenaline must be quite low. If the inward transport of an amine then increases the availability of the carrier (and the affinity of ^{3}H-noradrenaline to the carrier), outward transport may well be restricted by the low concentration of axoplasmic ^{3}H-noradrenaline. Hence, for any substantial outward transport of ^{3}H-noradrenaline, indirectly acting amines must not only be able to elicit facilitated exchange diffusion, they must also be able to "mobilize" vesicular ^{3}H-noradrenaline to produce a considerable increase in the axoplasmic concentration of ^{3}H-noradrenaline. If this proposal is accepted, then comparison of Figs. 3 and 7 indicates that amezinium, guanethidine and bethanidine are very poor "mobilizers" of vesicular ^{3}H-noradrenaline. Moreover, the "mobilizing" ability of tyramine appears to be greater than that of (+)-amphetamine (LANGELOH and TRENDELENBURG 1987).

As mentioned above, the NA/DOPEG ratio increases in the presence of the various indirectly acting amines. Hence it was of interest to plot the increase of the NA/DOPEG ratio against the ^{3}H-noradrenaline-releasing effect of the various concentrations of all indirectly acting amines (Fig. 8). Two conclusions are evident from Fig. 8. 1) For all amines the ratio NA/DOPEG increased with increasing rates of outward transport of ^{3}H-noradrenaline. 2) The compounds seem to fall into two groups. On the one hand, substrates of MAO (like (−)-noradrenaline, dopamine, (−)-phenylephrine, 5-HT and tyramine) as well as compounds that are neither substrates nor inhibitors of MAO (like guanethidine and (±)-metaraminol) produced relatively small increases of the ratio NA/DOPEG for any given increase in the release of ^{3}H-noradrenaline. It is likely that these seven compounds fail to block MAO (except for the very highest concentration of tyramine, to be discussed below). For the other compounds (i.e., amezinium, bretylium, bethanidine, (+)-amphetamine and debrisoquin), it is known that they block MAO. Hence, they can be expected to increase the ratio NA/DOPEG not only through the induction of an outward transport of ^{3}H-noradrenaline, but also through inhibition of MAO—and all their curves are above those of the first-mentioned group.

When the concentration of tyramine was increased from 1 to 10 mmol/l (Fig. 8, the last two points of the tyramine curve), there was a pronounced in-

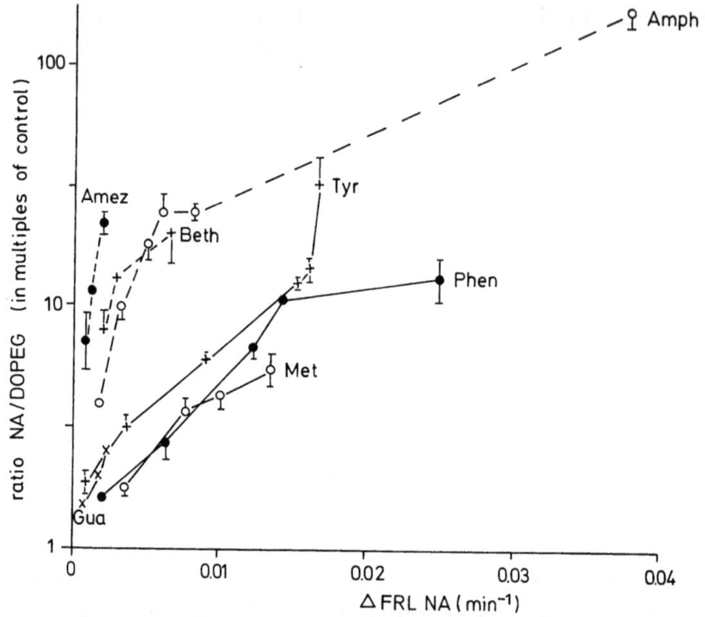

Fig. 8. The dependence of the increase in the NA/DOPEG ratio on the magnitude of the ^{3}H-noradrenaline-releasing effect of various indirectly acting amines. Same experiments as in Fig. 7. Ordinates: ratio NA/DOPEG (peak effect as multiple of controls, log scale); abscissae: magnitude of ^{3}H-noradrenaline-releasing effect (Δ FRL for ^{3}H-noradrenaline in min^{-1}). Shown are geometric means ± S.E. of 3 to 6 experiments per point. Abbreviations as in Fig. 7. (Langeloh and Trendelenburg 1987)

crease in the ratio NA/DOPEG without any corresponding increase in the release of ^{3}H-noradrenaline. As the K_m of tyramine for MAO A is slightly above 0.1 mmol/l (Tipton et al. 1982; Benedetti et al. 1983), and as tyramine is a very good substrate of the highly active neuronal MAO, it is highly likely that the inward transport of tyramine (by uptake$_1$) is unable to result in an axoplasmic concentration of this amine which saturates neuronal MAO. However, because of the pronounced lipophilicity of tyramine, an outside concentration of 10 mmol/l may well permit such massive diffusional influx of tyramine that neuronal MAO is partly saturated at this very high concentration. This would then explain the final steep increase of the tyramine curve in Fig. 8.

H. The Effects of Indirectly Acting Amines in the Presence of a Reserpine-like Compound

In the preceding section it was proposed that amezinium, guanethedine and bethanidine are poor releasers of ^{3}H-noradrenaline from intact adrenergic nerve endings, because they are unable to "mobilize" ^{3}H-noradrenaline from the vesicular store. If this interpretation is correct, then the ^{3}H-noradrenaline-releasing effect of these amines should be considerably increased when adrenergic nerve endings are exposed to a reserpine-like compound (which itself "mobilizes" vesicular ^{3}H-noradrenaline; see STUTE and TRENDELENBURG 1984).

Figure 9 shows the ^{3}H-noradrenaline-releasing effects of tyramine, amezinium and bethanidine in the absence and in the presence of the reserpine-like compound Ro 4-1284. The release of ^{3}H-noradrenaline by 10 µmol/l tyramine is not affected by the presence of Ro 4-1284 (Fig. 9a and d). However, for both amezinium (10 µmol/l, Fig. 9b and e) and bethanidine (1 000 µmol/l, Fig. 9c and f) (as well as for guanethidine, bretylium and debrisoquin, not shown) the ^{3}H-noradrenaline-releasing effect is about twice as high in the presence than in the absence of Ro 4-1284. Thus, rates of outward transport of ^{3}H-noradrenaline (induced by amezinium, bethanidine or guanethidine) are increased when vesicularly stored ^{3}H-noradrenaline is "mobilized" by the reserpine-like compound Ro 4-1284. It is reasonable to propose that the axoplasmic levels of ^{3}H-noradrenaline are increased by Ro 4-1284; hence, outward transport in the absence of Ro 4-1284 (but in the presence of any of these amines) is limited by the low substrate concentration at the inside of the axonal membrane.

Figure 9 shows an additional phenomenon of interest: in the presence of Ro 4-1284 amezinium and bethanidine lower the rate of efflux of tritium, but tyramine does not. This phenomenon is a consequence of the inhibition of MAO by amezinium and bethanidine (but not by tyramine), since it coincides with the very pronounced decline in the rate of formation of that highly lipophilic ^{3}H-metabolite (e.g. ^{3}H-DOPEG) which easily escapes from the nerve ending.

J. Factors Which Influence the ^{3}H-Noradrenaline-Releasing Effect of Indirectly Acting Amines

Various neuronal mechanisms can exert a minor or a major influence on the ^{3}H-noradrenaline releasing effects of indirectly acting amines.

I. Neuronal Uptake

As pointed out in preceding sections, it is an essential feature of indirectly acting amines that they are substrates of uptake$_1$. As a consequence, their ^{3}H-noradrenaline-releasing effects are largely determined by their K_m and V_{max} for uptake$_1$.

II. Inhibition of Neuronal MAO

LEITZ and STEFANO (1971) proposed that inhibition of neuronal MAO (by the indirectly acting amines) is an essential mechanism in their ^{3}H-noradrenaline-releasing effects (see also CLARKE 1980). This proposal was partly based on the assumption that good substrates of MAO (like tyramine or (−)-phenyl-

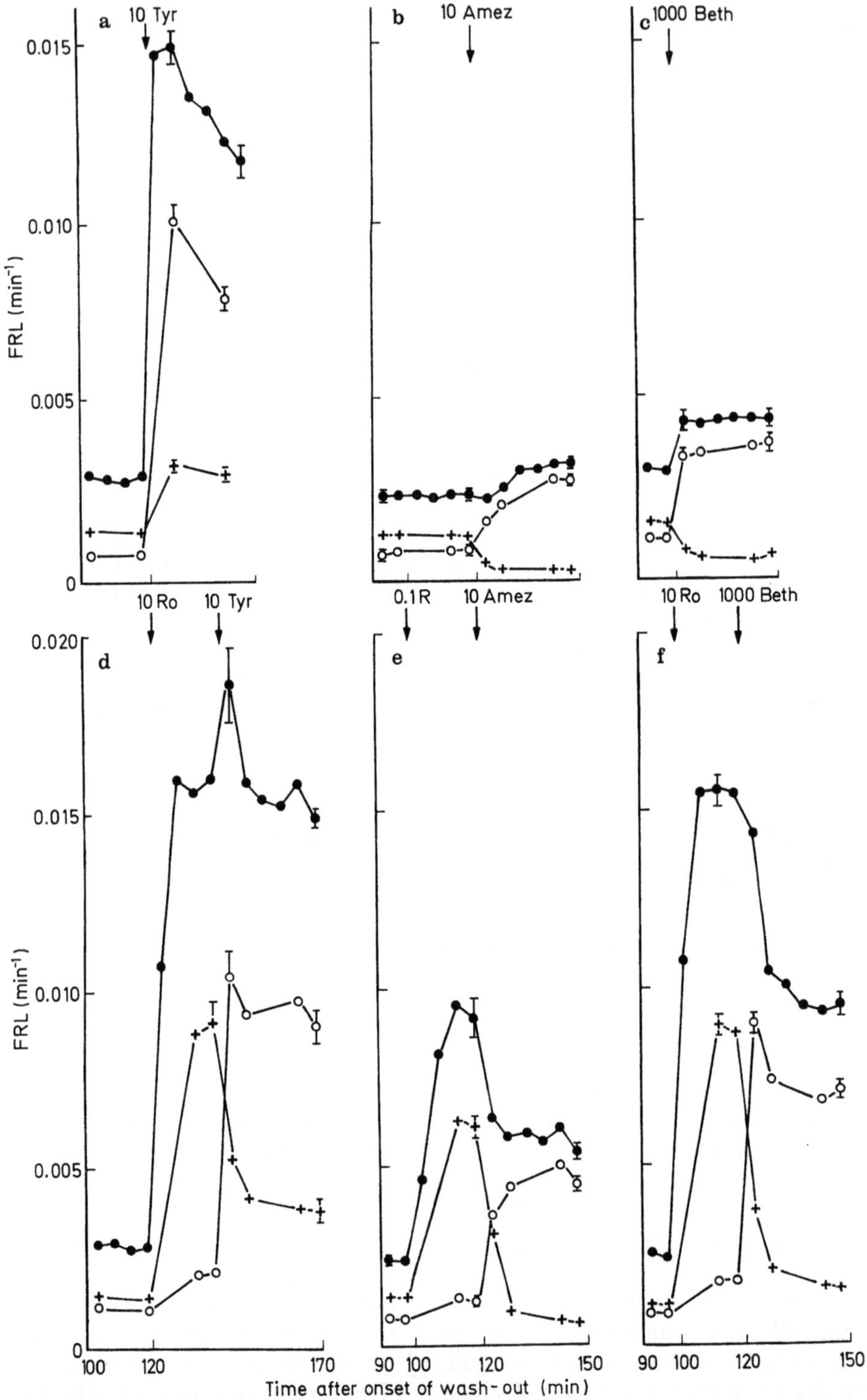
a
10 Tyr
b
10 Amez
c
1000 Beth
FRL (min⁻¹)
0.015
0.010
0.005
0
10 Ro
10 Tyr
0.1 R
10 Amez
10 Ro
1000 Beth
d
e
f
0.020
0.015
0.010
0.005
0
FRL (min⁻¹)
100
120
170
90
100
120
150
90
100
120
150
Time after onset of wash-out (min)

ephrine) are accumulated in the axoplasm to such a degree that they "inhibit" (i.e. saturate) neuronal MAO. However, as explained above, this is most unlikely, since neuronal MAO is an enzyme with a rather low affinity to its substrates (with K_m-values between 0.1 and 1 mmol/l for noradrenaline, tyramine etc.). Furthermore, the available evidence favours the view that the V_{max} for neuronal MAO appears to be very high, since the MAO activity within the varicosity is very high, in spite of the rather high K_m (see STEFANO and TRENDELENBURG 1984).

Furthermore, if a highly potent inhibitor of neuronal MAO (i.e. amezinium) is a very poor releaser of ^{3}H-noradrenaline (see Fig. 7), it is evident that inhibition of neuronal MAO is *not per se* an essential mechanism involved in the release of ^{3}H-noradrenaline. On the other hand, inhibition of neuronal MAO (in addition to an ability to "mobilize" vesicular ^{3}H-noradrenaline) would tend to increase the ^{3}H-noradrenaline releasing effect of indirectly acting amines (through an increase of the axoplasmic concentration of ^{3}H-noradrenaline).

III. The Existence of a Reserpine-like Effect for the Indirectly Acting Amine

Reserpine-like compounds increase the efflux of ^{3}H-DOPEG much more than that of ^{3}H-noradrenaline (see Fig. 1 b). Of all the indirectly acting amines studied here, only guanethidine had this effect, although only to a very moderate degree (as compared to the effects of Ro 4-1284 shown in Fig. 9). Although there is no doubt that any reserpine-induced mobilization of vesicular ^{3}H-noradrenaline can enhance the ^{3}H-noradrenaline releasing effects of indirectly acting amines (see Fig. 9), guanethidine is a very poor releaser of ^{3}H-noradrenaline (Fig. 7), in spite of its reserpine-like effects. A more detailed analysis of the time course of the effects of guanethidine revealed that the rather small ^{3}H-noradrenaline-releasing effect of this amine is seen immediately after the exposure of the tissue to guanethidine; however, the onset of the reserpine-like effects of guanethidine (indicated by an increase of the efflux of ^{3}H-DOPEG) is observed only after 20 to 30 min. Thus, the reserpine-like effect of

◄

Fig. 9 a–f. The influence of the reserpine-like compound Ro 4-1284 on the effects of tyramine (**a** and **d**), amezinium (**b** and **e**) and bethanidine (**c** and **f**). The adrenergic nerve endings of rat vasa deferentia were intact (but COMT was inhibited). They were loaded with ^{3}H-noradrenaline as in Fig. 7 and then washed out with amine-free solution. a, b and c: strips were exposed to 10 μmol/l tyramine (**a**), 10 μmol/l amezinium (**b**) or 1000 μmol/l bethanidine (**c**). **d**, **e** and **f**: strips were first exposed to Ro 4-1284 (0.1 μmol/l in **e**, 10 μmol/l in **d** and **f**), then to the three amines (in the continued presence of Ro 4-1284). Ordinates: fractional rate of loss (min^{-1}) (= rate of efflux/tritium content of tissue determined at that time); abscissae: time after onset of wash-out (in min). Shown are efflux curves for tritium (●——●), ^{3}H-noradrenaline (○——○) and ^{3}H-DOPEG (+——+), means (±S.E. as vertical bars) of 4 experiments each. Results taken from LANGELOH and TRENDELENBURG (1987)

guanethidine is developing too slowly to enhance the releasing effect of this amine. It is of interest to note that Muskus (1964) classified guanethidine as an agent that affected the "tyramine-sensitive pool of noradrenaline" more than reserpine did. As is evident from this discussion, guanethidine is both reserpine- and tyramine-like.

IV. Mobilization of Vesicular ^{3}H-Noradrenaline

This mechanism is of crucial importance in determining whether a substrate of uptake$_1$ is a "good" or a "poor" releaser of ^{3}H-noradrenaline (from intact adrenergic nerve endings). As pointed out in the text accompanying Figs. 1 and 9, it is highly likely that storage vesicles leak ^{3}H-noradrenaline (into the axoplasm) at a rather high rate. The highly efficient vesicular re-uptake compensates for most of this leakage, so that the rate of net leakage from the storage vesicles is low. In this situation, very substantial increases in the rate of net leakage can be brought about simply by inhibition of vesicular re-uptake. It is likely that indirectly acting amines "mobilize" vesicular noradrenaline by successfully competing for vesicular uptake (and thereby inhibiting the vesicular re-uptake of ^{3}H-noradrenaline). If this proposal is correct, then a good releaser (like tyramine) should have a high affinity to vesicular uptake and poor releasers (like amezinium, guanethidine and bethanidine) should have a very low (or no) affinity. In full agreement with this proposal Michalke et al. (1987) found the K_m for vesicular uptake to be low (2 to 15 µmol/l) for tyramine, 5-HT, noradrenaline and metaraminol, while that for bethanidine, amezinium and guanethidine was 200 µmol/l or higher. These findings support the view that the "mobilization" of vesicular noradrenaline by certain substrates of uptake$_1$ may well be brought about by saturation of the vesicular uptake mechanism by appropriate axoplasmic concentrations of high-affinity substrates of vesicular uptake.

Reserpine-like agents also inhibit this uptake mechanism. As explained in the text accompanying Fig. 1b, it is likely that Ro 4-1284 increases the efflux of tritium from the adrenergic nerve ending by increasing the rate of net leakage from the storage vesicles. The crucial difference between the reserpine-like compounds and the indirectly acting amines is found in the fact that most reserpine-like compounds enter the adrenergic nerve ending through passive diffusion across the axonal membrane, while the indirectly acting amines are transported into the adrenergic nerve endings.

When axoplasmic amine concentrations approach or even exceed 10 mmol/l, there is evidence for a massive net leakage of ^{3}H-noradrenaline from storage vesicles (see (+)-amphetamine in Fig. 7). It is likely that we now face a mechanism proposed by Johnson et al. (1982) and Phillips (1982): if very large amounts of a highly lipophilic indirectly acting amine enter the storage vesicles through passive diffusion, the pH-gradient (inside low) is reduced or even abolished. Since this reduces the protonation of the vesicular ^{3}H-noradrenaline, massive efflux occurs. However, the results presented here indicate that this mechanism does *not* come into play when the indirectly acting amines are mainly transported into adrenergic nerve endings.

In the discussion of Fig. 1 as well as in that of the "mobilization" of vesicular noradrenaline, the view has been expressed that there is always a net leakage of vesicular noradrenaline into the axoplasm (see Fig. 1) and that certain indirectly acting amines can greatly increase the rate of this net leakage (tyramine). It should be realized that this view is directly opposed to that of MARON et al. (1983). These authors proposed that the granular membrane is impermeable to intragranular noradrenaline or 5-HT; because of the low intragranular pH (see above) it constitutes a "pH-dependent kinetic barrier", so that the intragranular pH functions as a "gating mechanism". Unfortunately, this proposal is irreconcilable with the net leakage (from the storage vesicles into the axoplasm) that MUST take place in Fig. 1 (spontaneous efflux prior to the administration of any drug) to account for the spontaneous efflux of ^{3}H-noradrenaline and ^{3}H-DOPEG. Moreover, this proposal is also irreconcilable with any mobilization of vesicular ^{3}H-noradrenaline by *micro*molar concentrations of axoplasmic tyramine, since *milli*molar concentrations of this amine are needed for any appreciable increase in the intragranular pH (LANGER and BURGER, personal communication).

V. Exocytotic Release in the Absence of Extracellular Calcium

It has been demonstrated that a pronounced increase in the axoplasmic sodium concentration (induced by veratrine) is able to induce an exocytotic release of ^{3}H-noradrenaline even in the absence of extracellular calcium, presumably by mobilizing intracellular calcium. This conclusion was based on the observation that this veratrine-induced release was reduced by the presence of 1 μmol/l clonidine and enhanced by the presence of 3 μmol/l phentolamine (neocortical slices of rat brain, calcium-free medium; SCHOFFELMEER and MULDER 1983). However, this release mechanism does not seem to exist in the rat vas deferens, since 3 mmol/l ouabain (STUTE and TRENDELENBURG 1984) and 100 μmol/l veratridine plus 1 mmol/l ouabain (BÖNISCH and TRENDELENBURG 1987) induced a release of ^{3}H-noradrenaline that was modulated neither by 1 μmol/l clonidine nor by 3 μmol/l phentolamine (tissues from unpretreated rats, COMT inhibited—STUTE and TRENDELENBURG 1984; tissues from reserpine-pretreated rats, MAO and COMT inhibited—BÖNISCH and TRENDELENBURG 1987). It should be noted that, in the latter experiments, slightly more than 50% of the "neuronal ^{3}H-noradrenaline" was distributed into the storage vesicles (see Sect. C.I). Hence, it is unlikely that exocytotic release is obtained in rat vasa deferentia exposed to calcium-free medium.

VI. The Axoplasmic Compartment(s)

About two decades ago CROUT (1964) proposed that the axoplasm (and the storage vesicles) of the adrenergic nerve ending represent a multi-compartment system. For instance, if ^{3}H-noradrenaline leaks from superficially located storage vesicles, the diffusion distance to the axonal membrane is so short that any ^{3}H-noradrenaline molecule has a good chance of reaching the axonal membrane (and the carrier, if it is on the inside of the membrane).

However, when ^{3}H-noradrenaline leaks from centrally located storage vesicles, diffusion distances (and exposure to mitochondrial MAO) are such that the probalitity of any given molecule of ^{3}H-noradrenaline reaching the axonal membrane is very low. This proposal explains, for instance, why reserpine-like compounds lower the ratio NA/DOPEG (see Fig. 1b): when they increase the leakage of ^{3}H-noradrenaline from *all* (superficial and central) storage vesicles, a considerable proportion of the ^{3}H-noradrenaline is deaminated while diffusing towards the axonal membrane.

Thus, it is proposed that only the axoplasm (and the storage vesicles) in the immediate vicinity of the axonal membrane are of crucial importance for the ^{3}H-noradrenaline-releasing effects of indirectly acting amines. The following considerations support this proposal.

a) If the indirectly acting amine fails to block (or saturate) neuronal MAO, only that ^{3}H-noradrenaline, which leaks from storage vesicles in the vicinity of the axonal membrane, has a chance of reaching the axonal membrane.
b) When MAO is intact, a steep concentration gradient for indirectly acting amines like tyramine must exist within the adrenergic nerve ending. Hence, the most pronounced "mobilizing" effects of tyramine are likely to involve the storage vesicles that are in close proximity to the axonal membrane.
c) While the co-transport of Na^+ and Cl^- is also involved in the ^{3}H-noradrenaline-releasing effects of indirectly acting amines (see Sect. F), it is highly unlikely that enough Na^+ or Cl^- is co-transported to cause a general increase in the axoplasmic Na^+ or Cl^- concentration. However, it is feasible that the co-transport increases the Na^+ and Cl^- concentrations exclusively at the inside of the axonal membrane.

Even after inhibition of neuronal MAO, the axoplasm does not seem to represent a single distribution compartment for tyramine, presumably because a concentration gradient is generated by the vesicular uptake of tyramine. BRANDAO et al. (1981) exposed strips of the dog's saphenous vein (MAO, COMT and uptake$_2$ inhibited; preloaded with ^{3}H-noradrenaline and then washed out) to increasing concentrations of tyramine (0.49 to 3240 μmol/l) for 100 min. From the efflux curves for tritium the authors calculated the "bound fraction", i.e. that amount of tissue ^{3}H-noradrenaline that fails to contribute to the efflux of tritium (measured during 100 min). In the absence of tyramine, 85% of tissue ^{3}H-noradrenaline belonged to the "bound fraction" (i.e. failed to contribute to the efflux of tritium). In the presence of increasing concentrations of tyramine, the "bond fraction" declined more and more, and in the presence of the highest concentration of tyramine it failed to differ from zero.

The same dependence of the "bound fraction" on the concentrations of either (+)-amphetamine or tyramine was observed with vasa deferentia obtained from unpretreated rats (LANGELOH and TRENDELENBURG 1987): the "bound fraction" declined with increasing concentrations of these amines. Hence, it is false to speak of "tyramine-sensitive" and "tyramine-resistant" pools of ^{3}H-noradrenaline, since this is a function of the concentration of tyramine. Given high enough concentrations, tyramine can "mobilize" all the vesicular ^{3}H-noradrenaline.

Finally, tachyphylaxis to tyramine might be best interpreted by restriction of the discussion to the most superficially located storage vesicles of the adrenergic nerve ending. Repeated administrations of tyramine induce progressively decreasing responses, presumably because of the exhaustion of a crucial pool of ^{3}H-noradrenaline (or endogenous noradrenaline). Attempts to relate the development of tachyphylaxis to changes in *total* neuronal ^{3}H- (or endogenous) noradrenaline have not been very successful (see AXELROD et al. 1962; LEE et al. 1967). However, this relation is bound to be poor if tachyphylaxis is due exclusively to the exhaustion of the transmitter stores in the most superficially located storage vesicles. If one considers the fact that tachyphylaxis to tyramine is usually observed after moderate doses of tyramine, and if one takes into consideration that moderate doses of tyramine are unable to "mobilize" the transmitter from *all* storage vesicles of the adrenergic nerve endings, it is, indeed, highly likely that tachyphylaxis is due to the exhaustion of transmitter in the storage vesicles that are in the immediate neighbourhood of the axonal membrane. Moreover, since tachyphylaxis is well known to be easily and quickly abolished by an exposure of the tissue to exogenous noradrenaline, it is again likely that the most superficial layers of storage vesicles are involved; they are most easily re-loaded by this procedure.

K. The Release by Indirectly Acting Amines of Dopamine from Dopaminergic Nerve Endings

It is of interest to note that recent experiments by PARKER and CUBEDDU (1986a, b) revealed striking similarities with the results reported here for adrenergic nerve endings. After pretreatment of the animals with reserpine and pargyline (and after inhibition of COMT), equieffective concentrations (release of ^{3}H-dopamine) of six amines were strictly correlated with K_m-values (uptake into striatal slices) and $EC_{50} \gg K_m$. Moreover, evidence was presented that some of these amines are able to "mobilize" vesicularly stored ^{3}H-dopamine (subsequent to their inward transport), while very high concentrations of lipophilic amines (10 mmol/l β-phenylethylamine, for instance) seemed to be able to diffuse into the nerve endings in such amounts that the subsequent diffusion into the storage vesicles increased vesicular efflux (presumably through an increase of the pH inside the storage vesicles). Moreover, while some amines (like octopamine) were good mobilizers of vesicular dopamine, others (like β-phenylethylamine) appeared to lack this ability.

On the other hand, equally striking differences were reported. As pretreatment with reserpine prevents the synthesis of noradrenaline in adrenergic nerve endings, it is well known that it abolishes indirect effects mediated by adrenergic nerve endings (BURN and RAND 1958). However, this procedure not only fails to prevent the synthesis of dopamine, it actually increases it. As a consequence, there is enough dopamine in the axoplasm of dopaminergic nerve endings of reserpine-pretreated animals to permit a sustained release of dopamine in response to indirectly acting amines (like (+)-amphetamine).

Thus, for both types of nerve endings, the mechanism of action of indirectly acting amines is strikingly similar in that there is a) a multifactorial induction of outward transport of the transmitter and b) a "mobilization" of vesicularly stored transmitter, though the contribution by (b) may differ considerably from one substrate of the uptake mechanism to the next.

L. Conclusions

Currently available evidence strongly favours the view that indirectly acting sympathomimetic amines must a) be substrates of uptake$_1$ (in order to induce an outward transport of ^{3}H- (or endogenous) noradrenaline) and b) be able to "mobilize" vesicularly stored ^{3}H- (or endogenous) noradrenaline. Moreover, it is important to realize that substrates of uptake$_1$ induce an outward transport of ^{3}H- (or endogenous) noradrenaline not only through the mechanism of facilitated exchange diffusion (which amounts to an increase in the V_{max} for outward transport), but also via an increase in the affinity of axoplasmic noradrenaline to the inside carrier (as a consequence of the co-transport of Na^+ and Cl^-) and via inhibition of neuronal re-uptake. Because of this multifactorial mechanism, $K_m < EC_{50}$ for the release of ^{3}H-noradrenaline from the adrenergic nerve endings of tissues obtained from reserpine- and pargyline-pretreated animals. Moreover, the strict correlation between increasing K_m and increasing V_{max} (for uptake$_1$ of various substrates) leads to a limitation of the maximum releasing effects of high-affinity substrates of uptake$_1$ (again clearly seen in tissues obtained from reserpine- and pargyline-pretreated animals).

Acknowledgements. The work quoted here was supported by the Deutsche Forschungsgemeinschaft (SFB 176).

M. References

Axelrod J, Gordon E, Hertting G, Kopin IJ, Potter LT (1962) On the mechanism of tachyphylaxis to tyramine in the isolated rat heart. Brit J Pharmacol 19:56-63

Azzaro AJ, Ziance RJ, Rutledge CO (1974) The importance of neuronal uptake of amines for amphetamine-induced release of ^{3}H-norepinephrine from isolated brain tissue. J Pharmacol Exp Ther 189:110-118

Benedetti MS, Boucher T, Carlsson A, Fowler CJ (1983) Intestinal metabolism of tyramine by both forms of monoamine oxidase in the rat. Biochem Pharmacol 32:47-52

Bönisch H (1984) The transport of (+)-amphetamine by the neuronal noradrenaline carrier. Naunyn-Schmiedeberg's Arch Pharmacol 327:267-272

Bönisch H (1986) The role of co-transported sodium in the effect of indirectly acting sympathomimetic amines. Naunyn-Schmiedeberg's Arch Pharmacol 332: 135-141

Bönisch H, Rodrigues-Pereira E (1983) Uptake of ^{14}C-tyramine and release of extravesicular ^{3}H-noradrenaline in isolated perfused rabbit hearts. Naunyn-Schmiedeberg's Arch Pharmacol 323:233-244

Bönisch H, Trendelenburg U (1987) Veratridine-induced outward transport of ^{3}H-noradrenaline from adrenergic nerves of the rat vas deferens. Naunyn-Schmiedeberg's Arch Pharmacol 336:621-630

Bönisch H, Graefe K-H, Keller B (1983) Tetrodotoxin-sensitive and -resistant effects of veratridine on the noradrenergic neurone of the rat vas deferens. Naunyn-Schmiedeberg's Arch Pharmacol 324:264-270

Bönisch H, Fuchs G, Graefe K-H (1986) Sodium-dependence of the saturability of carrier-mediated noradrenaline efflux from noradrenergic neurones in the rat vas deferens. Naunyn-Schmiedeberg's Arch Pharmacol 332:131-134 (1986)

Brandao F, Rodrigues-Pereira E, Monteiro JG, Davidson R (1981) A kinetic study of the release of noradrenaline by tyramine. Naunyn-Schmiedeberg's Arch Pharmacol 318:83-87

Burn JH, Rand MJ (1958) The action of sympathomimetic amines in animals treated with reserpine. J Physiol (Lond) 144, 314-336

Cassis L, Ludwig J, Grohmann M, Trendelenburg U (1986) The effect of partial inhibition of monoamine oxidase on the steady-state rate of deamination of ^{3}H-catecholamines in two metabolizing system. Naunyn-Schmiedeberg's Arch Pharmacol 333:253-261

Clarke DE (1980) Amphetamine and monoamine oxidase inhibition: an old idea gains new acceptance. TIPS 1:312-313

Crout JR (1964) The uptake and release of ^{3}H-norepinephrine by the guinea-pig heart in vivo. Naunyn-Schmiedeberg's Arch Pharmacol 248:85-98

Friedrich U, Bönisch H (1986) The neuronal noradrenaline transport system of PC-12-cells: Kinetic analysis of the interaction between noradrenaline, Na^{+} and Cl^{-} in transport. Naunyn-Schmiedeberg's Arch Pharmacol 333:246-252

Henseling M, Trendelenburg U (1978) Stereoselectivity of the accumulation and metabolism of noradrenaline in rabbit aortic strips. Naunyn-Schmiedeberg's Arch Pharmacol 302:195-206

Henseling M, Rechtsteiner D, Trendelenburg U (1978) The influence of monoamine oxidase and catechol-O-methyl transferase on the distribution of ^{3}H-($\pm$)-noradrenaline in rabbit aortic strips. Naunyn-Schmiedeberg's Arch Pharmacol 302:181-194

Johnson RG, Scarpa A (1976) Internal pH of isolated chromaffin vesicles. J Biol Chem 251:2189-2191

Johnson RG, Carty SE, Hayflick S, Scarpa A (1982) Mechanism of accumulation of tyramine, metaraminol, and isoproterenol in isolated chromaffin granules and ghosts. Biochem Pharmacol 31:815-823

Langeloh A, Trendelenburg U (1987) The mechanism of the ^{3}H-noradrenaline releasing effect of various substrates of uptake$_1$: Role of monoamine oxidase and of vesicularly stored ^{3}H-noradrenaline. Naunyn-Schmiedebergs Arch Pharmacol 336:611-620

Langeloh A, Bönisch H, Trendelenburg U (1987) The mechanism of the ^{3}H-noradrenaline-releasing effect of various substrates of uptake$_1$: Multifactorial induction of outward transport. Naunyn-Schmiedeberg's Arch Pharmacol 336:602-610

Lee F-L, Weiner N, Trendelenburg U (1967) The uptake of tyramine and formation of octopamine in normal and tachyphylactic rat atria. J Pharmacol Exp Ther 155:211-222

Leitz FH, Stefano FJE (1971) The effect of tyramine, amphetamine and metaraminol

on the metabolic disposition of ^{3}H-norepinephrine released from the adrenergic neuron. J Pharmacol Exp Ther 178:464–473
Levi G, Roberts PJ, Raiteri M (1976) Release and exchange of neurotransmitters in synaptosomes: effects of the ionophore A23187 and of ouabain. Neurochem Res. 1:409–416
Liang NY, Rutledge CO (1982) Evidence for carrier-mediated efflux of dopamine from corpus striatum. Biochem Pharmacol 31:2479–2484
Lindmar R, Löffelholz K (1974) The neuronal efflux of noradrenaline: dependency on sodium and facilitation by ouabain. Naunyn-Schmiedeberg's Arch Pharmacol 284:93–100
Lindmar R, Löffelholz K, Muscholl E (1967) Unterschiede zwischen Tyramin und Dimethylphenylpiperazin in der Ca^{2+}-Abhängigkeit und im zeitlichen Verlauf der Noradrenalin-Freisetzung am isolierten Kaninchenherzen. Experientia (Basel) 23:933–934
Mack F, Bönisch H (1979) Dissociation constants and lipophilicity of catecholamines and related compounds. Naunyn-Schmiedeberg's Arch Pharmacol 310:1–9
Maron R, Stern Y, Kanner BI, Schuldiner S (1983) Functional asymmetry of the amine transporter from chromaffin granules. J Biol Chem 258:11476–11481
Michalke W, Langer R, Burger A (1987) Mobilization of biogenic amines from chromaffin-granule ghosts by indirectly acting sympathomimetic amines. Naunyn-Schmiedeberg's Arch Pharmacol 335:R81
Muscholl E (1966) Indirectly acting sympathomimetic amines. Pharmacol Rev 18:551–559
Muskus AJ (1964) Evidence for different sites of action of reserpine and guanethidine. Naunyn-Schmiedebergs Arch Pharmacol 248:498–513
Parker EM, Cubeddu LX (1986a) Effects of d-amphetamine and dopamine synthesis inhibitors on dopamine and acetylcholine neurotransmission in the striatum. I. Release in the absence of vesicular transmitter stores. J Pharmacol Exp Ther 237:179–192
Parker EM, Cubeddu LX (1986b) Effects of d-amphetamine and dopamine synthesis inhibitors on dopamine and acetylcholine neurotransmission in the striatum. II. Release in the presence of vesicular transmitter stores. J Pharmacol Exp Ther 237:193–203
Paton DM (1973) Evidence for carrier-mediated efflux of noradrenaline from the axoplasm of adrenergic nerves in rabbit atria. J Pharm Pharmacol 25:265–267
Paton DM (1981) Effect of cocaine on the efflux of noradrenaline, octopamine and metarminol in rabbit atria. IRCS Med Sci 9:128
Phillips JH (1982) Dynamic aspects of chromaffin granule structure. Neuroscience 7:1595–1609
Raiteri M, Levi G, Federico R (1974) d-Amphetamine and the release of ^{3}H-norepinephrine from synaptosomes. Europ J Pharmacol 28:237–240
Ross SB, Kelder D (1976) Effect of veratridine on the fluxes of ^{3}H-noradrenaline and ^{3}H-bretylium in the rat vas deferens in vitro. Naunyn-Schmiedeberg's Arch Pharmacol 295:183–189
Ross SB, Kelder D (1979) Release of ^{3}H-noradrenaline from rat vas deferens under various in vitro conditions. Acta physiol scand 105:338–349
Rutledge CO, Vollmer S (1979) Evidence for carrier mediated efflux of norepinephrine displaced by amphetamine. In: Usdin E, Kopin IE, Barchas J (eds) Catecholamines: Basic and Clinical Frontiers, vol 1. Pergamon Press, New York, pp 304–306
Sammet S, Graefe K-H (1979) Kinetic analysis of the interaction between noradrena-

line and Na^+ in neuronal uptake: kinetic evidence for co-transport. Naunyn-Schmiedeberg's Arch Pharmacol 309:99-107

Sanchez-Armass S, Orrego F (1977) A major role of chloride in (^3H)-noradrenaline transport by rat heart adrenergic nerves. Life Sci 20:1829-1838

Schömig E, Trendelenburg U (1987) Simulation of outward transport of neuronal ^3H-noradrenaline with the help of a two-compartment model. Naunyn-Schmiedeberg's Arch Pharmacol 336:631-640

Schoffelmeer ANM, Mulder AH (1983) ^3H-noradrenaline release from rat neocortical slices in the absence of extracellular Ca^{2+} and its presynaptic α_2-adrenergic modulation. A Study on the possible role of cyclic AMP. Naunyn-Schmiedeberg's Arch Pharmacol 323:188-192

Stefano JFE, Trendelenburg U (1984) Saturation of monoamine oxidase by intraneuronal noradrenaline accumulation. Naunyn-Schmiedeberg's Arch Pharmacol 328:135-141

Stute N, Trendelenburg U (1984) The outward transport of axoplasmic noradrenaline induced by a rise of the sodium concentration in the adrenergic nerve endings of the rat vas deferens. Naunyn-Schmiedeberg's Arch Pharmacol 327:124-132

Thoa NB, Wooten GF, Axelrod J, Kopin IJ (1975) On the mechanism of release of norepinephrine from symphathetic nerves induced by depolarizing agents and sympathomimetic drugs. Mol Pharmacol 11:10-18

Thoenen H, Hürlimann A, Haefely W (1968) Mechanism of amphetamine accumulation in the isolated perfused heart of the rat. J Pharm Pharmacol 20:1-11

Thoenen H, Hürlimann A, Haefely A (1969) Cation dependence of the noradrenaline-releasing action of tyramine. European J Pharmacol 6:29-37

Tipton KF, Fowler CJ, Houslay MD (1982) Specificities of the two forms of monoamine oxidase. In: Kamijo K, Usdin U, Nagatsu T (eds) Monoamine Oxidase—Basic and Clinical Frontiers Excerpta Medica, Amsterdam, pp 87-99

Trendelenburg U (1972) Classification of sympathomimetic amines. In: Blaschko H, Muscholl E (eds) Catecholamines. Springer, Berlin Heidelberg New York, pp 336-362 (Handbook of Experimental Pharmacology, vol 33)

Trendelenburg U (1978) Release induced by phenethylamines. In: Paton DM (ed) The Release of Catecholamines from Adrenergic Neurons. Pergamon Press, Oxford, pp 333-354

CHAPTER 6

The Extraneuronal Uptake and Metabolism of Catecholamines

U. TRENDELENBURG

A. Introduction

I. The Extraneuronal System

Twenty years ago IVERSEN (1965 a) discovered that, in the rat heart, noradrenaline is a substrate not only of the well-known neuronal uptake ($uptake_1$) but also of a second uptake mechanism ($uptake_2$) which clearly differed from $uptake_1$. The new uptake mechanism was first believed to be associated with adrenergic nerve endings (IVERSEN 1965 a), but fluorescence microscopy soon revealed that $uptake_2$ resulted in the accumulation of noradrenaline in myocardial cells (MALMFORS 1967; EHINGER and SPORRONG 1968; FARNEBO 1968); autoradiography later confirmed that myocardial cells are the main distribution compartment for isoprenaline perfused through the rat heart (AZEVEDO et al. 1983).

When discovered by IVERSEN (1965 a), $uptake_2$ in the rat heart was characterized by two findings. First, in contrast to $uptake_1$, $uptake_2$ had a low affinity for noradrenaline, its K_m being about 1000 times higher than that for $uptake_1$ (IVERSEN 1963, 1965 a). Second, at very low amine concentrations $uptake_2$ appeared not to function at all. From these two observations, the concept emerged that $uptake_2$ can play its role only at very high noradrenaline concentrations.

This concept had to be revised when it became apparent that $uptake_2$ is only the first step in a system that also involves intracellular COMT and MAO. At very low amine concentrations, there is no accumulation of noradrenaline inside the extraneuronal cells, because it is metabolized, primarily by intracellular COMT, but also by intracellular MAO. Indeed, it was then realized that the extraneuronal cells function mainly as an "extraneuronal O-methylating system". As the highly polar noradrenaline molecule is hardly able to diffuse through cell membranes, $uptake_2$ (as the first step) is an essential prerequisite for the intracellular metabolism of the amine (second step). Low outside concentrations of catecholamines (of up to about 1 µmol/l) mean also low inside concentrations; as the latter fail to saturate the intracellular enzymes, the whole system then functions as an "irreversible, metabolizing site of loss". However, high outside concentrations can lead to such high rates of inward transport that the intracellular enzymes are saturated. In that case, the whole system functions as a "reversible, accumulating site of loss". As IVERSEN (1965 a) did not inhibit COMT and MAO, he needed high concentrations of noradrenaline to induce an accumulation of noradrenaline inside the extraneuronal cells. However, the physiological function of the extraneuronal O-methylating system is likely to be restricted to those low extracellular noradrenaline concentrations at which uptake$_2$ is followed by highly efficient O-methylation (see TRENDELENBURG 1980).

Myocardial cells are the main sites of extraneuronal uptake and metabolism in the rat heart (see above). In an analogous manner, fluorescence microscopy (GILLESPIE and TOWART 1973) and autoradiography (BRANCO et al. 1981) revealed that smooth muscle cells have this function in blood vessels.

This review will deal mainly with uptake$_2$ and with the extraneuronal O-methylating system. It will concentrate on the evidence obtained with the rat heart, an organ that turned out to be especially useful for the analysis of this system. However, reference will be made to other organs, tissues or species, to document that uptake$_2$ and the extraneuronal O-methylating system of other organs and tissues have striking similarities with those of the rat heart. Moreover, this review aims at a comparison of the two noradrenaline-transporting mechanism, uptake$_1$ and uptake$_2$.

II. Other Non-neuronal System

In comparison with other uptake mechanisms, uptake$_2$ is characterized by a) preferring isoprenaline as a substrate (while this amine is a very poor substrate of uptake$_1$; IVERSEN 1967), b) being highly sensitive to various corticosteroids and O-methylated catecholamines (but not to the selective inhibitors of uptake$_1$, cocaine and desipramine) and c) being independent of the presence of extracellular Na^+ and Cl^- (while uptake$_1$ is crucially dependent on these ions).

Three other non-neuronal transport systems have been described; according to the above criteria, either they clearly differ from uptake$_2$ or they may well eventually turn out to differ. The three transport systems are as follows:

a) Catecholamines are taken up by, and are metabolized in, the endothelium of small vessels in the perfused lung. This uptake mechanism is highly sensitive to cocaine, it is dependent on the presence of extracellular NaCl, and isoprenaline is not the preferred catecholamine substrate (GILLIS 1976; BRYAN and O'DONNELL unpublished observations). Hence, this non-neuronal uptake mechanism resembles uptake$_1$ more than uptake$_2$, and it is associated with both, COMT and MAO (GILLIS 1976).
b) The gland epithelium in rabbit endometrium (KENNEDY and DE LA LANDE 1986) and the fibroblasts of the dental pulp of the rabbit (PARKER et al. 1987) both possess a cocaine-sensitive transport mechanism for noradrenaline that is resistant to inhibition by corticosteroids. The transport system in rabbit endometrium differs from uptake$_2$ also by being dependent on extracellular Na^+, but it resembles uptake$_2$ in having a rather high K_m for noradrenaline (78 µmol/l; KENNEDY and DE LA LANDE 1987). A definitive classification is not yet possible.
c) In the rat cerebral cortex there is a transport system which translocates isoprenaline (and other catecholamines; WILSON and TRENDELENBURG 1988)

as well as O-methylated catecholamines (HENDLEY et al. 1970; HENDLEY 1976) and is associated with COMT; this transport system is independent of extracellular NaCl (WILSON and TRENDELENBURG 1988) and, therefore, resembles uptake$_2$. However, its sensitivity to the selective inhibitors of uptake$_2$ (corticosterone and 3-O-methylisoprenaline) is very low (WILSON and TRENDELENBURG 1988). Hence, a definitive classification is at present impossible.
The present chapter does *not* deal with these "other non-neuronal transport mechanism", it concentrates on uptake$_2$.

B. Uptake$_2$

I. Inward Transport

1. Definition and Occurrence of Uptake$_2$

Uptake$_2$ is a saturable transport system characterized by a) a rather high K_m for various catecholamines (see below), b) a characteristic substrate spectrum (see below), and c) its sensitivity to three groups of inhibitors: O-methylated catecholamines (SALT 1972; MIREYLEES and FOSTER 1973; BÖNISCH et al. 1974), various corticosteroids (IVERSEN and SALT 1970; SALT 1972) and various β-haloalkylamines (IVERSEN et al. 1972). As corticosteroids have often been used as specific inhibitors of uptake$_2$, the term "corticosteroid-sensitive" will be used in this review, even if another inhibitor was actually used.

In the rat heart (and other tissues) both corticosterone and 3-O-methylisoprenaline (OMI) inhibit uptake$_2$ with a low K_i (about 2 μmol/l, BÖNISCH et al. 1974; GROHMANN and TRENDELENBURG 1984; see also O'DONNELL and REID 1984, for guinea-pig trachealis smooth muscle).

Uptake$_2$ has been identified in a variety of peripheral organs and tissues:

a) in myocardial cells of the rat and cat (IVERSEN 1965a; GRAEFE 1981),
b) in vascular smooth muscle (human umbilical artery: GULATI and SIVARAMAKRISHNA 1975; rat portal vein: HEDNER et al. 1981; rat coronary arterioles and venules: AZEVEDO et al. 1983; rabbit aorta: LEVIN 1974; rabbit coronary arteries: DE LA LANDE et al. 1974; rabbit ear artery: GILLESPIE and TOWART 1973; HEAD et al. 1980; cat coronary artery: CORNISH et al. 1978; dog saphenous vein: PAIVA and GUIMARAES 1978; dog mesenteric artery: GARRETT and BRANCO 1977; dog coronary artery: O'DONNELL and REID 1984).
c) in non-vascular smooth muscle (rabbit spleen: GILLESPIE and MUIR 1970; rat trachealis muscle: O'DONNELL and REID 1984; guinea-pig trachealis muscle: ANNING et al. 1979; GARLAND et al. 1981; guinea-pig uterus: ANNING et al. 1979; rabbit trachealis muscle: O'DONNELL and REID 1984; rabbit colon: GILLESPIE and PREUNER 1976; rabbit iris: PATIL and TRENDELENBURG 1982; rabbit uterus: KENNEDY et al. 1984; KENNEDY and DE LA LANDE 1984; cat nictitating membrane: GRAEFE and TRENDELENBURG 1974; cat

tracheal muscle: O'DONNELL and REID 1984; cultured bovine embryonic tracheal cells: POWIS 1973),

d) in additional tissues (rat submaxillary gland: ALMGREN and JONASON 1976; duodenal mucosa of rat: LANDSBERG 1976; brown fat of rat: CHINET and DURAND 1979).

For any given species, uptake$_2$ is not uniformely distributed in the various tissues. Although the reason for this heterogeneous distribution is unknown, it has been proposed that uptake$_2$ may be restricted to those tissues that have β_2-adrenoceptors (BRYAN et al. 1981). A few striking differences should be enumerated: in the guinea pig uptake$_2$ is prominent in the trachealis smooth muscle, but poorly developed in the heart (ANNING et al. 1979; BÖNISCH and TRENDELENBURG 1974); in the rabbit it is well developed in trachealis smooth muscle (O'DONNELL and REID 1984), aorta (LEVIN 1974) and uterus (KENNEDY et al. 1984), but very sparse in the heart (GRAEFE et al. 1978; GRAEFE 1981).

2. The Stereoselectivity of Uptake$_2$

For the rat heart, IVERSEN found no stereoselectivity for uptake$_2$ when he compared initial rates of uptake$_2$ of (−)- and (+)-noradrenaline (IVERSEN 1965a) and of (−)- and (±)-adrenaline (IVERSEN 1965b).

However, BARONE et al. (1983, 1985) found the O-methylation of (−)-isoprenaline in the rabbit aorta to exceed that of (+)-isoprenaline; this finding initiated a re-examination of the stereoselectivity of uptake$_2$. In three organs (guinea-pig trachealis smooth muscle, rat heart and rabbit aorta) stereoselectivity of uptake$_2$ was observed (BRYAN and O'DONNELL 1984a; GROHMANN and TRENDELENBURG 1984; HENSELING 1984); in all three organs the degree of stereoselective preference for the (−)-isomer declined in the order: isoprenaline > adrenaline > noradrenaline (with no stereoselectivity for noradrenaline

Table 1. Substrate specificity and stereoselectivity of uptake$_2$ of catecholamines

Tissue	dobutamine		isoprenaline		adrenaline		noradrenaline		dopamine
	(−)	(+)	(−)	(+)	(−)	(+)	(−)	(+)	
rat heart[a]	4.80	10.6[c]	29.8	125[c]	87.3	172[c]	224	173	498
guinea-pig trachealis smooth muscle[b]	–	–	67.7	331[c]	127	313[c]	207	334[c]	–

Shown are mean values (in µmol/l)

[a] IC_{50} values determined for unlabelled catecholamines (inhibiting initial rates of uptake of a tracer concentration of ^{3}H-(±)-isoprenaline) in rat perfused heart (GROHMANN and TRENDELENBURG 1984)

[b] K_m-values determined for unlabelled catecholamines (fluorescence histochemistry) in incubated tissues (BRYAN and O'DONNELL 1984a)

[c] $P < 0.05$ for comparison between the two stereoisomers of the same amine

in rat heart and rabbit aorta) (Table 1). However, it should be emphasized that the degree of stereoselectivity was small, even for isoprenaline (4- to 5-fold difference, see Table 1).

Another catecholamine, dobutamine, has its asymmetric carbon in the butyl substituent on the nitrogen. Stereoselectivity (favouring the (−)-isomer) was also observed for this agent (Table 1).

The occurrence of stereoselectivity for uptake$_2$ is in favour of the view that a carrier-mediated uptake process is involved.

3. The Substrate Spectrum of Uptake$_2$

K_m-values for uptake$_2$ can be obtained from an analysis of the saturability of the uptake of a substrate; this usually involves the use of radioactively labelled substrates which might be subject to an isotope effect (see TRENDELENBURG et al. 1983a). This problem is circumvented by the determination of that concentration of an unlabelled substrate which halves the initial rate of uptake of a marker of uptake$_2$ (IC_{50}). Such an IC_{50} equals K_m, provided one deals with transported substrates. IC_{50}-values were determined in the rat heart by BURGEN and IVERSEN (1965) and by GROHMANN and TRENDELENBURG (1984). For the details of the structure-action relations that emerged from these complementary studies, the reader is referred to the original publications. Here, it suffices to state that the structure-action relation characterizing uptake$_2$ is quite dissimilar from those known for either uptake$_1$ or the various subtypes of α- and β-adrenoceptors. Thus, in conjunction with stereoselectivity, the evidence is in favour of the view that uptake$_2$ is carrier-mediated.

For the catecholamines, the K_m typically increases in the order: (−)-dobutamine < (−)-isoprenaline < (−)-adrenaline < (−)-noradrenaline < dopamine (Table 1). It is legitimate to speak of a "K_m" for dobutamine, since GROHMANN and TRENDELENBURG (1985) provided evidence that this catecholamine is transported by uptake$_2$ (and O-methylated by COMT). It should be noted that there is a 100-fold difference between the K_m-values for dobutamine and dopamine (Table 1).

Comparison of the values obtained in the rat heart with those obtained by BRYAN and O'DONNELL (1984a) for guinea-pig trachealis smooth mucle shows a similar ranking order, but somewhat higher values for the latter than for the former (Table 1).

It is proposed that diffusion distances may explain this phenomenon. As the catecholamine diffuses into the extracellular space (either from the medium surrounding the incubated smooth muscle or from the capillaries of the perfused heart), uptake$_2$ generates a concentration gradient. Hence, cells at some distance from the surrounding medium (either in the centre of the incubated tissue or at some distance from the capillary) are exposed to a concentration of the catecholamine that is lower than that in the incubation medium or the perfusion fluid. Hence, if one wants to half-saturate uptake$_2$ *in the whole tissue*, the concentration in the incubation medium or perfusion fluid has to exceed the K_m for uptake$_2$ of each individual cell (see also GREEN 1976; EBNER and WAUD 1978; HENSELING 1983a). Moreover, if the average diffusion distances are short (as in a perfused tissue), this factor is bound to be much less important than in a tissue that involves considerable diffusion distances (as in an incubated tissue). Hence, it is very likely that the differences between the perfused heart and the incubated guinea-pig trachealis smooth muscle (Table 1) reflect diffusion distances rather than differences between different cell types.

Uptake$_2$ has a surprisingly broad substrate spectrum: it transports not only catecholamines and various phenethylamines (see above), it also transports a resorcinol (^{3}H-(±)-orciprenaline), an imidazoline derivative (^{3}H-clonidine), ^{3}H-histamine and ^{3}H-5-HT. Also rimiterol is a substrate of uptake$_2$ (and of COMT; BRAUN and O'DONNELL 1982). Apparently, uptake$_2$ is designed to remove a wide variety of substances from the extracellular space.

Interestingly enough, there was a positive correlation between the K_m and the V_{max} of uptake$_2$, when different substrates were studied (Fig. 1). For discussion of this finding, see Sect. B.II.2.

It is also of interest to note that ^{3}H-corticosterone and ^{3}H-(±)-fenoterol were not transported by uptake$_2$, although the unlabelled compounds were able to inhibit uptake$_2$ of ^{3}H-isoprenaline; hence, these two compounds have to be classified as non-transported inhibitors of uptake$_2$ (GROHMANN and TRENDELENBURG 1984).

Studies of uptake$_2$ of tritiated substrates are affected by the isotope effect introduced by the tritium substituent. This has been analysed for ^{3}H-(−)-nor-

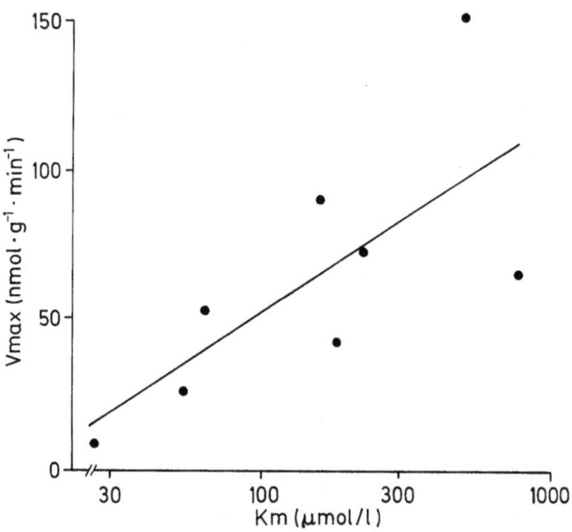

Fig. 1. The correlation between the V_{max} of uptake$_2$ and the K_m of uptake$_2$; rat perfused heart. Ordinate: V_{max} for uptake$_2$; abscissa: K_m for uptake$_2$ (log scale). The V_{max}-values for 8 ^{3}H-substrates of uptake$_2$ were taken from Table 6 of GROHMANN and TRENDELENBURG (1984), the K_m-values from Table 3 of the same publication. In ascending order of V_{max}: clonidine-(±)-isoprenaline-5-HT-tyramine-(±)-orciprenaline-(−)-noradrenaline-histamine-dopamine. Also shown is the regression line calculated for all points ($r = 0.7088$; $P < 0.02$). For details, see GROHMANN and TRENDELENBURG (1984)

adrenaline and ^{3}H-(−)-adrenaline (TRENDELENBURG et al. 1983a; GROHMANN and TRENDELENBURG 1983): the hindrance of $uptake_2$ increases with an increasing number of tritium substituents per catecholamine molecule. Hence, catecholamines with multiple tritium substituents are unsuitable for studies of $uptake_2$, but the isotope effect is small for catecholamines with just one tritium substituent ($uptake_2$ decreased by a factor of 1.4; TRENDELENBURG et al. 1983a).

Even a small isotope effect can lead to wrong conclusion. HENSELING and TRENDELENBURG (1978) reported a slightly higher rate of O-methylation of ^{14}C-(+)- than of ^{3}H-(−)-noradrenaline in the rabbit aorta, the MAO of which was not inhibited; they ascribed this to the stereoselective preference of MAO for the (−)-isomer of noradrenaline. However, since it is highly likely that the isotope effect is $^{3}H > {}^{14}C$, the reported stereoselectivity of the O-methylation of noradrenaline is likely to have been caused by the different labelling of the two isomers. As far as measurements of unlabelled (−)- and (+)-isoprenaline are concerned, there is no doubt that the O-methylation of low concentrations of the (−)-isomer exceeds that of the (+)-isomer (BARONE et al. 1985); and this difference is related to the stereoselective preference of $uptake_2$ for (−)-isoprenaline (Table 1).

4. The Distribution of ^{3}H-Catecholamines in the Rat Heart

Earlier studies of the distribution of ^{3}H-isoprenaline in the rat heart (BÖNISCH 1978) indicated that $uptake_2$ is responsible for the accumulation of this compound in two main compartments (compartments III and IV, characterized, in wash-out experiments, by half times for efflux of about 10 and 25 min, respectively) and a minor compartment II (characterized by a half time for efflux of about 4 min). Compartment I, on the other hand, represented the ^{3}H-isoprenaline distributed into the extracellular space and into the dead space of the apparatus.

A careful re-examination showed that, for unknown reasons, this view had to be revised. TRENDELENBURG et al. (1983b) found a single "main" distribution compartment (compartment III), while compartment IV was of very minor magnitude (as was compartment II). Recent experiments with ^{3}H-(−)-noradrenaline, ^{3}H-(±)-adrenaline and ^{3}H-dopamine (TRENDELENBURG 1985) showed a virtually identical pattern: the main distribution compartment is compartment III, although the characteristic half time for efflux differs from catecholamine to catecholamine (see below).

This revised distribution pattern of ^{3}H-catecholamines is in full agreement with an autoradiographic study of the corticosterone-sensitive distribution of ^{3}H-(±)-isoprenaline in the rat heart (AZEVEDO et al. 1983): about 80% of all grains were associated with the myocardial cells, just as about 80% of the amine distributed into the heart can be assigned to compartment III. Thus, we are justified in identifying compartment III as the myocardial cells. Moreover, a very high density of grains was found over the arterioles and venules of the heart; however, because of the sparsity of such vessels, only about 10% of all grains were associated with vascular smooth muscle. Quantitatively, this agrees well with the distribution of ^{3}H-(±)-isoprenaline into compartment II (TRENDELENBURG et al. 1983b).

Finally, if there is uptake$_2$ (i.e. corticosteroid-sensitive uptake) into both compartment II and compartment III, why do they differ with respect to the half time for efflux? As explained by AZEVEDO et al. (1983), if one compares two hypothetical cells that have identical membranes but which differ in size, then the surface/volume ratio becomes important. Small cells have a high surface/volume ratio, and they should lose their amine content much faster than large cells with a low surface/volume ratio. Since AZEVEDO et al. (1983) documented that the surface/volume ratio of the vascular smooth muscle cells is nearly 4 times higher than that of the myocardial cells, the difference between the half times for efflux from compartments II and III may reflect differences between surface/volume ratios rather than differences between two types of membranes.

5. The "Lipophilic Entry" of Agents into Extraneuronal Cells

When one studies corticosteroid-sensitive uptake$_2$, it is important to make sure that the results relate to the carrier-mediated uptake and that they are not distorted by the ability of agents with high lipid solubility to permeate through cell membranes by diffusion. Hence, "uptake$_2$" should, ideally, always refer to the difference "uptake in the absence of an inhibitor minus uptake in the presence of a fully effective inhibitor of uptake$_2$".

Among the catecholamines, lipid solubility increases in the order noradrenaline < adrenaline < dopamine < isoprenaline (MACK and BÖNISCH 1979). This is also true for "corticosteroid-resistant uptake" (GROHMANN and TRENDELENBURG 1984; BRYAN and O'DONNELL 1984a). However, this "lipophilic entrance" is rather minor for the polar catecholamines, as compared to much more lipophilic compounds like tyramine and especially clonidine (GROHMANN and TRENDELENBURG 1984). For clonidine, for instance, the "lipophilic entry" greatly exceeds corticosteroid-sensitive uptake$_2$ (measured after 2 min of perfusion of the rat heart with ^{3}H-clonidine; GROHMANN and TRENDELENBURG 1984). Also phenylephrine was found to be so lipid soluble that this interfered with quantitative determinations of uptake$_2$ (see RAWLOW et al. 1980).

II. Outward Transport

1. The Susceptibility of the Efflux of Catecholamines to Inhibition of Uptake$_2$

In the first study of uptake$_2$, IVERSEN (1965a) reported that inhibition of uptake$_2$ also causes inhibition of the efflux of catecholamines from extraneuronal tissues. This has been demonstrated for MN (rat heart: IVERSEN 1965a); OMI (rat heart: BÖNISCH et al. 1974; TRENDELENBURG 1985), corticosterone (rat heart: BÖNISCH et al. 1974; TRENDELENBURG et al. 1983b; rabbit aorta: ECKERT et al. 1976), phenoxybenzamine (rabbit aorta: ECKERT et al. 1976),

Table 2. Rate constants for the extraneuronal efflux of ^{3}H-catecholamines from the rat heart. Hearts were first loaded with 10 nmol/l ^{3}H-catecholamine for 20 min (COMT, MAO and uptake$_1$ inhibited; animals pretreated with reserpine), and then washed out with amine-free solution for 40 min, either in the absence or presence of 100 µmol/l OMI. Rate constants were determined from the slopes of the final efflux curves (i.e. for efflux from compartment III of TRENDELENBURG et al. 1983b). Shown are geometric means of rate constants (min^{-1}) with 95 % confidence limits in parentheses. The rate constant for total efflux (k_{out}) was determined in the absence of OMI, the rate constant for diffusional efflux ($k_{out\ diffusion}$) in the presence of 100 µmol/l OMI. Since $k_{out} = k_{out\ transport} + k_{out\ diffusion}$, $k_{out\ transport}$ was calculated as the difference between k_{out} and $k_{out\ diffusion}$. Also shown is k_{uptake} (the rate constant describing the inward transport by unsaturated uptake$_2$; taken from GROHMANN and TRENDELENBURG 1984)

^{3}H-catecholamine	n	k_{out}	$k_{out\ diffusion}$	$k_{out\ transport}$	k_{uptake}
(±)-isoprenaline	4	0.0513 (0.0487; 0.0541)	0.0176 (0.0151; 0.0206)	0.0337	0.4981
(±)-adrenaline	4	0.0532 (0.0475; 0.0593)	0.0123 (0.0059; 0.0255)	0.0409	0.6636
(−)-noradrenaline	6	0.0264 (0.0245; 0.0285)	0.0109 (0.0094; 0.0126)	0.0155	0.3266
dopamine	4	0.0368 (0.0332; 0.0408)	0.0267 (0.0184; 0.0388)	0.0101	0.3053

Note that a) $r = 0.973$ ($P < 0.05$) for the correlation between $k_{out\ transport}$ and k_{uptake}, and b) U-0521 (the inhibitor of COMT) was used at a concentration of 10 µmol/l, i.e., at a concentration that does not impair carrier-mediated fluxes (TRENDELENBURG 1985).

and also for high concentrations of U-0521 (3,4-dihydroxy-2methyl propiophenone, an inhibitor of COMT; rat heart: BÖNISCH et al. 1978; TRENDELENBURG 1985).

The inhibitory effect discussed here has the following characteristics: **a)** only part of the efflux of catecholamines is sensitive to this type of inhibition (TRENDELENBURG et al. 1983b; TRENDELENBURG 1985), and **b)** the IC_{50} for inhibition of efflux is virtually identical with the IC_{50} for inhibition of uptake$_2$ (which equals K_m, if the agent under discussion is a substrate of uptake$_2$, and K_i, if it is a non-transported inhibitor) (TRENDELENBURG et al. 1983b; TRENDELENBURG 1985). From this we may conclude that "total efflux" is due to two different mechanisms: carrier-mediated efflux (or outward transport) is sensitive to inhibitors of uptake$_2$, but diffusional efflux (due to the lipophilicity of the catecholamines) is resistant to these agents.

These two components of extraneuronal efflux were determined for four different ^{3}H-catecholamines in the rat heart (first loaded with the ^{3}H-compounds and then washed out with amine-free solution; TRENDELENBURG 1985). Table 2 shows the rate constants for the efflux from compartment III determined either in the absence (total efflux; k_{out}) or in the presence of a maximally effective concentration of OMI (= 50 times its IC_{50}; $k_{out\ diffusion}$ in

Table 2). The rate constants for the total efflux from compartment III (k_{out}) differed from catecholamine to catecholamine; they declined in the order: ^{3}H-($\pm$)-isoprenaline = ^{3}H-($\pm$)-adrenaline > ^{3}H-dopamine > ^{3}H-($-$)-noradrenaline.

In the presence of 100 μmol/l OMI, carrier-mediated efflux is virtually abolished. Table 2 (column $k_{out\ diffusion}$) shows that OMI greatly reduced the rate constants for efflux, though to a different extent for each catecholamine. This OMI-resistant efflux reflects the lipophilicity of the catecholamines.

The difference between the two rate constants (determined either in the absence or in the presence of OMI) then reflects the extent of carrier-mediated outward transport. As indicated by the values of the column "$k_{out\ transport}$" (Table 2) the ranking order for carrier-mediated efflux is ^{3}H-($\pm$)-adrenaline > ^{3}H-($\pm$)-isoprenaline > ^{3}H-($-$)-noradrenaline > ^{3}H-dopamine; it is of interest to note that this ranking order is identical with that for the rate constant determined for uptake$_2$ of the four catecholamines, determined for very low, non-saturating catecholamine concentrations. Thus, irrespective of whether we consider the inward or the outward transport of catecholamines, the uptake$_2$ carrier prefers adrenaline and isoprenaline to noradrenaline and dopamine. However, it should be noted that the rate constants for carrier-mediated efflux are only a small fraction of the rate constants for the inward transport by uptake$_2$ (i.e. 1/13 to 1/30; see GROHMANN and TRENDELENBURG 1984; k_{uptake} in Table 2).

2. The Interaction Between Different Substrates of Uptake$_2$

For the neuronal carrier (uptake$_1$) the phenomenon of "facilitated exchange diffusion" is well established: after loading of the axoplasm with "substrate A", the presence of "substrate B" in the extracellular space greatly accelerates the efflux of substrate A (PATON 1973). This increase in the efflux of substrate A in due to the increased availability of the carrier on the inside of the axonal membrane, as a consequence of the inward transport of substrate B.

Attempts were made to see whether a similar phenomenon can be observed for the extraneuronal system of the rat heart. Negative results were obtained with ^{3}H-catecholamines as substrate A and unlabelled isoprenaline or noradrenaline as substrate B (BÖNISCH et al. 1974; MEKANONTCHAI and TRENDELENBURG 1979). However, these negative results may well have been due to the fact that the concentrations of substrate B were too low (i.e. below the K_m for uptake$_2$).

Hence, in a further series of experiments rat hearts were loaded with ^{3}H-($\pm$)-isoprenaline (substrate A) and then washed out with amine-free solution. From the 30th min of wash-out onwards, either dopamine or 5-HT or clonidine or OMI were present in the wash-out solution at concentrations amounting to $2 \times K_m$. All four agents are substrates of uptake$_2$, even if (in the preceding sections) the O-methylated catecholamines were presented as "inhibitors of uptake$_2$". Uptake$_2$ of O-methylated catecholamines was demon-

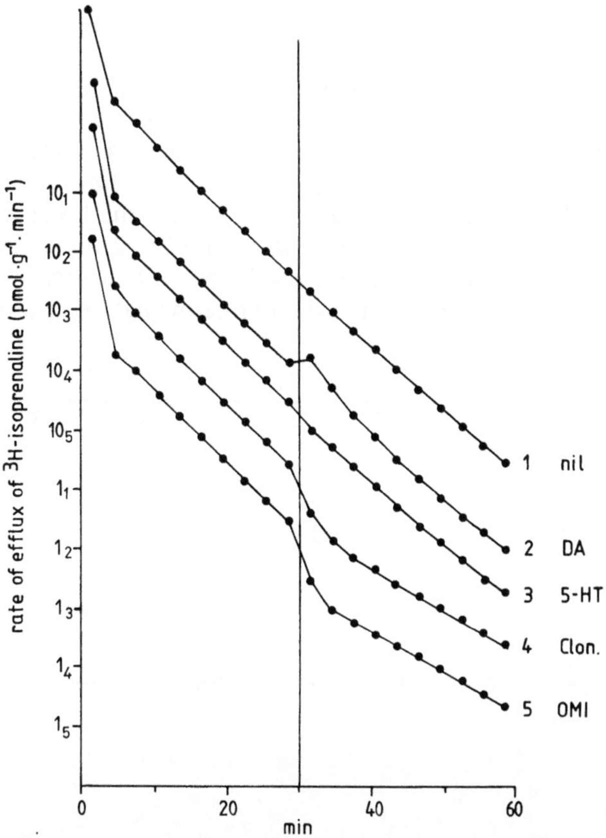

Fig. 2. The effect of four different unlabelled substrates of uptake$_2$ on the efflux of ^{3}H-(±)-isoprenaline from the extraneuronal compartment III of the rat heart. Reserpine- and pargyline-pretreated rats, COMT inhibited by the presence of 30 µmol/l U-0521. Hearts were first loaded by perfusion for 20 min with 40 nmol/l ^{3}H-(±)-isoprenaline and then washed out with amine-free solution for 30 min. From the 30th min of wash-out onwards the perfusion fluid either continued to be amine-free (curve 1; "nil") or it contained 1000 µmol/l dopamine (curve 2; "DA"), 350 µmol/l 5-HT (curve 3; "5-HT"), 50 µmol/l clonidine (curve 4; "Clon") or 4 µmol/l OMI (curve 5; "OMI"); these concentrations amount to "twice the K_m for uptake$_2$" (GROHMANN and TRENDELENBURG 1984). Ordinate: rate of efflux of ^{3}H-(±)-isoprenaline (in pmol·g^{-1}·min^{-1}; log scale); abscissa: time (in min) after onset of wash-out. Note that, for the sake of clarity, curves were displaced downward; hence, the rates "1_1 and 10_1" belong to curve 1, "1_2 and 10_2" to curve 2 etc. The vertical line indicates the time of the addition of the unlabelled substrates (curves 2 to 5). Shown are geometrical means of 5 experiments each. For the sake of clarity, standard errors were omitted. Taken from TRENDELENBURG (1985)

strated by SIMMONDS and GILLIS (1968) and UHLIG et al. (1976). Uptake$_2$ of dopamine, 5-HT and clonidine was demonstrated by GROHMANN and TRENDELENBURG (1984).

Figure 2 shows that the addition of dopamine elicited some minor degree of facilitated exchange diffusion: there was a transient increase in the rate of efflux of ^{3}H-isoprenaline. However, the addition of 5-HT to the perfusion fluid failed to affect the efflux of isoprenaline, while both clonidine and OMI clearly inhibited efflux (Fig. 2). Two facts should be pointed out in relation to these surprising results: a) the K_m for uptake$_2$ declines in the order dopamine > 5-HT > clonidine > OMI (GROHMANN and TRENDELENBURG 1984), and b) for 9 different ^{3}H-substrates of uptake$_2$ (excluding ^{3}H-orciprenaline), there was a close correlation between K_m and V_{max}: the lower K_m, the lower V_{max} (see Fig. 1, above). Thus, we can state from Fig. 2 that substrates of uptake$_2$ with a high V_{max} (dopamine) are able to elicit some facilitated exchange diffusion, that substrates of uptake$_2$ with an intermediate V_{max} (5-HT) are inef-

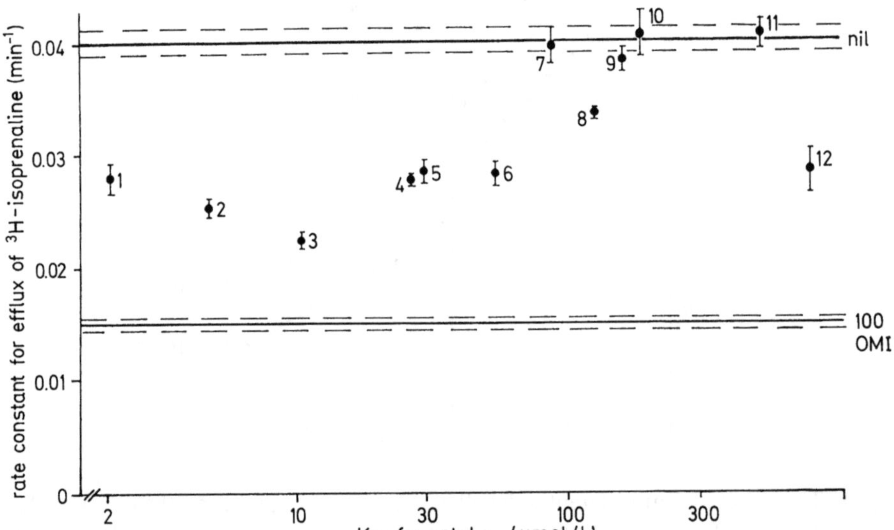

a

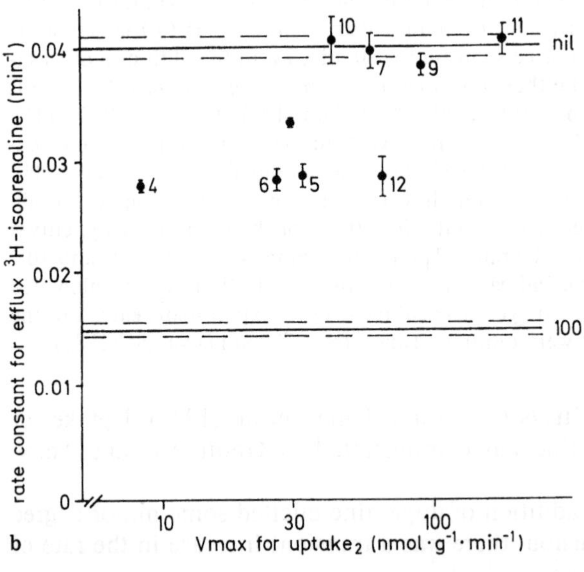

b

fective, and that substrates with a low V_{max} actually inhibit the efflux of ^{3}H-isoprenaline (clonidine, and presumably OMI).

In order to test the validity of this conclusion, further substrates of uptake$_2$ were studied in slightly modified experiments. Hearts were loaded as before, but the unlabelled substrates of uptake$_2$ (at concentrations equalling about 2 × K_m) were added to the wash-out solution from the end of the loading period onward. With this experimental design, no facilitated exchange diffusion can be detected, since it coincides with the initial steep fall of the efflux curve. Fig. 3 shows the magnitude of the inhibition of the efflux of ^{3}H-isoprenaline in dependence on the K_m (a) and on the V_{max} (b) for various substrates of uptake$_2$. As in the earlier study of GROHMANN and TRENDELENBURG (1984), (±)-orciprenaline failed to conform, but for all other substrates of uptake$_2$ it is evident that their ability to inhibit the efflux of ^{3}H-isoprenaline increases with decreasing V_{max} for uptake$_2$.

The extent of facilitated exchange diffusion (in the presence of dopamine) was small, especially when compared with the facilitated exchange diffusion elicited for uptake$_1$ (see BÖNISCH and TRENDELENBURG this volume Chap. 5). However, it should be borne in mind that the carrier of uptake$_1$ is highly asymmetrically distributed (SAMMET and GRAEFE 1979): most of the carrier molecules are on the outside of the axonal membrane (where the sodium concentration is high), very few are on the inside (where the sodium concentration is low). Hence, because of this asymmetrical distribution, the inward transport of substrate B can easily increase the number of carrier molecules on the inside of the axonal membrane by a factor of 10. If, on the other hand, the extraneuronal carrier were distributed more or less symmetrically, then even a fully saturating concentration of substrate B should only double the number of carriers available on the inside of the cell membrane. Thus, the rather minor degree of dopamine-induced facilitated exchange diffusion (Fig. 2) may well indicate a nearly symmetrical carrier distribution for uptake$_2$.

◀

Fig. 3 a, b. The dependence of the inhibitory effect of unlabelled substrates of uptake$_2$ (on the efflux of ^{3}H-(±)-isoprenaline from the extraneuronal compartment III of the rat heart) on the K_m for uptake$_2$ **(a)** and on the V_{max} for uptake$_2$ **(b)**. Experimental conditions as in Fig. 2, except that the unlabelled substrates were added to the perfusion fluid from the beginning of wash-out onwards. Shown are the rate constants for the efflux of ^{3}H-(±)-isoprenaline from compartment III in the absence (upper horizontal line) or in the presence of various unlabelled substrates of uptake$_2$, all at a concentration amounting to twice K_m for uptake$_2$. The lower horizontal line shows the rate constant for efflux in the presence of 100 μmol/l OMI (i.e. in the presence of a concentration of OMI which fully inhibits carrier-mediated outward transport of ^{3}H-(±)-isoprenaline). Note that not all substrates appearing in **(a)** also appear in **(b)**; while K_m is known for all, V_{max} is known for only some of the substrates of uptake$_2$. Ordinates: rate constant for efflux of ^{3}H-(±)-isoprenaline from compartment III; abscissae: K_m for uptake$_2$ (in **a**) or V_{max} for uptake$_2$ (in **b**), both with log scale. Shown are means ± S.E. for 3 to 6 observations per point. Note that unlabelled substrates tend to inhibit the efflux of ^{3}H-isoprenaline when either their K_m **(a)** or their V_{max} **(b)** is low. Taken from Trendelenburg (1985).

Two phenomena have to be explained: **a)** the substrate-dependence of V_{max} for uptake$_2$ and **b)** ability of substrates with high V_{max} to induce facilitated exchange diffusion, while substrates with low V_{max} inhibit the outward transport of ^{3}H-isoprenaline. The transport cycle can be broken down into at least four steps: **1)** association of the substrate with the outside carrier, **2)** translocation of the substrate-carrier complex, **3)** dissociation of the substrate from the inside carrier, and **4)** return of the unloaded carrier to the outside. If step 2 or 3 or 4 is rate-limiting for the whole transport cycle, then carrier accumulates inside the cell membrane or on the inside of the cell membrane, so that the availability of the carrier on the outside is restricted. Such a restriction of availability would restrict V_{max}. While it is easily possible to imagine the rate constant for step 2 or 3 to be subject to substrate-dependence, step 4 is most unlikely to be influenced by the substrate, since it involves the return of the *unloaded* carrier to the outside. Hence, it is highly likely that the substrate-dependence of V_{max} involves a rate-limiting and substrate-dependent rate constant for step 2 or 3.

This view also accounts for the fact that substrates with high V_{max} elicit facilitated exchange diffusion, while substrates with low V_{max} inhibit outward transport. Again, the accumulation of the loaded carrier, either inside the cell membrane (when step 2 is rate-limiting) or on the inside of the cell membrane (when step 3 is rate-limiting) should be very pronounced when V_{max} for uptake$_2$ is low; this should then reduce the availability of the free inside carrier for outward transport. A more detailed analysis of a transport model representing the whole transport cycle was presented by SCHÖMIG et al. (1988).

It is of considerable interest that uptake$_1$ exhibits an analogous dependence of the V_{max} on the K_m (see GRAEFE and BÖNISCH this volume, Chap. 4). Hence, the speculative explanation offered here may well also apply to uptake$_1$. However, there appears to exist a pronounced difference in so far as the free neuronal carrier is distributed very asymmetrically (outside $\gg$ inside), since it reflects the Na^+ concentration gradient (SAMMET and GRAEFE 1979). For uptake$_2$ the corresponding distribution is clearly less asymmetrical (see above). This difference may well account for the fact that high-affinity substrates of uptake$_1$ are able to release pre-stored ^{3}H-noradrenaline from adrenergic nerve endings (see BÖNISCH and TRENDELENBURG this volume, Chap. 5), while high-affinity substrates of uptake$_2$ tend to inhibit any outward transport from extraneuronal systems (see above).

III. The Effects of Various Ions on Uptake$_2$

While uptake$_1$ is absolutely dependent on the presence (in the extracellular space) of sodium and chloride (see GRAEFE and BÖNISCH this volume, Chap. 4), this is not true for uptake$_2$ (BÖNISCH et al. 1985). Uptake$_2$ is *not* affected by the acute omission of either Na^+ or Cl^- or Ca^{2+} from the perfusion fluid or incubation medium. However, uptake$_2$ is sensitive to alterations in the concentration of K^+ in the extracellular space: the increase in the concentration of K^+ from 2.7 to 15 mmol/l approximately halved the rate of uptake$_2$ of a tracer concentration of ^{3}H-($\pm$)-isoprenaline (rat heart: BÖNISCH et al. 1985;

LUDWIG et al. 1986; see there for additional references). The inhibitory effect of K^+ is due to a decrease of the V_{max} of uptake$_2$, not to any change in K_m (LUDWIG et al. 1986). Moreover, the same increase in the concentration of K^+ about doubled the rate constant for the efflux of ^{3}H-isoprenaline from hearts which had been preloaded first (BÖNISCH et al. 1985; GROHMANN 1988; see also KAUMANN et al. 1978).

Results of this kind indicate that either the potassium gradient or the membrane potential may somehow (directly or indirectly) provide the force that drives uptake$_2$.

ALMGREN and JONASON (1974) reported that the β-adrenoceptor antagonist propranolol increased the extraneuronal accumulation of ^{3}H-($\pm$)-isoprenaline in rat salivary gland slices. Since then it has been found that isoprenaline, by acting on β-adrenoceptors, depolarizes gland cells (PETERSEN 1976). Hence, it is possible that isoprenaline, by depolarizing gland cells, inhibits its own extraneuronal uptake, and that propranolol prevents depolarization. Recent experiments (TRENDELENBURG 1986) fully support this proposal. As the depolarizing β-effect of isoprenaline is unlikely to reduce the K^+ concentration gradient (PETERSEN 1976), it can be concluded that it is the membrane potential (and not the K^+ concentration gradient) which is able to modulate, perhaps to drive uptake$_2$. In this context it is of interest to note that depolarization also inhibits uptake$_1$, again through a reduction of V_{max} (UNGELL and GRAEFE 1987).

For the rabbit pulmonary artery, it is known that noradrenaline causes a depolarization of the smooth muscle cells that is mediated by activation of α_1-adrenoceptors (HAEUSLER 1978, 1983), while nicorandil elicits hyperpolarization (HAEUSLER and DE PEYER 1986). 1 µmol/l (−)-noradrenaline impaired the extraneuronal accumulation of ^{3}H-isoprenaline, an effect that was abolished by the presence of α_1-adrenoceptor antagonists. Nicorandil, on the other hand, enhanced the extraneuronal accumulation of ^{3}H-isoprenaline (in the absence of any (−)-noradrenaline; TRENDELENBURG 1987). Thus, also in this extraneuronal system depolarization hinders, and hyperpolarization facilitates the inward transport of ^{3}H-isoprenaline.

Depolarization of cells is also likely to provide the explanation for inhibition of uptake$_2$ by prolonged inhibition of the membrane Na^+, K^+-ATPase (rat heart and rabbit aorta: BÖNISCH et al. 1985; guinea-pig trachealis muscle: BRYAN and O'DONNELL 1984b) or by prolonged glucose deprivation plus either anoxia or dinitrophenol (rat submaxillary gland: HAMBERGER et al. 1967; ALMGREN and JONASON 1976; guinea-pig trachealis muscle: BRYAN and O'DONNELL 1984b).

Uptake$_2$ appears to be highly sensitive to changes in the resting membrane potential. For the smooth muscle of the rabbit pulmonary artery HAEUSLER (1978; 1983) found 1 µmol/l (−)-noradrenaline to depolarize the membrane potential by 15 mV; the same concentration reduced the OMI-sensitive accumulation of ^{3}H-isoprenaline by 44%. In view of this marked effect of relatively small changes in the resting membrane potential, one has to consider the possibility that the loaded extraneuronal carrier is a) positively charged and b) driven by the membrane potential.

C. The Extraneuronal Metabolizing Systems

I. The Extraneuronal O-Methylating-System

1. The Kinetic Constants

When uptake and intracellular COMT work hand in hand to dispose of extracellular catecholamines, it is of interest to relate the rate of metabolite formation to the *outside* ^{3}H-catecholamine concentrations ($S_{outside}$). Since saturable uptake is arranged in series with saturable COMT, the dependence of the rate of metabolite formation on $S_{outside}$ does *not* obey strict Michaelis-Menten kinetics (KURAHASHI et al. 1980). Until recently this fact had been neglected, and the dependence of the rate of metabolism on $S_{outside}$ was analysed with the help of methods that are based on the Michaelis-Menten equation.

Two constants are of interest: **a)** the "half-saturating outside concentration", i.e. that $S_{outside}$ which leads to half-saturation of the intracellular enzyme in steady state; and **b)** the V_{max} of the system (which equals V_{max} of the enzyme, if $V_{max\ enzyme} < V_{max\ uptake}$). It should be noted that in earlier publications the "half-saturating outside concentration" was also called the "K_m of the extraneuronal O-methylating system".

Extraneuronal O-methylating systems have been analysed in a variety of tissues. Table 3 shows that they are all characterized by having a low "half-saturating outside concentration" as well as a low V_{max}. This is surprising, since uptake$_2$ for catecholamines is characterized by a rather high K_m (see Sect. B.I.2) and a high V_{max} (Sect. B.I.4).

Table 3. Outside concentrations of labelled or unlabelled catecholamines that half saturate the intracellular COMT of extraneuronal 0-methylating systems

Organ	catecholamine	(μmol/l)	Reference
rat heart	^{3}H-(±)-isoprenaline	K_m[a] 2.9	BÖNISCH et al. (1974)
	^{3}H-(±)-isoprenaline	K_m[a] 3.3	BÖNISCH (1978)
	^{3}H-(−)-noradrenaline	K_m[a] 1.7	FIEBIG and TRENDELENBURG (1978b)
	(−)-dobutamine	HSOC[b] 0.93	GROHMANN and TRENDELENBURG (1985)
	(+)-dobutamine	HSOC[b] 1.04	GROHMANN and TRENDELENBURG (1985)
	(−)-isoprenaline	HSOC[b] 0.67[d]	GROHMANN and TRENDELENBURG (1985)
	(+)-isoprenaline	HSOC[b] 2.01[d]	GROHMANN and TRENDELENBURG (1985)
	(−)-adrenaline	HSOC[b] 0.76	GROHMANN and TRENDELENBURG (1985)
	(−)-noradrenaline	HSOC[b] 2.26	GROHMANN and TRENDELENBURG (1985)
	dopamine	HSOC[b] 2.73	GROHMANN and TRENDELENBURG (1985)

Table 3 (continued)

Organ	catecholamine	(μmol/l)		Reference
cat nictitating membrane	^{3}H-(−)-noradrenaline	K_m[a]	7.5	GRAEFE and TRENDELENBURG (1974)
	^{3}H-(±)-isoprenaline	K_m[a]	12.8	GRAEFE and TRENDELENBURG (1974)
rabbit aorta	^{3}H-(−)-noradrenaline	K_m[a]	3.7	HENSELING (1983a)
rat submaxillary gland	^{3}H-(±)-isoprenaline	K_m[a]	7.2	MAJOR et al. (1978)
guinea-pig trachea	^{3}H-(±)-isoprenaline	K_m[a]	6.1	GARLAND and MARTIN (1984)
	^{3}H-(−)-adrenaline	K_m[a]	8.7	GARLAND and MARTIN (1984)
	^{3}H-(−)-noradrenaline	K_m[a]	11.3	GARLAND and MARTIN (1984)
dog saphenous vein	(−)-isoprenaline	HSOC[c]	0.6[d]	PAIVA and GUIMARAES (1984)
	(±)-isoprenaline	HSOC[c]	1.1[d]	PAIVA and GUIMARAES (1984)
	(−)-adrenaline	HSOC[c]	0.7	PAIVA and GUIMARAES (1984)
	(−)-noradrenaline	HSOC[c]	1.4	PAIVA and GUIMARAES (1984)
	^{3}H-(±)-isoprenaline	K_m[a]	1.3	PAIVA and GUIMARAES (1984)
	^{3}H-(±)-adrenaline	K_m[a]	2.5	PAIVA and GUIMARAES (1984)
	^{3}H-(−)-noradrenaline	K_m[a]	3.4	PAIVA and GUIMARAES (1984)
rat brown fat	^{3}H-(±)-noradrenaline	K_m[a]	4.0	CHINET and DURAND (1979)

[a] K_m as calculated on the (slightly false) assumption that the dependence of the O-methylation on the outside catecholamine concentration obeys strict Michaelis-Menten kinetics. This assumption introduces a small error, as does the existence of an isotope effect.

[b] Half-saturating outside concentration after the correction presented by GROHMANN and TRENDELENBURG (1985).

[c] Half-saturating outside concentration without the correction presented by GROHMANN and TRENDELENBURG 1985).

[d] The difference between pairs of steroisomers is significant ($P < 0.05$)

The isotope effects at the level of the intracellular enzyme have to be considered. As far as COMT is concerned, the position of tritium on the ring impairs the O-methylation of ^{3}H-2,5,6-(−)-noradrenaline, while this is not true for tritium located on the side chain (i.e. in position 7 or 8; TRENDELENBURG et al. 1983a). This was demonstrated for the COMT of the intact rat heart as well as for homogenates of rat hearts.

2. The Revised Mathematical Model

KURAHASHI et al. (1980) presented a mathematical model which is able to simulate the extraneuronal uptake and O-methylation of catecholamines in the rat heart. Various comparisons of such simulated and of experimental results were presented by KURAHASHI et al. (1980) and TRENDELENBURG (1980). However, this model was based on two assumptions which, in the meantime, have been found to be false: **a)** at that time it was not realized that part of the efflux of catecholamines is carrier-mediated (see Sect. B.II.1), and **b)** k_{in} and k_{out} were assumed to be identical.

Figure 4 shows the revised model. The inward movement of the catecholamine is brought about either by carrier-mediated inward transport (characterized by the Michaelis-Menten constants $K_{m\ uptake}$ and $V_{max\ uptake}$) or by diffusion (characterized by the rate constant k_{in}). The outward movement of the catecholamine is either carrier-mediated or by diffusion; both components of efflux can be described together by k_{out} ($= k_{out\ diffusion} + k_{out\ transport}$). The use of $k_{out\ transport}$ implies that outward transport is not saturated. Indeed, no saturation of the efflux of ^{3}H-isoprenaline was observed when hearts were loaded with very high concentrations of this amine (TRENDELENBURG et al. 1983b).

Equation (1) of KURAHASHI et al. (1980) defined the steady-state tissue/medium ratio for the catecholamine, observed when the old model system is exposed to any outside concentrations of the amine ($S_{outside}$), including those that saturate either COMT or uptake$_2$ or both. For the revised model of Fig. 4, K. Kurahashi kindly derived the equivalent Eq. (1):

$$T/M_{st\text{-}st} = \frac{-(k_{out} \cdot K_{m\ enzyme} + V_{max\ enzyme} - v_{uptake} - k_{in} \cdot S_{outside})}{2\,k_{out} \cdot S_{outside}}$$

$$\pm \frac{\sqrt{(k_{out} \cdot K_{m\ enzyme} + V_{max\ enzyme} - v_{uptake} - k_{in} \cdot S_{outside})^2 + 4\,k_{out} \cdot K_{m\ enzyme}(v_{uptake} + k_{in} \cdot S_{outside})}}{2\,k_{out} \cdot S_{outside}} \quad (1)$$

where $K_{m\ enzyme}$ and $V_{max\ enzyme}$ represent the Michaelis-Menten constants of the intracompartmental enzyme, and where $v_{uptake} = (V_{max\ uptake} \cdot S_{outside})/(K_{m\ uptake} + S_{outside})$. As $T/M_{st\text{-}st} = S_{inside}/S_{outside}$, the intracellular catecholamine concentration can be calculated from Eq. (1)

Whenever experiments are carried out with such low $S_{outside}$ that neither the intracellular enzyme nor uptake$_2$ are saturated, Eq. (1) can be simplified. Under these experimental conditions,

$$T/M_{st\text{-}st} = \frac{k_{uptake} + k_{in}}{k_{enzyme} + k_{out}} \quad (2)$$

where k_{uptake} and k_{enzyme} equal V_{max}/K_m.

As many extraneuronal tissues contain not only COMT but also MAO (TRENDELENBURG 1980), and since the available evidence is consistent with a co-existence of both enzymes in the same compartment (see Sect. C.III), the corresponding equation is also of interest (but, again, valid only for non-saturating $S_{outside}$):

$$T/M_{st\text{-}st} = \frac{k_{uptake} + k_{in}}{k_{comt} + k_{mao} + k_{out}} \quad (3)$$

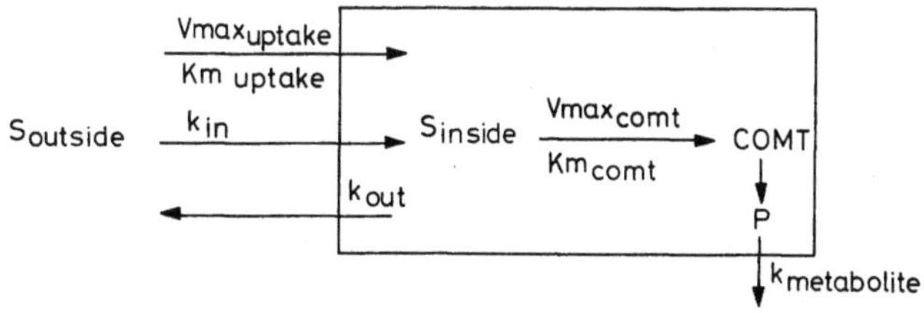

Fig. 4. Schematic representation of the mathematical model which simulates compartment III of the rat heart. The inward transport of isoprenaline is determined by $S_{outside}$ and the Michaelis-Menten constants for uptake$_2$. The passive influx of isoprenaline is determined by $S_{outside}$ and k_{in}. The outflow of isoprenaline is determined by S_{inside} and k_{out}, where k_{out} combines the diffusional outflow ($k_{out\ diffusion}$) with the carrier-mediated outward transport ($k_{out\ transport}$) (see Table 2). The intracellular COMT O-methylates isoprenaline at a rate that is dependent on S_{inside} and the Michaelis-Menten constants of the enzyme. The O-methylated metabolite leaves the cell at a rate dependent on the intracellular metabolite concentration (P) and the rate constant for the efflux of the metabolite ($k_{metabolite}$). This model has been adapted from that presented by KURAHASHI et al. (1980)

3. The Functional Characteristics

The functional properties of the extraneuronal O-methylating system can be deduced from calculations based on the mathematical model introduced by KURAHASHI et al. (1980). Fig. 5a shows the dependence of the steady-state tissue/medium ($T/M_{st\text{-}st}$) ratio on $S_{outside}$. The upper curve represents the extraneuronal O-methylating system after inhibition of COMT; it reflects the dependence of $T/M_{st\text{-}st}$ on $S_{outside}$ for a "pump and leak system" (WILBRANDT and ROSENBERG 1961): a high T/M_{st-st} is generated as long as uptake is not saturated; saturation of uptake (by increasing $S_{outside}$) leads to a decline of $T/M_{st\text{-}st}$, and no accumulation of the catecholamine (or a $T/M_{st\text{-}st}$ of unity) is obtained, when uptake is fully saturated. The lower curve in Fig. 5a represents the intact O-methylating system. At very low $S_{outside}$, $T/M_{st\text{-}st}$ is low (as an expression of the activity of COMT) and independent of $S_{outside}$ (as a consequence of the absence of any saturation of either uptake or enzyme). However, when $S_{outside}$ rises to levels which cause saturation of COMT, $T/M_{st\text{-}st}$ begins to rise: from here onward, the lower curve comes closer and closer to the upper curve of Fig. 5a.

The lower curve of Fig. 5a indicates that the extraneuronal O-methylating system can fulfil two very different functions, depending on $S_{outside}$: if it is exposed to *low* $S_{outside}$, it functions as an "irreversible, metabolizing site of loss" (as indicated by the low $T/M_{st\text{-}st}$), since uptake is followed by intracellular metabolism of the catecholamine; however, at *high* $S_{outside}$ the extraneuronal O-methylating system functions as a "reversible, accumulating site of loss" which accumulates rather than metabolizes the catecholamine. Moreover, the "half-saturating outside concentration" is that concentration below which the system functions predominantly as an irreversible site of loss, but above which it begins to function predominantly as a reversible site of loss. Thus,

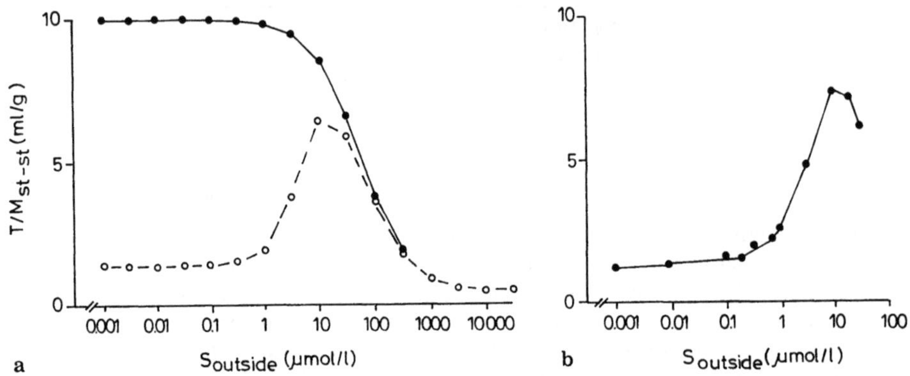

Fig. 5a, b. The dependence of the steady-state tissue/medium ratio for isoprenaline on the concentration of the amine. Shown are results obtained with the mathematical model described in Fig. 4a and results obtained in experiments with rat hearts perfused with ^{3}H-(±)-isoprenaline **(b)**. **a** The following constants were used for the model calculations: $K_{m\,uptake} = 53.6\,\mu mol/l$; $V_{max\,uptake} = 26.7\,nmol \cdot g^{-1} \cdot min^{-1}$; $K_{m\,comt} = 3.4\,\mu mol/l$; $V_{max\,comt} = 1.15\,nmol \cdot g^{-1} \cdot min^{-1}$; $k_{in} = 0.0161\,min^{-1}$; $k_{out} = 0.05134\,min^{-1}$. The constants were taken from GROHMANN and TRENDELENBURG (1984; 1985). ○---○: calculations for a system with an intact COMT; ●——●: calculations for a system with *no* COMT (i.e., for $V_{max\,comt} = 0$). **b** Rat hearts were perfused with various concentrations of ^{3}H-(±)-isoprenaline for 25 min, and tissue/medium ratios were determined at the end of the experiments (COMT intact). Taken from KURAHASHI et al. (1980). Note the close resemblance of the two curves (**a** and **b**) representing an O-methylating system with intact COMT

the "half-saturating outside concentration" is an important functional index of the extraneuronal O-methylating system.

Figure 5b shows that determination of $T/M_{st\text{-}st}$ for ^{3}H-(±)-isoprenaline, in rats hearts perfused with various concentrations of this catecholamine, yielded a curve that was virtually identical with the lower curve of Fig. 5a.

4. The Intracellular Enzyme

In systems like that presented in Fig. 4 the rate of O-methylation is largely determined by the rate of uptake. However, it is of interest to measure the activity of the intracellular enzyme, quite independently of the inward transport of the substrate. Such measurement are possible if the substrate concentration is so low that neither uptake$_2$ nor the intracellular enzyme are even partially saturated by catecholamine.

Under such conditions,

$$v_{st\text{-}st} = S_{inside} \cdot k_{comt} \tag{4}$$

where $v_{st\text{-}st}$ = steady-state rate of O-methylation, S_{inside} = intracellular amine concentration, and k_{comt} = rate constant describing the unsaturated intracellular COMT (= V_{max}/K_m). Thus, k_{comt} can be calculated from measurements of $v_{st\text{-}st}$ and S_{inside}. Rat hearts were perfused with 50 nmol/l ^{3}H-catecholamine un-

Table 4. Rate constants describing the unsaturated, extraneuronal COMT and MAO of the rat heart. All experiments in the presence of 30 μmol/l cocaine (to inhibit uptake$_1$); measurements of k_{comt} were made after pretreatment with pargyline (to block MAO), measurements of k_{mao} in the presence of 10 μmol/l U-0521 (to block COMT). Hearts were perfused with 50 nmol/l ^{3}H-catecholamine until the rate of appearance of the ^{3}H-metabolite(s) reached steady state. k_{enzyme} was then calculated from "steady-state rate of metabolism of ^{3}H-catecholamine in tissue", where "^{3}H-catecholamine in tissue" was corrected for extracellular space. Taken from GROHMANN (1987)

^{3}H-catecholamine	n	k_{comt} (min^{-1})	n	k_{mao} (min^{-1})
(±)-isoprenaline	5	0.779 ± 0.043	–	–
(−)-adrenaline	5	0.487 ± 0.031	4	0.053 ± 0.006
dopamine	5	0.312 ± 0.028	4	0.261 ± 0.007
(−)-noradrenaline	5	0.244 ± 0.017	4	0.107 ± 0.022

til the rate of appearance of the O-methylated ^{3}H-metabolites had reached a steady state (after inhibition of neuronal utpake and MAO). Table 4 shows that ^{3}H-(±)-isoprenaline and ^{3}H-(−)-adrenaline are preferred by COMT to ^{3}H-(−)-noradrenaline and ^{3}H-dopamine. It should be noted that a k_{comt} of 0.779 min^{-1} means that unsaturated COMT is able to O-methylate 77.9% of the intracellular ^{3}H-(±)-isoprenaline per minute.

It is of interest to note that estimates of k_{enzyme} (for an unsaturated, intracellular enzyme) are independent of the fractional size of the compartment under study (fractional size = compartment size as a fraction of total tissue). This is so, since k_{enzyme} simply denotes that fraction of S_{inside} which is metabolized per unit of time—a value that is independent of compartment size.

5. The Increase of S_{inside} Induced by Inhibition of the Intracellular Enzyme

It is a characteristic feature of all "metabolizing systems" that they are relatively resistant to inhibition of the intracellular enzyme. The reason for this relative resistance is evident form Eq. (2) (see Sect. C.I.2) which indicates that any reduction of k_{enzyme} leads to an increase in S_{inside} (provided $S_{outside}$ is too low to saturate either uptake or the intracellular enzyme). Because of this reciprocity of the changes in k_{enzyme} and S_{inside}, the decrease of the steady-state rate of metabolite formation is attenuated. This is also evident from Eq. (4).

Two conclusions can be drawn from these reciprocal changes of k_{enzyme} and S_{inside}. First, Eqs. (2) and (4) indicate that the most sensitive indicator of any decrease of k_{enzyme} is not a reduction of $v_{st\text{-}st}$ but rather the increase in S_{inside}. This postulate was verified in experiments with ^{3}H-catecholamines, the tritium substituent of which hindered either O-methylation by COMT (ring-labelled ^{3}H-catecholamines) or deamination by MAO (^{3}H-catecholamines labelled in position 8) (TRENDELENBURG et al. 1983a; GROHMANN et al. 1986).

This type of isotope effect results in a decrease of k_{comt} and k_{mao}, respectively. Second, in metabolizing systems enzyme inhibitors are less potent than indicated by their K_i for the intracellular enzyme. This was first demonstrated by BRYAN et al. (1983). The extent of this loss of potency of the enzyme inhibitor is directly proportional to the factor by which full inhibition of the enzyme increases S_{inside} (TRENDELENBURG, 1984b). Hence, the IC_{50} (i.e. that enzyme inhibitor concentration which halves the steady-state rate of O-methylation of a tracer concentration of a ^{3}H-catecholamine) exceeds K_i by the factor "S_{inside} after full enzyme inhibition/S_{inside} in the absence of the inhibitor".

Thus, in general terms, it can be stated that any metabolizing system is the more resistant to inhibition of the intracellular enzyme, the greater the factor by which inhibition of the enzyme increases S_{inside}. Inspection of Eq. (2) (Sect. C.I.2) indicates that this factor is greatest when k_{enzyme} is high and when k_{out} is low. It is of interest to note that k_{out} is rather high (relative to k_{comt}) for the extraneuronal O-methylating system; or in other words, this system is a rather "leaky" one. For the neuronal deaminating system, however, it is likely that k_{mao} (for neuronal MAO) is considerably higher than k_{comt} (for extraneuronal COMT), while k_{out} is lower for the neuronal than for the extraneuronal metabolizing system. Hence, inhibition of the intracellular enzyme is able to increase S_{inside} by a much greater factor inside the neuronal than inside the extraneuronal system. This explains the well-known failure of irreversible inhibitors of MAO to cause "full" inhibition of neuronal MAO (see HENSELING and TRENDELENBURG 1978; STEFANO and TRENDELENBURG 1984; CASSIS et al. 1986). It should be noted that this explanation does not involve any competition between the substrate (noradrenaline) and the irreversible enzyme inhibitor (pargyline). Whatever small percentage of MAO survives the exposure to pargyline, the rate at which it deaminates the amine is determined not only by the extent of inhibition, but also by S_{inside}. If S_{inside} rises by a factor of 50, so does the rate of deamination (provided saturation of the enzyme by S_{inside} is avoided).

6. The Importance of "Lipophilic Entry" of Catecholamines

If uptake$_2$ of the tissue under study is well developed, and if a very low substrate concentration is used, then the O-methylation of the substrate is easily and nearly completely prevented by inhibition of uptake$_2$ (see, for instance, Fig. 1 of UHLIG et al. 1974). However, this situation changes either when the tissue has a poorly developed uptake$_2$ (see FLEIG and TRENDELENBURG 1981, for experiments with guinea-pig papillary muscle) or when high catecholamine concentrations are used (see BRYAN et al. 1984, and BARONE et al. 1985). Under such conditions, the inhibitory effect of uptake$_2$-inhibitors is greatly reduced or lost. Two factors contribute to this finding. On the one hand, any increase in the concentration of the catecholamine leads to a proportional increase in the "lipophilic entry" of the catecholamine into the COMT-containing cells. On the other hand, the V_{max} of the intracellular COMT of extraneuronal O-methylating systems is low, much lower than the V_{max} for uptake$_2$. Hence, the diffusional influx of relatively modest amounts of a catecholamine (modest in comparison to the rate of inward transport) can suffice to generate relatively high rates of O-methylation that are then resistant to inhibitors of uptake$_2$.

This explanation for the inability of uptake$_2$ inhibitors to block the O-methylation of catecholamines offered at high concentrations represents a "one-compartment-system" with two modes of entry (transport and diffusion). It is too early to decide whether there are two COMT-containing compartments, as proposed by BARONE et al. (1985).

II. The Extraneuronal Deaminating System

As described in Sect. C.I.1 for the extraneuronal O-methylating system, "half-saturating outside concentrations" of various amines (known to be substrates of both uptake$_2$ and MAO) were determined in the perfused rat heart after inhibition of COMT (GROHMANN and TRENDELENBURG 1988). The results for five amines are shown in Table 5. Three catecholamines (noradrenaline, adrenaline and dopamine) are substrates of both enzymes. Comparison of the values (for rat heart) in Tables 3 and 5 shows that nearly 20 times higher concentrations of these three catecholamines were needed for half-saturation of the extraneuronal deaminating than for half-saturation of the corresponding O-methylating system. The interpretation of this discrepancy will be discussed in Sect. C.III. As far as noradrenaline is concerned, these recent results agree with those of an earlier study of the half-saturating outside concentrations of noradrenaline for both extraneuronal systems (FIEBIG and TRENDELENBURG 1978b).

Section C.I.3 described how k_{comt} was determined for unsaturated, extraneuronal COMT. In analogous experiments (rat heart, COMT and neuronal uptake inhibited) k_{mao} was determined for unsaturated, extraneuronal MAO (GROHMANN 1987). According to Table 4, for those three catecholamines which are substrates of both enzymes k_{comt} exceeds k_{mao}; or in other words, for these catecholamines the activity of extraneuronal COMT exceeds that of extraneuronal MAO. This is in agreement with the general impression that ex-

Table 5. The "half-saturating outside concentration" (HSOC) of various unlabelled amines; deaminating system of rat heart (vesicular uptake, neuronal uptake and COMT inhibited). Taken from GROHMANN and TRENDELENBURG (1988)

Amine	HSOC (μmol/1)
(−)-Adrenaline	14.6
Tyramine	21.1
Dopamine	40.4
(−)-Noradrenaline	41.7
5-Hydroxytryptamine	52.5

traneuronal COMT is the more important of the two extraneuronal enzymes, provided tissues are exposed to low concentrations of catecholamines. The ranking order of k_{comt}/k_{mao} (adrenaline > noradrenaline > dopamine) in Table 4 indicates that adrenaline is preferentially O-methylated in the extraneuronal cells of the heart, noradrenaline less so, while both enzymes can be expected to play similar roles in the extraneuronal metabolism of dopamine.

However, such statements are applicable only to such low catecholamine concentrations as do not saturate extraneuronal COMT. If concentrations are raised enough to saturate this enzyme, then the relative importance of yet unsaturated MAO increases—and deamination becomes predominant (see GILLESPIE and PREUNER 1976). When further increases in the concentration of catecholamines saturate also extraneuronal MAO, then the extraneuronal system ceases to function as an irreversible site of loss; it becomes an accumulating, reversible site of loss.

As stated above, earlier experiments already gave the general impression that MAO is the less important of the two extraneuronal enzymes. However, part of this impression was due to an experimental artifact. As pointed out by STARKE et al. (1980) and by GROHMANN et al. (1986), the deamination of noradrenaline is greatly hindered by the introduction of a tritium label in position 8 (i.e. on the α-carbon). Unfortunately, virtually all commercial samples of ^{3}H-(−)-noradrenaline that were sold as "^{3}H-7-(−)-noradrenaline" were contaminated by label in position 8. This is true up to the end of 1984. Because of this false labelling of (−)-noradrenaline, (−)-adrenaline and dopamine, the true activity of the extraneuronal MAO was underestimated. However, very recently, pure ^{3}H-7-(−)-noradrenaline, ^{3}H-7-(−)-adrenaline and ^{3}H-7-dopamine have become available — and it should be emphasized that the results presented in Table 4 were obtained with compounds labelled exclusively in position 7.

While the extraneuronal O-methylating system has been studied in a variety of tissues (see Table 3), the extraneuronal deaminating system has received less attention. However, while certain substrates of uptake$_2$ (i.e. the catecholamines) are good substrates of COMT, it should be realized that other substrates of uptake$_2$ are substrates of MAO only. Phenylephrine (RAWLOW et al. 1980) and 5-HT (PAIVA et al. 1984; GROHMANN and TRENDELENBURG 1984) are two examples.

Two subtypes of MAO are know to exist (MAO A und MAO B). CARAMONA and OSSWALD (1985) and CARAMONA et al. (1985) recently demonstrated for the extraneuronal metabolizing system of the dog saphenous vein that noradrenaline is deaminated exclusively by MAO A, while NMN appears to be a substrate of MAO B. However, as far as the rat heart is concerned, it appears to contain mainly MAO A as well as a semicarbazide-sensitive MAO, but no MAO B (LYLES and CALLINGHAM 1975).

III. The Co-existence of the Two Extraneuronal Enzymes

Are there separate sets of COMT- and MAO-containing extraneuronal cells—or are both enzymes co-existing in the same cells? Although this seems to be a simple question, it is difficult to answer.

The extraneuronal formation of O-methylated *and* deaminated metabolites of noradrenaline (i.e. of MOPEG and VMA) is no proof for the co-existence of both enzymes in the same cells. For instance, DOPEG formed in hypothetical MAO-containing cells might well be able to enter COMT-containing cells, since it is a highly lipophilic metabolite (TRENDELENBURG et al. 1979). In analogy, NMN may be formed in one cell type (containing COMT), in order to be deaminated in a second one (containing MAO).

When FIEBIG and TRENDELENBURG (1978a) determined the extraneuronal metabolism of ^{3}H-(−)-noradrenaline in the rat heart, they observed that inhibition of COMT caused a greater reduction of total metabolism than did inhibition of MAO. Such a result appeared to be in conflict with the idea that the two enzymes co-exist in the same cells, but this early study had certain imperfections, such as the use of a concentration of ^{3}H-(−)-noradrenaline which partly saturated COMT.

More recently, this question was studied again in the rat heart, with a tracer concentration of ^{3}H-(−)-noradrenaline which saturated neither COMT nor MAO (TRENDELENBURG 1984a). From steady-state rates of O-methylation (MAO inhibited) or deamination (COMT inhibited), initial rates of uptake of ^{3}H-(−)-noradrenaline into the extraneuronal O-methylating and extraneuronal deaminating systems respectively were calculated. As these initial rates of uptake were identical, it is highly likely that the two enzymes co-exist in the same extraneuronal cells.

If the two enzymes co-exist in the same cells, it is legitimate to assume that the catecholamine distributes into the same morphological compartment, when either MAO or COMT is inhibited. With this in mind, it is of interest to calculate a "pseudo-K_m" for either enzyme from "pseudo-$K_{m\,\mathrm{enzyme}}$" = $V_{\max\,\mathrm{enzyme}}/k_{\mathrm{enzyme}}$. The term "pseudo-$K_{m\,\mathrm{enzyme}}$" is used here to emphasize that only comparative values can be calculated, not the true K_m of the two enzymes. This is so, since the "fractional size" of the O-methylating and deaminating systems is unknown. "Fractional size" indicates that proportion of the total tissue of the rat heart that participates in the extraneuronal uptake and metabolism of the catecholamine; KURAHASHI et al. (1980) documented that "fractional size" must be known for any calculation of $K_{m\,\mathrm{enzyme}}$. From the $V_{\max\text{-values}}$ for the O-methylation and deamination of ^{3}H-(−)-noradrenaline (see FIEBIG and TRENDELENBURG 1978b), and from the k_{enzyme} values presented in Table 4, "pseudo-$K_{m\,\mathrm{enzyme}}$"-values are obtained which amount to about 10 μmol/l for COMT and about 250 μmol/l for MAO.

These values are surprising in so far as the pseudo-K_m for MAO is roughly the same as the K_m determined for MAO and (−)-noradrenaline in tissue homogenates (BENEDETTI et al. 1983), but pseudo-K_m for COMT is distinctly lower than the K_m of COMT for noradrenaline, as determined for purified soluble COMT (see GULDBERG and MARSDEN 1975). In this context, it is of interest to note that evidence is accumulating which indicates that COMT exists not only in the low-affinity soluble form, but also as a membrane-bound enzyme that is characterized by K_m-values for catecholamines in the low micromolar range (RIVETT and ROTH 1982; RIVETT et al. 1982, 1983).

HEAD et al. (1985) presented a very elegant analysis of the localization of COMT. It is based on the well-known need of COMT for S-adenosylmethionine as well as on the inhibition of COMT by calcium (GULDBERG and MARSDEN 1975). When ^{3}H-isoprenaline is added to segments of the rabbit aorta, the O-methylation of the amine is intracellular, since it is sensitive to inhibitors of uptake$_2$. However, on exposure of the tissure to ^{3}H-S-adenosylmethionine, the formation of ^{3}H-O-methyl-isoprenaline must take place in the extracellular space; this is also evident from the fact that a reduction of the extracellular calcium concentration increased the latter but not the former type of O-methylation. Thus, membrane-bound COMT appears to be in contact with both the intra- and the extracellular space. However, because of the high extracellular calcium concentration, and also because of the need of COMT for the intracellularly distributed cofactor, O-methylation is normally carried out by that membrane-bound COMT which faces the interior of the cell.

Further evidence for the existence of a high-affinity and membrane-bound COMT came from REID et al. (1986). From rabbit aorta they obtained a membrane-bound COMT characterized by a K_m of 0.9 µmol/l for isoprenaline ($V_{max} = 105\ \text{pmol} \cdot \text{g}^{-1} \cdot \text{min}^{-1}$), while soluble COMT was of low affinity ($K_m = 121$ µmol/l; $V_{max} = 174\ \text{pmol} \cdot \text{g}^{-1} \cdot \text{min}^{-1}$).

The proposal that the two extraneuronal enzymes co-exist in the same cells can be tested. As described in sects. C.I and C.II, half-saturating outside concentrations (HSOS's; of three catecholamines: (−)-noradrenaline, (−)-adrenaline and dopamine) were determined either after block of MAO (O-methylating system) or after block of COMT (deaminating system). As inhibition of either one of the enzymes should not affect uptake$_2$, the kinetic constants for uptake$_2$ (of any given catecholamine) should be the same in both types of experiments. Also k_{in} and k_{out} (for each catecholamine) should be the same in both types of experiments. Moreover, $V_{max\ comt}$ and $V_{max\ mao}$ have been determined experimentally (see GROHMANN and TRENDELENBURG 1985 for $V_{max\ comt}$, and GROHMANN and TRENDELENBURG 1988, for $V_{max\ mao}$). In order to use equation (1) (sect. C.I.2), we need $K_{m\ comt}$ and $K_{m\ mao}$. As discussed by GROHMANN (1987) a "pseudo-K_m" can be obtained as $V_{max\ enzyme}/k_{enzyme}$, where k_{enzyme} represents the rate constant characterizing the unsaturated intracellular enzyme. This value has to be called "pseudo-K_m", since its true value depends on the "fractional size" of the extraneuronal system (i.e. on what fraction of the perfused heart is participating in extraneuronal uptake and metabolism). However, if the two enzymes co-exist in identical cells, then "pseudo-K_m-values" can be used for calculations (that are based on the assumption that *all* cells of the heart participate).

With the help of Eq. (1), *T/M* ratios were calculated and then converted to Si. With the help of the Michaelis-Menten equation, rates of O-methylation and deamination were then calculated for a variety of outside concentrations (So), and from such calculations the "half-saturating outside concentrations"

Table 6. Comparison of experimental and calculated "half-saturating outside concentrations" (HSOC) for the extraneuronal 0-methylating and extraneuronal deaminating system; rat heart, vesicular and neuronal uptake inhibited, *one* extraneuronal enzyme inhibited. Shown are results for four catecholamines

Catecholamine	HSOC for 0-methylating system (μmol/l)		HSOC for deaminating system (μmol/l)	
	experimental	calculated	experimental	calculated
(−)-Adrenaline	0.76	0.57	14.6	9.9
(±)-Isoprenaline	1.1	1.0	–	–
(−)-Noradrenaline	2.3	1.6	41.7	35.8
Dopamine	2.7	2.2	40.4	32.6

Calculations are described in text. The following constants were used (in descending order of catecholamines): $K_{m\,uptake}$: 87.3, 53.6, 224, 498 μmol/l; $V_{max\,uptake}$: 59.3, 32.0, 47.7, 183 $nmol \cdot g^{-1} \cdot min^{-1}$; k_{in}: 0.0637, 0.0675, 0.0659, 0.0639 min^{-1}; k_{out}: 0.0533, 0.0553; 0.0265; 0.0531 min^{-1}; pseudo-$K_{m\,comt}$: 1.5, 1.6, 3.1, 3.5 μmol/l; $V_{max\,comt}$: 0.72, 1.15, 0.80, 1.13 $nmol \cdot g^{-1} \cdot min^{-1}$; pseudo-$K_{m\,mao}$: 84.0, -, 110, 84.0 μmol/l; $V_{max\,mao}$: 4.45, -, 11.8, 18.8 $nmol \cdot g^{-1} \cdot min^{-1}$. The constants were taken (or calculated) from TRENDELENBURG (1984a), GROHMANN and TRENDELENBURG (1984, 1985, 1988) and GROHMANN (1987)

of three catecholamines were calculated, for the extraneuronal O-methylating and the extraneuronal deaminating system. Table 6 shows the comparison of the calculated and the experimental values (as well as all the constants used in these calculations). The good agreement between experimental and calculated values clearly supports the view that the two intracellular enzymes coexist in the same extraneuronal cells.

It should be mentioned that the discrepancy between the HSOC's for the two systems has two reasons: **a)** as the K_m for MAO usually exceeds the K_m for COMT (see legend to Table 6), the HSOC for the deaminating system should exceed that for the O-methylating system; **b)** as the ratio $V_{max\,uptake}/V_{max\,enzyme}$ is lower for the deaminating than for the O-methylating system (see legend to Table 6), the HSOC for the deaminating system should exceed that for the O-methylating system (see TRENDELENBURG 1984b).

If Eq. (2) (sect. C.I.2) is changed to accomodate *two* intracellular enzymes (i.e. COMT and MAO), we obtain

$$T/M_{st\text{-}st} = \frac{k_{uptake} + k_{in}}{k_{comt} + k_{mao} + k_{out}} \tag{4}$$

provided neither $uptake_2$ nor COMT nor MAO is saturated. From this equation, it is obvious that inhibition of an intracellular enzyme with high activity (i.e. high k_{enzyme}) must increase the *T/M* ratio considerably more than inhibition of an intracellular enzyme with low activity (i.e. low k_{enzyme}). As for (−)-noradrenaline $k_{comt} \gg k_{mao}$ (see Table 4), the results presented in Fig. 6

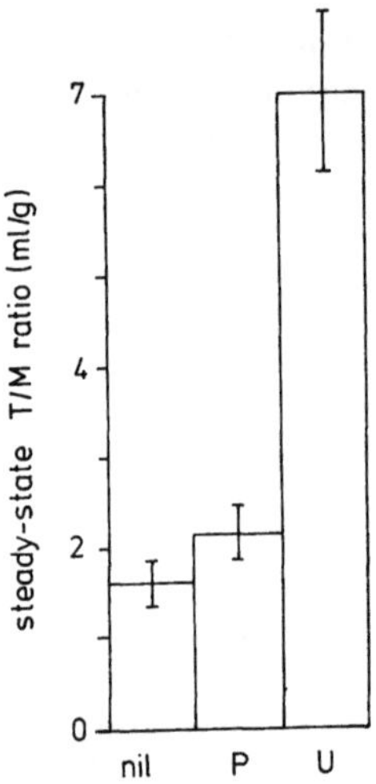

Fig. 6. Extraneuronal steady-state tissue/medium ratio for ^{3}H-(−)-noradrenaline in the perfused rat heart before and after inhibition of either MAO (P) or COMT (U). Hearts were perfused with 1 nmol/l ^{3}H-(−)-noradrenaline for 35 min (in the presence of 30 μmol/l cocaine, to block neuronal uptake), and the ^{3}H-noradrenaline content of the heart was determined. Tissue/medium ratios (in ml/g) were calculated after correction for the distribution of ^{3}H-noradrenaline into the extracellular space. "nil" — both enzymes intact; "P" — MAO inhibited by pretreatment of the rat with pargyline; "U" — COMT inhibited by the presence of 10 μmol/l U-0521. 5 experiments each; taken from TRENDELENBURG (1984a).

confirm the validity of this statement for the extraneuronal system of the rat heart: inhibition of COMT (by U-0521, U) increased the *T/M* ratio for ^{3}H-noradrenaline much more than did inhibition of MAO (by pargyline, P).

D. Supersensitivity to Catecholamines After Inhibition of the Extraneuronal O-Methylating System

If a site of loss is able to generate a concentration gradient from the incubation medium to the extracellular space in the centre of the tissue, inhibition of the site of loss increases the concentration of the agent in the biophase (i.e. in the extracellular space in close proximity to the receptors); supersensitivity to the substrate of the site of loss ensues.

It is typical for extraneuronal O-methylating systems that supersensitivity to catecholamines is induced when either uptake$_2$ or intracellular COMT is impaired. Moreover, the supersensitivity induced by *full* inhibition of one of the two mechanisms cannot be further increased by the additional inhibition of the second one (cat papillar muscle: KAUMANN 1972; cat nictitating membrane: GRAEFE and TRENDELENBURG 1974; pig bronchi: FOSTER et al. 1983; rabbit aorta: BARONE et al. 1987). In the majority of studies of this type of supersensitivity, U-0521 was used to block COMT. Hence, in the following the term "U-0521-induced supersensitivity" will be used.

The degree of supersensitivity that is induced by inhibition of a site of loss is dependent on several factors (see GUIMARAES et al. 1985; GUIMARAES and TRENDELENBURG 1985).

1) When an *unsaturated* site of loss is inhibited, the "maximally possible degree of supersensitivity" to any given agonist is induced. This "maximally possible degree of supersensitivity" is independent of the agonist concentration (provided it does not saturate the site of loss) and it is proportional to the rate constant describing the removal of the agonist from the extracellular space. For extraneuronal O-methylating systems, this rate constant is higher

for isoprenaline than for noradrenaline (PAIVA and GUIMARAES 1984; GROHMANN and TRENDELENBURG 1984); this is one reason why supersensitivity to isoprenaline usually exceeds that to noradrenaline.

2) If the potency of the agonist is low, the agonist concentration required to elicit measurable responses of the tissue may partly saturate the site of loss. Inhibition of a partially saturated site of loss then causes a degree of supersensitivity that is smaller than the "maximally possible degree of supersensitivity". The ratio "K_m of site of loss/EC_{50} of agonist" determines the extent of this reduction of the degree of supersensitivity: if the ratio is >10, then the site of loss is unsaturated, and the "maximally possible degree of supersensitivity" is obtained; if the ratio equals unity, the site of loss is halfsaturated by the agonist concentration, and only about half of the "maximally possible degree of supersensitivity" is obtained on inhibition of that site of loss; if the ratio is <0.1, the site of loss is virtually fully saturated (prior to its inhibition), and no supersensitivity can be expected to ensue from its inhibition (PAIVA and GUIMARAES 1984).

As isoprenaline has a high potency as a β_2-adrenoceptor agonist, while the potency of noradrenaline is much lower, and as the "half-saturating outside concentrations" for isoprenaline and noradrenaline and the extraneuronal O-methylating system are low (see section C.I.1), it follows that the ratio "K_m of site of loss/EC_{50} of agonist" is considerably higher for isoprenaline than for noradrenaline. Thus, there are two factors which together are responsible for the degree of U-0521-induced supersensitivity to be isoprenaline $\gg$ noradrenaline: the "maximally possible supersensitivity" is isoprenaline $>$ noradrenaline, and the ratio "K_m of site of loss/EC_{50}" also is isoprenaline $>$ noradrenaline (GUIMARAES and TRENDELENBURG 1985).

3) A third factor that must be taken into consideration is the thickness of the tissue under study. The term "concentration gradient" is in common use, but we have to realize that the degree of U-0521-induced supersensitivity depends not only on the concentration gradient (decline of agonist concentration in extracellular space per mm of diffusion distance towards the centre of the isolated tissue) but also on the diffusion distance. And the latter is greater in thick than in thin tissues. Hence, it is not surprising that GUIMARAES et al. (1974) reported that U-0521-induced supersensitivity to the beta-effects of isoprenaline was considerably greater in thick than in thin strips of the dog saphenous vein.

4) Finally, we have to consider the possibility that any given site of loss is in close proximity to one type of receptor, but rather distant from a second type of receptor. GUIMARAES and PAIVA (1981a, b) compared the cocaine- and the U-0521-induced degrees of supersensitivity to the α- and the β-effects of catecholamines (mainly adrenaline), and they did so in the same tissue (dog saphenous vein). While cocaine caused a greater degree of supersensitivity to the α- than to the β-effects of adrenaline, the opposite result was obtained on inhibition of COMT (in experiments in which the "other" site of loss was always inhibited). Hence, it is very likely that the α-adrenoceptors of this vascular smooth muscle are in close proximity to adrenergic nerve endings (and at quite some distance from the extraneuronal O-methylating system), while the

β-adrenoceptors are located in the neighbourhood of the extraneuronal O-methylating system (and rather far from the adrenergic nerve endings). The experimental results favour the view of a non-homogeneous distribution of the two receptor types.

Finally, it should be emphasized that the appearance of U-0521-induced supersensitivity is good evidence for the view that the extraneuronal O-methylating systems functions as a site of loss.

Any discussion concerning the effect of inhibition of a site of loss on the sensitivity of a tissue to a substrate of that site of loss is incomplete if the rate-limiting character of the diffusion of the substrate through the extracellular space is not mentioned. The ability of (uninhibited) sites of loss to generate a concentration gradient of the amine from the medium to the centre of the tissue (or even from the biophase to the site of loss, see point (4) above) is related to a discrepancy between the rate constant characterizing diffusion of the amine through the extracellular space and the rate constant characterizing the removal of the amine from the extracellular space by the site of loss: the larger the second rate constant (relative to the first), the greater the concentration gradient. If, for instance, the rate constant for diffusion through the extracellular space were very high, even an avid site of loss would be unable to generate a concentration gradient.

Diffusion through the extracellular space is poor, and this fact increases the probability that substrates of uptake$_2$ that diffuse through the extracellular space are actually taken up (and subsequently metabolized). Unfortunately, this difference cannot be quantified at the present time. However, if the results of GUIMARAES and PAIVA (1981a, b) indicate that the biophase of α-adrenoceptors is (on the average) closer to the neuronal deaminating than to the extraneuronal O-methylating site of loss (and vice versa for the biophase of β-adrenoceptors), then we must consider not only that concentration gradient that is generated between the medium and the centre of the incubated tissue, but also "minigradients" (superimposed onto the first-mentioned gradient) from any given biophase to the nearest site of loss.

The importance of concentration gradients is illustrated by an observation of HENSELING (1983b) with the rabbit aorta. Access of ^{3}H-(−)-noradrenaline was restricted to *one* side of an aortic strip (or ring), and the rate of ^{3}H-NMN formation was determined (after inhibition of uptake$_1$ and of MAO). Irrespective of whether the amine was offered from the intimal or the adventitial side of the tissue, the rate of ^{3}H-NMN-formation was the same. However, when ^{3}H-(−)-noradrenaline was offered to *both* sides of the tissue, the rate of ^{3}H-NMN-formation was nearly doubled. This observation clearly indicates that extraneuronal uptake and O-methylation generates such a steep concentration gradient within the extracellular space of the media that only a limited number of (superficial) layers of smooth muscle cells is responsible for most of the ^{3}H-NMN formation. Indeed, there may well be smooth muscle cells in the centre of the media that contribute very little to the O-methylation of ^{3}H-(−)-noradrenaline, irrespective of the side to which the amine is offered.

Quite different results were recently obtained by MAGARIBUCHI et al. (1987). In agreement with the observations described above they found an inhibitor of COMT (tropolone) to potentiate the positive chronotropic effect of isoprenaline on the perfused rat heart. However, when $uptake_2$ was inhibited as well (in addition to inhibition of COMT), responses to isoprenaline declined. Thus, in this series of experiments the sensitivity to isoprenaline increased when the accumulation of isoprenaline in the extraneuronal compartments increased (block of COMT), but it decreased again when the accumulation decreased (inhibition of COMT and $uptake_2$); hence, the authors regard it as possible that intracellular receptors may be involved in mediating at least part of the positive chronotropic response of the pacemaker to isoprenaline.

E. The Interaction Between the Neuronal and the Extraneuronal Sites of Loss

When noradrenaline is released from the adrenergic nerve endings of a tissue that has a very dense adrenergic innervation (cat's nictitating membrane), most of the metabolites are formed after neuronal re-uptake of the transmitter, while there is very little extraneuronal O-methylation (LUCHELLI-FORTIS and LANGER 1975). Unfortunately, this has not been studied in less densely innervated organs. The rabbit uterus might be highly suited for such a study, since the very pronounced growth of the uterus (under the influence of 17-β-oestradiol plus progesterone) goes hand in hand with a corresponding increase in the capacity of the extraneuronal O-methylating system, while the adrenergic innervation becomes less dense (KENNEDY et al. 1984).

After its release from adrenergic nerve endings, part of the transmitter diffuses into the extracellular space. Apparently, it is this transmitter which is then subject to extraneuronal uptake and metabolism. This is illustrated by observations by BELFRAGE et al. (1977). In a perfused skin flap, the vascular effects of noradrenaline released from adrenergic nerve endings were not affected by inhibition of COMT, presumably because the α-adrenoceptors of the vessels are so close to the adrenergic nerve endings that the extraneuronal site of loss has no influence. However, the response of the adipocytes of the tissue to the released noradrenaline was potentiated, presumably because the adipocytes are at some distance from the adrenergic nerve endings.

Thus, as far as noradrenaline released from nerve endings is concerned, the extraneuronal site of loss has the characteristics of a "second line of defense" (against the overspill of the transmitter into the circulation): it removes from the extracellular space that transmitter which escaped from neuronal re-uptake.

For circulating catecholamines, the situation is different, especially in those blood vessels which have an "asymmetric adrenergic innervation" (located at the adventitio-medial border). For circulating substrates of $uptake_2$ the extraneuronal site of loss represents an effective "barrier". As an example, the experiments of HENSELING (1983b) are of interest (see above).

Moreover, as the extraneuronal site of loss handles not only catecholamines, but also various phenethylamines, imidazoline derivatives, histamine and 5-HT (GROHMANN and TRENDELENBURG 1984), and as the extraneuronal system is able to deaminante 5-HT (PAIVA et al. 1984) and to N-methylate histamine (KALSNER 1970; 1975), the extraneuronal site of loss is somewhat reminiscent of a municipal garbage disposal system that is able to dispose of a large variety of exogenous and endogenous amines that appear in the extracellular space.

When the interaction between the neuronal and the extraneuronal sites of loss is considered for the rabbit aorta, we again have to consider the influence of the rate constant for the diffusion of the amines through the extracellular space. If strips (or rings) of rabbit aorta are incubated in the traditional way with a low, non-saturating concentration of ^{3}H-(−)-noradrenaline, 5 times more ^{3}H-(−)-noradrenaline metabolites are formed in the adrenergic nerve endings than in the extraneuronal tissue (i.e. the smooth muscle of the media). However, before this is interpreted as evidence that the rate constant for "neuronal uptake and deamination" greatly exceeds that for "extraneuronal uptake and O-methylation", a second observation should be considered. When a double chamber is used (see above; HENSELING 1983c), and when either uptake$_1$ or uptake$_2$ was inhibited, in each case about 60% of the ^{3}H-(−)-noradrenaline (offered to *one* side of the tissue) was metabolized on its way through the whole tissue. Hence, in this type of experiment the extraneuronal site of loss appeared to be as important as the neuronal one. The discrepancy between these two types of experiments is due to the fact that the rate constant for diffusion through the extracellular space is three times higher in the adventitia (i.e. around the adrenergic nerve endings) than in the media (i.e. in the neighbourhood of the extraneuronal site of loss) (HENSELING 1983c). Thus, the easy access of ^{3}H-(−)-noradrenaline to the adrenergic nerve endings explains why considerably more neuronal than extraneuronal metabolites are formed when the tissue is incubated in the traditional way.

While neuronal deamination of noradrenaline often predominates over extraneuronal O-methylation (see above), an inverse relation is seen for adrenaline (PAIVA and GUIMARAES 1978). Three factors combine to account for the predominant extraneuronal O-methylation of adrenaline: a) it is not as good a substrate of neuronal uptake, b) because of the N-methyl group, it is not as good a substrate of neuronal and extraneuronal MAO as is noradrenaline, and c) the rate constant for uptake$_2$ of adrenaline is higher than that for noradrenaline (PAIVA and GUIMARAES 1984; GROHMANN and TRENDELENBURG 1984).

Finally, it is of interest to note that the activity of the extraneuronal O-methylating system appears to be regulated by the adrenergic innervation. In the seventies, several attempts were made to see whether there is a neuronal COMT. In several organs, sympathetic denervation was found to cause a small decline of COMT activity determined in homogenates (JARROTT 1971; JARROTT and LANGER 1971). A small proportion of COMT activity appeared to be located in adrenergic nerve endings. However, more recent experiments lead to a different conclusion. The first hint came from a study of SALT and IVERSEN (1973) who found uptake$_2$ to be slightly reduced in the hearts of 6-hydroxydopamine-treated rats. More recently, BRANCO et al. (1984) denervated the dog saphenous vein and the rabbit ear artery and then determined the O-methylation of ^{3}H-isoprenaline. The observed reduction of the O-methylation of ^{3}H-isoprenaline in sympathetically denervated tissues went hand in hand with distinct morphological changes: the smooth muscle cells of the denervated tissues showed clear signs of de-differentiation. Hence, it is very likely that the sympathetic innervation has a trophic influence on the extraneuronal system responsible for the uptake and O-methylation of catecholamines. Moreover, a neuronal localization of COMT is highly unlikely.

F. Conclusions: A Comparison of Uptake$_2$ with Uptake$_1$

While uptake$_1$ is absolutely dependent on the presence (in the extracellular space) of Na^+ and Cl^-, uptake$_2$ shows no such dependence. While—under resting conditions—the carrier molecules of uptake$_1$ appear to be distributed asymmetrically, there is no evidence for any asymmetric distribution of the carrier of uptake$_2$. Both carriers can transport substrates in either direction. However, while a carrier-mediated outward transport of catecholamines by uptake$_2$ is easily demonstrable in any normal tissue, carrier-mediated outward transport by uptake$_1$ can be observed only under special conditions (i.e. either when a change in the sodium gradient changes the asymmetrical distribution of the carrier, or when the inward transport of a substrate increases the availability of the carrier on the inside). In terms of substrate specificity, there are clear-cut differences between uptake$_1$ and uptake$_2$. In terms of the kinetic constants for inward transport, uptake$_1$ tends to have a considerably lower K_m for the catecholamines than has uptake$_2$.

While the neuronal deaminating system and the extraneuronal O-methylating system have the same basic composition (i.e. an intracellular enzyme to which the substrate is transported by an uptake mechanism), they differ strikingly in various functional aspects. For the extraneuronal O-methylating system, $V_{max\,uptake} \gg V_{max\,enzyme}$. Consequently, this system functions as an irreversible site of loss only at low outside catecholamine concentrations; if exposed to concentrations in the micromolar range, the catecholamine is accumulated in the extraneuronal cells. Although V_{max} for neuronal MAO is unknown, all the evidence is compatible with the view that, for the neuronal deaminating system, $V_{max\,uptake} < V_{max\,enzyme}$. Hence, the nerve ending functions as an irreversible site of loss at virtually *all* outside concentrations.

As it is very likely that extraneuronal COMT and MAO co-exist in the same cells, the COMT of the intact extraneuronal cells appears to have a higher affinity to catecholamines than has been established for soluble COMT. Hence, it is likely that high-affinity, membrane-bound COMT is functionally important in intact cells.

Because metabolizing systems are comprised of two saturable processes arranged in series, they have certain complex kinetic characteristics. For instance, when $V_{max\,uptake} > V_{max\,enzyme}$, the "half-saturating outside concentration" delineates that concentration below which the system functions predominantly as an irreversible site of loss, but above which it functions as a reversible, accumulating site of loss. Moreover, any inhibition of the intracellular enzyme increases the steady-state accumulation of the catecholamine inside the system.

Acknowledgements. The author's work reported here was supported by the Deutsche Forschungsgemeinschaft (partly Tr. 96, partly SFB 176).

G. References

Almgren O, Jonason J (1974) On the role of the adrenergic receptors for extraneuronal amine uptake and retention in rat salivary glands in vitro. Naunyn-Schmiedeberg's Arch Pharmacol 283:21-35

Almgren O, Jonason J (1976) Extraneuronal amine transport in glandular tissue. In: Paton DM (ed) The Mechanism of Neuronal and Extraneuronal Transport of Catecholamines. Raven Press, New York, pp 299-311

Anning EN, Bryan LJ, O'Donnell SR (1979) The extraneuronal accumulation of isoprenaline in trachea and atria of guinea-pig and cat: a fluorescence histochemical study. Br J Pharmacol 65:175-182

Azevedo I, Bönisch H, Osswald W, Trendelenburg U (1983) Autoradiographic study of rat hearts perfused with ^{3}H-isoprenaline. Naunyn-Schmiedeberg's Arch Pharmacol 322:1-5

Barone S, Stitzel RE, Head RJ (1983) The vascular extraneuronal uptake (ENU) and O-methylation of isoproterenol (ISO) is a stereoselective process. Pharmacologist 25:267

Barone S, Stitzel RE, Head RJ (1985) The stereoselective O-methylation of isoprenaline in the isolated rabbit thoracic aorta. Naunyn-Schmiedeberg's Arch Pharmacol 329:9-17

Barone S, Panek D, Bennett L, Stitzel RE, Head RJ (1987) The influence of oestrogen and oestrogen metabolites on the sensitivity of the isolated rabbit aorta to catecholamines. Naunyn-Schmiedeberg's Arch Pharmacol 335:513-520

Belfrage E, Fredholm BB, Rosell S (1977) Effect of catechol-O-methyl-transferase (COMT) inhibition on the vascular and metabolic responses to noradrenaline, isoprenaline and sympathetic nerve stimulation in canine subcutaneous adipose tissue. Naunyn-Schmiedeberg's Arch Pharmacol 300:11-17

Benedetti MS, Boucher T, Fowler CJ (1983) The deamination of noradrenaline and 5-hydroxytryptamine by rat brain and heart monoamine oxidase and their inhibition by cimoxatone, toloxatone and MD 770222. Naunyn-Schmiedeberg's Arch Pharmacol 323:315-320

Bönisch H (1978) Further studies on the extraneuronal uptake and metabolism of isoprenaline in the perfused rat heart. Naunyn-Schmiedeberg's Arch Pharmacol 303:121-131

Bönisch H, Trendelenburg U (1974) Extraneuronal removal, accumulation and O-methylation of isoprenaline in the perfused heart. Naunyn-Schmiedeberg's Arch Pharmacol 283:191-218

Bönisch H, Uhlig W, Trendelenburg U (1974) Analysis of the compartments involved in the extraneuronal storage and metabolism of isoprenaline in the perfused heart. Naunyn-Schmiedeberg's Arch Pharmacol 283:223-244

Bönisch H, Graefe K-H, Trendelenburg U (1978) The determination of the rate constant for the efflux of an amine from efflux curves for amine and metabolite. Naunyn-Schmiedeberg's Arch Pharmacol 304:147-155

Bönisch H, Bryan LJ, Henseling M, O'Donnell SR, Stockmann P, Trendelenburg U (1985) The effect of various ions on uptake$_2$ of catecholamines. Naunyn-Schmiedeberg's Arch Pharmacol 328:407-416

Branco D, Azevedo I, Sarmento A, Osswald W (1981) The fate of isoprenaline in the isolated rabbit aorta. Radiochemical and morphologic observations. Naunyn-Schmiedeberg's Arch Pharmacol 316:120-125

Branco D, Teixeira AA, Azevedo I, Osswald W (1984) Structural and functional alterations caused at the extraneuronal level by sympathetic denervation of blood vessels. Naunyn-Schmiedeberg's Arch Pharmacol 326:302-312

Braun RA, O'Donnell SRO (1982) Some evidence for the extraneuronal uptake of rimiterol and its metabolism by catechol-O-methyltransferase in guinea-pig trachealis smooth muscle cells. J Pharm Pharmacol 34:668-671

Bryan LJ, O'Donnell SR (1984a) Stereoselectivity of extraneuronal uptake of catecholamines in guinea-pig trachealis smooth muscle cells. Br J Pharmacol 82:757-762

Bryan LJ, O'Donnell SR (1984b) Some factors affecting the extraneuronal accumulation of adrenaline in guinea-pig trachealis smooth muscle cells. Naunyn-Schmiedeberg's Arch Pharmacol 326:268-272

Bryan LJ, Cole JJ, O'Donnell SR, Wanstall JC (1981) A study designed to explore the hypothesis that beta-1 adrenoceptors are "innervated" receptors and beta-2 adrenoceptors are "hormonal" receptors. J Pharmacol exp Ther 216:395-400

Bryan LJ, Fleig H, Trendelenburg U (1983) A comparative study of the properties of the catechol-O-methyltransferase inhibitors, U-0521 and tropolone acetamide, in rat perfused heart. Naunyn-Schmiedeberg's Arch Pharmacol 322:6-19

Bryan LJ, O'Donnell SR, Trendelenburg U (1984) Kinetics of the O-methylating system for isoprenaline in the trachea and aorta of rabbit. Naunyn-Schmiedeberg's Arch Pharmacol 328:56-61

Burgen ASV, Iversen LL (1965) The inhibition of noradrenaline uptake by sympathomimetic amines in the rat isolated heart. Brit J Pharmacol 25:34-49

Caramona MM, Osswald W (1985) Effects of clorgyline and (−)-deprenyl on the deamination of normetanephrine and noradrenaline in strips and homogenates of the canine saphenous vein. Naunyn-Schmiedeberg's Arch Pharmacol 328:396-400

Caramona MM, Aráujo D, Brandao F (1985) Influence of MAO A and MAO B on the inactivation of noradrenaline in the saphenous vein of the dog. Naunyn-Schmiedeberg's Arch Pharmacol 328:401-406

Cassis L, Ludwig J, Grohmann M, Trendelenburg U (1986) The effect of partial inhibition of monoamine oxidase on the steady-state rate of deamination of ^{3}H-catecholamines in two metabolizing systems. Naunyn-Schmiedeberg's Arch Pharmacol 333:253-261

Chinet A, Durand J (1979) Control of the brown fat respiratory response to noradrenaline by catechol-O-methyltransferase. Biochem Pharmacol 28:1353-1361

Cornish EJ, Goldie RG, Miller RC (1978) Catecholamine uptake by isolated coronary arteries and atria of the kitten. Br J Pharmacol 63:445-456

De la Lande IS, Harvey JA, Holt S (1974) Response of the rabbit coronary arteries to autonomic agents. Blood Vessels 11:319-337

Ebner F, Waud DR (1978) The role of uptake of noradrenaline for its positive inotropic effect. Statistical evaluation. Naunyn-Schmiedeberg's Arch Pharmacol 303:1-6

Eckert E, Henseling M, Trendelenburg U (1976) The effect of inhibitors of extraneuronal uptake on the distribution of ^{3}H-($\pm$)-noradrenaline in nerve-free rabbit aortic strips. Naunyn-Schmiedeberg's Arch Pharmacol 293:115–127

Ehinger B, Sporrong B (1968) Neuronal and extraneuronal localization of noradrenaline in the rat heart after perfusion at high concentration. Experientia 24:265–266

Farnebo LO (1968) Histochemical studies on the uptake of noradrenaline in the perfused rat heart. Brit J Pharmacol 34:227P–228P

Fiebig ER, Trendelenburg U (1978a) The neuronal and extraneuronal uptake and metabolism of ^{3}H-($-$)-noradrenaline in the perfused rat heart. Naunyn-Schmiedeberg's Arch Pharmacol 303:21–35

Fiebig ER, Trendelenburg U (1978b) The kinetic constants for the extraneuronal uptake and metabolism of ^{3}H-($-$)-noradrenaline in the perfused rat heart. Naunyn-Schmiedeberg's Arch Pharmacol 303:37–45

Fleig H, Trendelenburg U (1981) The extraneuronal sites of loss for catecholamines in guinea-pig papillary muscle. Naunyn-Schmiedeberg's Arch Pharmacol 316:14–18

Fleig HA, Patil PN, Krieglstein GK (1987) The neuronal and extraneuronal uptake and metabolism of ^{3}H-($-$)-noradrenaline in the rabbit iris. Int Ophthalmol 10:15–22

Foster PS, Goldie RG, Paterson JW (1983) Effect of steroids on β-adrenoceptor-mediated relaxation of pig bronchus. Brit J Pharmacol 78:441–445

Garland LG, Martin GR (1984) Extraneuronal metabolism and supersensitivity to catecholamines in guinea pig trachea. In: Fleming WW, Langer SZ, Graefe KH, Weiner N, (eds) Neuronal and extraneuronal events in autonomic pharmacology. Raven Press, New York, pp 139–155

Garland LG, Marrion NV, Martin GR (1981) The extraneuronal O-methylation of ^{3}H-($\pm$)-isoprenaline by guinea-pig tracheal rings in vitro. Naunyn-Schmiedeberg's Arch Pharmacol 318:88–93

Garrett J, Branco D (1977) Uptake and metabolism of noradrenaline by the mesenteric arteries of the dog. Blood Vessels 14:43–54

Gillespie JS, Muir TC (1970) Species and tissue variation in extraneuronal and neuronal accumulation of noradrenaline. J Physiol (Lond.) 206:591–604

Gillespie JS, Preuner J (1976) Influence of iproniazid and 3′, 4′-dihydroxy-2-methylpropiophenone (U-0521) on uptake and metabolism of norepinephrine by rabbit colon. Naunyn-Schmiedeberg's Arch Pharmacol 293:R3

Gillespie JS, Towart R (1973) Uptake kinetics and ion requirements for extraneuronal uptake of noradrenaline by arterial smooth muscle and collagen. Brit J Pharmacol 47:556–567

Gillis CN (1976) Extraneuronal transport of noradrenaline in the lung. In: Paton DM (ed) The Mechanism of Neuronal and Extraneuronal Transport of Catecholamines. Raven Press, New York, pp 281–297

Graefe K-H (1981) The disposition of ^{3}H-($-$)-noradrenaline in the perfused cat and rabbit heart. Naunyn-Schmiedeberg's Arch Pharmacol 318:71–82

Graefe K-H, Trendelenburg U (1974) The effect of hydrocortisone on the sensitivity of the isolated nictitating membrane to catecholamines. Relationship to extraneuronal uptake and metabolism. Naunyn-Schmiedeberg's Arch Pharmacol 286:1–48

Graefe K-H, Bönisch H, Keller B (1978) Saturation kinetics of the adrenergic neurone uptake system in the perfused rabbit heart. A new method for determination of initial rates of amine uptake. Naunyn-Schmiedeberg's Arch Pharmacol 302:263–273

Green AL (1976) The kinetics of enzyme action and inhibition in intact tissues and tissue slices, with special reference to cholinesterase. J Pharm Pharmacol 28:265-274

Grohmann M (1987) The activity of the neuronal and extraneuronal catecholamine-metabolizing enzymes of the rat perfused heart. Naunyn-Schmiedeberg's Arch Pharmacol 336:139-147

Grohmann M (1988) The acceleration of the extraneuronal efflux of ^{3}H-(±)-isoprenaline induced by high extracellular potassium. Naunyn-Schmiedeberg's Arch Pharmacol (in press)

Grohmann M, Trendelenburg U (1983) The isotope effect of tritium in ^{3}H-(−)-adrenaline with very high specific activity. Naunyn-Schmiedeberg's Arch Pharmacol 324:233-234

Grohmann M, Trendelenburg (1984) The substrate specificity of uptake$_2$ in the rat heart. Naunyn-Schmiedeberg's Arch Pharmacol 328:164-173

Grohmann M, Trendelenburg U (1985) The handling of five catecholamines by the extraneuronal O-methylating system of the rat heart. Naunyn-Schmiedeberg's Arch Pharmacol 329:264-270

Grohmann M, Trendelenburg U (1988) The handling of five amines by the extraneuronal deaminating system of the rat heart. Naunyn-Schmiedeberg's Arch Pharmacol 337:159-163

Grohmann M, Henseling M, Cassis L, Trendelenburg U (1986) The errors introduced by a tritium label in position 8 of catecholamines. Naunyn-Schmiedeberg's Arch Pharmacol 332:34-42

Guimaraes S, Paiva MQ (1981a) Two distinct adrenoceptor-biophases in the vasculature: one for α- and the other for β-agonists. Naunyn-Schmiedeberg's Arch Pharmacol 316:195-199

Guimaraes S, Paiva QM (1981b) Two different biophases for adrenaline released by electrical stimulation or tyramine from the sympathetic nerve endings of the dog saphenous vein. Naunyn-Schmiedeberg's Arch Pharmacol 316:200-204

Guimareas S, Trendelenburg U (1985) Deviation supersensitivity and inhibition of saturable sites of loss. TIPS 6:371-374

Guimaraes S, Azevedo I, Cardoso W, Oliveira MC (1974) Relation between the amount of smooth muscle of venous tissue and the degree of supersensitivity to isoprenaline caused by inhibition of catechol-O-methyl transferase. Naunyn-Schmiedeberg's Arch Pharmacol 286:401-412

Guimaraes S, Paiva MQ, Moura D, Proenca J (1985) The saturability of a site of loss and the degree of supersensitivity to agonists which are substrates of this site of loss. Naunyn-Schmiedeberg's Arch Pharmacol 329:30-35

Gulati OD, Sivaramakrishna N (1975) Kinetics and some characteristics of uptake of noradrenaline by the human umbilical artery. Br J Pharmacol 53:152-154

Guldberg HC, Marsden CA (1975) Catechol-O-methyl transferase: Pharmacological aspects and physiological role. Pharmacol Rev, 27:135-206

Haeusler G (1978) Relationship between noradrenaline-induced depolarization and contraction in vascular smooth muscle. Blood Vessels 15:46-54

Haeusler G (1983) Contraction, membrane potential, and calcium fluxes in rabbit pulmonary arterial muscle. Fed Proc 42:263-268

Haeusler G, de Peyer J-E (1986) Role of membrane electrical events in endothelium and non-endothelium dependent vasodilatation. Naunyn-Schmiedeberg's Arch Pharmacol 332:R56

Hamberger B, Norberg K-A, Olson L (1967) Extraneuronal binding of catecholamines and 3,4-dihydroxyphenylalanine (dopa) in salivary glands. Acta physiol scand 69:1-12

Head RJ, de la Lande IS, Irvine RJ, Johnson SM (1980) The metabolism of isoproterenol in the rabbit ear artery. Blood Vessels 17:229-245

Head RJ, Irvine RJ, Barone S, Stitzel RE, de la Lande IS (1985) Nonintracellular, cell-associated O-methylation of isoproterenol in the isolated rabbit thoracic aorta. J Pharmacol Exp Ther 234:184-189

Hedner J, Almgren O, Jonason J, Lundberg D (1981) Neuronal and extraneuronal uptake of ^{3}H-noradrenaline in rat portal vein in vitro. Acta Physiol Scand 111:171-177

Hendley ED (1976) The mechanism of extraneuronal transport of catecholamines in the central nervous system. In: Paton DM (ed) The mechanism of Neuronal and Extraneuronal Transport of Catecholamines. Raven Press, New York, pp 313-324

Hendley ED, Taylor KM, Snyder SH (1970) ^{3}H-Normetanephrine uptake in rat brain slices. Relationship to extraneuronal accumulation of norepinephrine. Europ J Pharmacol 12:167-179

Henseling M (1983a) Kinetic constants for uptake and metabolism of ^{3}H-(−)-noradrenaline in rabbit aorta. Possible falsification of the constants by diffusion barriers within the vessel wall. Naunyn-Schmiedeberg's Arch Pharmacol 323:12-23

Henseling M (1983b) The influence of uptake$_2$ on the neuroeffector transmission in the isolated rabbit aorta. Naunyn-Schmiedeberg's Arch Pharmacol 324:99-107

Henseling M (1983c) The role of neuronal and extraneuronal uptake in the inactivation of ^{3}H-(−)-noradrenaline in the rabbit aorta determined by a method relating uptake with amine diffusion in the tissue. Naunyn-Schmiedeberg's Arch Pharmacol 324:163-168

Henseling M (1984) Stereoselectivity of extraneuronal uptake of catecholamines in rabbit aorta. Naunyn-Schmiedeberg's Arch Pharmacol 328:219-220

Henseling M, Trendelenburg U (1978) Stereoselectivity of the accumulation and metabolism of noradrenaline in rabbit aortic strips. Naunyn-Schmiedeberg's Arch Pharmacol 302:195-206

Iversen LL (1963) The uptake of noradrenaline by the isolated perfused rat heart. Brit J Pharmacol 21:523-537

Iversen LL (1965a) The uptake of catechol amines at high perfusion concentrations in the rat isolated heart: A novel catechol amine uptake process. Brit J Pharmacol 25:18-33

Iversen LL (1965b) The uptake of adrenaline by the rat isolated heart. Brit J Pharmacol 24:387-394

Iversen LL (1967) The uptake and storage of noradrenaline in sympathetic nerves, Cambridge University Press, Cambridge, England.

Iversen LL, Salt PJ (1970) Inhibition of catecholamine uptake$_2$ by steroids in the isolated rat heart. Brit J Pharmacol 40:528-530

Iversen LL, Salt PJ, Wilson HA (1972) Inhibition of catecholamine uptake in the isolated rat heart by haloalkylamines related to phenoxybenzamine. Brit J Pharmacol 46:647-657

Jarrott B (1971) Occurrence and properties of catechol-O-methyl transferase in adrenergic neurons. J Neurochem 18:17-27

Jarrott B and Langer SZ (1971) Changes in monoamine oxidase and catechol-O-methyl transferase activities after denervation of the nictitating membrane of the cat. J Physiol (Lond) 212:549-559

Kalsner S (1970) Effects of tyramine on responses to and inactivation of histamine in aortic strips. J Pharmacol exp Ther 175:489-495

Kalsner S (1975) Role of extraneuronal mechanisms in the termination of contractile responses to amines in vascular tissue. Brit J Pharmacol 53:267-277

Kaumann AJ (1970) Adrenergic receptors in heart muscle: relations among factors influencing the sensitivity of the cat papillary muscle to catecholamines. J Pharmacol exp Ther. 173:383-398

Kaumann AJ (1972) Potentiation of the effects of isoprenaline and noradrenaline by hydrocortisone in cat heart muscle. Naunyn-Schmiedeberg's Arch Pharmacol. 273:134-153

Kaumann AJ, Birnbaumer L, Wittmann R (1978) Heart β-adrenoceptors. In O'Malley B, Birnbaumer L (eds) Receptors and Hormone Action, vol III. Academic Press, New York, pp 133-177

Kennedy JA, de la Lande IS (1984) The effect of pregnancy on the metabolism of noradrenaline in reproductive organs of the rabbit. Naunyn-Schmiedeberg's Arch Pharmacol 326:143-147

Kennedy JA, de la Lande IS (1986) Effect of progesterone on the metabolism of noradrenaline in rabbit uterine endometrium and myometrium. Naunyn-Schmiedeberg's Arch Pharmacol 333:368-376

Kennedy JA, de la Lande IS (1987) Characteristics of the cocaine-sensitive accumulation and O-methylation of ^{3}H-(−)-noradrenaline by rabbit endometrium. Naunyn-Schmiedeberg's Arch Pharmacol 336:148-154

Kennedy JA, de la Lande IS, Morris RG (1984) Effect of ovarian steroids on the metabolism of noradrenaline in rabbit uterus. Naunyn-Schmiedeberg's Arch Pharmacol 326:132-142

Kurahashi K, Rawlow A, Trendelenburg U (1980) A mathematical model representing the extraneuronal O-methylating system of the perfused rat heart. Naunyn-Schmiedeberg's Arch Pharmacol 311:17-32

Landsberg L (1976) Extraneuronal uptake and metabolism of ^{3}H-L-norepinephrine by the rat duodenal mucosa. Biochem Pharmacol 25:729-731

Levin JA (1974) The uptake and metabolism of ^{3}H-l- and ^{3}H-dl-norepinephrine by intact rabbit aorta and by isolated adventitia and media. J Pharmacol Exp Ther 190:210-226

Luchelli-Fortis MA, Langer SZ (1975) Selective inhibition by hydrocortisone of ^{3}H-normetanephrine formation during ^{3}H-transmitter release elicited by nerve stimulation in the isolated nerve-muscle preparation of the cat nictitating membrane. Naunyn-Schmiedeberg's Arch Pharmacol 287:261-275

Ludwig J, Grohmann M, Trendelenburg U (1986) Inhibition by K^{+} of uptake$_2$ of ^{3}H-(±)-isoprenaline in the perfused rat heart. Naunyn-Schmiedeberg's Arch Pharmacol 334:393-396

Lyles GA, Callingham BA (1975) Evidence for a clorgyline-resistant monoamine metabolizing activity in the rat heart. J Pharm Pharmacol 27:682-691

Mack F, Bönisch H (1979) Dissociation constants and lipophilicity of catecholamines and related compounds. Naunyn-Schmiedeberg's Arch Pharmacol 310:1-9

Magaribuchi T, Hama T, Kurahashi K, Fujiwara M (1987) Effects of extraneuronal uptake inhibitors on the positive chronotropic response to isoprenaline and on the accumulation of isoprenaline in perfused rat heart after inhibition of catechol-O-methyl transferase. Naunyn-Schmiedeberg's Arch Pharmacol 335:123-128

Major H, Sauerwein I, Graefe K-H (1978) Kinetics of the uptake and metabolism of ^{3}H-(±)-isoprenaline in the rat submaxillary gland. Naunyn-Schmiedeberg's Arch Pharmacol 305:51-63

Malmfors T (1967) Fluorescent histochemical studies on the uptake, storage and release of catecholamines. Circ Res, Suppl III to vols XX and XXI, pp III-25-III-42

Mekanontchai R, Trendelenburg U (1979) The neuronal and extraneuronal distribution of ^{3}H-(−)-noradrenaline in the perfused rat heart. Naunyn-Schmiedeberg's Arch Pharmacol 308:199-210
Mireylees SE, Foster RW (1973) 3-Methoxyisoprenaline: a potent selective uptake$_2$ inhibitor. J Pharm Pharmacol 25:833-835
O'Donnell SR, Reid JJ (1984) Kinetic constants of isoprenaline and corticosterone for extraneuronal uptake in different cell types from various tissues. Naunyn-Schmiedeberg's Arch Pharmacol 325:54-61
Paiva MQ, Guimaraes S (1978) A comparative study of the uptake and metabolism of noradrenaline and adrenaline by the isolated saphenous vein of the dog. Naunyn-Schmiedeberg's Arch Pharmacol 303:221-228
Paiva MA, Guimaraes S (1984) The kinetic characteristics of the extraneuronal O-methylating system of the dog saphenous vein and the supersensitivity to catecholamines caused by its inhibition. Naunyn-Schmiedeberg's Arch Pharmacol 327:48-55
Paiva MQ, Caramona M, Osswald W (1984) Intra- and extraneuronal metabolism of 5-hydroxytryptamine in the isolated saphenous vein of the dog. Naunyn-Schmiedeberg's Arch Pharmacol 325:62-68
Parker DAS, de la Lande IS, Proctor C, Marino V, Lam NX, Parker I (1987) Cocaine-sensitive O-methylation of noradrenaline in dental pulp of the rabbit: comparison with the rabbit ear artery. Naunyn-Schmiedeberg's Arch Pharmacol 335:32-39
Patil PN, Trendelenburg U (1982) The extraneuronal uptake and metabolism of ^{3}H-isoprenaline in the rabbit iris. Naunyn-Schmiedeberg's Arch Pharmacol 318:158-165
Paton DM (1973) Evidence for carrier-mediated efflux of noradrenaline from the axoplasm of adrenergic nerves in rabbit atria. J Pharm Pharmacol 25:265-267
Petersen OH (1976) Electrophysiology of mammalian gland cells. Physiol Rev 56:535-577
Powis G (1973) The accumulation and metabolism of (−)-noradrenaline by cells in culture. Brit J Pharmacol 47:568-575
Rawlow A, Fleig H, Kurahashi K, Trendelenburg U (1980) The neuronal and extraneuronal uptake and deamination of ^{3}H-(−)-phenylephrine in the perfused rat heart. Naunyn-Schmiedeberg's Arch Pharmacol 314:237-247
Reid J, Stitzel RE, Head RJ (1986) Characterization of the O-methylation of catechol oestrogens by intact rabbit thoraric aorta and subcellular fractions thereof. Naunyn-Schmiedeberg's Arch Pharmacol 334:17-28
Rivett AJ, Roth JA (1982) Kinetic studies on the O-methylation of dopamine by human brain membrane-bound catechol O-methyltransferase. Biochem 21:1740-1742
Rivett AJ, Eddy BJ, Roth JA (1982) Contribution of sulfate conjugation, deamination and O-methylation to metabolism of dopamine and norepinephrine in human brain. J Neurochem 39:1009-1016
Rivett AJ, Francis A, Roth JA (1983) Localization of membrane-bound catechol-O-methyltransferase. J Neurochem 40:1494-1496
Salt PJ (1972) Inhibition of noradrenaline uptake$_2$ in the isolated rat heart by steroids, clonidine and methoxylated phenylethylamines. Europ J Pharmacol 20:329-340
Salt PJ, Iversen LL (1973) Catecholamine uptake sites in the rat heart after 6-hydroxydopamine treatment and in a genetically hypertensive strain. Naunyn-Schmiedeberg's Arch Pharmacol 279:381-386
Sammet S, Graefe K-H (1979) Kinetic analysis of the interaction between noradrenaline and Na^+ in neuronal uptake: kinetic evidence for co-transport. Naunyn-Schmiedeberg's Arch Pharmacol 309:99-107

Schömig E, Körber M, Bönisch K (1988) Kinetic evidence for a common binding site for substrates and inhibitors of the neuronal noradrenaline carrier. Naunyn-Schmiedeberg's Arch Pharmacol (in press)

Simmonds MA, Gillis CN (1968) Uptake of normetanephrine and norepinephrine by cocaine-treated rat heart. J Pharmacol Exp Ther 159:283-289

Starke K, Steppeler A, Zumstein A, Henseling M, Trendelenburg U (1980) False labelling of commercially available ^{3}H-catecholamines? Naunyn-Schmiedeberg's Arch Pharmacol 311:109-112

Stefano FJE, Trendelenburg U (1984) Saturation of monoamine oxidase by intraneuronal noradrenaline accumulation. Naunyn-Schmiedeberg's Arch Pharmacol 328:135-141

Trendelenburg U (1980) A kinetic analysis of extraneuronal uptake and metabolism of catecholamines. Rev Physiol Biochem Pharmacol 87:33-115

Trendelenburg U (1984a) The influence of inhibition of catechol-O-methyl transferase or of monoamine oxidase on the extraneuronal metabolism of ^{3}H-(−)-noradrenaline in the rat heart. Naunyn-Schmiedeberg's Arch Pharmacol 327:285-292

Trendelenburg U (1984b) Metabolizing systems. In: Fleming WW, Langer SZ, Graefe KH, Weiner N (eds) Neuronal and extraneuronal events in autonomic pharmacology. Raven Press, New York, pp 93-109

Trendelenburg U (1985) The outward transport of catecholamines by uptake$_2$ of the rat heart. Naunyn-Schmiedeberg's Arch Pharmacol 330:203-211

Trendelenburg U (1986) Modulation of uptake$_2$ of ^{3}H-(±)-isoprenaline by isoprenaline-induced depolarization of rat salivary gland cells. Naunyn-Schmiedeberg's Arch Pharmacol 334:388-392

Trendelenburg U (1987) The membrane potential of vascular smooth muscle appears to modulate uptake$_2$ of ^{3}H-isoprenaline. Naunyn-Schmiedeberg's Arch Pharmacol 336:33-36

Trendelenburg U, Bönisch H, Graefe K-H, Henseling M (1979) The rate constant for the efflux of metabolites of catecholamines and phenethylamines. Pharmacol Rev 31:179-203

Trendelenburg U, Stefano FJE, Grohmann M (1983a) The isotope effect of tritium in ^{3}H-noradrenaline. Naunyn-Schmiedeberg's Arch Pharmacol 323:128-140

Trendelenburg U, Fleig H, Bryan LJ, Bönisch H (1983b) The extraneuronal compartments for the distribution of isoprenaline in the rat heart. Naunyn-Schmiedeberg's Arch Pharmacol 324:169-179

Uhlig W, Bönisch H, Trendelenburg U (1974) The O-methylation of extraneuronally stored isoprenaline in the perfused heart. Naunyn-Schmiedeberg's Arch Pharmacol 283:245-261

Uhlig W, Fiebig R, Trendelenburg U (1976) The effect of corticosterone on the fluxes of ^{3}H-normetanephrine into and out of the extraneuronal compartments of the perfused rat heart. Naunyn-Schmiedeberg's Arch Pharmacol 295:45-50

Ungell A-L, Graefe K-H (1987) Failure of K^{+} to affect the potency of inhibitors of the neuronal noradrenaline carrier in the rat vas deferens. Naunyn-Schmiedeberg's Arch Pharmacol 335:250-254

Wilbrandt W, Rosenberg T (1961) The concept of carrier transport and its corrolaries in pharmacology. Pharmacol Rev 13:109-183

Wilson VG, Trendelenburg U (1988) The uptake and O-methylation of ^{3}H-(±)-isoprenaline in rat cerebral cortex slices. Naunyn-Schmiedeberg's Arch Pharmacol (in press)

CHAPTER 7

Catecholamine Receptors

B. B. WOLFE and P. B. MOLINOFF

A. Introduction

Specific receptors for drugs, hormones, and neurotransmitters are found on the surface membranes of many types of cells. Interaction of a receptor with an appropriate agonist results in a specific cellular response. This definition of a receptor is intended to distinguish receptors from binding sites for drugs that are not associated with a biologically meaningful function. For example, activation of β-adrenoceptors and at least some dopamine receptors leads to activation of the enzyme adenylate cyclase (ROBISON et al. 1971; KEBABIAN et al. 1972; PERKINS 1973). Stimulation of α-adrenoceptors in the central nervous system (CNS) also leads to an increase in the accumulation of cyclic AMP (PERKINS 1973), but this may be an indirect effect (PARTINGTON et al. 1980). In some systems, stimulation of α-adrenoceptors appears to reduce levels of cyclic AMP (BURNS and LANGLEY, 1975) by inhibiting adenylate cyclase activity (JAKOBS et al. 1976) while in other systems, it appears to increase the permeability of membranes to calcium (CHAN and EXTON 1977). The effect of α-adrenoceptor agonists on calcium flux has been associated with changes in the metabolism of phosphatidylinositol (MITCHELL 1975).

The availability of appropriate ligands radiolabelled with tritium or iodine has led to the development of quantitative assays for receptors for a number of neurotransmitters (see Fig. 1). The first successful assay of this type measured the nicotinic cholinoceptor present in the electric organs of fish and eels (MILEDI et al. 1971; CHANGEUX et al. 1971). Assays have now been developed for a variety of receptors including dopamine receptors and α- and β-adrenoceptors. The most widely used ligands for dopamine receptors are (^{3}H)-dopamine (BURT et al. 1975), (^{3}H)-spiroperidol (FIELDS et al. 1977), (^{3}H)-N-propylnorapomorphine (LEFF et al. 1981), and (^{3}H)-2-amino-6,7-dihydroxy-1,2,3,4-tetrahydronaphthalene (CREESE and SNYDER 1978). Other currently available ligands for dopamine receptors include (^{3}H)-haloperidol (BURT et al. 1975), (^{3}H)-domperidone (MARTRES et al. 1978; HUFF and MOLINOFF 1982), (^{3}H)-α-flupenthixol (CROSS and OWEN 1980), and (^{3}H)-apomorphine (SEEMAN et al. 1976). Assays for β-adrenoceptors are available using (^{125}I)-iodohydroxybenzylpindolol (AURBACH et al. 1974; HARDEN et al. 1976; MAGUIRE et al. 1976b), (^{125}I)-iodocyanopindolol (ENGEL et al. 1981), (^{125}I)-iodopindolol (BAROVSKY and BROOKER 1980), as well as the tritiated ligands (^{3}H)-propranolol (LEVITZKI et al. 1974), (^{3}H)-dihydroalprenolol (LEFKOWITZ et al. 1974), (^{3}H)-hydroxybenzylisoprenaline (LEFKOWITZ and WILLIAMS 1977;

Dopamine Receptors

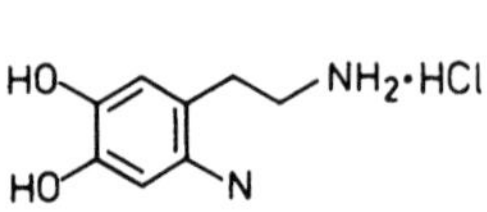

Dopamine

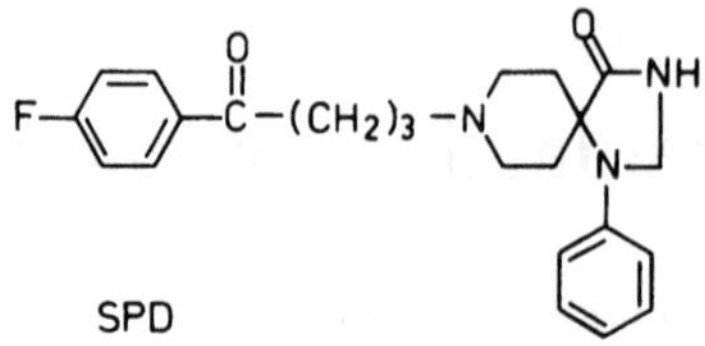

SPD

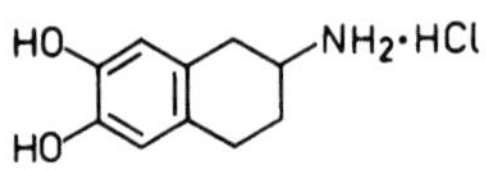

ADTN

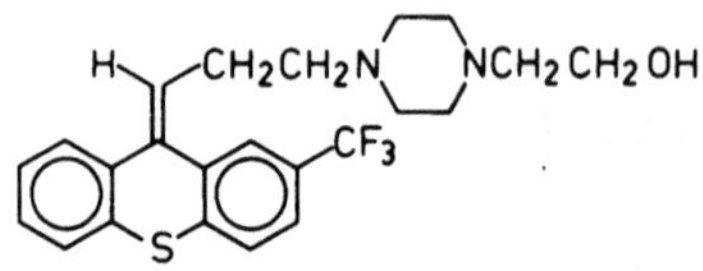

α-FPX

Beta-Adrenoceptors

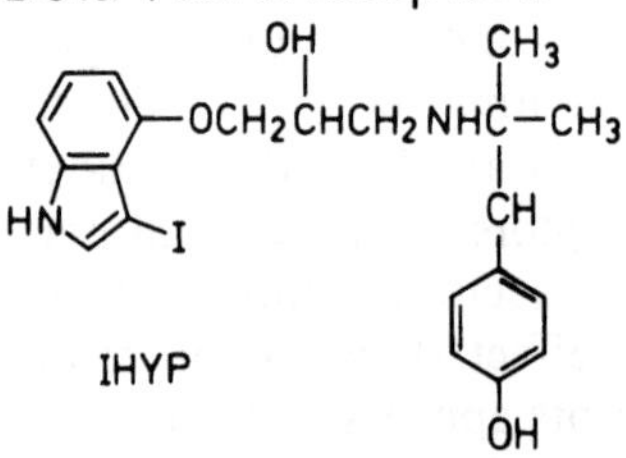

IHYP

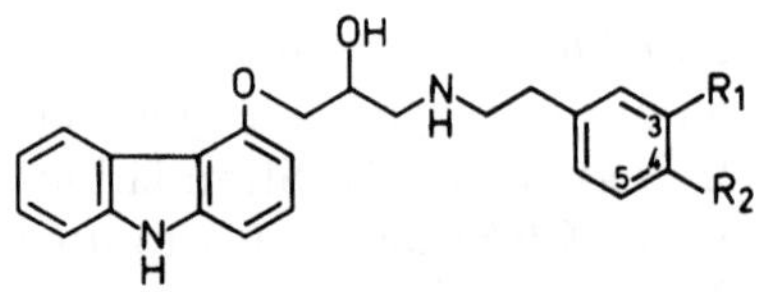

Carazolol

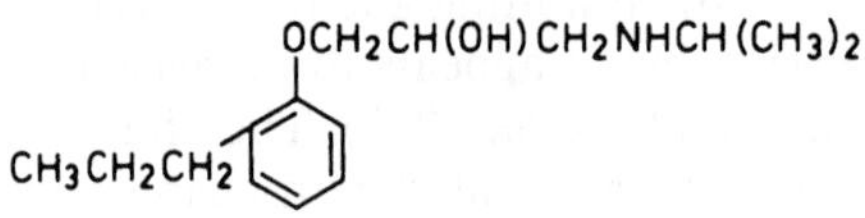

DHA

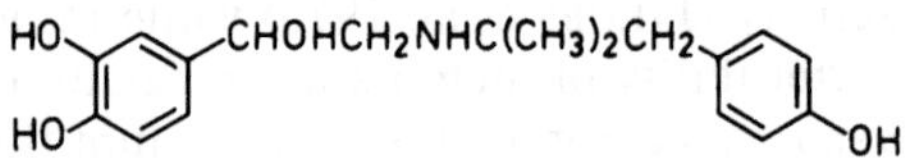

HBI

Alpha Adrenoceptors

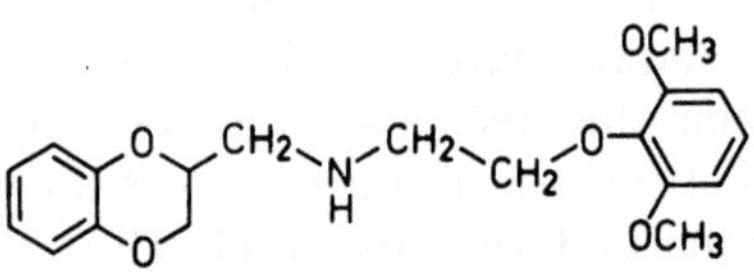

WB-4101

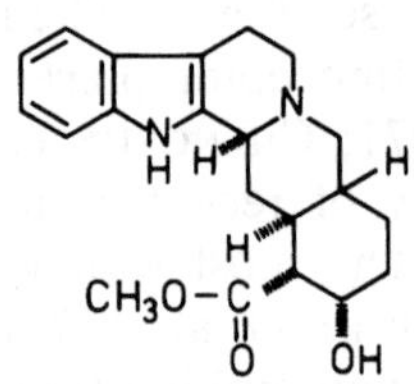

Yohimbine

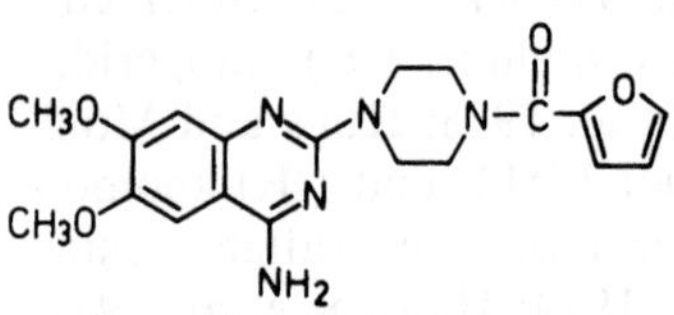

Prazosin

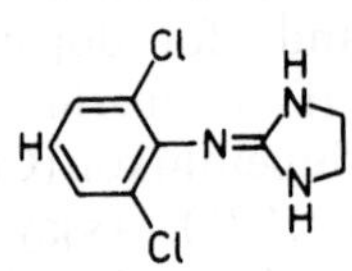

Clonidine

HEIDENREICH et al. 1980), and (^{3}H)-adrenaline (U'PRICHARD et al. 1978). *In vitro* assays for α-adrenoceptors are available using (^{3}H)-WB-4101 (2-(2',6'-dimethyloxy)phenoxyethylamino)methyl benzo-1,4-dioxane) (GREENBERG and SNYDER 1978), (^{3}H)-dihydroergocryptine (WILLIAMS and LEFKOWITZ 1976), (^{3}H)-yohimbine (DAIGUJI et al. 1981), (^{3}H)-rauwolscine (PERRY and U'PRICHARD 1981), (^{3}H)-prazosin (GREENGRASS and BREMNER 1979), (^{3}H)-adrenaline (U'PRICHARD and SNYDER 1977), and (^{3}H)-clonidine (U'PRICHARD et al. 1977). In many cases it is possible to use both radiolabelled agonists and antagonists to study neurotransmitter receptors *in vitro*. In view of the many differences in the way in which agonists and antagonists interact with receptors, the availability of both labelled agonists and antagonists has greatly facilitated our understanding of the molecular consequences of receptor occupancy. Since there are multiple subtypes of receptors for most of the known neurotransmitters, including noradrenaline and dopamine, it is important to develop methods for studying the physiological, pharmacological, and biochemical properties of these subtypes.

B. Relationship of Catecholamine Receptors to Effector Systems

I. Cyclic AMP Formation

SUTHERLAND and co-workers (1965; 1969) introduced the concept that adenosine 3',5'-monophosphate (cyclic AMP) serves as a second messenger for the actions of many hormones, including catecholamines acting at β-adrenoceptors (MURAD et al. 1962; SUTHERLAND et al. 1965; SUTHERLAND and ROBISON 1969). Thus, catecholamine-stimulated adenylate cyclase activity can serve as a biochemical measure of β-adrenoceptor function. α-Adrenoceptors do not usually act through increases in cyclic AMP levels (ROBISON et al. 1971). In the brain, however, activation of receptors with pharmacological characteristics similar to those of α-adrenoceptors in the periphery does result in increased intracellular levels of cyclic AMP (SCHULTZ and DALY 1973a, b; PERKINS and MOORE 1973b; DALY 1975). Thus, noradrenaline-stimulated cyclic AMP formation in the brain reflects stimulation of both α- and β-adrenoceptors. Subtypes of α-adrenoceptors have been shown to exist (see Section F below) and the relationship between subtypes of central α-adrenoceptors and

---

Fig. 1. Structures of principal ligands used to study catecholamine receptors

Abbreviations: ADTN, 2-amino-6,7-dihydroxy-1,2,3,4-tetrahydronaphthalene
DHA, dihydroalprenolol
α-FPX, α-flupenthixol
HBI, hydroxybenzylisoprenaline
IHYP, iodohydroxybenzylpindolol
SPD, spiroperidol

changes in cyclic AMP has been studied in several laboratories. DAVIS et al. (1978), using selective antagonists, showed that this increase is mediated through activation of α_1-adrenoceptors. However, incubation of slices of either hypothalamus or cerebral cortex in buffer containing inhibitors of prostaglandin synthesis such as indomethacin resulted in a decreased accumulation of cyclic AMP in response to noradrenaline (PARTINGTON et al. 1980). The decreased response to noradrenaline was primarily due to loss of the component mediated by α-adrenoceptors. Since the α-adrenoceptor-mediated component of the noradrenaline-stimulated accumulation of cyclic AMP is entirely dependent on the concentration of extracellular calcium, SCHWABE and DALY (1977) suggested that prostaglandins mediate α-adrenergic effects via a calcium-dependent mechanism. In this system an adenosine-induced accumulation of cyclic AMP is potentiated by noradrenaline. This effect was blocked by treatment with indomethacin (PARTINGTON et al. 1980).

The ability of catecholamines acting at α-adrenoceptors to increase cyclic AMP levels in the CNS is not observed in peripheral tissues. In fact, in many cases, activation of these receptors results in a decrease in basal or hormone-stimulated levels of cyclic AMP. This effect has been observed in liver (JARD et al. 1981), fat cells (GARCÍA-SÁINZ et al. 1980), platelets (SALZMAN and NERI 1969), and NG-108-15 neuroblastoma glioma hybrids (SABOL and NIRENBERG 1979). In contrast to studies of the CNS, where an effect on cyclic AMP accumulation is mediated by α_1-adrenoceptors, the inhibition of adenylate cyclase activity by catecholamines in peripheral tissues appears to be mediated through α_2-adrenoceptors (GARCÍA-SÁINZ et al. 1980).

Subtypes of dopamine receptors are also postulated to exist (see Sect. F below) and the subtype named D-1 is thought to mediate dopamine-stimulated increases in adenylate cyclase activity in homogenates or slices of retina (BROWN and MAKMAN 1972), as well as in sympathetic ganglia and some regions of the CNS (KEBABIAN et al. 1972). The effect is not due to stimulation of α- or β-adrenoceptors since in many cases isoprenaline is ineffective and non-catecholamines, including some agonists at α-adrenoceptors such as phenylephrine, are completely inactive. Furthermore, dopamine and its N-methyl analogue epinine are more potent agonists than noradrenaline or adrenaline (SHEPPARD and BURGHARDT 1974; MAKMAN et al. 1975; GREENGARD 1976; IVERSEN et al. 1976). In general, antagonists at β-adrenoceptors, such as propranolol, are ineffective inhibitors of dopamine-mediated adenylate cyclase stimulation, while α-adrenoceptor antagonists like phentolamine and phenoxybenzamine are effective only at concentrations above 20 μM (KEBABIAN and GREENGARD 1971; MAKMAN et al. 1975). These concentrations are considerably higher than those usually needed to block α-adrenoceptors. The most striking pharmacological property of dopamine-stimulated adenylate cyclase activity in the brain and retina is the potent inhibitory effect exerted by many of the known neuroleptics of the phenothiazine and thioxanthene classes (MAKMAN et al. 1975; IVERSEN et al. 1976). On the other hand, butyrophenones are potent neuroleptics but relatively weak inhibitors of dopamine-stimulated adenylate cyclase activity (JANSSEN 1965; KELLY and MILLER 1975). This appears to be a consequence of the existence of multiple types of dopamine receptors. Support for this suggestion comes from a report by MUNEMURA et al. (1980) who showed that dopamine inhibits adenylate cyclase activity in the intermediate lobe of the pituitary. This inhibition appears to be mediated by dopamine receptors that display a different pharmacological profile than the D-1 receptors described above. Thus, in the anterior pituitary, dopamine-inhibited adenylate cyclase activity is inhibited by low concentrations of butyrophenones and other drugs such as sulpiride thought to be selective for dopamine receptors that do not stimulate adenylate cyclase. These have been named D-2 receptors (KEBABIAN and CALNE 1979).

II. Calcium Flux and Phosphatidylinositol Breakdown

In addition to cyclic AMP, other small molecules have been proposed as second messengers which may mediate the effects of catecholamines. For example, cyclic GMP levels have been shown to be modulated by some α-adrenoceptors and this nucleotide has been proposed to act as a second messenger for these receptors (GOLDBERG et al. 1973). However, it has subsequently been shown that activation of guanylate cyclase activity is secondary to and dependent upon stimulation of calcium flux across the cell membrane (SCHULTZ et al. 1973). Thus, calcium has also been suggested as a second messenger for stimulation of these receptors. MICHELL (1975) has suggested, however, that even earlier events are responsible for transduction of receptor-mediated information across the cell membrane. Phosphatidylinositol, and its polyphosphorylated derivatives, relatively minor membrane constituents, have been shown to be hydrolyzed by a specific phospholipase C activated through stimulation of some α-adrenoceptors (MICHELL 1975; JOSEPH et al. 1984). This hydrolysis leads to the production of diacylglycerol and inositol phosphates and occurs independently of calcium flux. It has been suggested that the diglyceride can activate a protein kinase (TAKAI et al. 1979; KISHIMOTO et al. 1980) which results in phosphorylation of certain membrane proteins, ultimately leading to changes in the permeability of the membrane to calcium (BERRIDGE 1981). Alternatively, phosphatidic acid, which accumulates as a result of the breakdown of phosphatidylinositol, may function as a specific calcium ionophore (SALMON and HONEYMAN 1980; PUTNEY et al. 1980) involved in modulating a receptor-mediated calcium flux. Most recently, convincing evidence has been presented demonstrating that the initial breakdown product following receptor activation is inositol 1,4,5-trisphosphate (DOWNES 1983). This product comes from the hydrolysis of phosphatidylinositol 4,5-bisphosphate, a minor (1%) polyphosphoinositide that exists in the membrane. Inositol 1,4,5-trisphosphate has been shown to be a potent and specific agent for releasing calcium stored in the endoplasmic reticulum, thereby rapidly raising the internal concentration of free cellular calcium (JOSEPH et al. 1984; STREB et al. 1983).

GARCÍA-SÁINZ et al. (1980) have shown that the breakdown of phosphatidylinositol in hamster adipocytes is mediated by α_1-adrenoceptors since prazosin is a potent inhibitor of adrenaline-induced phosphatidylinositol turnover while yohimbine is a relatively weak antagonist of this effect. On the other hand, these authors demonstrated that the inhibition of cyclic AMP accumulation in this tissue was mediated by α_2-adrenoceptors since yohimbine potently blocked the effect of adrenaline while prazosin was relatively ineffective. Thus, the subtypes of α-adrenoceptors have been differentiated on the basis of their biochemical responses.

III. Other Effects of Catecholamines

Many of the effects of catecholamines acting through α, β-adrenoceptor or dopamine receptors have been shown to involve changes in adenylate cyclase activity, calcium flux, or phosphatidylinositol breakdown (see above). Some of the effects, however, involve mechanisms that are currently undefined. Changes in adenylate cyclase activity represent the only mechanisms clearly associated with effects mediated by dopamine receptors or β-adrenoceptors. However, it is unlikely that changes in adenylate cyclase activity are sufficient to explain all of the effects mediated by these classes of receptors. For example, dopamine acting at dopamine receptors on mammotrophs in the anterior pituitary inhibits the release of prolactin. This effect does not appear to involve changes in adenylate cyclase activity. Thus, administration of cholera toxin led neither to increases in cyclic AMP levels nor to inhibition of prolactin release (RAPPAPORT and GRANT 1974). Similarly, the autoreceptor on dopamine-containing nigrostriatal neurons regulates the activity of tyrosine hydroxylase (CARLSSON 1975) and cell firing (AGHAJANIAN and BUNNEY 1977), but it does not appear to involve changes in adenylate cyclase activity.

Although most effects of catecholamines on β_1- or β_2-adrenoceptors involve increases in cyclic AMP levels, there are some notable exceptions. For example, MAGUIRE and ERDOS (1980) have shown that the transport of magnesium into wild type S49 lymphoma cells is a saturable, carrier-mediated process. The transport of magnesium is inhibited by low concentrations of isoprenaline (10 nM), the effect of which is blocked by the β-adrenoceptor antagonist propranolol. Isoprenaline has been shown to inhibit magnesium influx in the kinase-deficient mutant of S49 lymphoma cells, suggesting that cyclic AMP does not mediate the effects of isoprenaline on magnesium transport. The finding that incubation of wild type, kinase-deficient, or UNC cells with 8-bromocyclic AMP, dibutyryl cyclic AMP, or cholera toxin had no effect on magnesium uptake was consistent with this conclusion. The lack of effect of isoprenaline on magnesium uptake in the cyc^- and UNC variants suggests that the effect requires a receptor capable of functionally interacting with a guanine nucleotide-binding regulatory protein (G_s). The specific way in which G_s affects magnesium transport remains, however, to be defined. It is also possible that an as yet unknown protein is required for the regulation of catecholamine-sensitive magnesium uptake (see SMIGEL et al. 1984).

Another effect of catecholamines mediated by β-adrenoceptors that appears not to involve increases in cyclic AMP content is the phenomenon of agonist-induced desensitization (see below). This phenomenon involves occupancy of receptors as evidenced by the fact that the changes in sensitivity are frequently receptor-specific. For instance, prior incubation of EH-118 astrocytoma cells with isoprenaline was associated with a decreased response to isoprenaline, but did not decrease sensitivity of the cells to prostaglandin E_1 (PGE_1), while incubation with PGE_1 affected the sensitivity of the tissue to PGE_1 but not to isoprenaline (JOHNSON et al. 1978). Furthermore, the effect of isoprenaline was blocked if the cells were coincubated with the antagonist sotalol. These findings suggest that the effect of isoprenaline is a consequence of occupancy of β-adrenoceptors. On the other hand, since both PGE_1 and isoprenaline increase cyclic AMP levels, the results suggest that changes in this nucleotide are not involved in the decreased sensitivity to isoprenaline.

In several systems it has been observed that the affinity of the receptor for agonists was much lower in studies carried out with intact cells compared to studies carried out with membranes prepared from the same cells (TERASAKI and BROOKER 1978; INSEL and STOOLMAN 1978; PITTMAN and MOLINOFF 1980). In contrast to results obtained in studies with agonists, the affinity of the receptor for antagonists was the same in membranes as in intact cells. The first evidence to suggest that changes in cyclic AMP levels were not involved came from the results of TERASAKI and BROOKER (1978), who suggested that the affinity of the receptor for some partial agonists was the same in broken and in intact cells. More recently, PITTMAN (1981) showed that there was a discrepancy of from 16- to 70-fold in the K_D of receptors on intact cells compared to the K_D of receptors on membranes. This discrepancy was seen with isoprenaline, Cc-25, fenoterol, adrenaline, and noradrenaline. On the other hand, the ratios of the K_D values of receptors on intact cells and on membranes for metaproterenol and terbutaline were 6.5 and 4.7, respectively, and there was no difference in the affinities of the receptors on intact and on broken cells for zinterol or salmefamol. Since all of these drugs were full agonists as measured by stimulation of adenylate cyclase activity in membranes and of cyclic AMP accumulation in whole cells, these findings suggest that changes in cyclic AMP levels are not involved in these agonist-specific effects. It is possible that some of these effects are mediated directly through a change in the conformation of the receptor or through the guanine nucleotide-binding regulatory protein.

Results obtained in studies with atypical agonists including pindolol and celiprolol have provided another example of a cyclic AMP-independent effect mediated by β-adrenergic receptors. Pindolol is a clinically used β-adrenergic receptor antagonist with intrinsic sympathomimetic activity (ISA) that does not enhance adenylate cyclase activity or cyclic AMP accumulation in L6 myoblasts (PITTMAN and MOLINOFF 1983), kitten heart (KAUMANN and BIRNBAUMER 1974), or rat adipocytes (YAMAMURA and RODBELL 1976). Celiprolol is a β_1-selective antagonist with ISA that does not stimulate adenylate cyclase activity in dog heart (PITTNER et al. 1983; SMITH and WOLF 1984). Despite the absence of observable stimulation of adenylate cyclase activity, pindolol and celiprolol have sympathomimetic activity in isolated tissues or in sympathectomized animals (KAUMANN and BLINKS 1980; FRISHMAN 1983; SMITH and WOLF 1984). Most of the effects of pindolol and celiprolol are inhibited by the antagonist propranolol, and they are thus thought to be mediated by β-adrenoceptors. In humans, the partial agonist activity of pindolol may account for the observation that pindolol, unlike pure antagonists, does not reduce resting cardiac rate or output in supine subjects (CARRUTHERS and TWUM-BARIMA 1981; AELLIG 1982; FRISHMAN 1983). In addition, whereas chronic administration of propranolol elevates the density of β-adrenoceptors on human lymphocytes (AARONS et al. 1980), chronic administration of pindolol results in a decrease in the density of receptors (HEDBERG et al. 1986).

A great deal of evidence has accumulated implicating cyclic AMP as the mediator of β-adrenoceptor effects on heart rate and contractility (DRUMMOND and HEMMINGS 1972; KUKOVETZ et al. 1975). However, some observations are inconsistent with this hypothesis. For example, HEDBERG and MATTSSON (1981) have shown that prenalterol acting via β-adrenoceptors was nearly as efficacious as isoprenaline at stimulating the force of contraction of isolated papillary muscles. On the other hand, prenalterol was only marginally effective in stimulating adenylate cyclase activity in homogenates of this tissue while isoprenaline markedly increased enzyme activity. In addition, HU and VENTER (1978) have shown that even though application of isoprenaline covalently attached to a polymer substrate resulted in increases in the rate and force of contraction of cat papillary muscles, no increase in cyclic AMP levels in this tissue could be demonstrated.

C. *In Vitro* Binding Assays

The most direct way to characterize a receptor is to study its interaction with a specific ligand. Through the use of *in vitro* binding assays the chemistry of the initial events involved in receptor-mediated responses has been directly investigated. In addition to characterizing the affinities of a variety of ligands for the receptor and providing a measure of the density of binding sites (Fig. 2), the use of radioligands makes it possible to determine the kinetics of ligand receptor association and dissociation. Furthermore, changes in the densities or properties of receptors occur and, in some cases, play a role in regulating the responsiveness of a tissue. These changes can be most easily examined with *in vitro* binding assays. Finally, by serving as a means through which receptors can be followed during solubilization, isolation, and reconstitution, *in vitro* assays may ultimately permit elucidation of the mechanisms through which receptor occupation leads to the biochemical responses that have been observed.

In any *in vitro* assay for a receptor it is necessary to quantify binding to the receptor as opposed to binding to other sites. Non-receptor binding may include hydrophobic and hydrophilic interactions with various membrane constituents and binding to specifiable non-receptor sites (MOLINOFF 1973). To measure specific binding, assays are routinely carried out in the presence of a competing agonist or antagonist at a concentration that will occupy essentially all of the receptor sites in the tissue (Fig. 2). In this experimental paradigm, specific binding of ligand to the receptor is equal to the amount of radioactivity bound in the absence of a competing ligand minus the amount

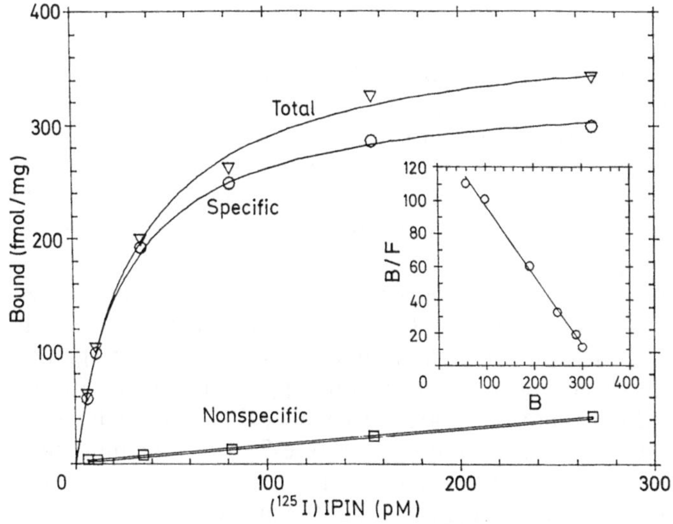

Fig. 2. Saturation analysis of binding of (^{125}I)-IPIN to membranes prepared from L6 myoblasts. Membranes were incubated with various concentrations of the radiolabelled ligand for 90 min at 25 °C. The amount of (^{125}I)-IPIN bound in the absence (▽) and the presence (□) of 50 μ*M* (−)-isoprenaline was determined at several concentrations of (^{125}I)-IPIN. The difference (○) is designated as specific binding. Inset: Scatchard transformation of the specific binding. The abscissa is fmol of specifically bound (^{125}I)-IPIN per mg of protein (B), and the ordinate is B divided by the concentration of free (^{125}I)-IPIN with units of 1/mg × 10^4. The K_D value was calculated to be 42 p*M* and the B_{max} was determined to be 331 fmol/mg of protein.

bound in its presence. In general, when assays are carried out under conditions such that the concentrations of the radioactive ligand and of the receptors are well below the K_D value for the radioactive ligand, then a competing ligand at a concentration which is one hundred times its K_D value will provide a satisfactory estimate of non-specific binding. This assumes that the Hill coefficient for binding is approximately 1. When the Hill coefficients of competing drugs are significantly less than 1, a higher concentration of a competing ligand is necessary to occupy 99% of the receptor sites in the tissue (Maguire et al. 1976a; see Weiland and Molinoff 1981). When the concentration of the radioactive ligand approaches or exceeds the K_D value of the ligand, a higher concentration of competing ligand must be used. It is often necessary, for example, to increase the concentration of the competing ligand when carrying out Scatchard analysis to determine the density of binding sites in a given tissue. The need for a quantitative definition of non-specific binding arises from the fact that too high a concentration of a competing ligand will not only block specific binding of the ligand to the receptor but may also compete for non-specific sites (Molinoff 1973; Sporn and Molinoff 1976; see Molinoff et al. 1981).

I. Assay of α-Adrenoceptors

Several ligands are available for the study of α-adrenoceptors. One of the first to be developed was the antagonist (^{3}H)-dihydroergocryptine (^{3}H-DHE). This compound has been reported to bind to α-adrenoceptors on rabbit uterine membranes. The binding is rapid and agonists and antagonists compete for the binding site with appropriate affinities. The expected stereoselectivity has been observed (WILLIAMS and LEFKOWITZ 1976). Problems with the specificity of binding of (^{3}H)-DHE limit its usefulness for the study of α-adrenoceptors. Ergot and its derivatives are well-known inhibitors of serotonin receptors and (^{3}H)-DHE has been reported to bind to dopamine receptors as well as to both subtypes of α-adrenoceptors. A second α-adrenoceptor antagonist, (^{3}H)-WB-4101, a derivative of methylbenzodioxan, appears to be a useful ligand for the study of α-adrenoceptors (U'PRICHARD et al. 1977). Although (^{3}H)-WB-4101 is relatively selective for α_1-adrenoceptors (see Sect. F), the binding of (^{3}H)-WB-4101 is inhibited by low concentrations of some antidepressants including amitriptyline. It has been suggested that the ability of tricyclic antidepressants to bind to α_1-adrenoceptors is related to their therapeutic effect (U'PRICHARD et al. 1978). Anticholinergic agents including benztropine, antihistaminics including promethazine, and a variety of antipsychotic drugs also inhibit the binding of (^{3}H)-WB-4101. (^{3}H)-Prazosin has been suggested to be more selective for α_1-adrenoceptors than (^{3}H)-WB-4101 (LYON and RANDALL 1980). Subsequently, an iodinated derivative of BE 2254 [2-(β(4-hydroxyphenyl)-ethyl aminomethyl) tetralone] has been introduced by ENGEL and HOYER (1981). The high selectivity and high specific activity of this ligand suggest that it is the ligand of choice to study α_1-adrenoceptors. The properties of this ligand make it possible to carry out assays with relatively small amounts of tissue.

Antagonists selective for α_2-adrenoceptors also exist. (^{3}H)-Yohimbine has been used to assay these receptors in platelets (DAIGUJI et al. 1981). This ligand does not appear to be useful for studying α_2-adrenoceptors in the CNS since the amount of nonspecific binding is extremely high. (^{3}H)-Rauwolscine, structurally similar to yohimbine, is highly selective for α_2-adrenoceptors and appears to be a useful ligand for studying these receptors in the CNS (PERRY and U'PRICHARD 1981).

In addition to α-adrenoceptor antagonists, agonists including (^{3}H)-clonidine (U'PRICHARD et al. 1977) and the catecholamines (^{3}H)-adrenaline and (^{3}H)-noradrenaline (PEROUTKA et al. 1978) have been used. Under the usual assay conditions, these ligands appear to bind predominantly to α_2-adrenoceptors. This does not necessarily reflect selectivity of these ligands for α_2-adrenoceptors. The observed results are thought to be a consequence of agonist-induced formation of a ternary complex involving agonist, receptor, and a guanine nucleotide-binding protein (see below). Agonists are less useful in terms of quantitation of receptors than are the antagonists discussed above because the interactions of agonists with receptors that act through adenylate cyclase are frequently complex. As discussed below, such interactions involve formation of a ternary complex of agonist, receptor, and a guanine nucleotide-binding regulatory protein. The formation of this complex complicates both kinetic and equilibrium determinations, and can result in curvilinear Scatchard plots (U'PRICHARD et al. 1977; see MOLINOFF et al. 1981).

II. Assay of β-Adrenoceptors

As noted above, a number of ligands have been developed for use as probes of β-adrenoceptors. In general, the results suggest that there are a finite number of receptors whose pharmacological specificity is consistent with that expected on the basis of earlier studies of the effects of catecholamines on physiological processes or on adenylate cyclase activity. β-Adrenoceptors show a marked degree of stereospecificity. The (−)-isomers of a variety of agonists and antagonists have a greater affinity for the receptor, whether determined by measuring adenylate cyclase activity or inhibition of radioligand binding, than do the corresponding (+)-isomers.

The earliest attempts to establish an *in vitro* binding assay for β-adrenoceptors utilized interaction of tritiated catecholamines with membrane preparations enriched with respect to catecholamine-sensitive adenylate cyclase activity. In general, binding was saturable and catecholamines were reasonably potent inhibitors of binding. Binding was also inhibited, however, by a variety of catechol-containing compounds that were not active at β-adrenoceptors. Furthermore, stereospecificity was not seen, bound radioactivity did not dissociate even after the addition of strong acid, and high concentrations of potent β-adrenoceptor antagonists, such as propranolol, had little effect on binding. Thus, workers in several laboratories concluded that the binding of catecholamines did not reflect binding to a β-adrenoceptor. The reaction appeared to involve a nonspecific, probably covalent, interaction of oxidized degradation products of catecholamines with unspecified tissue constituents (CUATRECASAS et al. 1974; MAGUIRE et al. 1974; WOLFE et al. 1974).

Although it has recently become possible to perform *in vitro* studies of β-adrenoceptors with tritium-labelled agonists (see below), the earliest successful assays for β-adrenoceptors involved antagonists. The major reason for this was that antagonists were available which had potencies greater by two to three orders of magnitude than those seen in studies with catecholamine agonists. The first successful assay for β-adrenoceptors was reported by LEVITZKI and his collaborators (1974). These investigators used turkey erythrocyte membranes and demonstrated specific binding of (^{3}H)-propranolol to β-adrenoceptors. Specific binding of (^{3}H)-propranolol to membranes prepared from chick cerebral cortex has also been reported (NAHORSKI 1976).

Lefkowitz and his co-workers (1974) reduced the carbon-carbon double bond of (−)-alprenolol with palladium and tritium gas to obtain (^{3}H)-dihydroalprenolol (^{3}H-DHA). The binding of this compound to β-adrenoceptors has been characterized in a variety of tissue preparations (ALEXANDER et al. 1975; BYLUND and SNYDER 1976; KEBABIAN et al. 1975b; WILLIAMS et al. 1976). A principal advantage of the use of (^{3}H)-DHA is that it is commercially available as a pure stereoisomer. It is also a relatively stable compound and most workers observe relatively low amounts of non-specific binding. The relatively high K_D value (about 1 nM) and the low specific activity of tritium-labelled compounds relative to that of (^{125}I)-labelled compounds render it difficult to use (^{3}H)-DHA when only a limited amount of tissue is available (see below).

Innis et al. (1979) have described a high-affinity β-adrenoceptor antagonist, carazolol, that can be labelled with tritium to relatively high specific activity. It is available as the active (−)-isomer. Because of its higher affinity and slower rate of dissociation, it appears to have a slight advantage as a radioligand over (^{3}H)-DHA. Lavin et al. (1981) have synthesized a radiolabelled azido derivative of carazolol which is photosensitive and appears to be a useful tool for irreversibly labelling β-adrenoceptors (see below).

Aurbach and his collaborators (1974) adopted a different approach to the study of β-adrenoceptors. They developed an iodinated ligand, (^{125}I)-iodohydroxybenzylpindolol (^{125}I-IHYP), which can be purified to a nearly theoretical specific activity of 2.2 Ci/µmol. Their expectation was that the hydroxyphenyl group would direct iodine onto one or both of the positions adjacent to the hydroxyl group. However, Bearer et al. (1980) carried out an extensive study of this radioligand and showed that the iodine is actually on the 3 position of the indole ring. Barovsky and Brooker (1980), recognizing that the hydroxyphenyl moiety was not required, demonstrated that pindolol itself would accept iodine. A major advantage of (^{125}I)-iodopindolol (^{125}I-IPIN) is that the percentage of specific binding with (^{125}I)-IPIN is higher than with (^{125}I)-IHYP. A further advantage of (^{125}I)-IPIN derives from the fact that the active (−)-isomer is available, whereas only the racemic mixture of hydroxybenzylpindolol is currently available.

Still another ligand that promises to be useful for the study of β-adrenoceptors is 2-cyanopindolol (Engel et al. 1981). The cyano substituent on the 2 position of the indole ring affords an increased stability to the compound compared to pindolol. Furthermore, although the affinity of β-adrenoceptors for pindolol is only half that for hydroxybenzylpindolol, the affinity of receptors for cyanopindolol is approximately 5 times that for pindolol. The percentage of specific binding observed with (^{125}I)-iodocyanopindolol is usually as high as that seen with (^{125}I)-IPIN. It is likely that iodocyanopindolol will be the ligand of choice for the study of β-adrenoceptors in cases where only a limited amount of tissue is available.

Several groups of workers have reinvestigated the binding of catecholamines to β-adrenoceptors. Malchoff and Marinetti (1976) studied the binding of (^{3}H)-isoprenaline to intact avian erythrocytes. Specific binding was inhibited by low concentrations of propranolol. These investigators were unable, however, to demonstrate stereospecific binding in preparations of broken cells. Lefkowitz and Williams (1977) studied the binding of (^{3}H)-hydroxybenzylisoprenaline (^{3}H-HBI) to β-adrenoceptors on membranes prepared from frog erythrocytes. The assays were carried out in the presence of high concentrations of ascorbic acid and catechol to suppress non-receptor binding. These investigators found it necessary to use a high concentration of membranes enriched with β-adrenoceptors. It now appears that this ligand is particularly useful in tissues such as the lung which have a relatively high density of β-adrenoceptors. The successful use of this ligand in mammalian tissues also requires inclusion of divalent cations (Heidenreich et al. 1980).

An increase in the high-affinity binding of (^{3}H)-HBI was also observed if membranes were washed extensively to minimize effects due to endogenous guanine nucleotides (see below). Finally, since the affinity of the β-adrenoceptor for agonists is increased as temperatures are reduced (WEILAND et al. 1979; 1980), studies of the binding of (^{3}H)-HBI are carried out at 20 °C rather than at 37 °C. The most striking finding of HEIDENREICH et al. (1980) was that Scatchard plots were curvilinear when assays were carried out in the absence of guanine nucleotides. This suggested a complex interaction of the agonist with the receptor. The total number of binding sites for (^{3}H)-HBI obtained by extrapolation of the curved Scatchard plot was the same as that determined in studies with (^{125}I)-IHYP. Furthermore, dissociation of (^{3}H)-HBI was the same whether induced by dilution or by the addition of excess agonist or antagonist. This suggested that the curvature was not due to negatively cooperative site-site interactions. Since curvilinear Scatchard plots were seen in tissues such as rat reticulocytes that appear to contain only a single subtype of receptor, the curvature was apparently not a consequence of the presence of multiple receptor subtypes. The authors concluded that the most likely explanation of the experimental data was a two-step (ternary complex) model of receptor/ligand interactions. A general model of this type has been developed by DE HAËN (1976) and by BOEYNEAMS and DUMONT (1977) and applied to the insulin receptor by JACOBS and CUATRECASAS (1976).

$$H + R \underset{}{\overset{K_a}{\rightleftharpoons}} H \cdot R$$

$$H \cdot R + E \underset{}{\overset{K_b}{\rightleftharpoons}} H \cdot R \cdot E$$

In these equations, K_a and K_b are equilibrium constants, H represents hormone, R a receptor, and E an effector molecule. This model can explain the curved Scatchard plots which describe the binding of an agonist such as (^{3}H)-HBI and the linear Scatchard plots that result from studies of the binding of a number of labelled antagonists. The most widely accepted hypothesis is that antagonists participate in only the first reaction while agonists participate in both reactions. DELEAN et al. (1980) have used this model to characterize the binding of agonists to β-adrenoceptors on frog erythrocyte membranes, in which the effector is likely to be the G_s protein. The model postulates that the initial step is the formation of a low-affinity hormone/receptor complex which is followed by formation of a high-affinity ternary complex of hormone, receptor, and G_s. Ternary complex formation appears to be required for the stimulation of adenylate cyclase activity (STADEL et al. 1980). High concentrations of guanine nucleotides are thought to destabilize the ternary complex. Thus, in the presence of guanine nucleotides, the ternary complex does not accumulate and the observed equilibrium constant is that of the initial binding reaction.

III. Assay of Dopamine Receptors

A number of tritiated ligands have been utilized for *in vitro* binding assays to determine the number and properties of dopamine receptors. In the first published studies, specific binding of (^{3}H)-dopamine was defined as the amount of (^{3}H)-dopamine bound in the absence of 1 μ*M* (−)-butaclamol minus the amount bound in the presence of 1 μ*M* (+)-butaclamol. The distribution of specific binding sites in the brain paralleled that of the dopaminergic innervation of various brain regions. Furthermore, the relative potencies of antipsychotic drugs in inhibiting the binding of (^{3}H)-dopamine in the caudate nucleus appeared to correlate with their clinical potencies (SEEMAN et al. 1975; BURT et al. 1975).

Other agonists that appear to offer some advantages over dopamine as radiolabelled ligands have been used in *in vitro* assays. The results obtained with (^{3}H)-apomorphine, utilized because of its slightly higher affinity, have been similar to those obtained in studies with (^{3}H)-dopamine (SEEMAN et al. 1976). Several investigators have studied binding of (^{3}H)-N-propylnorapomorphine. Under appropriate conditions, specific binding is observed (LEFF et al. 1981). However, the compound is very labile and appears to produce some irreversible reactions (ZAHNISER, HUFF and MOLINOFF unpublished observations). Furthermore, antagonists are relatively ineffective inhibitors of the binding of this compound. The compound (^{3}H)-2-amino-6,7-dihydroxy-1,2,3,4-tetrahydronaphthalene (^{3}H-ADTN), a rigid analogue of dopamine, has been used in *in vitro* studies of dopamine receptors (CREESE and SNYDER 1978; ZAHNISER et al. 1981).

In general, neuroleptics have a higher affinity for dopamine receptors than do catecholamine agonists. The distribution of binding sites for (^{3}H)-haloperidol in the brain has been shown to be similar to that for (^{3}H)-dopamine (SEEMAN et al. 1975; BURT et al. 1976). The affinity of receptors for (^{3}H)-spiroperidol was greater than for (^{3}H)-haloperidol and a higher percentage of specific binding was observed (FIELDS et al. 1977; LEYSEN et al. 1978a). It has been shown, however, that (^{3}H)-spiroperidol binds to 5-HT receptors as well as to dopamine receptors (LEYSEN et al. 1978b). Caution must therefore be exercised in interpreting data obtained with (^{3}H)-spiroperidol, especially in tissues such as the frontal cortex. It has been suggested that the agonist ADTN is a better choice than butaclamol as a competing ligand to define specific binding of (^{3}H)-spiroperidol to dopamine receptors (QUIK et al. 1978).

D. *In Vitro* Properties of Adrenoceptors

I. Effects of Guanine Nucleotides

1. Effects of Guanine Nucleotides on β-Adrenoceptors

Guanine nucleotides affect several components of the β-adrenoceptor/adenylate cyclase system. Both basal and hormone-sensitive adenylate cyclase activities are increased in the presence of GTP in a variety of systems (RODBELL et al. 1971; SMIGEL et al. 1984). GTP acts reversibly to increase the efficacy of catecholamines for the stimulation of adenylate cyclase activity. On the other hand, the action of the synthetic analogue guanyl-5′-yl imidodiphosphate (Gpp(NH)p) on adenylate cyclase activity is only slowly reversible (SCHRAMM and RODBELL 1975). This appears to be a consequence of the fact that Gpp(NH)p is resistant to hydrolysis.

Several laboratories have reported that the presence of GTP or Gpp(NH)p does not affect the K_D values of β-adrenoceptors for antagonists, as determined in direct binding studies (LEVITZKI et al. 1975; LEFKOWITZ et al. 1976b; BROWN et al. 1976). However, MAGUIRE et al. (1976a) and LEFKOWITZ et al. (1976b) have shown that guanine nucleotides affect the ability of agonists to inhibit the binding of (^{125}I)-IHYP and (^{3}H)-DHA to membranes of mammalian cell lines maintained in tissue culture or to membranes prepared from frog erythrocytes. In the presence of GTP, GDP, or Gpp(NH)p, the potency of agonists, as determined in direct binding studies, was decreased by approximately one order of magnitude (Fig. 3). In addition to changing the apparent affinity of β-adrenoceptors for agonists, a change in the Hill coefficient was observed (ROSS et al. 1977). In the absence of guanine nucleotides, Hill coefficients of 0.5-0.7 were observed. In the presence of guanine nucleotides including GTP or Gpp(NH)p, the Hill coefficient shifted towards 1. It is important to note that no changes in the affinities of antagonists were observed and that the Hill coefficients for antagonist inhibition of binding were approximately 1, either in the presence or absence of guanine nucleotides (Fig. 3).

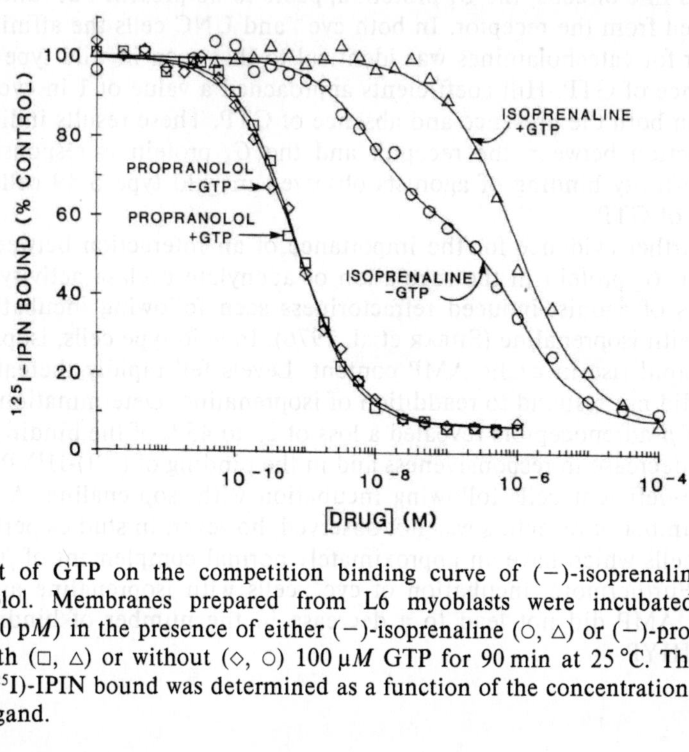

Fig. 3. Effect of GTP on the competition binding curve of (−)-isoprenaline and (−)-propranolol. Membranes prepared from L6 myoblasts were incubated with (^{125}I)-IPIN (50 p*M*) in the presence of either (−)-isoprenaline (○, △) or (−)-propranolol (□, ◇) with (□, △) or without (◇, ○) 100 μ*M* GTP for 90 min at 25 °C. The total amount of (^{125}I)-IPIN bound was determined as a function of the concentration of the competing ligand.

Ross et al. (1977) suggested that there is a negative heterotropic interaction between the binding of catecholamines to β-adrenoceptors and of nucleotides to a nucleotide binding site. WOLFE and HARDEN (1981) reported that guanine nucleotides increase the apparent affinity of antagonists for β-adrenoceptors on membranes prepared from L6 myoblasts. While this effect was large (2- to 5-fold) and easily demonstrable in membranes prepared from L6 myoblasts, it was found to be much smaller (50%) in other tissues such as rat lung and 132 1N1 astrocytoma cell membranes. It was suggested that the effect of guanine nucleotides on antagonist interactions is mediated by the same G_s protein that mediates the effect of GTP on the binding of agonists to β-adrenoceptors.

Guanine nucleotides have been reported to affect both the affinities of catecholamines for β-adrenoceptors and their Hill coefficients in rat heart, lung, and cerebellum (HEGSTRAND et al. 1979). They had no effect on the interaction of catecholamines with receptors on membranes prepared from cerebral cortex. The difference in the effect of guanine nucleotides on similarly prepared membranes from cerebellum and cortex may reflect differences in the efficiency of coupling between the receptor and the adenylate cyclase since hormone-sensitive adenylate cyclase activity is easily demonstrable in hypotonic homogenates of cerebellum but not of cortex. Ross et al. (1977) have shown that a mutant line of S-49 lymphoma cells that is missing or has a defective G_s protein (cyc^-) does not show a shift of agonist potency in the presence of GTP. Similarly, effects of GTP on the binding of agonists were not observed in an uncoupled mutant (UNC) of these cells (HAGA et al. 1977). In this line of cells, the G_s protein appears to be present but functionally uncoupled from the receptor. In both cyc^- and UNC cells the affinity of the receptor for catecholamines was identical to that seen in wild type cells in the presence of GTP. Hill coefficients approached a value of 1 in cyc^- and UNC cells in both the presence and absence of GTP. These results indicate that an interaction between the receptor and the G_s protein is responsible for the high-affinity binding of agonists observed in wild type S-49 cells in the absence of GTP.

Further evidence for the importance of an interaction between receptors and an G_s protein in the regulation of adenylate cyclase activity came from studies of agonist-induced refractoriness seen following incubation of S-49 cells with isoprenaline (SHEAR et al. 1976). In wild type cells, isoprenaline led to a rapid rise in cyclic AMP content. Levels fell rapidly thereafter and the cells did not respond to readdition of isoprenaline. Determination of the density of β-adrenoceptors revealed a loss of 25 to 45% of the binding sites. A similar decrease in responsiveness and in the binding of (^{125}I)-IHYP was seen in kinase-deficient cells following incubation with isoprenaline. A decrease in the number of receptors was not observed, however, in studies performed with cyc^- cells which have an approximately normal complement of β-adrenoceptors. Furthermore, incubation of cyc^- cells with isoprenaline and dibutyryl cyclic AMP did not lead to a decrease in the number of binding sites for (^{125}I)-IHYP.

2. Effects of Guanine Nucleotides on α-Adrenoceptors

An important question concerns the relationship between the guanine nucleotide-binding protein(s) involved in the stimulatory and inhibitory effects of catecholamines on adenylate cyclase activity (see SMIGEL et al. 1984). The decrease in adenylate cyclase activity mediated by α_2-adrenoceptors appears to involve a guanine nucleotide-binding protein since GTP is required for this effect (JAKOBS et al. 1978). Furthermore, guanine nucleotides have been shown to modulate the affinity of α_2-adrenoceptors for agonists (TSAI and LEFKOWITZ 1979; U'PRICHARD and SNYDER 1980; ROUOT et al. 1980; HOFFMAN et al. 1980; KAHN et al. 1982). TSAI and LEFKOWITZ (1979), using homogenates of human platelets which contain only α_2 receptors, showed that inclusion of GTP or Gpp(NH)p in binding assays resulted in an increase in the EC_{50} values of agonists for inhibition of the binding of the antagonist (^{3}H)-dihydroergocryptine (^{3}H-DHE). This effect was markedly potentiated in the presence of magnesium. Similarly, KAHN et al. (1982), using homogenates of NG-108 neuroblastoma glioma cells which contain only α_2 receptors, showed that while the binding of the antagonist (^{3}H)-yohimbine was not affected by guanine nucleotides, the ability of agonists to inhibit this reaction was decreased by high concentrations of GTP. U'PRICHARD and SNYDER (1980), ROUOT et al. (1980), and KAHN et al. (1982) examined the effects of guanine nucleotides directly on the binding of radiolabelled agonists and showed that high (10 μM-100 μM) concentrations of GTP resulted in a decrease in the amount of radioligand bound to α_2-adrenoceptors. Low (1 μM) concentrations of GTP, on the other hand, appeared to cause a small increase in the amount of labelled agonist bound to the α_2 receptor. Recently, the guanine nucleotide-binding protein involved in transducing information from α_2-adrenoceptors to adenylate cyclase has been purified and is called G_i since it modulates inhibition of adenylate cyclase activity (BOKOCH et al. 1984; SMIGEL et al. 1984).

The effects of guanine nucleotides on the affinity of α_1-adrenoceptors for agonists have also been investigated. HOFFMAN et al. (1980) studied the binding of (^{3}H)-DHE to rat liver plasma membranes and found that most of the receptors were of the α_1 subtype. Inclusion of GTP in the assay resulted in a small increase in the EC_{50} values for agonist inhibition of the binding of (^{3}H)-DHE. On the basis of a computer-assisted analysis, these authors concluded that this effect was a consequence of the small number of α_2 receptors present in the tissue. EL-REFAI and EXTON (1980) and EL-REFAI et al. (1979) studied the binding of (^{3}H)-adrenaline to rat liver plasma membranes. Since prazosin was a more potent inhibitor of the binding of (^{3}H)-adrenaline than was yohimbine, they concluded that the ligand was binding to α_1 receptors. GTP included in the assay caused a marked decrease in the amount of (^{3}H)-adrenaline bound. In contrast to the conclusions of HOFFMAN et al. (1980), these authors thus suggested that guanine nucleotides can modulate the affinity of α_1-adrenoceptors for agonists. Similarly, YAMADA et al. (1980b) examined the effect of Gpp(NH)p on the binding of the α_1 selective antagonist (^{3}H)-WB-4101 to α-adrenoceptors in rat heart and found that it caused a 4- to 5-fold increase in the EC_{50} value for adrenaline. No effect on the EC_{50} value for inhibition of the binding of (^{3}H)-WB-4101 by WB-4101 was observed. It seems possible that there is an as yet undefined guanine nucleotide binding protein involved in transducing information from the α_1-adrenoceptor to the phosphoinositide effector mechanism.

3. Effects of Guanine Nucleotides on Dopamine Receptors

GTP has effects on dopamine receptors that are similar in many ways to its effects on α- and β-adrenoceptors (ZAHNISER and MOLINOFF 1978; CREESE and SNYDER 1978). It decreases the ability of agonists and of the partial agonist apomorphine to compete for (^{3}H)-spiroperidol binding sites on rat striatal membranes. In the presence of 0.3 mM GTP, there was a 4- to 5-fold increase in the K_D values for agonists without any effect on the K_D values for antagonists. In contrast to the situation observed in studies of β-adrenoceptors, the Hill coefficients for agonists and at least some antagonists were significantly below 1 (0.5–0.8) in either the absence or presence of GTP. The Hill coefficients of antagonists were unaffected by inclusion of GTP while those of agonists were increased. They remained less than 1, however, even in the presence of GTP.

The effect of guanine nucleotides on the rate of dissociation of (^{3}H)-ADTN was investigated by ZAHNISER et al. (1980). In initial experiments performed at 37 °C, samples were equilibrated without nucleotide and the rate of dissociation of (^{3}H)-ADTN was determined. There was an increase in the rate of dissociation of (^{3}H)-ADTN when GTP was added along with the competing ligand. Thus, the agonist-specific decrease in affinity seen in the studies described above appeared to be due, at least partially, to an accelerated rate of dissociation of agonist from receptor. However, when assays were carried out at 37 °C in the presence of either ATP or GTP, binding of (^{3}H)-ADTN to striatal membranes did not appear to reach steady state but increased for approximately 5 min and then declined progressively (ZAHNISER et al. 1981). When assays were carried out at 20 °C, binding of (^{3}H)-ADTN reached equilibrium and was stable in either the absence or presence of purine nucleotides. Scatchard analysis of binding isotherms showed that the affinity of the binding sites for (^{3}H)-ADTN decreased markedly in the presence of either 0.3 mM ATP or GTP. An unexpected finding was that the density of binding sites increased approximately 5-fold when assays were carried out in the presence of either ATP or GTP. In contrast, the nonhydrolyzable purine nucleotide analogues Gpp(NH)p and adenosine imidodiphosphate did not affect the binding of (^{3}H)-ADTN.

II. Effects of Ions

1. Effects of Ions an α-Adrenoceptors

GREENBERG et al. (1978) reported that low concentrations (20 mM) of sodium and lithium specifically inhibited the binding of the agonists (^{3}H)-clonidine, (^{3}H)-adrenaline, and (^{3}H)-noradrenaline to α_2-adrenoceptors in calf brain. Potassium ion, on the other hand, was without effect at concentrations below

100 mM. The binding of the antagonists (^{3}H)-WB-4101 and (^{3}H)-DHE to α_1-adrenoceptors in calf brain was unaffected by monovalent cations. Similarly, TSAI and LEFKOWITZ (1978) examined the effects of mono- and divalent cations on agonist and antagonist inhibition curves of the binding of (^{3}H)-DHE to α_2-adenoceptors on rabbit platelets. The effects of ions appeared to be agonist-specific. Monovalent ions decreased the affinity of the receptor for agonists while divalent cations increased its affinity. No effects of mono- or divalent cations were observed on the binding of the antagonist (^{3}H)-DHE. ROUOT et al. (1980) examined the effects of divalent cations on the binding of (^{3}H)-clonidine to α_2-adrenoceptors in rat cerebral cortex. These authors reported that magnesium and manganese induced an increase in the amount of (^{3}H)-clonidine bound at equilibrium. Manganese appeared to be more potent as well as more efficacious. SUMMERS (1980) reported that the decreases in the binding of (^{3}H)-clonidine caused by sodium and potassium, as well as the increases in the binding of (^{3}H)-clonidine seen in the presence of magnesium and manganese were due to alterations in the apparent affinity of the α_2 receptor for the radioligand. On the other hand, GLOSSMANN and HORNUNG (1980) reported that the increase in the binding of (^{3}H)-clonidine seen in the presence of magnesium and manganese was due to both a change in the density of binding sites and their affinity for the radioligand.

2. Effects of Ions on β-Adrenoceptors

The effects of magnesium on the inhibition of the binding of (^{125}I)-IHYP by agonists and antagonists have been characterized (BIRD and MAGUIRE 1978). Magnesium resulted in an increase in the apparent affinity of β-adrenoceptors for agonists with no effect on the affinity of the receptor for antagonists. On the other hand, no effects of magnesium on the affinities of the receptor for agonists were seen in studies with variants of S49 cells which either lack or possess a defective G_s protein. Thus, it was suggested that the effect of magnesium is to promote the formation of a high-affinity ternary complex between an agonist, the receptor, and G_s. The presence of low concentrations of GTP, sufficient to destabilize the ternary complex, prevented the effects of magnesium (BIRD and MAGUIRE 1978). WILLIAMS et al. (1978) showed that magnesium, manganese, and calcium caused a 5-fold increase in the amount of the radiolabelled agonist (^{3}H)-HBI bound to frog erythrocyte membranes. These ions had no effect on the binding of the antagonist (^{3}H)-DHA to this same tissue. Strontium, barium, sodium, lithium, and lanthanum at the same low (1 m*M*) concentrations had no effect on either agonist or antagonist interactions with β-adrenoceptors. These authors suggested that the effect of magnesium was due to an increase in the affinity of the receptor for agonists since the agonist inhibition curves for the binding of (^{3}H)-DHA were shifted 20-fold to the left by magnesium. The effects of magnesium were lost following solubilization of the receptor in the detergent digitonin. Thus, these authors suggested that magnesium is promoting a high affinity hormone receptor complex that is dependent on an intact membrane. Since the magnitude of the GTP-induced shift in the concentration inhibition curves appeared to correlate with the intrinsic efficacy of the drugs, it was suggested that this high-affinity hormone receptor complex is a necessary intermediate in the activation of adenylate cyclase by catecholamines. HEIDENREICH et al. (1980, 1982) have performed studies of the effects of magnesium on the binding of the radiolabelled agonist (^{3}H)-HBI to β-adrenoceptors from rat lung, and have shown that the binding of (^{3}H)-HBI to the receptor appears to have both high- and low-affinity components. Furthermore, the B_{max} values for the slowly dissociable, high-affinity binding of (^{3}H)-HBI were increased in a dose-dependent fashion by magnesium. No effect on the apparent K_D value of the receptor for agonists was observed (HEIDENREICH et al. 1982). These authors suggested that the high-affinity component of the binding of (^{3}H)-HBI was a ternary complex consisting of hormone, receptor, and G_s protein, while the low-affinity binding was thought to represent the hormone/receptor complex.

3. Effects of Ions on Dopamine Receptors

Monovalent and divalent cations have marked effects on the binding of radiolabelled drugs to dopamine receptors in the striatum. ZAHNISER and MOLINOFF (1979) studied the interactions of agonists and antagonists with binding sites labelled with (^{3}H)-spiroperidol. Monovalent cations increased the K_D values for agonists but had no effect on the affinity of the receptor for antagonists. On the other hand, divalent cations decreased the K_D values for agonists while increasing the K_D values for antagonists. The presence of these ions did not appear to affect the total number of binding sites for (^{3}H)-spiroperidol. USDIN et al. (1980) reported somewhat different effects. These authors reported that inclusion of mono- or divalent cations or of EDTA resulted in an increased number of binding sites for (^{3}H)-spiroperidol. They reported no effects on K_D values. The binding of (^{3}H)-spiroperidol is, however, unstable in 50 mM Tris at times of incubation longer than 10 min (USDIN et al. 1980; ZAHNISER and MOLINOFF unpublished). Thus, since USDIN et al. (1980) used 50 mM Tris, some of the observed effects may be a consequence of the use of too high a concentration of buffer.

III. Energetics of the Interactions of Agonists and Antagonists with Catecholamine Receptors

1. Energetics of Interactions with β-Adrenoceptors

Turkey erythrocyte membranes have been used widely as a model system to study interactions of β-adrenoceptors with adenylate cyclase (JACOBS and CUATRECASAS 1976; MINNEMAN et al. 1980). It is also a convenient system with which to examine the effects of temperature on the binding of ligands since the kinetics of the binding of (^{125}I)-IHYP to turkey erythrocyte β-adrenoceptors are 6 to 10 times faster than the kinetics of binding to mammalian β-adrenoceptors (MINNEMAN et al. 1980). The incubation times required to reach equilibrium are therefore much shorter with turkey erythrocyte membranes than with mammalian tissues, even when assays are carried out at low temperatures.

The effects of temperature on the inhibition of the binding of (^{125}I)-IHYP to β-adrenoceptors by several drugs including (−)-propranolol and (−)-adrenaline were determined (WEILAND et al. 1979). A 16-fold increase in the affinity of the receptor for adrenaline was observed when the temperatore of the incubation was lowered from 37 °C to 1 °C. This difference was characteristic of other full agonists; the increases in affinities at 1 °C compared to 37 °C ranged up to 56-fold for noradrenaline. On the other hand, there was little change (2- to 3-fold) in the apparent affinity of the receptor for propranolol or other full antagonists. The temperature dependence of the binding of partial agonists was intermediate between that of full agonists and antagonists (WEILAND et al. 1979).

Van't Hoff plots, used to calculate the standard enthalpy change (ΔH^0), were linear over the temperature range studied, implying that the changes in enthalpy are independent of temperature. Both the enthalpy and the entropy changes (ΔS^0) of agonist binding were substantially different from those for antagonists. The binding of antagonists was associated with large increases in entropy and negligible changes in enthalpy. Such changes are characteristic of hydrophobic interactions. The most striking finding was that binding of full agonists resulted in decreases in entropy. These decreases are thermodynamically unfavorable and it is only because of large decreases in enthalpy that binding can occur. Thus, the binding of agonists is enthalpy-driven with marked net decreases in entropy, while the binding of antagonists is almost completely entropy-driven.

The changes in affinity did not result from changes in pH which occur when the temperature of a buffer is altered. Thus, the calculated K_D values were not significantly changed when determined at pH values between 6.8 and 7.8. Furthermore, the changes were reversible. For example, the affinity of the receptor for isoprenaline measured at 37 °C was unaffected by a 5-hour preincubation at 1 °C.

An excellent correlation was observed between the changes in entropy and the efficacies of a variety of drugs (WEILAND et al. 1979). Antagonists had zero efficacy, and their binding was associated with increases in entropy ranging from +13 to +42 entropy units (cal deg^{-1} mol^{-1}). Binding of full agonists resulted in changes in entropy of less than −12 entropy units, and the binding of partial agonists was associated with intermediate changes in entropy. The magnitude of the changes in enthalpy also correlated with the efficacies of agonists, partial agonists, and antagonists.

Although the turkey erythrocyte β-adrenoceptor is a useful model system, it has pharmacological and kinetic properties different from those of β-adrenoceptors in mammalian tissues (MINNEMAN et al. 1980). In view of the existence of subtypes of β-adrenoceptors in mammalian tissues, several mammalian tissues were examined. The tissues chosen included those containing predominantly β_1 receptors (rat heart and cerebral cortex) or β_2 receptors (rat lung and cerebellum), and those in which effects of GTP on agonist affinity are (heart, lung, and cerebellum) or are not (cerebral cortex) observed.

In studies of β-adrenoceptors on turkey erythrocytes (WEILAND et al. 1979), the affinities of agonists increased as the incubation temperature was decreased. Similar results were obtained with all four mammalian tissues studied (WEILAND et al. 1980). The temperature-dependent changes in the affinities of receptors for agonists were completely reversible. Guanine nucleotides such as GTP decreased the affinity of the receptors for agonists and increased the Hill coefficient for inhibition by agonists of the binding of (^{125}I)-IHYP in rat heart, lung, and cerebellum (HEGSTRAND et al. 1979). In these tissues the affinities of the receptor for agonists were less affected by temperature in the presence of GTP than in its absence. In the cerebral cortex, GTP had no effect on either the K_D values of the receptor for isoprenaline or adrenaline or on the temperture dependence of these values (WEILAND et al. 1980).

Calculation of thermodynamic parameters associated with the binding of ligands to β-adrenoceptors in mammalian tissues revealed that decreases in enthalpy provided most of the driving force for the binding of agonists (WEILAND et al. 1979). In most cases, the binding of agonists was associated with unfavorable decreases in entropy. In the heart, lung, and cerebellum, binding of agonists was associated with less negative changes in enthalpy and entropy in the presence of GTP than in its absence. The decreases in the affinity of the receptor for agonists caused by GTP were thus primarily due to less negative changes in the enthalpy of binding despite the effect of the changes in entropy. The thermodynamic changes associated with the binding of agonists to β-adrenoceptors in cerebral cortex were unaffected by the presence or absence of GTP.

The affinities of β-adrenoceptors for antagonists in mammalian tissues were relatively insensitive to changes in temperature. Binding of antagonists was largely entropy-driven although in some cases there was also a significant contribution from a decrease in enthalpy. GTP had no effect on the binding of antagonists at any temperature examined (WEILAND et al. 1980). The entropy-driven binding of antagonists to the β-adrenoceptor is thermodynamically similar to many other protein binding reactions in which information transfer is not involved (KLOTZ and URQUART 1949; SINGER and CAMPBELL 1955). These reactions may reflect hydrophobic binding, being driven by the increase in entropy which results when water molecules, ordered around hydrophobic components of the ligand and the binding site, are displaced (KAUZMANN 1959). Since binding reactions in general have large positive entropy components, it is reasonable to suggest that the initial interaction of agonists with the receptor also results in an increase in the entropy of the reaction. The observed net decrease in entropy associated with the binding of agonists may reflect a conformational change either of the receptor protein or of immediately surrounding membrane components. The calculated net decrease in entropy can be thought of as the sum of an increase in entropy associated with the initial ligand/receptor interaction plus a decrease in entropy resulting from conformational changes.

A model similar to that proposed for interactions of agonists with the nicotinic receptor (MAELICKE et al. 1977; MAGLEBY and STEVENS 1972a, b) can be used to describe the hypothesized agonist-induced isomerization of the β-adrenoceptor. According to this model the following sequence of reactions takes place:

$$H + R \overset{1}{\rightleftharpoons} H \cdot R \overset{2}{\rightleftharpoons} H \cdot R^*$$

In this model H is the ligand, R the receptor in its resting state, and R^* the agonist-induced (activated) state of the receptor. Antagonists would participate only in reaction 1, which is associated with little or no change in enthalpy and large increases in entropy. Agonists also participate in reaction 1; however, because of structural differences, agonists are able to induce a conformational change in the receptor to R^*. This conversion (reaction 2) appears to be associated with negative enthalpy and entropy components. A large decrease in enthalpy could compensate for an unfavorable decrease in entropy. This would result in a net negative free energy change for reaction 2. The extent of conversion to R^* determines the efficacy of a particular ligand. This model is consistent with the high degree of correlation that was observed between efficacy and the magnitude of the calculated changes in enthalpy and entropy (WEILAND et al. 1979).

Physicochemical interpretations of the entropy and enthalpy changes induced by agonists and the proposed conformational changes are necessarily speculative. However, a number of molecular events can be postulated to account for the thermodynamic changes observed. Increased hydrogen or ionic bonding within the protein or between the protein and other components of the system (solvent water or the polar groups of lipids, for example) could lead to decreases in enthalpy. The decreased entropy of the system could reflect either a more restricted conformation of the protein or the ordering of water molecules around newly exposed regions of the protein. The decrease in entropy could also reflect receptor aggregation or interaction of the receptor with another membrane component such as the G_s protein (LIMBIRD et al. 1980; LIMBIRD and LEFKOWITZ 1978). The possibility that agonists cause changes in the conformation of the receptor is supported by reports of an agonist-induced increase in the apparent size of the β-adrenoceptor (LIMBIRD and LEFKOWITZ 1978) and by the increased reactivity of the receptor toward N-ethylmaleimide following exposure to agonists (BOTTARI et al. 1979). An attractive hypothesis is that the decrease in entropy reflects an agonist-induced ordering of lipids surrounding the receptor. The effects of detergent solubilization on agonist affinity and of the polyene antibiotics, which are known to increase membrane fluidity, on the coupling of β-adrenoceptors with adenylate cyclase (HOWLETT et al. 1978; LIMBIRD and LEFKOWITZ 1976; PUCHWEIN et al. 1974) are consistent with this hypothesis. The recent report of the thermodynamics of receptormediated activation of adenylate cyclase by PGE_1 and by increased acyl chain ordering (SINENSKY et al. 1979) are also consistent with this speculation.

It is difficult to devise a unique physiochemical interpretation of the effects of GTP. The decrease in agonist affinity observed in the presence of GTP could be due to a less negative enthalpy change or to a more negative entropy change. Studies carried out in heart, lung, and cerebellum revealed that in the presence of GTP there were less negative changes in enthalpy and entropy.

The thermodynamic changes observed in studies of the binding of agonists in the absence of GTP may be a consequence of the formation of a ternary complex between hormone, receptor, and the G_s protein (LIMBIRD and LEFKOWITZ, 1978). Failure of this complex to accumulate in the presence of GTP may account for the more negative value of the change in entropy observed in the absence of GTP.

2. Energetics of Interactions with Dopamine Receptors

The characteristics of the binding of the dopamine receptor antagonist (^{3}H)-spiroperidol to rat striatal membranes were examined at six different incubation temperatures ranging from 1 °C to 37 °C (ZAHNISER and MOLINOFF 1983). Although the number of receptors labelled at each temperature was identical, the affinity of the receptor for (^{3}H)-spiroperidol decreased 10-fold as the incubation temperature was lowered. The thermodynamic parameters associated with the binding of (^{3}H)-spiroperidol were determined. The binding of (^{3}H)-spiroperidol was entropy-driven ($\Delta S^0 = +72.6$ cal/mol-deg), endothermic ($\Delta H^0 = +10.3$ kcal/mol), and exergonic ($\Delta G^0 = -11.0$ kcal/mol). Qualitatively similar results were observed in studies of (+)-butaclamol, another dopamine receptor antagonist. The binding of the agonists dopamine and ADTN to sites labelled by (^{3}H)-spiroperidol in the striatum also appeared to be entropy-driven ($\Delta S^0 = +30-40$ cal/mol-deg). The magnitude of the increase in entropy associated with the binding of agonists at dopamine receptors was only half that associated with the binding of an antagonist. Thus, agonists produce a smaller net increase in entropy than do antagonists. This may be due to the ability of agonists to induce a conformational change in the receptor as well as displace water from around the receptor. Unlike antagonists, the affinity of the receptor for agonists was not dependent on the temperature of the incubation. Thus, there was no change in the enthalpy associated with the interactions of agonists with the receptor, whether binding was studied at 8 °C or at 37 °C. In contrast to these results, NIEHOFF et al. (1980) have observed that the binding of the agonist (^{3}H)-N-propylnorapomorphine to rat striatal membranes was temperature-dependent and enthalpy-driven. This discrepancy may be explained by the fact that (^{3}H)-spiroperidol and (^{3}H)-N-propylnorapomorphine appear to label different populations of binding sites.

IV. Effects of Membrane Altering Agents

1. Effects on β-Adrenoceptors

The effects of specific lipid-perturbing agents on the sensitivity of adenylate cyclase to catecholamines have been investigated (LEFKOWITZ 1975; LIMBIRD and LEFKOWITZ 1976). Several agents, including phospholipases A, C, and D and amphotericin B, decrease the ability of catecholamines to stimulate adenylate cyclase activity. A dose-dependent decrease in hormone sensitivity was paralleled by a similar decrease in the specific binding of (^{3}H)-DHA. On the other hand, the polyene antibiotic filipin decreased the ability of catecholamines to stimulate adenylate cyclase activity without significantly altering binding of (^{3}H)-DHA. It has been suggested that filipin acts to "uncouple" receptors from catalytic sites. A decrease in the binding of (^{3}H)-DHA after treatment with heat, urea, or proteolytic enzymes suggests that the receptor is a protein or polypeptide. Furthermore, it was suggested that both hydrophobic and hydrophilic residues are necessary for the binding of (^{3}H)-DHA to β-adrenoceptors (LEFKOWITZ et al. 1976a).

As described above, the degree of coupling between the β-adrenoceptor and adenylate cyclase affects the properties of the binding of agonists to the receptor. Agents such as N-ethylmaleimide (NEM) have been reported to have several effects on the receptor/cyclase system. For example, ORLY and SCHRAMM (1976) used NEM to inactivate the adenylate cyclase of turkey erythrocytes in experiments in which cells containing catecholamine receptors were fused with cells that contained adenylate cyclase. In other experiments, NEM has been reported to prevent the loss of β-adrenoceptors induced by incubation with catecholamines (MUKHERJEE and LEFKOWITZ 1977). The slowly reversible binding of (^{3}H)-HBI to frog erythrocyte membranes was reduced by preincubation with NEM. However, the binding of the antagonist (^{3}H)-DHA was not affected (WILLIAMS and LEFKOWITZ 1977).

VAUQUELIN et al. (1980a, b) have shown that coincubation of agonists and NEM results in an apparent loss of binding sites for labelled β-adrenoceptor antagonists. No effect was observed when agonists alone, NEM alone, or antagonists plus NEM were incubated with membranes containing β-adrenoceptors. Guanine nucleotides were shown to prevent, but not reverse, the effect, implying a role for the ternary complex of hormone, receptor, and G_s protein. Additionally, these authors found that the rate of inactivation of binding sites was dependent both upon the agonist used and its concentration. The final number of receptors inactivated, however, was independent of the particular agonist used, with all agonists inactivating approximately 50 % of the receptors. The fact that the pseudo first-order rate constant of inactivation associated with each agonist was related to its ability to stimulate adenylate cyclase (efficacy) led the authors to suggest that two forms of the β-adrenoceptor exist, only one of which is involved in activation of adenylate cyclase. The authors suggested that NEM reacts with the HR complex involved with cyclase activation. VAUQUELIN and MAGUIRE (1980) extended these studies by examining the effects of coincubation of isoprenaline and NEM on radiolabelled antagonist binding sites in S49 lymphoma cells and variants of these cells. Using wild type cells, results similar to those obtained in studies of turkey erythrocyte membranes were observed, and only 65 % of the binding sites were affected by treatment with NEM/isoprenaline. On the other hand, cyc$^-$ and UNC cells that were missing the G_s protein or possessed a defective G_s protein were unaffected by pretreatment with NEM/isoprenaline, suggesting that coupling of the G_s protein to the receptor is necessary for the effect to occur. Similarly, only 65 % of the total number of receptors were affected by treatment of the β_d variant of S49 cells with NEM in the presence of an agonist. This mutant possesses 25 % of the normal density of β-adrenoceptors. These observations led VAUQUELIN and MAGUIRE (1980) to suggest that agonists induce a heterogeneity in the β-adrenoceptors that is not observed in studies of the interactions of antagonists with β-adrenoceptors. HEIDENREICH et al. (1982) have extended these observations by studying the effects of NEM directly on the binding of radiolabelled agonists. These authors demonstrated that the loss of binding sites for antagonists observed by VAUQUELIN et al. (1980 a, b) was due to an NEM-induced stabilization of a high-affinity, slowly reversible binding of an agonist. This high-affinity state was found to be only slowly reversible on addition of guanine nucleotides. These authors suggested that NEM stabilizes a ternary complex of agonist, receptor, and G_s protein. The maximal number of high-affinity binding sites formed in the presence of NEM represented approximately 40 % of the total number of receptors. This value is similar to the 40–60 % loss of (^{125}I)-IHYP binding sites seen after coincubation of agonist plus NEM. There are at least two explanations for the observation that only 40–60 % of the receptors are affected by incubation of membranes with NEM in the presence of an agonist: (1) the amount of functionally available G_s protein is stoichiometrically limiting; (2) some of the receptors can bind radioligands but are unable to interact in a productive way with the G_s protein. This NEM-induced increase in the binding of the agonist (^{3}H)-HBI was not observed in the presence of 1 mM Mg^{2+}. This is most likely due to the fact that Mg^{2+} already causes maximal formation of the high-affinity component of the binding of (^{3}H)-HBI (HEIDENREICH et al. 1982). Kinetic studies showed that in the absence of NEM, most of the bound (^{3}H)-HBI dissociated within 5 min after addition of Gpp(NH)p. After coincubation of membranes with (^{3}H)-HBI and NEM, however, binding of agonists was very slowly dissociable in the presence of Gpp(NH)p with a $t_{1/2}$ of over 3 hours.

In other experiments, KORNER et al. (1982) incubated membranes prepared from turkey erythrocytes with isoprenaline. This resulted in exposure of a specific sulfhydryl group. Addition of NEM or 5,5′-dithiobis-2-nitrobenzoic acid resulted in trapping of the agonist by the receptor and a consequent decrease in the density of binding sites for antagonists and in hormone-stimulated adenylate cyclase activity. Extended incubation at alkaline pH or the addition of high concentrations of GDP or GTP reactivated the receptor, with restoration of the ability of the receptor to bind radiolabelled antagonists and to mediate activation of adenylate cyclase. Since the reaction of NEM with sulfhydryl groups is usually irreversible, this finding suggests that NEM was not interacting with the receptor itself but with an associated component, most likely the G_s protein.

The molecular mechanism for the effect of NEM may involve alkylation of free sulfhydryl groups, resulting in stabilization of the ternary complex. Several reports support the existence of essential disulfide bonds on β-adrenoceptors. In the case of C6 glioma cells, pretreatment with dithiothreitol (DTT) resulted in a decrease in the affinity of β-adrenoceptors for both agonists and antagonists, whereas in turkey erythrocytes there was a decrease in the number of binding sites (LUCAS et al. 1977; VAUQUELIN et al. 1979). DTT also inactivated solubilized turkey erythrocyte receptors (VAUQUELIN et al. 1979). Preliminary data obtained in our laboratory indicate that pretreatment of rat lung membranes with DTT inactivates β-adrenoceptors as indicated by a decrease in the binding of both (^{125}I)-IHYP and (^{3}H)-HBI.

The results suggest that there is at least one essential disulfide bond on the receptor that must be intact for either agonists or antagonists to bind to the receptor. The hormone complex associates with the G_s protein, ultimately resulting in reduction of a disulfide bond or exposure of an existing sulfhydryl group. This free sulfhydryl group is then accessible to alkylation by NEM. After alkylation, the binding of an agonist is essentially irreversible since reoxidation of sulfhydryl groups is required for dissociation of the complex. This reaction sequence can account for the agonist-specific increase in high-affinity binding seen in the presence of NEM and the apparent lack of an effect of Gpp(NH)p on binding of agonists after treatment with NEM.

2. Effects on Dopamine Receptors

SUEN et al. (1980) observed a decrease in the specific binding of (^{3}H)-dopamine and a decrease in dopamine-stimulated adenylate cyclase activity following incubation of striatal membranes with NEM (SUEN et al. 1981). Preincubation of striatal membranes with NEM led to a decrease in the affinity of (^{3}H)-spiroperidol binding sites for agonists (HUFF and MOLINOFF 1981). This effect appeared to be similar to that seen in the presence of guanine nucleotides. Higher concentrations of NEM decreased the number of (^{3}H)-spiroperidol binding sites (HUFF and MOLINOFF 1981). These findings suggest that there are at least two reactive groups involved in the binding of ligands to dopamine receptors. One group appears to be required for high-affinity interactions of agonists. A second group, affected by higher concentrations of NEM, appears to be involved in the binding of antagonists.

Reduction of functional groups such as disulfide bonds with DTT decreased the specific binding of (^{3}H)-dopamine to dopamine receptors in the caudate with no change in the specific binding of (^{3}H)-spiroperidol (SUEN et al. 1980). This suggest that a disulfide bridge is necessary for the high-affinity binding of agonists to dopamine receptors.

There is a great deal of controversy with regard to the effects of ascorbate on dopamine receptors. Ascorbate and sodium metabisulfite have been reported to decrease the specific binding of (^{3}H)-ADTN to dopamine receptors without affecting the specific binding of (^{3}H)-spiroperidol (KAYAALP and NEFF 1980). In addition, ascorbate has been reported to inhibit dopamine-stimulated adenylate cyclase activity (THOMAS and ZEMP 1977). LESLIE et al. (1980) reported a decrease in the number of binding sites for (^{3}H)-haloperidol when receptors were measured in homogenates of guinea pig brain in the presence of ascorbate. Other investigators have shown that high concentrations of ascorbate must be maintained if one is to study the reversible binding of (^{3}H)-N-propylnorapomorphine to dopamine receptors (LEFF et al. 1981). We have recently observed that preincubation of rat striatal membranes with low concentrations of ascorbate decreases the affinity of the receptor for dopamine without affecting the binding of (^{3}H)-spiroperidol (HUFF and MOLINOFF unpublished observations).

E. Localization of Catecholamine Receptors

The complexity of a number of organs makes it difficult to interpret the results obtained when homogenates are assayed to determine the density of receptors. This problem is particularly acute in studies of the CNS where catecholamine receptors are thought to be associated with a variety of cellular elements. Microiontophoretic evidence has been obtained which shows that the electrical activity of some neuronal cells can be modified by the application of appropriate agonists (HOFFER et al. 1971). In addition to postsynaptic receptors, there are also receptors on nerve terminals which modulate the synthesis, release, or storage of transmitters (ROTH et al. 1975). Receptor-mediated control of glucose metabolism in the brain has been suggested to occur in glial cells (EDWARDS et al. 1974). Finally, receptor-mediated alterations in vascular resistance similar to those observed in the periphery may occur in the CNS, resulting in alterations in the distribution of blood flow. Receptors on different cell types may have distinct properties and thus be open to selective pharmacological manipulations. The possibility that different types of receptors exist on different types of cells is suggested by the existence of distinct regional localizations of β_1- and β_2-adrenoceptors in the mammalian brain (MINNEMAN et al. 1979d; RAINBOW et al. 1984).

Experiments have been carried out to investigate the distribution of α- and β-adrenoceptors in the vasculature. These experiments used strips of mesenteric artery and saphenous vein. The affinities of α- and β-adrenoceptors for adrenaline were similar in both tissues. GUIMARAES and PAIVA (1981a) compared the effect of inhibition of neuronal uptake by cocaine with that of inhibition of COMT by dihydroxy-2-methyl propiophenone (U-0521). The action of COMT is observed following extraneuronal uptake of a catecholamine into smooth muscle. In these experiments cocaine potentiated the effects of adrenaline on α-adrenoceptors more than on β-adrenoceptors. U-0521, on the other hand, potentiated the effects of adrenaline on β-adrenoceptors more than on α-adrenoceptors. The results suggest that there is not a homogenous distribution of adrenoceptors on smooth muscle. Thus, α-adrenoceptors may be situated close to presynaptic nerve endings while β-adrenoceptors may be located in close proximity to sites of extraneuronal uptake ($uptake_2$). Results consistent with this conclusion were obtained in studies of electrically stimulated strips of dog saphenous vein (GUIMARAES and PAIVA 1981b). In these experiments it was shown that the latency between the beginning of electical stimulation and the onset of the response was longer for responses mediated by β-adrenoceptors than for responses mediated by α-adrenoceptors.

β-Adrenoceptors are usually thought to be located on the plasma membranes of cells since catecholamine-stimulated adenylate cyclase activity is largely associated with such membranes (ROBISON et al. 1971; DAVOREN and SUTHERLAND 1963). Further evidence that β-adrenoceptors are located on the outer surface of cell membranes has come from experiments utilizing catecholamines immobilized on glass beads or on synthetic polymers (VENTER et al. 1973; VERLANDER et al. 1976).

The subcellular localization of β-adrenoceptors has been investigated using direct binding assays. WOLFE et al. (1976) reported that β-adrenoceptors copurified with fluoride-stimulated adenylate cyclase activity during subcellular fractionation of rat liver. WILLIAMS et al. (1976) showed that (^{3}H)-DHA binding sites are associated with a plasma membrane fraction obtained from rat adipocytes. Similarly, binding sites for (^{3}H)-DHA in rat cerebral cortex were shown to be associated with a synaptosomal fraction (BYLUND and SNYDER 1976; DAVIS and LEFKOWITZ 1976).

Some of the initial evidence for the existence of multiple subtypes of dopamine receptors came from studies comparing the subcellular distributions of dopamine-stimulated adenylate cyclase activity with those of binding sites for (^{3}H)-spiroperidol. CLEMENT-CORMIER and GEORGE (1979) reported that dopamine-stimulated adenylate cyclase activity was associated with the crude synaptosomal-plasma membrane fraction, while most of the binding sites for (^{3}H)-spiroperidol were associated with a microsomal fraction. NEAR and MAHLER (1981) also reported differences in the subcellular distribution of binding sites for (^{3}H)-spiroperidol and (^{3}H)-dopamine.

One way to circumvent the complexities that may arise because adrenoceptors are associated with both neuronal (SIGGINS et al. 1973; KEBABIAN et al. 1975a) and non-neuronal elements (DALY 1975; GILMAN and NIRENBERG 1971; CLARK and PERKINS 1971) is to use histological techniques, by which receptors can be labelled and visualized *in situ*. For example, YOUNG and KUHAR (1979; 1980) have demonstrated that (^{3}H)-WB-4101, when incubated with thin tissue slices from rat brain, will selectively label α_1-adrenoceptors. These receptors were visualized autoradiographically and were found in high concentrations in the olfactory bulb and the hippocampus while no specific binding could be demonstrated in white matter. Similarly, (^{3}H)-p-aminoclonidine has been utilized to autoradiographically localize α_2-adrenoceptors in rat brain (YOUNG and KUHAR 1979, 1980). The density of exposed photographic grains was found to be highest in the limbic system, the locus coeruleus, the nucleus of the solitary tract, and lamina II of the spinal cord.

Similar techniques have been utilized to examine β-adrenoceptors in relatively intact preparations. PALACIOS and KUHAR (1980) demonstrated propranolol-inhibitable autoradiographic labelling of β-adrenoceptors in rat cerebellum. These investigators showed that binding sites for (^{3}H)-DHA had a much higher density in the molecular and Purkinje cell layer than in the granular layer of the cerebellum. In addition, the density of β-adrenoceptors was found to be highest in the caudate-putamen and the cerebral cortex, in general agreement with radioligand binding studies using homogenates of brain regions (SPORN and MOLINOFF 1976).

Recently, RAINBOW et al. (1984) have developed an autoradiographic method of visualizing and quantifying the subtypes of β-adrenoceptors. Highly selective antagonists at either β_1- or β_2-adrenoceptors were used to inhibit the binding of (^{125}I)-IPIN at one receptor subtype but not the other.

This methodology allowed visualization and quantification at a resolution of approximately 100 μm, 10 to 100 times higher than had previously been possible. High levels of β_1-adrenoceptors were found in layers I and VI of the cerebral cortex, the cingulate cortex, the gelatinosis nucleus, CA1 of the hippocampus, the Islands of Calleja, and the dentate gyrus, while high levels of β_2-adrenoceptors were found in layer IV of the cerebral cortex, the molecular layer of the cerebellum, and the central, paraventricular, and caudal lateral posterior thalamic nuclei.

Using (^{3}H)-spiroperidol, MURRIN and KUHAR (1979) have developed a method for autoradiographically labelling binding sites for neuroleptics. The use of this ligand suffers from the fact that, in addition to labelling dopamine receptors, it may also label serotonin receptors. By incubating (^{3}H)-spiroperidol in the presence of pimpamperone, an antagonist at serotonin receptors, the binding of (^{3}H)-spiroperidol to 5-HT receptors was eliminated and the remainder of the pimozide-inhibitable binding appeared to be to dopamine receptors. In this study the (^{3}H)-spiroperidol was administered *in vivo* by intravenous infusion. One to two hours later, rats were killed and sections of the brain prepared for autoradiography. High densities of autoradiographic grains were found in brain areas known to have dopaminergic innervation including the olfactory tubercles, nucleus accumbens, nucleus caudate-putamen, and arcuate nucleus (KLEMM et al. 1979). MURRIN et al. (1979) studied the effects of various neurological lesions on the autoradiographic distribution of (^{3}H)-spiroperidol in rat brain. The intrastriatal administration of 6-hydroxydopamine or a median forebrain bundle lesion resulted in a significant increase in grain density in the caudate putamen. On the other hand, the intrastriatal administration of kainic acid resulted in a substantial decrease in grain density in the caudate putamen. Thus, it appears that the receptors visualized in this tissue are located postsynaptically mainly on neurons intrinsic to the caudate putamen.

MELAMED and his collaborators (1976) have described results obtained with a fluorescent compound, 9-aminoacridinpropranolol (9-AAP). This agent is a potent inhibitor of adrenaline-stimulated adenylate cyclase activity. Following its intravenous administration, a well-defined yellow fluorescent band was seen in the cerebellar cortex in the region of the Purkinje cell layer. The highest density of fluorescence was observed on the apical dendrites of Purkinje cells. On the other hand, the binding of 9-AAP was not inhibited by propranolol (HESS 1979) and much of the observed fluorescence appeared to be due to autofluorescence rather than to binding of the label to β-adrenoceptors. This ligand does not, therefore, appear to provide a means of visualizing β-adrenoceptors.

As noted above, catecholamine-mediated changes in cyclic nucleotide levels may occur both in neurons and in glia (HARDEN and MCCARTHY 1982). The *in vitro* culture of clonal cell lines derived from either neuronal or glial tumors of the nervous system has received increasing attention (AUGUSTI-TOCCO and SATO 1969; HAMPRECHT 1974). Hormone receptors controlling adenylate cyclase activity have been found in both types of tumor cells (CLARK and PERKINS 1971; GILMAN and NIRENBERG 1971; PRASAD and GILMER 1974; SABOL and NIRENBERG 1979). It is likely that receptor-controlled changes in cyclic nucleotide levels also take place in normal glial cells. Receptor-mediated increases in cyclic AMP may control carbohydrate metabolism in glial cells. Astrocytoma cells respond to noradrenaline with an increase in cyclic

AMP (GILMAN and NIRENBERG 1971), the conversion of glycogen phosphorylase from the "b" to the "a" form (BROWNING et al. 1974), and a decrease in glycogen content (PASSONNEAU and CRITES 1976). Furthermore, it has been suggested that glycogen is concentrated in normal cerebral astrocytes (NATHANSON 1977). Biogenic amine-mediated changes in cyclic AMP content in chick brain have also been linked with glycogenolysis (EDWARDS et al. 1974), while adenosine and isoprenaline, but not dopamine, appear to stimulate glycogenolysis in slices of rat striatum (WILKENING and MAKMAN 1976; 1977).

F. Multiplicity of Catecholamine Receptor Subtypes

On the basis of pharmacological, anatomical, and biochemical characteristics, it has been possible to subclassify receptors for a variety of neurotransmitters. Thus, the effects of acetylcholine can be attributed to the activation of either muscarinic or nicotinic receptors. Furthermore, nicotinic receptors on skeletal muscle have different properties from nicotinic receptors in ganglia. The existence of multiple types of adrenoceptors was first suggested by the results of DALE (1906). He showed that some of the effects of adrenaline or of sympathetic nerve stimulation were antagonized by ergot alkaloids while other responses were not affected. AHLQUIST (1948) divided adrenergic responses into two classes based on the effects elicited by a series of catecholamines. The first class of receptors, called α-adrenoceptors, was stimulated by agonists with the order of potency of adrenaline > noradrenaline > isoprenaline. Stimulation of these receptors led to contraction of the nictitating membrane, to vasoconstriction, and to contraction of smooth muscle in the uterus and ureter. The second class of receptors, called β-adrenoceptors, responded with the order of potency of isoprenaline > adrenaline > noradrenaline. Stimulation of these receptors resulted in increases in cardiac inotropy and chronotropy, in vasodilation, and in relaxation of uterine smooth muscle. This classification of adrenoceptors has been extended to include most of the effects mediated by catecholamines.

I. α-Adrenoceptor Subtypes

It has been known for some time that phenoxybenzamine is more potent in blocking postsynaptic α-adrenoceptors that mediate smooth muscle contraction than in blocking presynaptic α-adrenoceptors that regulate the release of noradrenaline during nerve stimulation (see LANGER 1977). This observation led to the suggestion that the postsynaptic α-adrenoceptor should be referred to as an α_1-adrenoceptor while the presynaptic receptor should be referred to as an α_2-adrenoceptor (LANGER 1974; BERTHELSEN and PETTINGER 1977). Snyder and his collaborators (PEROUTKA et al. 1978) have described two classes of α-adrenoceptors in the CNS. The receptor labelled by (^{3}H)-clonidine and by labelled catecholamines had a uniquely high affinity for agonists while the second class of receptor labelled by the antagonist (^{3}H)-WB-4101 had a lower affinity for agonists and a higher affinity for antagonists. The ligand

(^{3}H)-DHE appears to bind to both subclasses of α-adrenoceptor. In studies of the pharmacological specificity of the sites, the binding site for (^{3}H)-WB-4101 resembles the α_1-adrenoceptor while the binding site for (^{3}H)-clonidine resembles the α_2-adrenoceptor. As noted above, (^{3}H)-DHE appears to be a non-selective ligand binding to both classes of receptor. An important point with regard to the study of α-adrenoceptors is that there are drugs which appear to have a very high selectivity for α_1 or α_2 receptors. Thus, prazosin is 1,000–10,000 times more potent at α_1 than at α_2 receptors (U'PRICHARD et al. 1978a), while yohimbine is much more potent in blocking presynaptic than postsynaptic α-adrenoceptors (STARKE et al. 1975). Recent evidence suggests that post-synaptic α_2-adrenoceptors exist in both the CNS and the periphery (U'PRICHARD et al. 1979; RUFFOLO et al. 1981).

II. β-Adrenoceptor Subtypes

While many of the drugs which exist for distinguishing between the two subclasses of α-adrenoceptors act with both high affinity and selectivity, even the most selective drugs available for distinguishing between β_1- and β_2-adrenoceptors show only a limited difference in affinity (MINNEMAN et al. 1979d; see below).

The responses characteristic of β-adrenoceptors have been defined through the use of specific antagonists (FURCHGOTT 1972; see below). Thus, β-adrenoceptor-mediated responses have been divided into the subclasses β_1 and β_2 (LANDS et al., 1966; 1967). Adrenaline and noradrenaline are approximately equipotent activators of β_1-adrenoceptors, which are preferentially inhibited by practolol (DUNLOP and SHANKS 1968), metoprolol (PETRACK and CZERNICK 1973), and atenolol (HARMS 1976). Noradrenaline is much less potent than adrenaline in activating β_2-adrenoceptors. Congeners of methoxamine, including butoxamine and H35/25, are specific inhibitors at these receptors (LEVY 1966). Many antagonists, including propranolol, alprenolol, and pindolol, do not distinguish between β_1 and β_2 receptors.

The pharmacological specificity of adrenoceptor-mediated responses differs in the several systems that have been investigated carefully (see MINNEMAN et al. 1981a). In particular, differences have been reported between the potencies of drugs acting on β_2-adrenoceptors on vascular compared to tracheal smooth muscle. Differences of this type have led to the suggestion that there may be more than two subtypes of β-adrenoceptor (see AHLQUIST 1976). However, the potency of a drug determined *in vivo* or in an isolated organ preparation is not due only to the affinity of the drug for its receptor. For example, access of the drug to its receptor may vary due to the existence of diffusion barriers or to sequestration in membranes. Furthermore, the chemical or enzymatic degradation of a drug may vary between preparations. The densities of presynaptic neuronal and extrasynaptic non-neuronal uptake sites can also affect the amount of drug having access to the receptor. These factors can affect the relative specificity of a group of pharmacological agents. For example, differences in the apparent potencies of a number of drugs for stimulating or inhibiting tracheal as compared to vascular β_2-adrenoceptors have been shown to be due to differences in the extraneuronal uptake of the drugs and not to intrinsic differences in the properties of the receptors. Little or no selectivity was seen in the presence of inhibitors of extraneuronal uptake (O'DONNELL and WANSTALL 1976).

It has been suggested that β_1 and β_2 receptors may be present in the same organ (CARLSSON et al. 1972). In the dog heart both subtypes of β-adrenoceptor are involved in the inotropic and chronotropic effects of catecholamines. CARLSSON et al. (1972) suggested that β_2 receptors make a greater contribution to the control of heart rate than of contractile force. Similarly, using radioligand binding techniques, evidence has been obtained which suggests that two types of β-adrenoceptor coexist in a variety of rat tissues (MINNEMAN et al. 1979b). The affinities of a number of drugs for β_1 and β_2 receptors have been determined in several brain regions and peripheral organs of rats. The K_D value of a given drug for β_1 or β_2 receptors was the same in each of the tissues studied. Approximately 20 % of the β-adrenoceptors in rat heart have the pharmacological specificity of β_2 receptors. Similar results were obtained in studies of cat and guinea pig heart (HEDBERG et al. 1980). Approximately 25 % of the β-adrenoceptors in the atria were of the β_2 subtype. The cardiac ventricles contained essentially only β_1-adrenoceptors. These results may explain previous findings that β_2-selective drugs have more of an effect on chronotropy than on inotropy (CARLSSON et al. 1972). Similarly, approximately 15 % of the adrenoceptors in rat lung have the pharmacological specificity of β_1-adrenoceptors. The relative proportion of receptors with the pharmacological properties of β_1- or β_2-adrenoceptors was the same in a given tissue when determined with drugs whose affinities for the two types of receptors varied over a wide range. This suggests that there are only two classes of β-adrenoceptors in rat organs. If, for example, there were variable amounts of a third class of β-adrenoceptor, the apparent affinities of drugs for β_1- and β_2-adrenoceptors would be expected to vary. It is important to emphasize that these results were obtained both with β_1-selective and β_2-selective agonists and antagonists.

The strongest evidence that there are only two subtypes of β-adrenoceptors came from studies of tissues containing only one receptor subtype (MINNEMAN et al. 1979b). In these tissues, the inhibition of the specific binding of (^{125}I)-IHYP by β_1- and β_2-selective drugs resulted in linear Hofstee plots, suggesting that there is only a single class of binding sites. In the left ventricle of cat and guinea pig heart there appears to be a homogeneous population of β_1 receptors, whereas rat liver and cat soleus muscle contain only β_2 receptors (MINNEMAN et al. 1979b; HEDBERG et al. 1980). In these tissues it was possible to demonstrate directly that the interaction of each drug (including the radioligand and the selective drugs) with each of the subtypes of receptor followed the principles of mass action for a bimolecular interaction, yielding linear Hofstee plots and Hill coefficients of 1. In addition, it was possible to show that the affinities of each receptor for each drug in tissues that contain only one receptor subtype are the same as the affinities calculated from the computer-aided graphic analysis of biphasic Hofstee plots (MINNEMAN et al. 1979b, c).

In some non-mammalian tissues the kinetic properties and pharmacological specificity of β-adrenoceptors are distinct from those of mammalian β_1- or β_2-adrenoceptors. Although turkey erythrocytes contain an apparently homogeneous population of β-adrenoceptors, these receptors have major kinetic and pharmacological differences that distinguish them from mammalian β_1 or β_2 receptors (GIBSON et al. 1979; MINNEMAN et al. 1980). In terms of kinetics, the rate constants for association and dissociation of the radioligand (^{125}I)-IHYP were approximately 10 times greater in membranes prepared from turkey erythrocytes than in membranes prepared from mammalian tissues. Drugs selective for either β_1 or β_2 receptors had very different affinities for the receptor in turkey erythrocytes. On the other hand, drugs not selective for β_1 or β_2 receptors had the same affinities for the receptor on turkey erythrocytes as in mammalian tissues including rat heart and lung. The characteristics of the β-adrenoceptors in frog erythrocytes (WILLIAMS and LEFKOWITZ 1978) and heart (HANCOCK et al. 1979) also appear to be different from those observed in studies of mammalian β_1 or β_2 receptors (MINNEMAN et al. 1979b, c). Therefore, data obtained in studies with non-mammalian tissues should be extrapolated to mammalian tissues with caution.

The distribution of β-adrenoceptors in regions of the mammalian CNS has been examined extensively. Several groups of investigators (ALEXANDER et al. 1975; SPORN and MOLINOFF 1976; BYLUND and SNYDER 1976) have reported that β-adrenoceptors are distributed relatively homogeneously in various regions of the CNS. This is surprising because, if β-adrenoceptors are associated with noradrenergic neuronal function, one would expect them to be as asymmetrically distributed as is the noradrenergic innervation itself. Furthermore, the distribution of β-adrenoceptors in the CNS bears no apparent relationship to the catecholamine content of particular brain regions. Some regions with a high content of noradrenaline, such as the cerebral cortex, also have a high density of β receptors. Other regions, such as the caudate nucleus, with little or no detectable noradrenaline, also have a high density of receptors. Based on these findings, SPORN and MOLINOFF (1976) suggested that some of the β-adrenoceptors in the CNS might be associated with glia or blood vessels rather than with neuronal elements. The distribution of the two subtypes of β-adrenoceptors in regions of the mammalian CNS was determined and compared to the density of the noradrenergic innervation, the density of glial elements, and the degree of vascularization of various brain regions. The density of β_2-adrenoceptors in different brain regions varied only 2- to 3-fold. The highest density was in the cerebellum, while the lowest density was observed in the hippocampus and diencephalon. However, depending on the age of the animals, the ratio of the densities of β_1 receptors in cerebral cortex to β_1 receptors in cerebellum ranged up to approximately 50:1 (MINNEMAN et al. 1979b; PITTMAN et al. 1980a). These results suggested that β_1 receptors might be located on heterogeneously distributed populations of neurons, while β_2 receptors might be associated with more homogeneously distributed cellular elements such as glia or blood vessels. There was no correlation, however, between the density of β_1 receptors and the tissue content of noradrenaline. For example, the density of β_1 receptors was relatively high in the caudate which does not contain easily demonstrable noradrenaline or vesicular dopamine β-hydroxylase (SWANSON and HARTMAN 1975).

III. Dopamine Receptor Subtypes

Studies characterizing the pharmacological properties of dopamine receptors in a number of mammalian tissues have produced discrepant results (COOLS and VAN ROSSUM 1976; GOLDBERG et al. 1978). These discrepancies have led to the suggestion that dopamine receptors are either a single class of receptor molecule that can exist in two conformations (CREESE et al. 1975) or two distinct classes of receptors (COOLS and VAN ROSSUM 1976). The properties of receptors were markedly different, depending on whether they were determined by inhibition of the binding of labelled agonists or labelled antagonists (CREESE et al. 1975; BURT et al. 1976). In general, agonists were more potent inhibitors of the binding of (^{3}H)-dopamine than (^{3}H)-haloperidol while the affinities of the receptor for antagonists were greater if determined by inhibition of the binding of (^{3}H)-haloperidol than of (^{3}H)-dopamine.

Results of studies of the inhibition of the binding of radioligands to dopamine receptors by both agonists and antagonists are consistent with the existence of multiple classes of dopamine receptors. Thus, biphasic displacement curves are frequently observed (TITELER et al. 1978; CREESE et al. 1978). LEYSEN and LADURON (1977) showed that after differential centrifugation of homogenates of rat striatum, dopamine-sensitive adenylate cyclase activity was associated with a mitochondrial fraction while binding sites for (^{3}H)-haloperidol were found in a microsomal fraction, suggesting that the labelled butyrophenone does not bind to the dopamine receptor associated with the stimulation of adenylate cyclase.

KEBABIAN (1978) has summarized the evidence indicating that there are two major types of dopamine receptor, one of which (D-1) stimulates adenylate cyclase activity (see KEBABIAN and CALNE 1979). Dopamine-sensitive adenylate cyclase activity is present in the caudate nucleus and in a number of other tissues (KEBABIAN and GREENGARD 1971; KEBABIAN et al. 1972; BROWN and MAKMAN 1972; GALE et al. 1977). The effect of dopamine is inhibited by a variety of phenothiazines and butyrophenones and is distinguishable from effects due to other biogenic amines. Stimulation of adenylate cyclase activity by dopamine is correlated with and apparently involved in regulating the release of parathyroid hormone from isolated cells of the bovine parathyroid gland (BROWN et al. 1977). On the other hand, substantial evidence suggests that dopamine receptors occur on cells in the pituitary (CREESE et al. 1977b; CRONIN et al. 1978; MOWLES et al. 1978) where dopamine is involved in inhibiting the release of prolactin (MELTZER et al. 1977; CARON et al. 1978). This effect appears not to be mediated through increases in adenylate cyclase activity (SCHMIDT and HILL 1977). An example of a dopamine receptor that does not appear to act by changing adenylate cyclase activity is the autoreceptor on dopaminergic nigrostriatal neurons. These receptors are present both on nerve terminals, where they regulate the activity of tyrosine hydroxylase (CARLSSON 1975), and on dendrites of dopaminergic cell bodies in the substantia nigra, where they appear to regulate cell firing (Aghajanian and Bunney 1977).

There are differences between the pharmacological specificity of dopamine receptors that are involved in the stimulation of adenylate cyclase activity (D-1 receptors) and those that are not associated with increases in cyclic AMP levels (D-2 receptors). Apomorphine has been reported to be a partial agonist or a full antagonist at dopamine receptors that stimulate adenylate cyclase activity. These receptors are maximally stimulated by micromolar concentrations of dopamine and are inhibited by ergot alkaloids. In contrast, dopamine receptors that do not appear to be associated with stimulation of adenylate cyclase are maximally stimulated by nanomolar concentrations of dopamine and apomorphine. Ergot alkaloids are agonists at the dopamine receptor on mammotrophs but have little or no effect on striatal dopamine autoreceptors (Kebabian and Kebabian 1978). The antipsychotic agents molindone and sulpiride and the antiemetic metoclopramide are potent antagonists of D-2 receptors in the anterior pituitary but fail to inhibit dopamine-stimulated adenylate cyclase activity in the caudate.

Cote et al. (1982) have recently shown that at least some D-2 receptors are functionally linked to inhibition of adenylate cyclase activity via an inhibitory guanine nucleotide-binding regulatory subunit (G_i) (Smigel et al. 1984). This may be the explanation for the observation that guanine nucleotides regulate the affinity of agonists at D-2 receptors (Zahniser and Molinoff 1978).

Additional subtypes of dopamine receptors, designated D-3 and D-4, have been proposed based on results obtained in *in vitro* binding studies with labelled agonists. The putative D-3 receptor was reported to have a high affinity for agonists such as (^{3}H)-dopamine and a low affinity for butyrophenones such as spiroperidol (Titeler et al. 1979; Hamblin and Creese 1982). In contrast, the putative D-4 receptor was reported to have a high affinity for both agonists and butyrophenone neuroleptics (Seeman 1980). The interactions of agonists with receptors often involve multistep, multicompartment binding reactions that can resemble reactions observed in the presence of multiple classes of receptors even when there is only a single, homogeneous population of receptors in the tissue (see Minneman et al. 1981a). This can result in an erroneous conclusion that two independent subtypes of receptor exist (see Molinoff et al. 1981). Bacopoulos et al. (1983) have demonstrated that there is a significant correlation between the affinity of antagonists for the D-2 and the putative D-4 receptor. Futhermore, they reported that the affinity of the putative D-3 receptor for antagonists was significantly correlated with their potency as inhibitors of D-1 receptor-mediated stimulation of adenylate cyclase activity. These findings suggest that the proposed D-3 and D-4 subtypes represent different conformations or states of the D-1 and D-2 receptors, respectively, and not independent subtypes (Leff and Creese 1983).

It has been suggested that D-2 receptors in rat striatum can be further subdivided into D-2A and D-2B subtypes on the basis of *in vitro* binding studies with the antagonist (^{3}H)-spiroperidol (Huff and Molinoff 1982). Scatchard analysis of the binding of (^{3}H)-spiroperidol resulted in linear plots, suggesting that (^{3}H)-spiroperidol only labelled a single type of receptor in this tissue.

Studies of the inhibition of the binding of (^{3}H)-spiroperidol by a number of competing ligands gave results, however, that were not consistent with those expected from the interaction of these agents with a single class of binding sites. Hofstee plots of these data, analyzed by a computer-assisted iterative method, led to the conclusion that two distinct classes of binding sites, designated D-2A and D-2B, exist in the tissue. These sites were present in a ratio of approximately 1:3. Comparison of the properties of these striatal binding sites with those for (^{3}H)-spiroperidol in the frontal cortex, which contains both D-2 and 5-HT2 receptors, suggested that neither class of striatal sites is a receptor for 5-HT. In contrast to the linear Scatchard plots observed in the striatum, analysis of the binding of (^{3}H)-spiroperidol in the frontal cortex resulted in curvilinear plots, indicative of multiple classes of sites with different affinities for the radioligand. Moreover, both classes of sites in the striatum had a higher affinity for dopamine than 5-HT, whereas the large majority of sites in the frontal cortex had a higher affinity for 5-HT than for dopamine. Despite these observations, the possibility that (^{3}H)-spiroperidol labels a small population of 5-HT2 receptors in the striatum cannot be completely dismissed. It has been reported that the 5-HT2 antagonist (^{3}H)-ketanserin labels a small population of sites in the rat striatum (LEYSEN et al. 1982). If (^{3}H)-spiroperidol is labelling more than two populations of sites in the striatum, it will be very difficult to resolve and characterize these components using conventional radioligand binding techniques.

Another approach to investigating the multiplicity of dopamine receptor subtypes as well as dopamine-sensitive adenylate cyclase activity has utilized the neurotoxins 6-hydroxydopamine (6-OHDA) and kainic or ibotenic acid. Lesions of the nigrostriatal pathway produced by the administration of 6-OHDA have resulted in a marked depletion of dopamine from the caudate, reflecting degeneration of dopamine-containing nerve terminals. This treatment, however, did not result in a decrease in striatal dopamine-sensitive adenylate cyclase activity (MISHRA et al. 1974; KRUEGER et al. 1976). These results suggest that most of the dopamine-sensitive adenylate cyclase activity in the caudate is not associated with presynaptic terminals. Injection of kainic acid, a structurally rigid analogue of glutamate, or of ibotenic acid resulted in degeneration of cell bodies of neurons near the injection site but spared terminals and axons of afferent neurons (OLNEY et al. 1975). A single intrastriatal injection of kainic acid resulted in a rapid loss of neuronal perikarya and a marked decrease in the activities of the biosynthetic enzymes glutamic acid decarboxylase and choline acetyltransferase (COYLE and SCHWARCZ 1976; MCGEER and MCGEER 1976). Administration of kainic acid also resulted in proliferation of glial cells (MINNEMAN et al. 1978). Dopaminergic terminals in the striatum remained intact and levels of tyrosine hydroxylase were not affected (SCHWARCZ et al. 1978a). Biochemical changes similar to those seen following the administration of kainic acid have been observed in the brains of patients suffering from Huntington's Chorea, and the administration of kainic acid has been used as the basis for an animal model for this disease (COYLE and SCHWARCZ 1976; MCGEER and MCGEER 1976).

Following administration of kainic acid, there is a slow and relatively incomplete loss of neurotransmitter binding sites compared with the rapid loss of choline acetyl-transferase, glutamic acid decarboxylase and dopamine-sensitive adenylate cyclase activity. Two days post-injection, binding of (^{3}H)-haloperidol was decreased by only 20 %, whereas dopamine-sensitive adenylate cyclase activity was decreased by 85 % (SCHWARCZ et al. 1978a). One to three weeks after the injection of kainic acid, binding of (^{3}H)-haloperidol (SCHWARCZ et al. 1978a) and of (^{3}H)-spiroperidol (MINNEMAN et al. 1978) was reduced by 40 %-60 %. At this time dopamine-stimulated adenylate cyclase activity was virtually undetectable. FIELDS et al. (1978) have reported that binding of (^{3}H)-spiroperidol continued to decline over time so that seven weeks after the lesion only 25 % of the binding sites were found in the tissue. The time course of the loss of binding of (^{3}H)-quinuclidinyl benzilate to muscarinic receptors was very similar to that of binding of neuroleptics to dopamine receptors (FIELDS et al. 1978).

Two explanations have been proposed for the difference between the rate of decline of dopamine-sensitive adenylate cyclase activity and that of dopamine receptors. OLNEY and DE GUBAREFF (1978) utilized morphological techniques and reported that postsynaptic membranes are detectable for at least three weeks after treatment with kainic acid. These membranes may still have measurable binding sites that are not coupled to adenylate cyclase since the membranes are no longer attached to their original dendritic spine, shaft, or cell soma. An alternative explanation is that some of the labelled neuroleptic binding sites exist on the terminals of corticostriatal neurons (SCHWARCZ et al. 1978a, b; GARAU et al. 1978). Intrastriatal injection of kainic acid followed by cortical ablation led to a 70 % decrease in the binding of (^{3}H)-haloperidol within 5 days of the cortical ablation. These data support the idea that tritiated neuroleptics label two classes of dopamine receptor sites, one of which is on the soma of striatal cells and is sensitive to the effects of kainic acid, while the other is associated with cortical afferents and is resistant to the effects of kainic acid but is affected by cortical ablation.

IV. Assay of Receptor Subtypes

Pharmacological studies using intact preparations have provided much of the initial evidence with regard to the existence of receptor subtypes. There are, however, potential problems inherent in the use of such preparations. For example, drugs can be metabolically transformed between the time they are administered to a whole animal or added to a perfusion medium and the time they act on a receptor. There may be differences in the distribution of the administered drugs in the tissue. Drugs may be bound to different extents to plasma proteins, and the blood brain barrier or other diffusion barriers may differentially affect the ability of drugs to gain access to a given population of receptors. Enzymatic and chemical degradation can occur, and some drugs may be metabolically transformed into compounds that are either more or less active than the parent drug. There are also potential problems with tissue-specific drug removal. Uptake into both neuronal and extraneuronal cells can occur. These processes can lead to different pharmacological specificities when a series of agents is investigated even though the receptors are identical (see O'DONNELL and WANSTALL 1976). Finally, multiple receptor subtypes within a tissue or organ may markedly complicate determination of the pharmacological specificity of a response. For example, we now know that there are both β_1 and β_2 receptors in the hearts and lungs of a number of mammalian species (MINNEMAN et al. 1979 d). At least in the case of the heart, both β_1 and β_2 receptors mediate the chronotropic effects of catecholamines (CARLSSON et al. 1972; 1977). The degree of selectivity of a drug can affect the observed pharmacological response. The coexistence of receptor subtypes in a tissue can distort the pharmacological specificity observed in an intact preparation and lead to the erroneous conclusion that there are an ever-increasing number of receptor subtypes.

The development of direct biochemical assays for catecholamine receptors has circumvented many of the pharmacokinetic problems inherent in determining the affinity of a receptor for a drug in an intact tissue. The major advantage of the use of a broken cell preparation is that problems of selective uptake or unequal access of drugs to the receptor are minimized. Radioligand binding techniques can be used to monitor the initial drug/receptor interaction. In many cases, the initial interaction between the ligand and the receptor can be shown to be a simple bimolecular reaction. In this case quantitative characteristics of the reaction (rates of association and dissociation, equilibrium affinity constants) can be determined experimentally. In addition, experiments involving the use of radioligands allow quantitative comparisons of the affinities of various drugs for receptors in different tissues. Finally, the simplicity and technical ease of these experiments permit the collection of accurate data, suitable for computer-assisted quantitative analysis. Thus, radioligand binding assays can be very useful in characterizing receptor subtypes. On the other hand, it is important to recognize that most organs contain a variety of cell types. For example, mammalian brain contains glial cells and blood vessels as well as a variety of types of neurons. It is likely that different catecholamine receptors are associated with different cell types. It is also possible that some cells will contain multiple subtypes of a given transmitter receptor (see HOMBURGER et al. 1981).

Several requirements must be satisfied when binding of radioligands is used to study receptor subtypes. The basic technique involves monitoring the displacement of a radiolabelled compound (see Fig. 4), usually an antagonist, by subtype-selective, nonradioactive agonists or antagonists (see MOLINOFF et al. 1981). Alternatively, if a highly selective radioligand is available, the interaction of this compound with the receptor subtype in question can be monitored directly. In either case, the interactions of the labelled and/or nonradioactive drugs with the receptor must be shown to be reversible and competitive. It is also important that the amount of radioligand bound be small relative to the total amount of radioligand in the assay. Under these conditions, the concentration of the radioligand will not change during the course of the assay. Problems can arise if the gradual displacement of a radioligand from its binding sites results in a change in the concentration of free radioligand.

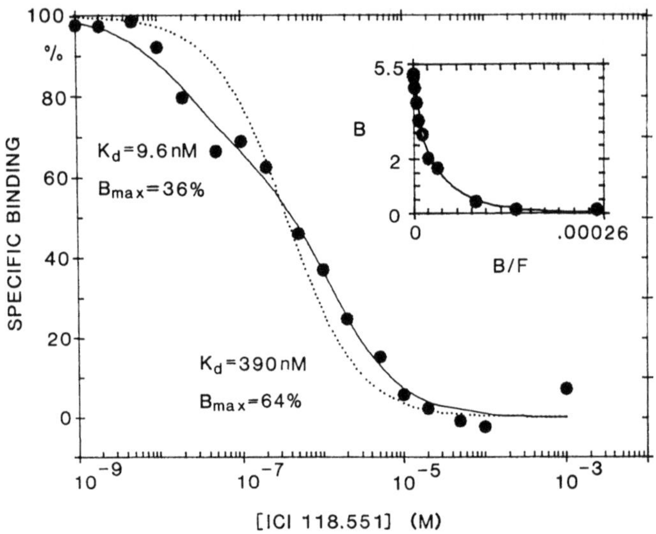

Fig. 4. Competition analysis using the selective drug ICI 118.551. Rat cortical membranes were incubated for 25 min at 37 °C with (^{125}I)-pindolol and increasing concentrations of ICI 118.551. The fit of the curve, assuming the presence of two types of receptors (solid line), was significantly better than that obtained when the data were fit to a one-site model (dotted line). Analysis of the two-site curve yielded values for the high-affinity ($K_H = 9.6$ nM) and low-affinity ($K_L = 390$ nM) sites of 36 % and 64 %, respectively. Inset: Hofstee transformation of the competition curve was not linear.

Additional complications may occur if the radioligand is a racemic mixture (BÜRGISSER et al. 1981). If the relative potencies of the two isomers are not known, it is difficult to carry out assays under conditions in which it is certain that only the active isomer binds to the receptor. In practice, assays should be carried out under conditions where less than 10 % of the ligand binds to the receptor. Since catecholamine receptors show a marked degree of stereoselectivity, this will usually eliminate binding of the inactive isomer (HARDEN et al. 1976; U'PRICHARD et al. 1977; BURT et al. 1976; see WEILAND and MOLINOFF 1981).

Any method of analyzing receptor subtypes is based on two important assumptions. The most critical assumption is that mass action principles are fully satisfied for the interactions of all ligands at all sites. This means that the radioligand must bind at each site with a Hill coefficient of 1 (i.e., linear Scatchard plot; see WEILAND and MOLINOFF 1981). Similarly, the binding of competing ligands to each site must be described by linear Hofstee plots with slope factors or Hill coefficients of 1. In practice, this assumption can be difficult to document. The second assumption concerns the number of subtypes assumed to be present. One means of testing this assumption involves a nonlinear least-squares analysis of binding isotherms. Such programs test the hypothesis that the data fit better to a model that includes two sites as compared to one site, or three sites as compared to two sites (HANCOCK et al. 1979). In practice, however, it is usually impossible to obtain sufficiently precise data to define more than two or at most three receptor subtypes. It takes 15–18 concentrations of a competing ligand to define two subtypes reliably and precisely (see MINNEMAN et al. 1981a). In studies of muscarinic cholinoceptors in the medulla/pons, BIRDSALL et al. (1980) used over 50 concentrations of competing ligand and were able to define three types of muscarinic receptor. Regardless of the number of concentrations of competing ligand used, it is probably not possible to define more than three receptor subtypes in one tissue using a single set of binding data.

Two basic approaches have been developed to utilize binding assays to quantitate receptor subtypes. The first approach uses a radioligand that is selective for one receptor subtype. If one has a radioligand that is 100% selective (i.e., binds to only one subtype), then it is only necessary to characterize the binding of the ligand to the receptor to define the properties of the subtype in question. For example, (^{3}H)-prazosin appears to be a selective radioligand for α_1 receptors while (^{3}H)-clonidine has been reported to be a selective radioligand for α_2 receptors (GREENGRASS and BREMNER 1979; U'PRICHARD et al. 1979). There are, however, relatively few cases in which a radioligand has been shown to be completely selective. Therefore, it is usually necessary to use a more involved mathematical analysis to determine the properties and densities of receptor subtypes present in a given tissue. When using a partially selective radioligand, one first determines a saturation curve. The data are then analyzed, often after transformation by the method of SCATCHARD (1949), which yields a curvilinear plot in the presence of more than one receptor subtype. A second approach to studying receptor subtypes involves the use of a selective competing ligand and a nonselective radioligand. A displacement curve for a selective competing drug is generated and the data are then analyzed after transformation by the method of EADIE (1952) and HOFSTEE (1952). This analysis again results in curvilinear plots in the presence of more than one receptor subtype (Fig. 4). In practice, the use of the second approach is greatly facilitated if both subtypes have the same affinity for the radioligand (see MINNEMAN and MOLINOFF 1980). It is important to perform studies using both selective agonists and selective antagonists, and the most credible results are obtained if a variety of drugs having markedly different structures is used. In most cases, studies with agonists will only be interpretable if assays are carried out in the presence of guanine nucleotides. The mathematical approach used to quantitate receptor subtypes assumes that the interactions of each of the drugs with each putative receptor subtype follow mass action principles (see MOLINOFF et al. 1981).

Evaluation of heterogeneous binding sites can be accomplished either by direct analysis of the log-dose response curve for inhibition of the binding of the radioligand by the drug, or by analysis after a linearizing transformation of the data into a form that will yield a straight line if a drug is interacting with a homogeneous population of receptor sites. This modified Scatchard (or Hofstee) analysis reveals an apparent curvature if more than one type of receptor, each with a different affinity for the drug, is present in the preparation. Several methods are now used to evaluate experimental data for the possible existence of multiple classes of binding sites. For direct analysis of non-linear, modified Scatchard (Hofstee) plots, the amount of ligand bound is plotted against the amount bound divided by the concentration of drug (see HOFSTEE 1952). An approach for analysis of such data was introduced by RUGG et al. (1978). Data were transformed by the Hofstee method and analyzed graphically to estimate the parameters of the two types of receptors. If drugs were available that were completely selective for each of the subtypes of receptors for a given transmitter, this method would yield accurate results. Unfortunately, at least for β-adrenoceptors, even the most selective drugs currently available have only limited selectivity. Use of this method, therefore, results in a significant underestimate of the number of high-affinity sites and an overestimate of the number of low-affinity sites (MINNEMAN and MOLINOFF 1980). A similar modified Scatchard (Hofstee) analysis was developed by MINNEMAN et al. (1979 b). These authors subjected transformed data to a computerized iterative analysis. This approach corrects each component for the contribution of the other component, resulting in an accurate analysis of theoretical data regardless of the degree of selectivity of the competing ligand (MINNEMAN et al. 1979 d). An advantage of this method is that it combines the intuitive simplicity of the Hofstee plot (where the characteristics of the two components can be easily visualized as the asymptotes of a biphasic curve) with a mathematically accurate analysis. The major limitation of the method derives from its use of unweighted linear regression. Furthermore, the transformation results in the propagation of errors onto both the abscissa and the ordinate. HANCOCK et al. (1979) have analyzed displacement curves directly using a computerized non-linear, least-squares, curve-fitting technique. Since this method uses untransformed data, errors are not expressed on the abscissa. For statistical reasons, this curve-fitting technique has advantages over the iterative graphic analysis described by MINNEMAN et al. (1979 d; see MOLINOFF et al. 1981). However, the method loses the intuitive simplicity inherent in the Hofstee analysis. The results of the computer analysis must therefore be taken at face value.

It is important to recognize that a number of other circumstances, in addition to the existence of receptor subtypes, will also result in curvilinear Hofstee or Scatchard plots (see MOLINOFF et al. 1981). The most common possible causes of such curvature include negatively cooperative interactions, ligand heterogeneity, and an incorrect definition of specific binding (see MOLINOFF et al. 1981). A two-step reaction involving three components can also generate curvilinear plots (DE HAËN 1976; BOEYNAEMS and DUMONT 1977; JACOBS and CUATRECASAS 1976; HEIDENREICH et al. 1980). According to this model, hormone or drug binds to the receptor to form a complex which then interacts with a third component in the membrane to form a ternary complex. This reaction scheme appears to apply to a large number of receptor systems, especially those that are coupled to activation or inhibition of adenylate cyclase activity. In these systems, the third component is thought to be the G_s or G_i protein that serves to couple the activated receptor to adenylate cyclase (see SMIGEL et al. 1984). The degree of curvature of Hofstee or Scatchard plots for a two-step/three-component system is dependent on the ratio of G_s protein to receptor. If there is a large excess of G_s relative to receptor, the Scatchard plot will be essentially linear. When this ratio is relatively low, a marked degree of curvature may result. It is important to point out that the interaction of the G_s protein with the agonist/receptor complex is a relatively high-affinity reaction. Furthermore, changing the ratios of the components of the system or the affinity of agonist-occupied receptor for G_s can change the degree of curvature. In any case, ternary complex formation, which is widely seen in systems that function through changes in the activity of adenylate cyclase, can generate markedly curvilinear Scatchard plots. Thus, with only a single type of receptor, a curvilinear Scatchard plot may result if studies are performed using ligands, most typically agonists, that induce formation of a ternary complex. In several systems examined in detail, GTP has been shown to prevent accumulation of the ternary complex and Scatchard plots are thus linear in the presence of GTP.

G. Regulation of the Sensitivity of Catecholamine Receptor Systems

I. Effects of Decreased Stimulation

The terms subsensitivity and supersensitivity have been used to describe changes in responsiveness to neurotransmitters that occur following various experimental manipulations. For example, removal of the neuronal input to an organ or nerve often results in an enhanced response to an appropriate neurotransmitter (CANNON and ROSENBLUETH 1949). The most detailed investigation of this phenomenon has involved studies of the vertebrate neuromuscular junction (AXELSSON and THESLEFF 1959; MILEDI 1960), but the phenomenon has also been observed in electrophysiological studies of the cardiac ganglion of the mudpuppy (ROPER, 1976) and of parasympathetic ganglion cells in frog heart (KUFFLER et al. 1971). Behavioral studies (UNGERSTEDT 1971a) and biochemical investigations carried out with preparations of the mammalian CNS (PALMER 1972; WEISS and STRADA 1972; HUANG et al. 1973; KALISKER et al. 1973; MISHRA et al. 1974) and of dog heart (PALMER et al. 1976) also provide support for the idea that denervation supersensitivity is a widespread phenomenon.

A number of cellular mechanisms have been shown to be involved in the supersensitivity that occurs following denervation. In vertebrate skeletal muscle, the increase in sensitivity involves a change in the number and distribution of receptors (MILEDI and POTTER 1971; BERG et al. 1972). Also contributing to an increase in the sensitivity of the tissue to acetylcholine is a marked decrease in the amount of acetylcholinesterase activity (HALL 1973; MCCONNELL and SIMPSON 1976). On the other hand, a decrease in the resting membrane potential and a consequent generalized increase in excitability have been reported in studies of postjunctional supersensitivity in the guinea pig vas deferens and in other preparations of smooth muscle (FLEMING et al. 1973; FLEMING and WESTFALL 1975).

Denervation of adrenergically innervated tissues leads to a supersensitive response to exogenous catecholamines (DISMUKES and DALY 1976; TRENDELENBURG 1963; 1966). This phenomenon, which has been extensively characterized in the peripheral sympathetic nervous system, includes at least two components. First, the loss of the presynaptic nerve terminal and its associated uptake system (IVERSEN 1967) results in a decreased capacity to inactivate catecholamines and in an increased concentration of amine at postsynaptic receptor sites. The result of this alteration is a shift to the left of the dose response curve for noradrenaline (KALISKER et al. 1973; SPORN et al. 1977). The second component of supersensitivity that occurs following denervation of sympathetically innervated organs develops with a slower time course and appears to reflect alterations at the level of the postsynaptic cell. This postsynaptic component has been clearly seen, for example, in studies of the effect of ganglionectomy on adenylate cyclase activity in the pineal gland (WEISS 1969).

The intraventricular administration of 6-OHDA leads to the destruction of nerve terminals and to a depletion of catecholamines from the brain (KALISKER et al. 1973; THOENEN and TRANZER 1968; URETSKY and IVERSEN 1970). PALMER (1972) and KALISKER et al. (1973) have studied the effects of the intraventricular injection of 6-OHDA on hormone-sensitive adenylate cyclase activity in rat cerebral cortex. In both studies, there was an enhanced response when slices were incubated with noradrenaline. KALISKER et el. (1973) described both a rapidly developing presynaptic effect and a more slowly developing increase in responsiveness that was attributed to postsynaptic phenomena. The latter effect developed during a time period (72–96 hours) when there was no further reduction in the degree of inhibition of the presynaptic uptake of (^{3}H)-noradrenaline. Furthermore, the administration of 6-OHDA led to an increased response to isoprenaline as well as to noradrenaline. The change in the efficacy of isoprenaline could not be explained by an alteration in the presynaptic uptake process (IVERSEN 1967).

SPORN et al. (1976; 1977) examined the development of supersensitivity in the rat cerebral cortex. The intraventricular administration of 6-OHDA resulted in the destruction of noradrenergic nerve terminals and a 90% decrease in noradrenaline content. The time courses of the effects of this treatment on catecholamine-stimulated cyclic 3′,5′-AMP accumulation in slices of cerebral cortex and on β-adrenoceptors were determined to investigate the mechanisms underlying supersensitivity in this system. The EC_{50} for stimulation of cyclic AMP accumulation by noradrenaline decreased within one day following 6-OHDA administration, and was probably due to the loss of the presynaptic uptake system for noradrenaline. The density of β-adrenoceptors, as determined by measuring the binding of the high-affinity β-adrenoceptor antagonist (^{125}I)-IHYP as well as maximal levels of catecholamine-stimulated cyclic AMP accumulation, increased with a slower time course, reaching peak levels between 8 and 16 days after treatment with 6-OHDA. The maximal increase in the density of β-adrenoceptors was 50%, while maximal levels of catecholamine-stimulated cyclic AMP accumulation in treated rats were approximately twice those measured in control animals. The K_D of (^{125}I)-IHYP for β-adrenoceptors and the EC_{50} of isoprenaline for stimulation of cyclic

AMP accumulation were unchanged following the administration of 6-OHDA, suggesting that the intrinsic properties of postsynaptic receptors were not affected by denervation. The administration of 6-OHDA did not affect fluoride-sensitive adenylate cyclase activity. The results are consistent with the idea that the supersensitivity to catecholamines that occurs in the cerebral cortex following 6-OHDA administration has both pre- and postsynaptic components. The slowly developing postsynaptic component can be at least partially explained by an increase in the density of β-adrenoceptors. Similarly, the administration of reserpine leads to a long-lasting depletion of catecholamines from both central and peripheral storage sites (SLOTKIN 1974). Cyclic AMP accumulation elicited by various concentrations of noradrenaline was increased by 40 to 60% in slices of cerebral cortex from animals pretreated with reserpine (DISMUKES and DALY 1974). The enhanced response was seen between 2 and 9 days after drug administration. A similar increase in the sensitivity of adenylate cyclase to catecholamines after reserpine administration was reported by WAGNER and PALMER (1975) and by PALMER et al. (1976). In the former study, an increase in adenylate cyclase activity was observed in brain slices and homogenates and in neuron-enriched fractions.

The effect of 6-OHDA on the ontogeny of the response to catecholamines has been investigated in rat cerebral cortex following the subcutaneous injection of 6-OHDA during the first 4 days after birth (PERKINS and MOORE 1973a; HARDEN et al. 1977b). This treatment caused a 60 to 70% inhibition of the development of the capacity of the cerebral cortex to take up (^{3}H)-noradrenaline (KALISKER et al. 1973). There was no change in the time course of development of sensitivity to either adenosine or noradrenaline (PERKINS and MOORE 1973a), but the maximal response to noradrenaline was increased 2- to 3-fold in slices from treated animals as compared to the responses observed in controls. The effect of 6-OHDA appeared to be specific in that there was no change in the magnitude of the response to adenosine (PERKINS and MOORE 1973a). The effect on the density and properties of β-adrenoceptors in rat cerebral cortex when newborn animals were treated with 6-OHDA has been determined (HARDEN et al. 1977b). The administration of 6-OHDA to rats on the 1st through the 4th day after birth produced a persistent 90 to 95% reduction in noradrenaline levels in the cerebral cortex. Isoprenaline-stimulated cyclic AMP accumulation and the specific binding of (^{125}I)-IHYP were barely detectable during the 1st week after birth in preparations from either control or 6-OHDA-treated animals. During the 2nd week, cyclic AMP accumulation in response to isoprenaline and the density of β-adrenoceptors increased markedly, reaching adult levels by day 16. Chemical sympathectomy did not affect the time course of development of β-adrenoceptors or the time course of development of the catecholamine-sensitive cyclic AMP generating system. Treatment with 6-OHDA had no effect on fluoride-stimulated adenylate cyclase activity in homogenates of cerebral cortex, on the basal conversion of (^{3}H)-ATP to (^{3}H)-cyclic AMP in cerebral cortical slices prelabelled with (^{3}H)-adenine, or on the affinities of (^{125}I)-IHYP or ($-$)-isoprenaline for β-adrenoceptors. However, a 40 to 65% increase in maximal levels of isoprenaline-stimulated cyclic AMP accumulation was observed in animals from 6 to

45 days of age. There was also a 45 to 75% increase in the density of β-adrenoceptors in the cerebral cortex of treated animals. These results (HARDEN et al. 1977b) suggest that the time course of development of postsynaptic β-adrenoceptors in rat cerebral cortex is not regulated by presynaptic noradrenergic nerve terminals. However, destruction of noradrenergic nerve terminals does result in β-adrenoceptor-mediated supersensitivity to catecholamines. In these experiments isoprenaline was used in preference to noradrenaline because isoprenaline stimulates only β-adrenoceptors while noradrenaline stimulates both α- and β-adrenoceptors and because isoprenaline is not a substrate for the neuronal uptake mechanism.

Similar experiments have been carried out in studies of α-adrenoceptors in rat cerebral cortex. Small but significant increases in the binding of (^{3}H)-WB-4101 were observed two weeks after 6-OHDA administration (U'PRICHARD et al. 1977). No significant change was observed in the total amount of (^{3}H)-clonidine bound, suggesting that the α_2-adrenoceptor in the CNS is not associated with presynaptic elements. However, the time course of this effect was not determined. It is possible that a loss of presynaptic binding sites for one or both of the ligands occurred but was masked by an increase in the density of postsynaptic receptors.

More recently, U'PRICHARD et al. (1979) have shown that the binding of (^{3}H)-clonidine to α_2-adrenoceptors is complex. Based on the existence of biphasic dissociation curves, a method of measuring both high-affinity and low-affinity components of the binding of this ligand was developed. The administration of 6-OHDA was shown to have a selective effect on the high-affinity component, which increased approximately 75% following drug treatment, while the low-affinity component increased by only 30%. The high-affinity component, by analogy to the β-adrenoceptor system, is thought to represent an agonist-induced ternary complex consisting of clonidine, the α_2 receptor, and a G_S protein which may or may not be identical to the one associated with the β-adrenoceptor. The low-affinity component of the binding of (^{3}H)-clonidine most likely represents the simple agonist/receptor complex. Thus, the results of U'PRICHARD et al. (1979) are consistent with the idea that the denervation induced by 6-OHDA may result in an increased availability or amount of the G_S protein. YAMADA et al. (1980a) have examined the effect of 6-OHDA administration on α_1-adrenoceptors in rat heart. These authors also described a regional variation in cardiac α_1 receptors. The density of receptors in the ventricles and septum was significantly greater than that found in the atria. In addition, the administration of 6-OHDA resulted in a significant increase in the density of α_1 receptors only in the ventricles and septum, indicating that these receptors probably receive noradrenergic input only in these regions of the heart.

Super- or subsensitive responses of dopamine receptors may play a role in the behavioral changes induced by many pharmacological agents (see UNGERSTEDT 1971a, b). The unilateral microinfusion of dopamine into the region of the caudate nucleus causes animals to circle in a direction away from the side of the infusion. This has been interpreted to mean that circling behavior is a function of a difference in the degree of stimulation of dopamine receptors on the two sides. Unilateral destruction of ascending nigrostriatal neurons by microinfusion of 6-OHDA (UNGERSTEDT 1971b) caused changes in posture, movement, and responses to drug administration. When apomorphine was administered to unilaterally lesioned animals, they rotated in a direction away from the side of the lesion, suggesting that the sensitivity of dopamine receptors was increased on the lesioned side.

An alternative to the intraventricular injection of 6-OHDA involves the production of unilateral lesions in specific locations in the CNS. MISHRA et al. (1974) have generated unilateral radiofrequency lesions and chemical lesions by the microinjection of 6-OHDA into the substantia nigra of rats.The sensitivity of postsynaptic dopamine receptors in the caudate was then determined. Both types of lesions led to an enhanced responsiveness of dopamine-sensitive adenylate cyclase. Interpretation of the results of this study was complicated by the fact that both manipulations resulted in an enhancement of dopamine-stimulated adenylate cyclase activity, while only the radiofrequency lesion resulted in a decrease in basal activity (MISHRA et al. 1974). The authors suggested that the findings were consistent with a selective increase in the dopamine-receptor component of the adenylate cyclase system or with an increase in the efficiency of coupling of the receptor to the catalytic component. On the other hand, KRUEGER et al. (1976) reported that stimulation of dopamine-sensitive adenylate cyclase in homogenates of the caudate nucleus was not affected by lesions of the ascending dopaminergic pathway. These findings led the authors to conclude that the enhanced ability of dopamine to increase cyclic AMP levels was the result of destruction of presynaptic dopaminergic terminals in the caudate nucleus. CREESE et al. (1977a) showed that the binding of (^{3}H)-haloperidol to dopamine receptors in the rat striatum was increased following unilateral lesions in the nigrostriatal dopaminergic pathway.

STAUNTON et al. (1981) reported phenomena consistent with postsynaptic supersensitivity which developed in the rat neostriatum following the destruction of dopaminergic afferent neurons. A gradual increase in the density of binding sites for (^{3}H)-spiroperidol occurred over a 2- to 3-week period. This increase was apparent only after the almost complete loss of dopamine-containing nerve terminals as measured by the depletion of endogenous dopamine. The properties of the receptor labelled by (^{3}H)-spiroperidol were not altered by denervation. Elimination of dopaminergic nerve endings in the neostriatum was accompanied by the gradual development of an increase in dopamine-sensitive adenylate cyclase activity in homogenates of the caudate ipsilateral to the lesion. The administration of apomorphine led to pronounced circling behavior. This effect occurred rapidly and was maximal within 3 days following destruction of dopaminergic neurons. Thus, the changes in receptor density did not appear to explain the drug-induced rotational behavior which develops after destruction of the dopamine-containing nigrostriatal pathway. This behavioral phenomenon clearly preceded the appearance of alterations in the receptors in the corpus striatum.

Changes in the magnitude of the response to stimulation of dopamine receptors have also been observed following the administration of neuroleptics

(JANSSEN 1967; CLEMENT-CORMIER et al. 1974). Chronic administration of haloperidol significantly depressed motor activity (MOORE and THORNBURG 1975). Subsequent withdrawal of this agent from the diet resulted in increased motor activity in treated mice compared to controls. After cessation of neuroleptic administration, enhanced responses to dopaminergic agoinsts have been reported (TARSY and BALDESSARINI 1973; GIANUTSOS et al. 1974; MOORE and THORNBURG 1975; EIBERGEN and CARLSON 1976; NOYCKY and ROTH 1977). Treatment with neuroleptics has also been shown to result in an increased density of binding sites for (^{3}H)-haloperidol and (^{3}H)-apomorphine (BURT et al. 1977; MULLER and SEEMAN 1977). PERT et al. (1978) showed that the concurrent administration of lithium carbonate and haloperidol prevented both the increase in the density of striatal dopamine receptors which normally occurs during the administration of haloperidol and the development of behavioral supersensitivity to the administration of apomorphine. Administration of lithium itself had no effect on the density of striatal dopamine receptors.

The effect of the chronic administration of lithium on the density of adrenoceptors has been examined by TREISER and KELLAR (1979), who showed that 5 weeks of oral administration of lithium resulted in a small (15%) but significant decrease in binding of ligands to β-adrenoceptors in rat cerebral cortex. When rats were treated with reserpine, the number of β-adrenoceptors also increased. However, the effect of reserpine on β-adrenoceptors was eliminated by the co-administration of lithium. The chronic administration of lithium had no effect on α-adrenoceptors in rat cerebral cortex. In a similar study, ROSENBLATT et al. (1979) showed that administration of lithium led to a decrease in the density of β-adrenoceptors in homogenates of whole rat brain. In these experiments, the chronic oral administration of lithium resulted in a 22% increase in the density of α_1-adrenoceptors. The discrepancy between these results and those of TRIESER and KELLER (1979) may be a consequence of the use of homogenates of whole brain as compared to cerebral cortex. ROSENBLATT et al. (1979) also studied the effects of the co-administration of imipramine and lithium. Chronic administration of imipramine resulted in a decrease in the density of β-adrenoceptors but did not affect binding to α_1-adrenoceptors in whole rat brain. The concomitant administration of lithium and imipramine did not affect this decrease in the density of β-adrenoceptors. Thus, it appears that lithium can prevent the increase in β-adrenoceptors caused by the administration of reserpine while it has no effect on the decrease in receptors caused by administration of imipramine. As noted above, the effect of lithium on α-adrenoceptors remains controversial.

II. Effects of Increased Stimulation

Catecholamine-stimulated cyclic AMP accumulation in slices from the limbic forebrain has been measured to study the effects of both increased and decreased availability of noradrenaline at the receptor (VETULANI et al. 1975; VETULANI and SULSER 1975). In general, an increased availability of noradrenaline led to a decreased responsiveness of the receptor/cyclase system, and a decreased availability led to an increase in responsiveness. Thus, an increase in the sensitivity of the system resulted from the administration of reserpine or 6-OHDA. Conversely, the chronic administration of monoamine oxidase inhibitors led to a decrease in the activity of a catecholamine-stimulated cyclic AMP-generating system (VETULANI and SULSER 1975). Effects of antidepressant drugs that do not elevate the level of monoamines in the brain have also been studied (VETULANI and SULSER 1975). Chronic administration of the antidepressants desipramine and iprindole reduced the responsiveness of the cyclic AMP-generating system to noradrenaline by approximately 70%. The effects of both monoamine oxidase inhibitors and antidepressants were seen after prolonged administration but not after short-term treatment. The authors suggested that the prolonged time course of the effects of monoamine oxidase inhibitors and antidepressants was consistent with the delayed therapeutic effects of these drugs, which often occur only after treatment for several weeks.

The effects of a variety of treatments that have in common an ability to increase the amount of noradrenaline having access to β-adrenoceptors in the cerebral cortex have been investigated (BANERJEE et al. 1977; MOLINOFF et al. 1978; WOLFE et al. 1978a). Treatment with the tricyclic antidepressant desipramine for 7 to 21 days resulted in a 35% to 45% decrease in the accumulation of cyclic AMP in response to a maximally effective concentration of (−)-isoprenaline in rat cerebral cortical slices. The EC_{50} for isoprenaline-stimulated cyclic AMP accumulation was not affected by desipramine administration. The diminution in responsiveness to catecholamines was accompanied by a 35% to 40% decrease in the density of β-adrenoceptors as measured by the binding of (^{125}I)-IHYP. Decreases in the accumulation of isoprenaline-sensitive cyclic AMP and in the density of β-adrenoceptors were temporally correlated, maximal decreases being observed after 5 to 7 days of treatment. Within 7 days after cessation of chronic desipramine treatment, the accumulation of isoprenaline-stimulated cyclic AMP and the density of β-adrenoceptors returned to normal. The role of presynaptic nerve terminals in mediating these phenomena was also investigated. Treatment of newborn rats with 6-OHDA prevented the development of noradrenergic nerve terminals in the cerebral cortex and blocked the effects of desipramine on the accumulation of cortical cyclic AMP and on the density of β-adrenoceptors. The administration of the β-adrenoceptor antagonist propranolol led to increases in maximal accumulation of isoprenaline-stimulated cyclic AMP and in the density of β-adrenoceptors in the rat cerebral cortex. The simultaneous administration of propanolol and desipramine prevented the effects of desipramine. Thus, the effect of desipramine appears to be mediated through an action of neuronally released noradrenaline at β-adrenoceptors. Chronic treatment with two other clinically effective antidepressants, pargyline and iprindole, led to effects similar to those observed after administration of desipramine. Pretreatment of neonates with 6-OHDA blocked the effect of iprindole on β-adrenoceptors, suggesting that even though its precise mechanism of action has not been defined, iprindole does appear to be acting presynaptically. Preincubation of cortical membranes with GTP prior to determining the density of β-adreno ceptors had no effect on the decreased number of receptors seen in desipramine treated animals. These results suggest that antidepressants, acting presynaptically, increase the concentration of transmitter at noradrenergic synapses and induce a compensatory decrease in the density of β-adrenoceptors.

The effects of amphetamine may be more complicated than those of the antidepressant drugs discussed above (see HUG 1972). Amphetamine causes the release of dopamine from dopaminergic nerve endings (MCKENZIE and SZERB 1968; CARR and MOORE 1970) and potentiates dopaminergic neurotransmission by competing with dopamine for reuptake sites on striatal dopaminergic terminals (COYLE and SNYDER 1969). Direct effects of amphetamine on dopamine receptors have also been reported (FELTZ and DECHAMPLAIN 1973). Interpretation of the biochemical effects of amphetamine is further complicated by the fact that amphetamine can inhibit monoamine oxidase (RUTLEDGE 1970).

In view of the multiplicity of effects observed after a single dose of amphetamine as well as after its chronic administration, it is not difficult to rationalize the disparate results of experiments in which binding sites and adenylate cyclase activity are measured after either acute or chronic treatment with this drug. MOBLEY et al. (1977) reported a decrease in the ability of catecholamines to stimulate cyclic AMP formation in the rat limbic forebrain following amphetamine administration. Conversely, BANERJEE et al. (1978) reported that amphetamine led to the development of β-adrenoceptor-mediated supersensitivity in whole mouse brain. In other experiments carried out with membranes prepared from cerebral cortices of animals treated for 9 days with amphetamine, decreases in cyclic AMP accumulation due to stimulation of both α- and β-adrenoceptors have been observed (WOLFE et al. 1978b). HOWLETT and NAHORSKI (1978) studied the effects of the chronic administration of amphetamine on β-adrenoceptors and dopamine receptors in rat brain. They reported no changes in β-adrenoceptors in the corpus striatum or the limbic forebrain but showed that the density of dopamine receptors in the striatum was decreased following 20 days of amphetamine administration. They also showed that this treatment resulted in a decrease in dopamine-stimulated cyclic AMP accumulation in the striatum.

III. Specific Alterations in Receptor Subtypes Caused by Changes in the Availability of Neurotransmitters

The density of β-adrenoceptors in the rat cerebral cortex is regulated by the amount of noradrenaline having access to receptors. The use of computer-aided graphic analysis of β-adrenoceptor subtypes made it possible to show that the cerebral cortex contains approximately 80% β_1 receptors and 20% β_2 receptors. To determine whether β-adrenoceptor subtypes are independently regulated in the cerebral cortex, the effects of altering the availability of noradrenaline on the density and properties of β_1 and β_2 receptors were investigated (MINNEMAN et al. 1979a; WOLFE et al. 1980). The monoamine oxidase inhibitor pargyline and the inhibitor of noradrenaline uptake desipramine were administered chronically. These treatments resulted in a 35% to 45% decrease in the density of β_1 receptors but had no effect on the density of β_2 receptors. Conversely, decreasing the availability of noradrenaline by chronic blockade of receptors with propanolol or destruction of noradrenergic nerve terminals with 6-OHDA resulted in a 60 % to 70 % increase in the density of β_1 receptors but had no effect on β_2 receptors. These data led to the suggestion that, in the rat cerebral cortex, β_1-adrenoceptors are primarily involved in neuronal function. β_2-Adrenoceptors, on the other hand, either do not have an endogenous input or are not regulated by changes in input. The β_2 receptor in the cerebral cortex may be located on non-neuronal cellular elements such as glia or blood vessels. In any case, these studies demonstrated for the first time that β receptor subtypes in the same tissue are independently regulated.

The cerebellum is an unusual tissue in the CNS in that the β-adrenoceptors in this tissue are primarily of the β_2 subtype. Only 5 % to 10 % of the total number of receptors are β_1-adrenoceptors (MINNEMAN et al. 1979d). Thus, in view of the results obtained in studies of receptors in the cerebral cortex (MINNEMAN et al. 1979a) and the fact that there is a well-characterized noradrenergic innervation of the Purkinje cells in the cerebellum, the effect of decreasing the availability of catecholamines on the density of β-receptor subtypes was investigated (WOLFE et al. 1982). The chronic administration of propranolol or depletion of cerebellar noradrenaline by the intracisternal administration of 6-OHDA resulted in a doubling in the number of β_1 receptors with no change in the density of β_2 receptors. These results suggest that either cerebellar β_2 receptors are not regulated by decreased availability of noradrenaline or that only β_1 receptors normally receive a neuronal input. The latter hypothesis is supported by the results of experiments in which neonatal rat pups were exposed to X-irradiation in such a manner as to selectively destroy the basket, granule, and stellate cells of the cerebellum while sparing the early-maturing Purkinje cells (MINNEMAN et al. 1981b). This treatment resulted in an 80 % to 85 % loss of cerebellar β_2-adrenoceptors. There was no effect on cerebellar β_1 receptors. These results, together with those described above, suggest that there are β_1-adrenoceptors associated with Purkinje neurons in the cerebellum and these receptors (which comprise only 5 % to 10 % of the total number of cerebellar β-receptors) may be the functionally important population of receptors in this tissue.

The caudate, in contrast to the cerebellum and the cerebral cortex, does not have a well-defined noradrenergic input. Endogenous levels of noradrenaline are low, and DBH, while present, has not been shown to be located in neuronal varicosities (SWANSON and HARTMAN 1975). On the other hand, the density of β-adrenoceptors in the caudate is similar to that found in the cerebral cortex and β_1 receptors comprise approximately 75 % of the total number of receptors (MINNEMAN et al. 1979d). The presence of receptors in the absence of an endogenous transmitter led to studies of the effects of desipramine, pargyline, and propranolol, as well as of bilateral adrenalectomy, on the density and properties of β-adrenoceptor subtypes in the caudate (MINNEMAN et al. 1982).The administration of pargyline and desipramine resulted in a 30 % to 40 % decrease in the density of β_1 receptors with no change in the density of β_2 receptors. These results were similar to those obtained in studies of cerebral cortex. The administration of propranolol also resulted in an increase in the density of β_1 receptors, but the magnitude of the change was small compared to that observed in studies of cerebellum and cerebral cortex. Since there is not a well-defined noradrenergic input to these receptors, the effect of removing the adrenal glands was examined. A small but significant increase in the density of β_1 receptors in the caudate but no effect on β_2 receptors was observed following adrenalectomy. This effect was restricted to the caudate in that no changes in either β_1 or β_2 receptors were observed in the cerebral cortex or cerebellum following adrenalectomy. A plausible explanation for these results is that some of the β_1 receptors in the caudate are located on blood vessels on the peripheral side of the blood-brain barrier. An alternative hypothesis can be developed to explain the effect of adrenalectomy on β-adrenoceptors in the caudate. Adrenal steroids may regulate the density of β-adrenoceptors in the caudate in a manner similar to that seen in studies of effects of steroids on β-adrenoceptors in the liver and lung (WOLFE et al. 1976; MANO et al. 1979; see below).

IV. Regulation of Responsiveness Not Mediated by Changes in the Density of Receptors

In many of the examples cited above, changes in the density of receptors have been shown to play a role in regulating the responsiveness of the receptor/cyclase system in rat cerebral cortex. In other situations, non-receptor mechanisms appear to be predominant. For example, reticulocytosis can be induced by administering phenylhydrazine to rats. Membranes prepared from reticulocytes have 25 times more isoprenaline-stimulated adenylate cyclase activity than do membranes from mature erythrocytes. The density of binding sites for (^{125}I)-IHYP was increased, however, by only about 30% in reticulocytes as compared to mature erythrocytes (BILEZIKIAN et al. 1977a; CHARNESS et al. 1976). As the reticulocytes matured, catecholamine-sensitive adenylate cyclase activity decreased rapidly but binding activity persisted. These results suggested that the β-adrenoceptor becomes uncoupled from the catalytic unit during differentiation of erythrocytes (BILEZIKIAN et al. 1977b).

The rat pineal has been extensively studied with regard to the control of the enzyme serotonin N-acetyltransferase (WEISSBACH et al. 1960). Catecholamines, acting through β-adrenoceptors, regulate the activity of this enzyme which is involved in the synthesis of melatonin. The responses of the pineal vary as a consequence of the prior level of neuronal activity. A period of stimulation results in a decrease in the response to subsequent stimulation, and a period without stimulation leads to a supersensitive response to subsequent stimulation. Several of the elements involved in the control, synthesis, and function of cyclic AMP change as a consequence of increased or decreased stimulation. The factors which change include the density of β-adrenoceptors (KEBABIAN et al. 1975b) and the amount of basal and hormone-sensitive adenylate cyclase activity (WEISS 1969). The pineal also shows super- and subsensitivity to the action of dibutyryl cyclic AMP (ROMERO and AXELROD 1975).

This agent bypasses the steps involved in regulating cyclic AMP levels and thus a step distal to cyclic AMP production must also be involved in the regulation of sensitivity. ZATZ and O'DEA (1976) showed that the activity of protein kinase was increased in supersensitive glands. There is also a diurnal cycle in phosphodiesterase activity which is elevated at night when the sensitivity of the transferase to induction is low (MINNEMAN and IVERSEN 1976). Thus, changes in phosphodiesterase activity may contribute to changes in end-organ sensitivity. The point to be emphasized with regard to the pineal is that sensitivity is regulated at multiple sites, thus providing an integrated system for the regulation of the sensitivity of the organ.

V. Effects of Hormones on Catecholamine Receptors and Responsiveness

The effects of altering the levels of a number of hormones on the ability of catecholamines to stimulate adenylate cyclase and on the density and properties of both α- and β-adrenoceptors have been studied in a variety of tissues. Alterations in the level of thyroid hormone have received a great deal of attention since it appears that the level of this hormone regulates the sensitivity of many organs, including the heart, to the effects of catecholamines. For example, when rats were made hypothyroid by feeding 6-propyl-2-thiouracil, β-adrenoceptor-stimulated adenylate cyclase activity from rat atria was diminished (BRODDE et al. 1980). Similar effects were observed in studies of adenylate cyclase activity from turkey erythrocytes (BILEZIKIAN and LOEB 1982) and rat adipocytes (MALBON et al. 1978). Effects of increased thyroid hormone on catecholamine-stimulated adenylate cyclase activity have been controversial in that an increase in activity was reported in cultured rat heart cells following 24 hours of exposure to 5 nM triiodothyronine (T_3) (TSAI and CHEN 1978), but no change was observed in studies of erythrocyte membranes prepared from turkeys that had been treated with L-thyroxine (600 µg/day–900 µg/day) for 2 months (BILEZIKIAN and LOEB 1982). The effects of thyroid status on the density and properties of β-adrenoceptors have also been examined. WILLIAMS et al. (1977) and STILES and LEFKOWITZ (1981) administered either T_3 or thyroxine to rats and reported that the density of β-adrenoceptors in the hearts of hyperthyroid rats was double that in the hearts of euthyroid rats. This increase in the density of receptors occurred in both male and female rats and could be induced by the administration of either T_3 or T_4. No changes in the K_D values for the radioligand (^{3}H)-DHA were observed. On the other hand, BILEZIKIAN and LOEB (1982), studying turkey erythrocytes, reported that hyperthyroidism does not result in an alteration in the density of β-adrenoceptors. Decreasing the level of thyroid hormone has been shown to result in a decrease in the density of β-adrenoceptors on turkey erythrocythes (BILEZIKIAN and LOEB 1982) and rat cardiac tissue (STILES and LEFKOWITZ 1981), but not to affect the density of these receptors on rat fat cells (MALBON et al. 1978). Thus, it appears that the effect of thyroid hormone on β-adrenoceptor density is tissue-dependent, and that hyperthyroidism results in increases in receptor density in the heart but not in erythrocytes.

Using membranes prepared from rat adipose tissue, MALBON (1980) showed that the EC_{50} value for isoprenaline inhibition of the binding of (^{3}H)-DHA to membranes prepared from euthyroid rats was increased 7-fold by the inclusion of Gpp(NH)p in the assay. Membranes prepared from either hypothyroid or hyperthyroid rats were insensitive to Gpp(NH)p. The EC_{50} value for isoprenaline using membranes prepared from hypothyroid rats was similar to that observed using membranes prepared from euthyroid rats assayed in the presence of Gpp(NH)p. This observation, coupled with the fact that stimulation of adenylate cyclase by adrenaline in membranes prepared from hypothyroid rats was reduced with no change in fluoride-stimulated activity, led the author to suggest that this system is similar to the UNC variant of the S49 mouse lymphoma cell (HAGA et al. 1977). Somewhat different results were obtained by STILES and LEFKOWITZ (1981) who investigated effects of hyperthyroidism. In these studies, Gpp(NH)p had similar effects on the affinity of cardiac β-adrenoceptors for isoprenaline in membranes prepared from euthyroid and hyperthyroid rats.

The effect of thyroid status on the density of α-adrenoceptors has also been examined. SHARMA and BANERJEE (1978) reported that the density of α-receptors in rat heart was decreased by the administration of T_3 to thyroidectomized rats. On the other hand, GARCÍA-SÁINZ et al. (1981) reported that the density of α-adrenoceptors in membranes prepared from adipose tissue of the hamster was unaffected by either hyperthyroidism or hypothyroidism.

The early experiments of SUTHERLAND and coworkers implicated cyclic AMP in the control of carbohydrate metabolism in the liver (see SUTHERLAND and ROBISON 1969). EXTON and HARPER (1975) have shown that both glucagon and adrenaline cause a rapid, dose-dependent increase in the levels of cyclic AMP in mammalian liver. Previously, BITENSKY et al. (1970) had shown that the magnitude of the adrenaline-induced stimulation of adenylate cyclase activity in rat liver was dependent upon the age, hormonal state, and sex of the animal. The response to adrenaline (a 3- to 4-fold increase in cyclic AMP levels) was fully developed in fetal rat liver 15 days after conception and remained constant until weaning (BITENSKY et al. 1970; BAR and HAHN 1971; BUTCHER and POTTER 1972). After weaning, the degree of stimulation began to fall until the maximal increase in cyclic AMP levels was only 1.5- to 2-fold in adult rat liver. Adenylate cyclase activity in liver obtained from female rats was more sensitive to adrenaline than that from male rats. The increased sensitivity to adrenaline in the liver from female rats was seen with weanlings, young adults, and adults (BITENSKY et al. 1970). The administration of meticortelone, testosterone, or ACTH greatly decreased the production of cyclic AMP which occurred in response to adrenaline in weanling rat liver. The administration of diethylstilbestrol had no effect while progesterone had an intermediate effect on adrenaline-sensitive adenylate cyclase activity (BITENSKY et al. 1970).

EXTON et al. (1972) have shown that the effect of adrenaline on gluconeogenesis and glycogenolysis in rat liver is reduced or abolished following adrenalectomy. Furthermore, livers from adrenalectomized rats were less sensitive to dibutyryl cyclic AMP in terms of stimulation of gluconeogenesis than were livers obtained from control rats. On the other hand, the response of rat liver adenylate cyclase to catecholamines was markedly increased by adrenalectomy or hypophysectomy (BITENSKY et al. 1970; EXTON et al. 1972; LERAY et al. 1973; WOLFE et al. 1976). This increase was reversed by the administration of cortisone. Diethylstilbestrol had no effect on catecholamine-sensitive adenylate cyclase activity in the liver of adrenalectomized rats. Adrenocorticotropic hormone reversed the effects of hypophysectomy but not the effects of adrenalectomy. The increase in responsiveness to catecholamines was not paralleled by an increase in the specific activity of fluoride-stimulated adenylate cyclase. This suggests that the increased responsiveness was not due to a change in the amount of enzyme.

The effects of adrenalectomy on the density and properties of β-adrenoceptors in the liver have been studied. Adrenalectomy did not change the kinetic properties of the binding of (^{125}I)-IHYP or the affinities of the receptor for agonists or antagonists (WOLFE et al. 1976): However, the concentration of β-adrenoceptors in liver obtained from adrenalectomized rats was significantly greater than in liver from control rats. This increase appears to explain the increase in hormone-sensitive adenylate cyclase activity seen in rat liver as a consequence of adrenalectomy. This effect on the density of β-adrenoceptors was prevented by the administration of cortisone. In another study, no change in the density of α-adrenoceptors was seen following adrenalectomy (GUELLAEN et al. 1978).

In contrast to the effects of glucocorticoids on rat liver, MANO et al. (1979) showed that adrenalectomy resulted in a 25% to 30% decrease in the density of β-adrenoceptors in rat lung while the chronic administration of hydrocortisone resulted in a 65% to 75% increase in the density of pulmonary β-adrenoceptors. The well-known observation that glucocorticoids potentiate the effects of adrenergic stimulation of airway smooth muscle may be explained by these results.

DAVIES and LEFKOWITZ (1981) examined the effects of the administration of cortisone to humans on the amount of agonist-induced high-affinity ternary complex that can be formed in membranes of neutrophils isolated from the blood of healthy volunteers. These authors, using computerized techniques for data analysis, found that administration of cortisone resulted in a substantial increase in the proportion of receptors in the high-affinity state, presumably the agonist-induced ternary complex. They suggested that this observation may partially explain the increase in catecholamine-stimulated adenylate cyclase activity seen following the administration of steroids.

The effects of the administration of estrogen and progesterone on α- and β-adrenoceptors in rabbit uterus have also been examined. In the non-pregnant uterus, stimulation of α-adrenoceptors causes contraction, while stimulation of β-adrenoceptors causes relaxation. Although both α- and β-adrenoceptors exist in this tissue, the administration of adrenaline causes relaxation (β-response). ROBERTS et al. (1977) reported that administration of progesterone affected neither α- nor β-adrenoceptors in this tissue. On the other hand, treatment with estrogen resulted in a 3-fold increase in the density of α-adrenoceptors. This change in receptor density was associated with a change in the response of the tissue to catecholamines from that characteristic of stimulation by β-adrenoceptors to that characteristic of stimulation by α-adrenoceptors. It is worth noting that even in the non-estrogen-treated animal, the density of α-adrenoceptors far exceeds that of β-adrenoceptors.

H. Agonist-Induced Desensitization

The interaction of neurotransmitters or hormones with an appropriate receptor often results in the rapid accumulation of cyclic AMP. The effects of many hormones, including catecholamines, are often transient, with a gradual decline in the rate of cyclic AMP synthesis occurring despite the continued presence of the agonist (KAKIUCHI and RALL 1968). In several systems, a decrease in the response of adenylate cyclase to catecholamines has appeared to accompany an apparent decrease in the number of β-adrenoceptors. This loss of receptors has been documented using both (^{125}I)-IHYP (SHEAR et al. 1976; JOHNSON et al. 1978) and (^{3}H)-DHA (MUKHERJEE et al. 1975; KEBABIAN et al. 1975b; MICKEY et al. 1975). Frequently, the magnitude of the decrease in responsiveness of adenylate cyclase to catecholamines exceeds that of the loss of β-adrenoceptors.

Perkins and coworkers (CLARK et al. 1975; SU et al. 1976a, b; JOHNSON et al. 1978) have studied hormone-induced refractoriness in a number of cell lines originally derived from a human astrocytoma. EH118MG astrocytoma cells showed both agonist-specific (homologous) and agonist-nonspecific (heterologous) desensitization (JOHNSON et al. 1978). Agonist-specific refractoriness was rapid in onset. Thus, only 20% to 30% of the normal response to isoprenaline was present following incubation of cells with 10 μM isoprenaline for 30 min. Although there was a 70% decrease in the response to isoprenaline, there was only a 17% decrease in the number of β-adrenoceptors. When cells were incubated with 10 μM PGE_1 for 2 hours and then challenged with isoprenaline, there was a 50% decrease in the amount of cyclic AMP that accumulated in the presence of 10 μM isoprenaline (JOHNSON et al. 1978). Under these conditions there was no change in the density of β-adrenoceptors. When cells that had been preincubated with PGE_1 for 2 hours were subsequently incubated with 10 μM isoprenaline for an additional 2 hours, a further 20% decrease in responsiveness to isoprenaline occurred. This loss was accompanied by a 20% decrease in the density of β-adrenoceptors. The studies with PGE_1 suggest that a significant fraction of the decrease in the response to isoprenaline can occur through mechanisms that do not involve β-adrenoceptors. These results and previous studies with cells in culture further suggest that desensitization to catecholamines may occur by several different mechanisms. Although the recovery of responsiveness after desensitization to catecholamines in human diploid fibroblasts in culture was dependent on protein synthesis (FRANKLIN et al. 1975), the desensitization process in these fibroblasts was not affected by inhibitors of protein synthesis. On the other hand, in C6 rat glioma cells, inhibition of protein synthesis affected the development of desensitization to the effects of noradrenaline but did not influence the recovery process after desensitization had taken place (DE VELLIS and BROCKER 1974). In studies carried out with astrocytoma cells, neither the onset nor the recovery of responsiveness following desensitization was affected by cycloheximide (SU et al. 1976a).

SHEAR et al. (1976) studied agonist-induced desensitization in several lines of S-49 mouse lymphoma cells including genetic variants with specific defects in the pathway of cyclic AMP formation and function. In wild type S-49 cells, cyclic AMP accumulation in response to isoprenaline reached a maximum within 30 min and then fell rapidly. Approximately 40% to 50% of the binding sites for (^{125}I)-IHYP were lost following incubation with isoprenaline, which suggested that changes in the density of β-adrenoceptors were involved in the desensitization process. In addition, exposure to isoprenaline led to a decrease in the density of receptors and a decrease in responsiveness to isoprenaline in protein kinase-deficient cells. Since an increase in phosphodiesterase activity was not seen following an increase in cyclic AMP levels in kinase-deficient cells, induction of phosphodiesterase activity does not appear to account for the refractoriness. Thus, cyclic AMP-dependent protein kinase does not appear to be required for either of these effects. When cyclic AMP content was increased by exposing cells to PGE_1 in the absence of isoprenaline, no decrease in either the density of binding sites for (^{125}I)-IHYP or the activity of isoprenaline-stimulated adenylate cyclase was seen. Also, when an S-49 clone that lacks hormone-sensitive adenylate cyclase activity but still possesses a normal complement of β-adrenoceptors was incubated with isoprenaline or with isoprenaline plus dibutyryl cyclic AMP, no decrease in the density of β-adrenoceptors was seen. Thus, receptor occupancy alone, or in combination with elevated cyclic AMP levels, is not sufficient to cause a decrease in responsiveness or in the density of β-adrenoceptors. It was concluded that in this system, adenlyate cyclase must be present for a loss of receptors to occur.

LEFKOWITZ and his colleagues have correlated decreases in postsynaptic responsiveness with alterations in the density of β-adrenoceptors. The chronic administration of catecholamines to frogs led to a decrease in adenylate cyclase activity in erythrocyte membranes in response to a subsequent *in vitro* challenge with isoprenaline (MUKHERJEE et al. 1975; 1976). A similar refractoriness to catecholamines was seen when erythrocytes were isolated and then incubated with isoprenaline *in vitro* (MICKEY et al. 1975; 1976) or if purified membranes from frog erythrocytes were incubated with catecholamines (MUKHERJEE and LEFKOWITZ 1976). Both *in vivo* and *in vitro* desensitization was accompanied by decreased numbers of β-adrenoceptors. Cycloheximide had no effect either on the desensitization process or on the subsequent reappearance of sensitivity (MUKHERJEE et al. 1976). Since the alterations in receptor density and catecholamine responsiveness appeared to be independent of protein synthesis, it was suggested that desensitization of the β-adrenoceptor/adenylate cyclase system involves inactivation and subsequent reactivation of existing receptor molecules.

In many cases the decrease in the number of functional β-adrenoceptors assayed by antagonist binding has been smaller than the decrease in catecholamine-stimulated adenylate cyclase activity (WESSELS et al. 1978). In some systems (JOHNSON et al. 1978) this discrepancy was partially due to a nonspecific or heterologous desensitization, while in the frog erythrocyte system, desensitization appeared to be entirely receptor-specific. In another series of experiments (SU et al. 1979; 1980), agonist-specific desensitization of human astrocytoma cells has been investigated. Incubation with isoprenaline resulted in a time-dependent, agonist-specific loss of responsiveness. At short periods of incubation, there was a large decrease in the cyclic AMP response to an isoprenaline challenge. This was not, however, accompanied by a decrease in the density of β-adrenoceptors. At longer times of incubation, the density of receptors decreased such that at 24 hours the decrease in receptors was similar to the decrease in responsiveness. Thus, it appears that the receptors become "uncoupled" at short times of incubation with agonists. The idea that this uncoupling occurs at the receptor level has been supported by two studies. HARDEN et al. (1980) have shown that short-term incubation of astrocytoma cells with isoprenaline results in a decrease in catecholamine-stimulated adenylate cyclase activity and a concomitant alteration in the sedimentation properties of particulate β-adrenoceptors. Thus, in control preparations, most of the receptors sediment in a single "heavy" peak on sucrose density gradients. These receptors exhibit the normal effects of GTP on agonist inhibition of the binding of antagonists (see above), implying that they are able to couple with the G_s protein. Catecholamine-stimulated adenylate cyclase activity co-sediments with this "heavy" peak. There is a much smaller, "light" peak of β-adrenoceptors in control preparations that is associated neither with catecholamine-stimulated adenylate cyclase activity nor with any GTP-induced effects on agonist inhibition of binding of antagonists. Short-term incubation of these cells with isoprenaline results in a marked change in the ratio of the receptors in the "heavy" and "light" peaks. Thus, after incubation with isoprenaline, approximately 50% to 60% of the receptors are in the "light" peak (compared to approximately 10% in the controls) and are "uncoupled" from the G_s protein. The rapid loss of catecholamine-stimulated cyclic AMP accumulation following incubation with isoprenaline is thought to be due to a functional alteration in the receptor. Similar results have been reported by STADEL et al. (1983a). Additional evidence for this hypothesis has come from the studies of GREEN and CLARK (1981) who demonstrated that incubation of adrenaline with S-49 mouse lymphoma cells of the cyc$^-$ variant, which is missing a functional G_s protein, results in a decrease in the ability to reconstitute catecholamine-sensitive adenylate cyclase activity following addition of exogenous G_s protein. This effect was receptor-specific since the ability to reconstitute PGE_1-stimulated adenylate cyclase activity was unaltered. Thus, the change in receptor properties induced by incubation with catecholamines is not dependent on the presence of a functional G_s protein and has no connection to adenylate cyclase. A similar conclusion was reached by IYENGAR et al. (1981), who showed that incubation of wild type S-49 cells with isoprenaline resulted in a decrease in isoprenaline-stimulated adenylate cyclase activity, with no effect on Gpp(NH)p-stimulated activity. Added G_s protein, obtained from untreated cells, did not restore sensitivity to isoprenaline. Only small losses in the density of β-adrenoceptors were detected. These data are consistent with the idea that incubation of cells with catecholamines results in a relatively rapid conversion of the receptors to a form that no longer interacts with the G_s protein nor activates adenylate cyclase but does retain its capacity to specifically bind antagonists. Under conditions where a significant proportion of the receptors are in this form, the loss of isoprenaline-stimulated adenylate cyclase activity will exceed the loss of binding sites (PERKINS 1981). Over an extended period of time this form of the receptor is converted to another form (perhaps degraded) that is capable neither of activating adenylate cyclase nor of binding radiolabelled antagonist. It is important to note that the above scheme applies to experiments in which whole cells were incubated with catecholamines.

Recently, STADEL et al. (1983b) showed that incubation of intact turkey erythrocytes with isoprenaline desensitized the cells to catecholamine stimulation and resulted in a phosphorylation of the β-adrenoceptor. The time course of this phosphorylation corresponded to the time course of the loss of isoprenaline-stimulated adenylate cyclase activity, suggesting that this phosphorylation is the mechanism by which the receptor is uncoupled from its effector system. The phosphorylated form of the receptor had slightly altered properties and was separated from normal receptors by SDS-polyacrylamide gel electrophoresis.

The possibility that internalization of β-adrenoceptors is responsible for the agonist-induced decrease in the density of receptors has been investigated by CHUANG and COSTA (1979). A protein that binds (^{3}H)-DHA was detected in the supernatant following homogenization of frog erythrocytes. The amount of binding to cytosolic receptors was increased when the erythrocytes were preincubated with isoprenaline. The EC_{50} values for the isoprenaline-induced increase in the binding of (^{3}H)-DHA in the cytosol and the decrease in this binding in membranes ranged between 60 and 90 nM. Furthermore, the increase in binding in the cytosol was blocked by exposure of the cells to the antagonist alprenolol. However, the magnitude of the observed decrease in membrane receptors far exceeded the magnitude of the increase in cytosolic receptors.

In a subsequent study, CHUANG et al. (1980) showed that the increase in soluble receptors did not occur when the preincubation was carried out at 0 °C. Incubation of cells with 2,4-dinitrophenol, an uncoupler of oxidative phosphorylation, lowered the amount of ligand bound to soluble β-adrenoceptors in both normal and treated erythrocytes. The amount of ligand bound to soluble receptors was also reduced if erythrocytes were treated with cordycepin, an inhibitor of ATP-dependent protein phosphorylation, with concanavalin A, a plant lectin that binds to membrane glycoproteins, and with methylamine, a drug that has been shown to inhibit the clustering and subsequent internalization of various hormone receptors. The results were interpreted to suggest that internalization of β-adrenoceptors occurs and is triggered by receptor clustering, possibly mediated by an increase in transglutaminase activity. The rank order of potency for inhibition of transglutaminase activity by methylamine and other agents was similar to that for prevention of receptor internalization (CHUANG 1981).

As reviewed by HARDEN (1983), stimulation of β-adrenergic receptors by agonists converts receptors to a form that does not couple to the components of the adenylate cyclase system. A decrease in catecholamine-stimulated adenylate cyclase activity is accompanied by a loss of GTP-sensitive high-affinity binding of agonists (TOEWS et al. 1985). The binding of some hydrophilic antagonists such as ^{3}H-CGP-12177 to intact cells is decreased when cells are incubated with agonists. This is followed by the appearance of receptors in a population of light vesicles that can be separated on density gradients from other markers used to identify plasma membranes (HERTEL et al. 1983a,b). The occurrence of sequestration is temporally correlated with a decrease in the affinity of the receptor for agonists (REYNOLDS et al. 1985). Receptors in light vesicles may be either recycled to or within the plasma membrane or may undergo lysosomal degradation. The precise relationship between these phenomena has not yet been determined. A plausible hypothesis is that the loss of GTP-sensitive binding of agonists, the diminished stimulation of adenylate cyclase, and the loss of binding of ^{3}H-CGP-12177 are different expressions of the same overall process, possibly related to phosphorylation of the receptor (SIBLEY et al. 1985). Sequestration of receptors in light vesicles may be related to the decreased affinity of receptors for agonists seen in studies with intact cells (PITTMAN and MOLINOFF 1980). Little is known about possible mechanisms of sequestration except that sequestration occurs in cyc^{-} S-49 lymphoma cells and thus does not require G_s (HOYER et al. 1984), and sequestration can be induced by phorbol esters (KASSIS et al. 1985). Sequestration can occur with or without prior loss of functional activity of receptors (uncoupling), and down-regulation of receptors can occur without prior sequestration (KASSIS et al. 1985; 1986; REYNOLDS and MOLINOFF 1986). Exposure of cells to atypical agonists like pindolol or celiprolol does not cause sequestration of receptors. Downregulation of receptors is observed in C_6 glioma cells and in wild-type but not cyc^{-} S-49 lymphoma cells (NEVE and MOLINOFF 1986; REYNOLDS and MOLINOFF 1986).

It is likely that some of the processes involved in desensitization of β-adrenoceptors are common to all catecholamine receptors or common to all receptors that couple to G proteins. The recent observation that there is a kinase that selectively phosphorylates β-adrenergic receptors in the presence of agonists provides a molecular explanation for the homologous or agonistspecific desensitization of at least β-adrenoceptors (BENOVIC et al. 1986). It is presently believed that heterologous desensitization is a result of cyclic AMP-dependent protein kinase-catalyzed phosphorylation of receptors (NAMBI et al. 1985).

Alterations in the density or properties of dopamine receptors following administration of dopamine agonists or drugs that increase the availability of dopamine to its receptors have also been studied. Intrastriatal administration of dopamine resulted in a decrease in the number of binding sites for (^{3}H)-spiroperidol (COSTALL and NAYLOR 1979). This phenomenon is consistent with observations of functional desensitization that occurs in other systems following long-term exposure to agonists (MUKHERJEE et al. 1975). On the other hand, the chronic administration of L-dopa (WILNER et al. 1980) or cocaine, which blocks the reuptake of catecholamines (TAYLOR et al. 1979), resulted in an increase in the density of binding sites for (^{3}H)-spiroperidol. An increase in the binding of (^{3}H)-apomorphine following incubation of membranes with dopamine and other agonists, including bromocriptine, lergotrile, and dihydroxyergotamine, has also been reported (McMANUS et al. 1978; ROBERTSON 1980). The increase in the binding sites for (^{3}H)-apomorphine was prevented by the presence of 1 μM ascorbate, and it has been suggested that the enhanced binding was a nonspecific reaction (LEFF et al. 1981). We have observed, however, a 50% increase in the number of binding sites for (^{3}H)-spiroperidol following *in vitro* incubation of rat striatal homogenates with dopamine (HUFF and MOLINOFF 1982). This increase occurred in the presence or absence of 200 μM ascorbate, and was inhibited by 2 μM (+)-butaclamol.

Agonist-induced desensitization of α-adrenoceptors has also been observed. Incubation of parotid slices with noradrenaline led to a marked decrease in the ability of noradrenaline to stimulate cyclic GMP accumulation. This decrease was not dependent on the accumulation of cyclic GMP since it was also observed in the absence of extracellular Ca^{2+} (HARPER and BROOKER 1977). STRITTMATTER et al. (1977) have shown that preincubation of parotid cells with adrenaline resulted in a 50% reduction in the amount of α-receptor-stimulated K^+ efflux from parotid cells. This phenomenon was not dependent on the presence of extracellular Ca^{2+} and the cells rapidly regained their responsiveness when incubated with K^+. The authors examined the binding of (^{3}H)-DHE to α-adrenoceptors following incubation with adrenaline. There was a 40% to 50% decrease in the specific binding of (^{3}H)-DHE which was rapidly reversed by addition of K^+. The EC_{50} values for the decrease in K^+ efflux and (^{3}H)-DHE binding were approximately 10 μM. The results suggest that the loss of receptors is responsible for the decrease in responsiveness but that the loss of receptors is not related to degradation of receptor protein molecules.

J. Ontogeny of Catecholamine Receptors in the Central Nervous System

The ontogeny of catecholamines has been studied both by measuring levels of individual amines and by use of the histofluorescence technique first described by FALCK and HILLARP (AGRAWAL et al. 1966; BREESE and TAYLOR 1972; LOIZOU 1972; COYLE and HENRY 1973). At 15 days of gestation both noradrenaline and dopamine can be detected in rat brain. At this time the concentrations of these amines are about 2% of adult levels while the specific activities of the biosynthetic enzymes are about 10% of those found in adult brain (see COYLE 1974). Beginning prior to birth and continuing for several weeks thereafter, the concentrations of the biosynthetic enzymes and their products, dopamine and noradrenaline, increase in a more or less parallel fashion (BREESE and TAYLOR 1972; LAMPRECHT and COYLE 1972). In addition to progressive increases in the activities of the biosynthetic enzymes during the first weeks after birth, there is also an increase in the activity of a specific uptake process for catecholamines (COYLE 1974). The development of adenylate cyclase activity has been determined in homogenates of rat cerebral cortex (PERKINS and MOORE 1973a). There was significant basal enzyme activity at birth. Enzyme activity increased gradually to approximately two and one-half times that found at birth by postnatal day 16 to 20. In contrast to the ontogenetic pattern of adenylate cyclase activity, hormone-sensitive cyclic AMP accumulation was essentially undetectable in neonatal rats. This has been reported for both noradrenaline- and dopamine-stimulated cyclic AMP accumulation, studied in the cerebral cortex and caudate, respectively (PERKINS and MOORE 1973a; COYLE and CAMPOCHIARO 1976). In both cases, responsiveness was first detected between 6 and 12 days after birth and reached adult levels over the ensuing 1 to 2 weeks. Incubation of slices of cerebral cortex with 100 μM adenosine led to an increased accumulation of cyclic AMP as early as the 5th day after birth (PERKINS and MOORE 1973a). The magnitude of the response to adenosine then increased gradually to maximal levels by day 15. Prior to the development of sensitivity to catecholamines, the response to adenosine was enhanced when the slices were incubated with both noradrenaline and adenosine. At a time when responses to adenosine and to catecholamines were fully developed, incubation with both noradrenaline and adenosine resulted in a greater response than that seen with either compound alone (PERKINS and MOORE 1973a).

The ontogenetic development of β-adrenoceptors in rat cerebral cortex has been investigated by HARDEN et al. (1977a). Binding sites for (^{125}I)-IHYP became detectable by day 5 to 8 and then increased rapidly, reaching adult levels by day 18. The time course of receptor development was thus correlated with that of catecholamine-stimulated cyclic AMP accumulation. The results were consistent with the idea that the development of receptors permits expression of hormone-stimulated adenylate cyclase activity. Although there was a marked increase in the density of β-adrenoceptors, there was no change in the affinities of the receptor for either agonists or antagonists as determined by studies of inhibition of the binding of (^{125}I)-IHYP. In another series of studies, the effect of the neonatal administration of 6-OHDA on the development of β-adrenoceptors was investigated (HARDEN et al. 1977b). This treatment led to the depletion of over 90% of the noradrenaline from the cerebral cortex. Although there was no change in the time course of receptor development, there was a 40% to 60% increase in the density of receptors in animals ranging in age from 6 days to adult. This was accompanied by a similar increase in the V_{max} for isoprenaline-stimulated cyclic AMP accumulation.

The development of β_1- and β_2-adrenoceptors was studied in rat cerebral cortex and cerebellum (PITTMAN et al. 1980a). In the cerebral cortex, which contains mostly β_1-adrenoceptors, the density of β-adrenoceptors increased sharply between postnatal days 10 and 21. The density of receptors decreased by approximately 25% in animals older than 7 weeks of age. The proportion of

β_1 and β_2 receptors was relatively constant in animals up to 18 months of age (see the following). The ontogeny of the two receptor subtypes thus paralleled the development of total β-adrenoceptors in the cerebral cortex.

The ontogeny of β-adrenoceptors in the cerebellum, which contains mainly β_2 receptors, was strikingly different from that observed in the cortex. Total cerebellar β receptor density exhibited a slow but steady increase from postnatal day 5 through day 42. The density of receptors then levelled off and remained constant until the animals were approximately 6 months of age. Unlike the results obtained in the cortex, the relative proportions of β_1 and β_2 receptors in the cerebellum changed markedly during development. Between postnatal days 8 and 13 approximately 18 % of the receptors were of the β_1 subtype. This proportion steadily decreased with age, and in 3- and 6-month-old animals only approximately 2 % to 5 % of the receptors were of the β_1 subtype. The results demonstrate that the two subtypes of β-adrenoceptors can have different developmental patterns in the same brain area, and that a single receptor subtype can follow different developmental patterns in different brain regions.

In addition to examining the development of receptor subtypes, the effects of aging on β-adrenoceptor subtypes in the cortex and cerebellum have been determined (PITTMAN et al. 1980b). The densities of β_1- and β_2-adrenoceptors were determined in homogenates of cerebral cortex and cerebellum of rats between 3 and 14 months of age. No change in either receptor population occurred in the cortex during this period. In the cerebellum, a 20 % to 25 % decrease in the density of β_2 receptors and a 350 % increase in the density of β_1 receptors occurred. The increase in β_1 receptors in the cerebellum may be the result of a decrease in the function of a noradrenergic projection from the locus coeruleus which forms synapses on cerebellar Purkinje cells, thereby causing a compensatory increase in receptors.

PARDO et al. (1977) have studied the development of dopamine receptors in the rat striatum by measuring the binding of (^{3}H)-haloperidol. Binding at birth was 10 % to 15 % of adult levels and showed little change until day 7, when it increased rapidly, more than doubling between days 7 and 14. The time course of receptor development was very different from that of (^{3}H)-dopamine uptake and endogenous dopamine content, which exhibited approximately linear increases from birth to day 28 (COYLE and CAMPOCHIARO 1976). Although the density of receptors was at adult levels by day 25, the presynaptic markers increased significantly between day 28 and adulthood. One biochemical marker whose development appeared to parallel that of the binding of (^{3}H)-haloperidol was endogenous acetylcholine. Tissue levels of acetylcholine at birth were approximately 20 % of adult levels. There was little change in acetylcholine content over the first 7 days after birth but adult levels were attained by day 28 (COYLE and YAMAMURA 1976). This is of particular interest in view of the pharmacological evidence which suggests that dopaminergic terminals may form synapses on cholinergic neurons in the corpus striatum (SETHY and VAN WOERT 1974; TRABUCCHI et al. 1975).

K. Solubilization, Purification, and Reconstitution of Catecholamine Receptor/Effector Systems

The β-adrenoceptor of frog erythrocytes has been solubilized with the plant glycoside digitonin (CARON and LEFKOWITZ 1976) and that in S49 lymphoma cells with Lubrol PX (HAGA et al. 1977). Following the use of digitonin, the receptor was still capable of stereospecifically binding β-adrenoceptor agonists and antagonists including (^{3}H)-DHA. Based on gel chromatography, the molecular weight was estimated to be 130,000 to 150,000 daltons.

The β-adrenoceptor in S49 cells was apparently labile after being exposed to detergents. HAGA and his collaborators (1977) were unable to demonstrate specific binding of (^{125}I)-IHYP or (^{3}H)-DHA to receptors solubilized with a variety of detergents including digitonin. However, because the rate of dissociation of (^{125}I)-IHYP is very slow ($k_2 = 0.044$/min at 30 °C), it was possible to prelabel receptors with (^{125}I)-IHYP prior to exposing the membranes to detergent. Under the conditions employed, 90% to 95% of the bound ligand appeared to be associated with the receptor prior to solubilization. In the same experiments, adenylate cyclase was irreversibly activated by incubation with Gpp(NH)p. Following solubilization, the hydrodynamic properties of the cyclase and the receptor were determined by gel filtration and by sucrose density gradient centrifugation. The two functions were resolved into distinct peaks by both techniques, suggesting that they are separate structures. The high partial specific volume of the receptor led to the suggestion that it may be an integral membrane constituent.

WITKIN and HARDEN (1981) developed an equilibrium binding assay for digitonin-solubilized β-adrenoceptors based on the use of (^{125}I)-iodopindolol. The binding of this ligand to solubilized receptors was inhibited in a stereoselective manner by a variety of agonists and antagonists. This approach allows receptors to be identified in soluble preparations from a variety of tissues.

Further evidence that the β-adrenoceptor and solubilized adenylate cyclase activity are discrete and independent entities came from the use of affinity chromatography (VAUQUELIN et al. 1977). A specific adsorbent was synthesized by coupling alprenolol to a spacer arm on agarose through a thio bond. Treatment with digitonin solubilized 20% to 30% of the adenylate cyclase and approximately the same percentage of the binding sites for (^{3}H)-DHA from turkey erythrocyte membranes. The properties of the solubilized receptors were similar to those of membrane-bound receptors. The receptor was then purified by column chromatography. Most of the adenylate cyclase activity was found in an eluate that did not contain receptor activity. The liquid binding activity was then eluted with 720 nM (±)-alprenolol in 1 M NaCl. The authors estimated that this procedure resulted in a 2000-fold purification of the receptor.

LEFKOWITZ and his colleagues have employed a number of techniques to purify the β-adrenoceptor from frog erythrocytes. Good results were obtained when a long-chain bisoxirane reagent was used to introduce a hydrophilic spacer arm on Sepharose 4B. Alprenolol covalently immobilized on Sepharose columns caused retention of digitonin-solubilized receptor (CARON et al. 1979) with only minimal retention of nonspecifically adsorbed protein. After washing the column with 100 mM NaCl, the receptor was eluted by addition of 1 mM (±)-isoprenaline. Agonists blocked retention by, and eluted receptor activity from, the column with the potency order expected of a β-adrenoceptor. The specific activity of material purified by two cycles of affinity chromatography was approximately 1,600 to 2,000 pmol/mg of protein. The use of Sepharose-alprenolol columns (CARON and LEFKOWITZ 1976; CARON et al. 1979) together with ion exchange chromatography on DEAE-Sepharose 6B-Cl resulted in a preparation of binding protein suitable for study with SDS-polyacrylamide gel electrophoresis. The specific activity of the preparation after four purification steps was 8,000 to 10,000 pmol of (^{3}H)-DHA binding sites/mg of protein, representing up to a 55,000-fold purification from crude frog erythrocyte membranes. Aliquots of partially purified receptor were iodinated with $Na^{125}I$ before a final Sepharose-alprenolol chromatography step and prior to sucrose density gradient centrifugation. Electrophoresis on SDS-polyacrylamide gels revealed two bands of iodinated material, one migrating with an apparent molecular size of 58,000 daltons and a more diffuse band in the lower molecular weight region of the gel. The 58,000-dalton-molecular-weight material was thought to correspond to the receptor since purified preparations bound (^{3}H)-DHA and other drugs with affinities and selectivity expected on the basis of results obtained with membrane-bound receptors.

Several groups of investigators have attempted to use chemically reactive site-directed ligands to study β-adrenoceptors. A bromoacetyl derivative of propranolol (ATLAS and LEVITZKI 1976) was reported to be a specific irreversible inhibitor of the β-adrenoceptor. This compound, prepared in tritiated form, was reported (ATLAS and LEVITZKI 1978) to label two proteins with molecular weights of 37,000 to 41,000 daltons on intact L6P myoblasts or on turkey erythrocytes. When cells were simultaneously incubated with hydroxybenzylpindolol and the (^{3}H)-affinity label, the labelling of the two proteins was partially prevented. In similar experiments, an azide derivative of propranolol (DARFLER and MARINETTI 1977) was reported to specifically block the binding of (^{3}H)-DHA to β-adrenoceptors on membranes prepared from turkey erythrocytes. These experiments were carried out in the dark and photolysis with this potential photoaffinity ligand was not attempted.

Three attempts to use photoaffinity labels to probe the properties of β-adrenoceptors have been reported. An azide derivative of the β-adrenoceptor antagonist acebutolol was reported to irreversibly inhibit β-adrenoceptor-stimulated adenylate cyclase activity of rat reticulocyte membranes after photolysis (WRENN and HOMCY 1980). RASHIDBAIGI and RUOHO (1981) prepared (^{125}I)-iodoazidobenzylpindolol with a specific activity of 1,300 Ci/mmol. When this compound was photolyzed in the presence of crude duck erythrocyte membranes and the membrane proteins subjected to SDS-polyacrylamide gel electrophoresis, specific labelling of two polypeptides was observed. These polypeptides, with molecular weights of 45,000 and 48,500 daltons, were photolabelled in a ratio of approximately 4:1. Another photoactive β-adrenoceptor antagonist, p-azidobenzylcarazolol, has been described by

LAVIN et al. (1981). This compound, labelled with tritium to a specific activity of 26 Ci/mmol, binds to the β-adrenoceptor of frog erythrocytes with high affinity ($K_D = 100$ pM). Photolysis resulted in the apparently covalent incorporation of radioactivity into a peptide with a molecular weight of 58,000 daltons. This peptide appears to be the same as that isolated by conventional purification techniques (SHORR et al. 1981). Other experiments involving polyacrylamide gel electrophoresis of (^{125}I)-p-azidobenzylcarazolol-labelled extracts of turkey erythrocyte membranes resulted in incorporation of label into two polypeptides of molecular weights 38,000 and 50,000 daltons. After incubating the erythrocytes with isoprenaline, the apparent molecular weights of the labelled peptides increased to 42,000 and 53,000 daltons. Thus, structural alterations in the receptor may accompany the desensitization process in turkey erythrocytes (STADEL et al. 1982).

As previously discussed, S49 lymphoma cells represent an important system with which to try to understand the mechanism of the regulation of catecholamine-sensitive adenylate cyclase. The availability of a variety of lines that are either protein kinase-deficient, G_s protein-deficient, or uncoupled makes this a uniquely valuable system (HAGA et al. 1977; SMIGEL et al. 1984). These lines have proved useful in reconstituting hormone-sensitive adenylate cyclase activity in membranes that have been depleted of one or more factors. Furthermore, it is likely that the various components will have to be properly integrated into a membrane structure to be able to express catecholamine-sensitive adenylate cyclase activity. FLEMING and ROSS (1980) have described the incorporation of the β-adrenoceptor into phospholipid vesicles. The receptors were solubilized from rat erythrocyte plasma membranes using digitonin. Solubilized receptors were reconstituted into vesicles by adding dimyristoylphosphatidylcholine and by removal of detergent. Sephadex chromatography was used to lower the digitonin concentration following the addition of excess lipid. This procedure restored the ability of the receptor to bind (^{125}I)-IHYP, a ligand that does not bind to detergent-solubilized receptors (see above). The binding of (^{125}I)-IHYP was saturable and was inhibited stereospecifically by agonists and antagonists at β-adrenoceptors.

A second approach to classifying the components of the adenylate cyclase system has been developed by ORLY and SCHRAMM (1976). These investigators used Sendai virus to fuse turkey erythrocytes in which the catalytic activity of adenylate cyclase had been inactivated by N-ethylmaleimide or by heat with Friend erythroleukemia cells that do not have detectable β-adrenoceptors. Cell ghosts of the fused preparations had adenylate cyclase activity which was markedly increased by isoprenaline. It was therefore concluded that the β-adrenoceptor of the turkey erythrocytes had become functionally coupled to the enzyme of the erythroleukemia cells.

EIMERL et al. (1980) solubilized the β-adrenoceptor of turkey erythrocytes with deoxycholate. After removing the detergent and adding phospholipid, the receptor was incorporated into the membrane of Friend erythroleukemia cells. The implanted β-adrenoceptor was efficiently coupled to the adenylate cyclase such that a 30-fold stimulation of adenylate cyclase activity occurred on incubating membranes with adrenaline. The concentration of adrenaline required for a half-maximal response was approximately the same as observed in the original erythrocyte membranes. In contrast to the use of covalent ligands, the use of reconstitution as an assay to follow receptor activity during purification would provide assurance that the component being purified is a functional receptor.

Recently CERIONE et al. (1983a, b) have demonstrated that the purified β-adrenoceptor can be functionally reconstituted into *Xenopus laevis* erythrocytes, which lack β-adrenoceptors. This fusion of a homogeneous, single polypeptide into these erythrocytes resulted in the ability of isoprenaline to markedly stimulate adenylate cyclase activity. Thus, it appears that the ligand recognition site and the site on the receptor that recognizes the effector mechanism (presumably the G_s protein) reside on the same polypeptide.

Several investigators have elicited antibodies that react with or cross-react with β-adrenoceptor. β-adrenoceptors purified from frog erythrocytes (STRADER et al. 1983), turkey erythrocytes (COURAD et al. 1981; 1983a, b), rat adipocytes (MOXHAM et al. 1986), guinea pig lung (MOXHAM et al. 1986), and A431 cells (STROSBERG et al. 1985) have been used to elicit antibodies. Only a few of these antibodies (MOXHAM et al. 1986) alter binding of ligands to the receptor. Strosberg and colleagues (GUILLET et al. 1984; 1985; STROSBERG et al. 1985) have elicited a monoclonal anti-idiotypic antibody that cross-reacts with β-adrenoceptor. The antibody interacts with native receptors in solution, labels a single band of the appropriate mobility on western blot analysis of crude membranes, and binds to cells that contain (A431 cells), but not to those that lack (rabbit lymphoma cells), β-adrenoceptor. The binding of this antibody to A431 cells was eliminated after incubation of the cells with isoprenaline, a procedure that causes loss of cellsurface receptors. Antibodies that will react with β-adrenoceptor have been obtained by immunizing animals with peptides corresponding to particular regions of the β-adrenoceptor (DIXON et al. 1986; 1987; YARDEN et al. 1986). Such anti-peptide antibodies have been shown to bind to solubilized, native receptor as well as to purified receptor on western blot analysis.

α_2-Adrenoceptors have been purified to homogeneity from human platelets (REGAN et al. 1986) and porcine brain (REPASKE et al. 1987). In both cases the receptor migrates as a single band on SDS polyacrylamide gels at approximately 64–65,000 daltons. The use of a novel affinity resin, yohimbine-agarose, allows purification with but a single chromatographic step (REPASKE et al. 1987). The ability of guanine nucleotides to regulate agonist affinity at the α_2-adrenoceptor is lost upon solubilization, but monovalent cations still regulate the affinity of the receptor for agonists.

Using purified α_2-adrenoceptors CERIONE et al. (1986) have functionally reconstituted the receptor with various guanine nucleotide-binding regulatory proteins (G-proteins) into phospholipid vesicles. In the presence of adrenaline, reconstituted receptor stimulated GTPase activity of several G-proteins. Maximal stimulation (1 mol of inorganic phosphate released per min per mol of G-protein) was observed when α_2-adrenoceptors were coupled with G_i and G_0. Much lower levels of stimulation were obtained with G_s or transducin.

This order of efficacy is not surprising since G_i is thought to be involved in biochemical events normally mediated by α_2-adrenoceptors, including inhibition of adenylate cyclase activity. Although the function of G_0 is unknown at the present time, one may speculate that its function is more similar to that of G_i than that of G_s.

Two experimental approaches have been used for solubilization of membrane-associated D-2 receptors. The first approach has been to solubilize the receptor using the detergent digitonin. After extraction of homogenates of caudate or pituitary tissue in 1% digitonin, 10–20% of the total membrane-associated receptors are found in the supernatant after centrifugation at 100,000 x g for 60 min. (LEFF and CREESE 1982; LUEDTKE and MOLINOFF 1987). A second approach for solubilization of D-2 receptors has been to extract tissue with a low concentration of a synthetic detergent in the presence of a high concentration of salt (HALL et al. 1983; LEW and GOLDSTEIN 1984). A potential advantage of the second approach is that synthetic detergents often have a higher critical micelle concentration than does digitonin and they can therefore be removed by dialysis. However, membrane fragments rather than soluble proteins may result when using the low-detergent/high-salt extraction protocol (LADURON and ILIEN 1982; LUEDTKE and MOLINOFF 1987).

Detergent extraction of caudate or pituitary tissue in the presence of agonists has resulted in solubilization of a form of the receptor with a high affinity for agonists (LEFF and CREESE 1984). In the pituitary, the guanine nucleotide-sensitive, high-affinity form of the receptor was shown to have a larger apparent molecular size than the low-affinity form based on the elution profile of pre-bound radioligand using size-exclusion HPLC (KILPATRICK and CARON 1984). Both the high-affinity form and the low-affinity form were retained on wheat germ agglutinin columns (KILPATRICK and CARON 1984).

Several attempts have been made to couple D-2-selective ligands to an insoluble matrix to construct affinity chromatography resins to be used for purifying the receptor. ANTONIAN et al. (1986) and RAMWANI and MISHRA (1986) constructed affinity resins using analogues of the butyrophenone haloperidol. RAMWANI and MISHRA (1986) reported a 2000-fold enrichment after solubilization with 0.2% cholate in the presence of 1.0 M NaCl and chromatographic separation on an epoxy-Sepharose CL6B-haloperidol column. ANTONIAN et al. (1986) synthesized a hemisuccinyl derivate of haloperidol for construction of an affinity resin for studies of canine caudate D-2 receptors solubilized with 1% digitonin.

^{3}H-SPD has greater specificity and higher affinity for D-2 receptors than does ^{3}H-haloperidol. Therefore, affinity resins constructed with derivatives of SPD might be expected to be more effective than those based on derivatives of haloperidol. SENOGLES et al. (1986) constructed an affinity resin using a carboxymethyloxime derivative of SPD for use in studies of preparations of bovine pituitary solubilized with digitonin. A 1000-fold enrichment of binding sites for ^{3}H-SPD was reported. A 5- to 10-fold increase in the calculated recovery of binding sites for ^{3}H-SPD was obtained when eluted proteins were reconstituted into vesicles. Reconstitution was required in these experiments because the affinity of the receptor for butyrophenones, including ^{3}H-SPD, decreases following solubilization and purification of the receptor (SENOGLES et al. 1986; LUEDTKE and MOLINOFF 1987).

Results of recent studies indicate that preparations of caudate tissue solubilized with digitonin are a source of D-1, as well as D-2, receptors. Studies of the binding of ^{3}H-SCH-23390 to receptors in solubilized preparations have been carried out. The same rank order of potency for the binding of agonists and antagonists exists as with D-1 receptors associated with membranes (NIZNIK et al. 1986). The observation of different elution profiles for the binding sites labeled with ^{3}H-SPD and ^{3}H-SCH-23390 strongly suggest that D-1 and D-2 receptors are distinct proteins (DUMBRILLE-ROSS et al. 1985).

Two reagents have been reported to bind irreversibly to the D-2 receptor. The first, azidoclebopride, was shown to be a photoactivated, irreversible inhibitor of the binding of ^{3}H-SPD to D-2 receptors in caudate tissue while D-1, 5-HT_2, benzodiazepine, α_1-, and β-adrenoceptors were not affected (WOUTERS et al. 1984; NIZNIK et al. 1985). Although selective for the D-2receptor, the usefulness of azidoclebopride is likely to be limited because tritiated or iodinated forms of the affinity label are not available. More recently, the synthesis of N-aminophenethylspiroperidol (NAPS) was described by AMLAIKY and CARON (1985). D-2 receptors have approximately the same affinity for NAPS as for SPD. Iodination followed by conversion of the arylamine to an arylazide generates a photoaffinty label that was used to selectively label a peptide with a molecular weight of 94,000 daltons. ^{125}I-NAPS can be used to study D-2 receptors but the ligand is relatively lipophilic and nonspecific binding is higher than with either ^{3}H-SPD or ^{125}I-iodobenzamide (BRUECKE et al. 1987). An N-aminophenyl derivative of SCH-23390 that can be iodinated with ^{125}I and used in cross-linking experiments designed to affinity-label D-1 receptors has also been described (AMLAIKY and CARON 1986). It appeared to cross-link to a protein of molecular weight 72,000.

L. Molecular Cloning of Catecholamine Receptors

Using the tools of molecular biology, the amino acid sequences of β-adrenoceptors in turkey erythrocytes (YARDEN et al. 1986), hamster lung DIXON et al. 1986), and human placenta and brain (KOBILKA et al. 1987a; CHUNG et al. 1987) have been determined. A striking structural feature of all the β-adrenoceptors that have been cloned and sequenced is their topographical orientation with respect to the membrane. Analysis of hydropathicity profiles suggests that there are seven hydrophobic regions, each of 20–25 amino acids. These are potentially membrane-spanning. Other structural features include a long C-terminal hydrophilic sequence, a somewhat shorter N-terminal hydrophilic sequence, and a long cytoplasmic loop between presumptive transmembrane segments V and VI (see Fig. 5).

The proposed structure of the β-adrenoceptor is strikingly similar in sequence and topography to that of rhodopsin (OVCHINNIKOV 1982; NATHANS and HOGNESS 1983) and the muscarinic acetylcholine receptor, whose cDNA has recently been cloned from porcine brain (KUBO et al. 1986 a) and heart

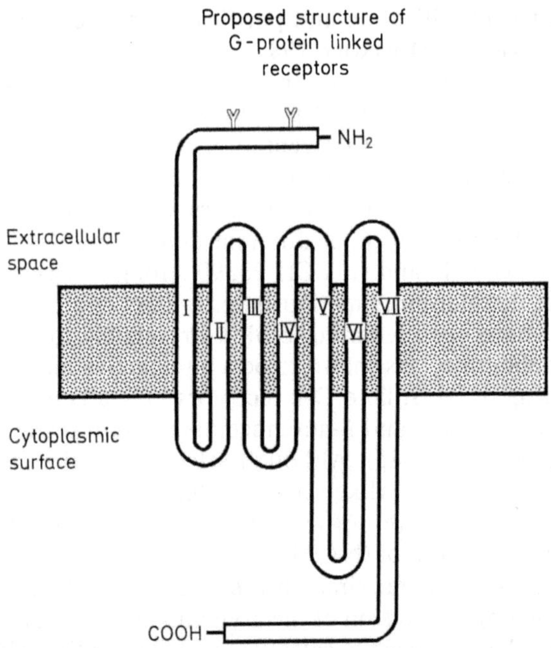

Fig. 5. Schematic diagram of the generic structure of G-protein-related receptors. The 7 membrane-spanning regions are indicated with roman numerals. Y denotes likely sites of glycosylation.

KUBO et al. 1986 b). Although these proteins mediate widely disparate biological effects, they show a high degree of homology. This is almost certainly related to the fact that, in each case, the immediate consequence of receptor activation is to promote an interaction between the receptor and a GTP-binding protein (see DOHLMAN et al. 1987).

Recently, a gene coding for the α_2-adrenoceptor has been cloned and sequenced (KOBILKA et al. 1987b). The protein sequence deduced from the DNA is fairly homologous to that of the β_2-adrenoceptor (DIXON et al. 1986) and shows the now typical seven membrane-spanning regions of high hydrophobicity. On the other hand, the third intracellular loop (i3) of the α_2-adrenoceptor is much longer than the corresponding region of the β_2-adrenoceptor. This feature of a long i3 loop is shared with the several muscarinic receptors that have been cloned and sequenced (BONNER et al. 1987). As with all of the other G-protein-coupled receptors, the α_2-adrenoceptor has two asparagine residues (10 and 14) that are likely to be sites of glycosylation. Using somatic cell hybridization analysis, it was shown that the gene coding for the human α_2-adrenoceptor resides on chromosome 10 (KOBILKA et al. 1987b).

M. Conclusion

It has been little more than a decade since the first successful *in vitro* binding assay for a catecholamine receptor was carried out. Enormous progress has been made over this relatively short time, and many of these receptors have been purified and the genes for some have been cloned. We now understand a great deal about the molecular properties of catecholamine receptors and the mechanisms by which they interact with and modulate the activity of effector systems. The tools of molecular biology promise that we will soon know the primary structure of all these molecules, and from this beginning we can try to understand something about the relationship between primary structure and the function of these receptors.

Knowledge of the molecular mechanisms involved in the effects of drugs on receptors is important, because receptors represent the molecular entity through which many pharmacologically active agents interact, and because a variety of disease processes may involve changes in the density of catecholamine receptors. It has also been shown that changes in catecholamine receptors occur as a consequence of the administration of agents widely used in the treatment of diseases ranging from hypertension to depression. An understanding of the molecular properties of catecholamine receptors may thus lead to significant improvement in our ability to treat these diseases.

Acknowledgment. We appreciate the excellent secretarial, editorial, and bibliographic assistance of Ms. Janet Malatesta and Ms. Lotte Gottschlich. During the time this article was being prepared the authors were supported by grants from the USPHS (NS 18479, NS 18591, AA 03527, and GM 31155). B.B. Wolfe is an Established Investigator of the American Heart Association.

N. References

Aarons RD, Nies AS, Gal J, Hegstrand LR, Molinoff PB (1980) Elevation of β-adrenergic receptor density of human lymphocytes after propranolol administration. J Clin Invest 65:949–957

Aellig WH (1982) Clinical pharmacology of pindolol. Am Heart J 104:346–356

Aghajanian GK, Bunney BS (1977) Dopamine "autoreceptors": pharmacological characterization by microiontophoretic single cell recording studies. Naunyn-Schmiedeberg's Arch Pharmacol 297:1–7

Agrawal HC, Glisson SN, Himwich WA (1966) Changes in monoamines of rat brain during postnatal ontogeny. Biochim Biophys Acta 130:511–513

Ahlquist RP (1948) A study of the adrenotropic receptors. Am J Physiol 153:586–600

Ahlquist RP (1976) Adrenergic beta-blocking agents. Prog Drug Res 20:27–42

Alexander RW, Williams LT, Lefkowitz RJ (1975) Identification of cardiac β-adrenergic receptors by (–)-(^{3}H) alprenolol binding. Proc Natl Acad Sci USA 72:1564–1568

Amlaiky N, Caron MG (1986) Identification of the D2-dopamine receptor binding subunit in several mammalian tissues and species by photoaffinity labeling. J Neurochem 47:196–204

Antonian L, Antonian E, Murphy RB, Schuster DI (1986) Studies on the use of a novel affinity matrix, sepharose amine-succinyl-amine haloperidol hemisuccinate, ASA-HHS, for purification of canine dopamine (D2) Receptor. Life Sci 38:1847-1858

Atlas D, Levitzki A (1978) An irreversible blocker for the β-adrenergic receptor. Biochem Biophys Res Commun 69:397-403

Atlas D, Levitzki A (1978) Tentative identification of β-adrenoreceptor subunits. Nature 272:370-371

Augusti-Tocco G, Sato G (1969) Establishment of functional clonal lines of neurons from mouse neuroblastoma, Proc Natl Acad Sci USA 64:311-315

Aurbach GD, Fedak SA, Woodard CJ, Palmer JS, Hauser D, Toxler F (1974) β-Adrenergic receptor: stereospecific interaction of iodinated β-blocking agent with high affinity site. Science 186:1223-1224

Axelsson J, Thesleff S (1959) A study of supersensitivity in denervated mammalian skeletal muscle. J Physiol (Lond) 147:178-193

Bacopoulos NG (1983) (^{3}H)Dopamine binds to D1 and D2 receptors in rat striatum. Eur J Pharmacol 87:353-356

Banerjee SP, Kung LS, Riggi SJ, Chanda SK (1977) Development of β-adrenergic receptor subsensitivity by antidepressants. Nature 268:455-456

Banerjee SP, Sharma VK, Kung LS, Chanda SK (1978) Amphetamine induces β-adrenergic receptor supersensitivity. Nature 271:380-381

Bär H-P, Hahn P (1971) Development of rat liver adenylcyclase. Can J Biochem 49:85-89

Barovsky K, Brooker G (1980) (−)-(^{125}I)-Iodopindolol, a new highly selective radioiodinated β-adrenergic receptor antagonist: measurement of β-receptors on intact rat astrocytoma cells. J Cyclic Nucleotide Res 6:297-307

Bearer CF, Knapp RD, Kaumann AJ, Swartz TL, Birnbaumer L (1980) Iodohydroxybenzylpindolol: preparation, purification, localization of its iodine to the indole ring, and characterization as a partial agonist. Mol Pharmacol 17:328-338

Benovic JL, Strasser RH, Caron MG, Lefkowitz RJ (1986) β-Adrenergic receptor kinase: identification of a novel protein kinase that phosphorylates the agonist-occupied form of the receptor. Proc Natl Acad Sci USA 83:2797-2801

Berg DK, Kelly RB, Sargent PB, Williamson P, Hall ZW (1972) Binding of α-bungarotoxin to acetylcholine receptors in mammalian muscle. Proc Natl Acad Sci USA 69:147-151

Berridge MJ (1981) Phosphatidylinositol hydrolysis: a multifunctional transducing mechanism. Mol Cell Endocrinol 24:115-140

Berthelsen S, Pettinger WA (1977) A functional basis for classification of α-adrenergic receptors. Life Sci 21:595-606

Bilezikian JP, Loeb JN (1982) Mechanisms of altered β-adrenergic responsiveness in the hyperthyroid and hypothyroid turkey erythrocyte. Life Sci 30:663-673

Bilezikian JP, Spiegel AM, Brown EM, Aurbach GD (1977a) Identification and persistence of β-adrenergic receptors during maturation of the rat reticulocyte. Mol Pharmacol 13:775-785

Bilezikian JP, Spiegel AM, Gammon DE, Aurbach GD (1977b) The role of guanyl nucleotides in the expression of catecholamine-responsive adenylate cyclase during maturation of the rat reticulocyte. Mol Pharmacol 13:786-795

Bird SJ, Maguire ME (1978) The agonist-specific effect of magnesium ion on binding by β-adrenergic receptors in s49 lymphoma cells. Interaction of gtp and magnesium in adenylate cyclase activation. J Biol Chem 253:8826-8834

Birdsall NJM, Hulme EC, Burgen A (1980) The character of the muscarinic receptors in different regions of the rat brain. Proc R Soc Lond Biol 207:1-12

Bitensky MW, Russell V, Blanco M (1970) Independent variation of glucagon and epinephrine responsive components of hepatic adenyl cyclase as a function of age, sex and steroid hormones. Endocrinology 86:154-159

Boeynaems JM, Dumont JE (1977) The two-step model of ligand-receptor interaction. Mol Cell Endocrinol 7:33-47

Bokoch GM, Katada T, Northup JK, Ui M, Gilman AG (1984) Purification and properties of the inhibitory guanine nucleotide-binding regulatory component of adenylate cyclase. J Biol Chem 259:3560-3567

Bonner TI, Buckley NJ, Young AC, Brann MR (1987) Identification of a family of muscarinic acetylcholine receptor genes. Science 237:527-532

Bottari S, Vauquelin G, Durieu OL, Klutchko C, Strosberg AD (1979) The β-adrenergic receptor of turkey erythrocyte membranes: conformational modification by β-adrenergic agonists. Biochem Biophys Res Commun 86:1311-1318

Breese GR, Traylor TD (1972) Developmental characteristics of brain catecholamines and tyrosine hydroxylase in the rat: effects of 6-Hydroxydopamine. Br J Pharmacol 44:210-222

Brodde O-E, Schümann H-J, Wagner J (1980) Decreased responsiveness of the adenylate cyclase system of left atria from hypothyroid rats. Mol Pharmacol 17:180-186

Brown EM, Fedak SA, Woodard CJ, Aurbach GD, Rodbard D (1976) β-Adrenergic receptor interactions: direct comparison of receptor interaction and biological activity. J Biol Chem 251:1239-1246

Brown EM, Caroll RJ, Aurbach GD (1977) Dopaminergic stimulation of cyclic AMP accumulation and parathyroid hormone release from dispersed bovine parathyroid cells. Proc Natl Acad Sci USA 74:4210-4213

Brown JH, Makman MH (1972) Stimulation by dopamine of adenylate cyclase in retinal homogenates and of adenosine-3':5'-cyclic monophosphate formation in intact retina. Proc Natl Acad Sci USA 69:539-543

Browning ET, Schwartz JP, Breckenridge BM (1974) Norepinephrine-sensitive properties of C-6 astrocytoma cells. Mol Pharmacol 10:162-174

Bruecke T, Tsai YF, Singhaniyom, McLellan C, Kung H, Chiueh CC (1987) Receptor binding properties of [i-125] labeled iodobenzamide (IBZM): a specific D2 dopamine receptor ligand for SPECT. Soc Neurosci Abstr 13:190

Bürgisser E, Hancock AA, Lefkowitz RJ, DeLean A (1981) Anomalous equilibrium binding properties of high-affinity racemic radioligands. Mol Pharmacol 19:205-216

Burns TW, Langley PE (1975) The effect of α- and β-adrenergic receptor stimulation on the adenylate cyclase activity of human adipocytes. J Cyclic Nucleotide Res 1:321-328

Burt DR, Enna SJ, Creese I, Snyder SH (1975) Dopamine receptor binding in the corpus striatum of mammalian brain. Proc Natl Acad Sci USA 72:4655-4659

Burt DR, Creese I, Snyder SH (1976) Properties of (^{3}H)-haloperidol and (^{3}H)-dopamine binding associated with dopamine receptors in calf brain membranes. Mol Pharmacol 12:800-812

Burt DR, Creese I, Snyder SH (1977) Antischizophrenic drugs: chronic treatment elevates dopamine receptor binding in brain. Science 196:326-328

Butcher FR, Potter VR (1972) Control of the adenosine 3',5'-monophosphate-adenyl cyclase system in the livers of developing rats. Cancer Res 32:2141-2147

Bylund DB, Snyder SH (1976) β-Adrenergic Receptor binding in membrane preparations from mammalian brain. Mol Pharmacol 12:568-580
Cannon WB, Rosenblueth A (1949) The supersensitivity of denervated structures. MacMillan, New York
Carlsson A (1975) Receptor-mediated control of dopamine metabolism. In: Usdin E, Bunney Jr WE (eds) Pre- and postsynaptic receptors, Marcel Dekker, New York, pp 49-65
Carlsson E, Åblad B, Brändström A, Carlsson B (1972) Differentiated blockade of the chronotropic effects of various adrenergic stimuli in the cat heart. Life Sci 11, Part I:953-958
Carlsson E, Dahlöf C-G, Hedberg A, Persson H, Tangstrand B (1977) Differentiation of cardiac chronotropic and inotropic effects of β-adrenoceptor agonists. Naunyn-Schmiedeberg's Arch Pharmacol 300:101-105
Caron MG, Lefkowitz RJ (1976) Solubilization and characterization of the β-adrenergic receptor binding sites of frog erythrocytes. J Biol Chem 251:2374-2384
Caron MG, Srinivasan Y, Pitha J, Kociolek K, Lefkowitz RJ (1979) Affinity chromatography of the β-adrenergic receptor. J Biol Chem 254:2923-2927
Carr LA, Moore KE (1970) Effects of amphetamine on the contents of norepinephrine and its metabolites in the effluent of perfused cerebral ventricles of the cat. Biochem Pharmacol 19:2361-2374
Carruthers SG, Twum-Barima Y (1981) Measurement of partial agonist activity of pindolol. Clin Pharmacol Ther 30:581-586
Cerione RA, Strulovici B, Benovic JL, Lefkowitz RJ, Caron MG (1983a) Pure β-adrenergic receptor: the single polypeptide confers catecholamine responsiveness to adenylate cyclase. Nature 306:562-566
Cerione RA, Strulovici B, Benovic JJ, Strader CD, Caron MG, Lefkowitz RJ (1983b) Reconstitution of β-adrenergic receptors in lipid vesicles: affinity chromatography-purified receptors confer catecholamine responsiveness on a heterologous adenylate cyclase system. Proc Natl Acad Sci USA 80:4899-4903
Cerione RA, Regan JW, Nakata H, Codina J, Benovic JL, Gierschik P, Somers RL, Spiegel AM, Birnbaumer L, Lefkowitz RJ, Caron MG (1986) Functional reconstitution of the α_2-adrenergic receptor with guanine nucleotide regulatory proteins in phospholipid vesicles. J Biol Chem 261:3901-3909
Chan TM, Exton JH (1977) β-Adrenergic-mediated accumulation of adenosine 3′:5′-monphosphate in calcium-depleted hepatocytes. J Biol Chem 252:8645-8651
Changeux J-P, Meunier J-C, Huchet M (1971) Studies on the cholinergic receptor protein of *electrophorus electricus*. I. An assay *in vitro* for the cholinergic receptor site and solubilization of the receptor protein from electric tissue. Mol Pharmacol 7:538-553
Charness ME, Bylund DB, Beckman BS, Hollenberg MD, Snyder SH (1976) Independent variation of β-adrenergic receptor binding and catecholamine-stimulated adenylate cyclase activity in rat erythrocytes. Life Sci 19:243-250
Chuang D-M (1981) Inhibitors of transglutaminase prevent agonist-mediated internalization of β-adrenergic receptors. J Biol Chem 256:8291-8293
Chuang D-M, Costa E (1979) Evidence for internalization of the recognition site of β-adrenergic receptors during receptor subsensitivity induced by (−)-isoproterenol. Proc Natl Acad Sci USA 76:3024-3028
Chuang D-M, Kinnier WJ, Farber L, Costa E (1980) A biochemical study of receptor internalization during β-adrenergic receptor desensitization in frog erythrocytes. Mol Pharmacol 18:348-355
Chung F-Z, Lentes K-U, Gocayne J, Fitzgerald M, Robinson D, Kerlavage AR, Fraser

CM, Venter JC (1987) Cloning and sequence analysis of the human brain β-adrenergic receptor: evolutionary relationship to rodent and avian β-receptors and porcine muscarinic receptors. FEBS Lett 211:200-206

Clark RB, Perkins JP (1971) Regulation of adenosine 3′:5′-cyclic monophosphate concentration in cultured human astrocytoma cells by catecholamines and histamine. Proc Natl Acad Sci USA 68:2757-2760

Clark RB, Su Y-F, Ortmann R, Cubeddu X, Johnson GL, Perkins JP (1975) Factors influencing the effect of hormones on the accumulation of cyclic AMP in cultured human astrocytoma cells. Metabolism 24:343-358

Clement-Cormier YC, George RJ (1979) Multiple dopamine binding sites: subcellular localization and biochemical characterization. J Neurochem 32:1061-1069

Clement-Cormier YC, Kebabian JW, Petzold GL, Greengard P (1974) Dopamine-sensitive adenylate cyclase in mammalian brain: a possible site of action of antipsychotic drugs. Proc Natl Acad Sci USA 71:1113-1117

Cools AR, van Rossum JM (1976) Excitation-mediating and inhibition-mediating dopamine-receptors: a new concept towards a better understanding of electrophysiological, biochemical, pharmacological, functional and clinical data. Psychopharmacol 45:243-254

Costall B, Naylor RJ (1979) Changes in dopamine receptor status after denervation or chronic receptor stimulation. Br J Pharmacol 66:492P-493P

Cote TE, Grewe CW, Tsurto K, Stoof JC, Eskay RL, Kebabian JW (1982) D-2 dopamine receptor-mediated inhibition of adenylate cyclase activity in the intermediate lobe of the rat pituitary gland requires guanosine 5′-triphosphate. Endocrinol 110:812-819

Courad P-O, Delavier-Klutchko C, Durieu-Trautmann O, Strosberg AD (1981) Antibodies raised against β-adrenergic receptors stimulate adenylate cyclase. Biochem Biophys Res Commun 99:1295-1302

Couraud P-O, Lü B-Z, Strosberg AD (1983a) Cyclical antiidiotypic response to antihormone antibodies due to neutralization by autologous anti-antiidiotype antibodies that bind hormone. J Exp Med 157:1369-1378

Couraud P-O, Lü B-Z, Schmutz A, Durieu-Trautmann O, Klutchko-Dalavier C, Hoebeke J, Strosberg AD (1983b) Immunological studies of β-adrenergic receptors. J Cell Biochem 21:187-193

Coyle JT (1974) Development of the central catecholaminergic neurons. In: Schmitt FO, Warden FG (eds) The neurosciences, third study program. The MIT Press, Cambridge MA USA, pp 877-884

Coyle JT, Campochiaro P (1976) Ontogenesis of dopaminergic-cholinergic interactions in the rat striatum: a neurochemical study. J Neurochem 27:673-678

Coyle JT, Henry D (1973) Catecholamines in fetal and newborn rat brain. J Neurochem 21:61-67

Coyle JT, Schwarcz R (1976) Lesion of striatal neurons with kainic acid provides a model for huntington's chorea. Nature 263:244-246

Coyle JT, Snyder SH (1969) Catecholamine uptake by synaptosomes in homogenates of rat brain: stereospecificity in different areas. J Pharmacol Exp Ther 170:221-231

Coyle JT, Yamamura HI (1976) neurochemical aspects of the ontogenesis of cholinergic neurones in the rat brain. Brain Res 118:429-440

Creese I, Snyder SH (1978) Dopamine receptor binding of ^{3}H-ADTN (2-amino-6,7-dihydroxy-1,2,3,4-tetrahydronaphthalene) regulated by guanyl nucleotides. Eur J Pharmacol 50:459-461

Creese I, Burt DR, Snyder SH (1975) Dopamine receptor binding: differentiation of agonist and antagonist states with ^{3}H-dopamine and ^{3}H-haloperidol. Life Sci 17:993-1002

Creese I, Burt DR, Snyder SH (1977a) Dopamine receptor binding enhancement accompanies lesion-induced behavioral supersensitivity. Science 197:596-598

Creese I, Schneider R, Snyder SH (1977b) ^{3}H-spiroperidol labels dopamine receptors in pituitary and brain. Eur J Pharmacol 46:377-381

Creese I, Prosser T, Snyder SH (1978) Dopamine receptor binding: specificity, localization and regulation by ions and guanyl nucleotides. Life Sci 23:495-500

Cronin MJ, Roberts JM, Weiner RI (1978) Dopamine and dihydroergocryptine binding to the anterior pituitary and other brain areas of the rat and sheep. Endocrinol 103:302-309

Cross AJ, Owen F (1980) Characteristics of ^{3}H-cis-flupenthixol binding to calf brain membranes. Eur J Pharmacol 65:341-347

Cuatrecasas P, Tell GPE, Sica V, Parikh I, Chang K-J (1974) Noradrenaline binding and the search for catecholamine receptors. Nature 247:92-97

Daiguji M, Meltzer HY, U'Prichard DC (1981) Human platelet α_2-adrenergic receptors: labelling with ^{3}H-yohimbine, a selective antagonist ligand. Life Sci 28:2705-2717

Dale HH (1906) On some physiological actions of ergot. J Physiol (Lond) 34:163-206

Daly J (1975) Role of cyclic nucleotides in the nervous system. In: Iversen LL, Iversen SD, Snyder SH (eds) Handbook of Psychopharmacology, vol 5, Plenum Press, New York, pp 47-130

Darfler FJ, Marinetti GV (1977) Synthesis of a photoaffinity probe for the β-adrenergic receptor. Biochem Biophys Res Commun 79:1-7

Davies AO, Lefkowitz RJ (1981) Agonist-promoted high affinity state of the β-adrenergic receptor in human neutrophils: modulation by corticosteroids. J Clin Endocrinol Metab 53:703-708

Davis JN, Lefkowitz RJ (1976) β-Adrenergic receptor binding: synaptic localization in rat brain. Brain Res 113:214-218

Davis JN, Arnett CD, Hoyler E, Stalvey LP, Daly JW, Skolnick P (1978) Brain α-adrenergic receptors: comparison of (^{3}H)-WB 4101 binding with norepinephrine-stimulated cyclic AMP accumulation in rat cerebral cortex. Brain Res 159:125-135

Davoren PR, Sutherland EW (1963) The cellular location of adenyl cyclase in the pigeon erythrocyte. J Biol Chem 238:3016-3023

De Haën C (1976) The non-stoichiometric floating receptor model for hormone sensitive adenylyl cyclase. J Theoret Biol 58:383-400

De Lean A, Stadel JM, Lefkowitz RJ (1980) A ternary complex model explains the agonist-specific binding properties of the adenylate cyclase-coupled β-adrenergic receptor. J Biol Chem 255:7108-7117

De Vellis J, Brooker G (1974) Reversal of catecholamine refractoriness by inhibitors of RNA and protein synthesis. Science 186:1221-1223

Dismukes K, Daly JW (1974) Norepinephrine-sensitive systems generating adenosine 3',5'-monophosphate: Increased responses in cerebral cortical slices from reserpine-treated rats. Mol Pharmacol 10:933-940

Dismukes RK, Daly JW (1976) Adaptive responses of brain cyclic AMP-generating systems to alterations in synaptic input. J Cyclic Nucleotide Res 2:321-336

Dixon RAF, Kobilka BK, Strader DJ, Benovic JL, Dohlman HG, Frielle T, Bolanowski MA, Bennett CD, Rands E, Diehl RE, Mumford RA, Slater EE, Sigal IS, Caron

MG, Lefkowitz RJ, Strader CD (1986) Cloning of the gene and cDNA for mammalian β-adrenergic receptor and homology with rhodopsin. Nature 321:75-79
Dixon RAF, Sigal IS, Rands E, Register RB, Candelore MR, Blake AD, Strader CD (1987) Ligand binding to the β-adrenergic receptor involves its rhodopsin-like core. Nature 326:73-77
Dohlman HG, Caron MG, Lefkowitz RJ (1987) A family of receptors coupled to guanine nucleotide regulatory proteins. Biochemistry 26:2657-2664
Downes CP (1983) Inositol phospholipids and neurotransmitter signalling mechanisms. Trends in Neuroscience 6:313-316
Drummond GI, Hemmings SJ (1972) Inotropic and chronotropic effects of dibutyryl cyclic AMP. Adv Cyclic Nucleotide Res 1:307-316
Dumbrille-Ross A, Niznik H, Seemann P (1985) Separation of dopamine D1 and D2 receptors. Eur J Pharmacol 110:151-152
Dunlop D, Shanks RG (1968) Selective blockade of adrenoceptive β-receptors in the heart. Br J Pharmacol Chemother 32:201-218
Eadie GS (1952) On the evaluation of the constants V_m and K_M in enzyme reactions. Science 116:688
Edwards C, Nahorski SR, Rogers KJ (1974) *In vivo* changes of cerebral cyclic adenosine 3′,5′-monophosphate induced by biogenic amines: association with phosphorylase activation. J Neurochem 22:565-572
Eibergen RD, Carlson KR (1976) Behavioral evidence for dopaminergic supersensitivity following chronic treatment with methadone or chlorpromazine in the guinea pig. Psychopharmacol 48:139-146
Eimerl S, Neufeld G, Korner M, Schramm M (1980) Functional implantation of a solubilized β-adrenergic receptor in the membrane of a cell. Proc Natl Acad Sci USA 77:760-764
El-Refai MF, Exton JH (1980) Subclassification of two types of α-adrenergic binding sites in rat liver. Eur J Pharmacol 62:201-204
El-Refai MF, Blackmore PF, Exton JH (1979) Evidence for two α-adrenergic binding sites in liver plasma membranes. J Biol Chem 254:4375-4386
Engel G, Hoyer D (1981) (^{125}I)-BE 2254, a new high affinity radioligand for α_1 adrenoceptors. Eur J Pharmacol 73:221-224
Engel G, Hoyer D, Berhold R, Wagner H (1981) (±)(^{125}Iodo)-cyanopindolol, a new ligand for β-adrenoceptors: identification and quantitation of subclasses of β-adrenoceptors in guinea pig. Naunyn-Schmiedeberg's Arch Pharmacol 317:277-285
Exton JH, Harper SC (1975) Role of cyclic AMP in the actions of catecholamines on hepatic carbohydrate metabolism. Adv Cyclic Nucleotide Res 5:519-532
Exton JH, Friedmann N, Wong EH-A, Brineaux JP, Corbin JD, Park CR (1972) Interaction of glucocorticoids with glucagon and epinephrine in the control of gluconeogenesis and glycogenolysis in liver and of lypolysis in adipose tissue. J Biol Chem 247:3579-3588
Feltz P, DeChamplain J (1973) The postsynaptic effect of amphetamine on striatal dopamine-sensitive neurones. In: Usdin E, Snyder SH (eds) Frontiers in catecholamine research, Pergamon Press, New York, pp 951-956
Fields JZ, Reisine TD, Yamamura HI (1977) Biochemical demonstration of dopaminergic receptors in rat and human brain using (^{3}H)-spiroperidol. Brain Res 136:578-584
Fields JZ, Reisine TD, Yamamura HI (1978) Loss of striatal dopaminergic receptors after intrastriatal kainic acid injection. Life Sci 23:569-574
Fleming JW, Ross EM (1980) Reconstitution of β-adrenergic receptors into phospho-

lipid vesicles: restoration of (^{125}I)-iodohydroxybenzylpindolol binding to digitonin-solubilized receptors. J Cyclic Nucleotide Res 6:407–419
Fleming WW, Westfall DP (1975) Altered resting membrane potential in the supersensitive vas deferens of the guinea pig. J Pharmacol Exp Ther 192:381–389
Fleming WW, McPhillips JJ, Westfall DP (1973) Postjunctional supersensitivity and subsensitivity of excitable tissues to drugs. Rev Physiol Biochem Pharmacol 68:55–119
Franklin TJ, Morris WP, Twose PA (1975) Desensitization of *beta* adrenergic receptors in human fibroblasts in tissue culture. Mol Pharmacol 11:485–491
Frishman WH (1983) Pindolol: A new β-adrenoceptor antagonist with partial agonist activity. N. Engl J Med 308:940–944
Furchgott RF (1972) The classification of adrenoceptors (adrenergic receptors). An evaluation from the standpoint of receptor theory. In: Blaschko H, Muscholl E (eds) *Catecholamines*, Springer, Berlin Heidelberg New York, pp 283–335 (Handbook of Experimental Pharmacology, vol 33)
Gale K, Guidotti A, Costa E (1977) Dopamine-sensitive adenylate cyclase: location in substantia nigra. Science 195:503–505
Garau L, Govoni S, Stefanini E, Trabucchi M, Spano PF (1978) Dopamine receptors: pharmacological and anatomical evidences indicate that two distinct dopamine receptor populations are present in rat striatum. Life Sci 23:1745–1750
García-Sáinz JA, Hoffman BB, Li S-Y, Lefkowitz RJ, Fain JN (1980) Role of alpha$_1$ adrenoceptors in the turnover of phosphatidylinositol and of alpha$_2$ adrenoceptors in the regulation of cyclic AMP accumulation in hamster adipocytes. Life Sci 27:953–961
García-Sáinz JA, Litosch I, Hoffman BB, Lefkowitz RJ, Fain JN (1981) Effect of thyroid status on α- and β-catecholamine responsiveness of hamster adipocytes. Biochim Biophys Acta 678:334–341
Gianutsos G, Drawbaugh RB, Hynes MD, Lal H (1974) Behavioral evidence for dopaminergic supersensitivity after chronic haloperidol. Life Sci 14:887–898
Gibson RE, Rzeszotarski WJ, Komai T, Reba RC, Eckelman WC (1979) Evaluation of beta adrenoceptor antagonist affinity and cardioselectivity by radioligand-receptor assay. J Pharmacol Exp Ther 209:153–161
Gilman AG, Nirenberg M (1971) Effect of catecholamines on the adenosine 3′:5′-cyclic monophosphate concentrations of clonal satellite cells of neurons. Proc Natl Acad Sci USA 68:2165–2168
Glossmann H, Hornung R (1980) *Alpha$_2$* adrenoceptors in rat brain. The divalent cation site. Naunyn-Schmiedeberg's Arch Pharmacol 314:101–109
Goldberg LL, Volkman PH, Kohli JD (1978) A comparison of the vascular dopamine receptor with other dopamine receptors. Ann Rev Pharmacol Toxicol 18:57–79
Goldberg ND, O'Dea RF, Haddox MK (1973) Cyclic GMP. Adv Cyclic Nucleotide Res 3:155–223
Green DA, Clark RB (1981) Adenylate cyclase coupling proteins are not essential for agonist-specific desensitization of lymphoma cells. J Biol Chem 256:2105–2108
Greenberg DA, Snyder SH (1978) Pharmacological properties of (^{3}H)dihydroergokryptine binding sites associated with *alpha* noradrenergic receptors in rat brain membranes. Mol Pharmacol 14:38–49
Greenberg DA, U'Prichard DC, Sheehan P, Snyder SH (1978) α-Noradrenergic receptors in the brain: differential effects of sodium on binding of (^{3}H)agonists and (^{3}H)antagonists. Brain Res 14:378–384
Greengard P (1976) The action of antipsychotic drugs on dopamine-stimulated adeny-

late cyclase activity. In: Sedvall G, Uvnäs B, Zotterman Y (eds) *Antipsychotic drugs: pharmacodynamics and pharmacokinetics* Pergamon Press, New York, pp 271-283

Greengrass P, Bremner R (1979) Binding characteristics of ^{3}H-prazosin to rat brain α-adrenergic receptors. Eur J Pharmacol 55:323-326

Guellaen G, Yates-Aggerbeck M, Vauquelin G, Strosberg D, Hanoune J (1978) Characterization with (^{3}H)dihydroergocryptine of the α-adrenergic receptor of the hepatic plasma membrane. Comparison with the β-adrenergic receptor in normal and adrenalectomized rats. J Biol Chem 253:1114-1120

Guillet J-G, Chamat S, Hoebeke J, Strosberg AD (1984) Production and detection of monoclonal anti-idiotype antibodies directed against a monoclonal anti-β-adrenergic ligand antibody. J Immunol Methods 74:163-171

Guillet J-G, Kaveri SV, Durieu O, Delavier C, Hoebeke J, Strosberg AD (1985) β-Adrenergic agonist activity of a monoclonal anti-idiotypic antibody. Proc Natl Acad Sci USA 82:1781-1784

Guimaraes S, Paiva MQ (1981a) Two distinct adrenoceptor-biophases in the vasculature: one for α- and the other for β-agonists. Naunyn-Schmiedeberg's Arch Pharmacol 316:195-199

Guimaraes S, Paiva MQ (1981b) Two different biophases for adrenaline released by electrical stimulation or tyramine from the sympathetic nerve endings of the dog saphenous vein. Naunyn-Schmiedeberg's Arch Pharmacol 316:200-204

Haga T, Ross EM, Anderson HJ, Gilman AG (1977) Adenylate cyclase permanently uncoupled from hormone receptors in a novel variant of S49 mouse lymphoma cells. Proc Natl Acad Sci USA 74:2016-2020

Hall JM, Frankham PA, Strange PG (1983) Use of cholate/sodium chloride for solubilisation of brain D2 dopamine receptors. J Neurochem 41:1526-1532

Hall ZW (1973) Multiple forms of acetylcholinesterase and their distribution in endplate and non-endplate regions of rat diaphragm muscle. J Neurobiol 4:343-361

Hamblin M, Creese I (1982) ^{3}H-dopamine binding to rat striatal D-2 and D-3 sites: enhancement by magnesium and inhibition by sodium. Life Sci 30:1587-1595

Hamprecht B (1974) Cell cultures as model systems for studying the biochemistry of differentiated functions of nerve cells. Jaenicke L (ed): *Biochemistry of Sensory Functions*, Springer-Verlag, New York, pp 391-423

Hancock AA, DeLean AL, Lefkowitz RJ (1979) Quantitative resolution of *beta*-adrenergic receptor subtypes by selective ligand binding: application of a computerized model fitting technique. Mol Pharmacol 16:1-9

Harden TK (1983) Agonist induced desensitization of the β-adrenergic receptor-linked adenylate cyclase. Pharmacol Rev 35:5-32

Harden TK, Cotton CU, Waldo GL, Lutton JK Perkins JP (1980). Catecholamine-induced alteration in sedimentation behavior of membrane bound β-adrenergic receptors. Science 210:441-443

Harden TK, McCarthy KD (1982) Identification of the beta adrenergic receptor subtype on astroglia purified from rat brain. J Pharmacol Exp Ther 222:600-605

Harden TK, Wolfe BB, Molinoff PB (1976) Binding of iodinated *beta* adrenergic antagonists to proteins derived from rat heart. Mol Pharmacol 12:1-15

Harden TK, Wolfe BB, Sporn JR, Perkins JP, Molinoff PB (1977a) Ontogeny of β-adrenergic receptors in rat cerebral cortex. Brain Res 125:99-108

Harden TK, Wolfe BB, Sporn JR, Poulos BK, Molinoff PB (1977b) Effects of 6-hydroxydopamine on the development of the *beta* adrenergic receptor/adenylate cyclase system in rat cerebral cortex. J Pharmacol Exp Ther 203:132-143

Harden TK, Cotton CU, Waldo GL, Lutton JK, Perkins JP (1980) Catecholamine-induced alteration in sedimentation behavior of membrane bound β-adrenergic receptors. Science 210:441-443

Harms HH (1976) Isoproterenol antagonism of cardioselective *beta* adrenergic receptor blocking agents: a comparative study of human and guinea-pig cardiac and bronchial *beta* adrenergic receptors. J Pharmacol Exp Ther 199:329-335

Harper JF, Brooker G (1979) Refractoriness to muscarinic and adrenergic agonists in the rat parotid: responses of adenosine and guanosine cyclic 3′,5′-monophosphates. Mol Pharmacol 13:1048-1059

Hedberg A, Mattsson H (1981) *Beta* adrenoceptor interaction of full and partial agonists in the cat heart and soleus muscle. J Pharmacol Exp Ther 219:798-808

Hedberg A, Minneman KP, Molinoff PB (1980) Differential distribution of *beta*-1 and *beta*-2 adrenergic receptors in cat and guinea-pig heart. J Pharmacol Exp Ther. 212:503-508

Hedberg A, Gerber JG, Nies AS, Wolfe BB, Molinoff PB (1986) Effects of pindolol and propranolol on beta adrenergic receptors on human lymphocytes. J Pharmacol Exp Ther 239:117-123

Hegstrand LR, Minneman KP, Molinoff PB (1979) Multiple effects of guanosine triphosphate on *beta* adrenergic receptors and adenylate cyclase activity in rat heart, lung and brain. J Pharmacol Exp Ther 210:215-221

Heidenreich KA, Weiland GA, Molinoff PB (1980) Characterization of radiolabeled agonist binding to β-adrenergic receptors in mammalian tissues. J Cyclic Nucleotide Res 6:217-230

Heidenreich KA, Weiland GA, Molinoff PB (1982) Effects of magnesium and *N*-ethylmaleimide on the binding of ^{3}H-hydroxybenzylisoproterenol to β-adrenergic receptors. J Biol Chem 257:804-810

Hertel C, Muller P, Portenier M, Staehelin M (1983a) Determination of the desensitization of β-adrenergic receptors by [^{3}H]-CGP-12177. Biochem J 216:669-674

Hertel C, Staehelin M, Perkins JP (1983b) Evidence for intracellular β-adrenergic receptors in membrane preparations from desensitized cells: binding of the hydrophilic ligand CGP-12177 only in the presence of alamethicin. J Cyclic Nucleotide Protein Phosphor Res 9:119-128

Hess A (1979) Visualization of beta-adrenergic receptor sites with fluorescent beta-adrenergic blocker probes—or autofluorescent granules? Brain Res 160:533-538

Hoffer BJ, Siggins GR, Bloom FE (1971) Studies on norepinephrine-containing afferents to purkinje cells of rat cerebellum. II. Sensitivity of purkinje cells to norepinephrine and related substances administered by microiontophoresis. Brain Res 25:523-534

Hoffman BB, Mullikin-Kilpatrick D, Lefkowitz RJ (1980) Heterogeneity of radioligand binding to α-adrenergic receptors. Analysis of guanine nucleotide regulation of agonist binding in relation to receptor subtypes. J Biol Chem 255:4645-4652

Hofstee HJ (1952) On the evaluation of the constants V_m and K_m in enzyme reactions. Science 116:329-331

Homburger V, Lucas M, Rosenbaum E, Vassent G, Bockaert J (1981) Presence of both $beta_1$ and $beta_2$-adrenergic receptors in a single cell type. Mol Pharmacol 20:463-469

Howlett DR, Nahorski SR (1978) Effect of acute and chronic amphetamine administration on β-adrenoceptors and dopamine receptors in rat corpus striatum and limbic forebrain. Br J Pharmacol 64:411P-412P

Howlett AC, Van Arsdale PM, Gilman AG (1978) Efficiency of coupling between the beta adrenergic receptor and adenylate cyclase. Mol Pharmacol 14:531-539

Hoyer D, Reynolds EE, Molinoff PB (1984) Agonist-induced changes in the properties of β-adrenergic receptors on intact S49 lymphoma cells: time-dependent changes in the affinity of the receptor for agonists. Mol Pharmacol 25:209-218

Hu EH, Venter JC (1978) Adenosine cyclic 3',5'-monophosphate concentrations during the positive inotropic response of cat cardiac muscle to polymeric immobilized isoproterenol. Mol Pharmacol 14:237-245

Huang M, Ho AKS, Daly JW (1973) Accumulation of adenosine cyclic 3',5'-monophosphate in rat cerebral cortical slices: stimulatory effect of *alpha* and *beta* adrenergic agents after treatment with 6-hydroxydopamine, 2,3,5-trihydroxyphenethylamine, and dihydroxytryptamines. Mol Pharmacol 9:711-717

Huff RM, Molinoff PB (1981) Effects of N-ethylmaleimide on ^{3}H-spiroperidol binding to rat striatal membranes. Soc Neurosci Abs 7:125

Huff RM, Molinoff PB (1982) Quantitative determination of dopamine receptor subtypes not linked to activation of adenylate cyclase in rat striatum. Proc Natl Acad Sci USA 79:7561-7565

Hug CC (1972) Characteristics and theories related to acute and chronic tolerance development. In: Mule SJ, Brill H (eds): *chemical and biological aspects of drug dependence*, CRC Press, Cleveland, pp 307-358

Innis RB, Correa FMA, Snyder SH (1979) Carazolol, an extremely potent β-adrenergic blocker: binding to β-receptors in brain membranes. Life Sci 24:2255-2264

Insel PA, Stoolman LM (1978) Radioligand binding to *beta* adrenergic receptors of intact cultured S49 cells. Mol Pharmacol 14:549-561

Iversen LL (1967) The uptake and storage of noradrenaline in sympathetic nerves, Cambridge University Press, Cambridge, U.K.

Iversen LL, Horn AS, Miller RJ (1976) Structure-activity relationships for interactions of agonist and antagonist drugs with dopamine-stimulated adenylate cyclase of rat brain—A model for CNS dopamine receptors? In: Sedvall G, Uvnäs B, Zotterman Y (eds) *Antipsychotic drugs: pharmacodynamics and pharmacokinetics*, Pergamon Press, New York pp 285-303

Iyengar R, Bhat MK, Riser ME, Birnbaumer L (1981) Receptor-specific desensitization of the S49 lymphoma cell adenylyl cyclase: unaltered behavior of the regulatory component. J Biol Chem 256:4810-4815

Jacobs S, Cuatrecasas P (1976) The mobile receptor hypothesis and "cooperativity" of hormone binding: application to insulin. Biochim Biophys Acta 433:482-495

Jakobs KH, Saur W, Schultz G (1976) Reduction of adenylate cyclase activity in lysates of human platelets by the alpha-adrenergic component of epinephrine. J Cyclic Nucleotide Res 2:381-392

Jakobs KH, Saur W, Schultz G (1978) Inhibition of platelet adenylate cyclase by epinephrine requires GTP. FEBS Lett 85:167-170

Janssen PAJ (1965) The evolution of the butyrophenones, haloperidol and trifluperidol, from meperidine-like 4-phenylpiperidines. Int Rev. Neurobiol 8:221-263

Janssen PAJ (1967) The pharmacology of haloperidol. Int J Neuropsychiatry 3 (Suppl 1):10-18

Jard S, Cantau B, Jakobs KH (1981) Angiotensin II and α-adrenergic agonists inhibit rat liver adenylate cyclase. J Biol Chem 256:2603-2606

Johnson GL, Wolfe BB, Harden TK, Molinoff PB, Perkins JP (1978) Role of β-adrenergic receptors in catecholamine-induced desensitization of adenylate cyclase in human astrocytoma cells. J Biol Chem 253:1472-1480

Joseph SK, Thomas AP, Williams RJ, Irvine RF, Williamson JR (1984) *Myo*-inositol 1,4,5-trisphosphate. A second messenger for the hormonal mobilization of intracellular Ca^{2+} in liver. J Biol Chem 259:3077-3081

Kahn DJ, Mitrius JC, U'Prichard DC (1982) *Alpha*$_2$-adrenergic receptors in neuroblastoma X glioma hybrid cells. Mol Pharmacol 21:17-26

Kakiuchi S, Rall TW (1968) The influence of chemical agents on the accumulation of adenosine 3′,5′-phosphate in slices of rabbit cerebellum. Mol Pharmacol 4:367-378

Kalisker A, Rutledge CO, Perkins JP (1973) Effect of nerve degeneration by 6-hydroxydopamine on catecholamine-stimulated adenosine 3′,5′-monophosphate formation in rat cerebral cortex. Mol Pharmacol 9:619-629

Kassis S, Zaremba T, Patel J, Fishman PH (1985) Phorbol esters and β-adrenergic agonists mediate desensitization of adenylate cyclase in rat glioma C6 cells by distinct mechanisms. J Biol Chem 260:8911-8917

Kassis S, Olasmaa M, Sullivan M, Fishman PH (1986) Desensitization of β-adrenergic receptor-coupled adenylate cyclase in cultured mammalian cells: receptor sequestration versus receptor function. J Biol Chem 261:12233-12237

Kaumann AJ, Birnbaumer L (1974) Studies on receptor-mediated activation of adenylyl cyclases. IV. Characteristics of the adrenergic receptor coupled to myocardial adenylyl cyclase: stereospecificity for ligands and determination of apparent affinity constants for β-blockers. J Biol Chem 249:7874-7885

Kaumann AJ, Blinks JR (1980) Stimulant and depressant effects of β-adrenoceptor blocking agents on isolated heart muscle. Naunyn-Schmiedeberg's Arch Pharmacol 311:205-218

Kauzmann W (1959) Some factors in the interpretation of protein denaturation. Adv Prot Chem 14:1-63

Kayaalp SO, Neff NH (1980) Differentiation by ascorbic acid of dopamine agonist and antagonist binding sites in striatum. Life Sci 26:1837-1841

Kebabian JW (1978) Multiple classes of dopamine receptors in mammalian central nervous system: the involvement of dopamine-sensitive adenylyl cyclase. Life Sci 23:479-484

Kebabian JW, Calne DB (1979) Multiple receptors for dopamine. Nature 277:93-96

Kebabian JW, Greengard P (1971) Dopamine-sensitive adenyl cyclase: possible role in synaptic transmission. Science 174:1346-1349

Kebabian JW, Kebabian PR (1978) Lergotrile and Lisuride: *in vivo* dopaminergic agonists which do not stimulate the presynaptic dopamine autoreceptor. Life Sci 23:2199-2204

Kebabian JW, Petzold GL, Greengard P (1972) Dopamine-sensitive adenylate cyclase in caudate nucleus of rat brain, and its similarity to the "dopamine receptor". Proc Natl Acad Sci USA 69:2145-2149

Kebabian JW, Bloom FE, Steiner AL, Greengard P (1975a) Neurotransmitters increase cyclic nucleotides in postganglionic neurons: immunocytochemical demonstration. Science 190:157-159

Kebabian JW, Zatz M, Romero JA, Axelrod J (1975 b) Rapid changes in rat pineal β-adrenergic receptor: alterations in l-(^{3}H)alprenolol binding and adenylate cyclase. Proc Natl Acad Sci USA 72:3735-3739

Kelly PH, Miller RJ (1975) The interaction of neuroleptic and muscarinic agents with central dopaminergic systems. Br J Pharmacol 54:115-121

Kilpatrick BF, Caron MG (1983) Agonist binding promotes a guanine nucleotide reversible increase in the apparent size of the bovine anterior pituitary dopamine receptors. J Biol Chem 258:13528-13532

Kilpatrick BF, Caron MG (1984) Dopamine receptor of the anterior pituitary gland: solubilization and characterization. Biochem Pharmacol 33:1981-1988

Kishimoto A, Takai Y, Mori T, Kikkawa U, Nishizuka Y (1980) Activation of calcium and phospholipid-dependent protein kinase by diacylglycerol, its possible relation to phosphatidylinositol turnover. J Biol Chem 255:2273-2276

Klemm N, Murrin LC, Kuhar MJ (1979) Neuroleptic and dopamine receptors: autoradiographic localization of (^{3}H)-spiperone in rat brain. Brain Res 169:1-9

Klotz IM, Urquhart JM (1949) The binding of organic ions by proteins. Effect of temperature. J Am Chem Soc 71:847-851

Kobilka BK, Dixon RAF, Frielle T, Dohlman HG, Bolanowski MA, Sigal IS, Yang-Feng TL, Francke U, Caron MG, Lefkowitz RJ (1987a) cDNA for human β_2-adrenergic receptor: a protein with multiple membrane-spanning domains and encoded by a gene whose chromosomal location is shared with that of the receptor for platelet-derived growth factor. Proc Natl Acad Sci USA 84:46-50

Kobilka BK, Matsui H, Kobilka TS, Yang-Feng TL, Francke U, Caron MG, Lefkowitz RJ, Regan JW (1987b) Cloning, sequencing, and expression of the gene coding for the human platelet α_2-adrenergic receptor. Science 238:650-656

Korner M, Gilon C, Schramm M (1982) Locking of hormone in the β-adrenergic receptor by attack on a sulfhydryl in an associated component. J Biol Chem 257:3389-3396

Krueger BK, Forn J, Walters JR, Roth RH, Greengard P (1976) Stimulation by dopamine of adenosine cyclic 3′,5′-monophosphate formation in rat caudate nucleus: effect of lesions of the nigro-neostriatal pathway. Mol Pharmacol 12:639-648

Kubo T, Fukuda K, Mikami A, Maeda A, Takahashi H, Mishina M, Haga T, Haga K, Ichiyama A, Kangawa K, Kojima M, Matsuo H, Hirose T, Numa S (1986a) Cloning, sequencing, and expression of complementary DNA encoding the muscarinic acetylcholine receptor. Nature 323:411-416

Kubo T, Maeda A, Sugimoto K, Akiba I, Mikami A, Takahashi H, Haga T, Haga K, Ichiyama A, Kangawa K, Matsuo H, Hirose T, Numa S (1986b) Primary structure of porcine cardiac muscarinic acetylcholine receptor deduced from the cDNA sequence. FEBS Lett 209:367-372

Kuffler SW, Dennis MJ, Harris AJ (1971) The development of chemosensitivity in extrasynaptic areas of the neuronal surface after denervation of parasympathetic ganglion cells in the heart of the frog. Proc R Soc Lond (Biol) 177:555-563

Kukovetz WR, Pöch G, Wurm A (1975) Quantitiative relations between cyclic AMP and contraction as affected by stimulators of adenylate cyclase and inhibitors of phosphodiesterase. Adv Cyclic Nucleotide Res 5:395-414

Laduron PM, Ilien B (1982) Solubilization of brain muscarinic, dopaminergic and serotonergic receptors: critical analysis. Biochem Pharmacol 31:2145-2151

Lamprecht F, Coyle JT (1972) DOPA Decarboxylase in the developing rat brain. Brain Res 41:503-506

Lands AM, Groblewski GE, Brown TG Jr (1966) Comparison of the action of isoproterenol and several related compounds on blood pressure, heart, and bronchioles. Arch Int Pharmacodyn Ther 161:68-75

Lands AM, Arnold A, McAuliff JP, Luduena FP, Brown TG Jr (1967) Differentiation of receptor systems activated by sympathomimetic amines. Nature 214:597-598

Langer SZ (1974) Presynaptic regulation of catecholamine release. Biochem Pharmacol 23:1793-1800

Langer SZ (1977) Presynaptic receptors and their role in the regulation of transmitter release. Br J Pharmacol 60:481-497

Lavin TN, Heald SJ, Jeffs PW, Shorr RGL, Lefkowitz RJ, Caron MG (1981) Photoaffinity labeling of the β-adrenergic receptor. J Biol Chem 256:11944-11950

Leff SE, Creese I (1982) Solubilization of D-2 dopamine receptors from canine caudate: agonist-occupation stabilizes guanine nucleotide sensitive receptor complexes. Biochem Biophys Res Commun 108:1150-1157

Leff SE, Creese I (1983) Dopaminergic D-3 binding sites are not presynaptic autoreceptors. Nature 306:586-589

Leff S, Sibley DR, Hamblin M, Creese I (1981) Ascorbic acid enables reversible dopamine receptor ^{3}H-agonist binding. Life Sci 29:2081-2090

Lefkowitz RJ (1975) Catecholamine stimulated myocardial adenylate cyclase: effects of phospholipase digestion and the role of membrane lipids. J Mol Cell Cardiol 7:27-37

Lefkowitz RJ, Williams LT (1977) Catecholamine binding to the β-adrenergic receptor. Proc Natl Acad Sci USA 74:515-519

Lefkowitz RJ, Mukherjee C, Coverstone M, Caron MG (1974) Stereospecific (^{3}H) (−)-alprenolol binding sites, β-adrenergic receptors and adenylate cyclase. Biochem Biophys Res Comm 60:703-709

Lefkowitz RJ, Limbird LE, Mukherjee C, Caron MG (1976a) The β-adrenergic receptor and adenylate cyclase. Biochim Biophys Acta 457:1-39

Lefkowitz RJ, Mullikin D, Caron MG (1976b) Regulation of β-adrenergic receptors by guanyl-5′-yl imidodiphosphate and other purine nucleotides. J Biol Chem 251:4686-4692

Leray F, Chambaut A-M, Perrenoud M-L, Hanoune J (1973) Adenylate-cyclase activity of rat-liver plasma membranes: hormonal stimulations and effect of adrenalectomy. Eur J Biochem 38:185-192

Leslie FM, Dunlap CE, III Cox BM (1980) Ascorbate decreases ligand binding to neurotransmitter receptors. J Neurochem 34:219-221

Levitzki A, Atlas D, Steer ML (1974) The binding characteristics and number of β-adrenergic receptors on the turkey erythrocyte. Proc Natl Acad Sci USA 71:2773-2776

Levitzki A, Sevilia N, Atlas D, Steer ML (1975) Ligand specificity and characteristics of the β-adrenergic receptor in turkey erythrocyte plasma membranes. J Mol Biol 97:35-53

Levy B (1966) The adrenergic blocking activity of N-*tert*-butylmethoxamine (butoxamine). J Pharmacol Exp Ther 151:413-422

Lew JY, Goldstein M (1978) Solubilization and characterization of striatal dopamine receptors. J Neurochem 42:1298-1305

Leysen J, Laduron P (1977) Differential distribution of opiate and neuroleptic receptors and dopamine-sensitive adenylate cyclase in rat brain. Life Sci 20:281-288

Leysen JE, Gommeren W, Laduron PM (1978a) Spiperone: A ligand of choice for neuroleptic receptors. 1. Kinetics and characteristics of *in vitro* binding. Biochem Pharmacol 27:307-316

Leysen JE, Niemegeers CJE, Tollenaere JP, Laduron PM (1978b) Serotonergic component of neuroleptic receptors. Nature 272:168-171

Leysen JE, Niemegeers CJE, Van Nueten JM, Laduron PM (1982) (^{3}H)ketanserin (R41 468), a selective ^{3}H-ligand for serotonin$_2$ receptor binding sites. Mol Pharmacol 21:301-314

Limbird LE, Lefkowitz RJ (1976) Adenylate cyclase-coupled *beta* adrenergic receptors: effect of membrane lipid-perturbing agents on receptor binding and enzyme stimulation by catecholamines. Mol Pharmacol 12:559-567

Limbird LE and Lefkowitz RJ (1978) Agonist-induced increase in apparent β-adrenergic receptor size. Proc Natl Acad Sci USA 75:228-232

Limbird LE, Gill DM, Lefkowitz RJ (1980) Agonist-promoted coupling of the β-adrenergic receptor with the guanine nucleotide regulatory protein of the adenylate cyclase system. Proc Natl Acad Sci USA 77:775-779

Loizou LA (1972) The postnatal ontogeny of monoamine-containing neurones in the central nervous system of the albino rat. Brain Res 40:395–418

Lucas M, Hanoune J, Bockaert J (1978) Chemical modification of the *beta*-adrenergic receptors coupled with adenylate cyclase by disulfide bridge-reducing agents. Mol Pharmacol 14:227-236

Luedtke R, Molinoff PB (1987) Characterization of binding sites for ^{3}H-spiroperidol. Biochem Pharmacol 36:3255-3264

Lyon TF, Randall WC (1980) Multiple Central WB-4101 Binding sites and the selectivity of prazosin. Life Sci 26:1121-1129

Maelicke A, Fulpius BW, Klett RP, Reich E (1977) Acetylcholine receptor: responses to drug binding. J Biol Chem 252:4811-4830

Magleby KL, Stevens CF (1972a) The effect of voltage on the time course of end-plate currents. J Physiol (Lond) 223:151-172

Magleby KL, Stevens CF (1972b) A quantitative description of end-plate currents. J Physiol (Lond) 223:173-197

Maguire ME, Erdös JJ (1980) Inhibition of magnesium uptake by β-adrenergic agonists and prostaglandin E_1 is not mediated by cyclic AMP. J Biol Chem 255:1030-1035

Maguire ME, Goldman PH, Gilman AG (1974) The reaction of (^{3}H)norepinephrine with particulate fractions of cells responsive to catecholamines. Mol Pharmacol 10:563-581

Maguire ME, Van Arsdale PM, Gilman AG (1976a) An agonist-specific effect of guanine nucleotides on binding to the *beta* adrenergic receptor. Mol Pharmacol 12:335-339

Maguire ME, Wiklund RA, Anderson HJ, Gilman AG (1976b) Binding of (^{125}I)iodohydroxybenzylpindolol to putative β-adrenergic receptors of rat glioma cells and other cell clones. J Biol Chem 251:1221-1231

Maguire ME, Ross EM, Gilman AG (1977) β-Adrenergic receptor: ligand binding properties and the interaction with adenylyl cyclase. Adv Cyclic Nucleotide Res 8:1-83

Makman MH, Brown JH, Mishra RK (1975) Cyclic AMP in retina and caudate nucleus: influence of dopamine and other agents. Adv Cyclic Nucleotide Res 5:661-679

Malbon CC (1980) The effects of thyroid status on the modulation of fat cell β-adrenergic receptor agonist affinity by guanine nucleotides. Mol Pharmacol 18:193-198

Malbon CC, Moreno FJ, Cabelli RJ, Fain JN (1978) Fat cell adenylate cyclase and β-adrenergic receptors in altered thyroid states. J Biol Chem 253:671-678

Malchoff CD, Marinetti GV (1976) Hormone action at the membrane level. V. Binding of ($\pm$)-(^{3}H)isoproterenol to intact chicken erythrocytes and erythrocyte ghosts. Biochim Biophys Acta 436:45-52

Mano K, Akbarzadeh A, Townley RG (1979) Effect of hydrocortisone on beta-adrenergic receptors in lung membranes. Life Sci 25:1925-1930

Martres M-P, Baudry M, Schwartz J-C (1978) Characterization of ^{3}H-domperidone binding on striatal dopamine receptors. Life Sci 23:1781-1784

McConnell MG, Simpson LL (1976) The role of acetylcholine receptors and acetylcholinesterase activity in the development of denervation supersensitivity. J Pharmacol Exp Ther 198:507-517

McGeer EG, McGeer PL (1970) Duplication of biochemical changes of huntington's chorea by intrastriatal injections of glutamic and kainic acids. Nature 263:517-519

McKenzie GM, Szerb JC (1968) The effect of dihydroxyphenylalanine, pheniprazine and dextroamphetamine on the *in vivo* release of dopamine from the caudate nucleus. J Pharmacol Exp Ther 162:302-308

McManus C, Hartley EJ, Seeman P (1978) Increased binding of (^{3}H)apomorphine in caudate membranes after dopamine pretreatment *in vitro*. J Pharm Pharmac 30:444-447

Melamed E, Lahav M, Atlas D (1976) Direct localisation of β-adrenoceptor sites in rat cerebellum by a new fluorescent analogue of propranolol. Nature 261:420-422
Meltzer HY, Fang VS, Simonovich M, Paul SM (1977) Effect of metabolites of chlorpromazine on plasma prolactin levels in male rats. Eur J Pharmacol 41:431-436
Michell RH (1975) Inositol phospholipids and cell surface receptor function. Biochim Biophys Acta 415:81-147
Mickey J, Tate R, Lefkowitz RJ (1975) Subsensitivity of adenylate cyclase and decreased β-adrenergic receptor binding after chronic exposure to (−)-isoproterenol *in vitro*. J Biol Chem 250:5727-5729
Mickey J, Tate R, Mullikin D, Lefkowitz RJ (1976) Regulation of adenylate cyclase-coupled *beta* adrenergic receptor binding sites by *beta* adrenergic catecholamines *in vitro*. Mol Pharmacol 12:409-419
Miledi R (1960) The acetylcholine sensitivity of frog muscle fibres after complete or partial denervation. J Physiol (Lond) 151:1-23
Miledi R, Molinoff P, Potter LT (1971) Isolation of the cholinergic receptor protein of *torpedo* electric tissue. Nature 229:554-557
Miledi R, Potter LT (1971) Acetylcholine receptors in muscle fibres. Nature 233:599-603
Minneman KP, Iversen LL (1976) Diurnal rhythm in rat pineal cyclic nucleotide phosphodiesterase activity. Nature 260:59-61
Minneman KP, Molinoff PB (1980) Classification and quantitation of β-adrenergic receptor subtypes. Biochem Pharmacol 29:1317-1323
Minneman KP, Quik M, Emson PC (1978) Receptor-linked cyclic AMP systems in rat neostriatum: differential localization revealed by kainic acid injection. Brain Res 151:507-521
Minneman KP, Dibner MD, Wolfe BB, Molinoff PB (1979a) β_1- and β_2-Adrenergic receptors in rat cerebral cortex are independently regulated. Science 204:866-868
Minneman KP, Hedberg A, Molinoff PB (1979b) Comparison of *beta* adrenergic receptor subtypes in mammalian tissues. J Pharmacol Exp Ther 211:502-508
Minneman KP, Hegstrand LR, Molinoff PB (1979c) The pharmacological specificity of *beta*-1 and *beta*-2 adrenergic receptors in rat heart and lung *In Vitro*. Mol Pharmacol 16:21-33
Minneman KP, Hegstrand LR, Molinoff PB (1979d) Simultaneous determination of *beta*-1 and *beta*-2 adrenergic receptors in tissues containing both receptor subtypes. Mol Pharmacol 16:34-36
Minneman KP, Weiland GA, Molinoff PB (1980) A comparison of the beta-adrenergic receptor of the turkey erythrocyte with mammalian *beta*$_1$ and *beta*$_2$ receptors. Mol Pharmacol 17:1-7
Minneman KP, Pittman RN, Molinoff PB (1981a) β-Adrenergic receptor subtypes: properties, distribution and regulation. Ann Rev Neurosci 4:419-461
Minneman KP, Pittman RN, Yeh HH, Woodward DJ, Wolfe BB, Molinoff PB (1981 b) Selective survival of β_1-adrenergic receptors in rat cerebellum following neonatal x-irradiation. Brain Res 209:25-34
Minneman KP, Wolfe BB, Molinoff PB (1982) Selective changes in the density of β_1-adrenergic receptors in rat striatum following chronic drug treatment and adrenalectomy. Brain Res 252:309-314
Mishra RK, Gardner EL, Katzman R, Makman MH (1974) Enhancement of dopamine-stimulated adenylate cyclase activity in rat caudate after lesions in substantia nigra: evidence for denervation supersensitivity. Proc Natl Acad Sci USA 71:3883-3887
Mobley PL, Smith HA, Sulser F (1977) Modification of the noradrenergic cyclic AMP

generating system in the rat limbic forebrain by amphetamine (A) and its hydroxylated metabolites. Fed Proc 36:319

Molinoff PB (1973) Methods of approach for the isolation of β-adrenergic receptors. In: Usdin E, Snyder SH (eds): *Frontiers in catecholamine research*, Pergamon Press, New York, pp 357-360

Molinoff PB, Sporn JR, Wolfe BB, Harden TK (1978) Regulation of β-adrenergic receptors in the cerebral cortex. Adv Cyclic Nucleotide Res 9:465-483

Molinoff PB, Wolfe BB, Weiland GA (1981) Quantitative analysis of drug-receptor interactions: II. Determination of the properties of receptor subtypes. Life Sci 29:427-443

Moore KE, Thornburg JE (1975) Drug-induced dopaminergic supersensitivity. Adv Neurol 9:93-104

Mowles TF, Burghardt B, Burghardt C, Charnecki A, Sheppard H (1978) The dopamine receptor of the rat mammotroph in cell culture as a model for drug action. Life Sci 22:2103-2112

Moxham CP, George ST, Graziano MP, Brandwein HJ, Malbon CC (1986) Mammalian β_1- and β_2-Adrenergic receptors: immunological and structural comparisons J Biol Chem 261: 14562-14570

Mukherjee C, Lefkowitz RJ (1976) Desensitization of β-adrenergic receptors by β-adrenergic agonists in a cell-free system: resensitization by guanosine 5′-(β, γ-imino) triphosphate and other purine nucleotides. Proc Natl Acad Sci USA 73:1494-1498

Mukherjee C, Lefkowitz RJ (1977) Regulation of *beta*-adrenergic receptors in isolated frog erythrocyte plasma membranes. Mol Pharmacol 13:291-303

Mukherjee C, Caron MG, Lefkowitz RJ (1975) Catecholamine-induced subsensitivity of adenylate cyclase associated with loss of β-adrenergic receptor binding sites. Proc Natl Acad Sci USA 72:1945-1949

Mukherjee C, Caron MG, Lefkowitz RJ (1976) Regulation of adenylate cyclase coupled β-adrenergic receptors by β-adrenergic catecholamines. Endocrinol 99:347-357

Muller P, Seeman P (1977) Brain neurstransmitter receptors after long-term haloperidol: dopamine, acetylcholine, serotonin, α-noradrenergic and naxolone receptors. Life Sc, 21:1751-1758

Munemura M, Cote TE, Tsuruta K, Eskay RL, Kebabian JW (1980) The dopamine receptor in the intermediate lobe of the rat pituitary gland: pharmacological characterization. Endocrinol 107:1676-1683

Murad F, Chi Y-M, Rall TW, Sutherland EW (1962) Adenyl cyclase. III. The effect of catecholamines and choline esters on the formation of adenosine 3′,5′-phosphate by preparations from cardiac muscle and liver. J Biol Chem 237:1233-1238

Murrin LC, Kuhar MJ (1979) Dopamine receptors in the rat frontal cortex: an autoradiographic study. Brain Res 177:279-285

Murrin LC, Gale K, Kuhar MJ (1979) Autoradiographic localization of neuroleptic and dopamine receptors in the caudate-putamen and substantia nigra: effects of lesions. Eur J Pharmacol 60:229-235

Nahorski SR (1976) Association of high affinity stereospecific binding of ^{3}H-propranolol to cerebral membranes with β adrenoceptors. Nature 259:488-489

Nambi P, Peters JR, Sibley DR, Lefkowitz RJ (1985) Desensitization of the turkey erythrocyte β-adrenergic receptor in a cell-free system: evidence that multiple protein kinases can phosphorylate and desensitize the receptor. J Biol Chem 260:2165-2171

Nathans J, Hogness DS (1983) Isolation, sequence analysis, and intron-exon arrangement of the gene encoding bovine rhodopsin. Cell 34:807-814

Nathanson, JA (1977) Cyclic nucleotides and nervous system function. Physiol Rev 57:157-256
Near JA, Mahler HR (1981) Dopamine receptors in subcellular fractions from bovine caudate: enrichment of (^{3}H)spiperone binding in a postsynaptic membrane fraction. J Neurochem 36:1142-1151
Neve KA, Molinoff PB (1986) Turnover of β_1- and β_2-adrenergic receptors after down-regulation or irreversible blockade. Mol Pharmacol 30:104-111
Niehoff DL, Palacios JM, Kuhar MJ (1980) Dopamine receptors: temperature effects on ^{3}H-spiperone and ^{3}H-N-n-propylnorapomorphine binding. Soc Neurosci Abs 6:252
Niznik HB, Guan JH, Neumeyer JL, Seeman P (1985) A photoaffinity ligand for dopamine D2 receptors: azidoclebopride. Mol Pharmacol 27:193-199
Niznik HB, Otsuka NY, Dumbrille-Ross A, Grigoriadis D, Tirpak A, Seeman P (1986) Dopamine D1 receptors characterized with (^{3}H)SCH 23390: solubilization of a guanine nucleotide-sensitive form of the receptor. J Biol Chem 261:8397-8406
Nowycky MC, Roth RH (1977) Presynaptic dopamine receptors. Naunyn-Schmiedeberg's Arch Pharmacol 300:247-254
O'Donnell SR, Wanstall JC (1976) The contribution of extraneuronal uptake to the trachea-blood vessel selectivity of β-adrenoceptor stimulants *in vitro* in guinea-pigs. Br J Pharmacol 57:369-373
Olney JW, de Gubareff T (1978) The fate of synaptic receptors in the kainate-lesioned striatum. Brain Res 140:340-343
Olney JW, Misra CH, de Gubareff T (1975) Cysteine-S-sulfate: brain damaging metabolite in sulfite oxidase deficency. J Neuropathol Exp Neurol 34:167-177
Orly J, Schramm M (1976) Coupling of catecholamine receptor from one cell with adenylate cyclase from another cell by cell fusion. Proc Natl Acad Sci USA 73:4410-4414
Ovchinnikov YA (1982) Rhodopsin and bacteriorhodopsin structure-function relationships. FEBS Lett 148:179-191
Palacios JM, Kuhar MJ (1980) Beta-adrenergic-receptor localization by light microscopic autoradiography. Science 208:1378-1380
Palmer DS, French SW, Narod ME (1976) Noradrenergic subsensitivity and supersensitivity of the cerebral cortex after reserpine treatment. J Pharmacol Exp Ther 196:167-171
Palmer GC (1972) Increased cyclic AMP response to norepinephrine in the rat brain following 6-hydroxydopamine. Neuropharmacol 11:145-149
Pardo JV, Creese J, Burt DR, Snyder SH (1977) Ontogenesis of dopamine receptor binding in the corpus striatum of the rat. Brain Res 125:376-382
Partington CR, Edwards MW, Daly JW (1980) Regulation of cyclic AMP formation in brain tissue by α-adrenergic receptors: requisite intermediacy of prostaglandins of the E series. Proc Natl Acad Sci USA 77:3024-3028
Passonneau JV, Crites SK (1976) Regulation of glycogen metabolism in astrocytoma and neuroblastoma cells in culture. J Biol Chem 251:2015-2022
Perkins JP (1973) Adenyl cyclase. Adv Cyclic Nucleotide Res 3:1-64
Perkins JP (1981) Catecholamine-induced modification of the functional state of β-adrenergic receptors. Trends in Pharmacol Sci 2:326-328
Perkins JP, Moore MM (1973a) Regulation of the adenosine cyclic 3′,5′-monophosphate content of rat cerebral cortex: ontogenetic development of the responsiveness to catecholamines and adenosine. Mol Pharmacol 9:774-782
Perkins JP, Moore MM (1973b) Characterization of the adrenergic receptors mediating a rise in cyclic 3′,5′-adenosine monophosphate in rat cerebral cortex. J Pharmacol Exp Ther 185:371-378

Peroutka SJ, Greenberg DA, U'Prichard DC, Snyder SH (1978) Regional variations in *alpha* adrenergic receptor interactions of (^{3}H)-dihydroergokryptine in calf brain: implications for a two-site model of *alpha* receptor function. Mol Pharmacol 14:403-412

Perry BD, U'Prichard DC (1981) ^{3}H-rauwolscine binding to α_2-adrenergic receptors in bovine brain. Soc Neurosci Abs 7:424

Pert A, Rosenblatt JE, Sivit C, Pert CB, Bunney WE Jr (1978) Long-term treatment with lithium prevents the development of dopamine receptor supersensitivity. Science 201:171-173

Petrack B, Czernik AJ (1976) Inhibition of isoproterenol activation of adenylate cyclase by metoprolol, oxprenolol, and the *para* isomer of oxprenolol. Mol Pharmacol 12:203-207

Pittman RN (1981) The pharmacology and regulation of β-adrenergic receptors on L-6 muscle cells. Thesis, University of Colorado

Pittman RN, Molinoff PB (1980) Interactions of agonists and antagonists with β-adrenergic receptors on intact L6 muscle cells. J Cyclic Nucleotide Res 6:421-435

Pittman RN, Molinoff PB (1983) Interactions of full and partial agonists with *beta*-adrenergic receptors on intact L6 muscle cells. Mol Pharmacol 24:398-408

Pittman RN, Minneman KP, Molinoff PB (1980a) Alterations in β_1- and β_2-adrenergic receptor density in the cerebellum of aging rats. J Neurochem 35:273-275

Pittman RN, Minneman KP, Molinoff PB (1980b) Ontogeny of β_1- and β_2-adrenergic receptors in rat cerebellum and cerebral cortex. Brain Res 188:357-368

Pittner H, Smith, Leibowitz M, Van Inweger RG, Wolf P (1983) investigations in the intrinsic sympathetic activity of the beta 1-adrenergic blocking agent celiprolol. Naunyn-Schmiedeberg's Arch Pharmacol 322:R41

Prasad KN, Gilmer KN (1974) Demonstration of dopamine-sensitive adenylate cyclase in malignant neuroblastoma cells and change in sensitivity of adenylate cyclase to catecholamines in "differentiated" cells. Proc Natl Acad Sci USA 71:2525-2529

Puchwein G, Pfeuffer T, Helmreich EJM (1974) Uncoupling of catecholamine activation of pigeon erythrocyte membrane adenylate cyclase by filipin. J Biol Chem 249:3232-3240

Putney JW Jr, Weiss SJ, Van De Walle CM, Haddas RA (1980) Is phosphatidic acid a calcium ionophore under neurohumoral control? Nature 284:345-347

Quik M, Iversen LL, Larder A, Mackay AVP (1978) Use of ADTN to define specific (^{3}H)-spiperone binding to receptors in brain. Nature 274:513-514

Rainbow TC, Parsons B, Wolfe BB (1984) Quantitative autoradiography of β_1- and β_2-adrenergic receptors in rat brain. Proc Natl Acad Sci USA 81:1585-1589

Ramwani J, Mishra RK (1986) Purification of bovine striatal dopamine D-2 receptor by affinity chromatography. J Biol Chem 261:8894-8898

Rappaport RS, Grant NH (1974) Growth hormone releasing factor of microbial origin. Nature 248:73-75

Rashidbaigi A, Ruoho AE (1981) Iodoazidobenzylpindolol, a photoaffinity probe for the β-adrenergic receptor. Proc Natl Acad Sci USA 78:1609-1613

Regan JW, Nakata H, DeMarinis RM, Caron MG, Lefkowitz RJ (1986) Purification and characterization of the human platelet α_2-adrenergic receptor. J Biol Chem 261:3894-3900

Repaske MG, Nunnari JM, Limbird LE (1987) Purification of the α_2-adrenergic receptor from porcine brain using a yohimbine-agarose affinity matrix. J Biol Chem 262:12381-12386

Reynolds EE, Molinoff PB (1986) Down regulation of β-adrenergic receptors in S49 lymphoma cells induced by atypical agonists. J Pharmacol Exp Ther 239:654-660

Reynolds EE, Hoyer D, Molinoff PB (1985) Agonist-induced changes in the properties of β-adrenergic receptors on intact S49 lymphoma cells: sequestration of receptors and desensitization of adenylate cyclase. In: Lefkowitz R, Lindenlaub E (eds): *Adrenergic receptors: molecular properties and therapeutic implications*. Schattauer, Stuttgart pp 505-530

Roberts JM, Insel PA, Goldfien RD, Goldfien A (1977) α Adrenoreceptors but not β adrenoreceptors increase in rabbit uterus with oestrogen. Nature 270:624-625

Robertson HA (1980) Stimulation of ^{3}H-apomorphine binding by dopamine and bromocriptine. Eur J Pharmacol 61:209-211

Robison GA, Butcher RW, Sutherland EW (1971) Cyclic AMP, Academic Press, New York

Rodbell M, Birnbaumer L, Pohl SL, Krans HMJ (1971) The glucagon-sensitive adenyl cyclase system in plasma membranes of rat liver. V. An obligatory role of guanyl nucleotides in glucagon action. J Biol Chem 246:1877-1882

Romero JA, Axelrod J (1975) Regulation of sensitivity to *beta*-adrenergic stimulation in induction of pineal *N*-acetyltransferase. Proc Natl Acad Sci USA 72:1661-1665

Roper S (1976) The acetylcholine sensitivity of the surface membrane of multiply-innervated parasympathetic ganglion cells in the mudpuppy before and after partial denervation. J Physiol (Lond) 254:455-473

Rosenblatt JE, Pert CB, Tallman JF, Pert A, Bunney WF Jr (1979) The effect of imipramine and lithium on α- and β-receptor binding in rat brain. Brain Res 160:186-191

Ross EM, Maguire ME, Sturgill TW, Biltonen RL, Gilman AG (1977) Relationship between the β-adrenergic receptor and adenylate cyclase. J Biol Chem 252:5761-5775

Roth RH, Walters JR, Murrin LC, Morgenroth VH III (1975) Dopamine neurons: role of impulse flow and pre-synaptic receptors in the regulation of tyrosine hydroxylase. In: Usdin E, Bunney WE Jr (eds) *Pre- and postsynaptic receptors*, Marcel Dekker, New York, pp 5-48

Rouot BM, U'Prichard DC, Snyder SH (1980) Multiple α_2-noradrenergic receptor sites in rat brain: selective regulation of high-affinity (^{3}H)clonidine binding by guanine nucleotides and divalent cations. J Neurochem 34:374-384

Ruffolo RR Jr, Waddell JE, Yaden EL (1981) Postsynaptic *alpha* adrenergic receptor subtypes differentiated by yohimbine in tissues from the rat. Existence of *alpha*-2 adrenergic receptors in rat aorta. J Pharmacol Exp Ther 217:235-240

Rugg EL, Barnett DB, Nahorski SR (1978) Coexistence of $beta_1$ and $beta_2$ adrenoceptors in mammalian lung: evidence from direct binding studies. Mol Pharmacol 14:996-1005

Rutledge CO (1970) The mechanisms by which amphetamine inhibits oxidative deamination of norepinephrine in brain. J Pharmacol Exp Ther 171:188-195

Sabol SL, Nirenberg M (1979) Regulation of adenylate cyclase of neuroblastoma X glioma hybrid cells by α-adrenergic receptors. J Biol Chem 254:1913-1920

Salmon DM, Honeyman TW (1980) Proposed mechanism of cholinergic action in smooth muscle. Nature 284:344-345

Salzman EW, Neri LL (1969) Cyclic 3',5'-adenosine monophosphate in human blood platelets. Nature 224:609-612

Scatchard G (1949) The attractions of proteins for small molecules and ions. Ann NY Acad Sci 51:660-672

Schmidt MJ, Hill LE (1977) Effects of ergots on adenylate cyclase activity in the corpus striatum and pituitary. Life Sci 20:789-798

Schramm M, Rodbell M (1975) A persistent active state of the adenylate cyclase system produced by the combined actions of isoproterenol and guanylyl imidodiphosphate in frog erythrocyte membranes. J Biol Chem 250:2232-2237

Schultz G, Hardman JG, Schultz K, Baird CE, Sutherland EW (1973) The importance of calcium ions for the regulation of guanosine 3′:5′-cyclic monophosphate levels. Proc Natl Acad Sci USA 70:3889-3893

Schultz J, Daly JW (1973a) Adenosine 3′,5′-monophosphate in guinea pig cerebral cortical slices: effects of α- and β-adrenergic agents, histamine, serotonin, and adenosine. J Neurochem 21:573-579

Schultz J, Daly JW (1973b) Accumulation of cyclic adenosine 3′,5′-monophosphate in cerebral cortical slices from fat and mouse: stimulatory effect of α- and β-adrenergic agents and adenosine. J Neurochem 21:1319-1326

Schwabe U, Daly JW (1977) The role of calcium ions in accumulations of cyclic adenosine monophosphate elicited by *alpha* and *beta* adrenergic agonists in rat brain slices. J Pharmacol Exp Ther 202:134-143

Schwarcz R, Creese I, Coyle JT, Snyder SH (1978a) Dopamine receptors localised on cerebral cortical afferents to rat corpus striatum. Nature 271:766-768

Schwarcz R, Fuxe K, Agnati LF, Gustafsson J-A (1978b) Effects of bromocriptine on (^{3}H)-spiroperidol binding sites in rat striatum. Evidence for actions of dopamine receptors not linked to adenylate cyclase. Life Sci 23:465-470

Seeman P (1980) Brain dopamine receptors. Pharmacol Rev 32:229-313

Seeman P, Chau-Wong M, Tedesco J, Wong K (1975) Brain receptors for antipsychotic drugs and dopamine: direct binding assays. Proc Natl Acad Sci USA 72:4376-4380

Seeman P, Lee T, Chau-Wong M, Tedesco J, Wong K (1976) Dopamine receptors in human and calf brains, using (^{3}H)-apomorphine and an antipsychotic drug. Proc Natl Acad Sci USA 73:4354-4358

Senogles SE, Amlaiky N, Johnson AL, Caron MG (1986) Affinity chromatography of the anterior pituitary D2-dopamine receptor. Biochemistry 25:749-753

Sethy VH, Van Woert MH (1974) Modification of striatal acetylcholine concentration by dopamine receptor agonists and antagonists. Res Comm Chem Pathol Pharmacol 8:13-28

Sharma VK, Banerjee SP (1978) α-Adrenergic receptor in rat heart: effects of thyroidectomy. J Biol Chem 253:5277-5279

Shear M, Insel PA, Melmon KL, Coffino P (1976) Agonist-specific refractoriness induced by isoproterenol: studies with mutant cells. J Biol Chem 251:7572-7576

Sheppard H, Burghardt CR (1974) Effect of tetrahydroisoquinoline derivatives on the adenylate cyclases of the caudate nucleus (dopamine-type) and erythrocyte (β-type) of the rat. Res Comm Chem Pathol Pharmacol 8:527-534

Shorr RGL, Lefkowitz RJ, Caron MG (1981) Purification of the β-adrenergic receptor. J Biol Chem 256:5820-5826

Sibley DR, Strasser RH, Caron MG, Lefkowitz RJ (1985) Homologous desensitization of adenylate cyclase is associated with phosphorylation of the β-adrenergic receptor. J Biol Chem 360:3883-3886

Siggins GR, Battenberg EF, Hoffer BJ, Bloom FE, Steiner AL (1973) Noradrenergic stimulation of cyclic adenosine monophosphate in rat purkinje neurons: an immunocytochemical study. Science 179:585-588

Sinensky M, Minneman KP, Molinoff PB (1979) Increased membrane acyl chain ordering activates adenylate cyclase. J Biol Chem 254:9135-9141

Singer SJ, Campbell DH (1955) Physical chemical studies of soluble antigen-antibody complexes. V. Thermodynamics of the reaction between ovalbumin and its rabbit antibodies. J Am Chem Soc 77:4851-4855

Slotkin TA (1974) Reserpine In: Simpson LL, Curtis DR (eds) *Neuropoisons*, Plenum Press, New York, pp 1-60

Smigel MD, Ross EM, Gilman AG (1984) Role of the β-adrenergic receptor in the regulation of adenylate cyclase. In: Elson EL, Frazier WH, Glaser L (eds) Cell membranes: methods and reviews, Plenum Publ, New York, pp 247–294

Smith RD, Wolf PS (1984) Celiprolol. New drugs ann.: Cardiovasc. Drugs 2:19–35

Sporn JR, Molinoff PB (1976) β-Adrenergic receptors in rat brain. J Cyclic Nucleotide Res 2:149–161

Sporn JR, Harden TK, Wolfe BB, Molinoff PB (1976) β-Adrenergic receptor involvement in 6-hydroxydopamine-induced supersensitivity in rat cerebral cortex. Science 194:624–626

Sporn JR, Wolfe BB, Harden TK, Molinoff PB (1977) Supersensitivity in rat cerebral cortex: pre- and postsynaptic effects of 6-hydroxydopamine at noradrenergic synapses. Mol Pharmacol 13:1170–1180

Stadel JM, DeLean A, Lefkowitz RJ (1980) A high affinity agonist β-adrenergic receptor complex is an intermediate for catecholamine stimulation of adenylate cyclase in turkey and frog erythrocyte membranes. J Biol Chem 255:1436–1441

Stadel JM, Nambi P, Lavin TN, Heald SL, Caron MG, Lefkowitz RJ (1982) Catecholamine-induced desensitization of turkey erythrocyte adenylate cyclase: structural alterations in the β-adrenergic receptor revealed by photoaffinity labeling. J Biol Chem 257:9242–9245

Stadel JM, Strulovici B, Nambi P, Lavin TN, Briggs MM, Caron MG, Lefkowitz RJ (1983a) Desensitization of the β-adrenergic receptor of frog erythrocytes. J Biol Chem 258:3032–3038

Stadel JM, Nambi P, Shorr RGL, Sawyer DF, Caron MG, Lefkowitz RJ (1983b) Catecholamine-induced desensitization of turkey erythrocyte adenylate cyclase is associated with phosphorylation of the β-adrenergic receptor. Proc Natl Acad Sci USA 80:3173–3177

Starke K, Borowski E, Endo T (1975) Preferential blockade of presynaptic α-adrenoceptors by yohimbine. Eur J Pharmacol 34:385–388

Staunton DA, Wolfe BB, Groves PM, Molinoff PB (1981) Dopamine receptor changes following destruction of the nigrostriatal pathway: lack of a relationship to rotational behavior. Brain Res 211:315–327

Stiles GL, Lefkowitz RJ (1981) Thyroid hormone modulation of agonist – beta-adrenergic receptor interactions in the rat heart. Life Sci 28:2529–2536

Strader CD, Pickel VM, Joh TH, Strohsacker, MW, Shorr RGL, Lefkowitz RJ, Caron MG (1983) Antibodies to the β-adrenergic receptor: attenuation of catecholamine-sensitive adenylate cyclase and demonstration of postsynaptic receptor localization in brain. Proc. Natl Acad Sci USA 80:1840–1844

Streb H, Irvine RF, Berridge MJ, Schulz I (1983) Release of Ca^{2+} from a nonmitochondrial intracellular store in pancreatic acinar cells by inositol-1,4,5-trisphosphate. Nature 306:67–69

Strittmatter WJ, Davis JN, Lefkowitz RJ (1977) α-adrenergic receptors in rat parotid cells. II. Desensitization of receptor binding sites and potassium release. J Biol Chem 252:5478–5482

Strosberg AD, Delavier-Klutchko C, Cervantes P, Guillet J-G, Schmutz A, Kaveri S, Durieu-Trautmann O, Hoebeke J (1985) Biochemical and immunological studies of β_1- and β_2-adrenergic receptors. In: Lefkowitz RJ, Lindenlaub E (eds) Adrenergic receptors: molecular properties and therapeutic implications, Schattauer Stuttgart, pp 53–81

Su Y-F, Cubeddu L, Perkins JP (1976a) Regulation of adenosine 3′:5′-monophosphate content of human astrocytoma cells: desensitization to catecholamines and prostaglandins. J Cyclic Nucleotides Res 2:257–270

Su Y-F, Johnson GL, Cubeddu L, Leichtling BH, Ortmann R, Perkins JP (1976b) Regulation of adenosine 3':5'-monophosphate content of human astrocytoma cells: mechanism of agonist-specific desensitization. J Cyclic Nucleotide Res 2:271-285

Su Y-F, Harden TK, Perkins JP (1979) Isoproterenol-induced desensitization of adenylate cyclase in human astrocytoma cells. J Biol Chem 254:38-41

Su Y-F, Harden TK, Perkins JP (1980) Catecholamine-specific desensitization of adenylate cyclase: evidence for a multistep process. J Biol Chem 255:7410-7419

Suen ET, Stefanini E, Clement-Cormier YC (1980) Evidence for essential thiol groups and disulfide bonds in agonist and antagonist binding to the dopamine receptor. Biochem Biophys Res Commun 96:953-960

Suen ET, Kwan PCK, Clement-Cormier YC (1981) Identification of an essential sulfhydryl group for the activation of dopamine-sensitive adenylate cyclase. Soc Neurosci Abstr 7:919

Summers RJ (1980) (^{3}H)-clonidine binding to α-adrenoceptors in membranes prepared from regions of guinea-pig kidney: alteration by monovalent and divalent cations. Br J Pharmacol 71:57-63

Sutherland EW, Robison GA (1969) The role of cyclic AMP in the control of carbohydrate metabolism. Diabetes 18:797-819

Sutherland EW, Oye I, Butcher RW (1965) The action of epinephrine and the role of the adenyl cyclase system in hormone action. Recent Prog Horm Res 21:623-646

Swanson LW, Hartman BK (1975) The central adrenergic system. An immunofluorescence study of the location of cell bodies and their efferent connections in the rat utilizing dopamine-β-hydroxylase as a marker. J Comp Neurol 163:467-506

Takai Y, Kishimoto A, Kikkawa U, Mori T, Nishizuka Y (1979) Unsaturated diacylglycerol as a possible messenger for the activation of calcium-activated, phospholipid-dependent protein kinase system. Biochem Biophys Res Commun 91:1218-1224

Tarsy D, Baldessarini RJ (1973) pharmacologically induced behavioural supersensitivity to apomorphine. Nature New Biol 245:262-263

Taylor DL, Ho BT, Fagan JD (1979) Increased dopamine receptor binding in rat brain by repeated cocaine injections. Commun Psychopharmacol 3:137-142

Terasaki WL, Brooker G (1978) (^{125}I)iodohydroxybenzylpindolol binding sites on intact rat glioma cells: evidence for β-adrenergic receptors of high coupling efficiency. J Biol Chem 253:5418-5425

Thoenen H, Tranzer JP (1968) Chemical sympathectomy by selective destruction of adrenergic nerve endings with 6-hydroxydopamine. Naunyn-Schmiedeberg's Arch Pharmacol Exp Pathol 261:271-288

Thomas TN, Zemp JW (1977) Inhibition of dopamine sensitive adenylate cyclase from rat brain striatal homogenates by ascorbic acid. J Neurochem 28:663-665

Titeler M, Weinreich P, Sinclair D, Seeman P (1978) Multiple receptors for brain dopamine. Proc Natl Acad Sci USA 75:1153-1156

Titeler M, List S, Seeman P (1979) High affinity dopamine receptors (D_3) in rat brain. Commun. Psychopharmacol 3:411-420

Toews ML, Waldo GL, Harden TK, Perkins JP (1985) Relationship between an altered membrane form and a low affinity form of the β-adrenergic receptor occurring during catecholamine-induced desensitization: evidence for receptor internalization. J Biol Chem 259:11844-11850

Trabucchi M, Cheney DL, Racagni G, Costa E (1975) *In vivo* inhibition of striatal acetylcholine turnover by L-dopa, apomorphine and (+)-amphetamine. Brain Res 85:130-134

Treiser S, Kellar KJ (1979) Lithium effects on adrenergic receptor supersensitivity in rat brain. Eur J Pharmacol 58:85-86

Trendelenburg U (1963) Supersensitivity and subsensitivity to sympathomimetic amines. Pharmacol Rev 15:225–276

Trendelenburg U (1966) Mechanisms of supersensitivity and subsensitivity to sympathomimetic amines. Pharmacol Rev 18:629–640

Tsai BS, Lefkowitz RJ (1978) Agonist-specific effects of monovalent and divalent cations on adenylate cyclase-coupled *alpha* adrenergic receptors in rabbit platelets. Mol Pharmacol 14:540–548

Tsai BS, Lefkowitz RJ (1979) Agonist-specific effects of guanine nucleotides on *alpha*-adrenergic receptors in human platelets. Mol Pharmacol 16:61–68

Tsai JS, Chen A (1978) Effect of L-triiodothyronine on (−)^{3}H-dihydroalprenolol binding and cyclic AMP response to (−)adrenaline in cultured heart cells. Nature 275:138–140

Ungerstedt U (1971a) Postsynaptic supersensitivity after 6-hydroxydopamine induced degeneration of the nigro-striatal dopamine system. Acta Physiol Scand 82 Suppl 367:69–93

Ungerstedt U (1971b) Stereotaxic mapping of the monoamine pathways in the rat brain. Acta Physiol Scand 82 Suppl 367:1–48

U'Prichard DC, Snyder SH (1977) Binding of ^{3}H-catecholamines to α-noradrenergic receptor sites in calf brain. J Biol Chem 252:6450–6463

U'Prichard DC, Snyder SH (1980) Interactions of divalent cations and guanine nucleotides at α_2-noradrenergic receptor binding sites in bovine brain mechanisms. J Neurochem 34:385–394

U'Prichard DC, Greenberg DA, Snyder SH (1977) binding characteristics of a radiolabeled agonist and antagonist at central nervous system *alpha* noradrenergic receptors. Mol Pharmacol 13:454–473

U'Prichard DC, Bylund DB, Snyder SH (1978) (±)-(^{3}H)epinephrine and (−)-(^{3}H)dihydroalprenolol binding to β_1- and β_2-noradrenergic receptors in brain, heart, and lung membranes. J Biol Chem 253:5090–5102

U'Prichard DC, Bechtel WD, Rouot BM, Snyder SH (1979) Multiple apparent *alpha*-noradrenergic receptor binding sites in rat brain: effect of 6-hydroxydopamine. Mol Pharmacol 16:47–60

Uretsky NJ, Iversen LL (1970) Effects of 6-hydroxydopamine on catecholamine containing neurones in the rat brain. J Neurochem 17:269–278

Usdin TB, Creese I, Snyder SH (1980) Regulation by cations of (^{3}H)spiroperidol binding associated with dopamine receptors of rat brain. J Neurochem 34:669–676

Vauquelin G, Maguire ME (1980) Inactivation of β-adrenergic receptors by *N*-ethylmaleimide in S49 lymphoma cells. Mol Pharmacol 18:362–369

Vauquelin G, Geynet P, Hanoune J, Strosberg AD (1977) Isolation of adenylate cyclase-free, β-adrenergic receptor from turkey erythrocyte membranes by affinity chromatography. Proc Natl Acad Sci USA 74:3710–3714

Vauquelin G, Bottari S, Kanarek L, Strosberg AD (1979) Evidence for essential disulfide bonds in β_1-adrenergic receptors of turkey erythrocyte membranes. J Biol Chem 254:4462–4469

Vauquelin G, Bottari S, Andre C, Jacobsson B, Strosberg AD (1980a) interaction between β-adrenergic receptors and guanine nucleotide sites in turkey erythrocyte membranes. Proc Natl Acad Sci USA 77:3801–3805

Vauquelin G, Bottari S, Strosberg AD (1980b) Inactivation of β-adrenergic receptors by *N*-ethylmaleimide: permissive role of β-adrenergic agents in relation to adenylate cyclase activation. Mol Pharmacol 17:163–171

Venter JC, Ross J Jr, Dixon JE, Mayer SE, Kaplan NO (1973) Immobilized catecholamine and cocaine effects on contractility of cardiac muscle. Proc Natl Acad Sci USA 70:1214–1217

Verlander MS, Venter JC, Goodman M, Kaplan NO, Saks B (1976) Biological activity of catecholamines covalently linked to synthetic polymers: proof of immobilized drug theory. Proc Natl Acad Sci USA 73:1009-1013

Vetulani J, Sulser F (1975) Action of various antidepressant treatments reduces reactivity of noradrenergic cyclic AMP-generating system in limbic forebrain. Nature 257:495-496

Vetulani J, Stawarz RJ, Blumberg JB, Sulser F (1975) Adaptive mechanisms in the norepinephrine (NE)-sensitive cyclic AMP generating system in the slices of the rat limbic forebrain (LFS). Fed Proc 34:265

Wagner HR, Palmer GC (1975) Increased sensitivity of rat brain adenylate cyclase to catecholamines following reserpine injections. Trans Am Soc Neurochem 6:80

Weiland GA, Molinoff PB (1981) Quantitative analysis of drug-receptor interactions: I. Determination of kinetic and equilibrium properties. Life Sci 29:313-330

Weiland GA, Minneman KP, Molinoff PB (1979) Fundamental difference between the molecular interactions of agonists and antagonists with the β-adrenergic receptor. Nature 281:114-117

Weiland GA, Minneman KP, Molinoff PB (1980) Thermodynamics of agonist and antagonist interactions with mammalian β-adrenergic receptors. Mol Pharmacol 18:341-347

Weiss B (1969) Effects of environmental lighting and chronic denervation on the activation of adenyl cyclase of rat pineal gland by norepinephrine and sodium fluoride. J Pharmacol Exp Ther 168:146-152

Weiss B, Strada SJ (1972) Neuroendocrine control of the cyclic AMP system of brain and pineal gland. Adv Cyclic Nucleotide Res 1:357-374

Weissbach H, Redfield BG, Axelrod J (1960) Biosynthesis of melatonin: enzymic conversion of serotonin to N-acetylserotonin. Biochim Biophys Acta 43:352-353

Wessels MR, Mullikin D, Lefkowitz RJ (1978) Differences between agonist and antagonist binding following β-adrenergic receptor desensitization. J Biol Chem 253:3371-3373

Wilkening D, Makman MH (1976) Stimulation of glycogenolysis in rat caudate nucleus slices by L-isopropylnorepinephrine, dibutyryl cyclic AMP and 2-chloroadenosine. J Neurochem 26:923-928

Wilkening D, Makman MH (1977) Activation of glycogen phosphorylase in rat caudate nucleus slices by L-isopropylnorepinephrine and dibutyryl cyclic AMP. J Neurochem 28:1001-1007

Williams LT, Lefkowitz RJ (1976) Alpha-adrenergic receptor identification by (^{3}H)dihydroergocryptine binding. Science 192:791-793

Williams LT, Lefkowitz RJ (1977) Slowly reversible binding of catecholamine to a nucleotide-sensitive state of the β-adrenergic receptor. J Biol Chem 252:7207-7213

Williams LT, Lefkowitz RJ (1978) Receptor binding studies in adrenergic pharmacology, New York, Raven Press

Williams LT, Jarett L, Lefkowitz RJ (1976) Adipocyte β-adrenergic receptors. Identification and subcellular localization by (−)-(^{3}H)dihydroalprenolol binding. J Biol Chem 251:3096-3104

Williams LT, Lefkowitz RJ, Watanabe AM, Hathaway DR, Besch HR Jr (1977) Thyroid hormone regulation of β-adrenergic receptor number. J Biol Chem 252:2787-2789

Williams LT, Mullikin D, Lefkowitz RJ (1978) Magnesium dependence of agonist binding to adenylate cyclase-coupled hormone receptors. J Biol Chem 253:2984-2989

Wilner KD, Butler IJ, Seifert WE, Clement-Cormier YC (1980) Biochemical alterations of dopamine receptor responses following chronic L-DOPA therapy. Biochem Pharmacol 29:701-706

Witkin KM, Harden TK (1981) A sensitive equilibrium binding assay for soluble β-adrenergic receptors. J Cyclic Nucleotide Res 7:235-246

Wolfe BB, Harden TK (1981) Guanine nucleotides modulate the affinity of antagonists at β-adrenergic receptors. J Cyclic Nucleotide Res 7:303-312

Wolfe BB, Zirrolli JA, Molinoff PB (1974) Binding of dl-(^{3}H)epinephrine to proteins of rat ventricular muscle: nonidentity with *beta* adrenergic receptors. Mol Pharmacol 10:582-596

Wolfe BB, Harden TK, Molinoff PB (1976) β-Adrenergic Receptors in rat liver: effects of adrenalectomy. Proc Natl Acad Sci USA 73:1343-1347

Wolfe BB, Harden TK, Sporn JR, Molinoff PB (1978a) Presynaptic modulation of *beta* adrenergic receptors in rat cerebral cortex after treatment with antidepressants. J Pharmacol Exp Ther 207:446-457

Wolfe BB, Staunton DA, Groves PM, Molinoff PB (1978b) Effects of chronic alterations in receptor stimulation on β-adrenergic receptors in rat brain. Soc Neurosci Abstr 4:526

Wolfe BB, Minneman KP, Dibner MD, Molinoff PB (1980) Regulation of β-adrenergic receptor subtypes in rat cerebellum, cerebral cortex and caudate. Soc Neurosci Abst 6:257

Wolfe BB, Minneman KP, Molinoff PB (1982) Selective increases in the density of cerebellar β_1-adrenergic receptors. Brain Res 234:474-479

Wouters W, Van Dun J, Laduron PM (1984) Photoaffinity labelling of dopamine receptors: synthesis and binding characteristics of azapride. Eur J Biochem 145:273-278

Wrenn SM Jr, Homcy CJ (1980) Photoaffinity label for the β-adrenergic receptor: synthesis and effects on isoproterenol-stimulated adenylate cyclase. Proc Natl Acad Sci USA 77:4449-4453

Yamada S, Yamamura HI, Roeske WR (1980a) Characterization of *alpha*-1 adrenergic receptors in the heart using (^{3}H)WB 4101: effect of 6-hydroxydopamine treatment. J Pharmacol Exp Ther 215:176-185

Yamada S, Yamamura HI, Roeske WR (1980b) The regulation of cardiac α_1-adrenergic receptors by guanine nucleotides and by muscarinic cholinergic agonists. Eur J Pharmacol 63:239-241

Yamamura H, Rodbell M (1976) Hydroxybenzylpindolol and hydroxybenzylpropranolol: partial *beta*-adrenergic agonists of adenylate cyclase in the rat adipocyte. Mol Pharmacol 12:693-700

Yarden Y, Rodriguez H, Wong SK-F, Brandt DR, May DC, Burnier J, Harkins RN, Chen EY, Ramachandran J, Ullrich A, Ross EM (1986) The avian β-adrenergic receptor: primary structure and membrane topology. Proc Natl Acad Sci USA 83:6795-6799

Young WS III, Kuhar MJ (1979) Noradrenergic α1 and α2 receptors: autoradiographic visualization. Eur J Pharmacol 59:317-319

Young WS III, Kuhar MJ (1980) Noradrenergic α1 and α2 receptors: light microscopic autoradiographic localization. Proc Natl Acad Sci USA 77:1696-1700

Zahniser NR, Molinoff PB (1978) Effect of guanine nucleotides on striatal dopamine Receptors. Nature 275:453-455

Zahniser NR, Molinoff PB (1979) GTP and cation effects on rat striatal dopaminergic and muscarinic cholinergic receptors. Soc Neurosci Abs 5:602

Zahniser NR, Molinoff PB (1983) Thermodynamic differences between agonist and antagonist interactions with binding sites for ^{3}H-spiroperidol in rat striatum. Mol Pharmacol 23:303-309

Zahniser NR, Heidenreich KA, Molinoff PB (1980) ATP- or GTP-induced increase in (^{3}H)-ADTN binding to rat striatal dopamine receptors is abolished by UV irradiation. Soc Neurosci Abstr 6:237

Zahniser NR, Heidenreich KA, Molinoff PB (1981) Binding of (^{3}H)amino-6,7-dihydroxy-1,2,3,4-tetrahydronaphthalene to rat striatal membranes: effects of purine nucleotides and ultraviolet irradiation. Mol Pharmacol 19:372–378

Zatz M, O'Dea RF (1976) Regulation of protein kinase in rat pineal: increased V_{max} in supersensitive glands. J Cyclic Nucleotide Res 2:427–439

CHAPTER 8

Presynaptic Receptors on Catecholamine Neurones

S. Z. LANGER and J. LEHMANN

A. Introduction and Definition of Terms

In 1963 CARLSSON and LINDQVIST observed that the administration of neuroleptic drugs caused an increase in the level of dopamine metabolites, suggesting that neuroleptics cause an increase in dopamine release. Subsequent confirmations and extensions of these results ultimately led to a general consensus ten years later favouring the dopamine hypothesis of schizophrenia (SNYDER 1972; MATTHYSE 1973; STEVENS 1973; MELTZER et al. 1974). The therapeutic effect of neuroleptics was thought to be due to the blockade of a postsynaptic dopamine receptor. The increase in dopamine turnover was ascribed to a receptor-mediated negative feedback system. The self-regulation of dopaminergic neurotransmission was generally assumed to be mediated by a long trans-synaptic feedback loop, i.e. by neurones postsynaptic to dopaminergic terminals, which relayed information over their long descending axons to dopaminergic cell bodies in the substantia nigra (see Fig. 1).

This traditional long-loop (trans-synaptic) feedback model has held a strong following in the past decade, even though a simpler explanation has been evolving in parallel. In 1971, FARNEBO and HAMBERGER (1971a) found that, in the isolated caudate slice, the release of ^{3}H-dopamine was increased by neuroleptics and decreased by the dopamine receptor agonist apomorphine, demonstrating that the self-regulatory dopamine receptors were present and functional in the slice. Subsequently KEHR et al. (1972) reported that the self-regulatory system modulating dopamine synthesis is still functional after transection of the ascending dopaminergic axons and the descending striatonigral axons, thus confirming that no long-loop feedback system involving trans-synaptic communication was necessary for dopamine receptor-mediated feedback to occur. In the early seventies, an analogous negative feedback mechanism modulating transmitter release was discovered for noradrenergic terminals in the peripheral nervous system (for reviews, see LANGER 1974, 1977, 1981). The fundamental aspects of this feedback system in the periphery were resolved much faster than for dopaminergic terminals in the CNS. In addition, because of the possibility of measuring simultaneously noradrenaline release and postsynaptic responses, the differences in the pharmacological properties of presynaptic α_2-autoadrenoceptors from the postsynaptic α_1-receptors were established rather early (LANGER 1974).

Looking back twenty years to the experiments of CARLSSON and LINDQVIST (1963) which germinated the dopamine hypothesis of schizophrenia, it ap-

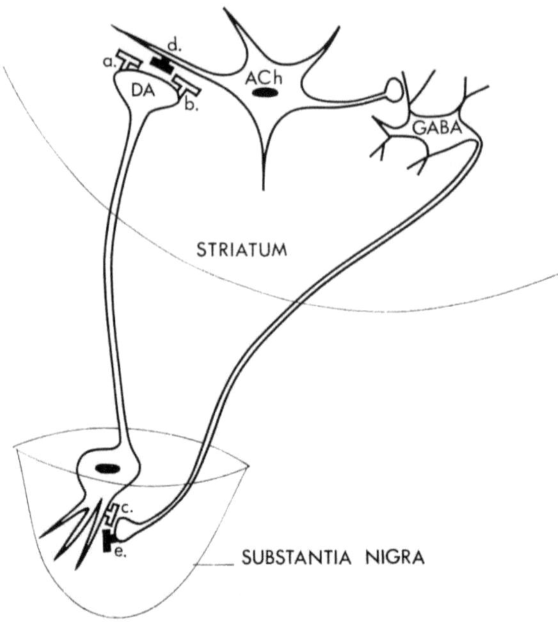

Fig. 1. Schematic representation of the localizations of postulated self-regulatory dopamine receptors. Self-regulatory dopamine receptors have been postulated to be localized to dopaminergic neurones, i.e., autoreceptors (a., b., and c.), or else to be localized postsynaptically to dopaminergic neurones (d.), or on gabaergic nerve terminals (e.), whereupon negative feedback occurs via a short or long neuronal loop. Presynaptic dopamine autoreceptors modulate depolarization-evoked release of dopamine (a.) and the synthesis of dopamine (b.). Somato-dendritic dopamine autoreceptors modulate the firing rate of dopaminergic neurones (c.). The postsynaptic dopamine receptors in the striatum (d.) are thought to be of D_1 and D_2 type, while the dopamine receptor on descending striato-nigral terminals (e.) is thought to be D_1. This simplified scheme of the biochemical neuroanatomy of the basal ganglia is adapted from models previously proposed by others (GROVES et al. 1975; WALTERS and ROTH 1976; COSTA et al. 1978).
Abbreviations: DA: dopamine, ACh: acetylcholine, GABA: γ-aminobutyric acid

pears today that these data probably reflect the interaction of neuroleptics with presynaptic dopamine autoreceptors. It is becoming apparent that presynaptic dopamine autoreceptors may play an important role in the pathology and pharmacotherapy of schizophrenia. A critical examination of the evidence for and against the existence of presynaptic dopamine receptors, their role in biological responses to dopamine receptor agonist and antagonist administration, and their possible role in schizophrenia, is the goal of the first part of this review.

To avoid confusion, definitions of several terms will be provided at the outset. *Presynaptic receptors* are considered to mean receptors which, in response to a neurotransmitter, hormone, or autacoid, modulate the synthesis and/or release of a neurotransmitter from the nerve terminal. By definition, presynaptic receptors that are involved physiologically in the modulation of neurotransmission respond to the endogenous receptor agonist which is present in the vicinity of the nerve terminals, without an involvement of receptors on dendrites or cell bodies which are the source of those nerve terminals. Presynaptic receptors may be divided into two categories: *Autoreceptors*, which modulate the synthesis and/or release of the same neurotransmitter to which they are sensitive, and *heteroreceptors*, which modulate the synthesis and/or release of a neurotransmitter different from the one to which they are sensitive.

A conceptual aspect of the definition of presynaptic receptors is that they are localized to the terminal whose neurotransmitter synthesis and/or release they modulate. At present it is very difficult to localize experimentally such presynaptic receptors to one side or the other of the 200 Å synaptic gap. The localization of presynaptic receptors to the terminals whose neurotransmitter synthesis and/or release they modulate, and the alternative second messenger hypotheses, will be a central concern of this review.

The dopamine autoreceptors modulating the electrical activity of dopaminergic neurones may be functionally related to the presynaptic dopamine autoreceptors modulating dopamine function in the dopaminergic terminals. However, referring to the dopamine autoreceptors modulating neuronal firing and located on the cell bodies or the dendrites of dopaminergic neurones as "presynaptic" invites confusion. Hence, we will refer to these dopamine autoreceptors as "somato-dendritic dopamine autoreceptors", since they modulate functions of the cell bodies ("somata") and dendrites of dopaminergic neurones.

B. The Dopamine Autoreceptor

I. Basic Functions of the Dopamine Autoreceptor

The term "dopamine autoreceptor" was introduced in 1975 by CARLSSON who suggested that its existence could explain *all* of the inhibitory actions of dopamine receptor agonists on 1) depolarization-evoked dopamine release, 2) dopamine synthesis, and 3) firing rate of dopaminergic neurones (see Fig. 2). The hypothesis as stated by CARLSSON (1975) that dopamine autoreceptors perform these three widely different basic functions provokes the questions: Are all these dopamine autoreceptors pharmacologically identical? Do they have the same cellular localization, possibly on dopaminergic neurones? Do they all function by one common mechanism, such as a calcium ion channel, or a common biochemical second messenger? We do not yet have the answers to these important questions. Despite its attractive and unifying simplicity, the hypothesis that the dopamine autoreceptors modulating the variety of functions of dopaminergic neurones are similar or identical must be approached with caution. In every experiment in which an effect is attributed to "dopamine autoreceptors", it is essential to ask *which* dopamine autoreceptors are

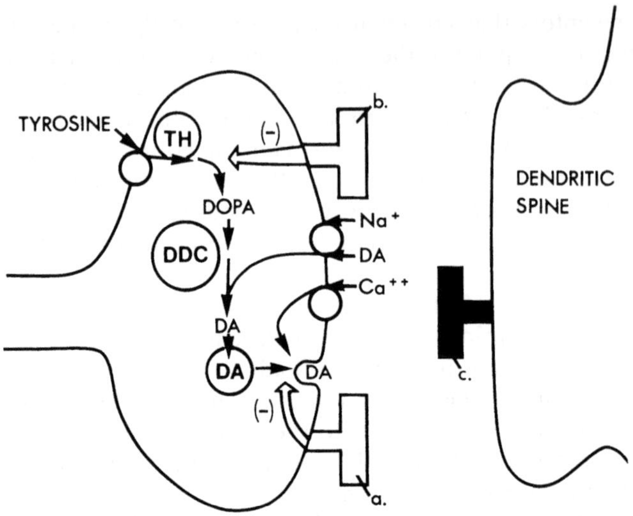

Fig. 2. The modulation of dopamine release and synthesis by inhibitory dopamine autoreceptors. The inhibitory dopamine autoreceptor modulating the release of dopamine (a.) exerts an effect only on calcium-dependent mechanisms of dopamine release. The depolarization-evoked release of endogenous dopamine, newly-synthesized dopamine, and recently taken up dopamine are all modulated by this dopamine autoreceptor. The inhibitory dopamine autoreceptor modulating the synthesis of dopamine (b.) intervenes at a point before the decarboxylation of dopa. The mechanisms and exact points of modulation for both of these inhibitory dopamine autoreceptors remain to be clarified. Shown presynaptically is also the dopamine transporter responsible for the Na^+-dependent uptake of dopamine (it can be labelled with ^{3}H-nomifensine or ^{3}H-cocaine). The postsynaptic dopamine receptor (c.) localized to dendritic spines upon which dopaminergic terminals synapse appears to be of D_2 type. The cellular or biochemical function of the D_2 receptor is not yet clarified, although it may be negatively coupled to adenylate cyclase.
Abbreviations: TH: tyrosine hydroxylase, DDC: dopa decarboxylase, DA: dopamine, DOPA: dihydroxyphenylalanine

involved: 1) autoreceptors modulating release of dopamine, 2) autoreceptors modulating synthesis of dopamine, or 3) somato-dendritic autoreceptors modulating electrical activity of dopaminergic neurones.

1. Modulation of Depolarization-Evoked Dopamine Release

a. The Origins of the Hypothesis that Dopamine Autoreceptors Modulate Dopamine Release

The first evidence supporting the existence of dopamine autoreceptors was furnished by FARNEBO and HAMBERGER (1971a), who demonstrated that the entire mechanism for dopamine receptor-mediated self-regulation of dopamine function existed in the striatal slice *in vitro*. When dopamine receptors are stimulated by apomorphine, the electrically evoked release of previously taken up ^{3}H-dopamine is reduced. Conversely, in the presence of neuroleptic dopamine receptor antagonists, ^{3}H-dopamine release is increased, suggesting that released endogenous dopamine stimulates the self-regulatory receptors.

The concept of a dopamine autoreceptor on the dopaminergic terminal, facing the synaptic gap, sensing the amount of dopamine released, and modulating the function of the dopaminergic terminal, is a simple and elegant concept, in analogy to negative feedback systems virtually ubiquitous in biochemical synthetic pathways. Nevertheless, this was not the hypothesis which FARNEBO and HAMBERGER (1971a) proposed to explain their results. They suggested rather that there was a "short feedback loop" involving a postsynaptic dopamine receptor on neurones or glia, the latter cells releasing an unidentified second messenger to act on an unidentified receptor situated on the dopaminergic terminals. In the final analysis, *there has to be some receptor on dopaminergic terminals*, since this is where the final effect on dopamine release is seen. The simpler model of an autoreceptor situated on the catecholaminergic terminal was first developed to explain the analogous inhibitory α_2-autoadrenoceptors in the peripheral nervous system (LANGER et al. 1971; ENERO et al. 1972; LANGER 1973, 1974). Later the presynaptic receptor concept was gradually appropriated to give rise to the hypothesis of a dopamine autoreceptor on dopaminergic terminals: FARNEBO and HAMBERGER (1973) proposed this hypothesis for dopaminergic terminals in the central nervous system, basing their argument on analogy to the peripheral noradrenergic system.

b. Seven Years of Controversy

Following the initial reports by FARNEBO and HAMBERGER (1971a, 1973), a number of laboratories failed to reproduce these original findings of an inhibitory dopamine autoreceptor modulating ^{3}H-dopamine release. SEEMAN and LEE (1975) reported the opposite effect from FARNEBO and HAMBERGER (1971a, 1973): neuroleptics caused a *decrease* in electrically evoked ^{3}H-dopamine release, while apomorphine was without effect. DISMUKES and MULDER (1977) carefully repeated these experiments under the conditions of both FARNEBO and HAMBERGER (1971a) and SEEMAN and LEE (1975) and essentially failed to reproduce the major findings of either laboratory. DISMUKES and MULDER (1977) did find that the apparent inhibition by neuroleptics of electrically evoked ^{3}H-dopamine release could be explained by an elevation of spontaneous outflow, an effect that was noted but not elaborated upon by SEEMAN and LEE (1975). DISMUKES and MULDER suggested that the apparent inhibition of electrically evoked ^{3}H-dopamine release observed by SEEMAN and LEE (1975) could be due to non-specific or local anaesthetic effects of the neuroleptics which were unrelated to dopamine receptors, and furthermore not specific to dopaminergic neurones. These views have subsequently been endorsed and supported by further experimental evidence (IVERSEN et al. 1976; MILLER and FRIEDHOFF 1979; SEEMAN 1981).

In the period 1973 to 1975 convincing evidence was obtained for the presence of presynaptic inhibitory dopamine receptors on peripheral noradrenergic nerves (LANGER et al. 1971; ENERO et al. 1972; LANGER 1973, 1974; FARAH and LANGER 1974; ENERO and LANGER 1975). Employing postganglionic electrical stimulation, ENERO and LANGER (1975) were able to monitor the response of the postsynaptic target organ, the cat nictitating membrane. Thus it was possible to select physiologically appropriate stimulation conditions. Lacking a measurable postsynaptic biological response to released endogenous dopamine, investigators studying the dopamine autoreceptor in the central nervous system suffered from a handicap in choosing appropriate stimulation conditions. This is probably one of the major reasons why the study of presynaptic receptors modulating noradrenaline release in the periphery advanced so much more rapidly in the past decade than did the study of central presynaptic receptors (LANGER 1974, 1977, 1981).

c. The Dopamine Autoreceptor Modulating ^{3}H-Dopamine Release Rediscovered

In retrospect, it is not so surprising that the groups trying to reproduce the results of FARNEBO and HAMBERGER (1971a) failed, but rather that FARNEBO and HAMBERGER succeeded in detecting the dopamine autoreceptor to begin with. FARNEBO and HAMBERGER (1971a) had employed a stimulation frequency of 10 Hz, while SEEMAN and LEE (1975) had stimulated at 25 Hz. From what is currently known about the frequency-dependence of autoreceptor function, in order to optimize conditions for autoreceptor modulation of transmitter release it is necessary to use a frequency in the physiological range of the firing of dopaminergic neurones, 0.5 to 8 Hz (see Sect. B.I.3).

The publication which reproduced the results of FARNEBO and HAMBERGER (1971a) employed electrical stimulation at a frequency of 3 Hz (STARKE et al. 1978). This group also reported the problem of adsorption of neuroleptic drugs to plastic tubing, which could contaminate subsequent experiments, possibly explaining the previous negative reports (SEEMAN and LEE 1975; DISMUKES and MULDER 1977). STARKE et al. (1978) found that the dopamine receptor agonists apomorphine and bromocriptine inhibited electrically evoked ^{3}H-dopamine release, and that the neuroleptic dopamine receptor antagonists chlorpromazine and haloperidol blocked this inhibition.

Taking into account the necessity 1) to stimulate the slice electrically with the same frequencies at which the dopaminergic neurones themselves fire and 2) to guard against inter-experiment contamination by neuroleptics, other groups (LANGER et al. 1980; STOOF et al. 1980) easily reproduced and extended the results of FARNEBO and HAMBERGER (1971a) and STARKE et al. (1978).

Prior to the report by STARKE et al. (1978) two other groups had reported evidence to support the existence of inhibitory dopamine autoreceptors modulating dopamine release from slices: **1)** the potassium-evoked release of ^{3}H-dopamine recently synthesized from ^{3}H-tyrosine is inhibited by dopamine, an effect antagonized by fluphenazine (WESTFALL et al. 1976) and **2)** the release of endogenous dopamine by electrical stimulation is decreased by apomorphine and increased by chlorpromazine (PLOTSKY et al. 1977). However, in both of these reports, the dopamine autoreceptor-mediated event may have been linked to modulation of synthesis and/or release, and not *exclusively* to modulation of transmitter release.

Analysis of the pattern of tritiated metabolites and dopamine released from striatal slices previously incubated with ^{3}H-dopamine has been performed in order to understand the mechanism by which dopamine autoreceptors modulate the release of dopamine (ZUMSTEIN et al. 1981). Haloperidol (10 n*M*), which increases by 38% the electrically evoked release of ^{3}H-dopamine, does not change the relative proportions of tritiated metabolites and dopamine released either during spontaneous outflow or under depolarizing conditions. Apomorphine (1 μ*M*), which reduces by 44% the electrically evoked release of ^{3}H-dopamine, reduces the fraction of total tritium released as ^{3}H-DOPAC and ^{3}H-HVA from the tissue under depolarizing conditions. These studies have confirmed that the amount of actual ^{3}H-dopamine released, and not a secondary metabolic process, is reduced by activation of the dopamine autoreceptor.

d. A Physiological Role for the Dopamine Autoreceptor?

As FARNEBO and HAMBERGER (1971a) reported, neuroleptic dopamine receptor antagonists increase the depolarization-evoked release of ^{3}H-dopamine from striatal slices. The interpretation offered for this phenomenon has been that the autoreceptor is stimulated by endogenous dopamine, creating a certain degree of receptor-mediated self-inhibition of release, which is eliminated by addition of the antagonist (REIMANN et al. 1979; LEHMANN et al. 1981). In favour of this hypothesis is the observation that the release-increasing effect of neuroleptics is directly proportional to the amount of ^{3}H-dopamine released as the frequency of stimulation (1, 3 or 6 Hz) is increased (LEHMANN and LANGER 1982c). Furthermore, when the endogenous dopamine levels are reduced by 65% by pretreatment with reserpine, the release-increasing effects of the dopamine autoreceptor antagonist sulpiride are dramatically decreased (LEHMANN and LANGER 1982c). The results of REIMANN et al. (1979) that dopamine itself is also an agonist at the dopamine autoreceptor support this hypothesis. It should be noted, however, that it is very difficult to obtain inhibition with *exogenous* dopamine itself as an agonist, requiring 1) the addition of uptake inhibitors to prevent the transport of exogenous dopamine into the nerve terminal, which otherwise decreases its concentration at the level of the dopamine autoreceptor, and also releases the preloaded ^{3}H-dopamine within the terminals, increasing the baseline of spontaneous outflow of radioactivity (REIMANN et al. 1979; ARBILLA et al. 1985); 2) the use of very low frequencies of stimulation (0.1 Hz) (REIMANN et al. 1979; ARBILLA et al. 1985); and 3) the reduction of the calcium concentration in the medium from 1.3 to 0.6 m*M* ARBILLA et al. 1982, 1985).

The increase in depolarization-evoked release caused by neuroleptics becomes much larger in magnitude when the brain slice is exposed to the inhibitor of dopamine uptake, nomifensine (LEHMANN and LANGER 1981). This is consistent with the hypothesis that nomifensine increases the concentration of dopamine in the synaptic cleft during the electrical stimulation, creating a greater activation of the dopamine autoreceptor by endogenous dopamine, the effect of which is antagonized by S-sulpiride (LEHMANN and LANGER 1981). These results also suggest that administration of dopamine re-uptake blockers *in vivo* will result in enhanced stimulation of the dopamine autoreceptor by endogenous neurotransmitter.

e. The Mechanism of Action of the Dopamine Autoreceptor Modulating Depolarization-Evoked Release of ^{3}H-Dopamine

Based on analogy to previous findings at presynaptic inhibitory α_2-autoadrenoceptors modulating noradrenaline release in the peripheral nervous system (LANGER 1981), the modulation of voltage-dependent calcium channels is a likely mechanism by which dopamine may inhibit transmitter release. While dopamine receptor agonists decrease and antagonists increase the calcium dependent release of ^{3}H-dopamine evoked by electrical stimulation (KAMAL et al. 1981) or elevated potassium (JACKISCH et al. 1980), the calcium-independent release of ^{3}H-dopamine evoked by amphetamine or tyramine is not subject to this modulation (KAMAL et al. 1981). These findings are supportive of a modulation of voltage-sensitive calcium channels by the dopamine autoreceptor, although other means of modulating calcium-dependent neurosecretion may also be possible. An essential role of voltage-dependent sodium channels in the action of the dopamine autoreceptor may be ruled out by the finding that tetrodotoxin does not modify the actions of dopamine receptor agonists or antagonists on potassium evoked ^{3}H-dopamine release (JACKISCH et al 1980).

An interesting model for the mechanism of action of presynaptic inhibitory α_2-autoadrenoceptors on noradrenergic terminals, involving **1)** the receptor-mediated reduction in conductance of voltage-dependent calcium channels and **2)** the facilitation of outward voltage-dependent potassium channels, has been described in detail (ALBERTS et al. 1981). At present, the data base given as support for these hypotheses is insufficient to draw clear conclusions. Important questions remain: a) what are the mechanisms of presynaptic receptor-mediated modulation of transmitter release, and b) are these mechanisms the same for inhibitory autoreceptors and presynaptic heteroreceptors in general? Another important question is whether a single mechanism for the dopamine autoreceptor can explain its three basic functions:

1) modulation of transmitter release,
2) modulation of dopamine synthesis,
3) modulation of the firing rate of dopaminergic neurones.

f. Are Autoreceptors Located on Dopaminergic Terminals or is There a Trans-synaptic Second Messenger?

We return now to the first hypothesis proposed by FARNEBO and HAMBERGER (1971a) that the local inhibition by dopamine receptor agonists of electrically evoked ^{3}H-dopamine release is a trans-synaptic and not an entirely presynaptic phenomenon. The concept of a second messenger released by the postsynaptic effector cell in proportion to its response to noradrenaline in the peripheral nervous system had gained support through the evidence presented by HEDQVIST (1970) that prostaglandins of the E series were released by these postsynaptic cells, and could influence noradrenaline release through an action on the terminals.

The possibility that prostaglandins mediate the inhibition of ^{3}H-dopamine release caused by dopamine receptor agonists must be considered unlikely. While PGE_2 may under certain conditions modulate the release of ^{3}H-noradrenaline from certain tissues (BERGSTRÖM et al. 1973; REIMANN et al. 1981), the electrically-evoked release of ^{3}H-dopamine from the striatum is not significantly modulated by prostaglandins, and a role for endogenous prostaglandins cannot be detected in experiments where the synthesis and release of endogenous prostaglandins is blocked by indomethacin (BERGSTRÖM et al. 1973; REIMANN et al. 1981).

It is obvious, however, that a great number of substances other than prostaglandins must be considered as candidates for the autoreceptor's hypothetical second messenger. Indeed, evidence for modulation of ^{3}H-dopamine release by a large number of other neurotransmitters has been presented (see Sect. C).

As an alternative to the hypothesis that autoreceptors are localized to the presynaptic terminal, the trans-synaptic model must supply a second messenger which fulfils certain criteria.

1) Dopamine release must be modulated not only by the exogenous, but also by the *endogenous*, second messenger.
2) The release of the second messenger must be modulated by dopamine receptor agonists and by *endogenous* dopamine itself.

It is in fact fairly easy to exclude a large number of neurotransmitters, hormones, and autacoids from roles as second messengers, as long as a receptor antagonist for the substance is available. Acetylcholine, GABA, L-aspartate, and L-glutamate, all substances reported to affect ^{3}H-dopamine release, have been excluded, since the dopamine autoreceptor still functions in the presence of antagonists of their receptors (LEHMANN and LANGER 1982b).

JACKISCH et al. (1980) found that the modulation by the dopamine autoreceptor of ^{3}H-dopamine release evoked by potassium persisted in the presence of tetrodotoxin, and suggested therefore that interneurones were not involved in this modulation. The belief that tetrodotoxin completely eliminates interneuronal communication is a common but incorrect assumption. Tetrodotoxin blocks voltage-dependent sodium channels which participate in action potential conduction along axons. There are, however, dendritic action potentials which are insensitive to tetrodotoxin, being mediated by voltage-dependent, tetrodotoxin-resistant calcium channels (LLINAS and HESS 1976). Such voltage-dependent calcium channels are involved in neurotransmitter release from nerve terminals, and probably also from dendrites (HATTORI et al. 1979; JOHNSON and PILAR 1980; MARTINEZ and ADLER-GRASCHINSKY 1980a, b). Thus, tetrodotoxin-resistant release of neurotransmitter can certainly occur from neural structures during depolarization by potassium. As JACKISCH et al. (1980) demonstrated, such release may be modulated by receptor-mediated processes. What is learned from the resistance to tetrodotoxin of dopamine autoreceptor-mediated effects is that these effects do not depend on axonal conduction in interneurones, nor on action potentials in dopaminergic terminals. Resistance to tetrodotoxin of dopamine autoreceptor-mediated effects does not, however, rule out the involvement of any neurotransmitter as a second messenger.

A case in point of the fallacy of considering tetrodotoxin as an eliminator of interneuronal function is to be found in the study of excitatory amino acid receptors modulating ^{3}H-dopamine release. It was found that L-glutamate evokes calcium-dependent release of ^{3}H-dopamine from slices of rat striatum (GIORGUIEFF et al. 1977a; ROBERTS and SHARIF 1978; ROBERTS and ANDERSON 1979). Such release was found to be insensitive to tetrodotoxin. However, after performing kainic acid lesions of the striatum, which destroys interneurones without harming afferent terminals (COYLE and SCHWARTZ 1976; MCGEER and MCGEER 1976), the effect of L-glutamate disappears (ROBERTS and ANDERSON 1979). Thus, it appears that interneurones are indeed involved in this response which is tetrodotoxin-resistant (see also Sect. C.III).

In this laboratory we have used kainic acid lesions to test the involvement of interneurones in dopamine autoreceptor-mediated effects. Although kainic acid has been found to be a useful tool for biochemical, anatomical and behavioural experiments (LEHMANN et al. 1979; SANBERG et al. 1979), the biological viability of slices from striatum previously lesioned with kainic acid is extremely poor. Despite our results that the dopamine autoreceptor continues to function after kainic acid lesions in the cat striatum (LEHMANN and LANGER, unpublished results), we feel that additional experiments are needed to clarify this point.

g. Autoreceptor Modulation of ^{3}H-Dopamine Release from Synaptosomes

The dopamine autoreceptor involved in the negative feedback modulation of depolarization-evoked ^{3}H-dopamine release may be localized to dopaminergic terminals, and involve a completely presynaptic mechanism. Alternatively, the dopamine receptor involved in this self-regulatory process may be localized to dendrites, terminals, or glial cells adjacent to dopaminergic terminals. The release of a second messenger from such postsynaptic structures might be modulated by the postsynaptic dopamine receptor; the second messenger may then interact with a specific receptor on dopaminergic terminals to modulate depolarization-evoked ^{3}H-dopamine release. These possibilities are difficult to exclude by experiments which rely upon pharmacological interventions. Indeed, the only solution would appear to be to physically separate these other structures from dopaminergic terminals and to test if they maintain the capacity to autoregulate the release of ^{3}H-dopamine via the autoreceptor.

This approach has been pursued using the synaptosome preparations. Resealed nerve terminals in sufficiently dilute suspension may be considered as isolated structures. In order to measure depolarization-evoked ^{3}H-dopamine release, such synaptosomal preparations may be superfused and depolarized with chemical stimuli such as elevated potassium or veratridine.

Two laboratories have attempted to study the modulation by the dopamine autoreceptor of the release of ^{3}H-dopamine from synaptosomes. One of these groups (WESTFALL et al. 1979b) reported positive results; methods and results were only sparsely described, however. Apomorphine (1 μM and 10 μM) inhibits the potassium (56 mM) evoked release of previously taken up ^{3}H-dopamine from rat striatal synaptosomes. The inhibition by 1 μM apomorphine (30%) is abolished by 1 μM chlorpromazine.

RAITERI et al. (1978, 1979) found that 1 μM apomorphine did not inhibit the release of previously taken up ^{3}H-dopamine from rat striatal synaptosomes, evoked either by veratridine or by potassium. Furthermore, the potassium (30 mM) or veratridine-evoked release of ^{3}H-dopamine synthesized from ^{3}H-tyrosine was not affected by 1 μM apomorphine. These workers concluded that dopamine autoreceptors modulating the release of ^{3}H-dopamine did not, therefore, exist on dopaminergic terminals. RAITERI et al. (1978, 1979) also failed to find evidence for dopamine autoreceptors modulating the synthesis of dopamine (see Sect. B.I.2).

Neither of these groups (WESTFALL et al. 1979b; RAITERI et al. 1978, 1979) studying ^{3}H-dopamine release from synaptosomes has reported testing the calcium dependence of release under their conditions. Previous workers have found that potassium-evoked ^{14}C-dopamine release from synaptosomes is *not* calcium-dependent (DE BELLEROCHE et al. 1976). As discussed above, it has been shown that the dopamine autoreceptor regulates only calcium-dependent ^{3}H-dopamine release (KAMAL et al. 1981).

In view of these considerations the relevance of synaptosomes to the physiology of normal dopaminergic terminals should be regarded with caution.

One of the more convincing arguments offered by RAITERI et al. (1978, 1979) in support of their conclusions is that presynaptic α-adrenoceptor modulation of ^{3}H-noradrenaline release from synaptosomes can be observed in their preparations. This supports their contention that the failure to observe dopamine autoreceptor modulation of ^{3}H-dopamine release is not an artefactual negative finding due to inappropriate techniques. The α-adrenoceptor modulating depolarization-evoked release of ^{3}H-noradrenaline from the cortex has been studied in a careful comparative study of the slice preparation and synaptosomes (MULDER et al. 1979). Noradrenaline (1 μM) inhibits the release of ^{3}H-noradrenaline by 63% from cortex slices, but by only 33% from cortical synaptosomes. Adrenaline (1 μM) inhibits the release of ^{3}H-noradrenaline by 62% from cortex slices, but by only 32% from cortical synaptosomes. It is seen that, although the α-adrenoceptor agonists adrenaline and noradrenaline continue to inhibit ^{3}H-noradrenaline release in synaptosomes, they are much less effective in doing so than in slices. It may be that presynaptic receptors are labile and may lose activity in the process of synaptosome preparation. The concept of a receptor-effector complex which is normally coupled but which may be uncoupled, as for the dopamine receptor coupled to the enzyme adenylate cyclase (KEBABIAN 1978), may also apply to the dopamine autoreceptor and explain the loss of activity of this receptor in synaptosomes. It is also possible that problems of experimental methods must still be resolved in order to reliably detect the dopamine autoreceptor modulating ^{3}H-dopamine release from synaptosomes, just as a great deal of searching was required to find the appropriate stimulation techniques to reliably detect the dopamine autoreceptor modulating ^{3}H-dopamine release from slices (see Sects. B.I.1.c and B.I.1.d).

h. Conclusions

The dopamine autoreceptor modulating ^{3}H-dopamine release evoked from striatal slices may be easily observed under conditions of stimulation which approximate the physiological frequency of firing of dopaminergic neurones. This dopamine autoreceptor is sensitive to low concentrations of dopamine receptor agonists and antagonists. The recent availability of dopamine receptor agonists and antagonists capable of distinguishing between D_1 and D_2 dopamine receptors has allowed us to establish that the autoreceptor is not of the D_1 (adenylate cyclase-linked) type (see Sect. B.VI.1). However, it is not clear at this time whether the dopamine autoreceptor is identical with the postsynaptic D_2 dopamine receptor. The dopamine autoreceptor modulates only the calcium-dependent depolarization evoked release of ^{3}H-dopamine. Although a second messenger trans-synaptic mechanism for the dopamine autoreceptor has been considered in the literature for more than ten years, no such possible second messenger has been found or proposed. At the same time, the localization to dopaminergic terminals of these release-modulating dopamine autoreceptors has not yet been convincingly established.

2. Synthesis of Dopamine, *In Vitro* and *In Vivo*

a. The Origins of the Hypothesis that Dopamine Autoreceptors Modulate the Synthesis of Dopamine

In 1972, KEHR et al. demonstrated that dopaminergic terminals contain receptors which modulate the synthesis of dopamine either in response to a dopamine receptor agonist, or in response to the dopamine which the terminals themselves release. In an elegant series of experiments, these authors showed that the increase in dopamine synthesis, caused by administration of the neuroleptic haloperidol, could be entirely explained by its blockade of these dopamine autoreceptors.

The modulation of dopamine synthesis by dopamine receptors appears to occur at the level of tyrosine hydroxylation, which is rate-limiting for the synthesis of dopamine. Dopa, which is formed from tyrosine, is normally rapidly decarboxylated to dopamine. In order to measure the rate of tyrosine hydroxylation *in vivo*, KEHR et al. (1972) administered a dopadecarboxylase inhibitor and measured the amount of dopa which accumulated during a given time interval. In the striatum, the terminal region of nigrostriatal dopaminergic neurones, the dopamine receptor agonist apomorphine decreased the accumulation of dopa, an effect that was blocked by haloperiodol. Haloperidol by itself caused an increase in dopa accumulation, an observation suggesting that the inhibitory dopamine receptor modulating dopamine synthesis was normally stimulated by released dopamine. Similar results were subsequently also found in the limbic forebrain, but not in the occipital cortex (KEHR et al. 1977).

In the intact brain, two different mechanisms could account for this phenomenon: either there is a neuronal feedback loop returning from the striatum to dopaminergic cell bodies in the substantia nigra, or else the dopaminergic nerve terminals are capable of responding to the stimulation of dopamine receptors in the striatum (Fig. 1). KEHR et al. (1972) performed coronal hemitransections of the brain between the substantia nigra and the striatum, eliminating all the possible neuronal feedback inputs to dopaminergic cells bodies, and separating the dopaminergic cell bodies from their axons and terminals. Under these conditions, impulse flow is lost, dopamine release is discontinued, and yet the accumulation of dopa is accelerated. KEHR et al. (1972) reasoned that this was because dopamine was no longer being released to stimulate the inhibitory dopamine receptor which otherwise reduces the rate of tyrosine hydroxylation. This was supported by their finding that haloperidol did not produce a further increase in dopa accumulation following hemitransection. The finding that, after hemitransection, apomorphine was still able to inhibit dopa accumulation, and that this action was antagonized by haloperidol, indicated that the inhibitory receptor-mediated autoregulation was still completely functional without the neuronal feedback loop involving striatonigral input upon dopaminergic cell bodies.

The conclusion reached by KEHR et al. (1972) was carefully formulated. Obviously, the response to these drugs was in the dopaminergic terminal, and could occur without cell body or impulse flow. The dopamine receptor must be localized such that it receives significant amounts of dopamine under normal conditions. Two alternative possibilities were proposed: either a) a postsynaptic dopamine receptor subsequently activates a trans-synaptic second messenger, or b) there is a dopamine receptor located on the dopaminergic terminal, facing the synaptic cleft.

b. The Localization of Synthesis Modulating Dopamine Autoreceptors to Dopaminergic Terminals

CHRISTIANSEN and SQUIRES (1974a) were the first to test these two alternative hypotheses regarding the localization of this dopamine receptor modulating dopamine synthesis. In synaptosomes, nerve terminals reseal, and in suspension can be considered as functioning in isolation and independent of other tissue components. When labelled tyrosine was used as precursor, the rate of labelled catechol formation in synaptosomes was decreased more than 50% in the presence of 0.6 μM apomorphine. Half of this inhibition was antagonized by 0.1 μM haloperidol. CHRISTIANSEN and SQUIRES concluded that the dopamine receptor-mediated inhibition of dopamine synthesis did occur in synaptosomes, strongly suggesting that the autoregulatory receptor was localized to dopaminergic terminals. Moreover, the potency of a number of neuroleptics in antagonizing the inhibition by apomorphine of dopamine synthesis positively correlated both with their therapeutic potency and with their ability to increase dopamine turnover *in vivo* (CHRISTIANSEN and SQUIRES 1974b). These authors were aware that the directly acting dopamine receptor agonist, apomorphine, was also a directly acting tyrosine hydroxylase inhibitor. Therefore the component of the inhibition by apomorphine of dopamine synthesis in synaptosomes that was not antagonized by neuroleptics was interpreted to be due to direct inhibition of the enzyme tyrosine hydroxylase.

IVERSEN et al. (1976) confirmed and extended the observations of CHRISTIANSEN and SQUIRES (1974a, b) by comparing the dopamine receptor inhibiting dopamine synthesis in synaptosomes with the dopamine-sensitive adenylate cyclase. The latter receptor was considered at the time as "the postsynaptic dopamine receptor". It is of interest to note that IVERSEN et al. (1976) cautiously noted the possibility of a pharmacological difference between the dopamine autoreceptor and the dopamine-stimulated adenylate cyclase: "The presynaptic 'autoreceptors' for dopamine on dopaminergic terminals may thus differ slightly in their pharmacological specificity from postsynaptic dopamine receptor sites." Today the dopamine autoreceptor is very easily distinguished from the dopamine-sensitive adenylate cyclase. We are now assuming the same careful attitude in pharmacologically distinguishing dopamine autoreceptors from the D_2 dopamine receptor, which iscurrently considered "the postsynaptic dopamine receptor" of physiological importance in endocrinological, neurological and psychiatric disorders (KEBABIAN and CALNE 1979).

IVERSEN et al. (1976) found that dopamine uptake inhibitors were effective in "antagonizing" the inhibition of dopamine synthesis by a number of catecholic dopamine receptor agonists, suggesting that these agonists were accumulated by the high-affinity dopamine uptake system and then inhibited the enzyme tyrosine hydroxylase directly. Most dopamine receptor antagonists are unable to completely antagonize the inhibition by apomorphine of dopamine synthesis. However, the concentrations of neuroleptic which are effective in antagonizing the effects of apomorphine are reasonably low: 16 n*M* for haloperidol, 35 n*M* for spiperone. IVERSEN et al. (1976) found that the pharmacological profile of the dopamine autoreceptor modulating dopamine synthesis correlated better with the therapeutic potency of neuroleptics, than did the pharmacological profile of the dopamine-stimulated adenylate cyclase. Nevertheless it was concluded at that time that the adenylate cyclase hat to be the important site of action of neuroleptics (IVERSEN et al. 1976). The avialability of a selective antagonist for the D_1 receptor, SCH 23390 (IORIO et al. 1983) has now added a powerful pharmacological tool in the characterization of dopamine receptor subtypes.

c. The Case Against Dopamine Autoreceptors Modulating the Synthesis of Dopamine

The challenge to the hypothesis stating that the inhibitory dopamine autoreceptor modulating dopamine synthesis was localized to dopaminergic terminals came from ZIVKOVIC et al. (1976), who argued that the acceleration of dopa accumulation caused by either axotomy or haloperidol was not due to a disinhibition phenomenon occurring at the dopamine autoreceptor, but rather to a direct activation of tyrosine hydroxylase, due to depolarization of the dopaminergic nerve terminal. The inhibition of dopamine synthesis by dopamine receptor agonists, they argued, was due to a direct inhibition of tyrosine hydroxylase by the catechol groups, which mimicked the end product inhibition by dopamine. While the arguments of ZIVKOVIC et al. (1976) were primarily directed toward the *in vivo* situation, analogous arguments are derived from experiments in superfused synaptosomes which suggested that whether by active uptake or passive diffusion, dopamine agonists with catechol moieties must enter the synaptosome in order to inhibit dopamine synthesis (CERRITO et al. 1981). In synaptosomes, for instance, the inhibitory effect of dopamine itself on dopamine synthesis is almost completely blocked by nomifensine (10 μ*M*) (CERRITO et al. 1981). In these studies in synaptosomes, however, the effects of dopamine receptor antagonists were not tested, and the successful receptor antagonism by neuroleptics found by the other laboratories working with synaptosomes were not explained by CERRITO et al. (1981).

The arguments raised by ZIVKOVIC et al. (1976) and CERRITO et al. (1981) prompted the testing of different dopamine receptor agonists which did not possess catechol groups. Bromocriptine and lergotrile were found to slightly diminish dopamine synthesis in synaptosomes (GOLDSTEIN et al. 1978; WESTFALL et al. 1979b). The modest inhibitory effect of these ergoline dopamine receptor agonists was ascribed to the fact that they are partial agonists, since they also antagonize the inhibition by the agonist apomorphine (GOLDSTEIN et al. 1978; WESTFALL et al. 1979b). This finding for bromocriptine is in agreement with studies on the dopamine autoreceptor modulating electrically evoked ^{3}H-dopamine release (LEHMANN and LANGER 1982c) and in locomotor sedation studies (COSTALL et al. 1981) where bromocriptine is also a partial agonist.

The non-catecholic dopamine receptor agonist 7-hydroxy-N,N-dipropyl-2-aminotetralin (7-HAT) does not inhibit directly the enzyme tyrosine hydroxylase in synaptosomes even at concentrations up to 1 m*M*, but this compound inhibits dopamine synthesis in micromolar concentrations (WAGGONER et al. 1980), and this inhibition by 7-HAT is antagonized by the neuroleptic fluphenazine (1 μ*M*) but not by the inhibitor of neuronal uptake of dopamine, benztropine (5 μ*M*). Thus, while direct inhibition of tyrosine hydroxylase by some dopamine receptor agonists accounts to some extent for the inhibition of dopamine synthesis in synaptosomes, dopamine receptor-mediated inhibition of dopamine synthesis in synaptosomes has been clearly established.

d. Conclusions

In summary, inhibitory dopamine autoreceptors modulating dopamine synthesis can definitely be detected *in vivo* and in synaptosomes, but it appears that in synaptosomes the receptor has lower affinity and possibly lower maximal inhibition than either *in vivo* or in slices. For instance, even though the inhibition of dopamine synthesis in synaptosomes is antagonized by inhibitors of dopamine uptake, the inhibition of dopamine synthesis by dopamine in slices is unaffected by 1 μ*M* benztropine, but blocked completely by 1 μ*M* fluphenazine (WESTFALL et al. 1976). The various laboratories that have detected the dopamine autoreceptor modulating synthesis of dopamine in synaptosomes have found different degrees of receptor-mediated inhibition by apomorphine compared to direct inhibition of the enzyme tyrosine hydroxylase: CHRISTIANSEN and SQUIRES (1974a) found that 50% of the inhibition by apomorphine could be antagonized by dopamine receptor antagonists. IVERSEN et al. (1976) found that the inhibition by low concentrations of apomorphine was completely antagonized by some dopamine receptor antagonists, while WAGGONER et al. (1980) failed to find any convincing antagonism by neuroleptics of the inhibition by apomorphine of dopamine synthesis. It would appear that different preparations of synaptosomes exhibit different degrees of modulation of synthesis via the dopamine autoreceptor. Methodological differences in the preparation of synaptosomes may account for the variability in results reported by various groups. The loss of dopamine autoreceptor function in synaptosomes, which may be related to uncoupling of the receptor from its effector and occur to a varying extent depending on preparation methods, has also been observed in neurotransmitter release studies (Sect. B.I.1.g).

The dopaminergic system of the striatum is unique, in that the modulation of neurotransmitter synthesis by autoreceptors has been extensively investigated and documented. To our knowledge, there are no equivalent studies on the modulation of the synthesis of 5-HT or noradrenaline by their respective autoreceptors, even though autoreceptors modulating the depolarization-evoked release of 5-HT and noradrenaline exist (LANGER 1981; LANGER and MORET 1982; see also Sect. D). The synthesis of noradrenaline seems in fact not to be modulated by presynaptic α-autoadrenoceptors (STEINBERG and KELLER 1978). Muscarinic autoreceptors modulating the synthesis of acetylcholine have been reported (BOWEN and MAREK 1982), but have not been investigated to the same extent as dopamine autoreceptors modulating dopamine synthesis.

3. Electrical Activity of Dopaminergic Neurones

a. The Origins of the Hypothesis that Dopamine Autoreceptors Modulate the Firing of Dopaminergic Neurones

The concept of a long-loop feedback mechanism postulates that postsynaptic dopamine receptor stimulation results in inhibition of the firing rate of dopaminergic cell bodies in the substantia nigra, via descending striato-nigral axons (Fig. 1). The anatomical demonstration of the striato-nigral projection was the original basis of the trans-synaptic long-loop feedback hypothesis (STEVENS 1973). At the same time as the first actual evidence was obtained to support the existence of the long-loop feedback, concurrent observations suggested the existence of inhibitory dopamine receptors on dopaminergic neurones (Aghajanian and Bunney 1973). When dopamine or the directly acting dopamine receptor agonist, apomorphine, is applied by iontophoresis into the substantia nigra, the firing rate of dopaminergic neurones is reduced. The opposite effect, an increase in firing rate of dopaminergic neurones, occurs upon intravenous admistration of haloperidol or chlorpromazine. This report by Aghajanian and Bunney (1973) is frequently cited as the discovery of the dopamine autoreceptor. The same authors contributed to a study on the inhibition of dopaminergic neurone firing by the indirectly acting dopamine agonist d-amphetamine, administered systemically (Bunney et al. 1973). This inhibition by d-amphetamine was also antagonized by systemic administration of neuroleptic dopamine receptor antagonists. In contrast to the publication by Aghajanian and Bunney (1973), Bunney et al. (1973) did not consider the possibility of a dopamine receptor located on dopaminergic neurones to explain the inhibitory effects of systemic amphetamine. Rather, a polysynaptic neuronal feedback via the striato-nigral projection was offered as an explanation for the depression of firing of dopaminergic neurones produced by amphetamine (see Fig. 1). As in other experimental approaches, these two alternative and mutually exclusive models tended to polarize scientific opinion.

a. The Debate over the Polysynaptic Feedback Loop vs. an Autoreceptor Mechanism for the Actions of Amphetamine

Elegant experiments by Groves et al. (1975) demonstrated that amphetamine infused into the striatum did not inhibit, but rather increased, the rate of firing of dopaminergic neurones in the substantia nigra. It was only when amphetamine was infused into the substantia nigra that it inhibited the firing of dopaminergic neurones. Likewise, haloperidol infused into the striatum caused a decrease in the firing rate of dopaminergic neurones, but when infused into the substantia nigra produced an increase in firing rate of dopaminergic neurones. Thus the existence of a poly- or monosynaptic striato-nigral feedback loop was confirmed by these workers; but it was a *positive* feedback loop! In other words, stimulation of postsynaptic dopamine receptors in the striatum produces an activation of the firing of dopaminergic neurones in the substantia nigra. This polysynaptic activation of dopaminergic neurone firing obviously could not explain the inhibition of dopaminergic neurone firing caused by systemic administration of dopamine receptor agonists. The *inhibitory* dopamine receptor modulating spike activity of dopaminergic neurones thus clearly appeared not to be localized to the striatum, but to the substantia nigra, and possibly localized to dopaminergic neurones (Fig. 1).

The study by Groves et al. (1975) was challenged, however, by that of Bunney and Aghajanian (1976), who presented results showing that amphetamine was very weak in depressing the firing rate of dopaminergic neurones when applied iontophoretically in the substantia nigra. In fact, amphetamine only decreased the firing rate of neurones when it also decreased the height of action potentials, the latter observation suggesting a nonspecific, possibly local anaesthetic action, but not a dopamine receptor-mediated effect. Furthermore, BUNNEY and AGHAJANIAN (1976) found that discrete lesions of the crus cerebri, which interrupt the descending striato-nigral axons while sparing dopaminergic nigrostriatal axons, eliminated the firing rate depression caused by systemically administered amphetamine.

The controversy was largely resolved in 1978 when BUNNEY and AGHAJANIAN made use of the recently discovered neurotoxin kainic acid. Kainic acid destroys neuronal perikarya and spares afferents and axons of passage when injected locally in the striatum. Following such lesions, BUNNEY and AGHAJANIAN (1978) performed dose-response studies on the inhibition of firing rate of dopaminergic neurones by intravenously administered amphetamine. They found a shift to the right in the dose-response curve to amphetamine after kainic acid lesions, the ID_{50} increasing 5-fold and the maximal inhibition decreasing by 40%. Their conclusion was therefore that autoreceptors in the substantia nigra may account for a significant proportion of firing rate inhibition by amphetamine administered i.v. particularly at higher doses (2 mg/kg and above), but that there was an important contribution of a striato-nigral feedback loop in the inhibitory effect of low doses of amphetamine on dopaminergic cell firing.

The relative contribution of trans-synaptic long-loop feedback, as opposed to a direct somato-dendritic action of amphetamine, has recently been re-examined by WANG (1981b). After transection of the diencephalon, WANG (1981b) found no change in the ID_{50} of amphetamine in inhibiting the spontaneous firing of dopamine cells in the ventral tegmental area. These results suggest that the inhibition of dopamine cell firing by amphetamine in the ventral tegmental area does not involve descending striato-nigral axons, and can be explained more readily by a local action in the ventral tegmental area, possibly by the involvement of dopamine autoreceptors.

c. Autoreceptor Mechanism for the Effects of Apomorphine and Dopamine on the Firing of Dopaminergic Neurones

The lively debate concerning the effects of amphetamine on dopaminergic neurones was conspicuously absent when the effects of directly acting dopamine receptor agonists were studied. Although the respective contributions of a polysynaptic feedback loop involving striato-nigral axons versus local negative feedback involving dopamine autoreceptors was difficult to resolve for the effects of amphetamine, no such difficulty was encountered in explaining the actions of apomorphine. The iontophoretic application of the directly acting dopamine receptor agonists apomorphine or dopamine in the substantia nigra reliably inhibited the firing of dopaminergic neurones (AGHAJANIAN and BUNNEY 1973). This effect of directly-acting dopamine receptor agonists was blocked by intravenously administered neuroleptics (AGHAJANIAN and BUNNEY 1973).

AGHAJANIAN and BUNNEY, who supported the importance of the long polysynaptic feedback loop for the inhibitory effects of amphetamine on dopaminergic cell firing, searched carefully for evidence of a similar polysynaptic mechanism of the inhibitory action of directly acting dopamine receptor agonists. While transection of the crus cerebri, which contains the descending striato-nigral axons, greatly reduces the effects of amphetamine, the same operation has no effect on the inhibition of dopaminergic cell firing by locally or intravenously administered apomorphine (BUNNEY and AGHAJANIAN 1976). Similarly, kainic acid lesions of the striatum, which also destroy the descending striato-nigral axons, and greatly reduce the inhibitory effect of amphetamine on dopaminergic cell firing, do not affect the inhibition exerted by apomorphine (BUNNEY and AGHAJANIAN 1978). Finally, intravenous administration of picrotoxin, which blocks GABA-receptor-mediated inhibition, diminishes substantially the inhibition of dopaminergic cell firing by amphetamine but not that by apomorphine (BUNNEY and AGHAJANIAN 1978). The failure of strychnine to modify the inhibition of cell firing by apomorphine excluded the possibility that a glycinergic interneurone might be involved in the local feedback inhibition mediated by dopamine receptor agonists (BUNNEY and AGHAJANIAN 1978). Thus, in this series of experiments, where the functional polysynaptic feedback loop was eliminated by three different procedures, the inhibition of dopaminergic cell firing by directly acting dopamine receptor agonists remained unaltered. This prompted AGHAJANIAN and BUNNEY (1978) to conclude that the existence of inhibitory dopamine autoreceptors on the somata and/or dendrites of dopaminergic neurones was highly probable: "Combined, these findings strongly suggest that the inhibitory effect of microiontophoretically applied dopamine on zona compacta neurones is mediated via an interaction of dopamine with autoreceptors on dopaminergic cell bodies or dendrites rather than through dopamine-induced release of GABA from afferent nerve terminals".

d. Pharmacological Aspects of the Dopamine Autoreceptor Modulating the Firing of Dopaminergic Neurones

Iontophoretically applied sulpiride stereoselectively antagonizes the inhibition by dopamine of the firing of dopaminergic neurones (WOODRUFF and PINNOCK 1981), while neither stereoisomer of sulpiride affects the inhibition by GABA of these neurones. Since sulpiride does not antagonize the stimulation by dopamine of adenylate cyclase (RUPNIAK et al. 1981), this finding eliminates the possibility that cAMP or the nigral adenylate cyclase localized to striatonigral terminals may be involved in the dopamine receptor-mediated inhibition of dopaminergic cell firing, contrary to what had been previously proposed (SPANO et al. 1976).

It appears that postsynaptic dopamine receptors in the striatum are not pharmacologically identical with the dopamine autoreceptors modulating the firing rate of dopaminergic neurones. Dopamine neurones are six to ten times more sensitive to iontophoretically applied dopamine or intravenously administered apomorphine than are the spontaneously active non-dopamine-containing neurones of the striatum (SKIRBOLL et al. 1979). This is consistent with other aspects of dopamine receptor function, notably behavioural (see

Sect. B.II.1), where dopamine autoreceptors are found to be more sensitive to agonists than are postsynaptic dopamine receptors.

e. Electrophysiological Aspects of the Function of the Dopamine Autoreceptor

By using statistical spike analysis, WILSON et al. (1979) were able to establish that a "trough" of inactivity follows the spike activity of dopaminergic neurones. It is this post-spike depression of electrical activity which is blocked by haloperidol. These workers suggested that haloperidol blocks the autoinhibition caused by action potential-evoked release of dopamine from nigral dendrites. The increase in probability of firing caused by haloperidol concerns a time period of 200–300 ms after an action potential (WILSON et al. 1979). The haloperidol-sensitive post-spike depression is associated with 1) a latency period of about 200 ms, and 2) a limited duration of 100 ms.

The phenomenon of post-action potential autoinhibition was studied more directly by NAKAMURA et al. (1979). Following kainic acid lesions of the striatum, pars compacta units were activated antidromically from the lesioned striatum. For a period of 240 ms (range 50–500 ms), inhibition of spontaneous spike activity was observed. These time parameters did not differ when the experiment was performed on the control, unlesioned side of the animal. Haloperidol (0.1 mg/kg i.v.), however, diminished the mean duration of post-antidromic action potential inhibition. Similar experiments performed on dopamine neurones in the ventral tegmental area (WANG 1981a) yielded results in general agreement with those of NAKAMURA et al. (1979). These results thus support the hypothesis of WILSON et al. (1979) that autoinhibition of firing of dopaminergic neurones via a dopamine receptor-mediated mechanism occurs maximally at approximately 250 ms following the occurrence of an action potential in dopaminergic neurones. Furthermore, this autoinhibition is dependent neither upon descending striatonigral axons nor upon neurones postsynaptic to dopaminergic terminals.

The baseline rate of firing of dopaminergic neurones plays a role in determining the effect of applied neuroleptic in increasing their firing rate. Dopaminergic neurones fire at frequencies relatively low (0.5 to 8 spikes/s) compared to other neurones in the CNS (GUYENET and AGHAJANIAN 1978). However, autocorrelation analysis suggests that self-inhibition does not occur significantly for those dopaminergic neurones firing at less than 3 spikes/s, and this is supported by the observation that haloperidol does not increase the firing rate of these relatively slowly firing units (WILSON et al. 1979). These data are in excellent agreement with results on the dopamine autoreceptor modulating ^{3}H-dopamine release evoked by electrical field stimulation in the striatum, where much greater disinhibition by neuroleptic dopamine receptor antagonists is observed at stimulation frequencies of 3 and 6 Hz than at lower frequencies of stimulation (see Sect. B.I.1). Results of recording unit activities in the substantia nigra of freely-moving cats have also confirmed the increase by neuroleptics and decrease by apomorphine of dopaminergic neurone firing frequency (TRULSON et al. 1981), supporting the concept that dopamine autoreceptor function can be observed under physiological circumstances, as well as under the most exigent experimental manipulations.

f. Conclusions

The localization to dopaminergic neurones of the inhibitory dopamine autoreceptor modulating the firing rate of dopaminergic neurones may be considered likely, but has not yet been rigorously demonstrated with electrophysiological methods. As noted above, several plausible second messengers have been eliminated by experimental test. The pharmacological similarity or identity of these dopamine autoreceptors modulating the electrical activity of dopaminergic neurones with dopamine autoreceptors modulating transmitter synthesis and release has not yet been established.

g. Questions for Future Electrophysiological Studies of Dopamine Autoreceptors

Among the wide variety of functions attributed to dopamine autoreceptors, dopamine autoreceptors on terminals or axons of dopaminergic neurones have also been reported to modulate the production and propagation of antidromic action potentials (GROVES et al. 1981).

The existence of postsynaptic receptors which modulate the membrane potential has been documented over a number of decades. There are, however, receptors which do not modulate the membrane potential of neurones directly, but rather modulate ion channels which are opened or closed by changes in membrane potential. These receptors modulating voltage-dependent currents carried by calcium, chloride, or potassium (PELLMAR 1981) share striking similarities with presynaptic receptors which modulate the *depolarization-evoked release* of neurotransmitter from nerve terminals, but have not effects of their own on *transmitter outflow from the resting terminal* (for review, see LANGER 1981). There may be an important similarity in the mechanisms of presynaptic dopamine autoreceptors, somatodendritic dopamine autoreceptors, and receptors which modulate voltage-dependent ion currents. This open question is an interesting area for future research in the field of presynaptic receptors.

II. Studies on the Functions of the Dopamine Autoreceptor

The dopamine receptor-mediated decrease in the dopamine metabolites HVA and DOPAC caused by the administration of apomorphine and other direct dopamine receptor agonists may be the result of any or all of the three basic functions attributed to dopamine autoreceptor activation: inhibition of release of dopamine, inhibition of synthesis of dopamine, or inhibition of firing of dopaminergic neurones.

In attempting to dissect out the multiple actions of the dopamine autoreceptor, it is necessary to use different experimental approaches. In order to measure release without also reflecting synthesis, the *in vitro* slice prelabelled with ^{3}H-dopamine is virtually obligatory. In order to measure synthesis *in vivo* the decarboxylation of dopa must be blocked, which is in itself a non-negligible pharmacological intervention disrupting the normal functioning of the brain. In electrophysiological studies, it is generally necessary to anaesthetize or immobilize the animal.

1. Locomotor Sedation

a. The Origin of Behavioural Models for Dopamine Autoreceptor Function

The ultimate goal of neuroscience is to explain behaviour in biological terms. A behavioural role for the dopamine autoreceptor was postulated on the basis of paradoxical observations first recorded in 1974: although moderate doses of the directly acting dopamine receptor agonist apomorphine increased locomotor activity in mice, considerably lower doses of apomorphine caused the opposite effect, namely a decrease in locomotor activity (PUECH et al. 1974; THORNBURG and MOORE 1974).

STRÖMBOM (1976) and DI CHIARA et al. (1976) considered the potential importance of this phenomenon in view of the theoretical basis for the dopamine autoreceptor laid down by AGHAJANIAN and BUNNEY (1973) and KEHR et al. (1972). STRÖMBOM and DI CHIARA et al. tested the hypothesis that the dopamine autoreceptor might mediate locomotor sedation by low doses of apomorphine. Three different behavioural responses are observed in response to different doses of apomorphine: a) at high doses, apomorphine induces stereotypy; b) at intermediate doses, locomotor activity is stimulated by apomorphine; c) at rather low doses, locomotor activity is decreased by apomorphine. The stereotypy and locomotor activation responses are clearly due to postsynaptic dopamine receptor activation, since these responses can be elicited by apomorphine even after pretreatment with reserpine (STRÖMBOM 1976).

The ED_{50} for apomorphine in stimulating locomotor activity in this model was found to be 0.28 mg/kg, which tends perhaps to be a low estimate because of the artificially low baseline of stimulation of dopamine receptors due to the depletion by reserpine of endogenous dopamine. The ED_{50} for apomorphine in decreasing locomotor activity is an order of magnitude lower, 0.035 mg/kg. Interesting enough, the ED_{50} of apomorphine in inhibiting dopa accumulation via the autoreceptor matches that of the sedative effect of apomorphine: STRÖMBOM (1976) found an ED_{50} of 0.05 mg/kg for apomorphine in decreasing dopa accumulation, indicating that this dopamine autoreceptor-mediated phenomenon occurred at the same low doses as the decrease in locomotor activity.

DI CHIARA et al. (1976) set out to determine whether the decrease in locomotor activity by low doses of apomorphine was mediated through a dopamine receptor, by antagonizing this behavioural response with neuroleptic dopamine receptor antagonists. This approach is difficult in practice because blockade by neuroleptics of postsynaptic dopamine receptors results in reduced locomotor activity. Nevertheless, DI CHIARA et al. (1976) sought and found low doses of neuroleptics just below those causing sedation, and found that several dopamine receptor antagonists—sulpiride, pimozide, haloperidol, droperidol, and benperidol—did antagonize the decrease in locomotor activity caused by apomorphine.

These behavioural studies were extended to other dopamine agonists, including the ergoline bromocriptine and the aminotetralin 2-(N,N-dipropyl)-5,6-dihydroxyaminotetralin, and the list of neuroleptics blocking these effects expanded (DI CHIARA et al. 1977b; COSTALL et al. 1981). An additional behavioural response ascribed to stimulation of the dopamine autoreceptor is the inhibition of spontaneous stereotyped climbing in mice by low doses of apomorphine (MARTRES et al. 1977), the opposite of the induction of stereotyped climbing produced by stimulation of postsynaptic dopamine receptors.

b. Criticisms of the Autoreceptor Hypothesis as the Mechanism of Locomotor Sedation Induced by Low Doses of Dopamine Receptor Agonists

COSTALL et al. (1981) presented data which suggest that a single dopamine receptor mediates both the locomotor sedation and the stereotyped climbing produced by dopamine receptor agonists. Despite the existence of a ten-fold lower ED_{50} for producing locomotor sedation than for inducing stereotyped climbing in mice, analysis of the potencies of a substantial number of dopamine receptor agonists in producing the two behavioural responses produces a striking positive correlation. The pharmacological properties of the proposed dopamine autoreceptor and the postsynaptic dopamine receptor mediating stereotyped climbing can thus be considered to be very similar. Moreover, these workers advanced the unorthodox hypothesis that low levels of stimulation of one dopamine receptor could produce the opposite behavioural effect as high levels of stimulation of the same receptor.

The pharmacological distinction between dopamine autoreceptors and postsynaptic dopamine receptors is currently under active investigation. Other methods, however, should be available for testing the involvement of the dopamine autoreceptor in the decrease in locomotor activity produced by low doses of apomorphine. It would appear simple, for instance, to chemically lesion the dopaminergic neurones with 6-hydroxydopamine and determine whether the decrease in locomotor activity response to low doses of apomorphine is still present. This strategy was, for instance, profitably applied to prove that the induction of stereotypy and locomotor activation are mediated by postsynaptic dopamine receptors in the caudate-putamen and nucleus accumbens of the rat (CREESE and IVERSEN 1973). Unfortunately, such a lesion produces itself a great decrease in the basal locomotor activity of rats (ROBERTS et al. 1975; VOITH 1980), and thus the appearance of a sedative effect of apomorphine would be difficult to detect, and a negative result would be difficult to interpret. In conclusion, the role of the dopamine autoreceptor as mediating the locomotor sedation caused by low doses of dopamine receptor agonists has not been conclusively proven, but there are no other seriously contending hypotheses to explain the same data.

2. "Turnover" of Dopamine

A large variety of strategies and experimental techniques have been used to measure dopamine turnover, and they have all resulted in very similar estimates of the basal rate of dopamine turnover (for reviews, see KORF 1981; SHARMAN 1981; see also MASSERANO et al. this volume, Chap. 20).

Dopamine turnover may be altered by activation or blockade of the dopamine autoreceptors modulating depolarization-evoked release, by changes in the rate of dopamine synthesis, or in the rate of firing of dopaminergic neurones.

a. Release of Newly Synthesized and Endogenous Dopamine

As early as 1970, it was observed that following acute *in vivo* administration to rats of the neuroleptic thioproperazine, there is *in vitro* an increased synthesis and efflux of ^{3}H-dopamine formed from ^{3}H-tyrosine in striatal slices (CHERAMY et al. 1970; BESSON et al. 1971). The potassium-evoked release of ^{3}H-dopamine formed from ^{3}H-tyrosine is increased by 1 μM fluphenazine (WESTFALL et al. 1976). These results were interpreted by the authors as "direct support for the concept that activation of presynaptic dopamine receptors located on dopaminergic terminals in the striatum of the rat results in an inhibition of synthesis and release of the transmitter" (WESTFALL et al. 1976). A subsequent investigation reported that the electrically evoked release of endogenous dopamine from rat striatal slices is decreased by apomorphine and increased by chlorpromazine (PLOTSKY et al. 1977).

Using the technique of perfusing the caudate nucleus with a push-pull cannula, BARTHOLINI et al. (1976) performed detailed studies on the increase in endogenous dopamine overflow caused by neuroleptics. Both the release of dopamine and the release of acetylcholine are increased by neuroleptic administration (BARTHOLINI et al. 1976). The push-pull cannula technique has recently been refined by the implantation of small diameter dialysis tubing, through which dopamine and other transmitters diffuse and are collected for analysis (UNGERSTEDT et al. 1982). Using this experimental technique, the release of dopamine *in vivo* has been shown to be decreased by apomorphine (UNGERSTEDT et al. 1982).

A very promising technique for studying dopamine turnover *in vivo* is now available; electrochemical detection of catechols *in vivo* offers not only a potentially rapid and relatively non-invasive method for studying catecholamine function, but also determines actual concentrations of catechols in the vicinity of the probe. For instance, in the nucleus accumbens DOPAC was found to have a concentration of 40 μM under normal conditions, which rises to 70 μM upon administration of haloperidol (BUDA et al. 1981).

b. Disappearance of Dopamine After α-Methyl-tyrosine Administration

Following the administration of a tyrosine hydroxylase inhibitor such as α-methyl-tyrosine, dopamine levels are no longer replenished by *de novo* synthesis, and through the normal impulse-coupled release from nerve terminals the dopamine levels gradually diminish (IVERSEN and GLOWINSKI 1966). As long as impulse activity continues in dopaminergic axons, neuroleptics accelerate the disappearance of dopamine (ANDÉN et al. 1971), while the dopamine receptor agonists apomorphine and piribedil decrease the rate of disappearance of dopamine following administration of α-methyl-tyrosine (GUDELSKY and MOORE 1976; DEMAREST and MOORE 1979). This experimental design precludes modulation of dopamine synthesis by the dopamine autoreceptor as the mechanism for the changes in rate of disappearance of dopamine, since tyrosine hydroxylase is already effectively inhibited. Autoreceptors may either modify the depolarization-evoked release of dopamine (see Sect. B.I.1) or the rate of firing of dopaminergic neurones (see Sect. B.I.3) in order to produce their effects in this model. It is not possible to exclude trans-synaptic modulation of firing rate of dopaminergic neurones as the mechanism responsible for changes in the rate of disappearance of dopamine following α-methyl-tyrosine administration: indeed, both baclofen and γ-butyrolactone, which act on receptors on dopamine cell bodies, also reduce the rate of dopamine disappearance in this model (DEMAREST and MOORE 1979).

CORRODI et al. (1972a) demonstrated histochemically that dopamine receptor agonists decelerated and antagonists accelerated the disappearance of dopamine following α-methyl-tyrosine. In agreement with later biochemical studies (GUDELSKY and MOORE 1976; DEMAREST and MOORE 1979) these early histochemical studies (CORROD; et al. 1972a) were the first to note the apparent absence of modulation by dopamine autoreceptor agonists or antagonists of the rate of dopamine disappearance in the median eminence following α-methyl-tyrosine administration.

c. The γ-Butyrolactone Model

The systemic administration of γ-hydroxybutyrate or its analog γ-butyrolactone eliminates action potentials in dopaminergic neurone cell bodies (ROTH et al. 1973) and causes an increase in dopamine levels and in the rate of dopamine synthesis (WALTERS and ROTH 1972; WALTERS et al. 1973). The effects of γ-hydroxybutyrate administration on dopamine metabolism appear identical to the effects of hemitransection (WALTERS et al. 1973) as studied by KEHR et al. (1972; see Sect. B.I.2.a). The γ-butyrolactone model offers a rapid and reliable experimental technique for studying dopamine autoreceptor function. WALTERS and ROTH (1976) used the γ-butyrolactone model to evaluate the relative importance of the dopamine autoreceptors vis-a-vis the postsynaptic dopamine receptors activating the postulated long loop feedback via the striato-nigral pathway (see Fig. 1).

The increase in dopamine levels (rather than dopa accumulation) by γ-butyrolactone or γ-hydroxybutyrate is also inhibited by apomorphine (HAND-

FORTH and SOURKES 1975; GIANUTSOS et al. 1976). The poor ability of pimozide to antagonize the action of apomorphine (GIANUTSOS et al. 1976) has been interpreted to mean that this neuroleptic is a selective postsynaptic dopamine receptor antagonist (ALANDER et al. 1980; MOORE 1981). Pimozide appears to require a relatively long time following injection to block dopamine receptors, but after a sufficiently long time it blocks pre- and postsynaptic dopamine receptors equally (MCMILLEN et al. 1980).

d. Measurement of Dopamine Metabolites

CARLSSON and LINDQVIST (1963) observed that neuroleptics administered with pargyline produced an increase in brain levels of 3-methoxytyramine. Currently it is generally agreed that the measurement of HVA and DOPAC levels (in the absence of monoamine oxidase inhibitors) are more suitable than 3-methoxytyramine as indices of dopamine release (WESTERINK 1979; WALDMEIER et al. 1981). HVA levels may be considered more reliable than DOPAC levels to indicate release of dopamine from dopaminergic terminals, since COMT is not thought to be present inside dopamine terminals. DOPAC and HVA levels increase in response to neuroleptic administration, and conversely decrease in response to administration of dopamine receptor agonists (ANDÉN et al. 1971; CORRODI et al. 1972b; JORI et al. 1974; WESTERINK 1979; WALDMEIER et al. 1981). In agreement with investigations of the dopamine autoreceptor modulating release of dopamine from slices (KAMAL et al. 1981), the amphetamine evoked increase in HVA is not modulated by dopamine receptor agonists (JORI et al. 1974).

α. The Involvement of Autoreceptors vs. the Long-Loop Feedback

The critical question is whether dopamine autoreceptors do indeed play a major role in modulating DOPAC and HVA levels, or whether postsynaptic dopamine receptors and the trans-synaptic long-loop feedback (Fig. 1) are more important. This question was addressed experimentally by employing kainic acid to lesion the striatal neurones (including the striato-nigral projections) without harming dopaminergic and other afferents to the striatum (DI CHIARA et al. 1977a). The postsynaptic dopamine receptors as well as the descending part of the putative long feedback loop are thus eliminated. The effectiveness of the lesion was verified not only histologically, but also by the complete disappearance of dopamine-stimulated adenylate cyclase in the striatum.

Under these conditions, administration of apomorphine decreases DOPAC levels in kainic acid lesioned striata by at least as much as in unlesioned striata, and the administration of haloperidol results in an increase in DOPAC levels that is clearly greater in kainic-acid lesioned than control striatum (DI CHIARA et al. 1977a). These findings were replicated by GARCIA-MUNOZ et al. (1977), who found that electrolytic lesions of the descending striatonigral projection did not alter the increase in DOPAC or HVA levels caused by haloperidol. The possibility of an interaction of dopaminergic terminals with glial elements or with striatal afferents such as the cortico-striatal projection cannot be ruled out. However, these results may be considered in favour of the hypothesis that dopamine autoreceptors modulating dopamine turnover are localized to dopaminergic neurones, and that the long striato-nigral feedback loop is of relatively little importance for the dopamine-receptor mediated regulation of dopamine turnover.

Di Chiara et al. (1978) found that kainic acid lesions of the striatum eliminated the activation of tyrosine hydroxylase (i.e. the reduction in K_m for cofactor) induced by administration of neuroleptics. This finding was replicated by Gale (1979), who found that electrolytic lesions of the descending striato-nigral axon bundle also eliminated the activation of striatal tyrosine hydroxylase. These results therefore suggest that the neuroleptic-induced activation of tyrosine hydroxylase is mediated by postsynaptic dopamine receptors via the striato-nigral feedback loop, a view consistent with results obtained using other experimental approaches (Zivkovic 1979).

The activation of tyrosine hydroxylase had previously been assumed to mediate the increased dopamine turnover caused by the administration of neuroleptics (Morgenroth et al. 1976; Gale 1979). Results from a number of laboratories now indicate that these two phenomena may be dissociated, as documented in the two preceding paragraphs. Neuroleptic-induced activation of tyrosine hydroxylase depends upon postsynaptic dopamine receptors and the long feedback loop. On the other hand, the neuroleptic-induced increase in dopamine metabolite levels depends on neither postsynaptic dopamine receptors, nor the striato-nigral feedback loop (see also Biggio et al. 1980).

β. Effects of Chronic Administration of Neuroleptics on Dopamine Metabolite Levels

As discussed in the previous section, the change in striatal levels of DOPAC and HVA induced by the acute administration of neuroleptics appears to be due predominantly to an interaction with dopamine autoreceptors, and not to involve postsynaptic dopamine receptors or the descending striato-nigral long feedback loop. The chronic administration of neuroleptics has been found to result in "tolerance" to the increase in DOPAC levels caused by neuroleptics, which may be observed in the striatum after 11 days or more of neuroleptic administration (Scatton 1977). Such tolerance requires a somewhat longer time to develop in the target nuclei of the mesolimbic dopaminergic neurones, namely the nucleus accumbens and olfactory tubercle, but after 40 days of treatment with neuroleptics, the tolerance which develops in these mesolimbic targets is greater in magnitude than the tolerance in the striatum (Scatton 1977). In contrast, "reverse tolerance" develops in the frontal cortex after 40 days of treatment with neuroleptics (Scatton 1977). In other words, acute administration of neuroleptics causes only a slight increase in HVA and DOPAC in the neocortex, but after repeated injections, the increase in HVA and DOPAC in neocortex increases in magnitude. These data have been confirmed and extended in primate brain (Bacopoulos et al. 1978).

Tolerance to the increase in DOPAC and HVA levels following chronic neuroleptic administration appears to reflect a supersensitivity of the dopamine receptor to agonists (BIGGIO et al. 1980). Fifteen days of twice daily haloperidol administration results in an enhanced decrease in striatal DOPAC levels by apomorphine (BIGGIO et al. 1980). The development of "tolerance" to neuroleptics and the development of supersensitivity of the dopamine autoreceptor to apomorphine, following chronic administration of neuroleptics, does not involve postsynaptic dopamine receptors or the striato-nigral long feedback loop, since these effects of chronic neuroleptic administration occur in identical fashion in animals previously lesioned intrastriatally with kainic acid (BIGGIO et al. 1980).

III. Receptor Binding Identification of the Dopamine Autoreceptor

1. Successes and Failures in Detecting Dopamine Autoreceptors with Receptor Binding Techniques

Based on the psychopharmacological studies of the decrease in locomotor activity induced by apomorphine and other dopamine receptor agonists, it would appear that the dopamine autoreceptor has a higher affinity for agonists than do the postsynaptic dopamine receptors. It is possible that a large receptor reserve exists for the dopamine autoreceptor (MELLER et al. 1986), which may explain why many dopamine receptor agonists are more potent in models on pre- versus postsynaptic dopamine receptors. This property of dopamine autoreceptors theoretically makes them an ideal target for direct identification with receptor binding techniques, which capitalize on the high affinity of recognition sites of receptors for the appropriate radiolabelled ligand. NAGY et al. (1978) were the first group to attempt to detect and establish the location of the dopamine autoreceptor. They produced specific lesions of the ascending dopaminergic nigrostriatal neurones by injecting the neurotoxin 6-hydroxydopamine into the axon bundle of this projection. After allowing time for anterograde and retrograde degeneration of the terminals and cell bodies of these neurones, ^{3}H-apomorphine, ^{3}H-haloperidol and ^{3}H-spiperone binding were studied in the striatum. ^{3}H-Apomorphine binding was decreased by more than 50 % in the striatum and by 75 % in the substantia nigra. The binding of the two labelled neuroleptic dopamine receptor antagonists, ^{3}H-haloperidol and ^{3}H-spiperone, was however increased in the striatum, with no change in ^{3}H-spiperone binding occurring in the substantia nigra. These results were explained by a preferential labelling of postsynaptic dopamine receptors by ^{3}H-spiperone, which increase in number following the lesion, and a preferential labelling of dopamine autoreceptors localized to dopaminergic terminals by ^{3}H-apomorphine, these sites being lost when the dopaminergic neurones degenerate.

These findings were not entirely confirmed by other groups (Table 1). For the direct demonstration of dopamine autoreceptors by ^{3}H-apomorphine binding, the first difficulties arose from CREESE and SNYDER (1979) who performed the same experiment but found the opposite result, namely that the binding of ^{3}H-apomorphine *increased* by 50 % in the striatum of 6-hydroxydopamine lesioned rats. CREESE and SNYDER (1979) explained the difference

Table 1. The effects of lesions of dopaminergic neurones by 6-hydroxydopamine injections into the ascending nigrostriatal axon bundle on the binding of the dopamine receptor antagonist ^{3}H-spiperone and the dopamine receptor agonist ^{3}H-apomorphine in substantia nigra and striatum of rats

		^{3}H-ligand binding, percent of control			
Region examined	^{3}H-ligand	NAGY et al. 1978	REISINE et al. 1979	CREESE and SNYDER 1979	MURRIN et al. 1979
Substantia nigra	^{3}H-apomorphine	24 %[c]	not tested	not tested	not tested
	^{3}H-spiperone	91 % (ns)	58 %[a]	not tested	52 %[a]
Striatum	^{3}H-apomorphine	44 %	not tested	152 %[d]	not tested
	^{3}H-spiperone	118 %[a]	118 %[a]	110 %[d]	126 %[b]

[a] $p < 0.05$
[b] $p < 0.01$
[c] $p < 0.001$
[d] statistical comparisons not performed
n.s.: no significant change

in results by suggesting that the binding of ^{3}H-apomorphine measured by NAGY et al. (1978) was too close to background to be meaningful. REISINE et al. (1979) attempted to reproduce the same results with the technique of displacing ^{3}H-spiperone binding with 10 μM 2-amino-6,7-dihydroxyaminotetralin, a dopamine receptor agonist, in parallel with the standard technique of displacing with a high concentration of a dopamine receptor antagonist, in this case, butaclamol. In this way it was thought that ^{3}H-spiperone might preferentially label a site with high affinity for agonists in the first case, and preferentially label a site with high affinity for antagonists in the second, these two sites possibly corresponding to the presynaptic dopamine autoreceptor and postsynaptic dopamine receptor, respectively. However, 6-hydroxydopamine lesions resulted in an increase in the number of both butaclamol- and 2-amino-6,7-dihydroxyaminotetralin-displaceable binding sites in the striatum, agreeing with the findings of CREESE and SNYDER (1979). Significantly, however, in the substantia nigra REISINE et al. (1979) found a decrease of 42 % in butaclamol-displaceable ^{3}H-spiperone binding after 6-hydroxydopamine lesions. This last finding was reproduced by MURRIN et al. (1979) who found a 48 % decrease in ^{3}H-spiperone binding sites in the region of dopaminergic cell bodies in the substantia nigra after 6-hydroxydopamine lesions, with the receptor binding autoradiographic technique.

The sum of these four reports (see Table 1) suggests that ^{3}H-spiperone binding sites in the substantia nigra may be found in appreciable concentration on dopaminergic neurones, where they may correspond to the dopamine autoreceptor. The detection of the dopamine autoreceptor in the striatum with receptor binding techniques remains, however, the topic of considerable controversy (see the following). The technique of studying receptor binding after lesions may produce rigorous evidence as to the localization of presynaptic receptors to specific cells, if those recognition sites are localized exclusively to specific neurone types, for instance, ^{3}H-imipramine binding sites on serotoninergic nerve terminals (SETTE et al. 1981), ^{3}H-desipramine binding sites on noradrenergic nerve terminals (RAISMAN et al. 1982), and ^{3}H-cocaine binding sites to the dopamine transporter in the striatum (PIMOULE et al. 1985; SCHOEMAKER et al. 1985).

2. The Proliferation of Dopamine Receptor Binding Sub-sites and Their Questionable Relationship to Functional Dopamine Receptors

As a result of the discrepancies noted above, as well as the differences in results obtained in various laboratories (see SEEMAN 1981 for review) and for several species (CREESE et al. 1979; SEEMAN 1981) using different labelled dopamine receptor agonists (SEEMAN 1981), an abundant literature rich in putative dopamine receptor subtypes has appeared (see COSTALL and NAYLOR 1981, for review). It should be pointed out that many of these so-called receptor subtypes may represent anomalous surface adsorption phenomena, rather than recognition sites of dopamine receptors (LEYSEN and GOMMEREN 1981). The possible functional significance of the four currently proposed dopamine "receptors" (SOKOLOFF et al. 1980a, b; SEEMAN 1981) has been completely neglected.

Of what is currently known about the inhibitory dopamine autoreceptor which modulates 1) electrically evoked ^{3}H-dopamine release *in vitro*, 2) the synthesis of dopamine *in vivo* and *in vitro*, 3) the electrophysiological activity of dopaminergic neurones, 4) the decrease in locomotor activity by dopamine receptor agonists, and 5) dopamine "turnover" *in vivo* and *in vitro*, two characteristics are all pervasive: 1) The dopamine autoreceptor is blocked by low concentrations or doses of neuroleptics. 2) The dopamine autoreceptor is sensitive to dopamine receptor agonists of aporphine, dihydroxyaminotetralin, and ergoline classes, probably at concentrations lower than those which stimulate postsynaptic dopamine receptors.

3. D_3: A Binding Site in Search of a Function

The D_3 site as defined by SOKOLOFF et al. (1980a, b) and SEEMAN (1981) has a generally lower sensitivity to dopamine receptor agonists and antagonists than do the postsynaptic dopamine receptor binding sites. In particular, sulpiride is inactive at the D_3 binding site (SOKOLOFF et al. 1980a, b), which is not consistent with findings on the pharmacology of the dopamine autoreceptor in its modulation of neurotransmitter release (see Sect. B.I.1), dopamine synthesis (Sect. B.I.2), electrophysiological activity of dopaminergic neurones (Sect. B.I.3), or locomotor activity (Sect. B.II.1). In studies of the decrease in locomotor activity by apomorphine, sulpiride is an effective antagonist of this action, mediated by dopamine autoreceptors (DI CHIARA et al. 1976; COSTALL et al. 1981). Sulpiride is also found to be an antagonist at the dopamine autoreceptor modulating ^{3}H-dopamine release from slices of the caudate with an

EC_{50} of 10 nM (LEHMANN et al. 1981). It is noteworthy that another laboratory found the opposite result of SOKOLOFF et al. (1980a, b) with respect to the potency of sulpiride at the dopamine autoreceptor identified with binding techniques, i.e. the ^{3}H-apomorphine binding site in the striatum which is lost following 6-hydroxydopamine lesions is a site with high affinity for sulpiride (FUJITA et al. 1980). The very low affinity of bromocriptine at the D_3 site also conflicts with investigations in which the agonist action of this compound at dopamine autoreceptors was examined (LEHMANN et al. 1983a).

The fact that phenoxybenzamine does not inactivate D_3 binding sites (HAMBLIN and CREESE 1980) but does inactivate all known functional dopamine receptors (LEHMANN and LANGER 1981) raises questions concerning the significance of D_3 binding sites. It is also of interest to note that all ^{3}H-apomorphine binding sites are inactivated by ascorbate (KAYAALP and NEEF 1980) at concentrations which do not substantially alter dopamine autoreceptor function in striatal slices (LEHMANN et al. 1981). As noted by SEEMAN (1981), no dopamine receptor identified by a biological response has been found which has the pharmacological properties of the D_3 binding site, and therefore it is not appropriate to refer to the D_3 binding site as a dopamine receptor.

4. Conclusions

The identification of dopamine autoreceptors by receptor binding techniques may have been accomplished in the substantia nigra in which dopamine receptors were labelled with ^{3}H-spiperone (REISINE et al. 1979; MURRIN et al. 1979), but the evidence cannot yet be considered compelling. The identification of dopamine autoreceptors by employing the potentially very powerful receptor binding techniques remains an important goal for the future. This pursuit should however be correlated with functional information on the dopamine autoreceptor.

IV. Dopamine Neurones Without Autoreceptors

The assumption that all dopaminergic neurones contain dopamine autoreceptors is not necessarily warranted. There are two brain regions in particular where it is suspected that dopamine autoreceptors may not be present, namely, the frontal cortex and the median eminence. It may be noted, furthermore, that a number of brain regions receiving dopaminergic innervation have not yet been tested for the presence of dopamine autoreceptors modulating the synthesis or the release of dopamine.

1. The Frontal Cortex

Using the γ-butyrolactone paradigm coupled with dopa-decarboxylase inhibition, BANNON et al. (1981) found results suggesting that dopamine autoreceptors modulating the rate of dopamine synthesis apparently do not exist in the frontal cortex. These workers emphasized that this finding does not permit extrapolation to the existence of dopamine autoreceptors modulating 1) depolarization-evoked dopamine release, or 2) the firing rate of the dopaminergic neuronal cell bodies from which these cortical dopaminergic terminals are derived. In fact, the acceleration of the disappearance of dopamine by haloperidol, following administration of α-methyl tyrosine, may suggest that these latter two classes of dopamine autoreceptor do modulate dopamine function in the frontal cortex (SCATTON 1977). It is noteworthy that in the frontal cortex, DOPAC levels increase to a much lesser extent than in the nucleus accumbens, caudate, or olfactory tubercle, in response to acute administration of neuroleptics (SCATTON 1977). The effect of chronic administration of neuroleptics on DOPAC levels in frontal cortex is also quite different from that in these other three regions (see before, SCATTON 1977). BANNON et al. (1982) have further documented the absence of inhibitory dopamine autoreceptors modulating dopamine synthesis in the frontal cortex, but did not reproduce the paradoxical time-dependent increase in dopamine turnover after chronic administration of neuroleptics found by SCATTON (1977). Also in disagreement with SCATTON (1977), BANNON et al. (1982) report that haloperidol does not accelerate the decline in dopamine levels following α-methyl-tyrosine administration. Thus the self-regulation of dopamine function in frontal cortex appears to be considerably different from that of other regions receiving nigrostriatal or mesolimbic dopamine innervation. Further studies are required to determine with certainty the existence and nature of self-regulatory dopamine receptors modulating dopamine function in the frontal cortex.

2. The Median Eminence

DEMAREST and MOORE (1979) reported extensive evidence suggesting that dopamine autoreceptors do not modulate dopamine function in the median eminence, unlike the nigrostriatal, mesolimbic and tuberohypophyseal dopamine systems they examined at the same time. Employing the experimental models of dopa accumulation, the modulation of dopamine levels after γ-butyrolactone administration, and the modulation of dopamine levels after α-methyl tyrosine administration, no effect of apomorphine was found in the median eminence, although the striatum, olfactory tubercle, and posterior pituitary responded in accordance with the classical dopamine autoreceptor hypothesis. In contrast, the dopaminergic system of the median eminence would appear to be modulated via a "trans-synaptic" hormonal feedback system (GUDELSKY et al. 1978), possibly mediated by prolactin. Dopamine synthesis in the median eminence is also apparently subject to end-product inhibition at the level of the tyrosine hydroxylase activity (DEMAREST and MOORE 1979).

V. The Role of Somatodendritic Dopamine Autoreceptors

While the electrophysiological evidence for the existence of somatodendritic dopamine autoreceptors modulating the firing rate of dopaminergic neurones in the substantia nigra and ventral tegmental area appears clear (see Sect. B.I.3), certain controversies exist concerning the function of dopamine autoreceptors in the cell body regions of dopaminergic neurones. The demonstration of release of dopamine *in vitro* and *in vivo* in the substantia nigra played a major role in the advancement of hypotheses concerning dendritic release of neurotransmitters in general (GEFFEN et al. 1976; NIEOULLON et al. 1977). Although the *in vitro* depolarization-evoked release of dopamine has since been demonstrated repeatedly in the substantia nigra (HEFTI and LICHTENSTEIGER 1978; ARBILLA and LANGER 1980; BEART and MCDONALD 1980; ACEVES and CUELLO 1981) and the ventral tegmental area (BEART and MCDONALD 1980), the modulation of such dopamine release by dopamine autoreceptors has never been reported. The possibility that ^{3}H-dopamine may be taken up, accumulated, and released from serotoninergic and noradrenergic terminals in the substantia nigra necessarily complicates such investigations.

In a preliminary report, the spontaneous release of ^{3}H-GABA in the rat substantia nigra was reported to be increased by dopamine, and this was offered as the trans-synaptic mechanism underlying the electrophysiological demonstration of dopamine autoreceptors (REUBI et al. 1977). Aside from the fact that electrophysiological studies have ruled out GABA receptors as participating in the dopamine autoreceptor effect (see Sect. B.I.3), subsequent workers have not reproduced the findings of REUBI et al. (BEART and MCDONALD 1980; ARBILLA et al. 1981b). Moreover, ARBILLA et al. (1981b) have reported an inhibition by apomorphine of potassium-evoked release of ^{3}H-GABA in the substantia nigra.

When administered systemically *in vivo*, the dopamine receptor agonist apomorphine decreases DOPAC and HVA levels in the substantia nigra, wherein reside the cell bodies and dendrites of dopaminergic neurones (WESTERINK and KORF 1976a, b; NICOLAOU 1980).

The neuroleptic haloperidol causes an increase in DOPAC and HVA levels in the substantia nigra (WESTERINK and KORF 1976a, b; ARGIOLAS et al. 1979; SCATTON 1979; NICOLAOU 1980). Some workers have found that DOPAC levels in the ventral tegmental area (containing dopamine perikarya which preferentially innervate nucleus accumbens, olfactory tubercle, and frontal cortex) also increase following haloperidol administration (SCATTON 1979; BEART and GUNDLACH 1980). On the other hand, ARGIOLAS et al. (1979) found no change in DOPAC levels in the ventral tegmental area following haloperidol administration. No attempts have yet been made to determine whether the dopamine receptor agonist- and antagonist-induced changes in these indices of dopamine turnover act via somatodendritic dopamine autoreceptors in the substantia nigra and ventral tegmental area, or alternatively via postsynaptic dopamine receptors in the forebrain and through the long feedback loop onto the dopaminergic neurones. For instance, it has not been determined what is the effect on nigral DOPAC and HVA of neuroleptics or dopamine receptor agonists injected into the striatum.

When dopamine, amphetamine, or benztropine is infused into the substantia nigra, the release of ^{3}H-dopamine (formed from ^{3}H-tyrosine) in the ipsilateral caudate is decreased. Infusion of neuroleptics into the substantia nigra produces the opposite effect (CHERAMY et al. 1977; NIEOULLON et al. 1979). Using an unusual method of assessing dopamine turnover, MAGGI et al. (1978) found that apomorphine injected intranigrally reduces 3-methoxytyramine levels in the striatum in rats. However, this reduction in 3-methoxytyramine in experiments in which MAO was not inhibited (MAGGI et al., 1978) is difficult to interpret (see WALDMEIER et al. 1981). A decrease in striatal HVA following intranigral injection of apomorphine has also been reported (WOLFARTH et al. 1978). Dopamine receptor-mediated turning behaviour after apomorphine injection into the substantia nigra is observed only if the dopaminergic neurones are lesioned (GLICK and CRANE 1978; KELLY and MOORE 1978; KOZLOWSKI et al. 1980).

VI. The Pharmacological Characteristics of the Dopamine Autoreceptor

The pharmacological characterization of dopamine receptors in general has been a controversial endeavor in the past decade. Several characterizations of dopamine receptors have appeared, only to be modified and superseded a year or two later by a new "definitive" characterization, owing to the development of new techniques and new drugs. Since the pharmacological characteristics of the dopamine autoreceptor must be described in the context of dopamine receptors in general, pertinent aspects of dopamine receptor pharmacology are briefly reviewed here (see also BARTHOLINI et al., WOLFE and MOLINOFF, Chaps. 16 and 7).

In 1975, IVERSEN presented evidence indicating that the dopamine-stimulated adenylate cyclase was *the* postsynaptic dopamine receptor of the brain and the key to the pathology and pharmacotherapy of schizophrenia. More recently the existence of two different types of dopamine receptors was suggested by the differences in potencies of agonists and antagonists in affecting adenylate cyclase (D_1 receptor) and the ^{3}H-neuroleptic binding site (D_2 receptor) (GARAU et al. 1978; KEBABIAN and CALNE 1979).

The distinction between D_1 and D_2 receptors was facilitated by the appearance of the benzamide neuroleptics, the best known of which is sulpiride. Subsequent investigation revealed that sulpiride stereoselectively displaces ^{3}H-neuroleptic binding only if sodium ions are present in the medium (THEODOROU et al. 1980; STEFANINI et al. 1980). Sulpiride does not antagonize the dopamine-stimulated adenylate cyclase, under any of the conditions tested, including high sodium concentrations (SPANO et al. 1979; RUPNIAK et al. 1981). Sulpiride would therefore appear to be inactive at D_1 receptors, but is a high-affinity antagonist at D_2 receptors.

Bulbocapnine antagonizes dopamine-sensitive adenylate cyclase with a K_i of 160 nM (IVERSEN 1975; SETLER et al. 1978) but is virtually inactive at displacing ^{3}H-haloperidol binding (BURT et al. 1976) or ^{3}H-spiperone binding (LEHMANN et al. 1983a). More recently, SCH 23390 became available as a selective antagonist for D_1 receptors (IORIO et al. 1983), as it is also tritiated as a ligand to label D_1-receptors (PIMOULE et al. 1985).

Recently a chemical family of dopamine receptor agonists of the ergolines family has been found to stimulate selectively the D_2 receptor (TSURUTA et al. 1981; RABEY et al. 1981). LY 141865 is the best known of these compounds.

Table 2. The pharmacological distinctions between D_1 and D_2 dopamine receptor subtypes

	Agonist	Antagonist
Selective for D_1	Fenoldopam (SKF 82526) SKF 38393	SCH 23390
Selective for D_2	Quinpirole (LY 141865)	S-sulpiride
Non selective for D_1 and D_2	Apomorphine Pergolide	Chlorpromazine Haloperidol

Some agonists and antagonists have been found to be highly selective for the dopamine-stimulated adenylate cyclase (D_1), and for dopamine receptors which are labelled by ^{3}H-neuroleptics (D_2). However, most dopamine receptor agonists and antagonists are active at both D_1 and D_2 dopamine receptors. Documentation is given in the text. Fenoldopam is SKF 82526. SK&F 38393 is 2,3,4,5-tetrahydro-7,8-dihydroxy-1-phenyl-1H-3-benzazepine. Quinpirole is LY 141865 [N-propyl tricyclic pyrazole].

Finally, the novel dopamine receptor agonist SK&F 38393 stimulates adenylate cyclase in the striatum via a dopamine receptor (SETLER et al. 1978), but does not displace ^{3}H-spiperone binding (LEHMANN et al. 1983a).

Thus, dopamine receptors can be distinguished into the two currently recognized types, D_1 and D_2, by selective agonists and antagonists (see Table 2). For the present, we are inclined to accept this two dopamine receptor subtypes model and to use the terms D_1 and D_2 defined in this way. It is necessary to note however that a *four* dopamine receptor model has been proposed (SOKOLOFF et al. 1980a, b; SEEMAN 1981). Also, an interaction of the D_2 receptor with adenylate cyclase has been proposed, whereby stimulation of D_2 receptors results in an inhibition of the adenylate cyclase which is activated by D_1 receptor stimulation (STOOF and KEBABIAN 1981).

Only in recent years have the pharmacological characteristics of the dopamine autoreceptor modulating electrically evoked ^{3}H-dopamine release come under intensive study, and this characterization cannot yet be considered complete. By analogy to α-adrenoreceptors, it may be that dopamine autoreceptors are pharmacologically different from postsynaptic dopamine receptors. The autoreceptor-mediated inhibition of electrically evoked ^{3}H-dopamine release from slices is a particularly suitable experimental system for studying the pharmacological properties of the dopamine autoreceptor, since the concentration of agonists and antagonists and other conditions can be completely controlled. The difficulty arises in that there is no comparable *in vitro* system for measuring responses mediated by postsynaptic dopamine receptors, unlike the noradrenergic system in the periphery.

1. *In Vitro* Pharmacological Characterization of Dopamine Autoreceptors

The dopamine autoreceptor modulating depolarization-evoked release of dopamine can be characterized pharmacologically *in vitro* with considerable precision. Other biological functions ascribed to dopamine autoreceptors are more difficult to characterize *in vitro*. The inhibition of dopamine synthesis mediated by dopamine receptor agonists can be demonstrated in synaptosomes, but a considerable proportion of such inhibition is ascribed to direct inhibition of the enzyme tyrosine hydroxylase, and not a receptor-mediated phenomenon (WAGGONER et al. 1980; see also Sect. B.I.1.g and B.I.2). The fraction of the inhibition of dopamine synthesis which is due to a direct effect on tyrosine hydroxylase varies with the particular dopamine receptor agonists employed. Somato-dendritic dopamine autoreceptors modulating the firing rate of dopaminergic neurones, on the other hand, have been extensively investigated in electrophysiological studies *in vivo*, but similar studies under *in vitro* conditions have not been carried out.

The dopamine autoreceptor stereospecifically recognizes antagonists. Similarly to postsynaptic dopamine receptors as assessed by ^{3}H-neuroleptic binding, S-butaclamol (commonly referred to as the "+" enantiomer) is active at nanomolar concentrations, while R-butaclamol is inactive even at 10 μM. Between S- and R-sulpiride, there is a ten- to one hundred-fold difference in EC_{50} at the dopamine autoreceptor (ARBILLA and LANGER 1981; DUBOCOVICH and WEINER 1981), while for ^{3}H-spiperone binding (SPANO et al. 1979) and for ^{3}H-haloperidol binding (GARAU et al. 1978) the stereospecificity is one hundred-fold or greater.

All known dopamine receptor agonists yet examined which are active at D_2 dopamine receptors inhibit electrically evoked ^{3}H-dopamine release from slices of the caudate nucleus of the cat (Table 3), the inhibition by all of these agonists being antagonized by S-sulpiride (LEHMANN and LANGER, unpublished data). The order of potency of these dopamine receptor agonists in inhibiting electrically evoked ^{3}H-dopamine release correlates quite closely to their potency in displacing ^{3}H-spiperone from its binding sites in membranes of the cat caudate (LEHMANN et al. 1983 a). The agonist specific for D_1 receptors, SK&F 38393, does not inhibit electrically evoked ^{3}H-dopamine release in concentrations up to 10 μM, and does not displace ^{3}H-spiperone binding in cat caudate (LEHMANN et al. 1983 a). As noted above, the agonist specific for the D_2 receptor, LY 141865, is active at the dopamine autoreceptor modulating electrically evoked ^{3}H-dopamine release from caudate slices.

On the basis of the effects of these dopamine receptor agonists, there is at least one difference between the D_2 receptor and the dopamine autoreceptor. Bromocriptine is 20-fold more potent than apomorphine at displacing ^{3}H-haloperidol binding, and 30 to 200 fold more potent than apomorphine at displacing ^{3}H-spiperone, from the D_2 receptor binding site (BURT et al. 1976; GOLDSTEIN et al. 1980; LEHMANN et al. 1983a). In contrast, apomorphine is more potent than bromocriptine as an agonist at the dopamine autoreceptor which modulated ^{3}H-dopamine release (Table 3). This comparison of the relative potencies of apomorphine and bromocriptine suggests that the D_2 dop-

Table 3. The potency and intrinsic activity of dopamine receptor agonists for the dopamine autoreceptor in vitro

Agonist	IC_{50}, nM (a)	Maximal inhibition (b)
Classic		
N-n-propylnorapomorphine	0.73	94 %
Pergolide	1.64	86 %
Apomorphine	13.5	87 %
Bromocriptine	30.0	50 %
Novel		
TL-99	4.5	97 %
M-7	7.0	96 %
LY 141865	43.0	90 %
SK & F 38393	Inactive	–

Classic and novel dopamine receptor agonists inhibit electrically-evoked ^{3}H-dopamine release from cat caudate slices (for experimental details see LEHMANN et al. 1981). The classic dopamine receptor agonists interact with both D_1 and D_2 receptors. Of the novel dopamine receptor agonists, SK & F 38393 interacts selectively with the D_1 receptor, while LY 141865 interacts selectively with the D_2 receptor (see Table 2). TL-99 (6,7-dihydroxy-2-(N,N-dimethyl)-aminotetralin) has been proposed as selective for dopamine autoreceptors. M-7 (5,6-dihydroxy-2-(N,N-dimethyl)-aminotetralin) is a rigid analogue of the α-trans-rotamer of dopamine, while TL-99 is a rigid analogue of the β-trans-rotamer of dopamine.
(a) IC_{50} is the concentration which produces half of the maximal inhibition of ^{3}H-dopamine release elicited by electrical stimulation.
(b) Maximal inhibition is the projected inhibition of electrically-evoked ^{3}H-dopamine release by a large concentration of agonist. Both maximal inhibition and IC_{50} were calculated by computer assisted fitting of data to sigmoid concentration-response curves as described by GOMENI and GOMENI (1980).

amine receptor, as measured by ^{3}H-neuroleptic binding (and reflecting primarily postsynaptic dopamine receptors, see Sect. B.III) may differ pharmacologically from the dopamine autoreceptor, which is thought to be presynaptic in location.

The pharmacological profile of the dopamine autoreceptor modulating electrically evoked ^{3}H-dopamine release with respect to antagonists has not yet been studied in great detail. Sufficient results have been obtained, however, to indicate that neuroleptics increase ^{3}H-dopamine release (by antagonizing the action of endogenous dopamine at the dopamine autoreceptor: LEHMANN and LANGER 1982c) with a pharmacological profile similar to that of their displacement of ^{3}H-spiperone binding to cat caudate homogenates (LEH-

Table 4. The potency and maximal increase caused by various dopamine receptor antagonists of electrically-evoked ^{3}H-dopamine release

Antagonist	EC_{50}, nM (a)	Maximal increase (b)
S-sulpiride	10	50 %
Chlorpromazine	30	65 %
7-Hydroxy-chlorpromazine	40	45 %
Clozapine	300	15 %
Bulbocapnine	Inactive	–
SCH 23390	Inactive	–

Dopamine receptor antagonists cause an increase in electrically-evoked ^{3}H-dopamine release by blocking the dopamine autoreceptor, which is otherwise stimulated by released endogenous dopamine (see LEHMANN and LANGER 1982c; for methods see LEHMANN and LANGER 1981). S-sulpiride is a selective D_2 receptor antagonist, while bulbocapnine and SCH 23390 are selective D_1 receptor antagonists. SCH 23390 is: R-(+)-8-chloro-2,3,4,5-tetrahydro-3-methyl-5-phenyl-1H-3-benzazepine-7-ol. Chlorpromazine and one of its active metabolites, 7-hydroxychlorpromazine, are thought to interact with both D_1 and D_2 receptors. In addition to its dopamine receptor blocking properties, clozapine has muscarinic antagonistic properties, and muscarinic receptor antagonists may cause a small decrease in ^{3}H-dopamine release (LEHMANN and LANGER 1982b).
(a) EC_{50} is the concentration of drug which produces half of the maximal increase in ^{3}H-dopamine release elicited by electrical stimulation.
(b) Maximal increase is defined as a plateau in the concentration-response curve, which is not significantly greater over a ten-fold increase in concentration of antagonist.

MANN et al. 1983a). Notably, S-sulpiride, which is inactive at D_1 receptors, is active at the dopamine autoreceptor modulating ^{3}H-dopamine release (Table 4). Moreover, the effect of sulpiride on the dopamine autoreceptor is stereospecific (ARBILLA and LANGER 1981). Bulbocapnine, which is inactive at D_2 receptors, is also inactive at dopamine autoreceptors modulating ^{3}H-dopamine release. More recently, we found that the D_1-selective antagonist SCH 23390 does not block the presynaptic dopamine autoreceptor modulating transmitter release (ARBILLA and LANGER, unpublished observations; Table 4).

It was recently reported that rapid desensitization to exogenous dopamine at the level of the autoreceptor develops in striatal slices (ARBILLA et al. 1985). This desensitization to exogenous dopamine is not observed for the D_2-receptor which modulated acetylcholine release in these striatal slices (ARBILLA et al. 1985).

2. *In Vivo* Pharmacological Characterization of Dopamine Autoreceptors

The characterization of dopamine receptors *in vivo* is a more complex task than the same endeavor *in vitro*. The bioavailability and metabolic fate of the administered drug, and the participation in the biological response of neurotransmitter receptors other than dopamine receptors, are two aspects which present greater complexity *in vivo* than *in vitro*.

Since the agonist which is specific for D_2 receptors, LY 141865, causes a decrease in DOPAC and HVA levels (Rabey et al. 1981), and since the antagonist specific for D_2 receptors S-sulpiride causes an increase in DOPAC and HVA levels (Spano et al. 1979; Jenner et al. 1980), it is probable that the dopamine receptor modulating dopamine turnover is of the D_2 subtype. Similarly, the somatodendritic dopamine autoreceptor modulating the firing rate of dopaminergic neurones has been shown to be antagonized stereospecifically by sulpiride (Woodruff and Pinnock 1981).

It is known that stereotypy evoked by high doses of dopamine receptor agonists is mediated by postsynaptic dopamine receptors (Voith 1980), while the decrease in locomotor activity caused by low doses of dopamine receptor agonists is thought to be due to an action at the dopamine autoreceptor (see Sect. B.II.1). These different responses to various doses of dopamine receptor agonists already suggest a pharmacological difference between the two receptors (postsynaptic and the autoreceptor). However, the similarity in the relative order of potency of dopamine receptor agonists for the two types of responses suggests that the two behaviours (i.e. decrease in locomotor activity and stereotypy) may be mediated by actions on very similar receptors (Costall et al. 1981).

A novel compound, 3-(3-hydroxyphenyl)-N-n-propylpiperidine (3-PPP) has been proposed as a selective dopamine autoreceptor agonist on the basis of *in vivo* investigations (Hjorth et al. 1981). In the models of 1) inhibition of dopa accumulation, 2) reduction of DOPAC levels, 3) retardation of dopamine disappearance after α-methyl tyrosine administration, and 4) reduction of locomotor activity, 3-PPP acts via a haloperidol-sensitive mechanism like other dopamine receptor agonists. However, unlike other dopamine receptor agonists, 3-PPP does not increase locomotor activity in reserpine-pretreated rats, induce stereotypy, or displace ^{3}H-spiperone binding (Hjorth et al. 1981; Martin et al. 1981). Costall et al. (1981) found that, while 3-PPP induces a decrease in locomotor activity, this effect is blocked by the α-adrenoceptor antagonist yohimbine, but not by the dopamine receptor antagonist spiperone. Furthermore, the compound 3-PPP does not inhibit electrically evoked ^{3}H-dopamine release from slices of rabbit caudate (Langer et al. 1983). Since the (+) and (−) stereoisomers of 3-PPP are now resolved, the *in vitro* studies on ^{3}H-dopamine and ^{3}H-acetylcholine release were repeated, the original studies having been carried out with racemic 3-PPP. The results obtained indicate that (+)3-PPP behaves as a dopamine receptor agonist, while (−)3-PPP has blocking properties on dopamine receptors (Arbilla and Langer 1984).

Similarly, 6,7-dihydroxy-2-(N,N-dimethyl)-aminotetralin (TL-99) has been proposed as an agonist selective for the dopamine autoreceptor (Goodale et al. 1980). TL-99 reduces locomotor activity and decreases DOPAC levels, but does not cause turning in rats with unilateral lesions of the nigrostriatal dopaminergic projection (GOODALE et al. 1980). MARTIN et al. (1981) found that TL-99 does consistently elicit turning in similarly lesioned rats. In reserpine-pretreated rats, TL-99 does stimulate locomotor activity, suggesting activation of postsynaptic dopamine receptors, but again the magnitude of such stimulation is markedly less than that produced by apomorphine (MARTIN et al. 1981). Finally, TL-99 is equipotent with apomorphine in displacing ^{3}H-spiperone binding (MARTIN et al. 1981; LEHMANN et al. 1983a). Overall, these results suggest that if TL-99 does distinguish between dopamine autoreceptors and postsynaptic dopamine receptors, it is by virtue of a reduced intrinsic activity at postsynaptic dopamine receptors, whereas TL-99 possesses full intrinsic activity at presynaptic dopamine autoreceptors (see Table 2).

Although it has been suggested that pimozide and clozapine are preferential postsynaptic dopamine receptor antagonists based on their low activity at autoreceptors in the γ-butyrolactone model, the basis for such a distinction is questionable (see Sect. B.II.2.c). It was, however, noted much earlier that thioridazine has a surprisingly small effect on CSF HVA levels in view of its therapeutic potency (MATTHYSE 1973). It has been suggested that these results reflect a difference in properties between striatal dopamine receptors and mesolimbic dopamine receptors (MATTHYSE 1973; LJUNGBERG and UNGERSTEDT 1978) rather than differences between dopamine autoreceptors and postsynaptic dopamine receptors.

It is apparent from this discussion that pharmacological differences between dopamine autoreceptors and postsynaptic dopamine receptors may exist, but that no compelling evidence for such a difference has yet been presented. There is, however, experimental evidence for a large receptor reserve at the level of the striatal dopamine autoreceptor (MELLER et al. 1986). The pharmacological profile presented by the D_2 receptor describes with reasonable accuracy the dopamine autoreceptor as well as the clinically relevant postsynaptic dopamine receptors which mediate behavioural effects of dopamine. As described in the introduction to Sect. B.V, dopamine receptor theory has been evolving and continues to evolve rapidly. The important question of whether the dopamine autoreceptor differs from postsynaptic dopamine receptors, in analogy to the α-adrenoceptors of the peripheral nervous system (LANGER 1974, 1981), remains to be resolved.

VII. Potential Clinical Relevance of Dopamine Autoreceptors

The dopamine hypothesis of schizophrenia was founded on observations that neuroleptics increase dopamine turnover via a dopamine receptor feedback mechanism (CARLSSON and LINDQVIST 1963). In the following decade the ability of neuroleptics to increase dopamine turnover by blocking self-regulatory dopamine receptors was documented in numerous laboratories, and gradually the idea emerged that dopamine receptors were involved in schizophrenia (MATTHYSE 1973). The ability of amphetamine to induce schizophrenic symptoms (SNYDER 1972) and the ability of neuroleptics to antagonize the dopamine receptor modulating prolactin release (MELTZER et al. 1974) provided further support for the dopamine hypothesis of schizophrenia.

These self-regulatory dopamine receptors were generally assumed to be localized postsynaptically, and to modulate dopamine turnover via long feedback loop neuronal circuits returning from the forebrain to the substantia nigra, since these neuronal projections were neuroanatomically identified (STEVENS 1973). However, in the decade since the dopamine hypothesis of schizophrenia was proposed, ample proof has appeared to document the involvement of presynaptic and somatodendritic dopamine autoreceptors as a principal mechanism for the increase in dopamine turnover caused by the administration of neuroleptics (see Sect. B.I.1–3, B.II.2.c and d). Postsynaptic dopamine receptors acting via the long neuronal feedback loop have been shown not to play an essential role in modulating dopamine turnover. There are, however, two exceptions: 1) Part of the reduction in firing rate of dopaminergic neurones by systemic amphetamine is mediated through striato-nigral axons via GABA receptors (BUNNEY and AGHAJANIAN 1978; but see WANG 1981b). 2) The activation of tyrosine hydroxylase by neuroleptics does depend on postsynaptic dopamine receptors and the long feedback loop (DI CHIARA et al. 1978). Nevertheless, the increase in dopamine turnover by neuroleptics has been shown to be independent of this postsynaptic dopamine receptor-mediated activation of tyrosine hydroxylase (see Sect. B.II.2.d).

It may appear paradoxical to postulate that schizophrenia is due to a pathologically high dopamine tone, when neuroleptics, which are used to treat the disease, increase the release of dopamine. It is currently still considered likely that the antipsychotic action of neuroleptics occurs by their blockade of postsynaptic dopamine receptors (HORNYKIEWICZ 1978; ANGRIST et al. 1980). The increase in dopamine release caused by neuroleptics acting at dopamine autoreceptors may act in opposition to this therapeutic action of neuroleptics, particularly during the first weeks of administration.

1. A Model for the Pre- and Postsynaptic Actions of Neuroleptics

The analysis of the two opposite effects of neuroleptics at dopaminergic synapses in the central nervous system is easier when compared with peripheral noradrenergic neurons.

Noradrenergic neuroeffector junctions in the peripheral nervous system may provide an analogous model for central dopaminergic synapses. For these synapses, 1) biological responses mediated by postsynaptic receptors *can* be measured, and 2) antagonists selective for pre- and postsynaptic receptors *are* known (for reviews, see LANGER 1977, 1981; LANGER et al. 1985). In the example shown in Fig. 3, the vasoconstriction in rabbit pulmonary artery is induced by noradrenaline released by nerve stimulation.

Presynaptic α_2-autoadrenoceptors, analogous to dopamine autoreceptors, inhibit the depolarization-evoked release of noradrenaline (LANGER 1974, 1977). These presynaptic inhibitory α_2-autoadrenoceptors are blocked selectively by yohimbine. Exposure to yohimbine results in an increase of noradrenaline release, and an increased postsynaptic response to nerve stimulation (Fig. 3).

The postsynaptic α_1-adrenoceptor (analogous to postsynaptic dopamine receptors), is blocked selectively by prazosin. Exposure to prazosin simply blocks the postsynaptic response to released noradrenaline (Fig. 3), since it does not modify the release of the neurotransmitter.

Phentolamine antagonizes both the presynaptic α_2-autoadrenoceptor and the postsynaptic α_1-adrenoceptor. The net response of the neuroeffector to phentolamine is the result of 1) increased noradrenaline release through blockade of the presynaptic α_2-autoadrenoceptor, and 2) decreased receptor activation by released noradrenaline through blockade of the postsynaptic α_1-adrenoceptor. While the net effect of phentolamine is to reduce the re-

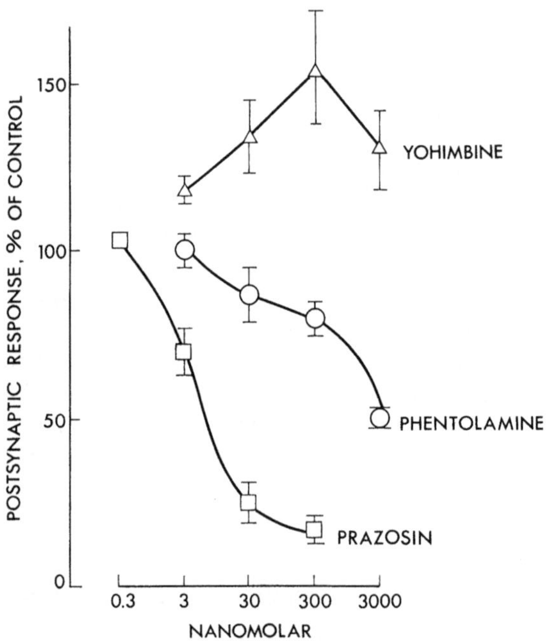

Fig. 3. The peripheral noradrenergic synapse: a potential model for central dopaminergic synapses. Noradrenaline released from nerve terminals interacts with postsynaptic receptors to cause a contraction of smooth muscle mediated by α_1-adrenoceptors (ordinate). α-Adrenoceptor blocking agents, added in the concentrations indicated (abscissa), modify the net response of the neuroeffector. When the presynaptic inhibitory α_2-noradrenergic autoreceptors are blocked selectively by yohimbine, the increased noradrenaline release causes a greater response. When the postsynaptic α_1-adrenoceptors are antagonized selectively by prazosin, the response to released noradrenaline is decreased and completely blocked. Phentolamine blocks both pre- and postsynaptic α-adrenoceptors simultaneously; with increasing concentrations, phentolamine gradually antagonizes the net response at the neuroeffector junction. Current neuroleptic dopamine receptor antagonists are thought to act analogous to phentolamine, i.e. neuroleptics block both pre and postsynaptic dopamine receptors. Figure taken from DAVEY (1980)

sponse to noradrenaline release, it does so with a flat slope, and does not block completely the postsynaptic response to nerve stimulation (Fig. 3). In fact, the effect of phentolamine resembles closely the algebraic sum of the effects of the selective antagonists, yohimbine and prazosin.

To the extent of our present knowledge, all neuroleptic dopamine receptor antagonists have approximately equal affinity for pre- and postsynaptic D_2 receptors, as discussed in Sect. B.V. The net effect of neuroleptics on dopamine synaptic function may therefore resemble the action of phentolamine on noradrenergic synapse function (Fig. 3). The availability of a selective antagonist for D_1 receptors, SCH 23390 (Iorio et al. 1983) offers a new and powerful research tool. It was recently reported that the binding of ^{3}H-SCH 23390 is unchanged in post-mortem brains from schizophrenic patients, while the binding of ^{3}H-spiroperidol to D_2-receptors was enhanced (Pimoule et al. 1985).

2. The Therapeutic Mechanism of Action of Neuroleptics

If indeed current neuroleptics are non-selective with respect to pre and postsynaptic D_2 receptors, one might justifiably ask how it is possible that neuroleptics are therapeutically effective at all. In point of fact, neuroleptics are not very effective anti-schizophrenic agents. They ameliorate the positive symptoms (hallucinations, delusions, and thought disorder) but do not help, and may worsen, the negative symptoms (affective flattening, poverty of speech, and loss of drive) associated with schizophrenia (Crow 1980; Mackay and Crow 1980; Angrist et al. 1980; Rifkin 1981; Johnson 1981; see also Lehmann and Langer 1982c).

In order to successfully treat a patient who suffers from an acute psychotic attack, neuroleptics must be administered for two to three weeks to achieve a therapeutic effect. There must be an explanation for this latency period in the therapeutic effect of neuroleptics.

Crow et al. (1980) performed an elegant investigation of the time course of therapeutic improvement as a result of four weeks administration of the neuroleptic α-flupenthixol (see Fig. 4). Since there is a non-negligible placebo effect (which is equivalent to the effect of administration of β-flupenthixol, the isomer of flupenthixol which is inactive at dopamine receptors), Crow et al. (1980) calculated "drug-specific clinical improvement" as the difference between the clinical rating for the α-flupenthixol group minus the control group.

Drug-specific clinical improvement occurs to a significant extent only between two and three weeks after the beginning of daily neuroleptic administration (Fig. 4). In contrast, the increase in serum prolactin levels caused by α-flupenthixol occurs rapidly. In fact, other studies have shown that serum prolactin increases within minutes following neuroleptic administration (Öhman et al. 1977). These data indicate that the delayed therapeutic effect of neuroleptics is unlikely to be due to the failure of the drugs to rapidly reach sufficiently high blood levels to antagonize dopamine receptors.

A possible second explanation is that neuroleptics do not cross the blood-brain barrier sufficiently rapidly to block *central* dopamine receptors. Some

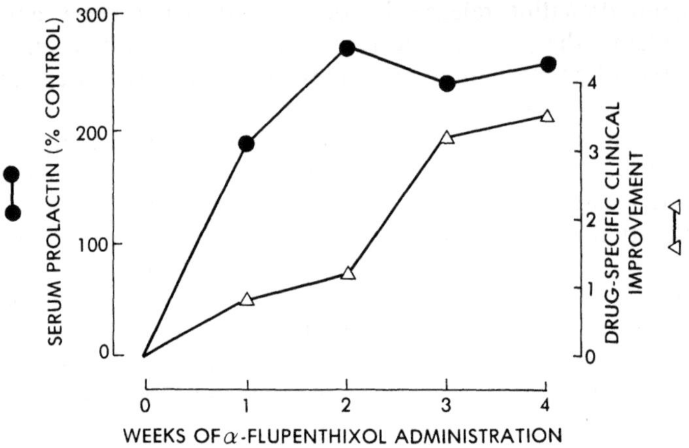

Fig. 4. The time course of changes in serum prolactin and of drug-specific clinical improvement of schizophrenic patients under medication. During the four weeks after initiation of daily α-flupenthixol administration (abscissa), the prolactin and therapeutic responses are clearly dissociated. Serum prolactin levels, expressed as per cent changes from control levels (ordinate, left) rise rapidly after beginning medication. The clinical improvement due to drugs (ordinate, right) is the improvement in clinical rating of the schizophrenic group treated with the active α-isomer, corrected for improvements also observed in groups receiving the inactive β-isomer or placebo. Data from CROW et al. (1980)

dopamine receptors modulating prolactin secretion are, after all, in the pituitary gland itself and not shielded by the blood-brain barrier. Experiments in rats show, however, that the increase in HVA and DOPAC levels in the brain occurs within minutes after the intraperitoneal administration of neuroleptics (WESTERINK and KORF 1976b; ÖHMAN et al. 1977; BUDA et al. 1981). Therefore it is unlikely that neuroleptics, which are in general lipophilic compounds, cross the blood-brain barrier with difficulty.

The hypothesis that neuroleptics ameliorate schizophrenic symptoms by blocking dopamine receptors therefore requires some revision: dopamine receptors are blocked upon acute administration of neuroleptics, and although there may be an acute sedative response, a significant therapeutic effect does not occur until after two to three weeks of neuroleptic administration. One explanation why neuroleptics do not have therapeutic benefit acutely may be that they have a tendency to increase dopaminergic neuro-transmission by their action at the autoreceptor, thus diminishing the effectiveness of their postsynaptic dopamine receptor blockade.

3. A Role of the Dopamine Autoreceptor in the Therapeutic Action of Neuroleptics?

How is it possible to explain the time lag between the beginning of neuroleptic administration and the onset of therapeutic effects? POST and GOODWIN (1975) presented data which may be the key to this puzzle. In Fig. 5, the time course of "clinical rating" of schizophrenic patients receiving daily administration of chlorpromazine is shown; the time course of this graph differs somewhat from that of Fig. 4, probably because the placebo effect has not been corrected for in Fig. 5 (see CROW et al. 1980). HVA levels in cerebrospinal fluid, reflecting primarily striatal dopamine turnover, increase as a result of neuroleptic administration (Fig. 5), as previously shown in animal studies (MATTHYSE 1973). "Tolerance" of the increase in HVA levels occurs (Fig. 5) in response to chronic neuroleptic administration, again as replicated in animal studies (SCATTON 1977).

The interesting feature of Fig. 5 is that the "tolerance" to increased HVA levels in CSF coincides with clinical improvement of schizophrenics in response to neuroleptic administration. The time course of the "tolerance" to chronic neuroleptic administration is also the time course of the development of supersensitivity of the dopamine autoreceptor. In investigations of chronic administration of neuroleptics in animals, supersensitivity has been demonstrated for dopamine autoreceptors modulating 1) depolarization-evoked release of dopamine from slices (ARBILLA et al 1981a; NOWAK et al. 1983), 2) dopamine accumulation following γ-butyrolactone administration NOWYCKY and ROTH 1977), and 3) DOPAC levels in striatum previously lesioned with kainic acid (BIGGIO et al. 1980). All three of these experimental approaches preclude an involvement of a trans-synaptic long loop feedback.

Significantly, tolerance does not develop to the increase in blood prolactin levels, with chronic administration of neuroleptics. This failure of tolerance to

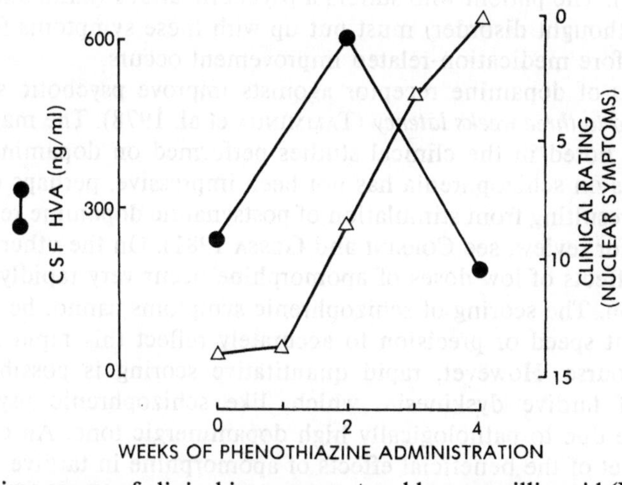

Fig. 5. The time course of clinical improvement and homovanillic acid (HVA) levels in cerebrospinal fluid (CSF) during phenothiazine administration in schizophrenics. As a function of time (abscissa) following the initiation of daily phenothiazine administration to schizophrenics, HVA levels in CSF (ordinate, left) was measured (data from POST and GOODWIN 1975; but see also GERLACH et al. 1975; SEDVALL et al. 1975). For comparison, the clinical improvement of patients receiving chlorpromazine (not corrected for placebo effect), are shown (data from PHILLIPSON et al. 1977). There seems to be a temporal correlation between clinical improvement and the development of "tolerance" to the HVA increase caused by neuroleptics

develop in the dopaminergic control of prolactin secretion may be linked to the absence of dopamine autoreceptors in the median eminence (see Sect. B.IV.2), wherein reside the dopamine terminals modulating prolactin release.

The time course of the therapeutic action of neuroleptics does not correlate, therefore, with their measurable blockade of pre- or postsynaptic receptors, but *does* correlate with their induction of supersensitivity of presynaptic dopamine autoreceptors which inhibit the release and the turnover of dopamine. It is tempting to take the step from correlation to causality, and suggest that the therapeutic mechanism of action of neuroleptics depends upon this induction of supersensitivity of dopamine autoreceptors (NOWAK et al. 1983). Such presynaptic supersensitivity would result in a diminished dopamine release and thus a decrease in the net response at the synapse (see LEHMANN and LANGER 1982c). Support for this hypothesis of the therapeutic mechanism of action of neuroleptics is derived from the therapeutic effect of low doses of dopamine receptor agonists in schizophrenia, the topic of the next section.

4. Agonists Acting at the Dopamine Autoreceptor: Potential as Therapeutic Agents

As previously noted, one of the drawbacks of neuroleptics in current use is the latency after beginning of administration until a therapeutic effect occurs (Figs. 4 and 5). The patient who suffers a psychotic attack (hallucinations, delusions, and thought disorder) must put up with these symptoms for at least two weeks before medication-related improvement occurs.

Low doses of dopamine receptor agonists improve psychotic symptoms, *without the two to three weeks latency* (TAMMINGA et al. 1978). The magnitude of improvement noted in the clinical studies performed on dopamine receptor agonist effects on schizophrenia has not been impressive, perhaps due to adverse effects resulting from stimulation of postsynaptic dopamine receptors by these drugs (for review, see CORSINI and GESSA 1981). On the other hand, the therapeutic effects of low doses of apomorphine occur very rapidly following administration. The scoring of schizophrenic symptoms cannot be performed with sufficient speed or precision to accurately reflect this rapid and short-lived time course. However, rapid quantitative scoring is possible for the symptoms of tardive dyskinesia, which, like schizophrenic psychosis, is thought to be due to pathologically high dopaminergic tone. An example of the rapid onset of the beneficial effects of apomorphine in tardive dyskinesia (CARROLL et al. 1977) is shown in Fig. 6.

In addition, since dopamine receptor agonists are effective in reducing dopamine turnover in humans receiving full therapeutic doses of neuroleptics (CUTLER et al. 1982), dopamine receptor agonists may be added to conventional neuroleptic therapy, to control occasional psychotic crises, or to improve therapeutic efficacy with a lower dose of neuroleptics.

The advantages of dopamine autoreceptor agonists as potential therapeutic agents for schizophrenia, tardive dyskinesia, and other diseases currently

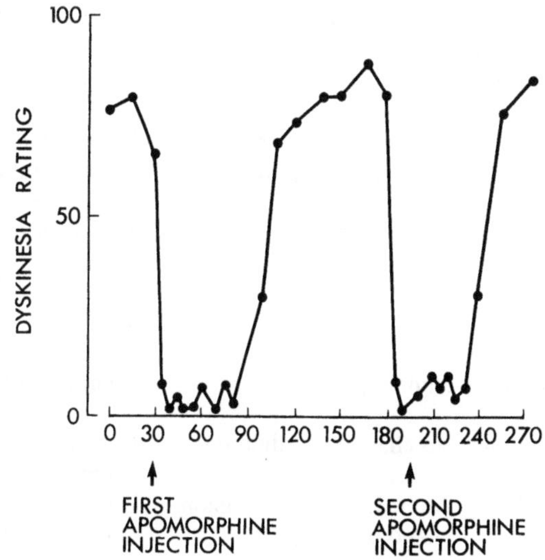

Fig. 6. The rapid and short-lasting therapeutic effect of apomorphine on symptoms of tardive dyskinesia. Within ten minutes after subcutaneous injection of 2–6 mg apomorphine, a dramatic improvement in the symptoms of tardive dyskinesia is observed. In this case report, it is clear that the therapeutic effect of apomorphine is also very short-lasting (about 1 h). From CARROLL et al. (1977). *Abscissa:* Time (in minutes). *Ordinate:* Clinical rating, severity of dyskinesic symptoms

treated with neuroleptic dopamine receptor antagonists must be considered in the light of the potential disadvantages of such medication (Table 5). The most important and obvious problem is that doses of dopamine receptor agonists slightly higher than those which produce therapeutic effects may stimulate postsynaptic dopamine receptors, and exacerbate rather than ameliorate the disease being treated. The importance of this particular problem depends on the degree of selectivity of the therapeutic agonist for the presynaptic as opposed to postsynaptic D_2 receptor, if such a pharmacological difference between these two receptors does indeed exist, as recently suggested (LEHMANN and LANGER 1982a).

The major advantage offered by selective presynaptic dopamine autoreceptor agonists is their efficacy upon acute administration. A question mark hangs over their advantages and disadvantages when chronically administered. To our knowledge, three weeks of 4 times daily administration of apomorphine is the longest clinical trial which has yet been performed in schizophrenic patients (HOLLISTER 1981). Since the therapeutic action of apomorphine lasts not longer than 60 minutes after injection (DAVIS et al. 1981), due to rapid metabolism of this drug, it is not well-suited for chronic clinical administration. It is hoped that in the near future, long-acting orally active agonists selective for the dopamine autoreceptor will begin to be tested in clinical trials. It is of interest to note that bromocriptine, administered for either 3 or 10 days produced no improvement in schizophrenic symptoms (C. A. TAM-

Table 5. Comparison of the advantages and disadvantages of antischizophrenic pharmacotherapies: the use of selective dopamine autoreceptor agonists vs. the use of current neuroleptic dopamine receptor antagonists

	Selective dopamine autoreceptor agonist	Neuroleptic dopamine receptor antagonist
Onset of therapeutic effect	Immediate	Two to three weeks
Treatment of tardive dyskynesia	Effective	Effective
Induction of tardive dyskinesia	?	Major side effect
Induction of extrapyramidal symptoms	None reported	Frequent side effect
Prolactin secretion	No change, or slight decrease	Induction of hyperprolactinemia
Cardiovascular side effects	Potential for hypotension and bradycardia	Hypotension, depending on α-blocking properties
Emesis	Potential obstacle	No problems
Development of tolerance	Open question	Not a major problem

Neuroleptic dopamine receptor antagonists have been used in the chronic treatment of schizophrenia for several decades, while dopamine receptor agonists remain primarily at the stage of clinical trials. Sufficiently selective dopamine autoreceptor agonists for routine clinical use have not yet appeared; however, some of their potential advantages and disadvantages can be predicted from existing clinical studies performed with some dopamine receptor agonists.

MINGA, personal communication). On the other hand, studies carried out with N-n-propylnorapomorphine show a significant clinical improvement in schizophrenia (TAMMINGA et al. 1981). It is possible that, unlike after neuroleptics, there will be less extrapyramidal Parkinson-like symptoms, acute dystonia, and tardive dyskinesia, when selective dopamine autoreceptor agonists are administered chronically (Table 5).

The hyperprolactinemia induced by conventional neuroleptics is likely not to be a problem with selective presynaptic dopamine autoreceptor agonists. In no case has an increase in prolactin levels been observed as a result of administration of dopamine receptor agonists, even at the low doses which ameliorate schizophrenic symptoms (TAMMINGA et al. 1977; MELTZER and SIMONOVIC 1981). These results may be related to the apparent lack of dopamine autoreceptors on tuberoinfundibular dopaminergic neurones (see Sect. B.IV).

Dopamine receptor agonists are emetic, and the dopamine receptor which mediates this response in dogs appears to be very sensitive to dopamine receptor agonists acting on the D_2 dopamine receptor. Emetic side effects of dopamine receptor agonists are perhaps the greatest obstacle to the therapeutic use of dopamine receptor effects, and tolerance develops rapidly to the emetic effects of dopamine receptor agonists administered to man; such examples are well documented for the dopamine receptor agonists pergolide and bromocriptine (CALNE et al. 1978; LEMBERGER and CRABTREE 1979) which are potent emetic drugs in the dog.

Dopamine receptor agonists are currently being considered as possible antihypertensive agents, an action ascribed to the inhibition of noradrenaline release via presynaptic inhibitory dopamine receptors on noradrenergic terminals in the periphery (ENERO and LANGER 1975; LANGER and DUBOCOVICH 1979; DUBOCOVICH and LANGER 1980; MASSINGHAM et al. 1980). In the schizophrenic patient who does not have hypertension, however, the orthostatic hypotension produced by dopamine receptor agonists acting on peripheral D_2 receptors may present a potential undesirable side effect. Yet, it is possible that the hypotensive effects of dopamine receptor agonists are more pronounced in hypertensive than in normotensive patients.

Two possible ways of avoiding the undesirable emetic and hypotensive effects of dopamine receptor agonists may be envisaged at present. First, it may be possible to develop selective presynaptic dopamine autoreceptor agonists which do not activate the dopamine receptors which induce emesis or hypotension; this would depend on the existence of a pharmacological difference between central dopamine autoreceptors and the peripheral presynaptic dopamine receptors in question, and it is not known whether such differences do in fact exist. Alternatively one may exploit the fact that the emesis-inducing hypotension are outside of the blood brain barrier. Dopamine receptor antagonists which essentially do not cross the blood-brain barrier such as domperidone (QUINN et al. 1981) can be effectively co-administered with dopamine receptor agonists to block their undesirable peripheral side effects.

The possibility even exists that a single drug may act as an antagonist at postsynaptic dopamine receptors and as an agonist at presynaptic dopamine autoreceptors. N-Chloroethyl-norapomorphine has exactly such properties, the postsynaptic antagonism being irreversible, while the inhibition of dopamine release by an agonistic action at the dopamine autoreceptor is reversible (LEHMANN and LANGER 1982a). Thus, N-chloroethyl-norapomorphine may serve as a prototype for the development of novel and effective antipsychotic agents.

C. Presynaptic Heteroreceptors Modulating Dopamine Function

The role of dopamine autoreceptors modulating the firing, synthesis and release of dopaminergic neurones has a longer history and has been studied in more detail than have presynaptic "heteroreceptors" (i.e. presynaptic receptors sensitive to neurotransmitters other than dopamine) which modulate dopamine release. The evidence for presynaptic heteroreceptors modulating dopamine release is derived from studies on potassium or electrically evoked release of dopamine, with very few reports existing on the possible role of presynaptic heteroreceptors which may modulate dopamine synthesis.

If heteroreceptors modulating the depolarization-evoked release of dopamine exist in the cell bodies or dendrites of dopaminergic neurones, these are not distinguishable with current techniques from conventional postsynaptic receptors on these neurones. The physiological relevance of these presynaptic heteroreceptors modulating dopamine release is probably associated with interactions between neurotransmitters at the level of transmitter release.

I. Cholinoceptors Modulating Dopamine Function

The possibility of a direct interaction of acetylcholine with dopaminergic nerve terminals is of particular importance because of the numerous dopamine-acetylcholine interactions which have been noted in the past decade in biochemical, animal psychopharmacological, and clinical studies (for review, see LLOYD 1978; see also BARTHOLINI et al., Chap. 16). Preliminary results from the MCGEERS' laboratory suggesting a direct synapse of dopaminergic terminals on immunohistochemically identified cholinergic dendrites (HATTORI et al. 1976) have been used to support the concept of dendro-axonic cholinergic neurotransmission onto dopaminergic terminals (HATTORI et al. 1979; MCGEER et al. 1979). However, MCGEERS' laboratory has withdrawn original immunohistochemical identification of the cholinergic neurone, and the neurones now thought to be the cholinergic interneurones of the striatum (LEHMANN and FIBIGER 1979; KIMURA et al. 1980) are probably not those receiving a synaptic input from dopaminergic terminals (LEHMANN and LANGER 1982b). Therefore, the existence of presynaptic cholinoceptors on dopaminergic terminals presents a considerable challenge for the interpretation of the possible physiological role of these receptors (see LEHMANN and LANGER 1982b).

The outflow of ^{3}H-dopamine newly synthesized from ^{3}H-tyrosine is doubled by exposure to 10 μM acetylcholine (GIORGUIEFF et al. 1977b). This effect is partially antagonized by the muscarinic antagonists atropine and scopolamine and the nicotinic antagonist pempidine. Oxotremorine produces a 60% increase in this basal release (GIORGUIEFF et al. 1977b), and nicotine more than doubles the basal release of ^{3}H-dopamine (GIORGUIEFF-CHESSELET et al. 1979c). The increase in basal release of ^{3}H-dopamine evoked by these cholinergic agonists is calcium-dependent, and the nicotinic but not the muscarinic component appears to be blocked by tetrodotoxin (GIORGUIEFF et al. 1977b; GIORGUIEFF-CHESSELET et al. 1979c). Therefore, nicotinic receptor stimulation appears to result in action potential-mediated, calcium-dependent release of ^{3}H-dopamine from dopaminergic terminals. However, the mechanism by which muscarinic agonists accelerate the outflow of ^{3}H-dopamine continuously synthesized from ^{3}H-tyrosine is less clear: if depolarization is the mechanism by which spontaneous release is increased, it is not via voltage-dependent, tetrodotoxin-sensitive sodium channels. While GIORGUIEFF et al. (1977b) demonstrated that muscarinic receptor antagonists alone do not modify the rate of synthesis of ^{3}H-dopamine, this information for muscarinic receptor *agonists* is curiously lacking.

The experimental model which has classically been utilized to investigate presynaptic receptors is the examination of modulation of depolarization-evoked transmitter release, rather than modulation of basal outflow as studied by GIORGUIEFF et al. (1977b). We investigated the presynaptic cholinergic modulation of electrically evoked release of previously taken up ^{3}H-dopamine in slices of cat caudate (LEHMANN and LANGER 1982b). The electrically evoked release of ^{3}H-dopamine is enhanced by the muscarinic receptor agonist, oxotremorine. The indirectly acting cholinomimetic agonist physostigmine also increases the electrically evoked release of ^{3}H-dopamine (LEHMANN and LANGER 1982b). Physostigmine has this effect by virtue of its inhibition of acetylcholinesterase, which normally degrades released endogenous acetylcholine. Thus, one possible interpretation is that endogenous acetylcholine released from the striatal cholinergic interneurones may attain high enough concentrations to interact with the muscarinic receptors modulating dopamine release, *if* acetylcholinesterase is inhibited. There is, however, no evidence that endogenous acetylcholine may stimulate nicotinic or muscarinic receptors modulating ^{3}H-dopamine release if acetylcholinesterase is not inhibited (LEHMANN and LANGER 1982b).

The increase in electrically evoked ^{3}H-dopamine release caused by oxotremorine or physostigmine is antagonized by the muscarinic receptor antagonist atropine (LEHMANN and LANGER 1982b). The observations that 1) the increase caused by physostigmine is *completely* blocked by atropine and that 2) d-tubocurarine has no apparent effect on ^{3}H-dopamine release in the presence of physostigmine (LEHMANN and LANGER 1982b) argues that even though nicotinic receptors may exist on dopaminergic terminals, the endogenous acetylcholine does not reach sufficiently high concentrations to stimulate these nicotinic receptors. Thus, muscarinic receptors would appear much more likely than nicotinic receptors to be involved in any physiological role of presynaptic cholinoceptors modulating dopamine function (LEHMANN and LANGER 1982b).

It may be noted that the nicotinic receptor antagonist mecamylamine (systemically administered) slows down the disappearance of dopamine in the striatum following α-methyl-tyrosine administration (AHTEE and KAAKKOLA 1978). Mecamylamine alone does not affect HVA levels, but if probenecid is co-administered, mecamylamine appears to slightly decrease HVA levels (AHTEE and KAAKKOLA 1978; NOSE and TAKEMOTO 1974). Arecoline and nicotine, nicotinic receptor agonists, increase HVA levels when injected systemically (NOSE and TAKEMOTO 1974). However, the increases elicited by physostigmine or the muscarinic receptor agonist oxotremorine are three times as large as those elicited by the nicotinic agonists (NOSE and TAKEMOTO 1974). In contrast to research on dopamine autoreceptors, efforts have not been made to determine whether receptors on dopaminergic terminals, as opposed to transsynaptic events, mediate the actions of cholinergic drugs on dopamine turnover *in vivo*.

A large number of investigators have attempted to detect cholinoceptors on dopaminergic terminals by performing 6-hydroxydopamine lesions and then measuring the binding of a variety of labelled muscarinic antagonists (KATO et al. 1978; DE BELLEROCHE et al. 1979; MCGEER et al. 1979; NOMURA et al 1979; REISINE et al. 1979; GURWITZ et al 1980; SUGA 1980) or of the labelled nicotinic antagonist, α-bungarotoxin (DE BELLEROCHE et al 1979; MCGEER et al. 1979). While α-bungarotoxin binding sites seem to decrease substantially in number following such lesions, the decreases in muscarinic binding sites are small or not statistically significant.

The failure to detect muscarinic receptors on dopaminergic terminals with the muscarinic receptor antagonist ^{3}H-quinuclidinyl benzilate may be explained by the low potency of this antagonist at the muscarinic receptors modulating ^{3}H-dopamine release (LEHMANN and LANGER 1982), as well as the high density of muscarinic receptors in the striatum on neuronal structures other than dopaminergic terminals. Recent data from MARCHI and RAITERI (1985) suggest that the muscarinic receptors which facilitate the release of ^{3}H-dopamine are of the M_1 subtype.

II. Opiate Receptors Modulating Dopamine Function

The potassium-evoked release of ^{3}H-dopamine in the striatum of the rat has been reported to be inhibited by morphine and β-endorphin, but not Met-enkephalin, an inhibition which is antagonized by naloxone (LOH et al. 1976). SUBRAMANIAN et al. (1977), however, reported that Met-enkephalin was effective in inhibiting potassium-evoked release of dopamine, an effect antagonized by naloxone. ARBILLA and LANGER (1978) found that neither morphine nor β-endorphin inhibits potassium-evoked release of ^{3}H-dopamine from the striatum, simultaneously demonstrating the validity of their experimental system by finding naloxone-sensitive receptor-mediated inhibition of potassium-evoked ^{3}H-noradrenaline release from rat cerebral cortex, by the opiate receptor agonists, morphine and β-endorphin.

These rather large discrepancies between laboratories remain essentially unexplained. The observations by MOLEMAN and BRUINVELS (1979) that systemically administered morphine accelerates the disappearance of dopamine following α-methyl-tyrosine administration only when the firing rate of dopaminergic neurones is retarded, may have some bearing on the discordant results on potassium-evoked ^{3}H-dopamine release: LOH et al. (1976) employed 53 m*M* potassium, SUBRAMANIAN et al. (1977) employed 40 m*M* potassium, and ARBILLA and LANGER (1978) employed 20 m*M* potassium as depolarizing stimuli. Presynaptic modulation of transmitter release can vary when different concentrations of potassium are used, and hence different degrees of depolarization are achieved. Therefore, these different experimental conditions may contribute to the different results obtained.

The possibility that there may be more than one type of opiate receptor on dopaminergic terminals is suggested by recent results from the laboratory of GLOWINSKI. The release of ^{3}H-dopamine continuously synthesized from ^{3}H-tyrosine in the cat caudate nucleus superfused by push-pull cannula is increased by local superfusion of either morphine or D-Ala^2-Metenkephalinamide, a synthetic enkephalin (CHESSELET et al. 1981). The action of morphine is antagonized by naloxone, while that of D-Ala^2-Metenkephalinamide is not. These workers therefore suggest that dopaminergic terminals possess both μ- and δ-type opiate receptors.

Adopting the experimental model used by KEHR et al. (1972) to suggest a presynaptic location of dopamine autoreceptors modulating dopamine synthesis, MOLEMAN and BRUINVELS (1979a) similarly suggested a presynaptic localization for opiate receptors modulating dopamine release. After acute electrolytic lesions in the diencephalon which sever nigrostriatal axons, administration of morphine accelerates the loss of dopamine which follows administration of α-methyl-tyrosine. These results indicate that this action of morphine is mediated by presynaptic receptors on dopaminergic terminals. Either morphine receptors are located on dopaminergic terminals, or else morphine receptors mediate the release of a second messenger, which acts on dopaminergic terminals. Morphine also elevates DOPAC and HVA levels in the striatum, nucleus accumbens, and olfactory tubercle of the intact rat (WESTERINK and KORF 1976a, b). The dopaminergic terminals in the cortex, however, would seem to be particularly sensitive to the acceleration by morphine of dopamine disappearance following α-methyl-tyrosine administration (MOLEMAN and BRUINVELS 1979b).

The opiate receptors which mediate an increase in dopamine metabolite levels, signalling an increase in dopamine turnover, have been suggested to be localized to dopaminergic terminals. BIGGIO et al. (1978) found that D-Ala^2-enkephalinamide, injected into the striatum, increases DOPAC levels, or, in animals pretreated with a dopa-decarboxylase inhibitor, increases dopa accumulation. These findings are reproduced even when the striatal neuronal perikarya are previously lesioned with kainic acid (BIGGIO et al. 1978). These results suggest that interneurones are not involved in the effect of enkephalins on dopamine turnover. The intraventricular injection of either Met-enkephalin or D-Ala^2-Met-enkephalinamide results in an increase in DOPAC and HVA levels in the striatum. Significantly, the effects of both of these enkephalins are antagonized by naloxone (BIGGIO et al. 1978). These results are in disagreement with the results of CHESSELET et al. (1981), who failed to antagonize the effects of D-Ala^2-Met-enkephalinamide with naloxone, and therefore proposed a second opiate receptor subtype on dopaminergic terminals.

The approach of performing 6-hydroxydopamine lesions followed by receptor-binding identification of opiate receptors has given positive results and marked reproducibility between laboratories. The binding of ^{3}H-Leu-enkephalin in the striatum is reduced by 29% and that of ^{3}H-naloxone reduced by 33% following 6-hydroxydopamine lesions (POLLARD et al. 1978). REISINE et al. (1979) similarly found a reduction in ^{3}H-naloxone binding of 28% in the striatum and 26% in the frontal cortex, but importantly, no change in ^{3}H-naloxone binding in the substantia nigra following 6-hydroxydopamine lesions. The binding of ^{3}H-naloxone is also reduced by 31% in the nucleus accumbens following local 6-hydroxydopamine lesions (POLLARD et al. 1977). By receptor autoradiography it is possible to discern two classes of opiate receptors which bind ^{3}H-diprenorphine: those which occur in "clusters" or patches, and those which do not. The lesion of dopaminergic afferents results in a preferential loss of the opiate receptors which are not clustered (MURRIN et al. 1980).

Although a great deal remains to be clarified regarding the pharmacological properties and physiological functions of opiate presynaptic receptors modulating dopamine function, the evidence is at least suggestive that such receptors exist. Several reports suggest that behavioural measurement of morphine-induced analgesia is reduced or abolished following 6-hydroxy-dopamine lesions of the substantia nigra (GROSSMAN et al. 1973; NAKAMURA et al. 1973; PRICE and FIBIGER 1975). On the other hand, enhancing dopamine function by administering the dopamine uptake inhibitor nomifensine has been shown to produce an apparent dopamine receptor mediated hyperalgesia.

III. Excitatory Amino Acid Receptors Modulating Dopamine Function

A major neuronal input to the striatum from the cortex utilizes an excitatory amino acid as neurotransmitter (McGEER et al. 1977). The question naturally arises whether dopamine function may be modulated via a direct interaction of excitatory amino acids with dopaminergic terminals. GIORGUIEFF et al. (1977a) found that L-glutamate but not D-glutamate increases the efflux of ^{3}H-dopamine continuously synthesized from ^{3}H-tyrosine, an effect which is tetrodotoxin-insensitive. ROBERTS and SHARIF (1978) found that L-glutamate increases the efflux of previously taken up ^{3}H-dopamine, again an effect which is tetrodotoxin-insensitive. In disagreement with GIORGUIEFF et al. (1977a), ROBERTS and co-workers found D-glutamate to be equipotent with L-glutamate in increasing ^{3}H-dopamine release (ROBERTS and SHARIF 1978; ROBERTS and ANDERSON 1979).

The findings of GIORGUIEFF et al. (1977a) merit further attention, however, since by virtue of their measurement of ^{3}H-dopamine synthesized from ^{3}H-tyrosine, the response to L-glutamate can be ascribed with certainty to dopaminergic terminals. The possibility exists that the increase in spontaneous release by L-glutamate does not reflect a receptor-mediated effect. The fact that D-glutamate is inactive in releasing ^{3}H-dopamine (GIORGUIEFF et al. 1977a) while D-glutamate is equipotent with L-glutamate in exciting neurones (WATKINS 1981) supports the proposition that the action of L-glutamate on the synthesis and release of ^{3}H-dopamine may *not be* mediated via a receptor. The effect of L-glutamate but not D-glutamate on ^{3}H-dopamine release may instead be a consequence of the metabolism of this amino acid, since L-glutamate but not D-glutamate enters the Krebs cycle. It is significant that GIORGUIEFF et al. (1977a) did not assess the effects of L-glutamate on the rate of ^{3}H-dopamine synthesis from ^{3}H-tyrosine.

In conclusion, there is as yet no firm support for the existence of an excitatory amino acid receptor which modulates dopamine function, situated either on dopaminergic terminals or acting indirectly via interneurones.

IV. GABA Receptors Modulating Dopamine Function

GABA in concentrations from 10 μM to 1 mM causes an increase in the basal release of ^{3}H-dopamine continuously synthesized from ^{3}H-tyrosine in striatal slices (GIORGUIEFF et al. 1978). These concentrations of GABA change neither the rate of ^{3}H-dopamine synthesis nor the reuptake of ^{3}H-dopamine

(GIORGUIEFF et al. 1978). The increase in basal release of ^{3}H-dopamine by GABA is mimicked by the GABA receptor agonists muscimol, lioresal, gammahydroxybutyrate, and 3-aminopropanesulfonate, and blocked by picrotoxin (GIORGUIEFF et al. 1978; GIORGUIEFF-CHESSELET et al. 1979a).

The GABA-mediated increase in ^{3}H-dopamine release is blocked by tetrodotoxin (GIORGUIEFF et al. 1978). For this reason, these authors concluded that the effect of GABA on dopaminergic terminal is mediated by an interneurone (GIORGUIEFF et al. 1978; GIORGUIEFF-CHESSELET et al. 1979a). As noted elsewhere in this review (see Sect. B.I.1.f), this conclusion is not warranted. Blockade by tetrodotoxin of the increase in the basal release of ^{3}H-dopamine means that action-potentials involving voltage-dependent sodium channels mediate the effect. These action potentials are as likely to be generated in dopaminergic terminals as in non-dopaminergic neural structures. The fact that veratridine-evoked neurotransmitter release from synaptosomes is also tetrodotoxin sensitive (BLAUSTEIN et al. 1972) illustrates the danger of equating tetrodotoxin sensitivity with the involvement of interneurones. GIORGUIEFF-CHESSELET et al. (1979a) observed that acetylcholine was not a mediator of the ^{3}H-dopamine release evoked by GABA.

Employing an *in vitro* model, STARR (1978, 1979) found that the basal release of previously taken up ^{3}H-dopamine is not modified by GABA (1.0 μM to 1 mM). However, the potassium-evoked release of ^{3}H-dopamine is increased by GABA (10 μM to 1 mM), this effect being blocked by picrotoxin but not bicuculline (STARR 1978). Subsequent work confirmed that GABA increased the potassium-evoked release of ^{3}H-dopamine but revealed that this effect was not sensitive to either picrotoxin or bicuculline (STARR 1979; ENNIS and COX 1981). STOOF et al. (1979) found that GABA (10 μM to 1 mM) increases the potassium-evoked release of ^{3}H-dopamine, the effect being antagonized by neither bicuculline nor picrotoxin. Other workers have reported that GABA does not increase potassium-evoked ^{3}H-dopamine release, but rather inhibits (BOWERY et al. 1980) or has no effect (MARTIN and MITCHELL 1979). The time of preincubation with GABA may be a crucial point in explaining these discrepancies.

It is important to recognize that the spontaneous release of ^{3}H-dopamine synthesized from ^{3}H-tyrosine, which is increased by GABA (GIORGUIEFF et al. 1978; GIORGUIEFF-CHESSELET et al. 1979a), may reflect a phenomenon different from the enhancement by GABA of potassium-evoked release of previously taken up ^{3}H-dopamine (STARR 1978, 1979; STOOF et al. 1979).

V. Other Presynaptic Heteroreceptors Modulating Dopamine Function

There is some evidence for a role of glycine increasing the basal release of ^{3}H-dopamine continuously formed from ^{3}H-tyrosine in striatal slices (GIORGUIEFF-CHESSELET et al. 1979b). The effect of glycine is partially antagonized by strychnine and blocked completely by tetrodotoxin or by the removal of calcium from the medium. There is as yet no convincing evidence that glycine is a neurotransmitter in the striatum (PYCOCK and KERWIN 1981).

Cox et al. (1981) reported inhibition of the potassium-evoked release of previously taken up ^{3}H-dopamine from striatal slices by 5-hydroxytryptamine and some related compounds. These results have not been reproduced in this laboratory in investigations of electrically evoked ^{3}H-dopamine release from rabbit caudate (ARBILLA and LANGER, unpublished observations).

A rigorous search for presynaptic histamine receptors modulating electrically evoked ^{3}H-dopamine release from rabbit caudate in this laboratory has not yielded positive results (NOWAK, ARBILLA and LANGER, unpublished results). Prostaglandins apparently do not modify depolarization evoked release of ^{3}H-dopamine (REIMANN et al. 1981).

It was recently reported that melatonin is a potent inhibitor of the calcium-dependent release of ^{3}H-dopamine in the rabbit and chicken retina (DUBOCOVICH 1983, 1984). These inhibitory effects of melatonin on ^{3}H-dopamine release from the rabbit and chicken retina are observed in the low nanomolar range (DUBOCOVICH 1983, 1984), but melatonin does not inhibit the release of ^{3}H-dopamine from striatal slices (DUBOCOVICH 1983). These interesting observations may provide an attractive experimental model for the study of melatonin receptors.

VI. The Physiological Relevance of Presynaptic Heteroreceptors

A frequently made, but nevertheless incorrect, assumption is that the existence of a presynaptic receptor implies a physiological function for that receptor. For presynaptic *autoreceptors*, such as the dopamine autoreceptor, access of the relevant neurotransmitter to the presynaptic receptor is obviously within the realm of possibility. For dopamine autoreceptors it is possible to demonstrate that the endogenous transmitter *does* in fact activate the presynaptic autoreceptor (by demonstrating the effect of receptor blocking agents on transmitter release) (LEHMANN and LANGER, 1982c). Likewise, it has been demonstrated that released endogenous noradrenaline does activate presynaptic α_2-adrenoceptors (ENERO and LANGER 1973; CUBEDDU and WEINER 1975).

However, the lack of physiological function of some presynaptic *heteroreceptors* has been well documented in the peripheral nervous system. For instance, presynaptic muscarinic receptors modulate the release of transmitter from noradrenergic terminals in rabbit ear artery (STEINSLAND et al. 1973), but the lack of effect of muscarinic antagonists alone, as well as the complete absence of cholinergic innervation (FLORENCE et al. 1981), indicate that these presynaptic muscarinic heteroreceptors do not serve a physiological function. Similar conclusions are drawn concerning the presynaptic dopamine receptors modulating noradrenaline release in the cat nictitating membrane (ENERO and LANGER 1975): convincing arguments can be found that dopamine does not reach high enough concentrations physiologically to stimulate these receptors, and in agreement with this view, the receptor antagonist, sulpiride, alone has no effect on noradrenaline release elicited by nerve stimulation (DUBOCOVICH and LANGER 1980).

The presynaptic receptors modulating depolarization-evoked release of ^{3}H-dopamine in the striatum appear to be exposed to concentrations of released endogenous acetylcholine sufficiently high to stimulate the muscarinic, but not nicotinic, receptors (LEHMANN and LANGER 1982b), when the inactivating enzyme acetylcholinesterase is inhibited. Therefore, these presynaptic muscarinic receptors may play a physiological role in modulating dopamine release in the striatum.

The possible role of endogenous GABA in stimulating presynaptic GABA receptors modulating depolarization-evoked ^{3}H-dopamine release has been suggested by the results of STOOF and MULDER (1977) showing that inhibition of GABA-transaminase (which normally catabolizes GABA) increases depolarization-evoked ^{3}H-dopamine release, presumably by potentiating the action of released endogenous GABA. Also, the effect of exogenous GABA on ^{3}H-dopamine release at different rostro-caudal levels within the caudate correlates with the levels of endogenous GABA and glutamic acid decarboxylase within the caudate (STOOF et al. 1979). The important piece of evidence which is lacking is the effect of a receptor antagonist at the presynaptic GABA heteroreceptor modulating ^{3}H-dopamine release. To be able to postulate an interaction between endogenous GABA and dopamine release, an antagonist to the presynaptic GABA heteroreceptor alone must be shown to affect ^{3}H-dopamine release.

Concerning the other putative presynaptic heteroreceptors modulating dopamine release, the question of the physiological relevance of these receptors has rarely been addressed experimentally. Once a given presynaptic heteroreceptor is clearly characterized pharmacologically, the next important question to be addressed concerns the physiological relevance of the receptor. This question may, in general, be addressed by determining whether the receptor antagonist alone has an action, which may be ascribed to blockade of receptor occupation by the naturally-occuring, endogenous neurotransmitter acting on the heteroreceptor.

D. Autoadrenoceptors

The first clues hinting at the existence of presynaptic α-autoadrenoceptors were obtained a quarter of a century ago, when BROWN and GILLESPIE (1957) observed an increase in noradrenaline release from the perfused cat spleen elicited by sympathetic nerve stimulation during exposure to phenoxybenzamine. Yet the concept of α_2-autoadrenoceptors modulating transmitter release developed only during the last decade (LANGER 1974). Several recent reviews (LANGER 1974, 1977, 1981; STARKE 1977, 1981; GILLESPIE 1980; RAND et al. 1980) attest to the great volume of research and theoretical progress pertaining to autoadrenoceptors. Here we will deal with selected aspects of autoadrenoceptors of the peripheral and central nervous system which will be compared with dopamine autoreceptors.

I. Functions of α_2-Autoadrenoceptors

In contrast to dopamine autoreceptors, autoadrenoceptors do not appear to modulate the rate of synthesis of noradrenaline in central (SALZMAN and ROTH 1980) or peripheral (STEINBERG and KELLER 1978) noradrenergic neurones, although this question appears not to have been explored in great depth. However, both the release of noradrenaline from nerve terminals and the electrical activity of noradrenergic neurones appear to be modulated by autoadrenoceptors of the α_2-subtype. Furthermore, there is some evidence to suggest the existence of a second pharmacological subtype of autoadrenoceptor, namely the β_2, which facilitates the release of the neurotransmitter.

1. Modulation of Noradrenaline Release

The increase in electrically evoked ^{3}H-noradrenaline overflow caused by phenoxybenzamine was first attributed to the inhibition of reuptake of noradrenaline caused by this drug (HERTTING 1965; IVERSEN 1965; IVERSEN and LANGER 1969). However, when neuronal uptake of catecholamines is blocked by cocaine or desipramine, subsequent exposure to phenoxybenzamine results in an even greater overflow of ^{3}H-noradrenaline (HÄGGENDAL 1970; FARNEBO and HAMBERGER 1970), in the same fashion that antagonists of dopamine autoreceptors have a greater effect in the presence of the dopamine uptake inhibitor, nomifensine (LEHMANN and LANGER 1981). In addition, in the isolated nerve-muscle preparation of the cat nictitating membrane, it was shown that phenoxybenzamine prevents the metabolism of ^{3}H-noradrenaline released by nerve stimulation, but this effect does not fully account for the large increase in transmitter overflow observed (LANGER 1970). Subsequently, it was demonstrated that phenoxybenzamine increases ^{3}H-noradrenaline release elicited by nerve stimulation, at concentrations too low to inhibit neuronal uptake of noradrenaline (ENERO et al. 1972). These findings clearly ruled out inhibition of catecholamine reuptake as the mechanism underlying the increase in depolarization evoked overflow caused by phenoxybenzamine. Furthermore, it was shown that phenoxybenzamine increases depolarization-evoked release of DBH, a biochemical marker of the amount of exocytosis (CUBEDDU et al. 1974), thereby suggesting that phenoxybenzamine increases exocytotic release of neurotransmitter.

The increase in depolarization-evoked overflow of ^{3}H-noradrenaline and of DBH depends upon the existence of endogenous noradrenaline stores. When such stores are depleted by reserpine pretreatment, the release-enhancing effects of phenoxybenzamine are greatly decreased (ENERO and LANGER 1973; CUBEDDU and WEINER 1975).

The second hypothesis which appeared to explain the effects of phenoxybenzamine involved the well known α-receptor blocking properties of the compound. This hypothesis was founded on the inverse relationship between postsynaptic responses and the presynaptic increase in transmitter overflow which was observed upon exposure to phenoxybenzamine. The local negative feedback mechanism linking postsynaptic responses and presynaptic transmitter release was postulated to be *trans-synaptic* (HÄGGENDAL 1969, 1970; HEDQVIST 1969, 1970; FARNEBO and HAMBERGER 1971b), although action on a presynaptic receptor was not excluded in these early theoretical formulations. The inhibitory action of α-adrenoceptor agonists on ^{3}H-noradrenaline overflow supported the receptor-mediated negative feedback model (STARKE 1972; LANGER et al. 1972).

An important change in orientation resulted when investigators proceeded to examine the heart, where the postsynaptic adrenoceptors are primarily of β subtype. According to the trans-synaptic hypothesis, in this tissue presynaptic modulation of ^{3}H-noradrenaline release should not occur. Nevertheless, it was clearly demonstrated that α-adrenoceptor antagonists increased ^{3}H-noradrenaline release in the guinea pig atrium (LANGER et al. 1971; MCCULLOCH et al. 1972) and rabbit and cat heart (STARKE 1972; FARAH and LANGER 1974).

The trans-synaptic hypothesis lost further ground and the presynaptic autoadrenoceptor hypothesis began to gain support, when the pharmacological differentiation between postsynaptic α_1-adrenoceptors and presynaptic α_2-adrenoceptors was established (LANGER 1974). Today it is known that the α_2-adrenoceptor (which was and is defined by its pharmacological properties, and *not* localization), occurs postsynaptically on several nonneuronal cell types (see LANGER 1981 for review). The distinction between α_1 and α_2-adrenoceptors remains useful in most tissues for distinguishing the responses mediated by postsynaptic adrenoceptors from the responses mediated by presynaptic release-modulating autoadrenoceptors. Finally, preliminary evidence has been reported to suggest differences in the pharmacological properties of pre- and postsynaptic α_2-adrenoceptors (HICKS 1981; DE JONGE et al. 1981).

In no case has the presynaptic response to drugs acting at α-adrenoceptors been shown to be causally linked to an effect on a postsynaptic α-adrenoceptor. Furthermore, while several trans-synaptic second messengers have been proposed, notably prostaglandins (HEDQVIST 1969, 1970), none of these have been found to adequately fulfil the role required by the trans-synaptic hypothesis. Nonetheless, considerable pains have been taken to account for the possibilities: 1) that there may be a postsynaptic α_2-adrenoceptor not linked to a known postsynaptic response, but involved in the modulation of noradrenaline release from nerves; 2) that there may be one or more second-messengers whose release is modulated by this silent α_2-adrenoceptor in all of the wide variety of tissues innervated by noradrenergic neurones; and 3) that α_2-adrenoceptors are not localized to the noradrenergic nerve terminal.

Atrophy of rat submandibullary gland following duct ligation does not affect the increase in potassium-evoked release of ^{3}H-noradrenaline caused by exposure to phentolamine (FILINGER et al. 1978). When sympathetic neurones in tissue culture are tested under conditions where intercellular synaptic contacts have not yet formed, α-adrenoceptor blockade increases depolarization-evoked release of ^{3}H-noradrenaline from these cells (VOGEL et al. 1972). Some receptor binding studies have suggested the localization of α_2-autoadrenoceptors to noradrenergic terminals in the heart, since the number of ^{3}H-dihydroergocriptine binding sites decreases following chemical sympathectomy (STORY et al. 1979). It is, however, important to note that the success of receptor-binding techniques in detecting the loss of α_2-autoadrenoceptors in the heart following lesion of the noradrenergic innervation (STORY et al. 1979) is the exception, rather than the rule. The heart contains some postsynaptic α_1-adrenoceptors (RAISMAN et al. 1979), even though the β-adrenoceptor is the primary postsynaptic receptor for noradrenaline in the heart. In the submaxillary gland, for instance, the same is not true. Postsynaptic α-adrenoceptors do play an important role in the submaxillary gland, and sympathetic denervation results in an increase in the B_{max} of ^{3}H-clonidine binding (PIMOULE et al. 1980). The inability to detect α_2-autoadrenoceptors in the submaxillary gland is neither due to a peculiarity of the tissue per se, nor to possible artefacts in the receptor binding technique, since the B_{max} of ^{3}H-desipramine, which labels noradrenaline uptake sites, does decrease dramatically following similar sympathetic denervation procedures (RAISMAN et al. 1982). Similar decreases in ^{3}H-desipramine binding to the noradrenergic transporter were obtained following sympathetic denervation of the rat heart (LANGER et al. 1984). In the central nervous system of the rat, no changes were obtained in either K_d or B_{max} of ^{3}H-idazoxan binding to α_2-adrenoceptors following chemical denervation of noradrenergic neurons with either DSP4 or 6-hydroxydopamine (PIMOULE et al. 1983). Thus, in general, it is as difficult to detect α_2-autoadrenoceptors as it is to detect dopamine autoreceptors with receptor binding techniques (see Sect. B.III).

α-Autoadrenoceptors also modulate the depolarization-evoked release of ^{3}H-noradrenaline in the brain (FARNEBO and HAMBERGER 1971b). These α-autoadrenoceptors resemble pharmacologically the α_2-subtype (TAUBE et al. 1977). Unlike the peripheral nervous system, the central nervous system lends itself to the preparation of "synaptosomes", resealed nerve endings, which retain some properties of intact nerve terminals. The synaptosome preparation presents the advantage that each nerve ending is effectively isolated from chemical influences of neighboring cells or cell fragments, in sufficiently dilute or superfused conditions. Autoadrenoceptors modulating depolarization-evoked release of ^{3}H-noradrenaline have been demonstrated in synaptosomal preparations of brain, strongly supporting the hypothesis that these autoreceptors are localized to noradrenergic terminals (MULDER et al. 1979). In contrast, analogous experiments designed to detect dopamine autoreceptors modulating ^{3}H-dopamine release in synaptosomes have led to controversial results (see Sect. B.I.1.g).

2. Modulation of the Electrical Activity of Noradrenergic Neurones

a. The Superior Cervical Ganglion

The noradrenergic neurones of the superior cervical ganglion would appear to present the ideal preparation for studying somato-dendritic autoreceptors. Since intracellular recording, bath application of drugs, and stimulation via pre- or postganglionic axons are all possible, the pharmacology and functions of somato-dendritic autoadrenoceptors are more accessible to investigation than is possible with somato-dendritic dopamine autoreceptors in the central nervous system.

Preganglionic stimulation of the isolated superior cervical ganglion results in the release of previously taken up ^{3}H-noradrenaline (MARTINEZ and ADLER-GRASCHINSKY 1980a). Thus dendritic release of noradrenaline would appear to occur in the superior cervical ganglion, and the released neurotransmitter should stimulate somato-dendritic autoadrenoceptors, if they exist. The possibility that somato-dendritic autoreceptors modulate the dendritic release of noradrenaline is indeed suggested by the decrease caused by α-adrenoceptor agonists, and conversely the increase caused by α-adrenoceptor antagonists, of ^{3}H-noradrenaline released from the isolated superior cervical ganglion elicited by preganglionic stimulation (MARTINEZ and ADLER-GRASCHINSKY 1980b). However, the involvement of dendro-axonic neurotransmission via α-adrenoceptors on cholinergic nerve terminals cannot presently be excluded as the mechanism underlying the results of MARTINEZ and ADLER-GRASCHINSKY (1980b), as argued by BROWN and CAULFIELD (1982).

α_2-adrenoceptor agonists cause hyperpolarization of noradrenergic cell bodies in the superior cervical ganglion (BROWN and CAULFIELD 1979). Yohimbine but not prazosin, blocks the hyperpolarization caused by noradrenaline; clonidine and oxymetazoline are potent agonists, while methoxamine is not (BROWN and CAULFIELD 1979). In its pharmacological profile, therefore, the somato-dendritic auto-adrenoceptor in the superior cervical ganglion strongly resembles presynaptic α_2-autoadrenoceptors which modulate depolarization-evoked noradrenaline release.

The resemblance between somato-dendritic and presynaptic autoadrenoceptors extends further, to the sensitivity to potassium and calcium ions. Like α_2-adrenoceptor-mediated inhibition of noradrenaline release (LANGER et al. 1975; MARSHALL et al. 1977; DISMUKES et al. 1977; ALBERTS et al. 1981, WEMER et al. 1981), α_2-adrenoceptor-mediated hyperpolarization is enhanced at low calcium concentrations and reduced in the presence of elevated potassium concentrations (BROWN and CAULFIELD 1979). As was suggested for dopamine autoreceptors, the possibility exists that the two categories, presynaptic and somato-dendritic α_2-autoadrenoceptors, may not only be pharmacologically identical, but may also function by a common ionic mechanism.

The key role of calcium conductance in the actions of α-autoadrenoceptors was highlighted again in the investigations of HORN and MCAFEE (1980). These workers identified three different voltage-dependent calcium currents: 1) the "shoulder" of action potentials; 2) the hyperpolarizing after potential; and 3) the action potential evoked in the presence of tetrodotoxin, called the "calcium spike". All three of these currents, which are calcium-dependent and activated by changes in membrane potential, are inhibited by α-adrenoceptor agonists (HORN and MCAFEE 1980). When the ganglion is continuously exposed to tetrodotoxin, and depolarization is evoked by passing current through an intracellular electrode, noradrenaline decreases the amplitude of the calcium spike; these experimental conditions are likely to exclude the effect of a trans-synaptic second messenger in the α-autoadrenoceptor-mediated response. HORN and MCAFEE raise the interesting possibility that the hyperpolarization caused by exposure to α_2-adrenoceptor agonists (BROWN and CAULFIELD 1979) may be due to modulation of voltage-dependent calcium currents.

b. The Locus Coeruleus

Iontophoretic application of noradrenaline in the locus coeruleus causes a maximal decrease of 70% to 80% in the firing rate of noradrenergic cell bodies (CEDERBAUM and AGHAJANIAN 1977). Other α-adrenoceptor agonists produce the same inhibition of firing rate, and by comparing the ejection currents required to produce inhibition, a rank order of potencies is found: clonidine $\gg$ α-methylnoradrenaline > adrenaline = noradrenaline $\gg$ phenylephrine (CEDERBAUM and AGHAJANIAN 1977). This relative order of potency correlates quite well with that of α_2-adrenoceptors of the peripheral nervous system (LANGER 1977, 1981).

The inhibition of firing of noradrenergic neurones is blocked by the α_2-adrenoceptor antagonist piperoxan applied iontophoretically, but neither by the β-adrenoceptor antagonist sotalol, nor by the neuroleptic dopamine receptor antagonist trifluoperazine (CEDERBAUM and AGHAJANIAN 1977). In contrast to these results, the inhibitory action of GABA on unit activity is unimpaired by the doses of piperoxan, systemically or iontophoretically applied, which effectively antagonize the α_2-adrenoceptor modulating the firing rate of noradrenergic neurones (CEDERBAUM and AGHAJANIAN 1976, 1977).

The threshold current required to evoke antidromic action potentials from the terminals of neurones in the locus coeruleus is subject to modulation by α-adrenoceptor agonists and antagonists applied in the region of the noradrenergic terminals (NAKAMURA et al. 1981). Therefore, similarly to dopaminergic neurones (see Sect. B.I.3.g), noradrenergic nerve terminals appear to possess α-autoadrenoceptors which modulate their electrophysiological excitability.

The firing rate of noradrenergic neurones of locus coeruleus, normally close to 1.1 spikes/second, increases to 7.5 spikes/second following systemic administration of piperoxan (CEDERBAUM and AGHAJANIAN 1976). In comparison, dopaminergic neurones of the substantia nigra, normally firing between 3 and 5 spikes/sec, increase their firing rate by only a factor of two or three in response to systemic or local administration of neuroleptics (BUNNEY et al. 1973; GROVES et al. 1975).

In contrast to the dopaminergic neurones of the substantia nigra, dendritic release of noradrenaline acting upon somato-dendritic autoadrenoceptors has not been proposed as a hypothesis to explain these electrophysiological data. Rather, the concept of axon collaterals of noradrenergic neurones which return to synapse upon noradrenergic dendrites has been proposed to explain these electrophysiological data (AGHAJANIAN et al. 1977). However, the possibility of dendritic release of noradrenaline and an action upon adjacent dendrites, or the same dendrite which releases the neurotransmitter, cannot be excluded by these data from AGHAJANIAN et al. (1977).

Anatomical data provide the major basis for hypotheses other than dendro-dendritic neurotransmission via somato-dendritic α_2-autoadrenoceptors. Golgi studies suggest the existence of recurrent collaterals arising from locus coeruleus neurones (SWANSON 1976). An input to the ventral locus coeruleus arising from the adrenaline-containing neurones in the medulla oblongata has also been suggested (HÖFKELT et al. 1974). Since adrenaline is as potent as noradrenaline at these α_2-adrenoceptors this ascending adrenergic pathway may certainly play a role in modulating the firing rate of noradrenergic neurones in the locus coeruleus. Curiously, however, recent immunohistochemical investigations at the light and electron microscopic levels have not furnished support for the existence of dopamine-β-hydroxylase-containing recurrent collaterals or afferent terminals which form synapses with the dendrites of noradrenergic neurones in the locus coeruleus (CIMARUSTI et al. 1979; GRZANNA and MOLLIVER 1980).

II. Comparisons Between α_2-Autoadrenoceptors and the Dopamine Autoreceptor

The α_2-autoadrenoceptor would appear to exercise a much greater degree of modulation over the release of noradrenaline, than does the dopamine autoreceptor over dopamine release. Antagonism of autoadrenoceptors by phentolamine results in a 285% increase in electrically-evoked ^{3}H-noradrenaline release from rat occipital cortex (PELAYO et al. 1980). By comparison, antagonism of dopamine autoreceptors by S-sulpiride results in only a 50% increase in electrically-evoked ^{3}H-dopamine release from cat caudate slices (LEHMANN and LANGER 1981). Inhibition of neuronal uptake of noradrenaline by cocaine and of dopamine uptake by nomifensine potentiate the enhancement by phentolamine of ^{3}H-noradrenaline release (HÄGGENDAL 1970; FARNEBO and HAMBERGER 1970) and by S-sulpiride of ^{3}H-dopamine release (LEHMANN and LANGER 1981). Nevertheless, even under conditions of inhibition of neuronal uptake, blockade of α_2-adrenoceptors produces a larger increase in noradrenaline release than that produced at the level of dopamine release by blockade of dopamine autoreceptors. These results suggest that released neurotransmitter activates the autoreceptor to a greater degree in noradrenergic neurones than in dopaminergic neurones.

While in a number of tissues, noradrenaline itself inhibits the release of ^{3}H-noradrenaline, the demonstration of inhibition of ^{3}H-dopamine release by dopamine itself requires special conditions, such as lowered calcium concentrations, very low frequencies of stimulation, and the pharmacological inhibition of dopamine uptake (REIMANN et al. 1979; ARBILLA et al. 1982, 1985). Nonetheless, the release-increasing effect of dopamine receptor antagonists depends upon the existence of endogenous dopamine stores (LEHMANN and LANGER 1982c), as has been demonstrated for the autoadrenoceptor modulating noradrenaline release (ENERO and LANGER 1973; CUBEDDU and WEINER 1975). Like the dopamine autoreceptor, the α-autoadrenoceptor modulates only calcium-dependent release of noradrenaline (STARKE and MONTEL 1974).

Auto-adrenoceptors modulating transmitter release have been shown to be "frequency-dependent": at higher frequencies (within limits) the facilitatory effect of α-adrenoceptor antagonists increases, while the inhibitory effect of α-adrenoceptor agonists decreases (LANGER et al. 1975; DUBOCOVICH and LANGER 1976). The dopamine autoreceptor is similarly frequency-dependent, and the nature of the frequency-dependence has been established in detail (LEHMANN and LANGER 1982c): with increasing frequency, the IC_{50} of the dopamine receptor agonist pergolide increases, and the maximal inhibition of ^{3}H-dopamine release decreases. The maximal facilitatory effect of the dopamine receptor antagonist S-sulpiride increases with increasing frequency of stimulation, between 1 and 6 Hz.

The localization of the α-autoadrenoceptor to noradrenergic neurones is well supported by both transmitter release and electrophysiological investigations, as noted in Sects. D.I.1 and D.I.2.a. This is in marked contrast to the case of the dopamine autoreceptor, where trans-synaptic models are endorsed by a number of workers. While it is possible that for dopamine neurones the negative feedback mechanism is trans-synaptic and for noradrenergic neurones it is exclusively presynaptic, a divergence of mechanisms to such a degree in these otherwise similar catecholaminergic neurones may be considered unlikely.

While α-autoadrenoceptors in the heart have been detected with receptor binding techniques and found to decrease following 6-hydroxydopamine lesions (STORY et al. 1979), neither central autoadrenoceptors nor central dopamine autoreceptors have been found to be particularly susceptible to an analogous experimental strategy (SKOLNICK et al. 1978; U'PRICHARD and SNYDER 1979; PIMOULE et al. 1983; see Sect. B.III).

As previously noted, autoadrenoceptors do not appear to modulate the synthesis of noradrenaline, in contrast to dopamine autoreceptors, which do modulate the synthesis of dopamine (see Sect. B.I.2).

III. Presynaptic β-Autoadrenoceptors on Peripheral Noradrenergic Neurones

Facilitation of the stimulation-evoked release of noradrenaline through presynaptic β-autoadrenoceptors was first reported by LANGER et al. (1974) and ADLER-GRASCHINSKY and LANGER (1974). The release of noradrenaline elicited by low frequency nerve stimulation is enhanced by low concentrations of isoprenaline and other β-adrenoceptor agonists in various tissues of several species, including man (ADLER-GRASCHINSKY and LANGER 1975; DAHLÖF et al. 1975; STJÄRNE and BRUNDIN 1975, 1976; CELUCH et al. 1978; GÖTHERT and HENTRICH 1985). The increase in the stimulation-evoked release of noradrenaline elicited by isoprenaline is completely antagonized by propranolol, but the β-adrenoceptor blocking agent does not reduce by itself the release of noradrenaline in most tissues. Yet, when peripheral noradrenergic nerves are loaded with adrenaline as false transmitter, instead of noradrenaline, blockade of β-adrenoceptors with propranolol becomes effective in reducing transmitter release induced by sympathetic nerve stimulation (GUIMARÃES et al. 1978; RAND et al. 1979).

The presynaptic facilitatory β-adrenoceptor is stereoselective both for the agonist, isoprenaline (CELUCH et al. 1978) as well as for the antagonist propranolol (DAHLÖF et al. 1980).

Under *in vitro* conditions, the experimental evidence indicates that the presynaptic facilitatory β-adrenoceptors are of the β_2-subtype (WESTFALL et al. 1979a; LANGER and GALZIN 1982). It is therefore possible that presynaptic β-adrenoceptors are mainly activated by increased levels of circulating adrenaline, rather than noradrenaline released from nerve terminals, to enhance noradrenergic neurotransmission. It is unlikely that circulating adrenaline actually does attain sufficiently high concentrations to stimulate presynaptic β_2-adrenoceptors (MAJEWSKI et al. 1982).

An alternative model for the physiological activation of presynaptic β_2-adrenoceptors has been evolving in recent years. Adrenaline released from the adrenal medulla is taken up from the circulation by noradrenergic terminals (LANGER and VOGT 1971). Adrenaline taken up in this fashion is later released by noradrenergic nerve terminals, upon depolarization (MAJEWSKI et al. 1982). Thus, adrenaline may act as a *cotransmitter* in noradrenergic nerve terminals. Following incubation *in vitro* with adrenaline, the stimulation-evoked release of ^{3}H-noradrenaline is increased by the cotransmitter, via activation of presynaptic facilitatory β-adrenoceptors (MAJEWSKI et al. 1982). The model that circulating adrenaline is taken up by noradrenergic terminals, released as a cotransmitter, and facilitates noradrenaline release by activating presynaptic β-adrenoceptors, has received experimental support in clinical studies. Moreover, this sequence of events may play a key role in essential hypertension. When adrenaline is administered i.v. to volunteers, diastolic and mean blood pressure fall (CLUTTER et al. 1980; FITZGERALD et al. 1980). However, some time after such an infusion of adrenaline has ended, there is an *increase* in diastolic and mean blood pressure as well as in heart rate (BROWN and MACQUIN 1981). These long lasting effects of adrenaline are not observed in patients treated with desipramine to inhibit neuronal uptake (M. BROWN, personal communication).

It is possible that presynaptic β_2-adrenoceptors are normally "silent" (i.e. noradrenaline release *is not* decreased by β_2-adrenoceptor antagonists) under resting conditions, but are "active" (i.e. noradrenaline release *is* decreased by β_2-adrenoceptor antagonists) following physiological elevation of plasma adrenaline levels. Aside from offering an additional explanation for the effectiveness of β-adrenoceptor antagonists in treating essential hypertension, this new hypothesis potentially offers insights for novel and better antihypertensive therapy.

E. Presynaptic Dopamine Receptors Modulating Noradrenaline Release

Dopamine was shown to inhibit the depolarization-evoked release of ^{3}H-noradrenaline in the cat nictitating membrane and perfused spleen (LANGER 1973). Dopamine causes this inhibition not through α_2-autoadrenoceptors, which, however, *may* be activated by sufficiently high concentrations of dopamine (DUVAL et al. 1985), but through presynaptic dopamine receptors, which are blocked by chlorpromazine and pimozide but not by phentolamine (ENERO and LANGER 1975). Subsequent investigations have confirmed the existence of inhibitory presynaptic dopamine receptors modulating the release of noradrenaline in rabbit heart (FUDER and MUSCHOLL 1978) and cat spleen (DUBOCOVICH and LANGER 1980). These presynaptic dopamine receptors are antagonized stereoselectively by sulpiride (DUBOCOVICH and LANGER 1980), and can be demonstrated under *in vivo* experimental conditions (MASSINGHAM et al. 1980; SHEPPERSON et al. 1982).

Postsynaptic dopamine receptors also exist in certain vascular smooth muscles, notably the renal and mesenteric beds (GOLDBERG et al. 1978; SHEPPERSON et al. 1982). These postsynaptic vascular dopamine receptors pharmacologically resemble the D_1 receptor of the central nervous system (see Sect. B.VI), and are in fact called DA-1 receptors in the periphery (GOLDBERG et al. 1978). The DA-1 receptor is sensitive to the agonist SK&F 38393 and the antagonist bulbocapnine (MASSINGHAM et al. 1980; GOLDBERG and KOHLI 1981; SHEPPERSON et al. 1982), and SCH 23390 (DUVAL and LANGER, unpublished observations). S-sulpiride appears to be a rather weak antagonist at DA-1 receptors, where it is equipotent with the R-isomer (SHEPPERSON et al. 1982). On the other hand, only S-sulpiride (and not the R-isomer) is a potent antagonist at the presynaptic dopamine receptor which inhibits noradrenaline release in the periphery (DUBOCOVICH and LANGER 1980; SHEPPERSON et al. 1982). At this time, therefore, the presynaptic dopamine receptors modulating noradrenaline release in the peripheral nervous system appear to resemble central D_2 dopamine receptors, and in fact are termed DA-2 receptors in the periphery (see also STEINSLAND and HIEBLE 1978). The clearly distinct properties of these peripheral pre- and postsynaptic receptors on the basis of agonist and antagonist selectivities are summarized in Table 6.

It is currently not clear whether any dopaminergic neurones exist in the periphery, with the majority of the evidence being negative (but see CHAPMAN et al. 1980). The physiological role, if any, of DA-1 receptors is for the moment obscure.

Both of the subtypes of peripheral dopamine receptors, the DA-1 and DA-2, mediate a reduction in blood pressure upon administration of dopamine receptor agonists. Yet, the most important contribution to the hypotension produced by dopamine receptor agonists originates from the stimulation of presynaptic inhibitory dopamine receptors of peripheral noradrenergic nerves. In support of this view it was shown that DA-2 receptor agonists like N,N,-di-n-propyldopamine, propylbutyldopamine and pergolide produce de-

Table 6. The relative selectivities of agonists and antagonists for dopamine receptor subtypes in the periphery

	RELATIVE SELECTIVITY OF AGONISTS
DA_1	SKF 38393, Fenoldopam (SKF 82526)
$DA_1 = DA_2$	Dopamine
DA_2	Apomorphine < Pergolide < PBDA < DPDA < LY 141865 (quinpirole)
	RELATIVE SELECTIVITY OF ANTAGONISTS
DA_1	R-Sulpiride < Bulbocapnine < < < SCH 23390
$DA_1 = DA_2$	(+)-Butaclamol
DA_2	Pimozide < Haloperidol < Domperidone < S-sulpiride

The results shown were obtained under *in vitro* and *in vivo* conditions at peripheral noradrenergic neuroeffector junctions in several species.
DA-1: The postsynaptic dopamine receptor mediating vasodilatation in vascular smooth muscle (renal and mesenteric vascular beds).
DA-2: The presynaptic dopamine receptor mediating the inhibition of noradrenaline release elicited by postganglionic sympathetic stimulation.
SK&F 38393 is 2,3,4,5-tetrahydro-7,8-dihydroxy-1-phenyl-1H-3-benzazepine.
Fenoldopam (SKF 82526)
LY 141865 (quinpirole): N-propyl tricyclic pyrazole
PBDA: N-n-propyl-N-n-butyldopamine
DPDA: N,N-di-n-propyldopamine
SCH 23390: R-(+)-8-chloro-2,3,4,5-tetrahydro-3-methyl-5-phenyl-1H-3-benzazepine-7-ol

creases in blood pressure, and that these effects are selectively blocked by low doses of S-sulpiride or domperidone (MASSINGHAM et al. 1980; BARRETT and LOKHANDWALA 1982; W.H. FENNELL, personal communication).

The presynaptic DA-2 receptors modulating noradrenaline release do not seem to play a physiological role. The absence of an increase in ^{3}H-noradrenaline release or end organ responses to sympathetic stimulation upon exposure to antagonists of these presynaptic DA-2 receptors (ENERO and LANGER 1975; DUBOCOVICH and LANGER 1980; SHEPPERSON et al. 1982) supports this concept, and is furthermore in sharp contrast to the effects of dopamine receptor antagonists in the central nervous system (see Sect. B, particularly B.I).

Release of ^{3}H-noradrenaline from central noradrenergic neurones in the rabbit hypothalamus and cerebral cortex appears to be modulated by presynaptic dopamine receptors (GALZIN et al. 1982; GALZIN and LANGER 1985). Apomorphine and pergolide are active agonists at this presynaptic dopamine receptor, while bromocriptine is apparently a partial agonist (GALZIN et al. 1982). Butaclamol and sulpiride antagonize stereoselectively the inhibition of ^{3}H-noradrenaline release by these dopamine receptor agonists (GALZIN et al. 1982). Thus the presynaptic dopamine receptor modulating ^{3}H-noradrenaline release in the brain would appear to be D_2-like. Whether this presynaptic receptor is identical to or different from D_2 receptors or dopamine autoreceptors in other respects remains to be determined. The fact that neither S-sulpiride nor (+)butaclamol enhances ^{3}H-noradrenaline release evoked by electrical stimulation, in concentrations which block the effects of the dopamine receptor agonists, supports the view that inhibitory dopamine receptors do not play a physiological role in modulating central noradrenergic neurotransmission (GALZIN et al. 1982).

Dopamine itself has been found to be inactive in reducing electrically evoked ^{3}H-noradrenaline release in the hypothalamus when the time of exposure of the tissues to dopamine is 20 minutes (GALZIN et al. 1982). Yet, when the concentration of calcium and the time of exposure to dopamine are reduced, dopamine does inhibit the electrically evoked release of ^{3}H-noradrenaline in the rabbit hypothalamus (GALZIN and LANGER 1985). These observations suggest that rapid desensitization may develop for the presynaptic dopamine receptor that modulates noradrenaline release in the central nervous system when dopamine itself is employed as the agonist. Dopamine itself slows the firing rate of noradrenergic neurones in the locus coeruleus, but this effect is not blocked by the dopamine receptor antagonist trifluoperazine, and is thought rather to be mediated via α_2-adrenoceptors (CEDERBAUM and AGHAJANIAN 1977).

It is of interest that M7, a drug possessing both α_2-adrenoceptor agonist (SHEPPERSON and LANGER 1981) and dopamine agonist properties (Table 3), inhibits the electrically evoked release of ^{3}H-noradrenaline in the rabbit hypothalamus. This inhibitory effect of M7 is partly blocked by S-sulpiride or yohimbine alone, but when both yohimbine and S-sulpiride are added together, the inhibition by M7 of ^{3}H-noradrenaline release is completely antagonized (GALZIN and LANGER 1985). These results indicate that M7 stimulates both α_2-autoadrenoceptors and dopamine receptors in the noradrenergic nerve terminals of the rabbit hypothalamus.

The curious situation exists that noradrenergic terminals possess a presynaptic receptor for dopamine, although dopamine itself does not appear to be a physiological substrate for this receptor. The converse situation does not exist: the depolarization-evoked release of ^{3}H-dopamine from cat caudate is not modulated by α-adrenoceptor agonists or antagonists (LEHMANN and LANGER, unpublished observations). The presynaptic dopamine receptor modulating ^{3}H-noradrenaline release may be related to the postulated interaction of dopamine with noradrenergic cell bodies in the superior cervical ganglion (GREENGARD and KEBABIAN 1974). However, the dopamine receptor involved at noradrenergic cell bodies has been suggested to be a dopamine-stimulated adenylate cyclase, i.e. probably of type D_1 (GREENGARD and KEBABIAN 1974). The role of dopamine-containing, small, intensely fluorescent neurosecretory cells in the superior cervical ganglion and the existence of dopamine receptors in sympathetic ganglia has more recently been severely questioned (DUN 1980; BROWN and CAULFIELD 1982).

A final possible answer to the question why noradrenergic neurones possess dopamine presynaptic receptors when they are not physiologically functional, is that they may be a phylogenetic vestige of these catecholamine neurones, antedating the evolution of DBH.

F. Conclusions

Inhibitory autoreceptors modulate the fundamental processes of depolarization-evoked neurotransmitter release and electrical activity of both dopaminergic and noradrenergic neurones. The modulation of these processes seems to be intimately related with voltage-dependent calcium channels, in the terminal as well as the cell body region of both dopaminergic and noradrenergic neurones. In addition, the modulation of the synthesis of dopamine by autoreceptors has been extensively documented, whereas modulation of noradrenaline synthesis by autoadrenoceptors has not been reported.

It appears likely, though by no means certain, that both dopamine autoreceptors and α-autoadrenoceptors are activated by released neurotransmitter under physiological conditions. Although this is a negative feedback system, it is not necessarily true that autoreceptors play a homeostatic role. The modulation of neurotransmitter release by autoreceptor agonists is more effective at low frequencies that at high frequencies of stimulation. This frequency dependence may indicate that low levels of neurotransmitter release are lowered further, and high levels of neurotransmitter are affected relatively little by autoreceptor activation, producing an increase in the "signal-to-noise" ratio of neurotransmission.

Presynaptic autoreceptors and heteroreceptors do not always play a physiological role. Presynaptic heteroreceptors in particular often appear unlikely to be activated by their endogenous neurotransmitters, and are sometimes referred to as "pharmacological receptors". If indeed certain presynaptic receptors can be identified as not playing physiological roles, these may be particularly valuable as targets of therapeutic drug design, since their activation by drugs would not disrupt a normal physiological process.

Where presynaptic receptors do play a physiological role, it is important to determine what changes occur in their properties and functions in disease states, such as schizophrenia (for the dopamine autoreceptor) and essential hypertension (for α- and β-autoadrenoceptors). Likewise, the effect of chronic administration of therapeutic drugs on these presynaptic receptors is essential for the rational design of new drugs. Since, for some neurotransmitters, such as noradrenaline, there are differences between the pre- and postsynaptic receptors, it is possible to selectively modify neurotransmitter release with drugs which act preferentially on the presynaptic receptor.

Presynaptic autoreceptors on catecholamine neurones have been studied intensively for more than two decades. Presynaptic receptors appear to function also on serotoninergic, cholinergic, and possibly neurones utilizing amino acids as neurotransmitters. It would appear that these regulatory systems are not accidents of nature, but rather sufficiently important in the process of neurotransmission to have evolved for various neurotransmitters.

G. References

Aceves J, Cuello AC (1981) Dopamine release induced by electrical stimulation of microdissected caudate-putamen and substantia nigra of the rat brain. Neuroscience 6:2069–2075

Adler-Graschinsky E, Langer SZ (1974) Possible role of a β-adrenoceptor regulating noradrenaline release by nerve stimulation. Acta Physiol Latinoamer 24:185P

Adler-Graschinsky E, Langer SZ (1975) Possible role of a β-adrenoceptor in the regulation of noradrenaline release by nerve stimulation. Br J Pharmacol 53:43–50

Aghajanian GK, Bunney BS (1973) Central dopaminergic neurons: neurophysiological identification and responses to drugs In: Usdin E, Snyder SH (eds) Frontiers in Catecholamine Research, Pergamon Press Oxford, pp 643–648

Aghajanian GK, Cederbaum JM, Wang RY (1977) Evidence for norepinephrine mediated collateral inhibition of locus coeruleus neurons. Brain Res 136:570–577

Ahtee L, Kaakkola S (1978) Effect of mecamylamine on the fate of dopamine in striatal and mesolimbic areas of rat brain: interaction with morphine and haloperidol. Br J Pharmacol 62:213–218

Alander T, Andén N-E, Grabowska-Andén M (1980) Metoclopramide and sulpiride as selective blocking agents of pre- and postsynaptic dopamine receptors. Naunyn-Schmiedeberg's Arch Pharmacol 312:145–150

Alberts P, Bartfai T, Stjärne L (1981) Site(s) and ionic basis of α-autoinhibition and facilitation of ^{3}H-noradrenaline secretion in guinea pig vas deferens. J Physiol (Lond.) 312:297–334

Andén NE, Corrodi H, Fuxe K, Ungerstedt U (1971) Importance of nervous impulse flow for the neuroleptic induced increase in amine turnover in central dopamine neurons. Eur J Pharmacol 15:193–199

Angrist B, Rotrosen J, Gershon S. (1980) Differential effects of amphetamine and neuroleptics on negative vs. positive symptoms in schizophrenia. Psychopharmacol 72:17–19

Arbilla S, Langer SZ (1978) Morphine and β-endorphin inhibit release of noradrenaline from cerebral cortex but not of dopamine from rat striatum. Nature 271:559–561

Arbilla S, Langer SZ (1980) Influence of monoamine oxidase inhibition on the release of ^{3}H-dopamine elicited by potassium and by amphetamine from the rat substantia nigra and corpus striatum. Naunyn-Schmiedeberg's Arch Pharmacol 311:45–52

Arbilla S, Langer SZ (1981) Stereoselectivity of presynaptic autoreceptors modulating dopamine release. Eur J Pharmacol 76:345–351

Arbilla S, Langer SZ (1984) Differential effects of the stereoisomers of 3PPP on dopaminergic and cholinergic neurotransmission in superfused slices of the corpus striatum. Naunyn-Schmiedeberg's Arch Pharmacol 327:6–13

Arbilla S, Galzin AM, Langer SZ, Nowak JZ (1981a) Supersensitivity of central dopamine autoreceptors following chronic haloperidol treatments in the rabbit. Br J Pharmacol 74:901P

Arbilla S, Kamal LA, Langer SZ (1981b) Inhibition by apomorphine of the potassium-evoked release of ^{3}H-γ-aminobutyric acid from the rat substantia nigra *in vitro*. Br J Pharmacol 74:389–397

Arbilla S, Langer SZ, Nowak JZ (1982) Influence of calcium on the autoreceptor-mediated inhibition of ^{3}H-dopamine release from the rabbit caudate nucleus. Br J Pharmacol 77:359P

Arbilla S, Nowak JZ, Langer SZ (1985) Rapid desensitization of presynaptic dopamine autoreceptors during exposure to exogenous dopamine. Brain Res 337:11–17

Argiolas A, Fadda F, Melis MR, Gessa GL (1979) Haloperidol increases DOPAC in the substantia nigra but not in the ventral tegmental area. Life Sci 24:2279-2284

Bacopoulos NG, Bustos G, Redmond DE, Baulu J, Roth RH (1978) Regional sensitivity of primate brain dopaminergic neurons to haloperidol: alterations following chronic treatment. Brain Res 157:396-407

Bannon MJ, Michaud RL, Roth RH (1981) Mesocortical dopamine neurons. Lack of autoreceptors modulating dopamine synthesis. Mol Pharmacol 11:270-275

Bannon MJ, Reinhard JF, Bunney EB, Roth RH (1982) Unique response to antipsychotic drugs is due to absence of terminal autoreceptors in mesocortical dopamine neurones. Nature 296:444-446

Barrett RJ, Lokhandwala MF (1982) Dopaminergic inhibition of cardiac sympathetic nerve function by pergolide. Eur J Pharmacol 77:79-83

Bartholini G, Stadler H, Gadea-Ciria M, Lloyd KG (1976) The use of the push-pull cannula to estimate the dynamics of acetylcholine and catecholamines within various brain areas. Neuropharmacol 15:515-519

Beart PM, Gundlach AL (1980) 3,4-Dihydroxyphenylacetic acid (DOPAC) and the mesolimbic dopaminergic pathway: drug effects and evidence for somatodendritic mechanisms. Br J Pharmacol 69:241-247

Beart PM, McDonald D (1980) Neurochemical studies of the mesolimbic dopaminergic pathway: somatodendritic mechanisms and GABAergic neurones in the rat ventral tegmentum. J Neurochem 34:1622-1629

Bergström S, Farnebo LO, Fuxe K (1973) Effect of prostaglandin E_2 on central and peripheral catecholamine neurons. Eur J Pharmacol 21:362-368

Besson JM, Cheramy A, Glowinski J (1971) Effects of some psychotropic drugs on dopamine synthesis in rat striatum. J Pharmacol Exp Ther 177:196-205

Biggio G, Casu M, Corda MG, Di Bello C, Gessa GL (1978) Stimulation of dopamine synthesis in caudate nucleus by intrastriatal enkephalins and antagonism by naloxone. Science 200:522-554

Biggio G, Casu M, Klimek V, Gessa GL (1980) Dopamine synthesis: tolerance to haloperidol and supersensitivity to apomorphine depend on presynaptic receptors. In: Cattabeni F, Racagni G, Spano PF, Costa E (eds) Long-Term Effects of Neuroleptics, Raven Press New York, pp 17-22

Blaustein MP, Johnson EM, Needleman P (1972) Calcium-dependent norepinephrine release from presynaptic nerve endings in vitro. Proc Nat Acad Sci (USA) 69:2237-2240

Bowen DM, Marek KL (1982) Evidence for the pharmacological similarity between the central muscarinic autoreceptor and postsynaptic muscarinic receptors. Br J Pharmacol 75:367-372

Bowery NG, Hill DR, Hudson AL, Doble A, Middlemiss DN, Shaw J, Turnbell M (1980) (−) Baclofen decreases neurotransmitter release in the mammalian CNS by an action at a novel GABA receptor. Nature 283:92-94

Brown DA, Caulfield MP (1979) Hyperpolarizing "α_2"-adrenoceptors in rat sympathetic ganglia. Br J Pharmacol 65:435-445

Brown DA, Caulfield MP (1982) Adrenoceptors in ganglia. In: Kunos G (ed) Adrenoceptors and Catecholamine Action. Wiley & Sons, Chichester, England

Brown GL, Gillespie JS (1957) The output of sympathetic transmitter from the spleen of the cat. J Physiol (Lond) 138:81-102

Brown MJ, Macquin I (1981) Is adrenaline the cause of essential hypertension? The Lancet II:1079-1082

Buda M, Gonon F, Cespuglio R, Jouvet M, Francois J (1981) In vivo electrochemical detection of catechols in several dopaminergic brain regions of anaesthetized rats. Eur J Pharmacol 73:61-68

Bunney BS, Aghajanian GK (1976) D-Amphetamine-induced inhibition of central dopaminergic neurons: mediation by a striatonigral feedback pathway. Science 192:391–393
Bunney BS, Aghananian GK (1978) D-Amphetamine-induced depression of central dopamine neurons: evidence for mediation by both autoreceptors and a striato-nigral feedback pathway. Naunyn-Schmiedeberg's Arch Pharmacol 304:255–261
Bunney BS, Walters JR, Roth RH, Aghajanian K (1973) Dopaminergic neurons: effect of antipsychotic drugs and amphetamine on single cell activity. J Pharmacol Exp Ther 185:560–571
Burt DR, Creese I, Snyder SH (1976) Properties of ^{3}H-haloperidol and ^{3}H-dopamine binding associated with dopamine receptors in calf brain membranes. Mol Pharmacol 12:800–812
Calne DB, Plotkin C, Williams AC, Nutt JG, Neophytides A, Teychenne PF (1978) Long-term treatment of parkinsonism with bromocriptine. Lancet I:735–738
Carlsson A (1975) Receptor-mediated control of dopamine metabolism. In: Usdin E, Bunney WE (eds) Pre and Postsynaptic Receptors. Marcel Dekker Inc New York, pp 49–65
Carlsson A, Lindqvist M (1963) Effect of chlorpromazine or haloperidol on formation of 3-methoxytyramine and normetanephrine in mouse brain. Acta Pharmacol Toxicol 20:140–144
Carroll BJ, Curtis GC, Kokmen E (1977) Paradoxical response to dopamine agonists in tardive dyskinesia. Am J Psychiatry 134:785–789
Cederbaum JM, Aghajanian GK (1976) Noradrenergic neurons of the locus coeruleus: inhibition by epinephrine and activation by the α-antagonist piperoxane. Brain Res 112:413–419
Cederbaum JM, Aghajanian GK (1977) Catecholamine receptors on locus coeruleus neurons: pharmacological characterization. Eur J Pharmacol 44:375–385
Celuch SM, Dubocovich ML, Langer SZ (1978) Stimulation of presynaptic β-adrenoceptors enhances ^{3}H-noradrenaline release during nerve stimulation in the perfused cat spleen. Br J Pharmacol 63:97–108
Cerrito F, Maura G, Raiteri M (1981) Do presynaptic dopamine autoreceptors exist? In: Gessa GL, Corsini GU (eds) Apomorphine and Other Dopaminomimetics, Vol I. Raven Press New York, pp 123–132
Chapman BJ, Horn NM, Munday KA, Robertson MJ (1980) The actions of dopamine and of sulpiride on regional blood flows in the rat kidney. J Physiol (Lond) 298:437–452
Cheramy A, Besson MJ, Glowinski J (1970) Increased release of dopamine from striatal dopaminergic terminals in the rat after treatment with a neuroleptic: thioproperazine. Eur J Pharmacol 10:206–214
Cheramy A, Nieoullon A, Michelot R, Glowinski J (1977) Effects of intranigral application of dopamine and substance P on the *in vivo* release of newly synthesized ^{3}H-dopamine in the ipsilateral caudate nucleus of the cat. Neurosci. Letters 4:105–109
Chesselet MF, Chéramy A, Reisine TD, Glowinski J (1981) Morphine and δ-opiate agonists locally stimulate *in vivo* dopamine release in cat caudate nucleus. Nature 192:320–322
Christiansen J, Squires RF (1974a) Antagonistic effects of apomorphine and haloperidol on rat striatal synaptosomal tyrosine hydroxylase. J Pharm Pharmacol 26:367–369
Christiansen J, Squires RF (1974b) Antagonistic effects of neuroleptics and apomorphine on synaptosomal tyrosine hydroxylase in vitro. J Pharm Pharmac 26:742–743

Cimarusti DL, Saito K, Vaughn JE, Barber R, Roberts E, Thomas PE (1979) Immunocytochemical localization of dopamine-β-hydroxylase in rat locus coeruleus and hypothalamus. Brain Res 162:55-67

Clutter WE, Bier DM, Shah SD, Cryer PE (1980) Epinephrine plasma metabolic clearance rates and physiologic thresholds for metabolic and haemodynamic actions in man. J Clin Invest 66:94-101

Corrodi H, Farnebo LO, Fuxe K, Hamberger B, Ungerstedt U (1972a) ET495 and brain catecholamine mechanisms: evidence for stimulation of dopamine receptors. Eur J Pharmacol 20:195-204

Corrodi H, Fuxe K, Lidbrink P (1972b) Interaction between cholinergic and catecholaminergic neurones in rat brain. Brain Res 43:397-416

Corsini GU, Gessa GL (1981) Apomorphine and Other Dopaminomimetics, Vol 2: Clinical Pharmacology. Raven Press New York, 1981.

Costa E, Cheney DL, Mao CC, Moroni F (1978) Action of antischizophrenic drugs on the metabolism of γ-aminobutyric acid and acetylcholine in globus pallidus, striatum, and n. accumbens. Fed Proc 37:2408-2414

Costall B, Naylor RJ (1981) The hypotheses of different dopamine receptor mechanisms. Life Sci 28:215-229

Costall B, Lim SK, Naylor RJ (1981) Characterisation of the mechanisms by which purported dopamine agonists reduce spontaneous locomotor activity of mice. Eur J Pharmacol 73:175-188

Cox B, Ennis C, Kemp JD (1981) Inhibitory 5-hydroxytryptamine receptors located on dopamine nerve terminals in the rat striatum. Br J Pharmacol 72:491P

Coyle JT, Schwartz R (1976) Lesion of striatal neurones with kainic acid provides a model for Huntington's chorea. Nature (Lond) 263:244-246

Creese I, Iversen SD (1973) Blockage of amphetamine induced motor stimulation and stereotypy in the adult rat following neonatal treatment with 6-hydroxydopamine. Brain Res 55:369-382

Creese I, Snyder SH (1979) Nigrostriatal lesions enhance striatal ^{3}H-apomorphine and ^{3}H-spiroperidol binding. Eur J Pharmacol 56:277-281

Creese I, Stewart K, Snyder SH (1979) Species variations in dopamine receptor binding. Eur J Pharmacol 60:55-66

Crow TJ (1980) Molecular pathology of schizophrenia: more than one disease process? Br Med J 280:66-68

Crow TJ, Cross AJ, Johnstone EC, Longden A, Owen F, Ridley RM (1980) Time course of the antipsychotic effect in schizophrenia and some changes in postmortem brain and their relation to neuroleptic medication. In: Cattabeni F, Racagni G, Spano PF, Costa E (eds) Longterm effects of Neuroleptics. Raven Press New York, pp 495-503

Cubeddu LX, Weiner N (1975) Nerve stimulation-mediated overflow of norepinephrine and dopamine-β-hydroxylase. III. Effects of norepinephrine depletion on the alpha-presynaptic regulation of release. J Pharmacol Exp Ther 192:1-14

Cubeddu LX, Barnes EM, Langer SZ, Weiner N (1974) Release of norepinephrine and dopamine-β-hydroxylase by nerve stimulation. I. Role of neuronal and extraneuronal uptake and of alpha-presynaptic receptors. J Pharmacol Exp Ther 190:431-450

Cutler NR, Jeste DV, Karoum F, Wyatt RJ (1982) Low-dose apomorphine reduces homovanillic acid concentration in schizophrenic patients. Life Sci 30:753-756

Dahlöf C, Åblad B, Borg KO, Ek L, Waldeck B (1975) Prejunctional inhibition of adrenergic nervous vasomotor control due to β-receptor blockade, In: Almgren O, Carlsson A, Engel J (eds) Chemical Tools in Catecholamine Research, vol. II. North-Holland Publishing Company Amsterdam, pp 201-210

Dahlöf C, Ljung B, Åblad B (1980) Pre- and postjunctional beta-adrenoceptor mediated effects on transmitter release and effector response in the isolated rat portal vein. Acta Physiol Scand 108:39–47

Davey MJ (1980) Aspects of the pharmacology of prazosin. Med J Aust, Spec. Suppl. 2:4–8

Davis JM, Tamminga C, Schaeffer MH, Smith RC (1981) Effects of apomorphine on schizophrenia. In: Corsini GU, Gessa GL (eds) Apomorphine and Other Dopaminomimetics, vol. 2. Raven Press New York, pp 45–48

De Belleroche JS, Bradford HF, Jones DG (1976) A study of the metabolism and release of dopamine and amino acids from nerve endings isolated from sheep corpus striatum. J Neurochem 26:561–571

De Belleroche J, Luqmani Y, Bradford HF (1979) Evidence for presynaptic cholinergic receptors on dopaminergic terminals: degeneration studies with 6-hydroxydopamine. Neuroscience Letters 11:209–213

De Jonge A, Santing PN, Timmermans PBMWM, van Zwieten PA (1979) A comparison of peripheral pre- and postsynaptic α_2-adrenoceptors using meta-substituted imidazolines. J Auton Pharmacol 1:377–383

Demarest KT, Moore KE (1979) Comparison of dopamine synthesis regulation in the terminals of nigrostriatal, mesolimbic, tuberoinfundibular and tuberohypophyseal neurons. J Neural Transm 46:263–277

Di Chiara G, Porceddu ML, Vargiu L, Stefanini E, Gessa GL (1976) Evidence for dopamine receptors mediating sedation in the mouse brain. Nature 264:564–567

Di Chiara G, Porceddu ML, Spano PF, Gessa GL (1977a) Haloperidol increases and apomorphine decreases dopamine metabolism after destruction of striatal dopamine-sensitive adenylate cyclase by kainic acid. Brain Res 130:374–382

Di Chiara G, Porceddu ML, Vargiu L, Stefanini E, Gessa GL (1977b) Evidence for selective and long-lasting stimulation of "regulatory" dopamine receptors by bromocriptine (CB 154). Naunyn-Schmiedeberg's Arch Pharmacol 300:239–245

Di Chiara G, Onali PL, Tissari AH, Porceddu ML, Morelli M, Gessa GL (1978) Destruction of postsynatic dopamine receptor prevents neuroleptic-induced activation of striatal tyrosine hydroxylase but not dopamine synthesis stimulation. Life Sci 23:691–696

Dismukes K, Mulder AH (1977) Effects of neuroleptics on release of ^{3}H-dopamine from slices of rat corpus striatum. Naunyn-Schmiedeberg's Arch Pharmacol 297:23–29

Dismukes K, de Boer AA, Mulder AH (1977) On the mechanism of α-receptor mediated modulation of ^{3}H-noradrenaline release from slices of cat brain. Naunyn-Schmiedeberg's Arch Pharmacol 299:115–122

Dubocovich ML (1983) Melatonin is a potent modulator of dopamine release in the retina. Nature 306:782–784

Dubocovich ML (1984) N-acetyltryptamine antagonizes the melatonin-induced inhibition of [^{3}H]-dopamine release from retina. Eur J Pharmacol 105:193–194

Dubocovich ML, Langer SZ (1976) Influence of the frequency of nerve stimulation on the metabolism of ^{3}H-norepinephrine released from the perfused cat spleen: differences observed during and after the period of stimulation. J Pharmacol Exp Ther 198:83–101

Dubocovich ML, Langer SZ (1980) Dopamine and α-adrenoceptor agonists inhibit neurotransmission in the cat spleen through different presynaptic receptors. J Pharmacol Exp Ther 212:144–152

Dubocovich ML, Weiner N (1981) Modulation of the stimulation evoked release of ^{3}H-dopamine in the rabbit retina. J Pharmacol Exp Ther 219:701–707

Dun NJ (1980) Ganglionic transmission: electrophysiology and pharmacology. Fed Proc 39:2981–2289

Duval N, Hicks PE, Langer SZ (1985) Dopamine preferentially stimulates postsynaptic α_2-adrenoceptors in the femoral vascular bed, but α_1-adrenoceptors in the renal vascular bed of the anaesthetised dog. Eur J Pharmacol 108:265–272

Enero MA, Langer SZ (1973) Influence of reserpine-induced depletion of noradrenaline on the negative feedback mechanism for transmitter release during nerve stimulation. Br J Pharmacol 49:214–225

Enero MA, Langer SZ (1975) Inhibition by dopamine of ^{3}H-noradrenaline release elicited by nerve stimulation in the isolated cat's nictitating membrane. Naunyn-Schmiedeberg's Arch Pharmacol 289:179–203

Enero MA, Langer SZ, Rothlin RP, Stefano FJE (1972) The role of the alpha receptor in regulating noradrenaline overflow by nerve stimulation. Br J Pharmacol 44:672–689

Ennis C, Cox B (1981) GABA enhancement of ^{3}H-dopamine release from rat striatum: dependence on slice size. Eur J Pharmacol 70:417–420

Farah MB, Langer SZ (1974) Protection by phentolamine against the effects of phenoxybenzamine on transmitter release elicited by nerve stimulation in the perfused cat heart. Br J Pharmacol 52:549–557

Farnebo LO, Hamberger B (1970) Effects of desipramine, phentolamine and phenoxybenzamine on the release of noradrenaline from isolated tissues. J Pharm Pharmacol 22:855–857

Farnebo LO, Hamberger B (1971a) Drug-induced changes in the release of ^{3}H-monoamines from field stimulated rat brain slices. Acta Physiol Scand Suppl 371:35–44

Farnebo LO, Hamberger B (1971b) Drug-induced changes in the release of ^{3}H-noradrenaline from field stimulated rat iris. Br J Pharmacol 43:97–106

Farnebo LO, Hamberger B (1973) Catecholamine release and receptors in brain slices. In: Usdin E, Snyder SH (eds) Frontiers in Catecholamine Research. Pergamon Press Oxford, pp 589–593

Filinger EJ, Langer SZ, Perec CJ, Stefano FJE (1978) Evidence for the presynaptic location of the α-adrenoceptors which regulate noradrenaline release in the rat submaxillary gland. Naunyn-Schmiedeberg's Arch Pharmacol 304:21–26

Fitzgerald GA, Barnes PJ, Hamilton CA, Dollery CT (1980) Circulating adrenaline and blood pressure: the metabolic effects of infused adrenaline in man. Eur J Clin Invest 10:401–406

Florence VM, Hume WR, Matsunaga ML (1981) Choline acetyltransferase in the sympathetic nerves of the rabbit ear artery. Br J Pharmacol 73:721–724

Fuder H, Muscholl E (1978) The effect of dopamine on the overflow of endogenous noradrenaline from the perfused rabbit heart evoked by sympathetic nerve stimulation. Naunyn-Schmiedeberg's Arch Pharmacol 305:109–115

Fujita N, Saito K, Hirata A, Iwatsubo K, Noguchi Y, Yoshida H (1980) Effects of dopaminergic agonists and antagonists on ^{3}H-apomorphine binding to striatal membranes: sulpiride lack of interactions with positive cooperative ^{3}H-apomorphine binding. Brain Res 199:335–342

Gale K (1979) Pre- vs post-synaptic control of tyrosine hydroxylase activity: the "long and short" of nigrostriatal feedback loops. In: Langer SZ, Starke K, Dubocovich ML (eds) Presynaptic Receptors. Pergamon Press Oxford, pp 179–183

Galzin AM, Langer SZ (1985) Inhibition by M7 of noradrenergic neurotransmission in the rabbit hypothalamus: role of α_2-adrenoceptors and of dopamine receptors. J Pharmacol Exp Ther 233:459–465

Galzin AM, Dubocovich ML, Langer SZ (1982) Presynaptic inhibition by dopamine receptor agonists of noradrenergic neurotransmission in the rabbit hypothalamus. J Pharmacol Exp Ther 221:461-471

Garau L, Govoni S, Stefanini E, Trabucchi M, Spano PF (1978) Dopamine receptors: pharmacological and anatomical evidences indicate that two distinct dopamine receptor populations are present in rat striatum. Life Sci 23:1745-1750

Garcia-Munoz M, Nicolaou NM, Tulloch IF, Wright AK, Arbuthnott GW (1977) Feedback loop or output pathway in striato-nigral fibres? Nature 265:363-365

Geffen LB, Jessell TM, Cuello AC, Iversen LL (1976) Release of dopamine from dendrites in rat substantia nigra. Nature 260:258-260

Gerlach J, Thorsen K, Fog R (1975) Extrapyramidal reactions and amine metabolites in cerebrospinal fluid during haloperidol and clozapine treatment of schizophrenic patients. Psychopharmacol 40:341-350

Gianutsos G, Thornburg JE, Moore KE (1976) Differential actions of dopamine agonists and antagonists on the γ-butyrolactone-induced increase in mouse brain dopamine. Psychopharmacol 50:225-229

Gillespie JS (1980) Presynaptic receptors in the autonomic nervous system. In: Szekeres L (eds) Adrenergic Activators and Inhibitors. Springer Berlin Heidelberg New York, pp 353-425 (Handbook of Experimental Pharmacology, vol 54, Part I)

Giorguieff MF, Kemel ML, Glowinski J (1977a) Presynaptic effect of L-glutamic acid on the release of dopamine in rat striatal slices. Neuroscience Letters 6:73-78

Giorguieff MF, Le Floc'h ML, Glowinski J, Besson MJ (1977b) Involvement of cholinergic presynaptic receptors of nicotinic and muscarinic types in the control of the spontaneous release of dopamine from striatal dopaminergic terminals in the rat. J Pharmacol Exp Ther 200:535-544

Giorguieff MF, Kemel ML, Glowinski J, Besson MJ (1978) Stimulation of dopamine release by GABA in rat striatal slices. Brain Res 139:115-130

Giorguieff-Chesselet MF, Kemel ML, Wandscheer D, Glowinski J (1979a) Attempts to localize the GABA receptors involved in the GABA-induced release of newly-synthesized ^{3}H-dopamine in rat striatal slices. Brain Res 175:383-386

Giorguieff-Chesselet MF, Kemel D, Wandscheer D, Glowinski J (1979b) Glycine stimulates the spontaneous release of newly synthesized ^{3}H-dopamine in rat striatal slices. Eur J Pharmacol 60:101-104

Giorguieff-Chesselet MF, Kemel ML, Wandscheer D, Glowinski J (1979c) Regulation of dopamine release by presynaptic nicotinic receptors in rat striatal slices: effect of nicotine in a low concentration. Life Sci 25:1257-1262

Glick SD, Crane LA (1978) Function of nigral dopamine receptors? Brain Res 156:360-381

Goldberg LI, Kohli JD (1981) Agonists and antagonists of peripheral pre- and post-synaptic dopamine receptors: clinical implications. In Gessa GL, Corsini GU (eds) Apomorphine and Other Dopaminomimetics, vol 1. Raven Press New York, pp 273-284

Goldberg LI, Volkman PH and Kohli JD (1978) A comparison of the vascular dopamine with other dopamine receptors. Ann Rev Pharmacol Toxicol 18:57-79

Goldstein M, Lew JY, Nakamura S, Battista AF, Lieberman A, Fuxe K (1978) Dopaminephilic properties of ergot alkaloids. Fed Proc 37:2202-2206

Goldstein M, Lieberman A, Lew JY, Asano T, Rosenfeld MR, Makman MH (1980) Interaction of pergolide with central dopaminergic receptors. Proc Natl Acad Sci (USA) 77:3725-3728

Gomeni R, Gomeni C (1980) A conversational graphic program for the analysis of the sigmoid curve. Computers Biomed Res 13:489-499

Goodale DB, Rusterholz DB, Long JP, Flynn JR, Walsh B, Cannon JG, Lee T (1980) Neurochemical and behavioral evidence for a selective presynaptic dopamine receptor agonist. Science 210:1141-1143

Göthert M, Hentrich F (1985) Identification of presynaptic β_2-adrenoeptors on the sympathetic nerve fibres of the human pulmonary artery. Br J Pharmacol 85:933-941

Greengard P, Kebabian JW (1974) Role of cyclic AMP in synaptic transmission in the mammalian peripheral nervous system. Fed Proc 33:1059-1067

Grossman W, Jurna I, Nell T, Theres C (1973) The dependence of the antinociceptive effect of morphine and other analgesic agents on spinal motor activity after central monoamine depletion. Eur J Pharmacol 24:67-77

Groves PM, Wilson CJ, Young SJ, Rebec GV (1975) Self-inhibition by dopaminergic neurons. Science 190:522-529

Groves PM, Fenster GA, Tepper JM, Nakamura S, Young SJ (1981) Changes in dopaminergic terminal excitability induced by amphetamine and haloperidol. Brain Res 221:425-431

Grzanna R, Molliver ME (1980) The locus coeruleus in the rat: an immunohistochemical delineation. Neuroscience 5:21-40

Gudelsky GA, Moore KE (1976) Differential effects on dopamine concentrations and rates of turnover in the median eminence, olfactory tubercle and corpus striatum. J Neural Transm 38:95-105

Gudelsky GA, Annunziato L, Moore KE (1978) Localization of the site of the haloperidol-induced, prolactin-mediated increase of dopamine turnover in the median eminence: studies in rats with complete hypothalamic deafferentations. J Neural Transm 42:181-192

Guimarães S, Brandão F, Paiva MQ (1978) A study of the adrenoceptor-mediated feedback mechanisms by using adrenaline as a false transmitter. Naunyn-Schmiedeberg's Arch Pharmacol 305:185-188

Gurwitz D, Kloog Y, Ergozi Y, Sokolovsky M (1980) Central muscarinic receptor degeneration following 6-hydroxydopamine lesion in mice. Life Sci 26:79-84

Guyenet PG, Aghajanian GK (1978) Antidromic identification of dopaminergic and other output neurons of the rat substantia nigra. Brain Res 150:69-84, 305:185-188

Häggendal J (1969) On release of transmitter from adrenergic nerve terminals at nerve activity. Acta Physiol Scand (Suppl) 330:29

Häggendal J (1970) Some further aspects on the release of the adrenergic transmitter. In: Kronberg G, Schümann HJ (eds) New Aspects of Storage and Release Mechanisms of Catecholamines. Springer Berlin Heidelberg New York, pp 100-109

Hamblin M, Creese I (1980) Phenoxybenzamine discriminates multiple dopamine receptors. Eur J Pharmacol 65:119-121

Handforth A, Sourkes TL (1975) Inhibition by dopamine agonists of dopamine accumulation following γ-hydroxybutyrate treatment. Eur J Pharmacol 34:311-319

Hattori T, Singh VK, McGeer EG, McGeer PL (1976) Immunohistochemical localization of choline acetyltransferase containing neostriatal neurons and their relationship with dopaminergic synapses. Brain Res 102:164-173

Hattori T, McGeer PL, McGeer EG (1979) Dendro axonic neurotransmission. II. Morphological sites for the synthesis, binding and release of neurotransmitters in dopaminergic dendrites and cholinergic dendrites in the neostriatum. Brain Res 170:71-83

Hedqvist P (1969) Antagonism between prostaglandin E_2 and phenoxybenzamine on noradrenaline release from the cat spleen. Acta Physiol Scand 76:383-384
Hedqvist P (1970) Studies on the effect of prostaglandins E_1 and E_2 on the sympathetic neuromuscular transmission in some animal tissues. Acta Physiol Scand Suppl 345:1-40
Hefti F, Lichtensteiger W (1978) Dendritic dopamine: studies on the release of endogenous dopamine from subcellular particles derived from dendrites of rat nigrostriatal neurons. Neurosci Letters 10:65-70
Hertting G (1965) Effects of drugs and sympathetic denervation on noradrenaline uptake and binding in animal tissues. In: Douglas WW, Carlsson A (eds) Pharmacology and Cholinergic and Adrenergic Transmission. Pergamon Press Oxford, pp 277-288
Hicks PE (1981) Antagonism of pre and postsynaptic α-adrenoceptors by BE 2254 (Heat) and prazosin. J Auton Pharmacol 1:391-397
Hjorth S, Carlsson A, Wikstrom H, Lindberg P, Sanchez D, Hacksell U, Arvidsson LE, Svensson U, Nilsson JLG (1981) 3-PPP, a new centrally acting DA-receptor agonist with selectivity for autoreceptors. Life Sci 28:1225-1238
Hökfelt T, Fuxe K, Goldstein M, Johansson O (1974) Immunohistochemical evidence for the existence of adrenaline neurons in the rat brain. Brain Res 66:235-251
Hollister LE (1981) Experiences with dopamine agonists in depression and schizophrenia. In: Corsini GU, Gessa GL (eds) Apomorphine and Other Dopaminomimetics, vol 2. Raven Press New York, pp 57-64
Horn JP, McAfee DA (1980) Alpha-adrenergic inhibition of calcium-dependent potentials in rat sympathetic neurones. J Physiol (Lond) 301:191-204
Hornykiewicz O (1978) Psychopharmacological implications of dopamine and dopamine antagonists: a critical evaluation of current evidence. Neuroscience 3:773-783
Iorio CL, Barnett A, Leitz FH, Houser VP, Korduba CA (1983) SCH 23390, a potential benzazepine antipsychotic with unique interactions on dopaminergic systems. J Pharmacol Exp Ther 226:462-468
Iversen LL (1965) The inhibition of noradrenaline uptake by drugs. Adv Drug Res 2:5-23
Iversen LL (1975) Dopamine receptors in the brain. Science 188:1084-1089
Iversen LL, Glowinski J (1966) Regional studies of catecholamines in the rat brain. II. Rate of turnover in various brain regions. J Neurochem 13:671-682
Iversen LL, Langer SZ (1969) Effects of phenoxybenzamine on the uptake and metabolism of noradrenaline in the rat heart and vas deferens. Br J Pharmacol 37:627-637
Iversen LL, Rogawski MA, Miller RJ (1976) Comparison of the effects of neuroleptic drugs on pre- and postsynaptic dopaminergic mechanisms in the rat striatum. Mol Pharmacol 12:251-262
Jackisch R, Zumstein A, Hertting G, Starke K (1980) Interneurones are probably not involved in the presynaptic dopaminergic control of dopamine release in rabbit caudate nucleus. Naunyn-Schmiedeberg's Arch Pharmacol 314:129-133
Jenner P, Clow A, Reavill C, Theodorou A, Marsden CD (1980) Stereoselective actions of substituted benzamide drugs on cerebral dopamine mechanisms. J Pharm Pharmacol 32:39-44
Johnson DA, Pilar G (1980) The release of acetylcholine from postganglionic cell bodies in response to depolarization. J Physiol (Lond) 229:605-619
Johnson DAW (1981) Depressions in schizophrenia: some observations on prevalence, etiology and treatment. Acta Psychiat Scand 63 (Suppl. 291):137-143

Jori A, Cerchetti G, Dolfini E, Monti E, Garattini S (1974) Effect of piribedil and one of its metabolites on the concentration of homovanillic acid in the rat brain. Eur J Pharmacol 27:245-248

Kamal L, Arbilla S, Langer SZ (1981) Presynaptic modulation of the release of dopamine from the rabbit caudate nucleus: differences between electrical stimulation, amphetamine and tyramine. J Pharmacol Exp Ther 216:592-598

Kato G, Carson S, Kemel ML, Glowinski J, Giorguieff MF (1978) Changes in striatal specific ^{3}H-atropine binding after unilateral 6-hydroxydopamine lesions of nigrostriatal dopaminergic neurones. Life Sci 22:1607-1614

Kayaalp SO, Neff NH (1980) Differentiation by ascorbic acid of dopamine agonist and antagonist binding sites in striatum. Life Sci 26:1837-1841

Kebabian JW (1978) Venom of Russell's viper uncouples the dopamine receptor from striatal adenylyl cyclase. Brain Res 144:194-198

Kebabian JW, Calne DB (1979) Multiple receptors for dopamine. Nature 277:93-96

Kehr W, Carlsson A, Lindqvist M, Magnusson T, Atack C (1972) Evidence for a receptor-mediated feedback control of striatal tyrosine hydroxylase activity. J Pharm Pharmacol 24:744-747

Kehr W, Carlsson A, Lindqvist M (1977) Catecholamine synthesis in rat brain after axotomy: interaction between apomophine and haloperidol. Naunyn-Schmiedeberg's Arch Pharmacol 297:111-117

Kelly PH, Moore KE (1978) Dopamine concentrations in the rat brain following injections into the substantia nigra of baclofen, γ-aminobutyric acid, γ-hydroxybutyric acid, apomorphine and amphetamine. Neuropharmacol 17:169-174

Kimura H, McGeer PL, Peng F, McGeer EG (1980) Choline acetyltransferase-containing neurons in rodent brain demonstrated by immunohistochemistry. Science 208:1057-1059

Korf J (1981) Turnover of neurotransmitters in the brain: an introduction. In: Pycock CJ, Taberner PV (eds) Central Neurotransmitter Turnover. Croom Helm London, pp 1-19

Kozlowski MR, Sawyer S, Marshall JF (1980) Behavioural effects and supersensitivity following nigral dopamine receptor stimulation. Nature 287:52-54

Langer SZ (1970) The metabolism of ^{3}H-noradrenaline released by electrical stimulation from the nictitating membrane of the cat and from the vas deferens of the rat. J Physiol (Lond) 208:515-546

Langer SZ (1973) The regulation of transmitter release elicited by nerve stimulation through a presynaptic feedback mechanism. In: Usdin E, Snyder S (eds) Frontiers in Catecholamine Research. Pergamon New-York pp 543-549

Langer SZ (1974) Presynaptic regulation of catecholamine release. Biochem Pharmacol 23:1793-1800

Langer SZ (1977) Presynaptic receptors and their role in the regulation of transmitter release. Br J Pharmac 60:481-497

Langer SZ (1981) Presynaptic regulation of the release of catecholamines. Pharmacol Rev 32:337-362

Langer SZ, Dubocovich ML (1979) Physiological and pharmacological role of the regulation of noradrenaline release by presynaptic dopamine receptors in the peripheral nervous system. In: Imbs JL, Schwartz J (eds) Peripheral dopaminergic receptors. Pergamon Press, pp 233-246

Langer SZ, Galzin AM (1982) Importance physiologique et pharmacologique des récepteurs β-présynaptiques dans la modulation de la libération de noradrénaline. Thérapie 37:523-532

Langer SZ, Moret C (1982) Citalopram antagonizes the stimulation by LSD of presyn-

aptic inhibitory serotonin autoreceptors in the rat hypothalamus. J Pharmacol Exp Ther 222:220-226
Langer SZ, Vogt M (1971) Noradrenaline release from the isolated muscles of the nictitating membrane of the cat. J Physiol (Lond) 214:159-171
Langer SZ, Adler E, Enero MA, Stefano FJE (1971) The role of the alpha receptor in regulating noradrenaline overflow by nerve stimulation. Proc XXVth Int Congr Physiol Sci, Munich, p 335
Langer SZ, Enero MA, Adler-Graschinsky E, Stefano FJE (1972) The role of the α-receptor in the regulation of transmitter overflow elicited by stimulation. Vth International Congress of Pharmacology, San Francisco, p 5
Langer SZ, Adler-Graschinsky E, Enero MA (1974) Positive feedback mechanism for the regulation of noradrenaline released by nerve stimulation. Jerusalem Satellite Symposia, XXVth International Congress of Physiological Sciences, p 81, Israel Physiological and Pharmacological Society, Jerusalem
Langer SZ, Dubocovich ML, Celuch SM (1975) Prejunctional regulatory mechanism for noradrenaline release elicited by nerve stimulation. In: Almgren O, Carlsson A, Engel J (eds) Chemical Tools in Catecholamine Research, vol II. North-Holland Amsterdam, pp 183-191
Langer SZ, Arbilla S, Kamal L (1980) Autoregulation of noradrenaline and dopamine release through presynaptic receptors. In: Littauer UZ et al. (eds) Neurotransmitters and their Receptors. Wiley and Sons Chichester, pp 7-21
Langer SZ, Arbilla S, Kamal L, Cantrill R (1983) Peripheral and central dopamine receptors modulating the release of neurotransmitters. In: Carlsson A, Nilsson JLG (eds) Proceedings of the Symposium on Dopamine Receptor agonists. Swedish Pharmaceutical Press Stockholm, pp 108-117
Langer SZ, Tahraoui L, Raisman R, Arbilla S, Najar M, Dedek J (1984) ^{3}H desipramine labels a site associated with the neuronal uptake of noradrenaline in the peripheral and central nervous system. In: Fleming W et al. (eds) Neuronal and extraneuronal events in autonomic pharmacology. Raven Press New York, pp 37-49
Langer SZ, Duval N, Massingham R (1985) Pharmacological und therapeutic significance of alpha-adrenoceptor subtype. J Cardiovasc Pharmacol 7:S1-S8
Lehmann J, Fibiger HC (1979) Acetylcholinesterase and the cholinergic neuron. Life Sci 25:1939-1947
Lehmann J, Langer SZ (1981) Phenoxybenzamine blocks dopamine autoreceptors irreversibly: implications for multiple dopamine receptor hypotheses. Eur J Pharmacol 75:247-254
Lehmann J, Langer SZ (1982a) Dopamine autoreceptors differ pharmacologically from postsynaptic dopamine receptors: effects of (−) N (2-chloroethyl)-norapomorphine. Eur J Pharmacol 77:85-86
Lehmann J, Langer SZ (1982b) Muscarinic receptors on dopamine terminals in the cat caudate nucleus: neuromodulation of ^{3}H-dopamine release *in vitro* by endogenous acetylcholine. Brain Res 248:61-69
Lehmann J, Langer SZ (1982c) The pharmacological distinction between central pre- and postsynaptic dopamine receptors: implications for the pathology and therapy of schizophrenia. In: Kohsaka M, Shohmori T, Tsukuda Y, Woodruff GN (eds) Advances in Dopamine Research. Pergamon Press Oxford, pp 25-39
Lehmann J, Fibiger HC, Butcher LL (1979) The localization of acetylcholinesterase in the corpus striatum and substantia nigra of the rat following kainic acid lesions of the corpus striatum: a biochemical and histochemical study. Neuroscience 4:217-225

Lehmann J, Arbilla S, Langer SZ (1981) Dopamine receptor mediated inhibition by pergolide of electrically-evoked ^{3}H-dopamine release from striatal slices of cat and rat: slight effect of ascorbate. Naunyn-Schmiedeberg's Arch Pharmacol 317:31-35

Lehmann J, Briley M, Langer SZ (1983a) Characterization of classical and novel dopamine receptor agonists at dopamine autoreceptors and at ^{3}H-spiperone binding sites *in vitro*. Eur J Pharmacol 88:11-26

Lehmann J, Smith RV, Langer SZ (1983b) Stereoisomers of apomorphine differ in affinity and intrinsic activity at presynaptic dopamine receptors modulating [^{3}H]-dopamine and [^{3}H]-acetylcholine release in slices of cat caudate. Eur J Pharmacol 88:81-88

Lemberger L, Crabtree RE (1979) Pharmacological effects in man of a potent, long-acting dopamine receptor agonist. Science 205:1151-1153

Leysen JE, Gommeren W (1981) Optimal conditions for ^{3}H-apomorphine binding and anomalous binding of ^{3}H-apomorphine and ^{3}H-spiperone to rat striatal membranes: involvement of surface phenomena versus multiple binding sites. J Neurochem 36:201-219

Ljungberg T, Ungerstedt U (1978) Classification of neuroleptic drugs according to their ability to inhibit apomorphine-induced locomotion and gnawing: evidence for two different mechanisms of action. Psychopharmacol 56:239-247

Llinas R, Hess R (1976) Tetrodotoxin-resistant dendritic spikes in avian Purkinje cells. Proc Natl Acad Sci (USA) 73:2520-2523

Lloyd KG (1978) Neurotransmitter interactions related to central dopamine neurons. In: Youdim MBH Novenberg W, Sharman DS, Lagmato SR (eds) Essays in Neurochemistry and Neuropharmacology John Wiley and Sons, Chichester, vol 3. pp 129-207

Loh HH, Brase DA, Sampath-Khanna S, Mar JB, Way EL, Li CH (1976) β-Endorphin *in vitro* inhibition of striatal dopamine release. Nature 264:567-568

Mackay AVP, Crow TJ (1980) Positive and negative schizophrenic symptoms and the role of dopamine. Br J Psychiat 137:379-386

Maggi A, Bruno F, Cattabeni F, Groppetti A, Parenti M, Racagni G (1978) Apomorphine-induced release: role of dopaminergic receptors in substantia nigra. Brain Res 145:180-184

Majewski H, Tung L-H, Rand MJ (1982) Adrenaline activation of prejunctional β-adrenoceptors and hypertension. J Cardiovasc Pharmacol 4:99-106

Marchi M, Raiteri M (1985) Differential antagonism by dicyclomine, pirenzepine and secoverine at muscarinic receptor subtypes in the rat frontal cortex. Eur J Pharmacol 107:287-288

Marshall I, Nasmyth PA, Shepperson NB (1977) The relationship between presynaptic α-adrenoceptors, stimulation frequency and calcium. Br J Pharmac 61:128P

Martin GE, Haubrich DR, Williams M (1981) Pharmacological properties of the putative dopamine autoreceptor agonists 3-PPP and TL-99. Eur J Pharmacol 76:15-23

Martin IL, Mitchell PR (1979) Diazepam facilitates the potassium-stimulated release of ^{3}H-dopamine from rat striatal tissue. Br J Pharmacol 66:107P

Martinez AE, Adler-Graschinsky E (1980a) Release of norepinephrine induced by preganglionic stimulation of the isolated superior cervical ganglion of the cat. J Pharmacol Exp Ther 212:527-532

Martinez AE, Adler-Graschinsky E (1980b) Modulatory role of α-adrenoceptors on the release of ^{3}H-norepinephrine elicited by preganglionic stimulation of the cat superior cervical ganglion. J Pharmacol Exp Ther 212:533-535

Martres MP, Costentin J, Baudry M, Marcais H, Protais P, Schwartz JC (1977) Long-term changes in the sensitivity of pre- and postsynaptic dopamine receptors in mouse striatum evidenced by behavioural and biochemical studies. Brain Res 136:319-337

Massingham R, Dubocovich ML, Langer SZ (1980) The role of presynaptic receptors in the cardiovascular actions of N,N-di-n-propyldopamine in the cat and dog. Naunyn-Schmiedeberg's Arch Pharmacol 314:17-28

Matthyse S (1973) Antipsychotic drug actions: a clue to the neuropathology of schizophrenia? Fed Proc 32:200-205

McCulloch MW, Rand MJ, Story DF (1972) Inhibition of ^{3}H-noradrenaline release from sympathetic nerves of guinea-pig atria by a presynaptic α-adrenoceptor mechanism. Br J Pharmac 46:523-524P

McGeer EG, McGeer PL (1976) Duplication of biochemical changes of Huntington's chorea by intrastriatal injections of glutamic and kainic acids. Nature (Lond) 263:517-519

McGeer PL, McGeer EG, Scherer U, Singh K (1977) A glutamatergic corticostriatal path? Brain Res 128:369-373

McGeer PL, McGeer EG, Innanen VT (1979) Dendro-axonic transmission. I. Evidence from receptor binding of dopaminergic and cholinergic agents. Brain Res 169:433-441

McMillen BA, German DC, Sanghera MK, Warnack W, Shore PA (1980) Pimozide: delayed onset of action at rat striatal pre- and postsynaptic dopamine receptors. J Pharmacol Exp Ther 215:150-155

Meller E, Helmer-Matyjek E, Bohmaker K, Adler CH, Friedhoff AJ, Goldstein M (1986) Receptor reserve at striatal dopamine autoreceptors: implications for selectivity of dopamine agonists. Eur J Pharmacol 123:311-314

Meltzer HY, Simonovic M (1981) Effect of 3-PPP, a putative dopamine autoreceptor agonist, on rat serum prolactin levels. Life Sci 29:99-105

Meltzer HY, Sachar EJ, Frantz AG (1974) Serum prolactin levels in unmedicated schizophrenic patients. Arch Gen Psychiatry 31:564-569

Miller JC, Friedhoff AJ (1979) Effects of haloperidol and apomorphine on the K^+-depolarized overflow of ^{3}H-dopamine from rat striatal slices. Biochem Pharmacol 28:688-690

Moleman P, Bruinvels J (1979a) Morphine-induced striatal dopamine efflux depends on the activity of nigrostriatal dopamine neurones. Nature 281:686-687

Moleman P, Bruinvels J (1979b) Effect of morphine on dopaminergic neurons in the rat basal forebrain and striatum. J Neural Transm 46:225-237

Moore KE (1981) Dyskinesia: animal experimental correlates. Acta Psychiat Scand 63 (Suppl 291):88-99

Morgenroth VH, Walters JR, Roth RH (1976) Dopaminergic neurons—alteration in the kinetic properties of tyrosine hydroxylase after cessation of impulse flow. Biochem Pharmacol 25:655-661

Mulder AH, Wemer J, De Langen CDJ (1979) Presynaptic receptor-mediated inhibition of noradrenaline release from brain slices and synaptosomes by noradrenaline and adrenaline. In: Langer SZ, Starke K, Dubocovich ML (eds) Presynaptic Receptors. Pergamon Press Oxford, pp 219-224

Murrin LC, Gale K, Kuhar M (1979) Autoradiographic localization of neuroleptic and dopamine receptors in the caudate putamen and substantia nigra: effects of lesions. Eur J Pharmacol 60:229-235

Murrin LC, Coyle JT, Kuhar MJ (1980) Striatal opiate receptors: pre- and postsynaptic localization. Life Sci 27:1175-1183

Nagy JI, Lee T, Seeman P, Fibiger HC (1978) Direct evidence for presynaptic and postsynaptic dopamine receptors in brain. Nature 274:278-281

Nakamura K, Kuntzman R, Maggio AC, Augulic V, Conney AH (1973) Influence of 6-hydroxydopamine on the effect of morphine on the tail-flick latency. Psychopharmacol 31:117-189
Nakamura S, Iwatsubo K, Tsai CT, Iwana K (1979) Neuronal activity of the substantia nigra (pars compacta) after injection of kainic acid into the caudate nucleus. Exp Neurol 66:682-691
Nakamura S, Tepper JM, Young SJ, Groves PM (1981) Neurophysiological consequences of presynaptic receptor activation: changes in noradrenergic terminal excitability. Brain Res 226:155-170
Nicolaou NM (1980) Acute and chronic effects of neuroleptics and acute effects of amphetamine on dopamine turnover in corpus striatum and substantia nigra of the rat brain. Eur J Pharmacol 64:123-132
Nieoullon A, Chéramy A and Glowinski J (1977) Release of dopamine *in vivo* from cat substantia nigra. Nature 266:375-377
Nieoullon A, Chéramy A, Leviel V, Glowinski J (1979) Effects of the unilateral nigral application of dopaminergic drugs on the *in vivo* release of dopamine in the two caudate nuclei of the cat. Eur J Pharmacol 53:289-296
Nomura Y, Kajiyama H, Nakata Y, Segawa T (1979) Muscarinic cholinergic binding in striatal and mesolimbic areas: reduction by 6-hydroxydopamine. Eur J Pharmacol 58:125-131
Nose T, Takemoto H (1974) Effect of oxotremorine on homovanillic acid concentration in the striatum of the rat. Eur J Pharmacol 25:51-55
Nowak JZ, Arbilla S, Galzin AM, Langer SZ (1983) Changes in sensitivity of release modulating dopamine autoreceptors following chronic treatment with haloperidol. J Pharmacol Exp Ther 226:558-564
Nowycky MC, Roth RH (1977) Presynaptic dopamine receptors. Development of supersensitivity following treatment with fluphenazine decanoate. Naunyn-Schmiedeberg's Arch Pharmacol 300:247-254
Öhman R, Larsson M, Nilsson IM, Engel J, Carlsson A (1977) Neurometabolic and behavioural effects of haloperidol in relation to drug levels in serum and brain. Naunyn-Schmiedeberg's Arch Pharmacol 299:105-114
Pelayo F, Dubocovich M, Langer SZ (1980) Inhibition of neuronal uptake reduces the presynaptic effects of clonidine but not of α-methyl-noradrenaline on the stimulation-evoked release of ^{3}H-noradrenaline from rat occipital cortex slices. Eur J Pharmacol 64:143-155
Pellmar TC (1981) Transmitter control of voltage-dependent currents. Life Sci 28:2199-2205
Phillipson OT, McKeown JM, Baker MJ, Healy AF (1977) Correlation between plasma chlorpromazine and its metabolites and clinical ratings in patients with acute relapse of schizophrenic and paranoid psychosis. Br J Psychiat 131:172-184
Pimoule C, Briley MS, Langer SZ (1980) Short-term surgical denervation increases ^{3}H-clonidine binding in rat salivary gland. Eur J Pharmacol 63:85-87
Pimoule C, Scatton B and Langer SZ (1983) [^{3}H]-RX 781094: a new antagonist ligand labels α-adrenoceptors in the rat brain cortex. Eur J Pharmacol 95:79-85, 1983
Pimoule C, Schoemaker H, Reynolds GP, Langer SZ (1985) [^{3}H]-SCH 23390 labeled D_1 dopamine receptors are unchanged in schizophrenia and Parkinson's disease. Eur J Pharmacol 114:235-237
Plotsky PM, Wightman RM, Chey W, Adams RN (1977) Liquid chromatographic analysis of endogenous catecholamine released from brain slices. Science 197:904-906
Pollard H, Llorens C, Bonnet JJ, Costentin J, Schwartz JC (1977) Opiate receptors on mesolimbic dopaminergic neurons. Neurosci Letters 7:295-299

Pollard H, Llorens C, Schwartz JC, Gros C, Dray F (1978) Localization of opiate receptors and enkephalins in the rat striatum in relationship with the nigrostriatal dopaminergic systems: lesion studies. Brain Res 151:392–398
Post RM, Goodwin FK (1975) Time-dependent effects of phenothiazines on dopamine turnover in psychiatric patients. Science 190:488–489
Price MT, Fibiger HC (1975) Ascending catecholamine systems and morphine analgesia. Brain Res 99:189–193
Puech AJ, Simon P, Chermat R, Boissier JR (1974) Profil neuropsychopharmacologique de l'apomorphine. J Pharmacol (Paris) 2:241–254
Pycock CJ, Kerwin RW (1981) The status of glycine as a supraspinal neurotransmitter. Life Sci 28:2679–2686
Quinn N, Illas A, Lhermitte F, Agid Y (1981) Bromocriptine and domperidone in the treatment of Parkinson disease. Neurology 31:662–667
Rabey JM, Passeltiner P, Markey K, Asano T, Goldstein M (1981) Stimulation of pre- and postsynaptic dopamine receptors by an ergoline and by a partial ergoline. Brain Res 225:347–356
Raisman R, Briley M, Langer SZ (1979) Specific labeling of postsynaptic α_1-adrenoceptors in rat heart ventricle by ^{3}H-WB 4101. Naunyn-Schmiedeberg's Arch Pharmacol 307:223–226
Raisman R, Sette M, Pimoule C, Briley M, Langer SZ (1982) High affinity ^{3}H-desipramine binding in the peripheral and central nervous system: a specific site associated with the neuronal uptake of noradrenaline. Eur J Pharmacol 78:345–351
Raiteri M, Cervoni AM, Del Carmine AM, Levi G (1978) Do presynaptic autoreceptors control dopamine release? Nature 274:706–708
Raiteri M, Cervoni AM, Del Carmine R, Levi G (1979) Lack of presynaptic autoreceptors controlling dopamine release in striatal synaptosomes. In: Langer SZ, Starke K, Dubocovich ML (eds) Presynaptic Receptors. Pergamon Press Oxford, pp 225–230
Rand MJ, Majewski H, McCulloch MW, Story DF (1979) An adrenaline-mediated positive feedback loop in sympathetic transmission and its possible role in hypertension. In: Langer SZ, Starke K, Dubocovich ML (eds) Presynaptic Receptors. Pergamon Press Oxford, pp 263–269
Rand MJ, McCulloch MW, Story DF (1980) Catecholamine receptors on nerve terminals. In Szekeres L (ed) Adrenergic Activators and Inhibitors. Springer Berlin Heidelberg New York, pp 223–266 Handbook of Experimental Pharmacology vol 54, Part I
Reimann W, Zumstein A, Jackisch R, Starke K, Hertting G (1979) Effect of extracellular dopamine on the release of dopamine in the rabbit caudate nucleus: evidence for a dopaminergic feedback inhibition. Naunyn-Schmiedeberg's Arch Pharmacol 306:53–60
Reimann W, Steinhauer HB, Hedler L, Starke K (1981) Effect of prostaglandins D_2, E_2 and $F_{2\alpha}$ on catecholamine release from slices of rat and rabbit brain. Eur J Pharmacol 69:421–427
Reisine TD, Nagy JI, Fibiger HC, Yamamura HI (1979) Localization of dopamine receptors in rat brain. Brain Res 169:209–214
Reubi JC, Iversen LL, Jessell TM (1977) Dopamine selectively increases ^{3}H-GABA release from slices of rat substantia nigra *in vitro*. Nature 268:652–654
Rifkin A (1981) The risks of long-term neuroleptic treatment of schizophrenia: especially depression and akinesia. Acta Psychiat Scand 63 (Suppl 291):129–134
Roberts DCS, Zis AP, Fibiger HC (1975) Ascending catecholamine pathways and amphetamine-induced locomotor activity: importance of dopamine and apparent non-involvement of norepinephrine. Brain Res 93:441–454

Roberts PJ, Anderson SD (1979) Stimulatory effects of L-glutamate and related amino acids on ^{3}H-dopamine release from rat striatum: an *in vitro* model for glutamate actions. J Neurochem 32:1539-1545
Roberts PJ, Sharif NA (1978) Effects of L-glutamate and related amino acids upon the release of ^{3}H-dopamine from rat striatal slices. Brain Res 157:391-395
Roth RH, Walters JR, Aghajanian GK (1973) Effect of impulse flow on the release and synthesis of dopamine in the rat striatum. In Usdin E, Snyder SH (eds) Frontiers in Catecholamine Research. Pergamon Press New York, pp 567-574
Rupniak NMJ, Jenner P, Marsden CD (1981) The absence of sodium ions does not explain the failure of sulpiride to inhibit, in vitro, rat striatal dopamine-sensitive adenylate cyclase. J Pharm Pharmacol 33:602-603
Salzman PM, Roth RH (1980) Poststimulation catecholamine synthesis and tyrosine hydroxylase activation in central noradrenergic neurons. II. Depolarized hippocampal slices. J Pharmacol Exp Ther 212:74-84
Sanberg PR, Lehmann J, Fibiger HC (1979) Sedative effects of apomorphine on an animal model of Huntington's disease. Arch Neurology 36:349-350
Scatton B (1977) Differential regional development of tolerance to increase in dopamine turnover upon repeated neuroleptic administration. Eur J Pharmacol 46:363-369
Scatton B (1979) Acute and subacute effects of haloperidol on DOPAC levels in the substantia nigra and ventral tegmental area of rat brain. Eur J Pharmacol 56:183-184
Schoemaker H, Pimoule C, Arbilla S, Scatton B, Javoy-Agid F, Langer SZ (1985) Sodium-dependent [^{3}H]cocaine binding associated with dopamine uptake sites in the rat striatum and human putamen: decrease after dopaminergic denervation and in Parkinson's disease. Naunyn-Schmiedeberg's Arch Pharmacol 329:227-235
Sedvall G, Aflredson G, Bjerkenstedt L, Eneroth P, Fryö B, Härnryd C, Swahn CG, Wiesel FA, Wode-Helgodt B (1975) Selective effects of psychoactive drugs on levels of monoamine metabolites and prolactin in cerebrospinal fluid of psychiatric patients. Proc Sixth Internatl Cong Pharmacol (Helsinki) 3:255-267
Seeman P (1981) Brain dopamine receptors. Pharmacol Rev 32:229-313
Seeman P, Lee T (1975) Antipsychotic drugs: direct correlation between clinical potency and presynaptic action on dopamine neurons. Science 188:1217-1219
Setler PE, Sarau HM, Zirkle CL, Saunders HL (1978) The central effects of a novel dopamine agonist. Eur J Pharmacol 50:419-430
Sette M, Raisman R, Briley M, Langer SZ (1981) Localisation of tricyclic antidepressant binding sites on serotonin nerve terminals. J Neurochem 37:40-42
Sharman DF (1981) The turnover of catecholamines. In: Pycock CJ, Taberner PV (eds) Central Neurotransmitter Turnover. Croom Helm London, pp 20-58
Shepperson NB, Langer SZ (1981) The effects of the 2-amino-tetrahydronaphthalene derivative M7, a selective α_2-adrenoceptor agonist *in vitro*. Naunyn-Schmiedeberg's Arch Pharmacol 318:10-13
Shepperson NB, Duval N, Massingham R, Langer SZ (1982) Differential blocking effects of several dopamine receptor antagonists for peripheral pre- and postsynaptic dopamine receptors in the anaesthetized dog. J Pharmacol Exp Ther 221:753-761
Skirboll LR, Grace AA, Bunney BS (1979) Dopamine auto- and postsynaptic receptors: electrophysiological evidence for differential sensitivity to dopamine agonists. Science 206:80-82
Skolnick P, Stalven LP, Daly JW, Hoyler E, Davis JN (1978) Binding of α- and β-adrenergic ligands to cerebral cortical membranes: effects of 6-hydroxydopamine treatment and relationship to the responsiveness of cyclic AMP-generating systems in two rat strains. Eur J Pharmacol 47:201-210

Snyder SH (1972) Catecholamines in the brain as mediators of amphetamine psychosis. Arch Gen Psychiatry 27:169-179

Sokoloff P, Martres MP, Schwartz JC (1980a) ^{3}H-Apomorphine labels both dopamine postsynaptic receptors and autoreceptors. Nature 288:283-286

Sokoloff P, Martres MP, Schwartz JC (1980b) Three classes of dopamine receptor (D-2, D-3, D-4) identified by binding studies with ^{3}H-apomorphine and ^{3}H-domperidone. Naunyn-Schmiedeberg's Arch Pharmacol 315:89-102

Spano PF, DiChiara G, Tonon GC, Trabucchi M (1976) A dopamine-stimulated adenylate cyclase in rat substantia nigra. J Neurochem 27:1565-1568

Spano PF, Stefanini E, Trabucchi M, Fresia P (1979) Stereospecific interaction of sulpiride with striatal and non-striatal dopamine receptors. In: Spano PF, Trabucchi M, Corsini G, Gessa GL (eds) Sulpiride and other Benzamides. Italian Brain Research Foundation Press Milan

Starke K (1972) Alpha-sympathomimetic inhibition of adrenergic and cholinergic transmission in the rabbit heart. Naunyn-Schmiedeberg's Arch Pharmacol 274:18-45

Starke K (1977) Regulation of noradrenaline release by presynaptic receptor systems. Rev Physiol Biochem Pharmacol 77:1-124

Starke K (1981) Presynaptic receptors. Ann Rev Pharmacol Toxicol 21:7-30

Starke K, Montel H (1974) Influence of drugs with affinity for α-adrenoceptors on noradrenaline release by potassium, tyramine and dimethyl-phenylpiperazinium. Eur J Pharmacol 27:273-280

Starke K, Reimann W, Zumstein A, Hertting G (1978) Effect of dopamine receptor agonists and antagonists on release of dopamine in the rabbit caudate nucleus *in vitro*. Naunyn-Schmiedeberg's Arch Pharmacol 305:27-36

Starr MS (1978) GABA potentiates potassium-stimulated ^{3}H-dopamine release from rat substantia nigra and corpus striatum. Eur J Pharmacol 48:325-328

Starr MS (1979) GABA-mediated potentiation of amine release from nigrostriatal dopamine neurons *in vitro*. Eur J Pharmacol 53:215-226

Stefanini E, Marchisio AM, Devoto P, Vernaleone F, Collu R, Spano PF (1980) Sodium-dependent interaction of benzamides with dopamine receptors. Brain Res 198:229-223

Steinberg MI, Keller CE (1978) Enhanced catecholamine synthesis in isolated rat superior cervical ganglia caused by nerve stimulation: dissociation between ganglionic transmission and catecholamine synthesis. J Pharmacol Exp Ther 204:384-399

Steinsland OS, Hieble JP (1978) Dopaminergic inhibition of adrenergic neurotransmission as a model for studies on dopamine receptor mechanisms. Science 199:443-445

Steinsland OS, Furchgott RF, Kirpekar SM (1973) Inhibition of adrenergic neurotransmission by parasympathomimetics in the rabbit ear artery. J Pharmacol Exp Ther 184:346-356

Stevens JR (1973) An anatomy of schizophrenia? Arch Gen Psychiatry 29:177-189

Stjärne L, Brundin J (1975) Dual adrenoceptor-mediated control of noradrenaline secretion from human vasoconstrictor nerves: facilitation by β-receptors and inhibition by α-receptors. Acta Physiol Scand 94:139-141

Stjärne L, Brundin J (1976) β_2-Adrenoceptors facilitating noradrenaline secretion from human vasoconstrictor nerves. Acta Physiol Scand 97:88-93

Stoof JC, Kebabian JW (1981) Opposing roles for D-1 and D-2 dopamine receptors in efflux of cyclic AMP from rat neostriatum. Nature 294:366-368

Stoof JC, Mulder AH (1977) Increased dopamine release from rat striatal slices by inhibitors of GABA-aminotransferase. Eur J Pharmacol 46:177-180

Stoof JC, Den Breejen EJS, Mulder AH (1979) GABA modulates the release of dopamine and acetylcholine from rat caudate nucleus slices. Eur J Pharmacol 57:35-42

Stoof JC, Horn AS, Mulder AH (1980) Simultaneous demonstration of the activation of presynaptic dopamine autoreceptors and postsynaptic dopamine receptors in vitro by N,N-dipropyl-5,6-ADTN. Brain Res 196:276-281

Story DF, Briley MS, Langer SZ (1979) The effects of chemical sympathectomy with 6-hydroxydopamine on α-adrenoceptor and muscarinic cholinoceptor binding in rat heart ventricle. Eur J Pharmacol 57:423-426

Strömbom U (1976) Catecholamine receptor agonists. Effects on motor activity and rate of tyrosine hydroxylation in mouse brain. Naunyn-Schmiedeberg's Arch Pharmacol 292:167-176

Subramanian N, Mitznegg P, Sprügel W, Domschke W, Domschke S, Wünsch E, Demling L (1977) Influence of enkephalin on K^+-evoked efflux of putative neurotransmitters in rat brain. Selective inhibition of acetylcholine and dopamine release. Naunyn-Schmiedeberg's Arch Pharmacol 299:163-165

Suga M (1980) Effect of long-term L-DOPA administration on the dopaminergic and cholinergic (muscarinic) receptors of striatum in 6-hydroxydopamine lesioned rats. Life Sci 27:877-882

Swanson LW (1976) The locus coeruleus: a cytoarchitectonic, Golgi and immunohistochemical study in the albino rat. Brain Res 110:39-56

Tamminga CA, Smith RC, Pandey G, Frohman LA, Davis JM (1977) A neuroendocrine study of supersensitivity in tardive dyskinesia. Arch Gen Psychiat 34:1199-1203

Tamminga CA, Schaffer MH, Smith RC, Davis JM (1978) Schizophrenic symptoms improve with apomorphine. Science 200:567-568

Tamminga CA, DeFraites EG, Gotts MD, Chase TN (1981) Apomorphine and N-n-propyl-norapomorphine in the treatment of schizophrenia. In: Corsini GU, Gessa GL (eds) Apomorphine and Other Dopaminomimetics, vol 2. Raven Press New York, pp 49-55

Taube HD, Starke K, Borowski E (1977) Presynaptic receptor systems on the noradrenergic neurones of rat brain. Naunyn-Schmiedeberg's Arch Pharmacol 299:123-141

Theodorou AE, Hall MD, Jenner P, Marsden CD (1980) Cation regulation differentiates binding of ^{3}H-sulpiride and ^{3}H-spiperone to rat striatal preparations. J Pharm Pharmacol 32:441-444

Thornburg JE, Moore KE (1974) A comparison of effects of apomorphine and ET 495 on locomotor activity and circling behaviour in mice. Neuropharmacol 13:189-197

Trulson ME, Preussler DW, Howell GA (1981) Activity of substantia nigra units across the sleep-waking cycle in freely moving cats. Neuroscience Letters 26:183-188

Tsuruta K, Frey EA, Grewe CW, Cote TE, Eskay RL, Kebabian JW (1981) Evidence that LY-141865 specifically stimulates the D-2 dopamine receptor. Nature 292:463-465

Ungerstedt U, Herrera-Marschitz M, Jungnelius U, Stahle L, Tossman U, Zetterström T (1982) Dopamine synaptic mechanisms reflected in studies combining behavioural recording and brain dialysis. In: Kohsaka M, Shohmori T, Tsukuda Y, Woodruff GN (eds) Advances in Dopamine Research. Pergamon Press Oxford

U'Prichard DC, Snyder SH (1979) Distinct α-noradrenergic receptors differentiated by binding and physiological relationships. Life Sci 24:79-88

Vogel SA, Silberstein SD, Berv KR, Kopin IJ (1972) Stimulation-induced release of

norepinephrine from rat superior cervical ganglion in vitro. Eur J Pharmacol 20:308-311
Voith K (1980) Supersensitivity to apomorphine in experimentally induced hypokinesia and drug-induced modifications of the apomorphine response. Psychopharmacol 79:247-254
Waggoner WG, McDermed J, Leighton HJ (1980) Presynaptic regulation of tyrosine hydroxylase activity in rat striatal synaptosomes by dopamine analogs. Mol Pharmacol 18:91-99
Waldmeier PC, Lauber J, Blum W, Richter WJ (1981) 3-Methoxytyramine: its suitability as an indicator of synaptic dopamine release. Naunyn-Schmiedeberg's Arch Pharmacol 315:219-225
Walters JR, Roth RH (1972) Effect of γ-hydroxybutyrate on dopamine and dopamine metabolites in the rat striatum. Biochem Pharmacol 21:2111-2121
Walters JR, Roth RH (1976) Dopaminergic neurons: an in vivo system for measuring drug interactions with presynaptic receptors. Naunyn-Schmiedeberg's Arch Pharmacol 296:5-14
Walters JR, Roth RH, Aghajanian GK (1973) Dopaminergic neurons: similar biochemical and histochemical effects of γ-hydroxybutyrate and acute lesions of the nigro-neostriatal pathway. J Pharmacol Exp Ther 186:630-639
Wang RY (1981a) Dopaminergic neurons in the rat ventral tegmental area. II. Evidence for autoregulation. Brain Res Rev 3:141-151
Wang RY (1981b) Dopaminergic neurons in the rat ventral tegmental area III. Effects of D- and L-amphetamine. Brain Res Rev 3:153-165
Watkins JC (1981) Pharmacology of excitatory amino acid receptors. In: Roberts PJ, Storm-Mathisen J, Johnston GAR (eds) Glutamate: Transmitter in the Central Nervous System. Wiley and Sons Chichester, pp 1-24
Wemer J, Schoffelmeer ANM, Mulder AH (1981) Studies on the role of Na^+, K^+ and Cl^- ion permeabilities in K^+-evoked release of ^3H-noradrenaline from rat brain slices and synaptosomes and in its presynaptic α-adrenergic modulation. Naunyn-Schmiedeberg's Arch Pharmacol 317:103-109
Westerink BHC (1979) Further studies on the sequence of dopamine metabolism in the rat brain. Eur J Pharmacol 56:313-322
Westerink BHC, Korf J (1976a) Comparison of effects of drugs on dopamine metabolism in the substantia nigra and the corpus striatum of rat brain. Eur J Pharmacol 40:131-136
Westerink BHC, Korf J (1976b) Regional rat brain levels of 3,4-dihydroxyphenylacetic acid and homovanillic acid: concurrent fluorometric measurement and influence of drugs. Eur J Pharmacol 38:281-291
Westfall TC, Besson MJ, Giorguieff MF, Glowinski J (1976) The role of presynaptic receptors in the release and synthesis of ^3H-dopamine by slices of rat striatum. Naunyn-Schmiedeberg's Arch Pharmacol 292:279-287
Westfall TC, Peach MJ, Titermary V (1979a) Enhancement of the electrically induced release of norepinephrine from the rat portal vein: mediation by β_2-adrenoceptors. Eur J Pharmacol 58:67-74
Westfall TC, Perkins NA, Paul C (1979b) Role of presynaptic receptors in the synthesis and release of dopamine in the mammalian central nervous system. In: Langer SZ, Starke K, Dubocovich ML (eds) Presynaptic Receptors. Pergamon Press Oxford, pp 243-248
Wilson CJ, Fenster GA, Young SJ, Groves PM (1979) Haloperidol-induced alteration of post-firing inhibition in dopaminergic neurons of rat substantia nigra. Brain Res 179:165-170

Wolfarth S, Dulska E, Golembiowska-Nikitin K, Vetulani J (1978) A role of the polysynaptic system of substantia nigra in the cholinergic dopaminergic equilibrium in the central nervous system. Naunyn-Schmiedeberg's Arch Pharmacol 302:123-131

Woodruff GN, Pinnock RD (1981) Some central actions of ADTN—a potent dopamine receptor agonist. In: Gessa GL, Corsini GU (eds) Apomorphine and Other Dopaminomimetics, vol 1. Raven Press New York, pp 241-251

Zivkovic B (1979) The involvement of dopamine postsynaptic and autoreceptors in the regulation of dopamine synthesis. In: Langer SZ, Starke K, Dubocovich ML (eds) Presynaptic Receptors. Pergamon Press Oxford, pp 249-253

Zivkovic B, Guidotti A, Costa E (1976) Cyclic AMP content and regulation of tyrosine-3-mono-oxygenase in rat striatum. J Cyclic Nucl Res 2:1-10

Zumstein A, Karduck W, Starke K (1981) Pathways of dopamine metabolism in the rabbit caudate nucleus in vitro. Naunyn-Schmiedeberg's Arch Pharmacol 316:205-217

CHAPTER 9

Adaptive Supersensitivity

W. FLEMING and D.P. WESTFALL

A. Introduction

Over the years, the term supersensitivity has been applied to several phenomena and it was coined long before any of the underlying mechanisms were understood. Given the variety of phenomena to which the term has been applied, FLEMING et al. (1973) offered the following definition and explanation of supersensitivity: "Supersensitivity may be defined as the phenomenon in which the amount of a substance required to produce a given biological response is less than 'normal', i.e. the dose-response curve is shifted to the left. In a few instances there may also be an increase in the maximum response to a drug. However, this increase in maximal response is not a regular occurrence in supersensitivity. Likewise, supersensitivity may, but need not, be associated with a change in the slope of the dose-response curve. Thus, the one consistent sign of supersensitivity is the shift of the dose-response curve to the left."

As the mechanisms responsible for changes in sensitivity began to be unraveled, it became clear that the term supersensitivity encompassed at least two entirely separate types of phenomena (FLEMING 1975). The first type (deviation supersensitivity) is the result of an alteration in the disposition of a drug or transmitter in the biological system. In this case, the leftward displacement of the dose-response curve for an agonist is the result of the agonist achieving, in the biophase, a greater than normal fraction of the dose or concentration administered. There is no change in the relationship between the stimulus (i.e. a given concentration of the agonist in the biophase) and the cellular response. Essentially, deviation supersensitivity incorporates those examples of supersensitivity which result from elimination of some transport and/or metabolic site of loss of the agonist. The second type (adaptive supersensitivity), which, until recently, has been most commonly termed postjunctional or nondeviation supersensitivity (FLEMING 1976), is the result of a change in the quantitative relationship between the stimulus (drug or transmitter) and the cellular response.

Factors producing the deviation type of supersensitivity in tissues with an adrenergic innervation include inhibition of the neuronal uptake of noradrenaline (for example, by cocaine or desipramine), destruction of the nerve terminals into which the uptake occurs and the inhibition of extraneuronal uptake or metabolism of catecholamines (TRENDELENBURG 1966; TRENDELENBURG and GRAEFE 1975; TRENDELENBURG 1986). Although the term deviation supersensitivity has not generally been applied to cholinergically innervated tissues, there is a clear analogy in the shift of the dose-response curve of acetylcholine to the left as a consequence of the inhibition of cholinesterase or the loss of neuronal cholinesterase, for example by denervation (WESTFALL et al. 1974; MCCONNELL and SIMPSON 1976). It will be obvious that deviation supersensitivity is as highly specific as was the site of loss before its inhibition. The dose-response curve is shifted only in the case of agonists which are substrates for the transport or enzymatic process which has been inhibited or lost.

Adaptive supersensitivity is usually induced by a prolonged interruption or reduction in transmission across a neuroeffector junction (FLEMING et al. 1973). The effector cell may be a muscle cell, gland cell, another neuron, etc. Thus, the term "junctional" is used in a broad sense. Adaptive supersensitivity is generally less specific than deviation supersensitivity. That is, the sensitivity is increased to a wider variety of agonists, as will be discussed later.

Chronic enhancement of neuroeffector transmission has been demonstrated to induce subsensitivity of effector systems (FLEMING et al. 1973; FLEMING 1976). Although much less experimental work has been done with adaptive subsensitivity than with supersensitivity, there are clear examples of the same effector becoming supersensitive after chronic reduction in neurotransmission and subsensitive after chronic enhancement of neurotransmission (EMMELIN 1964; BITO et al. 1971; OVERSTREET et al. 1973; RUSSELL and OVERSTREET 1987). It has been suggested that supersensitivity and subsensitivity are both expressions of cellular homeostatic mechanisms by which excitable cells compensate for chronic changes in the stimulus they receive (FLEMING et al. 1973).

This hypothesis of cellular adaptation is consistent with the growing body of evidence that tolerance and withdrawal reactions to some drugs are expressions of cellular adaptations to the chronic effects of the drugs which quantitatively alter the neuroeffector relationship. Such explanations have received support in reference to narcotics (COLLIER 1966; SCHULZ and GOLDSTEIN 1973; GIANUTSOS et al. 1974; SCHULZ and HERZ 1976; JOHNSON et al. 1978) and ethanol (FRENCH et al. 1975; ENGEL and LILJEQUIST 1976; TABAKOFF and HOFFMAN 1979, TABAKOFF et al. 1979). An extensive review of tolerance and withdrawal is, however, beyond the scope of this chapter.

Adaptive supersensitivity has been demonstrated in several types of effector cells, including skeletal muscle, smooth muscle, cardiac muscle, exocrine glands, the pineal gland and neurons (FLEMING 1976). Furthermore, it has been produced by interference with several different types of neurotransmission processes, including adrenergic, cholinergic, dopaminergic (see review by FLEMING 1976) and serotoninergic (KLAWANS et al. 1975; TRULSON et al. 1976; CARRUBA et al. 1979; WANG et al. 1979). It would be impossible to adequately discuss adaptive supersensitivity in terms of only adrenergically innervated effector systems. Consequently, this chapter is not limited to those systems.

There have been a number of extensive reviews of sensitivity pheomena. This chapter will make frequent use of earlier reviews to provide background information, will concentrate primarily on the original literature from the mid 1970's to 1987 and will emphasize progress in understanding the cellular mechanisms of supersensitivity. For the reader who wishes to examine in detail the earlier literature on sensitivity, the following reviews and symposia are suggested: CANNON and ROSENBLUETH (1949), THESLEFF (1960), EMMELIN (1961), GUTMANN and HONIK (1963), TRENDELENBURG (1963), SHARPLESS (1964), EMMELIN (1965), GUTH (1968), FLEMING et al. (1973), DRACHMAN (1974), CALNE et al. (1975), FLEMING (1976), WESTFALL (1981), FLEMING (1984).

B. The Induction of Adaptive Supersensitivity

1. Experimental Procedures Which Have Been Used to Produce Supersensitivity

The procedures which are recognized as effective in producing adaptive supersensitivity have in common the ability to chronically interrupt or markedly reduce neurotransmission to effector cells. It should be understood that the interference of transmission must be maintained for a prolonged period, generally several days or more, in order to induce supersensitivity. Consequently, pharmacologic procedures used to produce the phenomenon usually involve chronic administration of a drug. It will be recognized that the experimental procedure chosen will depend upon several factors such as surgical accessibility of the nerves, type of neurotransmission (e.g. adrenergic vs. cholinergic), importance of reflexes in the normal function of the effector cells, etc. The following classification of procedures inducing adaptive supersensitivity is based upon the list in Fleming et al. (1973).

I. Surgical interruption of the nerves.
   A. Destruction of the neurons which directly innervate the effector cells ("denervation").
   B. Destruction of neurons "upstream" in the chain of activation of the effector cells.
      1) Preganglionic denervation (decentralization) of autonomically innervated cells.
      2) Sensory denervation of cells (iris sphincter, salivary glands) which are activated primarily by reflex.
II. Chemical denervation (immunosympathectomy, 6-hydroxydopamine).
III. Chronic pharmacological inhibition of neurotransmission "upstream", such as with ganglionic blockade.
IV. Chronic depletion of transmitter.
V Chronic inhibition of transmitter release.
   A. Adrenergic neuron blockers, such as guanethidine.
   B. Botulinum toxin (cholinergic neurons).
VI. Chronic inhibition of transmission at the receptors on the effector cells.
VII. Chronic alteration of sensory stimuli such as light (pineal gland, iris spincter).

In some tissues, for example the nictitating membrane of the cat, the vas deferens and atrium of the guinea pig, the ability of two or more procedures to produce adaptive supersensitivity have been compared (FLEMING et al. 1973; FLEMING 1984). When care is taken to use optimal conditions for each experimental procedure, the different procedures produce qualititively and quantitatively similar supersensitivity in the same effector organ. In other words, if neurontransmission is chronically interrupted, the supersensitivity which appears in the effector cells is independent of the method used to interrupt transmission. Some of the procedures which produce adaptive supersensitivity also cause a deviation type of supersensitivity. Clearly, it is important to be able to measure the two phenomena separately. This is generally accomplished by taking advantage of the greater specificity of deviation types of supersensitivity. Destruction of adrenergic neurons produces a rapidly developing (usually 1-2 days) deviation supersensitivity, due to loss of neuronal uptake of certain amines, and a more slowly developing adaptive supersensitivity. Once the adaptive supersensitivity has developed, the leftward shift of the dose-response curve of noradrenaline (a good substrate for neuronal uptake) is a reflection of both types of supersensitivity. Conversely, the dose-response curve for methoxamine is shifted to a lesser degree and reflects only the adaptive supersensitivity because methoxamine is not a substrate for the neuronal amine transport system (TRENDELENBURG et al. 1970; WESTFALL and FEDAN 1975). In denervated skeletal muscle, the sensitivity is increased more to acetylcholine, a substrate for cholinesterase, than it is to carbachol, which is not a substrate for the enzyme (MCCONNELL and SIMPSON 1976). The shift of the dose-response curve of carbachol reflects the adaptive supersensitivity alone.

The use of drugs to interfere with neurotransmission in order to induce adaptive supersensitivity has serious pitfalls which a number of investigators have failed to appreciate. The choice of the proper dose and schedule of treatment for each species and effector organ is critical. For instance, a very small precentage of the total store of noradrenaline is adequate to maintain neurotransmission (see for example, ANTONACCIO and SMITH 1974) and thus the dose and schedule of pretreatment with reserpine must be adequate for nearly complete depletion of the transmitter. On the other hand, large doses of reserpine and prolonged treatment can lead to important effects other than depletion, including toxic effects on effector cells (see, for example, IWAYAMA et al. 1973). For further references and discussion of the use of reserpine to induce adaptive supersensitivity, see FLEMING et al. (1973) and FLEMING (1984).

In those situations in which the preganglionic autonomic nerves are easily reached and separated from surrounding tissues, decentralization is a particularly desirable means of producing adaptive supersensitivity. It does not produce any form of deviation supersensitivity, does not create the side effects which are unavoidable with chronic treatment with drugs and allows one to interrupt transmission to a relatively discrete area, leaving the neurotransmission to other organs intact.

II. Effector Cell Activity (Use—Disuse) Versus Trophic Substances as Factors Regulating Sensitivity

It is evident that some aspect of the neuroeffector relationship has a suppressing effect on the sensitivity of the effector cells. Removal of that factor, whatever it may be, allows sensitivity to increase. For nearly as long as the effect of denervation on sensitivity has been recognized, there has been speculation about, and, more recently, intense investigation of, the factor or factors which control sensitivity. The factor could be the level of activity of the effector cells themselves, in which case the role of the nerve is only secondary as the physiologic modulator of that cellular activity. If the controlling factor is the target cell activity, one must define which specific cellular activity is critical. For example, in a muscle cell, is the frequency of contraction, the frequency of action potentials, the frequency of subthreshold junction potentials, some aspect of transmembrane ion movement or the intracellular concentration of calcium ions (to name a few possibilities) the ultimate factor signaling the cell to adjust its sensitivity? On the other hand, sensitivity may be regulated by a specific trophic substance released from the nerve. Such a trophic factor could be the transmitter itself or a separate substance. If there is a trophic factor other than the transmitter, it might be released in intimate association with the transmitter or independently. Gathering clear evidence for or against these various possibilities has been exceedingly difficult.

1. Skeletal Muscle

The cellular changes which occur in association with denervation in skeletal muscle are very complex and include an increase in density of extrajunctional cholinoceptors, a partial membrane depolarization, altered sodium channels, the appearance of fibrillations and changes in the binding of calcium (FLEMING 1976). The interpretation of studies of trophic factors in skeletal muscle is complicated by the use by various authors of different indices of denervation supersensitivity as the end point to be measured. The problem is compounded by the fact that these different cellular changes appear at different times after denervation (see for example, ALBUQUERQUE et al. 1971; REDFERN and THESLEFF 1971) and may even be influenced to different degrees by separate trophic factors (FISCHBACH and ROBBINS 1971; GILLIATT et al. 1978).

The evidence is solid that electrical activity of the skeletal muscle cells themselves is important in regulating the sensitivity of the cells (LÖMO 1976). Chronic electrical stimulation of muscle *in vivo* inhibits the appearance of cellular changes associated with supersensitivity or reverses these changes after they have appeared (DRACHMAN and WITZKE 1972; LÖMO and ROSENTHAL 1972; LÖMO and SLATER 1978). Cultured embryonic chick skeletal muscle cells are very sensitive to acetylcholine applied by iontophoresis, presumably because they are not innervated. COHEN and FISCHBACH (1973) found that chronic electrical stimulation of cultured cells induced a decrease in their sensitivity to acetylcholine and a reduction in receptor binding. Of particular interest is the finding that chronic stimulation at voltages either below or above the threshold for contraction of denervated muscles of the rat was equally effective in supressing the development of supersensitivity (GRUENER et al. 1974). Furthermore, procedures which interrupt nerve transmission without eliminating minature end plate potentials (mepps) result in denervation-like changes which are quantitatively less than caused by denervation (JOHNS and THESLEFF 1961; LÖMO and ROSENTHAL 1972; CANGIANO et al. 1977). STANLEY and DRACHMAN (1979) compared the partial depolarization (the earliest change associated with supersensitivity) of cells after either denervation or nerve block produced by repeated injection of tetrodotoxin (TTX) into the epineural space. TTX blocks the passage of nerve impulses but not the spontaneous release of transmitter which is responsible for mepps. Denervation or TTX caused equal degrees of depolarization, about 15 mV, but the rate of development was slower after TTX (7 days) than after denervation (2 days). It would thus appear that fluctuations in membrane potential and/or transmembrane movements of ions less than those which occur with action potentials do suppress sensitivity in skeletal muscle cells. However, LÖMO (1976) has challanged this conclusion.

Botulinum toxin is a highly specific inhibitor of the release of acetylcholine. SIMPSON (1977) reported that chronic poisoning of the rat hemidiaphragm with botulinum toxin produced increases in sensitivity to acetylcholine and in cholinoceptor binding (radiolabelled α-bungarotoxin) which were similar in magnitude to those induced by denervation. In contrast, PESTRONK et al. (1976) found less of an increase in receptor binding produced by chronic botulinum poisoning than by denervation. However, the botulinum treatment employed by the latter authors reduced, but did not abolish, mepps, indicating that a small amount of spontaneous release of acetylcholine continued in the presence of the toxin. Indeed, MATHERS and THESLEFF (1978) also found that botulinum toxin caused a lesser magnitude of denervation-like changes than did denervation itself. However, when they combined botulinum toxin to inhibit release of acetylcholine with α-bungarotoxin to antagonize irreversibly any released acetylcholine at the receptor sites, the effects were equal to denervation.

It is tempting to conclude that the only factors controlling sensitivity in skeletal muscle are the activity of the muscle and/or a direct trophic influence of acetylcholine. Unfortunately, one cannot be certain that botulinum toxin does not inhibit the release of some unknown trophic substance from the nerve in addition to acetylcholine (GUTH 1968).

The results of experiments with colchicine or vinblastine have provided evidence for a trophic factor (not the transmitter) which reaches the nerve terminal by axonal transport (HOFMANN and THESLEFF 1972; ALBUQUERQUE et al. 1974; SELLIN and MCARDLE 1977a). Colchicine or vinblastine is able to inhibit axonal transport without interfering markedly with the passage of nerve impulses or the release of transmitter. Colchicine or vinblastine, applied either by epineural injection or placed around the nerves in silicone cuffs, causes the appearance of extrajunctional receptors and other cellular changes associated with denervation supersensitivity. Unfortunately, the interpretation of these results is not entirely unambiguous in view of the evidence of CAN-

GIANO and FRIED (1977) that colchicine can cause denervation-like changes *without* inhibiting axonal transport. It is also interesting to note that, whatever trophic substances are released by nerve, they must not diffuse very far. Denervated muscle fibers develop typical signs of denervation even though they are adjacent to fibers which are innervated (TIEDT et al. 1977).

2. Smooth Muscle

Adrenergic denervation of the nictitating membrane of the cat causes supersensitivity to several unrelated agonists, including sympathomimetic and parasympathomimetic agents (TRENDELENBURG 1963). Regeneration of the adrenergic innervation or reinnvervation by a *cholinergic* nerve leads to a restoration of the normal sensitivity (SIMEONE 1937; LUCO and VERA 1964). This indicates that, if a neurotrophic factor is involved, it is a substance common to adrenergic and cholinergic neurons. If the trophic factor is the transmitter, another transmitter which is atypical for that tissue may readily substitute for it.

The supersensitivity which develops subsequent to suppression of autonomic innervation can be prevented or reversed by the chronic administration of drugs which stimulate the effector cells. For example, the supersensitivity of salivary glands which is induced by parasympathetic decentralization can be reversed by chronic administration of any of several secretogogues, including pilocarpine, carbachol and adrenaline (EMMELIN 1965). The development of denervation supersensitivity in sweat glands is inhibited by chronic administration of pilocarpine (REAS and TRENDELENBURG 1967) and supersensitivity of the guinea-pig ileum following suppression of parasympathetic function can be prevented by chronic administration of carbachol (FLEMING 1968). It thus appears that the transmitter per se may not be the factor which is normally suppressing sensitivity, since its effect can be replaced by other agonists acting through the same or even different receptors. The results of EMMELIN (1965), in particular, argue against any important role for a specific trophic substance and in favor of the effector cells' own level of activity (use-disuse) as the factor regulating sensitivity.

This interpretation is complicated by results of experiments with the adrenergically innervated guinea-pig vas deferens. In that organ, decentralization or chronic treatment of the animals with reserpine leads to adaptive supersensitivity which is typical of that seen in smooth muscles in general. That is, the dose-response curves of several unrelated agonists are shifted to the left without a change in maximum (WESTFALL 1970a). However, denervation, in addition to causing shifts of the dose-response curves for the same agonists, also increases the maximal response (WESTFALL et al. 1972). This effect is not secondary to loss of neuronal uptake because it is not mimicked by cocaine. The increased maximal response appears to be the result of improved electrical coupling among the smooth muscle cells (see Sect. D.IV). There are two possible explanations for the appearance of this unusual effect with denervation but not with decentralization or depletion of transmitter. One possibility is that the transmitter is functioning as a trophic substance to regulate cell-to-cell coupling and is so potent in this regard that it is able to continue to suppress coupling even in the low concentration released in tissues which are decentralized or markedly depleted of transmitter. The alternative explanation is that cell-to-cell coupling is controlled by a separate trophic factor in the postganglionic neuron, the storage and release of which is unaffected by decentralization or depletion of transmitter.

GOTO et al. (1979) applied colchicine to the hypogastric plexus of the rat and, two weeks later, observed denervation-like supersensitivity of the vas deferens, including increased maximal responses to noradrenaline and methacholine. The authors were able to rule out any direct effects of colchicine on the muscle cells. The colchicine treatment did cause some reduction in the total amount of noradrenaline in the tissue and in neurotransmission. However, these reductions were less than generally required to produce adaptive supersensitivity, leading the authors to conclude that control of sensitivity does involve a trophic substance which is transported along the axon. A subsequent study using intracellular recording (GOTO 1980) found that spontaneous junction potentials could be observed in virtually every smooth muscle cell impaled in the colchicine pretreated group. Although the frequency and amplitude were somewhat reduced, these results support the contention of the author that neurotransmission was not abolished by the colchicine.

GOTO (1983) has also found that chlorpromazine, trifluperizine and local anesthetics applied to the hypogastric plexus produce supersensitivity identical to that associated with denervation minus its prejunctional (deviation) component. The peak change is four days after application of the drug to the nerve. This effect is not duplicated by acute application of the drugs to the smooth muscle in the organ bath. Interestingly, the capacity of the phenothiazines and local anesthetics to induce supersensitivity in this manner was relatively proportional to their potency in antagonizing calmodulin. GOTO (1983) suggested that the supersensitivity was the result of a combination of membrane-stabilizing effects of the drugs on the nerves and interference with calmodulin-dependent axonal transport. The hypothesis is supported by his finding that junctional potentials and action potentials could be elicited in the smooth muscle by nerve stimulation, but higher voltage or longer duration pulses were required.

As concluded by DRACHMAN (1974) for skeletal muscle, one cannot rule out the possibility of multiple factors additively regulating sensitivity in effector systems innervated by autonomic nerves. In both cases, there is evidence in favor of effector cell activity being at least partially responsible for controlling sensitivity in the same cells. For a very useful discussion of the problems involved in investigating and interpreting trophic relationships, see GUTMANN (1976).

C. Characteristics of Adaptive Supersensitivity

I. Temporal Aspects

There is, generally, a delay in the appearance of adaptive supersensitivity. The suppression of contact between transmitter and effector cells must be maintained for a period, usually of one to several days, before the supersensitivity begins to develop (FLEMING et al. 1973). This is in contrast to deviation supersensitivity which is always coincident with the loss or inhibition of the critical transport or enzymatic process.

Skeletal muscle is particularly intriguing in this regard because denervation supersensitivity is related to several cellular changes which appear and progress at different rates in the same muscle. For example, ALBUQUERQUE et al. (1971) reported that the partial depolarization of the cells of the extensor digitorum longus muscle of the rat began as early as 2 hours after denervation, when the site of denervation was very close (less than 3.5 mm from the end plate). This was well before the loss of spontaneous transmitter release, as judged by the cessation of mepps, which occurred 10–12 hours post-denervation (ALBUQUERQUE et al. 1971; DESHPANDE et al. 1976). The depolarization was maximal after about 48 hours. In contrast, the signs of extrajunctional spread of receptors did not begin until 24 hours after denervervation and was still progressing 48 hours after denervation. An increase in membrane resistance of the muscle cells began about 72 hours after denervation. The changes were delayed if the denervation site was further from the muscle cell. The authors concluded that depolarization is the earliest sign of denervation changes, is not regulated by spontaneous release of transmitter and that more than one trophic factor is involved in regulation of these cellular characteristics. In most experiments with skeletal muscle, the nerve is cut or crushed at a considerable distance from the muscle. Under these more typical conditions, supersensitivity requires several days to develop in mammals and much longer in amphibians (THESLEFF 1960).

In effector tissues with autonomic innervation, the duration of the delay in onset and maximal development of supersensitivity is characteristic of the effector cells involved and, if optimal conditions are used, independent of the experimental procedures chosen to induce supersensitivity. The time-course of the development of supersensitivity in smooth muscle and cardiac muscle has been extensively covered in previous reviews (FLEMING et al. 1973; FLEMING 1976). Suffice it to say that, in some smooth muscles, such as the aorta (HUDGINS and FLEMING 1966) and the saphenous artery (ABEL et al. 1981), supersensitivity is well developed within 3 days whereas, in the nictitating membrane, adaptive supersensitivity does not become maximal until 2–4 weeks (FLEMING 1963; LANGER et al. 1967).

Adaptive supersensitivity appears to be a very long-lasting phenomenon, but reversible if reinnervation is allowed to occur (for early literature, see CANNON and ROSENBLEUTH 1949). EKSTRÖM and EMMELIN (1974) produced supersensitivity in the parotid gland of the cat by cutting the auriculo-temporal nerve. Supersensitivity developed rapidly over the first few days and remained high for about 2½ months. At that time, evidence of reinnervation appeared and sensitivity began to decline. However, after 7 months, reinnervation was still not complete and sensitivity had not returned completely to the control level. NEIDLE (1950) removed the ciliary ganglion in the cat and followed changes in the sensitivity of the iris to pilocarpine. Supersensitivity was maximal at 5 to 8 days and remained constant until 35 days after denervation, at which time a gradual decline in sensitivity began. Unfortunately, the author did not have evidence as to whether any reinnervation had occurred. COLASANTI and HOOVER (1979) found that supersensitivity of the cat iris, which was well developed at 21 days after ciliary ganglionectomy, was absent two years later, even though cholinergic reinnervation had not occurred.

II. Agonist Specificity

Denervation supersensitivity in skeletal muscle is generally regarded as specific for agonists acting via nicotinic cholinoceptors. This is not suprising in view of the magnitude of the increase in sensitivity to acetylcholine (see, for example, THESLEFF 1960). Although part of this effect is due to loss of cholinesterase, the magnitude of supersensitivity is very large also to those cholinomimetics which are not substrates for cholinesterase (MCCONNELL and SIMPSON 1976). However, there are modest increases in responsiveness of denervated skeletal muscle to a variety of agonists including potassium ion, histamine, 5-hydroxytryptamine, caffeine and bradykinin (FLEMING et al. 1973). It is quite clear that denervated skeletal muscle presents a large and specific supersensitivity to nicotinic-type agonists superimposed upon a modest nonspecific supersensitivity. The modest nonspecific supersensitivity is consistent with the very basic changes in cellular physiology, such as partial depolarization and altered calcium dynamics, which occur consequent to denervation.

In contrast, a specific component of adaptive supersensitivity is conspicuously absent in smooth muscle. This fact has been extensively documented (FLEMING et al. 1973). Briefly, adaptive supersensitivity in smooth muscle is consistently characterized by increases in sensitivity of similar magnitude to several unrelated agonists acting on separate receptor systems and to ions such as potassium and barium which do not act via receptors.

Adaptive supersensitivity is relatively specific in the heart. Chronotropic and inotropic supersensitivity to catecholamines is readily demonstrated in a variety of species (see review by FLEMING 1984). Some experiments have also indicated supersensitivity to calcium and some have not. However, there consistently is no change in sensitivity to agents such as histamine, aminophylline or parasympathomimetics.

Although the recent literature is filled with examples of putative adaptive supersensitivity in the CNS (see FLEMING 1976, and Sect. D.I.5), evidence relating to the specificity of the increase in sensitivity for various agonists is sparse. The technical problems in determining which agonists, other than the transmitter and drugs which mimic it, will affect a single type of neuron are enormous. The best tool available is microiontophoresis and even with this method the problems of determining which agonists act on which neurons require a level of sophistication greater than is frequently applied (BLOOM 1974). The technical problems are compounded by the misconception of many investigators that adaptive supersensitivity *must* be specific (see, for example, SEGAL 1977). The assumption seems to be based on a superficial knowledge of the literature on adaptive supersensitivity and an emphasis on the *older* literature on skeletal muscle. Among the points to be stressed in this chapter, and already stressed in earlier reviews (FLEMING et al. 1973; FLEMING 1976) are (1) skeletal muscle is not necessarily a good model for supersensitivity in other tissues, (2) adaptive supersensitivity is not truly specific in skeletal muscle and (3) since activation of a receptor is only the first step in a chain of physiologic events leading to response, there is no reason, *a priori*, that supersensitivity *must* be the result of changes in receptors or be specific for agonists acting through one receptor type.

Some of the difficulties associated with determining the specificity of supersensitivity to drugs which induce behavioral responses may be illustrated from a study by KLAWANS et al. (1977). These investigators used two procedures to induce supersensitivity in guinea pigs. Chronic treatment of the animals with dopamine-receptor antagonists produced supersensitivity of stereotyped behavior to dopamine-like agonists but no change in sensitivity to stereotyped behavior induced by activation of 5-HT receptors. Chronic antagonism at 5-HT receptors produced the opposite effect. Since the dopamine and 5-HT receptors are associated with distinctly different stereotyped behaviors and presumably, therefore, different neurons, KLAWANS et al. (1977) concluded that the supersensitivity is confined to those neurons to which neurotransmission is inhibited. This is a valid and important conclusion. However, KLAWANS et al. (1977) also concluded that the supersensitivity is specific for the *receptors* at which the chronic antagonism occurred. This is a conclusion which exceeds the available data. Considering that 5-HT agonists do not cause dopamine-like stereotypy, there may be no 5-HT receptors on the neurons responsible for that behavior. The determination of specificity of supersensitivity in a single type of neuron requires, first, the identification of the receptor systems existing in *those* neurons and then determination of whether or not sensitivity is increased to two or more agonists acting on those neurons via different receptors.

A few investigators have studied the effects of two different classes of agonists, applied by iontophoresis, in relation to supersensitivity in the central nervous system. SEGAL (1977) examined the effects of aspartate and acetylcholine on hippocampal neurons of rats. Three groups of rats were used: controls, rats with the commissural path (which contains neurons in which aspartate may be the transmitter) severed and rats in which the septo-hippocampal path (cholinergic) was severed. The author found a decrease in threshold to aspartate but not acetylcholine in hippocampal neurons of decommissurized rats and a decrease in threshold to acetylcholine but not aspartate in neurons of rats with septal lesions. Unfortunately, the author was unable to establish that the supersensitivity to aspartate was not of the deviation type and did not even consider the possibility that the supersensitivity to acetylcholine was of the deviation type which, indeed, it appears to be (BIRD and AGHAJANIAN 1975).

Supersensitivity of caudate neurons after repeated administration of haloperidol (a dopamine receptor antagonist) has been investigated by iontophoresis (YARBROUGH 1975; SKIRBOLL and BUNNEY 1979). There were indications of increased sensitivity of the caudate neurons to dopamine and apomorphine (a dopamine-like agonist which is not subject to neuronal uptake) but not to GABA. Thus, within the limits of the agonists tested, there seemed to be an adaptive supersensitivity specific to dopamine-like agonists. A surprising finding was that the dose of haloperidol used did not antagonize iontophoretic dopamine, making it difficult to understand how haloperidol induces supersensitivity in the preparation (SKIRBOLL and BUNNEY 1979).

WANG et al. (1979), in a very extensive and thorough study, have presented evidence of nonspecific supersensitivity in central neurons. Destruction of serotoninergic neurons of forebrain cells of rats was accomplished by injection of 5,7-dihydroxytryptamine into the lateral ventricle or the ascending pathway in the ventromedial tegmentum. Sensitivity to agonists was tested by iontophoresis. Supersensitivity to 5-HT appeared within 24 hours and was maximal after about 7 days. During the first two days the increase in sensitivity was specific for 5-HT, the period during which histological studies showed the serotoninergic neurons to be disappearing. Subsequently, the cells became supersensitive to lysergic acid diethylamide (a 5-HT-like agonist which is not a substrate for transport into serotoninergic neurons), noradrenaline and GABA.

D. Possible Mechanisms for Changes in Sensitivity of Effector Cells and Evidence Supporting or Opposing these Mechanisms

I. Changes in Receptors

1. Skeletal Muscle and Application of Receptor Theory

The concept that supersensitivity can result from an increase in the number of receptors has been clearly established for denervated skeletal muscle (see reviews by THESLEFF 1960; 1974: GUTMANN and HONICK 1963; GUTH 1968; DRACHMAN 1974; and FLEMING 1976). Denervation of skeletal muscle produces an enormous increase in sensitivity to acetylcholine, nicotine and other nicotinic agonists (THESLEFF 1960), and there is excellent evidence that this is due in large part to the spread of cholinoceptors outward from the endplate, where the receptors are confined in normally innervated muscle, to render the entire muscle fiber responsive to cholinoceptor agonists.

The initial experimental evidence leading to the conclusion of "receptor spread" as an underlying mechanism for supersensitivity was based on a combination of intracellular electrical recording and localized iontophoretic application of acetylcholine (AXELSSON and THESLEFF 1959; MILEDI 1960). This technique enabled the area of the muscle responsive to acetylcholine to be "mapped". This approach is not readily adaptable to the study of receptors in other types of excitable tissues. In smooth muscle, for example, cells are considerably smaller than those of skeletal muscle which makes it extremely difficult to "map" the transmitter-sensitive areas via the combination of intracellular recording and iontophoresis. In addition, because autonomic effectors such as smooth muscle are variably and diffusely innervated *en passage* (BURNSTOCK 1970) rather than having a specialized endplate, it is uncertain whether receptors are restricted to a localized area.

The development in recent years of methods to measure the extent of binding of a radioactive ligand with high affinity for a specific type of receptor has provided a new approach to estimating receptor density. In the case of skeletal muscle, the compound employed is α-bungarotoxin, a potent nicotinic cholinoceptor antagonist. After chronic denervation of skeletal muscle, the number of $[^{125}I]\alpha$-bungarotoxin binding sites is markedly increased, indicating an expansion of the cholinoceptor population (COLQUHOUN et al. 1974; MCCONNELL and SIMPSON 1976; Ko et al. 1977). The fact that this approach yielded results similar to those of the electrophysiological studies establishes an important precedent for the use of ligand binding assays as an indication of receptor density in other tissues which exhibit supersensitivity.

It is not difficult to accept the concept of "receptor spread" as a factor contributing to supersensitivity of skeletal muscle. In innervated muscle, acetylcholine affects only the small endplate region because of the localization of receptors at the neuromuscular junction. After denervation, because of the appearance of extrajunctional receptors, the entire surface of the muscle fiber becomes responsive, i.e. a much larger surface area is available for acetylcholine to interact with. This explanation for supersensitivity, an expansion of the receptor population, is frequently assumed to be responsible for supersensitivity of other tissues. It seems appropriate, therefore, from the standpoint of "classical drug-receptor theory", to consider whether an increase in receptor number can, in fact, provide a theoretical basis for supersensitivity.

Calculations concerning the effect of irreversible receptor inhibition on the position of concentration-response curves have been derived previously (FURCHGOTT 1966). Similar calculations can be used to quantify changes in the EC_{50}, and maximal response to, an agonist after a receptor population has been increased. However, the following assumption must be included with those made by FURCHGOTT (1966): all "new" receptors are identical to the original receptors, coupled to their proper effectors and fully functional.

Under equilibrium conditions, the law of mass action dictates that the concentrations of the agonist, receptor, and agonist-receptor complex are related to the dissociation constant (K_A) by the following Eq.

$$\frac{[RA]}{[R_t]} = \frac{[A]}{[A] + K_A} \tag{1}$$

where $[RA]$, $[R_t]$ and $[A]$ represent the concentrations of agonist-receptor complex, total receptors and free agonist, respectively. As proposed by STEPHEN-

SON (1956), the interaction of an agonist with free receptors produces a stimulus (S) which is proportional to the efficacy (e) of the agent and the fractional receptor occupation ($[RA/R_t]$):

$$S = ue\frac{[RA]}{[R_t]} \tag{2}$$

The term u is analogous to FURCHGOTT's q (1966) and is used to relate changes in $[R_t]$ to changes in e. Since u varies in direct proportion to $[R_t]$ (FURCHGOTT 1966), a treatment which doubles the initial $[R_t]$ will increase u from 1.0 to 2.0. The relative response (E_A/E_m) of an effector to the formation of the agonist-receptor complex is a function of the stimulus (S) generated (STEPHENSON 1956):

$$\frac{E_A}{E_m} = f(s) = f\left(ue\frac{[RA]}{[R_t]}\right) \tag{3}$$

where E_A is the response measured to a single concentration of a particular agonist and E_m is the potential maximal response of the tissue when S approaches very high values (i.e. the response to an agonist of maximal efficacy). An arbitrary scale for stimulus can be set such that $S = 1$ when relative response $(E_A/E_m) = 0.5$ (STEPHENSON 1956). In this case, E_A/E_m is a rectangular hyperbolic function of S (FURCHGOTT 1966):

$$\frac{E_A}{E_m} = \frac{S}{1+S} \tag{4}$$

Such a relationship is represented by the action of α-adrenoceptor agonists on vascular smooth muscle (BESSE and FURCHGOTT 1976; RUFFOLO et al. 1979).

In the first example, the effect of a 50% increase in receptor population on the concentration-response curve to an agonist ($K_A = 10^{-7}$) with a high efficacy ($e = 100$) will be determined. Before receptor proliferation, $u = 1.0$. Using the above values for u and e in Eq. (2), $S = 100$ when 100% receptor occupation is achieved (i.e., $[RA]/[R_t] = 1.0$). Substituting 100 for S in Eq. (4): $E_A/E_m = 100/(1 + 100) = 0.99$. Thus, 99% of the theoretical maximum can be attained with 100% receptor occupancy. The relative response at the EC_{50} is $0.99/2 = 0.495$. Using this value for E_A/E_m in Eq. (4), S is found to equal 0.98. The fractional receptor occupancy necessary to obtain an S of 0.98 can be determined from Eq. (2). Substituting the values for S, u and e, fractional receptor occupation at the EC_{50} is found to equal 0.0098. The EC_{50}, obtained by solving for $[A]$ in Eq. (1), equals 9.9×10^{-10} M.

A 50% increase in receptor population will raise u from 1.0 to 1.5. Using this new value, a series of calculations identical to those above can be performed to determine the effect of this receptor change on the concentration-response curve for the same agonist. The maximum response (i.e. E_A/E_m at 100% receptor occupancy) after receptor proliferation is slightly increased (0.3%) while the EC_{50} is reduced to 6.64×10^{-10} M. In this case, the shift of the concentration-response curve induced by the change in receptor population is equal to 1.49-fold. Further calculations reveal that the maximum shift obtained by a 50% increase in $[R_t]$ approaches 1.5-fold as e approaches infinity. Our equations predict that, if supersensitivity is due solely to an increase in the concentration of receptors, the ratio of EC_{50} (control cells): EC_{50} (supersensitive cells) will never exceed the ratio of [receptors] (supersensitive cells): [receptors] (control cells).

Similar calculations can be performed for partial agonists. Before a change in $[R_t]$, the maximum response to an agonist with an e equal to 1.0 is 50% of E_m. After a 50% increase in $[R_t]$, the maximum response to the agonist is elevated by 20% and the EC_{50} reduced by 20% (i.e. a 1.25-fold shift of the concentration-response curve). For agonists with very low efficacies ($e \leqq 0.1$), the maximum response increases proportionally with $[R_t]$ while the EC_{50} is essentialy unaffected by a 50% change in receptor number. Thus, depending on the efficacy of the agonist used, an increase in the number of receptor sites can decrease the EC_{50}, enhance the maximum response, or both.

The relevance of the above assumptions and calculations can be tested using published data for skeletal muscle. Part of the supersensitivity of denervated skeletal muscle is the consequence of the loss of cholinesterase which accompanies denervation (McConnell and Simpson 1976). The adaptive component of the supersensitivity can be determined from the shift of the mean dose-response curve of a nonhydrolyzable cholinomimetic such as carbachol. McConnell and Simpson (1976) reported a 50-fold shift of the curve for carbachol in the denervated rat hemidiaphragm. That shift in the carbachol curve would be the multiple consequence of all changes in muscle, including spread of receptors, partial depolarization and altered calcium mobilization (Fleming 1976). The contribution of non-receptor changes in the diaphragm is relatively small, of the order of 2-fold (see, for example, Fleming 1971). The contribution of receptor spread to the shift in the carbachol curve may be estimated to be 25-fold (50 divided by 2). McConnell and Simpson (1976) measured binding of α-bungarotoxin and obtained evidence that the cholinoceptors in the hemidiaphragm were increased 20-fold by denervation. From the above equations, the predicted decrease in the EC_{50} is 19.8-fold, which agrees well with the estimated shift of 25-fold.

2. Salivary Glands

Work by Emmelin and colleagues (see reviews by Emmelin 1961, 1965) on the sensitivity of salivary glands of cats has been important to our current understanding of the regulation of effector sensitivity. Emmelin established that chronic interruption of the excitatory innervation to the salivary glands, either surgically (denervation or decentralization) or pharmacologically with drugs such as atropine (to chronically block muscarinic cholinoceptors), reserpine (to deplete noradrenaline from adrenergic nerves) or ganglion blocking agents, produces a nonspecific increase in the secretory response to drugs.

Supersensitivity of salivary glands of other species has also been demonstrated. Two glands in particular, the parotid and submaximallary gland of the rat, have been studied with the specific goal of determining whether an increase in receptor number is responsible for the supersensitivity that they exhibit. Ten days after parasympathetic denervation or ten days after administration of chlorisondamine, a ganglionic cholinoceptor antagonist, the volume of saliva produced by the parotid gland in response to methacholine, a cholinoceptor agonist, is enhanced (ALM and EKSTRÖM 1976; EKSTRÖM and LINDMARK 1978). This effector, therefore, exhibits the classical characteristics of adaptive supersensitivity.

Using the rat parotid gland as a model of supersensitivity, TALAMO et al. (1979) investigated the influence of parasympathetic denervation on the number of muscarinic cholinoceptors as estimated by the binding of the radio-labelled muscarinic antagonist (^{3}H)-quinuclidinyl-benzilate (^{3}H-QNB). Rather than an increase in the magnitude of ^{3}H-QNB binding, TALAMO et al. (1979) observed a significant decrease. The largest part of the decrease was already present three days postoperatively and, therefore, appeared to coincide with the loss of parasympathetic nerve terminals (which was monitored by assaying choline acetyltransferase activity). The absence of any increase in ^{3}H-QNB binding sites at longer periods after denervation (6 to 16 days) suggested that an increase in muscarinic cholinoceptors does not account for the supersensitivity of this gland to cholinoceptor agonists.

The rat submaxillary gland also exhibits the characteristics of postjunctional supersensitivity. The production of saliva in response to agonists is enhanced after chronic ganglion blockade (EKSTRÖM and LINDMARK 1978) or after sympathetic denervation produced either surgically by removal of the superior cervical ganglion or chemically by treatment with 6-hydroxydopamine (PEREC et al. 1973). Responses other than production of saliva are also enhanced. For example, MARTINEZ and QUISSELL (1977) showed that K^+ release from submaximally gland slices induced by noradrenaline and carbachol was greater than normal when slices were prepared from animals pretreated with reserpine for seven days.

The possible role of receptor alterations in the phenomenon of supersensitivity of the submaxillary gland was investigated by ARNETT and DAVIS (1979); the isoprenaline-induced accumulation of cAMP in dispersed cells was markedly increased (600%) 14 days after sympathetic denervation. The number of β-adrenoceptors, as determined by Scatchard analysis of the binding of [^{3}H]-dihydroalprenolol (^{3}H-DHA), a β-adrenoceptor antagonist, was also increased by denervation, but only by 125%. ARNETT and DAVIS (1979) also examined the influence of denervation on an α-adrenoceptor-mediated response (adrenaline-induced K^+ release) of dispersed submaxillary gland cells, and quantified the number of α-adrenoceptors by analysis of the binding of [^{3}H]-dihydroergocryptine (^{3}H-DHE), an α-adrenoceptor antagonist. Curiously, membranes prepared from glands exhibited a 50% increase in ^{3}H-DHE binding sites after denervation but the K^+-releasing effect of adrenaline from dispersed cells was not enhanced.

Hedlund et al. (1983) studied binding to muscarinic receptors (N-methyl-4-piperidinyl benzilate ^{3}H-4-NMPB) and to receptors for vasoactive intestinal polypeptide (^{125}I-VIP) in rat salivary glands (which type of salivary glands not indicated). Chronic administration of atropine (20 mg/kg/day for 14 days) doubled the B_{max} for specific binding to each of the receptor types. On the other hand, 14 days after postganglionic denervation, there was a 60% increase in ^{3}H-4-NMPB binding but *no* change in ^{125}I-VIP binding. Hedlund and co-workers did not measure sensitivity changes per se.

It seems probable that up- and down-regulation of receptors does occur in salivary glands. However, conflicting results and lack of quantitative correlations between changes in dose-response curves and changes in specific binding make it difficult to asses the contribution of changes in receptors to supersensitivity in salivary glands.

3. Smooth Muscle

We have long maintained that changes in receptors are probably *not* the primary mechanism underlying supersensitivity in smooth muscle (FLEMING et al. 1973; FLEMING 1976). The basis of this is the marked non-specifity of the supersensitivity in smooth muscle. That is, the sensitivity is regular increased, often by very similar magnitudes, to a variety of unrelated agonists, including ions such as potassium and barium which do not act via receptors. Ligand-binding assays have generally supported this concept.

Parasympathetic denervation of the cat iris, by removal of the ciliary ganglion, results in a prominent (10- to 20-fold) increase in sensitivity to pilocarpine and carbachol (BITO and DAWSON 1970). Employing the same technique for producing supersensitivity, SACHS et al. (1979) evaluated the status of the muscarinic cholinoceptors by determining the binding characteristics of radio-labelled N-methyl-4-piperidyl benzilate (^{3}H-4-NMPB), a potent muscarinic cholinoceptor antagonist. At a time when the iris was supersensitive to pilocarpine, SACHS et al. (1979) found no change in the dissociation constant of ^{3}H-4-NMPB or in the maximum number of binding sites for this ligand. KARLSEN (1978) also found no change in the binding of [^{3}H]QNB to muscarinic receptors in the denervated guinea-pig iris, but did not measure sensitivity.

Removal of the superior cervical ganglion produces a sympathetic denervation of the rabbit iris and supersensitivity to the mydriatic effects of catecholamines (COLASANTI et al. 1978). Dilation of the pupil occurs via an α-adrenoceptor-mediated contraction of the dilator muscle. PAGE and NEUFELD (1978) studied the characteristics of the binding of the α-adrenoceptor antagonist ^{3}H-DHE to membranes prepared from control and chronically sympathectomized irides. There was no increase in the affinity or density of ^{3}H-DHE binding sites. Curiously, the denervated irises exhibited a modest increase (20%) in the number of β-adrenoceptors, as judged by Scatchard analysis of ^{3}H-DHA binding. The possible influence this change would have on the sensitivity of the iris dilator muscle is unclear and probably unimportant.

Adrenergic denervation or chronic pretreatment with reserpine induces non-specific supersensitivity of the guinea-pig vas deferens (WESTFALL et al. 1972). Denervation, however, does not change the affinity or maximum binding of [^{3}H]-2[(2′-6′-dimethoxy)phenoxyethylaminomethyl] benzodioxan ([^{3}H] WB-4101), an α-adrenoceptor-binding ligand (HATA et al. 1980; COWAN et al. 1985). Likewise, chronic pretreatment with reserpine does not have any effect on the binding of [^{3}H] WB-4101 (COWAN et al. 1985). HATA et al. (1980) did report that adrenergic denervation increased ^{3}H-QNB binding by 43% and attempted to relate that effect to a specific supersensitivity of the vas deferens to cholinomimetics which was demonstrable only at 20 °C. Considering that the supersensitivity is very non-specific at 37 °C, the increase in cholinoceptor binding is modest in magnitude and there are much better explanations for supersensitivity in the vas deferens (see Sects. D.II.2 and D.III.1.c), it seems unlikely that changes in the density of receptors have a significant role in supersensitivity in the guinea-pig vas deferens.

Similarly, changes in α_1-adrenoceptors do not explain supersensitivity in the rat vas deferens. ABEL et al. (1985) demonstrated adaptive supersensitivity to noradrenaline and phenylephrine in the chronically denervated rat vas deferens without any changes in the binding of [^{125}I] Be-2254 (2-[β-4-hydroxyphenyl)-ethylaminomethyl]-tetralone). Using BE-2254 as well as [^{3}H] rauwolscine, the same laboratory has established that there are no measurable changes in ligand binding to α_1- or α_2-adrenoceptors in supersensitive caudal arteries of rats (NASSERI et al. 1985). Thus investigations in several different smooth muscles in several different species have failed to detect any connection between adaptive supersensitivity and changes in receptor density.

4. Cardiac Muscle

Chronotropic, inotropic and arrhythmogenic supersensitivity of the adaptive type has been observed in hearts or cardiac preparations of humans, dogs, cats, guinea pigs, rabbits and rats. The supersensitivity has been induced by several of the possible procedures listed in Sect. B.I, including postganglionic surgical denervation, 6-hydroxydopamine, immunosympathectomy, chronic administration of reserpine and chronic administration of the β-adrenoceptor antagonist, propranolol. In many of the studies, the supersensitivity has been relatively specific for catecholamines. The reader interested in details and further references to the above is referred to the review by FLEMING (1984).

There have been conflicting results regarding the association of receptor changes with supersensitivity in cardiac muscle. Withdrawal of propanolol from rats following chronic administration (10 mg/kg every 8 hrs for 14 days) is associated with supersensitivity (2-fold) to isoprenaline's chronotropic effects in atria and inotropic effects in strips of ventricle (TENNER 1983). This same treatment has been reported to double the maximum binding of the β-adrenoceptor ligand, ^{3}H-DHA, to membrane preparations of rat ventricle (GLAUBIGER and LEFKOWITZ 1977). Based upon the equations presented above, a doubling of receptor density would account for a 2-fold shift of the concentration-response curve. BAKER and POTTER (1980) were unable to obtain the same results when they used homogenates of whole hearts. BAKER and POTTER (1980) suggested that the cardiac microsomes of GLAUBIGER and LEFKOWITZ (1977) probably contained only 5–10% of the β-adrenoceptors of homogenates and that the earlier investigation may have uncovered a small propranolol-sensitive pool of receptors obscured in the total receptor population.

Chronic treatment with guanethidine is associated with depletion of noradrenaline and increased binding (65%) of ^{3}H-DHA to membrane fractions of ventricular muscle in rats (GLAUBIGER et al. 1978) and rabbits (BOBIK et al. 1980). Although GLAUBIGER et al. (1978) also reported a 23% increase in cAMP accumulation in their preparations from guanethidine-pretreated rats, neither group investigated chronotropic or inotropic sensitivity.

TENNER et al. (1982) pretreated rabbits with reserpine (0.3 mg/kg/day for 7 days). This treatment induced a seven-fold increase in sensitivity to isoprenaline in the papillary muscle. They also found a small (30%) increase in the maximum binding of ^{3}H-DHA to membrane fractions. TENNER et al. (1982) concluded that an increase in receptors may contribute to the supersensitivity but could not fully account for it (see also Sect. D.II.1.b). Indeed, if the change in receptor density is uniform, a 30% increase would cause only a 1.3-fold increase in sensitivity.

TORPHY et al. (1982) investigated the effects of pretreatment with reserpine on the sensitivity to drugs and the characteristics of the β-adrenoceptors of the guinea-pig right atrium. Treatment of guinea pigs for seven days with reserpine (0.1 mg/kg/day) produces chronotropic and inotropic supersensitivity which is specific for catecholamines. Ligand-binding studies utilizing [^{125}I]-iodohydroxybenzylpindolol (^{125}I-HYP), a β-adrenoceptor antagonist, revealed that neither the number nor the affinity of β-adrenoceptors in right atrial membrane preparations was altered by either short- or long-term pretreatment with reserpine. HAWTHORN and BROADLEY (1982) produced inotropic supersensitivity in guinea-pig papillary muscle using the same dose schedule of pretreatment with reserpine as TORPHY et al. (1982) as well as larger doses for shorter durations. The sensitivity was increased to isoprenaline but not histamine or calcium. Using ^{3}H-DHA as the β-adrenoceptor ligand, HAWTHORN and BROADLEY (1982) observed no changes in the affinity or maximum binding of the ligand to membrane fractions of the muscle.

In regard to cardiac muscle, it is concluded that up-regulation of β-adrenoceptors probably contributes to adaptive supersensitivity under some conditions or in some species. However, chronic pretreatment of guinea pigs with reserpine induces a catecholamine-specific supersensitivity which is not associated with measurable changes in binding to membrane β-adrenoceptors. Whether these differences are a function of species or the procedures used to induce supersensitivity needs to be determined. A new technique for inducing adaptive supersensitivity in both the rat and the guinea-pig heart by surgical sympathectomy, developed by Dr. GOTO in our laboratory (GOTO et al. 1985), presents a promising new approach to the problem.

5. Pineal Gland

The pineal gland is an adrenergically innervated effector which exhibits alterations in sensitivity as a consequence of changes in the intensity of its excitatory innervation (AXELROD 1974). Denervation of the pineal gland by bilateral superior cervical ganglionectomy or 6-hydroxydopamine-pretreatment or pretreatment of animals with reserpine results in an increased responsiveness of β-adrenoceptor-mediated increases in cAMP accumulation, adenylate cyclase activity and induction of 5-HT-N-acetyltransferase (an enzyme involved in the synthesis of the hormone melatonin) (WEISS and COSTA 1967; WEISS 1969; DEGUCHI and AXELROD 1973; STRADA and WEISS 1974).

Alterations in the environmental conditions in which animals are housed will also influence sensitivity. For example, interrupting the normal light-dark cycle by keeping animals in continuous light produces supersensitivity to β-adrenoceptor agonists in the pineal gland (DEGUCHI and AXELROD 1973; STRADA and WEISS 1974). Presumably this occurs because of a suppression of the release of noradrenaline from the adrenergic nerves which innervate the gland (AXELROD 1974). Conversely, animals kept in continuous dark exhibit a loss of responsiveness to β-adrenoceptor agonists. Alterations in the sensitivity of the pineal gland occur rapidly and exhibit a diurnal rhythm, sensitivity being highest at the end of a period of light and lowest at the end of a period of darkness (AXELROD 1974). KEBABIAN et al. (1975) reported that alterations in the sensitivity of the pineal gland, which either occur during the day-night cycle or are experimentally induced, were accompanied by changes in the number of binding sites for ^{3}H-DHA.

6. Peripheral and Central Nervous System

There has been extensive investigation of the sensitivity of postganglionic neurons of the superior cervical ganglion following chronic interruption of the preganglionic neurons. However, the results of experiments in the 1960s were conflicting and involve numerous methodological problems (FLEMING et al. 1973). DUN et al. (1976a) circumvented the earlier methodological problems by studying single neurons of the ganglion by means of intracellular recording. They demonstrated that the denervated neurons are subsensitive to the depolarizing actions of nicotinic cholinomimetic agonists and supersensitive to the depolarizing actions of muscarinic cholinomimetic agonists. These changes in sensitivity occur without appreciable changes in passive or active membrane properties (DUN et al. 1976b) or binding of the muscarinic cholinoceptor ligand ^{3}H-QNB (BURT 1978) to the ganglion cells. Unfortunately, the experiments of BURT (1978) were done in rats while those of DUN et al. (1976a, 1976b) were in rabbits.

Distinct parallels to the phenomenon of supersensitivity of peripheral tissues occur in the central nervous system. There has been much speculation that certain pathological conditions and the phenomena of drug tolerance and dependence are related to altered sensitivity of groups of central neurons (see, for example, STAVRAKY 1961; COLLIER 1966; SHARPLESS 1969; SCHWARTZ et al. 1978; OVERSTREET and YAMAMURA 1979). In their minireview, OVERSTREET and YAMAMURA (1979) point out that, although changes in receptor binding in the CNS have been shown after chronic treatment of animals with a variety of drugs, few studies have established a good quantitative association between such changes and tolerance to the drugs. One of the most persistent problems is the lack of parallel studies of sensitivity and ligand binding under identical conditions.

Surgical procedures and chronic drug administration have been shown to produce phenomena in the central nervous system which almost certainly are examples of what we have come to identify as adaptive supersensitivity. The same or similar procedures have been shown to lead to changes in the maximum binding of receptor-specific ligands. The number of papers relating to CNS supersensitivity and/or receptor changes are legion and beyond the scope of this chapter. There is little question that interfering with neurotransmission in the central nervous system can lead to altered sensitivity *and* altered receptor binding in the effector neurons (SEEMAN 1980). The ability of neurotransmitters in the CNS to "up-or-down-regulate" the receptors upon which they act is strongly supported. The question to be adressed here is, do the changes in receptors fully account for the changes in sensitivity?

A system in which relevant quantitative information is particularly bountiful, is the dopaminergic system, especially of the nigrostriatal pathway. Unilateral destruction of dopaminergic nigrostriatal pathways in animals leads to a unique rotating behavioral response to dopamine-like agonists (UNGERSTEDT 1971), due to supersensitivity of the denervated neurons.

The degree of supersensitivity in the rotational model is difficult to assess because the altered response consists of a behavior for which there is no quantitative control, i.e. unlesioned animals do not exhibit the contralateral turning behavior. However, the work of THORNBURG and MOORE (1975) does provide an approximation of the enhanced sensitivity. These authors determined dose-response curves for apomorphine after 6-hydroxydopamine-induced nigrostriatal lesions in mice. Two days after the lesion, apomorphine produced a dose-dependent contralateral rotation by the animals, that is, by two days there is some degree of supersensitivity. By 20 to 30 days after the lesion, the dose-response curve for apomorphine was shifted to the left of the curve at two days by 20 fold. The magnitude of the supersensitivity was therefore at least 20-fold. Subsequent work by MARSHALL and UNGERSTEDT (1977) and UNGERSTEDT et al. (1978) indicate the sensitivity change could be even greater. SCHWARCZ et al. (1979) used kainic acid lesions to unilaterally destroy dopamine receptors and allow "control" rotational determinations via receptors on the untouched side of the nigrostriatal pathway. By comparing dose-response curves for rotation induced by agonists acting upon dopamine receptors in the kainic acid-lesioned animals and in 6-hydroxydopamine-lesioned animals, SCHWARCZ et al. (1979) found that supersensitivity was associated with an increase in the slope of the dose-response curve and a 5-fold shift at the ED_{50} of apomorphine.

SEEMAN (1980) has reviewed studies of dopamine-receptor binding induced by 6-hydroxydopamine in the striatum. Most of the increases in binding ranged from 20–60% with an average of about 40%. SEEMAN used the equation below to estimate the change in sensitivity that such an increase in receptor density might produce.

$$f = \frac{R_{\text{occ}}}{R_{\text{tot}}} = \frac{D}{D + K_{\text{D}}}$$

In this equation, f = the fraction of receptors occupied, R_{tot} = the total receptor pool, D = the concentration of the agonist and K_D is the dissociation constant for the agonist. SEEMAN's calculations indicated that a 40% increase in the concentration of the receptors would increase the slope of the dose-response curve and result in a 4-fold shift *at the* ED_{90}. SEEMAN (1980) concluded that this represented a good fit with the measurements made by SCHWARCZ et al. (1979, vide supra).

There are several weaknesses in the above analysis. **1)** The equation used does indeed predict a change in slope with an increase in the concentration of receptors, which makes it incorrect to compare the actual difference measured at the ED_{50} with a predicted shift at the ED_{90}. The equation would predict, on the basis of a 40% increase in the concentration of receptors, a 1.3-fold shift at the ED_{50} as compared with the measured shift of 5-fold, not a good correlation. **2)** SEEMAN (1980) compared the results of his calculations to the dose-response curves obtained by SCHWARCZ et al. (1979) for apomorphine but ignored the data in the same paper for another dopamine-receptor agonist, elymoclavine, for which the shift at the ED_{50} is 30-fold. **3)** The equation of SEEMAN (1980) is not consistent with more modern developments in receptor theory because it does not account for the concept of efficacy (STEPHENSON 1956; FURCHGOTT 1966).

The more appropriate equations, based upon those of STEPHENSON (1956) and FURCHGOTT (1966), were given in the introduction to Sect. D.I. These equations predict an increase in slope with an increase in the concentration of receptors for agonists with low efficacy but not for agonists with high efficacy. Note that the measurements of SCHWARCZ et al. (1979) show an increase in the slope of the dose-response curve of apmorphine, but not for elymoclavine. Furthermore, our equations predict that, if supersensitivity is due solely to an increase in the concentration of receptors, the ratio of EC_{50} (control cells): EC_{50} (supersensitive cells) will never exceed the ratio of [receptors] (supersensitivite cells): [receptors] (control cells). In other words, a 40% increase in receptors would cause a shift at the EC_{50} of no more than 1.4-fold. It is clear that neither SEEMAN's equations nor ours would predict shifts of dose-response curves of even 5-fold at the EC_{50}, let alone shifts of 20 to 40-fold.

There are several possible explanations for the failure of such calculations to predict the large changes in sensitivity which are actually observed.

1) The distribution of agonists within the tissue is altered in such a manner as to increase access to the receptors. This certainly could apply to dopamine because of the loss of neurons and their associated uptake of dopamine. However, there is no known reason why any alteration should occur in the distribution of, for example, apomorphine.

2) There could be very large increases in receptors at very limited but important sites on the responding neurons, such that the total increase in receptors is modest. There is no evidence known to this reviewer which would either support or reject such an hypothesis. It is clearly *not* what happens to receptors in denervated skeletal muscle, where mapping of receptors is possible.

3) There could be other cellular changes which contribute to the alteration

in sensitivity, independent of increases in receptors. These might include altered coupling of recepors to second messengers, changes in membrane potential, changes in the Na^+, K^+ pump, changes in ion channels, etc. Since, as discussed in subsequent sections of this chapter, there is extensive evidence that such changes do contribute to supersensitivity in other tissues, notably skeletal muscle and smooth muscle, there is no reason to reject this possibility in central neurons until it has been thoroughly tested.

II. Changes in Electrophysiologic Characteristics

Membrane potential (E_m) in excitable cells is determined primarily by the permeability to, and the electrochemical gradients of, potassium, sodium and chloride ions across the cell membrane. In terms of the "resting" E_m, potassium ions play a particularly important role. That is, potassium conductance in the "resting" state contributes a sizable constant outward positive current which is not balanced by an equal inward current of the less permeant sodium ions.

The Na^+, K^+ pump is essential for the long-term maintenance of resting E_m because it maintains the normal gradients of Na^+ and K^+ as they gradually run downhill. The pump may be either electroneutral, in which case it maintains a one-to-one exchange of Na^+ for K^+ across the membrane, or it may be electrogenic, in which case there is an unequal exchange, for example $3\,Na^+$ for $2\,K^+$. The pump can be electroneutral under one set of conditions and electrogenic under other conditions. When the pump is electrogenic, it creates a positive outward current across the resistance of the cell membrane and, therefore, contributes a few millivolts directly to E_m ($E = IR$). Thus, in the presence of an electrogenic pump, $E_m = E_{ec} + I_p R_m$ where E_{ec} is the electrochemical potential, I_p is the current contributed by the electrogenic pump and R_m is membrane resistance.

A basic understanding of the contributions of ionic gradients, permeability and electrogenic pumping are important to the discussion of supersensitivity as it relates to cellular electrophysiology. For this purpose, the above explanations are adequate. However, for additional references and a more detailed discussion, especially of electrogenic pumping, see the reviews by THOMAS (1972) and FLEMING (1980).

1. Skeletal Muscle

Denervation of skeletal muscle leads to a depolarization of 15 mV or more (see FLEMING et al. 1973, for early references). This depolarization has attracted considerable interest because it is the earliest detectable change in the muscle after denervation (see Sect. C.I). The depolarization is particularly marked during the first few days after denervation, declines slightly by the seventh day and persists at a constant level thereafter (SELLIN and THESLEFF 1980).

Studies of ion conductances are consistent with changes in permeability to ions being at least partly responsible for the depolarization in denervated skeletal muscle. In the absence of changes in the ionic gradients, changes in conductance reflect changes in permeability of a similar direction. LORKOVIC and TOMANEK (1977) found a marked decrease in chloride conductance (g_{Cl}) and an increase in potassium conductance (g_K) in denervated muscle. Sodium permeability (P_{Na}) has been reported to be increased in denervated skeletal muscle (CREESE et al. 1968; ROBBINS 1977). The impact of changes in permeability upon membrane potential depends upon the equilibrium potential for each ion. From a knowledge of equilibrium potentials in skeletal muscle, it is predicted that changes in P_{Cl} would have little effect, an increase in P_K would hyperpolarize and an increase in P_{Na} would depolarize. Since the ratio of P_{Na}/P_K is increased more than two-fold by denervation (ROBBINS 1977), one would expect some degree of depolarization to be the net result.

Several studies support the hypothesis that a reduction in electrogenic pumping is a contributing factor to the partial depolarization of denervated skeletal muscle. The support comes from experiments with cardiac glycosides, specific inhibitors of the Na^+, K^+ pump (THOMAS 1972). Briefly, ouabain causes a rapid depolarization, presumably due to inhibition of the electrogenic pump, in control skeletal muscle (bringing its E_m to approximately the level of denervated muscle) but produces little or no depolarization in denervated muscle (LOCKE and SOLOMON 1967; MCARDLE and ALBUQUERQUE 1975; SELLIN and MCARDLE 1977b). When denervation was produced by crushing the nerve (SELLIN and MCARDLE 1977b), reinnervation began at 9 days, at which time both the membrane potential and the depolarizing effect of ouabain began to recover, following the same time course back to control values. BRAY et al. (1976) reported that the difference in membrane potential between control and denervated pieces of rat diaphragm was reduced or abolished by catecholamines or dibutyryl cyclic AMP. These authors also found that ouabain (1-5 mM) produced a partial depolarization of control muscle cells or denervated cels exposed to catecholamines or dibutyryl cyclic AMP but not of denervated cells in the absence of such substances. BRAY et al. (1976) concluded that denervation caused a reduction in electrogenic pumping which could be stimulated and restored by catecholamines or dibutyryl cyclic AMP.

ROBBINS (1977) noted that all of the evidence favoring the hypothesis that a reduction of electrogenic pumping contributes to the depolarization in denervated rat skeletal muscle is dependent upon the assumption that the depolarizing action of ouabain is the consequence of inhibition of the Na^+, K^+ pump. To counter that assumption, ROBBINS (1977) demonstrated that (1) decreasing $[K^+]$ in the Krebs' solution altered E_m of normal cells to the precise amount to be expected if E_m were entirely dependent upon the permeability and gradients of ions, i.e. as if the pump were electroneutral and (2) ouabain ($10^{-4}M$) altered P_{Na}. The issue is certainly cloudy. The mammalian species generally chosen for investigation of denervation of skeletal muscle is the rat, a species which requires unusually high concentration of cardiac glycosides to inhibit the Na^+, K^+ pump. The use of high concentrations obviously increases the possibility of the drug producing nonspecific effects, such as changes in permeability, and complicates the use of ouabain as an inhibitor of the pump. More recent work, reviewed by CLAUSEN (1986) indicates that there is a decrease in the density of Na^+, K^+ pump sites after denervation or immobilization. However, this decrease begins after the partial depolarization and, therefore, cannot be the cause of the depolarization.

Several other changes in electrophysiologic measurements have been documented in denervated skeletal muscle (see FLEMING et al. 1973, for references). These include the appearance of fibrillation potentials, increases in membrane resistance, capacitance, time constant and length constant, decreases in the rate of rise and recovery of the action potential and the development of resistance of the action potential to inhibition by tetrodotoxin. How these changes relate to each other and to supersensitivity is not clear in all instances.

The partial depolarization may contribute to the nonspecific aspect of denervation supersensitivity in skeletal muscle (e.g. supersensitivity to potassium ion), as it does in smooth muscle (see below). The development of fibrillations is another early event following denervation; THESLEFF and WARD (1975) have presented evidence that fibrillations are intimately related to the partial depolarization.

On the other hand, the decrease in the rate of rise of the action potential is independent of the partial depolarization because (1) the change in the action potential begins later and (2), when control and denervated muscle cells are both hyperpolarized to -90 mV by the passage of current, the rate of rise of the action potential in the denervated cells is only about half of the value in the control cells (THESLEFF 1973). THESLEFF (1973) concluded that denervation must reduce "either the number or the efficiency of the membrane channels responsible for the action current."

The slowing of the rate of rise of the action potential, the development of resistance of the action potential to tetrodotoxin (TTX) and the increased density of extrajunctional receptors are intimately associated (THESLEFF 1973, 1974). Skeletal muscle has at least two types of sodium channel. One (the "fast" sodium channel) is sensitive to inhibition by TTX and the other (the "slow" sodium channel) is resistant to TTX. In normal muscle, the action potential is associated with the entry of sodium via TTX-sensitive channels. After denervation, the action potential remains dependent upon extracellular sodium but becomes resistant to TTX. The evidence has been interpreted to indicate that denervation causes a persistent reduction of fast Na^+ channels and an increase of slow Na^+ channels (COLQUHOUN et al. 1974; SELLIN and THESLEFF 1980). The decrease in the rate of rise of the action potential, resistance to TTX and the increase in density of extrajunctional receptors all appear near the endplate with the same delay after denervation and spread toward the tendon ends of the muscle cell at the same rate. All three changes are prevented by the inhibitor of RNA synthesis, actinomycin D, but are not abolished by it once the changes are established. It is important to note, however, that the increase in cholinoceptors, as estimated from the binding of ^{125}I-iodo-α-bungarotoxin, is far greater than the estimated increase of slow Na^+ channels (Colquhoun et al. 1974).

2. Smooth Muscle

The most complete investigation of the relationship between membrane potential and adaptive supersensitivity in smooth muscle has been in the vas deferens of the guinea pig. Adaptive supersensitivity is induced in that organ by denervation, decentralization or chronic administration of reserpine (WESTFALL 1970a; WESTFALL et al. 1972). The sensitivity is increased to α-adrenoceptor agonists, muscarinic agonists, histamine and potassium ion. Supersensitivity appears on the fourth day and thereafter is maintained at a constant level (WESTFALL 1970a; FLEMING and WESTFALL 1975). Both decentralization and denervation cause a depolarization of the membranes of the smooth muscle cells of 8–10 mV and this depolarization follows the same time course as does supersensitivity (FLEMING and WESTFALL 1975).

The depolarization of the guinea-pig vas deferens occurs without a change in the threshold potential at which action potentials are elicited (GOTO et al. 1978). Decentralization does not cause a significant change in the magnitude of the junction potentials elicited in the smooth muscle by single maximal electric shocks delivered to the nerves by transmural stimulation (FLEMING and WESTFALL 1975). Thus, assuming no change in the amount of transmitter released per impulse, it seems that a given amount of transmitter produces similar amounts of depolarization in control and supersensitive cells. This supports the conclusion that adaptive supersensitivity in the guinea-pig vas deferens is related to the reduced difference between the resting membrane potential and the threshold of activation. This difference is a mean of 22 mV in control and 14 mV in denervated cells, a reduction of approximately 40 % (GOTO et al. 1978).

The Na^+, K^+-pump is electrogenic in smooth muscle cells of vasa deferentia from control guinea pigs and contributes directly to the resting potential (see FLEMING 1980). URQUILLA et al. (1978) used procedures which alter either the electrochemical potential or the electrogenic pump potential to explore the relationship between E_m and sensitivity. Control cells were acutely depolarized 8-10 mV by ouabain ($10^{-5}M$), by doubling the concentration of potassium $[K^+]_0$ in the bathing medium or by decreasing $[K]_0$ to 50 % of normal. Ouabain and low $[K]_0$ are presumed to depolarize by inhibiting the electrogenic pump; high $[K]_0$ depolarizes by decreasing the K^+ gradient across the membrane (see FLEMING 1980 for a more detailed discussion). All three of these procedures increase the sensitivity to methoxamine and histamine to a degree similar to denervation or decentralization (URQUILLA et al. 1978). Thus, the depolarization produced by denervation or decentralization is adequate to account for the non-specific supersensitivity.

The depolarization caused by the acute exposure of the vas deferens to ouabain and that caused by chronic denervation are not additive. That is, these two procedures produce similar amounts of depolarization separately (8-10 mV) but ouabain induces no further depolarization when added to the bathing medium of denervated vasa deferentia (GERTHOFFER et al. 1979). This suggests that the electrogenic contribution to E_m in control guinea-pig vasa deferentia is approximately 8-10 mV and that the depolarization associated with denervation results from the virtual elimination of that contribution. The decrease in membrane potential and electrogenic pumping in the supersensitive guinea-pig vas deferens correlates with evidence that there is a reduction in the number of Na^+, K^+ pump sites on the smooth muscle membrane as indicated by ^3H-ouabain binding (WONG et al. 1981). Denervation (6-hydroxydopamine), decentralization or treatment with reserpine for one day had no effect on the binding. Since the degeneration of the adrenergic nerve terminals is virtually complete 24 hours after denervation (WESTFALL, unpublished observation), these nerve terminals must contribute a neglible amount of the total pump sites in the tissue. However, 5-7 days after denervation, decentralization or the beginning of daily administration of reserpine, there is a 20-40 % decrease in the number of ^3H-ouabain-binding sites.

The apparent decline in the density of Na^+, K^+ pump sites in supersensitive smooth muscle cells of the guinea-pig vas deferens is also consistent with enzymatic studies in subcellular fractions of the same tissue (GERTHOFFER et al. 1979). Denervation, decontralization, or chronic treatment with reserpine produced a decrease in the maximum activity of Na^+, K^+ATPase, the biochemical correlate of the Na^+, K^+ pump. The decline in maximum activity of the enzyme coincided in time with the appearance of nonspecific supersensitivity. By analyzing the kinetics of the reaction, GERTHOFFER et al. (1979) established that denervation or decentralization caused little, if any, change in the K_m for ATP, $K_{0.5}$ values for Na^+ or K^+ activation or the IC_{50} for ouabain. However, both procedures resulted in significant reductions in the V_{max} of the enzymatic reaction. The results of GERTHOFFER et al. (1979) and WONG et al. (1981) indicate a decrease of 20-40 % in maximum activity of the enzyme and in the number of pump sites.

BOWEN and McDONOUGH (1987) have established that the regulation of Na^+, K^+ pump sites in cultured canine kidney cells occurs via prolonged changes in the intracellular Na^+:K^+ ratio. This regulation involves alterations in the mRNA. If such a mechanism also exists in smooth muscle, it might provide a means for the down regulation of the Na^+, K^+ pump in adaptive supersensitivity. It is reasonable to expect that, initially, a decrease in the intracellular Na^+:K^+ ratio would follow the interruption of stimulatory innervation. This could cause a decrease in RNA for the pump and a decrease in pump sites.

Collectively, the above evidence gathered with electrophysiological and biochemical methods suggests that chronic interference with adrenergic transmission in the vas deferens of the guinea pig leads to a decline in the maximum activity of the Na^+, K^+ pump. This reduced maximum activity appears to be associated with an alteration in the ratio of Na^+-K^+ exchange such that the pump shifts from electrogenicity to electroneutrality. The withdrawal of electrogenic pumping represents the loss of a factor contributing to E_m and results in a partial depolarization. The partial depolarization decreases the amount of additional change in E_m needed to initiate action potentials and thus produces nonspecific supersensitivity. Note, however, that partial depolarization will also enhance the responses of smooth muscle to an agonist acting via pharmacomechanical coupling (MULVANY et al. 1982).

As will be discussed later, supersensitivity occurs in the guinea-pig vas deferens without any measurable changes in intracellular Na^+ or K^+. This is not incompatible with a modest decline in *maximum* pumping capacity or a change from electrogenic to electroneutral pumping. It is likely that, under resting conditions, the pump is working in the steep portion of the activation curve, not at maximum. A switch from electrogenic to electroneutral Na^+-K^+ exchange would cause a modest increase in $[Na^+]_i$. This would stimulate the pump to extrude Na^+ at a slightly greater rate, compensating for the tendency for Na^+ to accumulate intracellularly. The net changes in $[Na^+]_i$ and $[K^+]_i$ may be too small to detect with available techniques.

SHARMA and BANERJEE (1977, 1979) have studied ouabain binding in several organs of the cat after destruction of adrenergic nerves with 6-hydroxydopamine. These authors reported decreases of 50 to 90 % in ouabain binding in, for example, salivary glands, heart, nictitating membrane and vas deferens. Adaptive supersensitivity has been demonstrated in three of these organs in the cat (heart, salivary glands, nictitating membrane; see FLEMING et al. 1973, for references). SHARMA and BANERJEE (1979) concluded that the loss of Na^+, K^+ pump sites reflects the loss of those sites associated with the degenerating adrenergic neurons. That conclusion seems improbable. Considering the very small portion of the total tissue mass represented by the nerves and the importance of the Na^+, K^+ pump to all excitable cells, it is very doubtful that 50-90 % of the pump sites in these tissues reside in the nerves. For example, based on electron micrographs, the ratio of nerve mass to muscle mass in the vas deferens of the rat, a densely innervated smooth muscle, has been estimated to be 1:200 (WESTFALL et al. 1975). On the other hand, if the pump sites in the neurons constitute an insignificant fraction of the pump sites in the whole organ, one must postulate that pump sites were lost from the muscle cells just as in the guinea-pig vas deferens. The changes reported by SHARMA and BANERJEE (1977, 1979) are so large, however, that it would be comforting to have them confirmed.

There is little or no detectable change in ion content of smooth muscles associated with supersensitivity. Marked loss of sodium, potassium and calcium from vascular tissues of animals chronically treated with reserpine in some early reports was subsequently found to be associated only with doses of reserpine greater than those required to produce supersensitivity (see FLEMING et al. 1973 for references). In the vas deferens of the guinea pig, supersensitivity induced by either decentralization or chronic pretreatment with reserpine develops without any significant changes in total tissue Na^+, K^+, Mg^{2+}, Ca^{2+} or Cl^-; in intracellular Na^+, K^+, or Cl^-; or in ^{42}K or ^{36}Cl efflux (WESTFALL 1970b).

URQUILLA et al. (1980) confirmed the lack of any significant change in steady-state ^{42}K or ^{36}Cl turnover associated with chronic treatment with reserpine or denervation in the guinea-pig vas deferens. This supports the conclusion that the partial depolarization of supersensitive cells of the vas deferens is due to a reduction in electrogenic pumping rather than a change in permeability to ions. URQUILLA et al. (1980) did report that denervation or chronic administration of reserpine enhanced the fractional exchange of ^{42}K and ^{36}Cl induced by methoxamine or furthrethonium. The authors concluded that, in addition to the role of partial depolarization, supersensitivity may also involve an enhanced increase in ionic permeabilities per unit of receptor activation.

A partial depolarization may contribute to adaptive supersensitivity also in vascular smooth muscle cells. Treatment of rats with 6-hydroxydopamine induces adaptive supersensitivity in the portal vein which is maximally developed in three days. The smooth muscle cells undergo a depolarization of 4 mV simultaneously with the appearance of supersensitivity (APRIGLIANO and HERMSMEYER 1977).

Treatment of rabbits for three days with reserpine produces adaptive supersensitivity and a mean depolarization of 4 mV in the smooth muscle cells of the saphenous artery (ABEL et al. 1981). Twenty-four hours after a single dose of reserpine there was virtually complete depletion of transmitter but no change from control values of sensitivity or E_m. The threshold membrane potential at which potassium elicited contractions of the saphenous artery was not changed by the pretreatment. Ouabain, $3 \times 10^{-6} M$, produced a rapid depolarization (approximately 5 mV) of the smooth muscle cells and non-specific supersensitivity in saphenous arteries from control animals or animals pretreated with reserpine one day before the experiment, but had no effect on arteries from rabbits which had been treated with reserpine for three days (ABEL et al. 1981). Thus, as in the guinea-pig vas deferens, there appears to be a temporal association among a decline in electrogenic pumping, partial depolarization and adaptive supersensitivity in the rabbit saphenous artery.

The only smooth muscle preparation thus far investigated which does not fit the above pattern is the vas deferens of the rat. Under normal resting conditions the Na^+, K^+ pump in this tissue is not electrogenic. Thus, ouabain, in a concentration which does inhibit the Na^+, K^+ ATPase (FEDAN et al. 1978a) does not acutely depolarize the rat vas deferens nor does it alter the tissue's sensitivity (URQUILLA et al. 1978). Chronic denervation (LEE et al. 1975) or the local application of colchicine to the hypogastric plexus (GOTO et al. 1979) does produce adaptive supersensitivity, but it is not associated with a partial depolarization (GOTO et al. 1978; GOTO 1980).

Even in the rat vas deferens, there are some electrophysiologic correlations with supersensitivity. The threshold potential at which action potentials are elicited becomes more negative, i.e. closer to resting potential, after denervation (GOTO et al. 1978) or after the application of colchicine to the hypogastric plexus (GOTO 1980). Thus, there is an interesting contrast between the guinea-pig and the rat vas deferens. In the former, supersensitivity is related to the resting potential coming closer to threshold. In the latter, the threshold comes closer to the resting potential. The end result is the same in both tissues: less stimulus is required to elicit an action potential and, therefore, a contraction. However, two entirely different mechanisms appear to be involved, a reduction in electrogenic pumping vs. a change in membrane properties, possibly due to a decrease in the binding of Ca^{2+} to the membrane (GOTO et al. 1978).

It is also of interest that the rat vas deferens becomes spontaneously active after chronic denervation or chronic treatment of the rats with reserpine (LEE et al. 1975). The appearance of spontaneous contractions is temporally correlated with adaptive supersensitivity. The magnitude of spontaneous contractions is directly related to the sensitivity to noradrenaline and to the completeness of denervation. The evidence indicates that the spontaneous activity is probably myogenic, perhaps a function of the altered electrical threshold. Unfortunately, we have no data on membrane potential during such spontaneous activity in the rat vas deferens because (a) it is difficult to maintain impalements in cells of contracting smooth muscle and (b) the spontaneous contractions are greatest shortly after removing the tissue from the animal and disappear rapidly. Spontaneous activity is also a characteristic of the supersensitive nictitating membrane (see FLEMING et al. 1973).

3. Cardiac Muscle

Adaptive supersensitivity to the chronotropic, arrhythmic and inotropic effects of catecholamines and, in some instances, calcium on hearts of several species has been demonstrated after chronic treatment with reserpine. These species include the cat (FLEMING and TRENDELENBURG 1961; FLEMING 1962), the dog (WESTFALL and FLEMING 1968a), the rabbit (TENNER et al. 1978) and the guinea pig (WESTFALL and FLEMING 1968b; ANTONNACIO and SMITH 1974; TAYLOR et al. 1976; TORPHY et al. 1982). The sensitivity is not increased to histamine, theophylline or acetylcholine (WESTFALL and FLEMING 1968a; TORPHY et al. 1982). The supersensitivity is not associated with any change in binding of radio-ligands to β-adrenoceptors in the guinea-pig heart (TORPHY et al. 1982; HAWTHORN and BROADLEY 1982). Thus, there is a relatively specific increase in sensitivity which must be explained by a change in cellular function beyond the receptor.

TAYLOR et al. (1976) investigated the membrane potential in right atrial cells of perfused hearts and isolated right atria. In both preparations, there is a reduction of the maximum diastolic potential (4 mV) after chronic pretreatment with reserpine. The depolarization and supersensitivity occur simultaneously. SCHULZ et al. (1984) confirmed the finding of a modest depolarization and determined that it occurred in epicardial, but not endocardial, cells of the right atrium. A partial depolarization in nonpacemaker cells could increase the probability of ectopic pacemakers arising (NOBLE 1979) and thus account for the supersensitivity to the arrhythmic effects of noradrenaline. It will be important, therefore, to examine membrane potential in supersensitive ventricular muscle.

Electrophysiological experiments in right atrial nonpacemaker cells (SCHULZ et al. 1984) have provided evidence that (1) there is little or no contribution of electrogenic pumping to maximum diastolic potential at "resting" rate, (2) the cells are capable of electrogenic pumping when they are sodium loaded, and (3) chronic treatment with reserpine does depress the electrogenic pumping induced by sodium loading. If the depression of electrogenic pumping by chronic treatment with reserpine is associated with depression of the maximum capacity of the Na^+, K^+ pump, as it is in smooth muscle (GERTHOFFER et al. 1979), this effect could contribute to the inotropic supersensitivity induced by treatment with reserpine by enhancing the normal transient increase in $[Na^+]_i$ during each action potential. AKERA and BRODY (1977) have proposed that such an increased sodium transient would enhance Na^+-Ca^{2+} exchange leading to transient rises in $[Ca^{2+}]_i$ during each action potential.

SCHULZ et al. (1984) also investigated the potential role of electrogenic pumping in chronotropic supersensitivity induced by chronic treatment with reserpine in guinea-pig right atria. Extensive examination of the electrophysiology of pacemaker cells and the interactions among chronic treatment with reserpine, acute exposure to ouabain and acute exposure to agonists (isoprenaline and histamine) led to the following conclusions regarding these cells. (1) In contrast to rabbit atria (COURTNEY and SOKOLOVE 1979), electrogenic pumping does not have a significant role in the electrophysiology of pacemaker cells at basal rate; (2) electrogenic pumping does not have a significant role in the regulation of chronotropic sensitivity; (3) the catecholamine-specific chronotropic supersensitivity induced by chronic treatment with reserpine cannot be related to any clear pattern of electrophysiologic changes. In other words, all of the alterations in the electrophysiology of supersensitive pacemaker cells induced by isoprenaline are similar to the alterations induced by isoprenaline in control cells but simply require lower concentrations of the agonist. In the absence of measurable changes in ligand binding to β-adrenoceptors (see Sect. D.I.4), the above electrophysiological findings suggest a change in the coupling of the receptors to a second messenger or to ion channels (SCHULZ et al. 1984).

4. Central Nervous System

To our knowledge, no work has been done on the membrane potential of central neurons relative to supersensitivity. However, chronic administration of neuroleptics (WESTFALL et al. 1981) and chemical denervation (SWANN et al. 1982) have been shown to cause significant changes in ^3H-ouabain binding, suggesting alterations in the Na^+, K^+ pump. This is a potentially exciting direction of research that deserves more attention.

III. The Role of "Second Messengers" in Supersensitivity

The evidence discussed in Sects. D.I and D.II indicates that, in some instances, alterations in receptors and electrophysiological properties can contribute to adaptive supersensitivity. It is clear, however, that other changes in effector cells must also contribute to supersensitivity. For example, the evidence is firm that adaptive supersensitivity of the vas deferens of the guinea pig involves a decrease in resting membrane potential such that drug-induced contractions due to depolarization of the cell membrane will be enhanced. Altered electrical properties cannot be the only change, however, because completely depolarized vasa deferentia (by a K-rich, Na-free bathing solution) contract in response to drugs and also exhibit adaptive supersensitivity (WESTFALL 1977). This membrane potential-independent component of the supersensitivity would appear to result from an increased efficiency of cellular signaling systems.

As stated by BERRIDGE (1975) "calcium and the cyclic nucleotides (cAMP and cGMP) are the main components of an internal signaling system which regulates the activity of most cells. The primary function of these intracellular signals is to mediate the cell's response to a wide range of external stimuli. The plasma membrane usually acts as a transducer where the incoming signals are received and transformed into these internal signals (second messengers) which are ultimately responsible for adjusting cellular activity". Because certain aspects of supersensitivity phenomena cannot be credited to a change at the level of the "transducer", i.e. cell membrane electrical properties or receptors, alterations in second messenger systems are likely to be involved.

In this section, the potential involvement of second messengers in supersensitivity will be reviewed. While possible changes in calcium homeostasis and cyclic nucleotides will be discussed separately, it should be borne in mind that these substances interdigitate in several ways in mediating cellular events.

1. Calcium

a. Skeletal Muscle

GUTMANN and SANDOW (1965) found that caffeine caused contraction of denervated skeletal muscle but not normally innervated muscle. They suggested that caffeine has a calcium-releasing effect on the sarcoplasmic reticulum of denervated muscle, an effect absent in normal muscle. ISAACSON and SANDOW (1967) subsequently confirmed that the translocation of calcium under the influence of caffeine was much greater after denervation. Furthermore, calcium accumulation into denervated muscle (HOWELL et al. 1966) and into microsomal preparations of denervated muscle (BRODY 1966) is enhanced. Along with increased sensitivity, denervated skeletal muscle develops a characteristic contracture in response to acetylcholine. LÜLLMANN et al. (1974), by studying the interaction of acetylcholine, caffeine and varying concentrations of calcium in the medium, provided evidence that the contracture is the consequence of calcium release from the sarcoplasmic reticulum. EVANS and SMITH (1976) studied the nature of the catecholamine-induced contracture of chronically denervated skeletal muscle. The permeability of skeletal muscle, as measured by ^{45}Ca uptake, was increased by denervation. Catecholamines, which had no effect on ^{45}Ca-influx in innervated muscle, substantially increased ^{45}Ca entry into denervated muscle.

It thus appears that denervation of skeletal muscle results in several changes in calcium dynamics which may contribute to supersensitivity to drugs. The permeability of the cell membrane to calcium is enhanced by denervation. In addition, there may be a larger amount of calcium in sarcoplasmic reticular stores and the ease by which this calcium is mobilized is enhanced.

b. Cardiac Muscle

TENNER et al. (1978) reported that pretreatment of rabbits with reserpine (0.1 mg/kg/day for 7 days) produces a 3.6-fold supersensitivity of left atrial preparations to the inotropic effects of noradrenaline. There is evidence, although somewhat indirect, for an involvement of changes in calcium dynamics in this phenomenon. TENNER et al. (1978) found that atria from reserpine pretreated animals exhibited larger rested-state contractions than atria from untreated animals. The rested-state contraction, first described by WOOD et al. (1969), refers to the magnitude of the first myocardial contraction which follows a long period of quiescence. It is believed that the rested-state contraction is a measure of the degree of activation of the myocardial cell and thus a measure of the relative amount of calcium released to the myofilaments (LANGER et al. 1975). While the rested-state contraction was larger after pretreatment with reserpine, the ability of noradrenaline to augment the contraction was apparently not altered. TENNER et al. (1978) suggested, therefore, that reserpine induces supersensitivity to the inotropic effects of noradrenaline, not because of an altered ability of noradrenaline to release "activator" calcium, but because of a greater availability of releasable calcium.

c. Smooth Muscle

Much of the evidence suggesting that calcium is involved in the supersensitivity of smooth muscle has been obtained with vascular smooth muscle preparations rendered supersensitivity by pretreatment of animals with reserpine. In selecting evidence, it is particularly important to exclude studies which used high doses of reserpine because of the marked effects of reserpine, independent of transmitter depletion, on ion balance (FLEMING et al. 1973; FLEMING 1984).

Pretreatment with reserpine produces a nonspecific supersensitivity of rabbit aortic strips (HUDGINS and FLEMING 1966) without significantly altering either total tissue calcium content or ^{45}Ca tissue space (HUDGINS and HARRIS 1970; GARRETT and CARRIER 1971; CARRIER 1975; DEFELICE and JOINER 1976). There is evidence, however, for an alteration in the characteristics of binding of calicum to various tissue "compartments". Analysis of the efflux of ^{45}Ca from aortic strips indicates that there may be a reduced affinity (HUDGINS and HARRIS 1970) or number (DEFELICE and JOINER 1976) of superficial, or membrane, binding sites for calcium after pretreatment with reserpine. The amount of calcium which is retained by the reserpine-treated tissue when incubated in a calcium-free solution is larger than normal (HUDGINS and HARRIS 1970; GARRETT and CARRIER 1971; CARRIER and JUREVICS 1973) suggesting an increase in a more firmly-bound, possibly intracellular, pool of calcium.

FEDAN et al. (1978b; 1979) examined the influence of pretreatment with reserpine on calcium uptake into subcellular fractions of rabbit aorta. Oxalate-facilitated uptake of calcium into microsomal fractions was significantly enhanced by pretreatment with reserpine as was the spontaneous release of calcium from these extra-mitochondrial organelles. These observations suggest that there is a larger and more easily mobilized calcium fraction. Further studies by FEDAN et al. (1980) implicated the involvement of an unidentified cytosolic factor in this phenomenon. It was found that a factor, present in the 105,000 g supernatant fraction, enhanced calcium uptake to a greater extent into microsomes prepared from tissues from reserpine-pretreated animals.

FEDAN et al. (1980) suggested that the cytosolic factor may be similar to the activator of 3′:5′-cyclic nucleotide phosphodiesterase (calcium-dependent regulator, CDR or calmodulin; CHEUNG 1970, KAKIUCHI and YAMAZAKI 1970, CHEUNG et al. 1978), since calmodulin is known to stimulate calcium transport into microsomal preparations of cardiac muscle (KATZ and REMTULLA 1978), erythrocyte plasma membranes (BOND and CLOUGH 1973); FARRANCE and VINCENZI 1977; HINDS et al. 1978) and a phosphodiesterase activating factor, probably calmodulin, enhances smooth muscle calcium transport (KROEGER et al. 1977). Indeed, recent studies have shown that purified calmodulin does enhance calcium uptake into smooth muscle microsomes (HOGABOOM and FEDAN 1981).

The evidence is accumulating that calmodulin plays an essential role in smooth muscle contractility. For example, calmodulin directly mediates the calcium-dependent activation of contraction by virtue of its interaction with myosin light chain kinase (see reviews by ADELSTEIN 1980; and MRWA and HARTSHORNE 1980) and may also modulate contractility by its effects on calcium transport. One might speculate, then, that supersensitivity is related to a greater effect of calmodulin on calcium binding and transport such that higher concentrations of calcium are present and more easily released upon the appropiate stimulation. Recent work by RAMOS et al. (1984) has produced additional evidence of a role of calcium in supersensitivity. The concentration-response curve for calcium was shifted to the left without a change in maximum in skinned fibers of chronically denervated guinea-pig vasa deferentia in relation to skinned fibers from control vasa deferentia. This suggests a change in the interaction of calcium with the contracile apparatus. Calmodulin is one of several components of the contractile machinery that may be involved in this altered relationship.

d. Central Nervous System

As discussed in Sect. D.I, the supersensitivity of striatal neurons to dopamine does not appear to be accounted for entirely by a proliferation of receptors for dopamine. There is evidence for another alteration which is related indirectly to calcium homeostasis. Procedures which cause supersensitivity of striatal neurons, such as chronic treatment with dopamine receptor antagonists like haloperidol (GNEGY et al. 1977a, b) or denervation of the corpus striatum (LUCHELLI et al. 1978) cause a significant increase in the striatal concentration of calmodulin. Calmodulin is known to activate a specific form of phosphodiesterase and is also known to modulate the activity of brain adenylate cyclase (BROSTROM et al. 1975; CHEUNG et al. 1975). GNEGY et al. (1977a, b) have proposed that supersensitivity of striatal adenylate cyclase to dopamine agonists, and by implication the behavioral supersensitivity, is related to the higher concentration of calmodulin in striatal membranes.

2. Cyclic Nucleotides

As discussed earlier in this chapter, there is evidence of enhanced sensitivity of the adenylate cyclase-cAMP system in several effectors including the pineal gland, submaxillary gland and brain. There is also evidence that activation of adenylate cyclase by noradrenaline is enhanced in chronically denervated dog hearts (PALMER et al. 1975). A number of factors need to be considered as possibly contributing to the enhanced responsiveness to drugs of the adenylate cyclase-cAMP system. 1) The number of receptors which mediate the action of the drug on adenylate cyclase is increased. 2) There is an alteration in the coupling between the receptor site and the adenylate cyclase. 3) The activity of phosphodiesterase is altered. 4) There is a greater amount of adenylate cyclase or of the regulatory protein. 5) There are changes in the amount of substrate for adenylate cyclase. 6) Changes in the activity of the system are subservient to another change in the cell's physiology such as ion permeability. Unfortunately the experimental support for these possibilities is often equivocal or scarce.

The first three possibilities have already been addressed to some degree. An increase in the number of β-adrenoceptors or dopamine receptors in the brain, as determined by ligand binding experiments, has been shown to occur under certain circumstances but, as discussed in Sect. D.I, there is a question of whether the magnitude of this change is sufficient to fully account for the enhanced responsiveness. The postulate of GNEGY et al. (1977a), presented above, concerning an enhanced efficiency of coupling between the regulatory protein and the dopamine-sensitive adenylate cyclase due to the actions of calmodulin, is attractive but requires confirmation. A change in the level of cAMP secondarily to an alteration in phosphodiesterase activity has little direct support. In fact, there is evidence that changes in phosphodiesterase activity do not account for the enhanced cAMP accumulation of the supersensitive pineal gland (WEISS and STRADA 1972).

The fourth possibility, an increase in the amount of adenylate cyclase, has been reported to occur in several systems including supersensitive striatal neurons (GNEGY et al. 1977a, b). This does not seem to be a universal finding, however, and there are reports that, in the dopamine-sensitive system, the amount of adenylate cyclase does not increase even though there is evidence for behavorial supersensitivity (VON VOIGTLANDER et al. 1975).

The fifth possibility, that changes in cAMP may occur as a consequence of changes in the concentration of its precursor has received little attention. If the activity of adenylate cyclase were to be assessed by measuring the conversion of endogenous ATP to cAMP, any increased activity could be the result of increased substrate (ATP) rather than a change in the enzyme. It is known that, in smooth muscle, procedures which induce supersensitivity cause an increase in endogenous levels of ATP (WESTFALL et al. 1975). Furthermore, alterations in the level of ATP, produced by denervation of the vas deferens, are accompanied by parallel changes in the level of cAMP (G. BROOKER, T. C. WESTFALL and D.P. WESTFALL unpublished observations). SKOLNICK and DALY (1975) have drawn attention to the importance of substrate levels in this phenomenon.

In the absence of evidence to the contrary, the sixth suggestion, that enhanced responses of the adenylate cyclase-cAMP system may be due to changes in such parameters as ion permeability, is a viable possiblity. The relationship between adenylate cyclase, on the one hand, and resting membrane potential, ion permeability and calcium dynamics on the other hand, has not been clearly delineated.

Changes in the guanylate cyclase-cGMP system have also been reported to occur in supersensitive effectors. HSU et al. (1976) reported that decentralization of the guinea-pig vas deferens resulted in a greater increase in cGMP levels in response to methacholine. As reviewed by SCHULTZ et al. (1973), the elevation of cGMP in the vas deferens is secondary to an increase in the intracellular free calcium concentration. The enhancement of the methacholine-induced cGMP accumulation by decentralization, therefore, appears to be a consequence, rather than a cause, of the supersensitive state.

3. Phosphoinositides

Noradrenaline causes a concentration-dependent incorporation of radiolabeled inorganic phosphorus into phosphotidic acid in the rat vas deferens and this effect is enhanced by chronic denervation (TAKENAWA et al. 1983). AKHTAR and ABDEL-LATIF (1986) demonstrated that chronic sympathetic denervation (superior cervical ganglionectomy) was associated with supersensitivity to α_1-adrenoreceptor-mediated contraction and accumulation of myo-inositol triphosphate in the rabbit iris dilator muscle. Subsequent work established that, under similar conditions, arachidonate release and prostaglandin synthesis were also enhanced. Much work needs to be done to assess the significance of these findings in regard to adaptive supersensitivity. The investigations just cited were restricted to the effects of α-adrenoreceptor agonists. As discussed in this chapter and other reviews (FLEMING et al. 1973; FLEMING 1976) adaptive supersensitivity in smooth muscle is nonspecific, the sensitivity increasing by similar magnitudes, to a variety of agonists, including histamine and potassium ion, which are *not* known to act through the phosphoinositol system.

IV. Cell-to-Cell Coupling

In general, adaptive supersensitivity is not believed to result from morphological changes in effectors (FLEMING 1976). There is one example, however, where a morphological alteration may contribute to a change in the pattern of response. Denervation of the vas deferens produces, in addition to a leftward shift of the dose-response curves for stimulants, an increase in the maximal response of the tissue to drugs. This occurs in both the guinea-pig (WESTFALL et al. 1972) and rat (KASUYA et al. 1969; LEE et al. 1975; GOTO et al. 1979), after denervation or the local application of colchicine to the hypogastric plexus but not after decentralization or pretreatment with reserpine. Because these four procedures produce equivalent leftward shifts of dose-response curves, the increase in maximum response appears to result from a mechanism which is independent of that process which causes supersensitivity.

It has been suggested that the increased maximum response after denervation occurs as a result of an improvement in electrical coupling between cells (WESTFALL et al. 1972; LEE et al. 1975) which may well lead to an enhanced synchronization of an agonist-induced contraction. Under control conditions a stimulant drug probably does not directly stimulate all muscle cells simultaneously because of the limitations of a finite diffusion rate through the layers of muscle. Synchrony would be aided if electrical transmission from cell-to-cell were enhanced.

There is considerable evidence to support this suggestion. GOTO et al. (1976) have shown that denervation of the rat vas deferens enhanced the propagation of a drug-induced contraction and that this was accompanied by a decrease in the electrical resistance among smooth muscle cells. The space constant, a measure of the spread of electrical current among cells, of both the rat and guinea-pig vas deferens is significanly increased by denervation or application of colchicine to the hypogastric plexus (GOTO et al. 1978; GOTO 1980). These changes are consistent with, and explained by, the finding that denervation of the vas deferens increased the incidence of gap junctions (nexuses), the morphological structures believed to be involved in low electrical resistance pathways between smooth muscle cells (WESTFALL et al. 1977). Pretreatment with reserpine, which produces supersensitivity but no increase in maximum response, lacks this effect (WESTFALL et al. 1977).

E. Summary and Conclusions

Adaptive supersensitivity is a compensatory phenomenon which occurs in most types of excitable cells, including all types of muscle, exocrine glands, the pineal gland and neurons. These cells adapt to long term (generally days or weeks) changes in the neural input which they receive. The controlling factor or factors regulating effector cell sensitivity may include some or all of the following: (1) some aspect of the cellular activity, (2) a trophic effect of the neurotransmitter or (3) a trophic substance, other than the transmitter, from the nerve.

The cellular changes which make alterations in sensitivity possible are multiple, including changes in density of receptors, membrane potential, electrogenic Na^+, K^+ transport, ion channels and the function of second messengers, specifically calcium and cyclic AMP. Alterations of the synthesis and/or breakdown of specific proteins (receptors, Na^+, K^+-ATPase, adenylate cyclase, calmodulin) and other substances (ATP) have been implicated in specific instances. Two or more cellular changes may contribute to supersensitivity in any one type of effector cell and which one (or ones) is (are) more important varies from one type of effector to another. Determining which precise points in the cascade of cellular events leading to a response are the cause, and which the consequences, of supersensitivity has been difficult and the literature has, at times, become confused because of over-simplification of this cascade of events.

An additional problem has been the natural desire among many investigators to find a single unifying cellular explanation for supersensitivity in all types of cells. It is absolutely clear that there is no such single explanation. Finally, there has been a lack of consideration of whether or not the magnitude of measured changes in cellular function are adequate to fully explain supersensitivity. The following is a summary of the state of understanding of the cellular bases for adaptive supersensitivity in different types of effector organs.

I. Skeletal Muscle

There is a specific and marked increase in sensitivity to nicotinic cholinomimetics in denervated skeletal muscle. This is primarily the consequence of a dramatic increase in the density of nicotinic cholinoceptors in extrajunctional areas of the muscle fibers. The specific supersensitivity to cholinomimetics is superimposed upon a very non-specific supersensitivity of lesser magnitude. The latter is the result of complex ionic-electrophysiologic changes as well as altered calcium dynamics.

II. Smooth Muscle

Adaptive supersensitivity in smooth muscle is characterized by a moderate but very nonspecific increase in sensitivity. Changes in receptors are apparently of little, if any, consequence. In some smooth muscles, an important factor for the development of supersensitivity is a partial depolarization consequent to a reduction in the density of Na^+-K^+ pump sites and altered stoichiometry of the Na^+-K^+ exchange. In the rat vas deferens, a change in the threshold for initiation of action potentials is a significant factor. Also contributing to supersensitivity in smooth muscle is an alteration in calcium dynamics, possibly related to the function of calmodulin. There is no evidence that direct changes in the adenylate cyclase-cyclic AMP system are important to supersensitivity in smooth muscle.

In the vas deferens of the rat and guinea pig, denervation causes an increase in the maximum response to drugs. This increase in maximum can be separated from the more conventional supersensitivity in the same tissues and is due to improved electrical coupling among the smooth muscle cells.

III. Cardiac Muscle

In heart, adaptive supersensitivity is relatively specific for catecholamines. Although changes in receptors may contribute to cardiac supersensitivity in some species, supersensitivity occurs without any changes in either the affinity or the density of β-adrenoceptors in the guinea-pig heart. Changes in membrane potential, the Na^+, K^+ pump, calcium dynamics and the adenylate cyclase system all may be implicated in cardiac supersensitivity. However, the relationships are still far from clear.

IV. Exocrine Glands

Supersensitivity in salivary glands is relatively non-specific. There are reports in the literature of increases in the density of various types of receptors in supersensitive glands. However, the quantitative correlation between supersensitivity and increased concentration of receptors is poor. The possible role of other cellular changes in supersensitivity in exocrine glands has been relatively unexplored.

V. Pineal Gland

Supersensitivity in this organ is observed as increases in the formation of cyclic AMP and, subsequently, the induction of 5-HT-N-acetyltransferase in response to β-sympathomimetic agents. Evidence indicates that an increase in the density of β-adrenoceptors contributes to the supersensitivity. An increase in adenylate cyclase may also be a factor. Other possible cellular changes have not been investigated.

VI. Central Nervous System

Many reports have indicated increased binding of radioactive ligands to adrenoceptors and to dopamine receptors associated with supersensitivity in the central nervous system. However, the quantitative fit is not good. It appears that, although increases in receptor density probably do contribute to supersensitivity in the CNS, other factors may also be involved. Much more work is needed to determine definitely what the role of calcium, calmodulin and/or adenylate cyclase may be in neuronal supersensitivity. Thus far, the question of a change in electrophysiologic characteristics in supersensitive neurons has not been addressed, primarily because of technical problems. Considering that the purpose of neurons is to conduct electrical impluses, the "uncharted waters" of ionic-electrophysiologic changes relative to altered sensitivity in the nervous system represent a particularly intriguing challenge.

Acknowledgements. The research by the authors on supersensitivity has been supported in part by grants from the National Institutes of Health, NB 03034, NS 08300, GM 29840 and 2 T32 GM 07039. The authors are grateful to Dr. Theodore Torphy and Robert Furchgott for their advice and assistance in developing the quantitative analysis of the relationship between EC_{50} values and changes in receptor population.

F. References

Abel PW, Urquilla PR, Goto K, Westfall DP, Robinson RL, Fleming WW (1981) Chronic reserpine treatment alters sensitivity and membrane potential of the rabbit saphenous artery. J Pharmacol Exp Ther 217:430–439

Adelstein RS (1980) Symposium on phosphorylation of muscle contractile proteins: Introduction. Fedn Proc Fedn Am Socs Exp Biol 39:1544–1546

Akera T, Brody TM (1977) The role of Na^+, K^+-ATPase in the inotropic action of digitalis. Pharmacol Rev 29:187-220

Akhtar, R-A, Abdel-Latif AA (1986) Surgical sympathetic denervation increases α_1-adrenoceptor-mediated accumulation of myo-inositol triphosphate and muscle contraction in rabbit iris dilator smooth muscle. J Neurochem 46:96-104

Albuquerque EX, Schuh FT, Kauffman FC (1971) Early membrane depolarization of the fast mammalian muscle after denervation. Pflügers Arch ges Physiol 328:36-50

Albuquerque EX, Warnick JE, Sansone FM, Onur R (1974) The effects of vinblastine and colchicine on neural regulation of muscle. Ann NY Acad Sci 228:224-243

Alm P, Ekström J (1976) Cholinergic nerves of unknown origin in the parotid glands of rats. Arch oral Biol 21:417-421

Antonaccio MJ, Smith CB (1974) Effects of chronic pretreatment with small doses of reserpine upon adrenergic nerve function. J Pharmacol Exp Ther 188:654-667

Aprigliano O, Hermsmeyer K (1977) Trophic influence of the sympathetic nervous system on the rat portal vein. Circulation Res 41:198-206

Arnett CO, Davis JA (1979) Denervation-induced changes in alpha- and beta-adrenergic receptors of the rat submandibular gland. J Pharmacol Exp Ther 211:394-400

Axelrod J (1974) The pineal gland: a neurochemical transducer. Science 184:1341-1348

Axelsson J, Thesleff S (1959) A study of supersensitivity in denervated mammalian skeletal muscle. J (Physiol) Lond 147:178-193

Baker SP, Potter LT (1980) Effect of propranolol on β-adrenoceptors in rat hearts. Br J Pharmacol 68:8-10

Berridge MS (1975) The interaction of cyclic nucleotides and calcium in the control of cellular activity. Adv Cyclic Nucleotide Res 6:1-96

Besse JC, Furchgott RF (1976) Dissociation constants and relative efficacies of agonists acting on alpha adrenergic receptors in rabbit aorta. J Pharmacol Exp. Ther. 197:66-78

Bird SJ, Aghajanian GK (1975) Denervation supersensitivity in the cholinergic septo-hippocampal pathway: a microiontophoretic study. Brain Res 100:355-370

Bito LZ, Dawson MS (1970) The site and mechanism of the control of cholinergic sensitivity. J Pharmacol Exp Ther 175:673-684

Bito LZ, Dawson MJ, Petrinovic L (1971) Cholinergic sensitivity: normal variability as a function of stimulus background. Science 172:583-585

Bloom FE (1974) To spritz or not to spritz: the doubtful value of aimless iontophoresis. Life Sci 14:1819-1834

Bobik A, Korner P, Carson V, Oliver JR (1980) Cardiac β-adrenoceptors and adenyl cyclase activity in rabbit heart during conditions of altered sympathetic activity. Circ Res 46, (Suppl I), I-43 to 1-44

Bond GH, Clough DL (1973) A soluble protein activator of (Mg^{2+} + Ca^{2+})-dependent ATPase in human red cell membranes. Biochim Biophys Acta 323:592-599

Bowen JW, McDonough A (1987) Pretranslational regulation of Na-K-ATPase in cultured canine kidney cells by low K^+. Amer J Physiol 252:C179-C189

Bray JJ, Hawken MJ, Hubbard JI, Pockett S, Wilson L (1976) The membrane potential of rat diaphragm muscle fibres and the effect of denervation. J Physiol (Lond) 225:651-667

Brody IA (1966) Relaxing factor in denervated muscle: a possible explanation for fibrillations. Amer J Physiol 211:1277-1280

Brostrom CO, Huang Y-C, Breckenridge B McL, Wolff DS (1975) Identification of a calcium-binding protein as a calcium-dependent regulator of brain adenylate cyclase. Proc Nat Acad Sci 72:64–68

Burnstock G (1970) Structure of smooth muscle and its innervation. In: Bülbring E, Brading AF, Jones AW, Tomita T (eds) Smooth muscle. Williams and Wilkins Baltimore, pp 1–69

Burt DR (1978) Muscarinic receptor binding in rat sympathetic ganglia is unaffected by denervation. Brain Res 143:573–579

Calne D, Chase TN, Barbeau A (1975) Dopaminergic mechanisms. Adv Neurology, Vol 9. Raven Press New York

Cangiano A, Fried TA (1977) The production of denervation-like changes in rat muscle by colchicine, without interference with axonal transport or muscle activity. J Physiol (Lond) 265:63–84

Cangiano A, Lutzemberger L, Nicotra L (1977) Non-equivalence of impulse blockade and denervation in the production of membrane changes in rat skeletal muscle. J Physiol (Lond) 273:691–706

Cannon WB, Rosenblueth A (1949) The supersensitivity of denervated structures. Macmillan New York

Carrier O (1975) Role of calcium in postjunctional supersensitivity. Fedn Proc Fedn Am Socs Exp Biol 34:1975–1980

Carrier O, Jurevics HA (1973) The role of calcium in "nonspecific" supersensitivity of vascular muscle. J Pharmacol Exp Ther 184:81–94

Carruba MO, Nistico G, Mantegazza P (1979) Evidence for a receptor supersensitivity following impairment of central serotoninergic activity in the rabbit. Naunyn-Schmiedeberg's Arch Pharmacol 309:125–129

Cheung WY (1970) Cyclic 3′,5′-Nucleotide phosphodiesterase. Biochem Biophys Res Comm 38:533–538

Cheung WY, Bradham LS, Lynch TJ, Lin YM, Tallant EA (1975) Protein activator of cyclic 3′:5′-nucleotide phosphodiesterase of bovine or rat brain also activates its adenylate cyclase. Biochem Biophys Res Comm 66:1055–1062

Cheung WY, Lynch TJ, Wallace RW (1978) An endogenous Ca^{2+}-dependent activator protein of brain adenylate cyclase and cyclic nucleotide phosphodiesterase. Adv Cyclic Nucleotide Res 9:233–251

Clausen T (1986) Regulation of active $Na^+ - K^+$ transport in skeletal muscle. Physiol Rev 66:542–580

Cohen SA, Fischbach GD (1973) Regulation of muscle acetylcholine sensitivity by muscle activity in cell culture. Science 181:76–78

Colasanti BK, Hoover DB (1979) Loss of supersensitivity of the cat eye to carbachol at prolonged periods after ciliary ganglionectomy. Fedn Proc Fedn Am Socs Exp Biol 38:276

Colasanti BK, Chiu P, Trotter RR (1978) Adrenergic and cholinergic drug effects on rabbit eyes after sympathetic denervation. Europ J Pharmacol 47:311–318

Collier HOJ (1966) Tolerance: physical dependence and receptors. Adv Drug Res 3:171–188

Colquhoun D, Rang HP, Ritchie JM (1974) The binding of tetrodotoxin and α-bungarotoxin to normal and denervated mammalian muscle. J Physiol (Lond) 240:199–226

Courtney KR, Sokolove PG (1979) Importance of electrogenic sodium pump in normal and overdriven sinoatrial pacemaker. J Mol Cell Cardiol 11:787–794

Cowan FF, Wong SK, Westfall DP, Fleming WW (1985) Effect of postganglionic denervation and pretreatment with reserpine on alpha adrenoceptors of the guinea-pig vas deferens. Pharmacology 30:289–295

Creese R, El-Shafie AL, Vrbova G (1968) Sodium movements in denervated muscle and the effects of antimycin A. J Physiol (Lond) 197:279-294

Defelice A, Joiner P (1976) Effects of reserpine on tissue calcium and contractility of rat and rabbit aorta. Canad J Physiol Pharmacol 54:520-528

Deguchi T, Axelrod J (1973) Supersensitivity and subsensitivity of the β-adrenergic receptor in pineal gland regulated by catecholamine transmitter. Proc Nat Acad Sci 70:2411-2414

Deshpande SS, Albuquerque EX, Guth L (1976) Neurotrophic regulation of prejunctional and postjunctioual membrane at the mammalian motor endplate. Exp Neurol 53:151-165

Drachman DB (1974) The role of acetylcholine as a neurotrophic transmitter. Ann NY Acad Sci 228:160-176

Drachman DB, Witzke F (1972) Trophic regulation of acetylcholine sensitivity of muscle: effect of electrical stimulation. Science 176:314-316

Dun N, Nishi S, Karczmar AG (1976a) Alteration in nicotinic and muscarinic responses of rabbit superior cervical ganglion cells after chronic preganglionic denervation. Neuropharmacology 15:211-218

Dun N, Nishi S, Karczmar AG (1976b) Electrical properties of denervated mammalian sympathetic ganglion cells. Neuropharmacology 15:219-223

Ekström J, Emmelin N (1974) Reinnervation of the denervated parotid gland of the cat. Q J Exp Physiol 59:1-9

Ekström J, Lindmark B (1978) Choline acetyltransferase activity in post ganglionic parasympathetic nerves after "pharmacological decentralization". Acta Pharmacol Toxicol 43:103-110

Emmelin N (1961) Supersensitivity following "pharmacological denervation." Pharmacol Rev 13:17-37

Emmelin N (1964) Action of acetylcholine on the responsiveness of effector cells. Experientia 15:275

Emmelin N (1965) Action of transmitters on the responsiveness of effector cells. Experientia 21:57-65

Engel J, Liljequist S (1976) The effect of long-term ethanol treatment on the sensitivity of dopamine receptors in the nucleus accumbens. Psychopharmacologia 49:253-257

Evans RH, Smith JW (1976) The effect of catecholamines on the influx of calcium and the development of tension in denervated mouse diaphragm muscle. Br J Pharmacol 58:109-116

Farrance ML, Vincenzi FF (1977) Enhancement of (Ca^{2+} + Mg^{2+})-ATPase activity of human erythrocyte membranes by hemolysis in isosmotic imidazole buffer. II. Dependence on calcium and a cytoplasmic activator. Biochim Biophys Acta 471:59-66

Fedan JS, Westfall DP, Fleming WW (1978a) Species differences in sodium-potassium adenosine triphosphatase activity in the smooth muscle of the guinea pig and rat vas deferens. J Pharmacol Exp Ther 207:356-363

Fedan JS, Westfall DP, Fleming WW (1978b) Ca^{2+} uptake by subcellular fractions of rabbit aorta: effect of treatment of animals with reserpine. The Pharmacologist 20:226

Fedan JS, Westfall DP, Fleming WW (1979) Ca^{2+} release and supernatant effects on Ca^{2+} uptake by rabbit aortic microsomes: effect of treatment of animals with reserpine. Fedn. Proc. Fedn. Am. Socs Exp. Biol. 38, 760

Fedan JS, Westfall DP, Fleming WW (1980) Potentiation by supernatant of Ca^{2+} uptake in rabbit aortic microsomes: effect of pretreatment of animals with reserpine. Life Sci. 26:469-474

Fischbach GD, Robbins N (1971) Effect of chronic disuse of rat soleus neuromuscular junctions on postsynaptic membrane. J Neurophysiology 34:562-569
Fleming WW (1962) Supersensitivity of the cat heart to catecholamine-induced arrhythmias following reserpine pretreatment. Proc Soc Exp Biol Med 3:484-486
Fleming WW (1963) A comparative study of supersensitivity to norepinephrine and acetylcholine produced by denervation, decentralization and reserpine J Pharmacol Exp Ther 141:173-179
Fleming WW (1968) Nonspecific supersensitivity of the guinea-pig ileum produced by chronic ganglion blockade. J Pharmacol Exp Ther 162:277-285
Fleming WW (1971) Supersensitivity of the denervated rat diaphragm to potassium: a comparison with supersensitivity in other tissues. J Pharmacol Exp Ther 176:160-166
Fleming WW (1975) Supersensitivity in smooth muscle. Introduction and historical perspective. Fedn Proc Fedn Am Socs Exp Biol 34:1969-1970
Fleming WW (1976) Variable sensitivity of excitable cells: possible mechanisms and biological significance. Rev Neuroscience 2:43-90
Fleming WW (1980) The electrogenic Na^+, K^+-pump in smooth muscle: physiologic and pharmacologic significance. Ann Rev Pharmacol 20:129-149
Fleming WW (1984) A review of postjunctional supersensitivity in cardiac muscle. In: Fleming WW, Graefe K-H, Langer SZ, Weiner N (eds) Neuronal and Extraneuronal Events in Autonomic Pharmacology. Raven Press New York pp 205-219
Fleming WW, Trendelenburg U (1961) The development of supersensitivity to norepinephrine after pretreatment with reserpine. J Pharmacol Exp Ther 133:41-51
Fleming WW, Westfall DP (1975) Altered resting membrane potential in the supersensitive vas deferens of the guinea-pig. J Pharmacol Exp Ther 192:381-389
Fleming WW, McPhillips JJ, Westfall DP (1973) Postjunctional supersensitivity and subsensitivity of excitable tissues to drugs. Rev Physiol Biochem Pharmacol 68:56-119
French SW, Palmer DS, Narod ME, Reid PE, Ramey CW (1975) Noradrenergic sensitivity of the cerebral cortex after chronic ethanol ingestion and withdrawal. J Pharmacol Exp Ther 194:319-326
Furchgott RF (1966) The use of β-haloalkylamines in the differentiation of receptors and in the determination of dissociation constants of receptor-agonist complexes. In: Harper NJ, Simmons AB (eds) Advances in Drug Research, 3. Academic Press London, pp 21-55
Garrett RL, Carrier O (1971) Alteration of extracellular calcium dependence in vascular tissue by reserpine. Europ J Pharmacol 13:306-311
Gerthoffer WT, Fedan JS, Westfall DP, Goto K, Fleming WW (1979) Involvement of the sodium-potassium pump in the mechanism of postjunctional supersensitivity of the vas deferens of the guinea-pig. J Pharmacol Exp Ther 210:27-36
Gianutsos G, Hynes MD, Puri SK, Drawbaugh RB, Lal H (1974) Effect of apomorphine and nigrostriatal lesions on aggression and striatal dopamine turnover during morphine withdrawal: evidence for dopaminergic supersensitivity in protracted abstinence. Psychopharmacologia 34:37-44
Gilliatt RW, Westgaard RH, Williams IR (1978) Extrajunctional acetylcholine sensitivity of inactive muscle fibres in the baboon during prolonged nerve pressure block. J Physiol (Lond) 280:499-514
Glaubiger G, Lefkowitz RJ (1977) Elevated beta-adrenergic receptor number after chronic propranolol treatment. Biochim Biophys Acta 78:720-725
Glaubiger G, Tsai BS, Lefkowitz RJ, Weiss B, Johnson EM (1978) Chronic guanethidine treatment increases cardiac β-adrenergic receptors. Nature 273:240-242

Gnegy ME, Luchelli A, Costa E (1977a) Correlation between drug-induced supersensitivity of dopamine dependent striatal mechanisms and the increase in striatal content of the Ca^{2+} regulated protein activator of cAMP phosphodiesterase. Naunyn-Schmiedeberg's Arch Pharmacol 301:121-127

Gnegy M, Lislinov P, Costa E (1977b) Participation of an endogenous Ca^{++}-binding protein activator in the development of drug-induced supersensitivity of striatal dopamine receptors. J Pharmacol Exp Ther 202:558-564

Goto K (1980) Electrophysiological analysis of colchicine-induced supersensitivity in the rat vas deferens. J Physiol (Lond) 308:465-477

Goto K (1983) Postjunctional supersensitivity of the smooth muscle of the rat vas deferens induced by calmodulin-antagonizing drugs applied locally to the hypogastric plexus. J Pharmacol Exp Ther 224:231-238

Goto K, Masuda Y, Kasuya Y (1976) The effect of denervation in the synchronization of contraction of the rat vas deferens. Europ J Pharmacol 36:395-404

Goto K, Westfall DP, Fleming WW (1978) Denervation-induced changes in electrophysiologic parameters of the smooth muscle of the guinea-pig and rat vas deferens. J Pharmacol Exp Ther 204:325-333

Goto K, Masaki T, Saito A, Kasuya Y (1979) Denervation-like supersensitivity in the rat vas deferens induced by local application of colchicine to the hypogastric plexus. J Pharmacol Exp Ther 209:376-381

Goto K, Longhurst PA, Cassis LA, Head RJ, Taylor DA, Rice PJ, Fleming WW (1985) Surgical sympathectomy of the heart in rodents and its effect on sensitivity to agonists. J Pharmacol Exp Ther 234:280-287

Gruener R, Baumbach N, Coffee D (1974) Reduction of denervation supersensitivity of muscle by submechanical threshold stimulation. Nature, Lond 248:68-69

Guth L (1968) "Trophic" influences of nerve on muscle. Physiol Rev 48:645-681

Gutmann E (1976) Problems in differentiating trophic relationships between nerve and muscle cells. In: Thesleff S (ed) Motor Innervation of Muscle. Academic Press London, pp 323-343

Gutmann E, Honik P (1963) The effect of use and disuse on neuromuscular functions. Elsevier Amsterdam

Gutmann E, Sandow A (1965) Caffeine-induced contracture and potentiation of contraction in normal and denervated rat muscle. Life Sci 4:1149-1156

Hata F, Takeyasu K, Morikawa Y, Lai R-T, Ishida H, Yoshida H (1980) Specific changes in the cholinergic system in guinea-pig vas deferens after denervation. J Pharmacol Exp Ther 215:716-722

Hawthorn MH, Broadley KJ (1982) β-Adrenoceptor ligand binding and supersensitivity to isoprenaline of ventricular muscle after chronic reserpine pretreatment. Naunyn-Schmiedeberg's Arch. Pharmacol 320:240-245

Hedlund B, Abens J, Barfai T (1983) Vasoactive intestinal polypeptide and muscarinic receptors: supersensitivity induced by long-term atropine treatment. Science 220:519-521

Hinds TR, Larsen FL, Vincenzi FF (1978) Plasma membrane Ca^{2+} transport: stimulation by soluble proteins. Biochem Biophys Res Comm 81:455-461

Hofmann WW, Thesleff S (1972) Studies on the trophic influence of nerve on skeletal muscle. Europ J Pharmacol 20:256-260

Hogaboom GK, Fedan JS (1981) Calmodulin stimulation of calcium uptake and (Ca^{2+}, Mg^{2+})-ATPase activities in microsomes from canine tracheal smooth muscle. Biochem Biophys Res Comm 99:737-744

Howell JN, Fairhurst AS, Jenden DJ (1966) Alterations on the calcium accumulating ability of striated muscle following denervation. Life Sci 5:439-446

Hsu CY, Leighton HJ, Westfall TC, Brooker G (1976) Enhancement of methacholine-stimulated guanosine 3′:5′-cyclic monophosphate formation in supersensitive guinea-pig vasa deferentia. J Cyclic Nucleotide Res 2:359–364

Hudgins PM, Fleming WW (1966) A relatively nonspecific supersensitivity in aortic strips resulting from pretreatment with reserpine. J Pharmacol Exp Ther 153:70–80

Hudgins PM, Harris TM (1970) Further studies on the effects of reserpine pretreatment on rabbit aorta: calcium and histologic changes. J Pharmacol Exp Ther 175:609–618

Isaacson A, Sandow A (1967) Caffeine effects on radiocalcium movement in normal and denervated rat skeletal muscle. J Pharmacol Exp Ther 155:376–388

Iwayama T, Fleming WW, Burnstock G (1973) Ultrastructure of mitochondria in atrial muscle associated with depression and supersensitivity produced by reserpine. J Pharmacol Exp Ther 184:95–105

Johns TR, Thesleff S (1961) Effects of motor inactivation on the chemical sensitivity of skeletal muscle. Acta Physiol Scand 51:136–141

Johnson SM, Westfall DP, Howard SA, Fleming WW (1978) Sensitivities of the isolated ileal longitudinal smooth muscle-myenteric plexus and hypogastric nerve-vas deferens of the guinea pig after chronic morphine pellet implantation. J Pharmacol Exp Ther 204:54–66

Kakiuchi S, Yamazaki R (1970) Calcium dependent phosphodiesterase activity and its activating factor from brain. Biochem Biophys Res Comm 41:1104–1110

Karlsen RL (1978) Muscarinic receptor binding and the effect of atropine on the guinea-pig iris. Exp Eye Res 27:577–581

Kasuya Y, Goto K, Hashimoto H, Watanabe H, Munakata H, Watanabe M (1969) Nonspecific denervation supersensitivity in the rat vas deferens “in vitro”. Europ J Pharmacol 8:177–184

Katz S, Remtulla MA (1978) Phosphodiesterase protein activator stimulates calcium transport in cardiac microsomal preparations enriched in sarcoplasmic reticulum. Biochem Biophys Res Comm 83:1373–1379

Kebabian JW, Zatz M, Romero JA, Axelrod J (1975) Rapid changes in rat pineal β-adrenergic receptor: alterations in 1-[^{3}H]alprenolol binding and adenylate cyclase. Proc Natn Acad Sci 72:3735–3739

Klawans HL, d'Amico DS, Patel BC (1975) Behavioral supersensitivity to 5-hydroxytryptophan induced by chronic methysergide pretreatment. Psychopharmacologia 44:297–300

Klawans HL, d'Amico DJ, Nausieda PA, Weiner WS (1977) The specificity of neuroleptic- and methysergide-induced behavioral hypersensitivity. Psychopharmacology 55:49–52

Ko PK, Anderson MJ, Cohen MW (1977) Denervated skeletal muscle fibers develop discrete patches of high acetylcholine receptor density. Science 196:540–542

Kroeger EA, Teo TS, Ho H, Wang JH (1977) Relaxants, cyclic adenosine 3′:5′-monophosphate, and calcium metabolism in smooth muscle. In: Stephens NL (ed) The Biochemistry of Smooth Muscle. University Park Press Baltimore

Langer GA, Serena SD, Nudd LM (1975) Localization of contractile-dependent calcium: comparison of manganese and verapamil in cardiac and skeletal muscle. Amer J Physiol 229:1003–1008

Langer SZ, Draskoczy PR, Trendelenburg U (1967) Time course of the development of supersensitivity to various amines in the nictitating membrane of the pithed cat after denervation or decentralization. J Pharmacol Exp Ther 157:255–273

Lee TJ-F, Westfall DP, Fleming WW (1975) The correlation between spontaneous con-

tractions and postjunctional supersensitivity of the smooth muscle of the rat vas deferens. J Pharmacol Exp Ther 192:136–148
Locke S, Solomon HC (1967) Relation of resting potential of rat gastrocnemius and soleus muscles to innervation, activity and the Na-K pump. J Exp Zool 166:377–386
Lömo T (1976) The role of activity in the control of membrane and contractile properties of skeletal muscle. In: Thesleff S (ed) Motor Innervation of Muscle. Academic Press London, pp 289–321
Lömo T, Rosenthal J (1972) Control of ACH sensitivity by muscle activity in the rat. J Physiol (Lond) 221:493–513
Lömo T, Slater CR (1978) Control of acetylcholine sensitivity and synapse formation by muscle activity. J Physiol (Lond) 275:391–402
Lorkovic H, Tomanek RJ (1977) Potassium and chloride conductances in normal and denervated rat muscles. Amer J Physiol 232:C109–C114
Luchelli A, Guidotti A, Costa E (1978) Striatal content of Ca^{2+}-dependent regulator protein and dopaminergic receptor function. Brain Res 155:130–135
Luco SV, Vera C (1964) Sensitivity to acetylcholine of the nicititating membrane reinnervated by cholinergic fibers. Acta Physiol Lat Amer 14:289–294
Lüllmann H, Preuner J, Sunano S (1974) On the interaction of acetylcholine, caffeine and altered Ca-concentrations upon the excitation-contraction coupling in chronically denervated skeletal muscle. Pflügers Arch 352:279–290
Marshall JF, Ungerstedt U (1977) Supersensitivity to apomorphine following destruction of the ascending dopamine neurons: quantification using the rotational model. Europ J Pharmacol 41:361–367
Martinez JR, Quissel DO (1977) Potassium release from the rat submaxillary gland in vitro. III. Effects of pretreatment with reserpine. J Pharmacol Exp Ther 201:206–217
Mathers DA, Thesleff S (1978) Studies on neurotrophic regulation of murine skeletal muscle. J Physiol (Lond) 282:105–114
McArdle JJ, Albuquerque EX (1975) Effects of ouabain on denervated and dystrophic muscles of the mouse. Exp Neurol 47:353–356
McConnell MG, Simpson LL (1976) The role of acetylcholine receptors and acetylcholinesterase activity in the development of denervation supersensitivity. J Pharmacol Exp Ther 198:507–517
Miledi R (1960) Junctional and extra-junctional acetylcholine receptors in skeletal muscle fibers. J Physiol (Lond) 151:24–30
Mrwa U, Hartshorne DJ (1980) Symposium on phosphorylation of muscle contractile proteins: phosphorylation of smooth muscle myosin and myosin light chains. Fedn Proc Fedn Am Socs Exp Biol 39:1564–1568
Mulvany MJ, Nilsson H, Flatman JA (1982) Role of membrane potential in the response of rat small mesenteric arteries to exogenous noradrenaline stimulation. J Physiol (Lond) 332:363–373
Neidle EA (1950) Pilocarpine sensitization in the parasympathetically denervated pupil of the cat. Amer J Physiol 160:467–473
Noble D (1979) The initiation of the heartbeat. Oxford: Clarendon Press
Overstreet DH, Kozar MD, Lynch GS (1973) Reduced hypothermic effects of cholinomimetic agents following chronic anticholinesterase treatment. Neuropharmacology 12:1017–1032
Overstreet DH, Yamamura HI (1979) Receptor alterations and drug tolerance. Life Sci 25:1865–1878
Page ED, Neufeld AH (1978) Characterization of α- and β-adrenergic receptors in

membranes prepared from rabbit iris before and after development of supersensitivity. Biochem Pharmacol 27:953-958
Palmer GC, Spurgeon HA, Priola DV (1975) Involvement of adenylate cyclase in mechanisms of denervation supersensitivity following surgical denervation of the dog heart. J Cyclic Nucleotide Res 1:89-95
Perec CJ, Stefano FJE, Barrio Rendo ME (1973) Long-lasting supersensitivity after 6-hydroxydopamine in the submaxillary gland of the rat. J Pharmacol Exp Ther 186:220-229
Pestronk A, Drachman DB, Griffin JW (1976) Effect of botulinum toxin on trophic regulation of acetylcholine receptors. Nature 264:787-789
Ramos K, Gerthoffer WT, Westfall DP (1984) Atypical calcium sensitivity of chemically skinned smooth muscle of the guinea pig vas deferens. Proc West Pharmacol Soc 27:387-389
Reas HW, Trendelenburg U (1967) Changes in the sensitivity of the sweat glands of the cat after denervation. J Pharmacol Exp Ther 156:126-136
Redfern P, Thesleff S (1971) Action potential generation in denervated rat skeletal muscle. I. Quantitative aspects. Acta Physiol Scand 81:557-564
Robbins N (1977) Cation movements in normal and short term denervated rat fast twitch muscle. J Physiol (Lond) 271:605-646
Ruffolo RRJR, Rosing EL, Waddell JR (1979) Receptor interactions of imidazolines. I. Affinity and efficacy for alpha adrenergic receptors in rat aorta. J Pharmacol Exp Ther 209:429-436
Russell RW, Overstreet DH (1987) Mechanisms underlying sensitivity to organophosphorus anticholinesterase compounds. Prog Neurobiol 28:97-129
Sachs DI, Kloog Y, Korezyn AD, Heron DS, Sokolusk VM (1979) Denervation, supersensitivity and muscarinic receptors in the cat iris. Biochem Pharmacol 28:1513-1518
Schulz JC, Fleming WW, Westfall DP, Millecchia R (1984) Cellular potentials, electrogenic sodium pumping and sensitivity in guinea-pig atria. J Pharmacol Exp Ther, 231:181-188
Schulz R, Goldstein A (1973) Morphine tolerance and supersensitivity to 5-hydroxytryptamine in the myenteric plexus of the guinea-pig. Nature 244:168-170
Schulz R, Herz A (1976) Aspects of opiate dependence in the myenteric plexus of the guinea-pig. Life Sci 19:1117-1128
Schultz G, Hardman JG, Schultz K, Baird CE, Sutherland EW (1973) The importance of calcium ions for the regulation of guanosine 3′:5′-cyclic monophosphate levels. Proc Natn Acad Sci 70:3889-3893
Schwarcz R, Fuxe K, Agnati LF, Hökfelt T, Coyle JT (1979) Rotational behavior in rats with unilateral striatal kainic acid lesions: a behavioral model for studies on intact dopamine receptors. Brain Res 170:485-495
Schwartz JC, Costentin J, Martires MP, Protais P, Baudry M (1978) Modulation of receptor mechanisms in the CNS: hyper- and hyposensitivity to catecholamines. Neuropharmacology 17:665-685
Segal M (1977) Supersensitivity of hippocampal neurons to acidic amino acids in decomissurized rats. Brain Res 119:476-479
Seeman P (1980) Brain dopamine receptors. Pharmacol Rev 32:229-313
Sellin LC, McArdle JJ (1977a) Colchicine blocks neurotrophic regulation of the resting membrane potential in reinnervating skeletal muscle. Exp Neurol 55:483-492
Sellin LC, McArdle JJ (1977b) Effect of ouabain on reinnervating mammalian skeletal muscle. Europ J Pharmacol 41:337-340

Sellin LC, Thesleff S (1980) Alterations in membrane electrical properties during long-term denervation of rat skeletal muscle. Acta Physiol Scand 108:243-246

Sharma VK, Banerjee SP (1977) The effect of 6-hydroxydopamine on specific ^{3}H-ouabain binding to some sympathetically innervated organs of the cat. Mol Pharmacol 13:796-804

Sharma VK, Banerjee SP (1979) Regeneration of [^{3}H]ouabain binding to (Na^+-K^+)-ATPase in chemically sympathectomized cat peripheral organs. Mol Pharmacol 15:35-42

Sharpless SK (1964) Reorganization of function in the nervous system — use and disuse. Ann Rev Physiol 26:357-388

Sharpless SK (1969) Isolated and deafferented neurons: disuse supersensitivity. In: Jasper RH, Ward AA, Pope A (eds) Basic mechanisms of the epilepsies Little Brown Boston, pp 329-348

Simeone FA (1937) The effect of regeneration of the nerve supply on the sensitivity of the denervated nictitating membrane to adrenine. Amer J Physiol 120:466-474

Simpson LL (1977) The effects of acute and chronic botulinum toxin treatment on receptor number, receptor distribution and tissue sensitivity in rat diaphragm. J Pharmacol Exp Ther 200:343-351

Skirboll LR, Bunney BS (1979) The effects of acute and chronic haloperidol treatment on spontaneously firing neurons in the caudate nucleus of the rat. Life Sci 26:1419-1434

Skolnick P, Daly JW (1975) Stimulation of adenosine 3′,5′-monophosphate formation in rat cerebral cortical slices by methoxamine interaction with an alpha-adrenergic receptor. J Pharmacol Exp Ther 193:549-558

Stanley EF Drachman DB (1979) Effect of disuse on the resting membrane potential of skeletal muscle. Exp Neurol 64:231-234

Stavraky GW (1961) Supersensitivity following lesions of the nervous system. University of Toronto Press Toronto

Stephenson RP (1956) A modification of receptor theory. Brit J Pharmacol 11:379-393

Strada SJ, Weiss B (1974) Increased response to catecholamines of the cyclic AMP system of rat pineal gland induced by decreased sympathetic activity. Arch Biochem Biophys 160:197-204

Swann AC, Grant SJ, Mass JW (1982) Brain (Na^+, K^+)-ATPase and noradrenergic activity: effects of hyper-innervation and denervation on high-affinity ouabain binding. J Neurochem 38:836-839

Tabakoff B, Hoffman PL (1979) Development of functional dependence on ethanol in dopaminergic systems. J Pharmacol Exp Ther 208:216-222

Tabakoff B, Munoz-Marcus M, Fields JZ (1979) Chronic ethanol feeding produces an increase in muscarinic cholinergic receptors in mouse brain. Life Sci 25:2173-2180

Takenawa T, Masaki T, Goto K (1983) Increase in norepinephrine-induced formation of phosphatidic acid in rat vas deferens after denervation. J Biochem 93:303-306

Talamo BR, Adler SC, Burt DR (1979) Parasympathetic denervation decreases muscarinic receptor binding in rat parotid. Life Sci 24:1573-1580

Taylor DA, Westfall DP, de Moraes S, Fleming WW (1976) The effect of pretreatment with reserpine on the diastolic potential of guinea-pig atrial cells. Naunyn-Schmiedeberg's Arch Pharmacol 293:81-87

Tenner TE (1983) Propranolol withdrawal supersensitivity in rat cardiovascular tissue, in vitro. Europ J Pharmacol 92:91-97

Tenner TE, McNeill JH, Carrier O (1978) The role of calcium in supersensitivity to the inotropic effects of norepinephrine. Europ J Pharmacol 50:359-367
Tenner TE, Mukherjee A, Hester RK (1982) Reserpine-induced supersensitivity and the proliferation of cardiac β-adrenoceptors. Europ J Pharmacol 77:61-65
Thesleff S (1960) Effects of motor innervation on the chemical sensitivity of skeletal muscle. Physiol Rev 40:734-752
Thesleff S (1973) Functional properties of receptors in striated muscle. In: Rang HP (ed) Drug Receptors. University Park Press Baltimore, pp 121-133
Thesleff S (1974) Physiological effects of denervation of muscle. Ann NY Acad Sci 228:89-103
Thesleff S, Ward MR (1975) Studies on the mechanism of fibrillation potentials in denervated muscle. J Physiol (Lond) 244:313-323
Thomas RC (1972) Electrogenic sodium pump in nerve and muscle cells. Physiol Rev 52:563-594
Thornburg JE, Moore KE (1975) Supersensitivity to dopamine agonists following unilateral 6-hydroxydopamine-induced striatal lesions in mice. J Pharmacol Exp Ther 192:42-49
Tiedt TN, Albuquerque EX, Guth L (1977) Degenerating nerve fiber products do not alter physiological properties of adjacent innervated skeletal muscle fibers. Science 198:839-841
Torphy TJ, Westfall DP, Fleming WW (1982) Effect of reserpine pretreatment on mechanical responsiveness and [^{125}I]-iodohydroxybenzylpindolol binding sites in the guinea-pig right atrium. J Pharmacol Exp Ther 223:332-341
Trendelenburg U (1963) Supersensitivity and subsensitivity to sympathomimetic amines. Pharmac Rev 15:225-276
Trendelenburg U (1966) Mechanisms of supersensitivity and subsensitivity to sympathomimetic amines. Pharmacol Rev 18:629-640
Trendelenburg U (1986) The metabolizing systems involved in the inactivation of catecholamines. Naunyn-Schmiedeberg's Arch. Pharmacol 332:201-207
Trendelenburg U, Graefe K-H (1975) Symposium on supersensitivity in smooth muscle: supersensitivity to catecholamines after impairment of extraneuronal uptake or catechol-O-methyl transferase. Fedn Proc Fedn Am Soc Exp Biol 34:1971-1974
Trendelenburg U, Maxwell RA, Pluchino S (1970) Methoxamine as a tool to assess the importance of intraneuronal uptake of l-norepinephrine in the cat's nictitating membrane. J Pharmacol Exp Ther 172:91-99
Trulson ME, Eubanks EE, Jacobs BL (1976) Behavioral evidence for supersensitivity following destruction of central serotonergic nerve terminals by 5,7-dihydroxytryptamine. J Pharmacol Exp Ther 198:23-32
Ungerstedt U (1971) Postsynaptic supersensitivity after 6-hydroxydopamine induced degeneration of nigro-striatal dopamine system. Acta Physiol Scand 82 Suppl 367:69-93
Ungerstedt U, Ljünberg T, Schultz W (1978) Dopamine receptor mechanisms: behavioral and electrophysiological studies. Adv Biochem Psychopharmacol 19:311-321
Urquilla PR, Westfall DP, Goto K, Fleming WW (1978) The effects of ouabain and alterations in potassium concentration on the sensitivity to drugs and the membrane potential of the smooth muscle of the guinea-pig and rat vas deferens. J Pharmacol Exp Ther 207:347-355
Urquilla PR, Jones AW, Fleming WW (1980) Effects of sympathetic denervation or chronic reserpine on potassium (^{42}K) and chloride (^{36}C1) efflux from guinea pig vas deferens. Europ J Pharmacol 66:11-19

Von Voigtlander PF, Losey EG, Triezenberg HJ (1975) Increased sensibility to dopaminergic agents after chronic neuroleptic treatments. J Pharmacol Exp Ther 193:88-94

Wang RY, Montigny C de, Gold BI, Roth RH, Aghajanian GK (1979) Denervation supersensitivity to serotonin in rat forebrain: single cell studies. Brain Res 178:479-497

Weiss B (1969) Effects of environmental lighting and chronic denervation on the activation of adenyl cyclase of rat pineal gland by norepinephrine and sodium fluoride. J Pharmacol Exp Ther 168:146-152

Weiss B, Costa E (1967) Adenyl cyclase activity in rat pineal gland: effects of chronic denervation and norepinephrine. Science 156:1750-1752

Weiss B, Strada SJ (1972) Neuroendocrine control of the cyclic AMP system of brain and pineal gland. Adv Cyclic Nucleotide Res 1:357-374

Westfall DP (1970a) Nonspecific supersensitivity of the guinea-pig vas deferens produced by decentralization and reserpine treatment. Br J Pharmacol 39:110-120

Westfall DP (1970b) The effect of reserpine treatment and decentralization on the ion distribution in the vas deferens of the guinea-pig. Br J Pharmacol 39:121-127

Westfall DP (1977) The effects of denervation, cocaine, 6-hydroxydopamine and reserpine on the characteristics of drug-induced contractions of the depolarized smooth muscle of the rat and guinea-pig vas deferens. J Pharmacol Exp Ther 201:267-275

Westfall DP (1981) Supersensitivity of smooth muscle. In: Bülbring E, Brading A, Jones F, Tomita T (eds) Smooth Muscle: An assessment of current knowledge. Arnold Ltd London, pp 285-310

Westfall DP, Fedan JS (1975) The effect of pretreatment with 6-hydroxydopamine on the norepinephrine concentration and sensitivity of the rat vas deferens. Europ J Pharmacol 33:413-417

Westfall DP, Fleming WW (1968a) Sensitivity changes in the dog heart to norepinephrine, calcium and aminophylline resulting from pretreatment with reserpine. J Pharmacol Exp Ther 159:98-106

Westfall DP, Fleming WW (1968b) The sensitivity of the guinea-pig pacemaker to norepinephrine and calcium after pretreatment with reserpine. J Pharmacol Exp Ther 164:259-269

Westfall DP, McClure DC, Fleming WW (1972) The effects of denervation, decentralization and cocaine on the response of the smooth muscle of the guinea-pig vas deferens to various drugs. J Pharmacol Exp Ther 181:328-338

Westfall DP, McPhillips JJ, Foley DJ (1974) Inhibition of cholinesterase activity after postganglionic denervation of the rat vas deferens: evidence for prejunctional supersensitivity to acetylcholine. J Pharmacol Exp Ther 189:493-498

Westfall DP, Goto K, Stitzel RE, Fedan JS, Fleming WW (1975) Effects of various denervation techniques on the ATP of the rat vas deferens. Europ J Pharmacol 34:397-400

Westfall DP, Millecchia LL, Lee TJ-F, Corey SP, Smith DJ, Fleming WW (1977) Effect of denervation and reserpine on nexuses in the rat vas deferens. Europ J Pharmacol 41:239-242

Westfall DP, Wong SK, Fleming WW (1981) An increase in the number of Na-K pump sites in corpus striatum following chronic treatment of rats with neuroleptics. Abst VIII International Congress of Pharmacology, p 474

Wong SK, Westfall DP, Fedan JS, Fleming WW (1981) The involvement of the sodium-potassium pump in postjunctional supersensitivity of the guinea-pig vas deferens as assessed by [^{3}H] ouabain binding. J Pharmacol Exp Ther 219:163-169

Wood EH, Heppner RL, Weidmann J (1969) I. Positive and negative effects of constant electric currents or current pulses applied during cardiac action potentials. II. Hypothesis: calcium movements, excitation-contraction coupling and inotropic effects. Circ Res 24:409–445

Yarbrough GG (1975) Supersensitivity of caudate neurons after repeated administration of haloperidol. Europ J Pharmacol 31:367–369

Yousufzai SYK, Abdel-Latif AA (1986) alpha$_1$-Adrenergic receptor induced subsensitivity and supersensitivity in rabbit iris-ciliary body. Invest Ophthalmol Vis Sci 28:409–419

Subject Index

Please note that a uniform (and "English") terminology was used in this volume and subject index.

MIX
Papier aus verantwortungsvollen Quellen
Paper from responsible sources
FSC® C105338

If you have any concerns about our products,
you can contact us on
ProductSafety@springernature.com

In case Publisher is established outside the EU,
the EU authorized representative is:
Springer Nature Customer Service Center GmbH
Europaplatz 3, 69115 Heidelberg, Germany

Printed by Libri Plureos GmbH
in Hamburg, Germany